T5-BAM-415

MANTLE METASOMATISM and ALKALINE MAGMATISM

Edited by
Ellen Mullen Morris
and
Jill Dill Pasteris

215

Mantle Metasomatism
and
Alkaline Magmatism

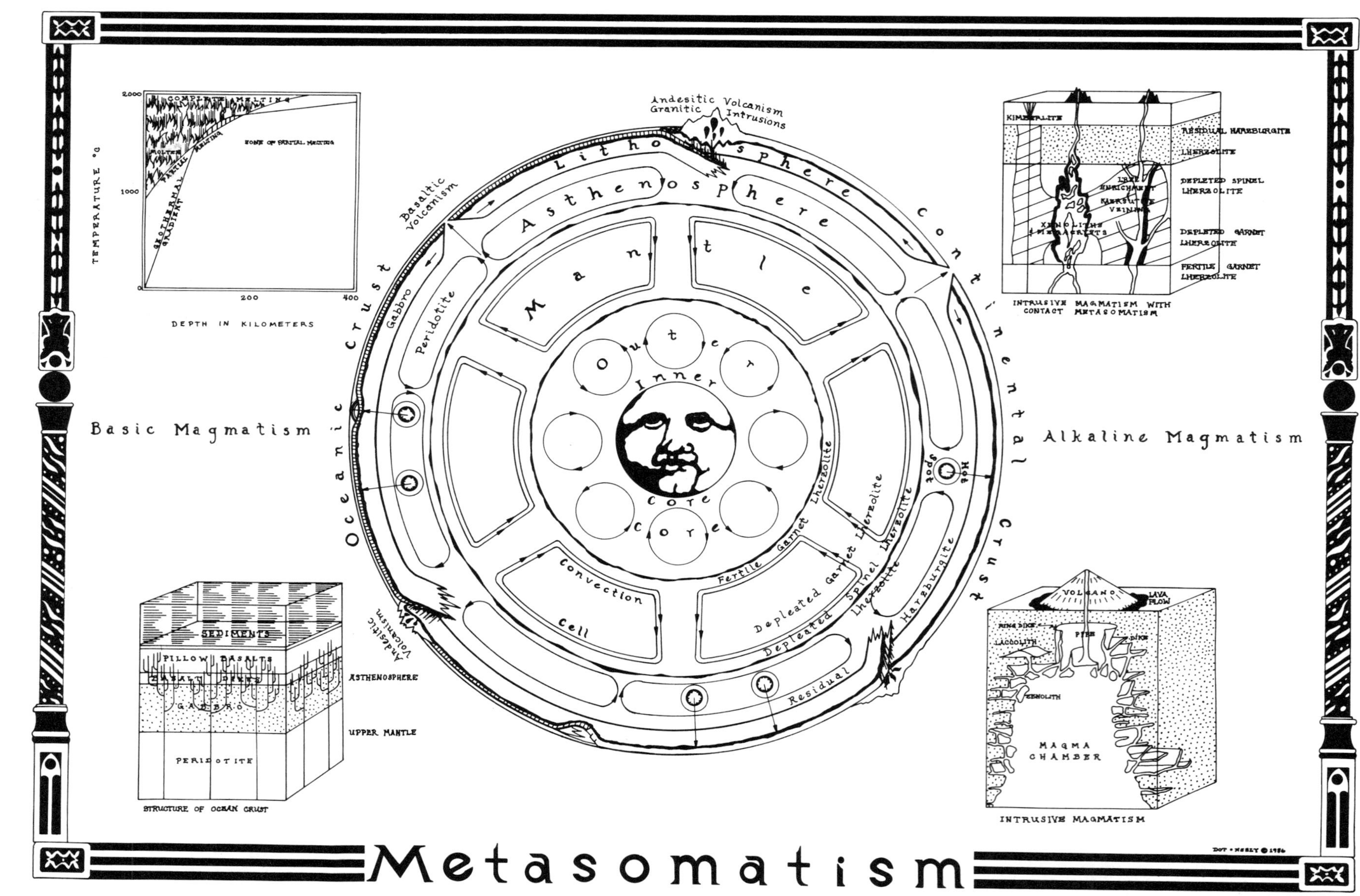
Metasomatism
Basic Magmatism
Alkaline Magmatism
Andesitic Volcanism
Granitic Intrusions
Basaltic Volcanism
Andesitic Volcanism
Lithosphere
Asthenosphere
Mantle
Outer
Inner
core
core
Oceanic Crust
continental Crust
Gabbro
Peridotite
Hot Spot
Residual
Depleated Harzburgite
Depleated Spinel Lherzolite
Depleated Garnet Lherzolite
Fertile Garnet Lherzolite
Lherzolite
Convection
Cell
TEMPERATURE °C
2000
1000
0
200
400
DEPTH IN KILOMETERS
COMPLETE MELTING
MOLTEN
PARTIAL MELTING
ZONE OF PARTIAL MELTING
GEOTHERMAL GRADIENT
SEDIMENTS
PILLOW BASALTS
BASALT DIKES
GABBRO
PERIDOTITE
ASTHENOSPHERE
UPPER MANTLE
STRUCTURE OF OCEAN CRUST
KIMBERLITE
RESIDUAL HARZBURGITE
LHERZOLITE
DEPLETED SPINEL LHERZOLITE
DEPLETED GARNET LHERZOLITE
FERTILE GARNET LHERZOLITE
XENOLITHS & MEGACRYSTS
INTRUSIVE MAGMATISM WITH CONTACT METASOMATISM
VOLCANO
LAVA FLOW
RING DIKE
LACCOLITH
PIPE
DIKE
XENOLITH
MAGMA CHAMBER
INTRUSIVE MAGMATISM
DOT + NEELY © 1986

Mantle Metasomatism and Alkaline Magmatism

Edited by

Ellen Mullen Morris
Department of Geology
Sul Ross State University
Alpine, Texas 79832

Jill Dill Pasteris
Department of Earth and Planetary Sciences
Washington University
St. Louis, Missouri 63130

215

1987

Published by The Geological Society of America, Inc.
3300 Penrose Place, P.O. Box 9140, Boulder, Colorado 80301

Printed in U.S.A.

GSA Books Science Editor Campbell Craddock

Library of Congress Cataloging-in-Publication Data

Mantle metasomatism and alkaline magmatism / edited by Ellen Mullen
Morris, Jill Dill Pasteris.
p. cm. -- (Special paper / Geological Society of America :
215)
Papers from a Symposium on Alkalic Rocks and Kimberlites, held at
the Geological Society of America South-Central Section meeting,
Apr. 15-16, 1985, Fayetteville, Ark.
Includes bibliographies and index.
ISBN 0-8137-2215-2
1. Metasomatism (Mineralogy)--Congresses. 2. Earth--Mantle-
-Congresses. 3. Alkalic igneous rocks--Congresses. I. Morris,
Ellen Mullen, 1948– . II. Pasteris, Jill Dill, 1952– .
III. Geological Society of America. South-Central Section.
IV. Symposium on Alkalic Rocks and Kimberlites (1985 : Fayetteville,
Ark.) V. Series: Special paper (Geological Society of America) ;
215.
QE364.2.M4M36 1987 87-17390
552'.1--dc19 CIP

Cover photo: Photomicrograph of harzburgite with image of olivine gabbro superimposed. Rocks from the Canyon Mountain complex ophiolite, eastern Oregon. Crossed nicols; area shown is 2 mm × 2.6 mm. (Photo by Ellen M. Morris)

Contents

Preface

The papers in this volume were all presented at the Symposium on Alkalic Rocks and Kimberlites, held at the Geological Society of America South-Central Section meeting, April 15–16, 1985, in Fayetteville, Arkansas. This two-day symposium included a total of 55 papers dealing with mantle metasomatism and alkaline magmatism. The four symposium sessions focused upon mantle metasomatism and the origin of alkaline magmas, kimberlites and related rocks, alkalic rocks in oceanic settings, and alkalic rocks in continental settings. Perhaps the most noteworthy outgrowth of this symposium was a heightened awareness that alkaline magmatism is not restricted to any single scenario, but may occur in virtually all tectonic and petrologic settings.

The papers in this volume provide a representative cross section of the Fayetteville symposium. The meeting was designed to bring together researchers from all aspects of alkaline rock petrology to provide better insight into the complex diversity of alkalic systems, and thus to better understand the entire problem rather than just one element. The resulting publication reflects this intent. Here, as in the symposium papers, we initially focus on mantle processes which preceed and accompany alkaline magmatism, then examine kimberlitic and oceanic systems, track across the continental margin, and finally emerge onto the continent itself. Abstracts of all papers presented at the symposium and not published in full here are included in an appendix at the end of this volume so that the broad scope of mantle metasomatism and alkaline magmatism evident at the meeting may be preserved. Readers should keep in mind that the papers in this volume for the most part reflect the authors' views as of mid-1985.

Numerous petrologists not directly involved in the Symposium on Alkalic Rocks and Kimberlites gave unselfishly of their time in order to review manuscripts for this volume. The editors and authors appreciate and thank the following individuals for their thoughtful and prompt reviews: Jun Abrajano, Calvin Barnes, Barbara Barriero, Nabil Boctor, Arthur Boettcher, Douglas Brookins, Bruce Doe, Ian Duncan, Susan Eriksson, Kenneth Foland, Fred Frey, Kiyota Futa, Michael Garcia, William C. Hart, James Hawkins, Carter Hearn, Jr., Glenn Himmelberg, Michael J. Howard, Anthony Irving, Robert Jacobi, William P. Leeman, Gail Mahood, Anthony Mariano, Ian McCallum, V. Rama Murthy, Bjorn Mysen, Donald Parker, Michael Perfit, Martin Prinz, Malcolm Ross, Jean-Guy Schilling, Robert Stern, John Stormer, Lawrence Taylor, Allan Treiman, David Vanko, and Michael Walawander.

The index for this volume was compiled by Patricia Sheahan, and the editors gratefully acknowledge her work.

Support for the Symposium on Alkalic Rocks and Kimberlites was provided by the National Science Foundation (Grant EAR 8500390) and the South-Central Section of the Geological Society of America. These grants contributed greatly to the symposium's value.

Finally, this volume, as well as the symposium from which it is derived, would not have been possible without the encouragement and faith of the late Robert C. Morris, chairman of the South-Central Section of the Geological Society of America. We dedicate this work to his memory.

Ellen Mullen Morris
Jill Dill Pasteris

Geological Society of America
Special Paper 215
1987

Prologue

Ellen Mullen Morris*
Department of Geology, University of Arkansas, Fayetteville, Arkansas 72701
Jill Dill Pasteris
Department of Earth and Planetary Sciences, McDonnell Center for the Space Sciences, Washington University, Box 1169, St. Louis, Missouri 63130

The papers presented at the Symposium on Alkalic Rocks and Kimberlites overwhelmingly demonstrate that alkaline magmatism is the result of complex, multistage processes, and occurs in a variety of settings. Mantle metasomatism is required for most alkalic rocks, from kimberlites to comendites, but the alkalinity of many magmas is also enhanced by fractionation and crustal assimilation. Rocks that are mineralogically and geochemically considered to be alkaline are found not only in environments now universally recognized as sites of alkaline magmatism—such as continental rifts, seamounts, and transforms—but are also associated with island arcs and ophiolites. The following summary of symposium proceedings condenses and highlights the most important new understandings in alkalic rock petrology that evolved at this meeting; papers by authors mentioned herein are included elsewhere in this volume.

The keynote address by Barry Dawson summarized past research on kimberlites, xenoliths and related rocks, and posed questions for future investigation. Dawson emphasized that, although metasomatism is closely associated with kimberlitic magmatism, it is not necessarily contemporaneous; he cited trace element and isotopic evidence that metasomatism related to the Cretaceous (80 to 95 Ma) kimberlites of the Kimberley District, Republic of South Africa, occurred 150 to 200 Ma ago at a depth of about 100 km. Most of the long-held "facts" regarding kimberlites are subjects of present controversy: kimberlites are *not* found only on stable cratons (cf. the Argyle pipe in the mobile belt of northwest Australia), diamonds are not found *exclusively* in kimberlites (cf. the diamondiferous lamprophyres of Australia and Arkansas), and diamonds are most certainly *not* precipitated from their kimberlite or lamproite hosts (cf. Richardson's dating of a garnet inclusion in diamond at 3,200 Ma. Suggested new fields of investigation include the age and source of diamonds, alternate host rocks, and the relation between metasomatism and kimberlite eruption.

Mantle metasomatism is a complex and diverse process that may occur in discrete stages and is usually related to magmatic injection. There seems to be a specific chemical chronology to these injections (early K, later Fe, Ti), as they are derived from increasingly lower-pressure regimes over time (Wilshire). Metasomatism is considered a necessary precursor and cause of alkaline magmatism because geochemical models indicate that the sources of the most primitive alkaline magmas must be a selectively LIL-enriched, previously depleted mantle peridotite. Mantle metasomatism may also be a consequence of alkaline magmatism because not all melt is removed from the mantle source region. However, even trace-element modeling and Sr and Nd isotope systematics cannot distinguish this cause-and-effect relationship because the residual fluid derived from a hydrous melt or basaltic magma would have a composition similar to the metasomatizing agent (Roden).

There are several distinct styles and processes of metasomatism that affect mantle rocks. Patent metasomatism is the most obvious, with resulting minerals occurring in veins or as new growth within the mantle rock. However, not all newly recrystallized minerals in mantle peridotite are the product of metasomatism. For example, on the basis of mineral-xenolith compositional relationships, Dromgoole and Pasteris demonstrate that the formation of sulfides is lithologically controlled. Sulfide inclusions in spinel lherzolite were derived from an immiscible sulfide liquid that formed during partial melting of the same mantle rock.

Cryptic metasomatism is more subtle than patent metasomatism and is reflected by changes in mineral and whole-rock chemical composition. Nielson and Noller document patent metasomatism on a microscopic scale, with the development of minute solid and fluid inclusions in metasomatized olivine and clinopyroxenes. However, such metasomatism might result from the reaction of peridotite wallrock with a residual hydrous fluid from a basaltic melt (Roden).

Hydrous phases such as phlogopite and amphibole are of great importance to mantle metasomatism. In considering the trace- and minor-element contents of hydrous minerals in mantle assemblages, Kempton concludes that only hydrous peridotites are likely source rocks for alkalic melts, because such sources contribute substantial LIL, Fe, Al, and alkali enrichment, and at the same time provide H_2O, CO_2, and other essential volatile

*Present address: Department of Geology, Sul Ross State University, Box C130, Alpine, Texas 79832.

fluxes. Certainly, the increase in LREE/HREE in mantle clinopyroxenes correlates strongly with the development of hydrous phases in the peridotite host (Nielson and Noller).

Although the presence of hydrous phases seems essential to metasomatism, the metasomatizing fluid is not necessarily hydrous. In fact, the constraints of *P-T* stability and element solubility rule out carbonic fluids and silicic melts, as well as hydrous fluids as essential metasomatizing agents for many mantle-derived rocks (Meen). Metasomatism may occur when an ultra-alkalic melt, derived from a hydrated, carbonated peridotite, rises and intersects the peridotite-H_2O-CO_2 solidus. This ultra-alkalic melt would react with mantle wallrock, transforming its mineralogy to more alkali-rich compositions. The exact pressure at which the solidus was intersected and the specific composition of the ultra-alkalic melt would control the final mineralogy and determine whether a fluid phase would form.

Both the mineralogy and geochemistry of metasomatized mantle samples provide conflicting evidence of the timing and relationships of metasomatism and magmatism. The mineral compositions of shallow (<100 km) mantle samples from the Colorado-Wyoming kimberlite district suggest that infertile peridotite was fertilized metasomatically long before the kimberlite eruptions (Eggler and others). Isotopic evidence (Nd and Sr) from the 25-Ma Westland dike swarm of New Zealand indicates a Paleozoic mantle enrichment (Barreiro and Cooper). However, calculation of diffusion rates for Fe and Ti enrichment in rims of garnets in garnet peridotites from both a potassic minette of the Thumb (Colorado Plateau) and a kimberlite of Lesotho indicate that their parent lherzolites were fertilized by metasomatism only shortly before eruption (Smith). Similarly, REE data suggest that mantle metasomatism occurred just prior to the generation of some carbonatites (Bell and Blenkinsop).

This near-synchroneity of enrichment and eruption may be fortuitous, or the rate of magma generation and eruption following metasomatism may be governed by as-yet unquantified factors such as the amount and composition of metasomatism, the *P-T* of the mantle target, and the availability of a fracture/conduit system leading toward the surface. Pressure release may be a mechanism for the generation of alkalic magmas (Lameyre), with melting of the mantle source ending as fractures fill with fluid.

Some xenoliths that are nearly contemporaneous (<4 Ma) with their hosts apparently were derived from the kimberlites themselves. Rb/Sr and Nd isotopic systematics of mineral separates from glimmerite, MARID, and PKP xenoliths from Kimberley, Republic of South Africa, indicate that glimmerites are derived from pegmatites of group I kimberlites, whereas the MARID and PKP suites are related to mixing of fluid components of group I and group II kimberlites (Jones and others). These Kimberley xenolith suites are also unusual because they are rather low-temperature assemblages that are intensely sheared. The simple truth, recognized by most authors, is that models developed for one kimberlite or alkalic rock province cannot necessarily be extended entirely to other locales.

Carbonatites have their source in the mantle, and may be the result of mixed and/or multiple metasomatic events. Isotopic evidence from Sr and Nd systems indicates that carbonatites from Ontario and Quebec were derived from an LIL-depleted mantle source that differentiated from a "bulk-earth" composition mantle at about 3,000 Ma and that has remained coupled to the crust ever since (Bell and Blenkinsop). Similar isotopic and trace-element evidence (Sr, Nd, Pb) indicates that carbonatites, lamprophyres, and related rocks of South Island, New Zealand, had a mantle source that also experienced early (Paleozoic) depletion and later selective LIL enrichment (Barreiro and Cooper). More than one enrichment event may be involved in carbonatite genesis, and carbonatites may not be the result of alkali metasomatism. Treiman notes that many carbonatites associated with alkaline complexes such as Oka and Magnet Cove have minimal K and Na content; he suggests that alkali metasomatism and carbonate segregation may be separate processes that frequently occur together. Fenitization is not associated with low-alkali carbonatites. However, it is common and in places rampant in the alkaline carbonatites of northeast Paraguay as is discussed in the paper by Mariano and Drueker.

Alnoites are rare ultra-alkalic rocks that may be closely related to kimberlites. Hearn presents major and trace-element evidence of a compositional continuum from kimberlite to monticellite-peridotite to alnoite, and suggests that olivine fractionation controls the gradation from kimberlite to alnoite with decreasing whole-rock MgO. Such a close relationship would make alnoites such as those newly discovered in Paraguay (Meyer and Villar) an exciting prospecting target for diamonds.

The magmatic evolution of kimberlites may involve processes such as magma mixing and fractionation, which are regularly recognized in more common igneous rock types. Work by Shervais and Taylor on the differences in major- and trace-element contents of low-Ti and high-Ti garnet megacryst suites from the same rock indicate that two batches of magma were mixed in the low-velocity zone before eruption of the hybrid kimberlite melt. Kimberlite spinels follow four different evolutionary trends, and possibly may be used to distinguish among kimberlite types (Mitchell). Spinels in lamproites are not as distinctive, and may simply determine the degree of evolution of any particular lamproite.

New data on a number of newly discovered and/or reinvestigated kimberlitic rocks were presented at the symposium session entitled "Kimberlites and Related Rocks." These include Kansas localities that Mansker and others, characterize as Winkler-type (crater facies), Stockdale-type (diatreme facies), and Bala-type (hypabyssal facies). This latter type produces a very strong magnetic anomaly. These authors suggest that if all the kimberlites in the Kansas province tapped the same source, then strong devolatization through the crater facies would have permitted the contemporaneous intrusion of more volatile-poor facies in the surrounding area. In contrast to the Kansas kimberlites, which occur in pipe-like bodies, Kansas lamproites at Rose Dome and Silver City occur predominantly as sills, which apparently were localized along deep-seated faults (Berendsen and others). Lamproites with more pipe-like morphologies occur to the south at the Twin Knobs pipe in Arkansas, several kilometers southeast of the well-known Prairie Creek complex. This lamproite contains tuff, breccia, and hypabyssal facies of olivine lamproite. Megacrysts, mantle xenoliths, and diamonds have been recovered from the pipe (Waldman and others). All that contains diamonds, however, is not necessarily of kimberlitic heritage. Bergman and others presented geologic and geochemical evidence that strongly suggests that the diamondiferous Pamali Breccia of Indonesia is neither a kimberlite breccia pipe nor a lamproite, but instead a conglomerate derived from the pyroxenites and serpentinites of the underlying Bobaris ophiolite. The source of the diamonds remains unknown, but Bergman suggests that the harzburgite of the ophiolite might be a likely candidate.

Alkaline basaltic magmas are the product of some combination of a metasomatized mantle source, fractionation, and possibly some crustal assimilation. The Sr and Nd systematics of suboceanic mantle determined by Neal and Davidson indicate an isotopically and mineralogically zoned mantle in which the shallower spinel lherzolite is more affected by metasomatism and LIL enrichment than the deeper garnet and garnet-spinel lherzolite. Evidence from the basal portion of ophiolite cumulates suggests that early oceanic magmas may in fact have an alkaline character due to small amounts of partial melting in the MORB source or mantle heterogeneity (Evans).

Alkalic lavas of oceanic hot spots (Speiss Ridge, Dick and Bryan; Walvis Ridge, Dietrich and Carmen) seem to be derived from multiple, heterogeneous, metasomatized mantle sources. Two distinct series of lavas are identifiable at each hot spot. Possible metasomatism by early subduction related to Iapetus closure is suggested for a source of metasomatizing fluids at the Walvis Ridge. Alkalic rocks of the Kerguelen Archipelago also have bimodal (silica undersaturated/silica oversaturated) trends, which may be explained by fractionation modeling (Beaux and Giret). Experimental work by Sack and Carmichael indicates that the ankaramite-to-trachyte series lavas of the East African Rift have undergone substantial low-pressure fractionation, and that polybaric fractionation commonly has affected alkaline basaltic magmas of diverse settings.

Alkaline magmatism has been recognized increasingly in island arc settings. Perfit reported that alkalic rocks from the southwest Pacific arcs include high-K basalts, and rare high-Na, high-Ti basalt, as well as strongly undersaturated basanite, tephrite, and phonolite. Most of the island arc alkalic suite is low in TiO_2 (<1.2 wt %) and other high-field–strength cations, but has large abundances of LIL. Characteristically, arc alkalics lack the marked HREE depletion of continental and hot spot alkaline basalts. These alkalic rocks, as well as a similar back-arc alkalic series from Trans Pecos, Texas (Price and others), are related to extensional and/or strike-slip tectonics in an older arc and may be derived from small degrees of partial melting in the enriched subarc mantle. In contrast, boninites, which are among the earliest propducts of arc volcanism, also have alkalic characteristics. According to Bloomer, boninites display an increase in Ti/Zr, Y, Ba, and REE, which is consistent with their derivation from a depleted MORB-type source that has been enriched in REE and Zr. Later arc tholeiites are much less enriched in REE and LREE, suggesting that the early boninites removed much of the LIL elements from the arc source material. Hence, alkalic rocks associated with arcs seem to be of two generations, very early and late, and result from somewhat different processes.

There is a close similarity between the sources and overall geochemistry of continental and oceanic alkaline basalts. Eiche reports that alkaline rocks of the northern Canadian cordillera follow fractionation and enrichment trends very similar to oceanic alkalic basalts. Fitton notes that the Cameroon line, a within-plate alkalic province that includes volcanic centers on both the African continent and in the south Atlantic, has nearly indistinguishable isotopic and trace-element signatures or evident evolution among alkaline rocks of this province erupted over period of 65 m.y. in continental and oceanic settings. Their geochemistry indicates derivation from a relatively homogeneous mantle streaked with old LIL-enriched material. The contrast between recent alkaline volcanism in the East Pacific (Clarion Island), and the Late Eocene alkalic Trans-Pecos province of Texas is somewhat greater, but not as stark as might be thought. Both provinces have sodic enrichment trends; the Trans-Pecos province also has a potassic trend. Compositional trends for Rb, Sc, Ti, and REE are similar, and in both provinces, silica depletion increases markedly with fractionation. However, the Trans-Pecos province shows evidence of crustal assimilation and enrichment, and a somewhat different fractionation trend, which can be accounted for by its thicker continental crust (Nelson and Nelson).

The complexity of the Trans-Pecos province is further explained by changes in petrology and sources throughout time. Price and others note that early (Eocene) magmatism of the Trans-Pecos varied from calc-alkaline in Mexico near the east-dipping subduction zone, to alkalic in the Trans-Pecos region, and that overall the rocks developed in a compressional regime. Two magma series—one critically undersaturated basalts and the other a peralkaline series that included rhyolites—coexisted in the early magmatism, and to some extent the two mixed to produce hybrid magmas. A second episode of alkalic magmatism in late Oligocene to early Miocene time was related to early extension of the Texas portion of the Basin and Range province.

Alkalic magmatism that is commonly considered continental, or associated with old continental rifting, seems to show little crustal assimilation or involvement. The data on the Balcones Texas province as reported by Barker and others indicate an age more compatible with—but slightly younger than—the lamproite-to-syenitic rocks of the Arkansas alkalic province to the northeast, which is in near-alignment. The Balcones rocks are strongly bimodal, with nephelinite to alkaline basalt and phonolite as the two types. Partial fusion of mantle peridotite generated the mafic rocks; the phonolite is probably the result of fractionation. There is little evidence for crustal involvement.

Similarly, there is little evidence for crustal involvement in the syenites of the Arkansas alkalic province, according to Pb isotope data (Tilton). This finding is substantiated by the presence of Fo_{70} olivine in primitive syenites of Granite Mountain, and the low Ba content of K-Na feldspars in these rocks (Morris). Potassic metasomatism, followed by Na-Fe-Ti metasomatism, seems to be responsible for changes in some mineral compositions, rather than simple fractionation. Magma mixing was demonstrated for similar rocks in the Highwood Mountains of Montana, based on oscillatory zoning in clinopyroxenes (O'Brien and others).

In summary, the most important and widely agreed-on conclusions from the Symposium on Alkalic Rocks and Kimberlites are that: (1) alkalic rocks are present in nearly every tectonic environment; (2) mantle metasomatism is important to their genesis; and (3) there is little or no crustal involvement in their production and evolution, regardless of their setting. There are multitudes of debatable and conflicting details such as the precise timing and relation of metasomatism and magmatism, the source(s) of metasomatizing fluids, the relation between different alkalic magma series erupted in the same setting, and the role of fractionation in increasing alkalinity. As these questions are further resolved, new problems will undoubtedly arise to take their place. We should all look forward to them.

Printed in U.S.A.

Geological Society of America
Special Paper 215
1987

The mantle redox state; An unfinished story?

Gene C. Ulmer, David E. Grandstaff, David Weiss,* Mark A. Moats, and Tom J. Buntin*
Geology Department†, Temple University, Philadelphia, Pennsylvania 19122
David P. Gold
Department of Geosciences, The Pennsylvanian State University, University Park, Pennsylvania 16802
Christopher J. Hatton
Institute for Geological Research on the Bushveld Complex, Pretoria, South Africa
Arnold Kadik
Vernadsky Institute of Geochemistry and Analytical Chemistry, Academy of Science, Moscow, Union of Soviet Socialist Republics
Richard A. Koseluk
Pennzoil Exploration and Production Company, P.O. Box 13410, The Hague, 2501 E.K., Netherlands
Matthias Rosenhauer
Mineralogisch-Petrologisches Institut, Goldschmidtstrasse #1, 3400 Göttingen, Federal Republic of Germany

ABSTRACT

We review mantle redox models and present new data for xenolithic and megacrystic intrinsic oxygen fugacity (IOF) studies from varied geologic settings. The roles of fluid inclusions, carbon, autoreduction, auto-oxidation, Ti^{3+}, crystal defects, disproportionation, metasomatism, and gravity are examined with regard to their possible influences on redox data.

In several IOF studies, the geothermometric determination for a multimineral sample yields very close temperature concordance with totally different geothermometric techniques; these specific studies mandate some confidence in the IOF fO_2 values obtained. Also of high confidence are the IOF data that come from nearly flawless, gem-quality megacrysts (GQ) of various silicates. Both types of these high-confidence IOF data indicate that the redox state of the mantle is nearer the wüstite-iron (WI) buffer than the quartz-fayalite-magnetite (QFM) buffer.

The fO_2 data of the ilmenite-containing xenolithic assemblages that have either been calculated (Eggler, 1983), IOF-measured (Arculus and others, 1984), or gas mixture-equilibrated (McMahon, 1984) have all shown reasonable overlap in fO_2-T values, and all indicate that these samples come from a mantle region more like (QFM). The ilmenitic xenoliths subjected to IOF analysis in our laboratory have exhibited auto-oxidation and therefore are difficult to interpret unequivocally.

The high-confidence IOF data do argue strongly for mantle redox inhomogeneity, but much more work is needed to establish whether there is a general systematic redox decrease with depth into the mantle, on which the observed redox inhomogeneity is superimposed.

*Present addresses: Weiss, Corning Ceramic Research, Sullivan Park, Corning, New York 14831; Buntin, Department of Environmental Resources, 1875 New Hope Street, Norristown, Pennsylvania 19401.

†More than five years of data accumulations obtained in the Intrinsic Oxygen Fugacity Laboratory at Temple University are presented in this paper.

INTRODUCTION

Values for the redox (reduction-oxidation) state of the earth's mantle may be obtained from models of its chemical and mineralogic properties, from thermochemical calculations of equilibrium assemblages, and from oxygen fugacity measurements (either gas mixtures or intrinsic) from mantle xenolith minerals. This chapter summarizes results of these three methods and attempts to reconcile some of the differences among them.

Models of the earth's mantle are classically integrated from and reiterated against planetary mechanical-density constraints, meteoritic clues, seismic-velocity information, cosmochemical data, experimental petrologic-crystal chemical data, and chemical phobias and affinities in the Goldschmidt tradition. Thus the hypothesized redox in the earth's interior was both the "by-product" and the "hostage" of the model utilized. Only within the last decade, with the expansion of interest in xenoliths, has research directly focused on the mantle redox state become a primary goal of some investigations.

Indeed, as recently as the late 1970s, Ringwood (1979) inferred mantle redox in his model that required decarburization of the earth as a whole because upper mantle Fe^{2+}/Fe^{3+} ratios and volcanic C-H-O gas compositions are too oxidized to be in equilibrium with pure metallic core phases. He used the increased miscibility in the system Fe-O above 2,000°C to infer oxygen in the core to satisfy density constraints. In Ringwood's model, diamond was the "odd bit of left-over carbon from this decarburization process with the bulk of the carbon tied up in Paleozoic limestones, which saved us from being like Venus" (A. E. Ringwood, personal communication, 1980). By comparison, Sato, in a series of papers (1977, 1978a, b) invoked carbon as one of the important low-density elements in the core. He has interpreted the presence of carbon in rocks produced in deep-seated plutons, as well as the increase in fO_2 prior to volcanic eruptions, as evidence that the oxidation state of the crust is higher than that of the mantle, and that as mantle magmas rise they are oxidized by the shallower, more oxidized crustal conditions. These two models are at the extreme opposite ends of a range of mantle models too numerous to review in this chapter, but the origin of the dichotomy of thought regarding redox potential in the mantle is exemplified by these two cases.

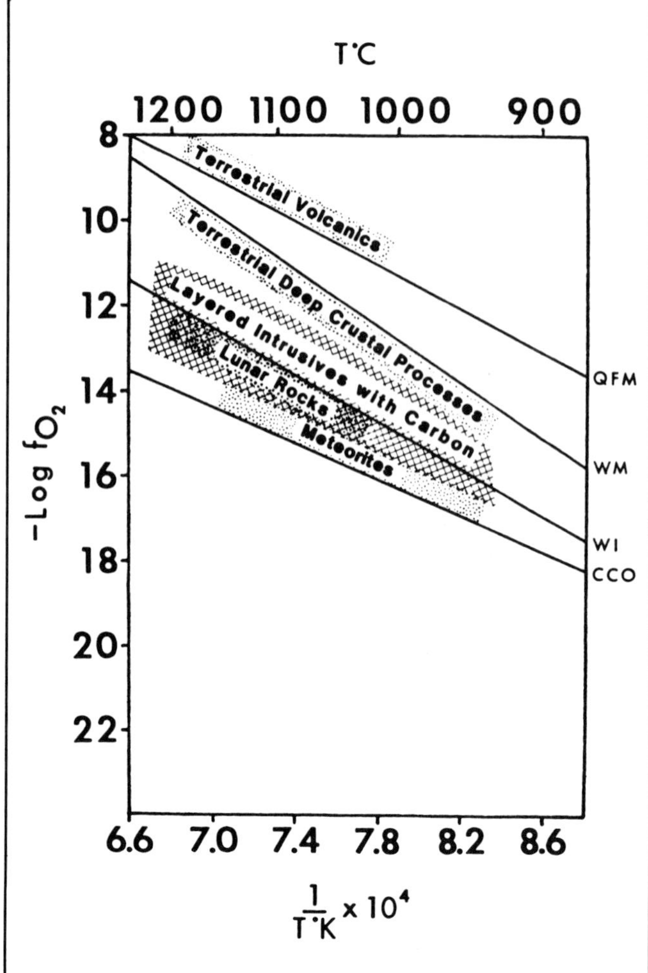

Figure 1. Review of intrinsic oxygen fugacity data (IOF) available for samples other than mantle xenoliths (based on Ulmer, 1984, with modifications). Quartz-fayalite-magnetite (QFM) buffer, wüsite-magnetite (WM) buffer, and wüsite-iron (WI) buffer all plotted at 1 bar from data of Eugster and Wones (1962), whereas graphite-CO_2 (CCO) buffer plotted at 1 bar from data of French and Eugster (1965). For simplicity, some overlaps and exceptions to general envelopes not shown.

The availability of mantle xenoliths since the 1973 International Kimberlite Conference has led to expanded research on mantle samples (cf. Nixon, 1973; Meyer, 1977; Boyd and Meyer, 1979a, b). Simultaneously, the development of the intrinsic oxygen fugacity (IOF) laboratories, begun with Sato's (1965) pioneering work, had progressed with studies by Ulmer and others (1976) and by Arculus and Delano (1980). The petrogenetic role of oxygen fugacity, fO_2, had been heightened by Osborn's (1962) restudy of the Bowen and Fenner differentiation trends, resulting in a renewed interest in the role of fO_2. Thus, interest intensified for some measurements of the mantle redox conditions.

When Haggerty reviewed the role of fO_2 on igneous processes in 1976, he examined more than 400 literature references; most of these studies had been performed using the Buddington and Lindsley fO_2-geothermometer based on iron-titanium oxide equilibria (1964 or later improved versions). Haggerty's 1976 review showed that, regardless of rock type, the results of most studies were within one order of magnitude of the 1 bar quartz-fayalite-magnetite buffer (QFM) of Eugster and Wones (1962). At that time, the newer IOF technique had very few published results. However, Haggerty's review pointed out that the then-available IOF studies did show that plutonic samples were consistently more reduced than the envelope around the (QFM) buffer.

This review also provided a hint of IOF data indicating that the deeper the sample, the more reduced it was. Meanwhile,

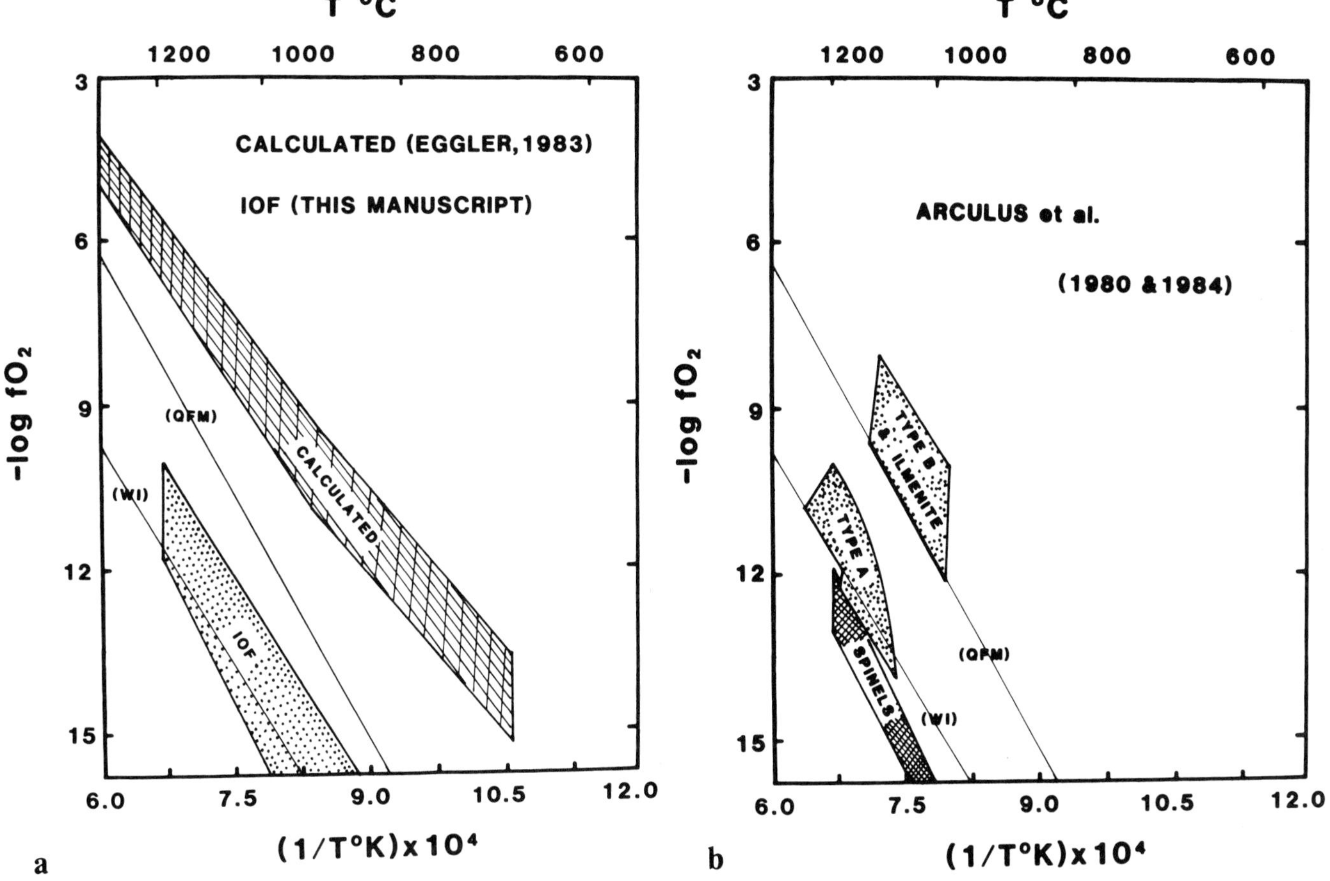

Figure 2. a. Comparison of thermodynamically calculated values for ilmenite-silicate assemblages (Eggler, 1983) with IOF data of high-confidence presented and discussed in this chapter. The (QFM) and (WI) buffers are shown at 1 bar. Sample localities are widely diverse, but can be obtained from literature cited. b. Comparison of ilmenitic megacryst data with Type I and Type II (Frey and Prinz, 1978) silicate data and spinel data from IOF studies by Arculus and Delano (1980) and Arculus and others (1984). (QFM) and (WI) buffers are shown at 1 bar. Sample localities are widely diverse but can be obtained from original literature cited.

meteorites and lunar rocks were also being examined by the IOF technique and consistently showing values reduced much more than terrestrial crustal processes. The results of many IOF studies are compared (Fig. 1, after Ulmer, 1984) with the 1-atm values of the standard buffers (QFM), wüstite-magnetite (WM), wüstite-iron (WI), all from Eugster and Wones (1962), and carbon-carbon dioxide (CCO) from French and Eugster (1965).

Thus, the stage was set for much more quantitatively focused redox research on mantle samples. From the mid-1970s on, there have been both fO_2 measurements and thermodynamic calculations of the redox memory of mantle xenoliths and megacrysts. The dichotomy in mantle redox conditions that resulted from comparisons of the earlier qualitative models has, if anything, become more pronounced with the attempts at quantification utilizing thermodynamic fO_2-geothermometry or IOF measurements. This dichotomy for mantle samples is summarized in Figures 2a and 2b: in both graphs the upper, or more oxidized, envelope surrounds the data for the ilmenitic xenoliths, and a lower envelope surrounds data for silicate xenoliths. More specifically, in Figure 2a the upper envelope is composed of 40 xenoliths for which the fO_2 relationships were calculated thermodynamically from electron microprobe analyses (Eggler, 1983). The lower envelope encloses data from IOF measurements made at the U.S. Geological Survey at Reston (Sato's Laboratory) and at Temple University (Ulmer's laboratory) on about 40 mineral fractions or whole-rock samples of xenoliths. Note that in both figures the 1-bar values of the (WI) and (QFM) buffers are given for reference. (The pressure effect on these buffers is discussed in a later section and also in Fig. 10.) There is a difference of at least three orders of magnitude between the data envelopes of these two diagrams.

Arculus and others (1984) recognized the redox separation of

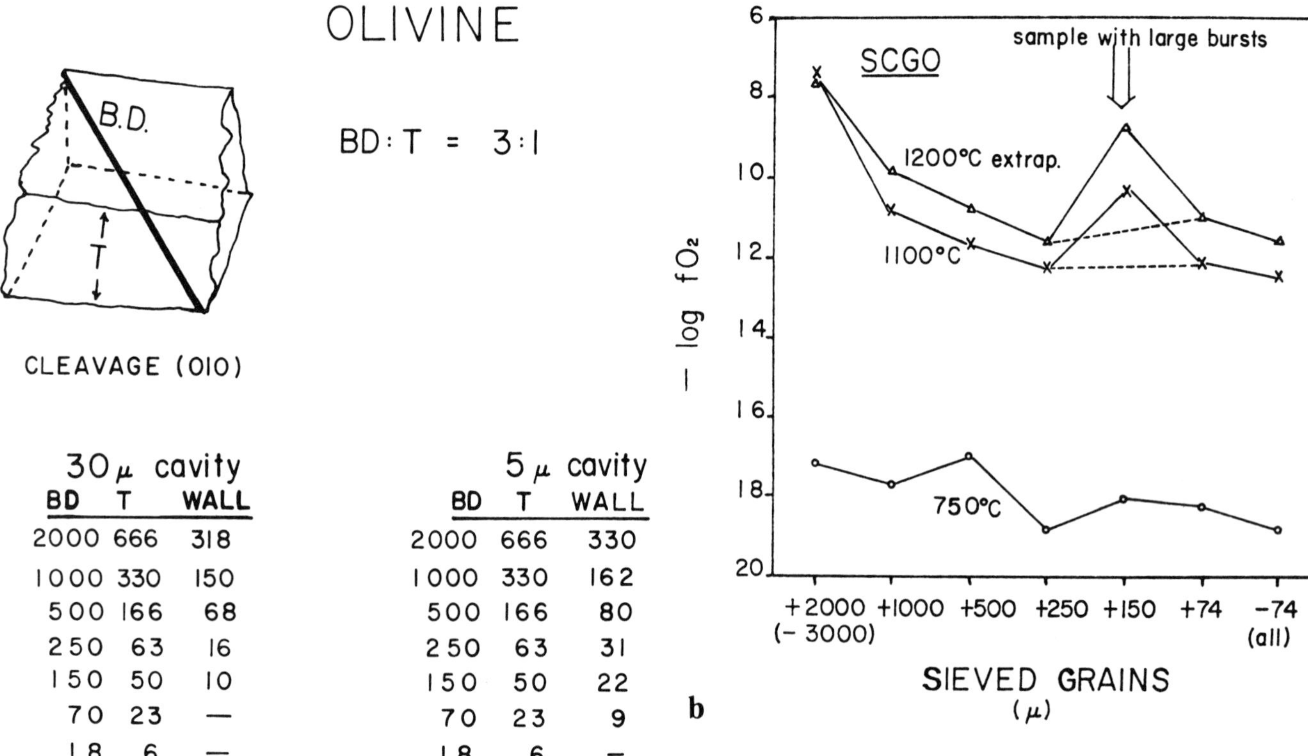

Figure 3. a. Schematic of olivine cleavage influence on retention of fluid inclusion cavities centrally located with a grain of body diagonal (BD) and thickness (T) as shown. Olivine studied seemed to fragment on crushing into grains with BD:T ratio of about 3:1. Measurements in microns; WALL dimensions listed are minimum wall thicknesses remaining for centrally located spherical cavity. Note that grinding to sizes as small as sieve sizes (BD) of 70 microns will not free all centrally located 5 micron spherical fluid inclusions. b. Effect of grain size on determined IOF value at various temperatures for Type I San Carlos gem olivine (Fo_{92}). Note that as grain size decreases, influence of decrepitation of CO_2 inclusions becomes less. See text for further details.

Type A and B xenoliths and discussed some possible reasons for it. The goal of our group effort is to examine the whole problem of the dichotomy in mantle redox data in order to understand the possible difference between measurement and calculation techniques, the various IOF data types that have nonexplicit interpretations, the various types of IOF data that have explicit interpretations, and the role of any inherited source xenolith inhomogeneity.

FACTORS AND ASSUMPTIONS INVOLVED IN MANTLE REDOX RESEARCHES

The IOF technique

Long and detailed discussions of the experimental technique for intrinsic oxygen fugacity measurements were provided by Sato (1971), Sato and Valenza (1980), Arculus and Delano (1981), and Elliott and others (1982). It is sufficient to note that the many pitfalls of the method have been remedied. We believe that all IOF data in Figure 2 are free of mistakes in experimental technique, although there are several idiosyncrasies of mantle samples that make them uniquely difficult to examine by IOF techniques; the resulting problems are worth discussing.

Fluid inclusions. The xenolithic silicate minerals have been studied in great detail for the presence of fluid inclusions (e.g., Roedder, 1965, 1984). The influence of fluid inclusions on IOF data was studied in our laboratory on two types of olivine-rich xenolithic nodules collected from the basanite flows at San Carlos Mesa in Arizona. Both Type I olivine (Fo_{92}) and Type II olivine (Fo_{77}) (classification of Frey and Prinz, 1978) were found to contain fluid inclusions. These inclusions can have profound effects on the IOF data collected for these olivines.

The predominant cleavage for olivine is (010), as shown in Figure 3a. If olivine is crushed to various grain sizes in preparation for IOF work, it need not release all of the fluid inclusions, even if grain sizes of <70 microns are used, as is shown in the figure. Note that the grain size is effectively controlled by the body diagonal (BD) of the grain, and note that cleavage typically produces grains whose thickness (T) is about one-third of the BD. A centrally located fluid cavity smaller than 5 microns will not be released at a grain size of 70 microns; this makes it imperative to prove that the material is inclusion-free before doing the IOF work.

A study of San Carlos megacryst olivine (SCGO) of Type I demonstrates the effect of fluid inclusions. The data show that the measured fO_2 is strongly dependent on the amount of fluid inclusions. In Figure 3b, grain size is plotted against measured fO_2. As expected, the smaller grains have released a larger proportion of their inclusions. However, as grain size decreases, the influence or effectiveness of CO_2-release becomes less, with the measured fO_2 values starting to plateau at grain sizes smaller than 250 microns. At high temperature there is a spread of four orders of magnitude in measured fO_2, depending on the grain size used. The results portrayed in Figure 4 suggest that with extreme care in selecting each grain of the -150 ± 74 micron size sample ("hand picked," thus avoiding inclusions by microscopic examination of each grain), a sensibly coherent and relatively strongly reduced IOF plot is obtained for the xenolithic olivine, as compared with a similar grain-sized fraction of the matrix basanite (Koseluk and others, 1979). The lack of intersection of these two IOF plots agrees with the hand-sample observation of the olivine being xenolithic, i.e., not in fO_2-T equilibrium with the basanitic matrix.[1]

The linearity of the hand-picked SCGO sample (Fig. 4) was possible because of the careful exclusion (done optically) of fluid inclusions. In contrast, grains containing fluid inclusions (Fig. 3b) show oxidized values. Roedder (1965), for this very same locality (then called Rice Station at Globe, Arizona), has proven that the fluid inclusions are nearly pure CO_2 and that at least some of the inclusions formed at as much as 5 kbar of total pressure. It could be that these CO_2 inclusions are exsolved from the host phase; i.e., the CO_2 is a component that belongs in mantle IOF measurements. However, this does not seem plausible for reasons discussed by such authors as Pasteris (1987).

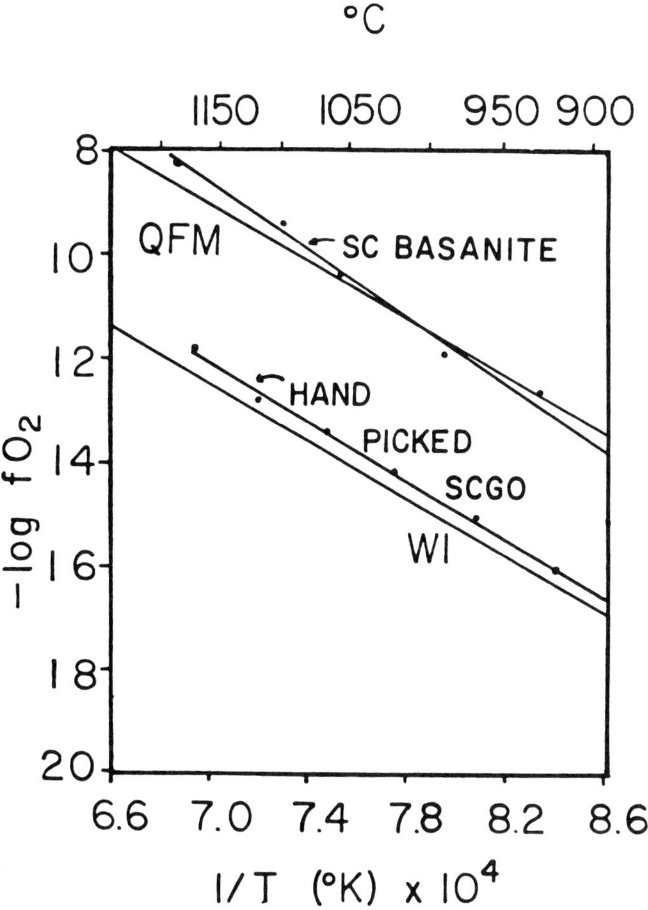

Figure 4. IOF data for Type I hand-picked San Carlos gem olivine (SCGO) compared to host basanitic basalt whose eruption brought these peridotitic nodules to surface. Lack of IOF intersection agrees with hand sample appearance of the olivine being xenolithic. Hand picking greatly reduces the fluid inclusion content.

If a sample containing a number of these fluid inclusions is placed in the stagnant, argon-filled IOF cell and allowed to decrepitate, the CO_2 released into the 2-ml cell volume could dominate the fO_2 measurement by forming a CO_2-argon gas mixture. This mixture, depending on both the pCO_2 and the temperature, will either react with the silicate to produce a new, "blended" signature for the assemblage (solid + gas), or the CO_2-argon mixture will not react with the silicate but be read itself as the "new" unknown. In either case, the original (i.e., intrinsic) oxygen fugacity has not been read. The exact behavior of such a burst is shown in Figure 5 for Type I olivine. As this sample (+150 microns with many inclusions) was first heated, inclusion bursting occurred below 800°C. Two critical temperatures, T* and T** are discussed below. From about 800°C to T*, no further decrepitation occurred, and the CO_2-argon mixture became the material being read by the IOF technique. At T*, reaction of the CO_2-argon gas mixture with the olivine began to occur and continued to T**. Above T** the IOF technique is reading the re-equilibrated olivine-CO_2-argon gas mixture at 1 atm, i.e., the total pressure of the IOF test.

The following statements generalize what Figure 5 illustrates:

1. Fluid inclusions may burst by decrepitation during IOF heating (the acronym for this condition is DICEY, which stands for demonstrable inclusion contribution: erroneous yield).

2. At some T and at some partial pressure of the fluid inclusion species, a reaction with the solid sample may occur; T* will be the symbol used for this temperature throughout the rest of this chapter (also referred to as DICEY data: see 1 above).

3. At some temperature higher than T*, the fluid species–

[1]Many lines of evidence support disequilibria: from the lack of reaction rims to the isotopic systematics between Nd-Sm, Rb-Sr in kimberlitic xenoliths (Richardson and others, 1984), as well as Pb-U in Cenozoic volcanic xenoliths (Menzies and others, 1983).

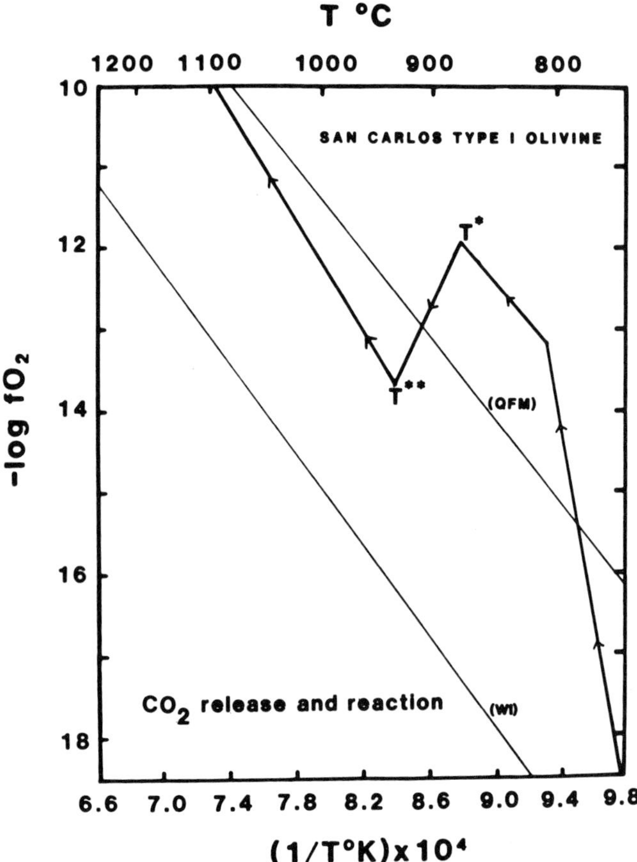

Figure 5. IOF data for San Carlos Type I olivine without selection to avoid fluid inclusions. CO_2 released by decrepitation while heating to 800°C mixes with argon in IOF cell between 800°C and T*. At T*, CO_2 begins reacting with olivine and continues to react until T**. From 800°C to T*, IOF technique is reading argon-CO_2 gas mixture in stagnant IOF cell. From T** upward, IOF technique is reading re-equilibration between released CO_2 and olivine. Text explains why these are called DICEY data.

argon mixture may react with the solid and come to a new equilibrium, but the IOF plot above this temperature will be recording a new re-equilibrated fO_2-T signature; this higher temperature is hereafter designated T** (also referred to as DICEY data; see 1 above).

4. The partial pressure of the fluid inclusion species may not be sufficient for fluid-solid reaction; i.e., the species population concentration may not be kinetically effective for solid-gas reactions in the time frame of the IOF studies (usually not more than 30 hr); in this case, no T* nor T** slope break appears in the data (the acronym for this condition is KRIPPTIC, which stands for kinetically retarded inclusion partial pressure that imposes caution).

5. Even if the partial pressure of the fluid does not cause reaction with the solids, the fixed volume of the IOF argon-filled cell can be overpowered by the fluid of the inclusion, and the IOF data are not representative of the mantle (also falls into the category of KRIPPTIC data; see 4 above).

6. Even without the obvious problem of T* or T** slope breaks in an IOF pattern, the proof that the IOF data are not merely a reading of a fluid-argon mixture at one bar in the cell remains with the researcher. (DICEY and KRIPPTIC effects must be avoided by the researcher.)

The data in Figures 6 and 7 further illustrate these six points. Several Type II composite xenoliths from San Carlos were used for these studies (Weiss and others, 1985); these xenoliths are unusually fresh peridotitic xenoliths with pyroxenitic layers. The objective was to separate silicates and oxides from the xenolith to prepare monomineralic separates from discrete layers. These samples were to be used for IOF geothermometry utilizing the mineral intersect technique (Sato, 1965). The data in Figure 6 are for several mineral separates: Ol = olivine Fo_{77}; Sp = spinels (magnetite-hercynite-spinel solid solution); and Cpx = clinopyroxene (aluminous augite). The separated materials were split and analyzed for IOF both at Temple University and at Sato's laboratory (USGS, Reston). Obviously no mineral intersects were found in the data from either laboratory despite the fact that these grains were in intimate contact in the hand samples. It could be argued that the pyroxene layer was not in equilibrium with the peridotitic host matrix, but the spinel within the peridotite also does not intersect its coexisting olivine in the IOF plots. All of the fO_2-T data are statistically subparallel, but are also similar in slope to CO_2-CO gas mixtures, as calculated by Deines and others (1974).

To interpret these results, it was necessary to estimate by point counting the number of fluid inclusions in the olivines involved. It was optically estimated that, in a 25-mg sample of Type II olivine-separate used for the IOF runs, there were between 10^6 and 10^7 fluid inclusions of the 1- to 5-micron diameter size. As observed from the calculated data in Figure 7, the influence of the decrepitation of this number of inclusions on the 2-ml argon-filled volume of the IOF cell (at 1,000°C and 1 atm total pressure) would be the creation of a log pCO_2 of –3.5 to –4.0 within the IOF cell.

The data in Figure 6, when examined in the light of the six points discussed above, are therefore considered to be KRIPPTIC data. To prove that the KRIPPTIC condition was involved, Type II olivine from these xenoliths was subjected to thermogravimetric analyses (TGA) at 1,100°C under controlled argon–CO_2 gas mixing (Darken and Gurry, 1945). The TGA results show that the olivine at 1,100°C does not react with argon–CO_2 mixtures until the log pCO_2 reaches higher than –4.0. Thus, the inference is that the subparallel, nonintersecting data in Figure 6 are indeed KRIPPTIC data; i.e., the IOF data are reading the CO_2–argon mixtures. Thus, the data are not re-equilibrated silicate values, and neither are they the mantle redox memory.

By comparison, the data in Figures 3a and 4 are for Type I olivine that was hand-picked from a nonlayered nodule. Point counting in the original bulk olivine showed fluid inclusions at

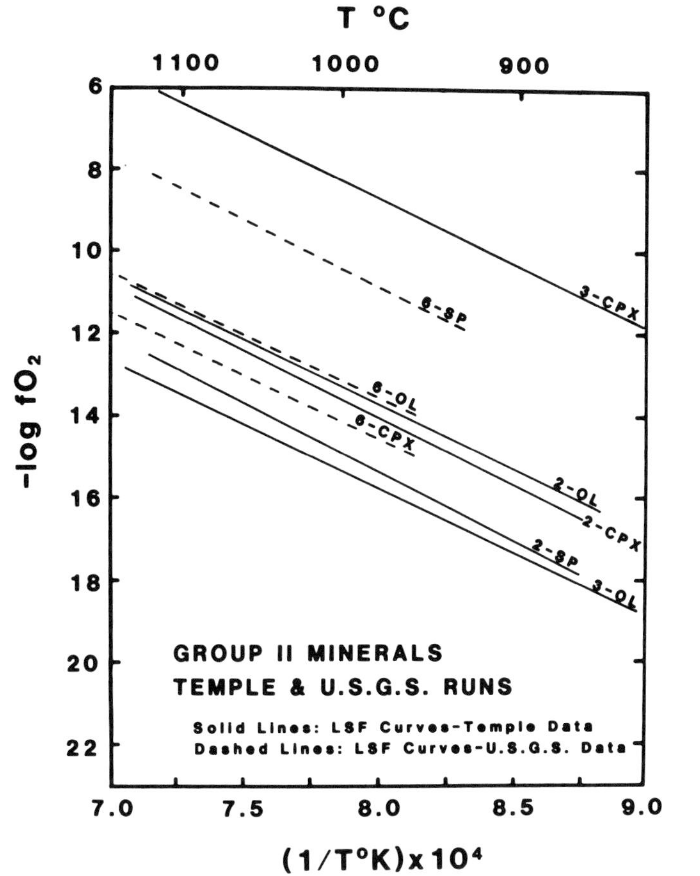

Figure 6. Minerals from Type II composite xenoliths from San Carlos run both at U.S. Geological Survey, Reston, and at Temple. No mineral intersects were found; least-squares fit (LSF) slopes are all subparallel to each other and to CO_2/CO equilibria. Each line was obtained from more than a dozen points. Text discussion concludes that these data are only reading the CO_2 released from fluid inclusions. The CO_2-argon mixtures formed in IOF cell have too small a pCO_2 to react with olivine as happened between T* and T** in IOF study shown in Figure 5. Text explains why these are called KRIPPTIC data. Reproduced from Weiss (1984).

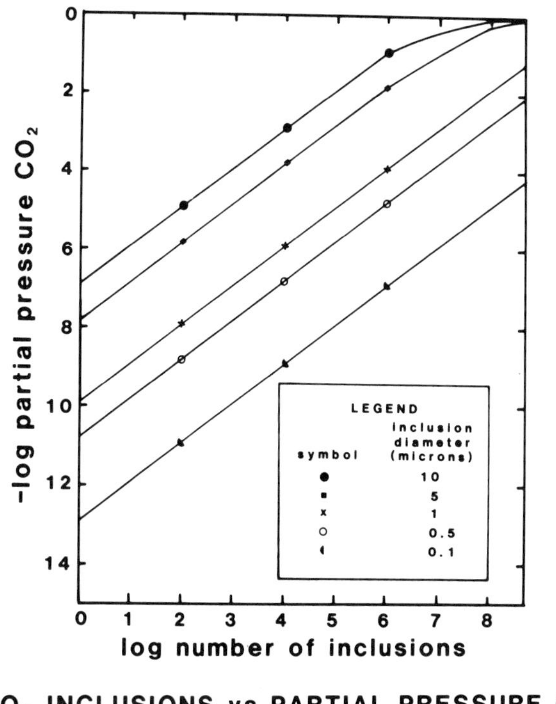

Figure 7. Calculated pCO_2 that will result at 1,000°C by release of given number of CO_2 fluid inclusions (at 5 kb) of given size into 2 ml of argon-filled dead space in typical IOF cell. Text points out that for San Carlos Type II olivine, $-\log pCO_2$ can be as much as 4 for 25 mg of sample being used in IOF work of Figure 6. Reproduced from Weiss (1984).

least two orders of magnitude less than the Type II olivine from layered nodules (Fig. 6). Even for the bulk Type I olivine, examination of Figure 7 leads to the conclusion that the log pCO_2 would be less than –6 at 1,000°C and 1 atm in the IOF cell. However, hand-picking of this bulk olivine removed many of the grains continuing inclusions, and crushing to –74 microns still further eliminated larger inclusions. Because the data in Figure 3b indicate that the fO_2 became less dependent on grain size at –74 microns, and because any inclusion contribution into log pCO_2 can be estimated to be <-10, the IOF data in Figure 4 can be entered for consideration as the 1 bar memory of mantle redox conditions, as further discussed below.

The CO and CH_4 components of fluid inclusions in mantle xenoliths are reviewed by Roedder (1984). No systematic investigation of their effects nor of the partial pressures that would produce re-equilibration with their matrix solids have yet been worked out. In every mantle peridotite that was examined from about 20 geologic settings, the recent work of Mathez and others (1984) confirms the presence of carbon, most of which exists in CO_2-rich fluid inclusions. Hence, the questions of re-equilibration of fluid with the sample or fluid overpowering the stagnant IOF cell gas must be addressed by researchers to insure the validity of their IOF work.

Elemental carbon. The effect of elemental carbon on IOF results must also be examined in the light of reported inorganic, noncarbonate carbon in xenoliths (e.g., Hatton and Gurney, 1977; Freund and others, 1980; Pasteris, 1981; Duba and Shankland, 1982; Ulmer, 1983; Ulmer and others, 1985a, b).

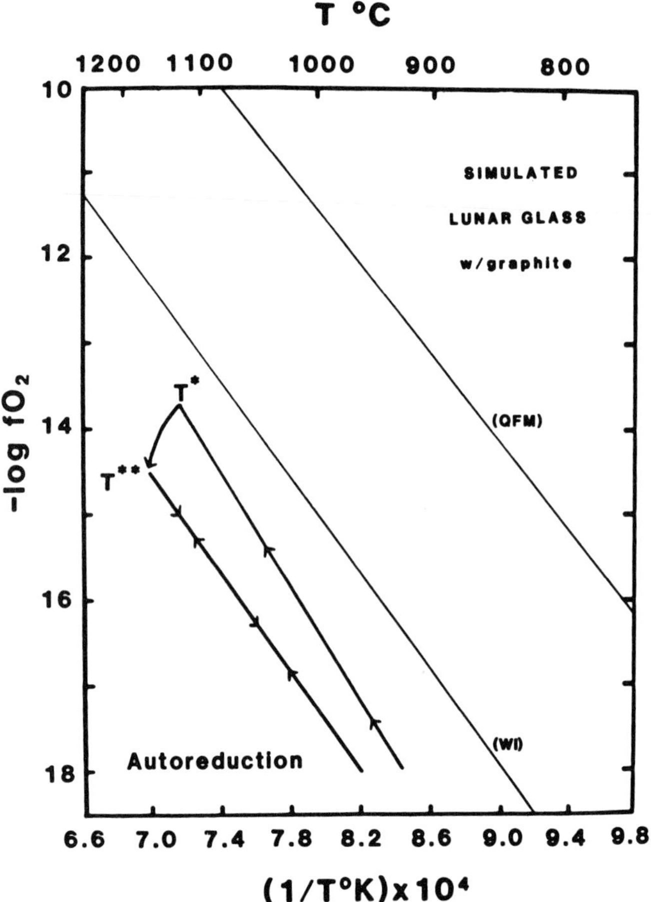

Figure 8. Typical pattern of known graphite-caused autoreduction in low Fe^{3+} synthetic lunar glass. This test was a Temple University duplication of data obtained by Sato in his laboratory at Reston. T* and T** explained in text.

In an intralaboratory comparison test of synthetic low–Fe^{3+} lunar glass (Sample CG 68415) prepared by Minkin and others (1976) and spiked with 50 ppmw of well-crystallized synthetic graphite, both the U.S. Geological Survey's (Sato's) and Temple's laboratories obtained very similar results. Temple's results are shown in Figure 8. In this sample with known content of carbon, the initial IOF pattern indicates that the glass and the carbon remain unreacted until a temperature, T*, is reached. At this temperature (in this case, 1,120°C) the carbon begins to react with the solid sample, generating CO, which in turn causes the IOF signal between T* and T** to drop to more reduced values[2]. As the CO comes to equilibrium with the glass and carbon reaction is complete, all subsequent IOF determinations on this sample, even with up- and down-temperature reheatings, produce no new IOF slopes, but track on the reduced fO_2-T slope shown. This behavior has been termed "autoreduction" and was first reported by Sato and Valenza in some of the Skaergaard plutonic rocks (1980). To generalize, samples low in ferric iron and containing carbon may be expected to re-equilibrate at some temperature (T*), with the production of a CO–CO_2 gas mixture rich in CO, with the subsequent more reduced fO_2-T signature for the solid sample.

However, if the original sample containing carbon has a high ferric iron content [(WM) or (QFM)], the resulting CO–CO_2 gas mixture may be rich in CO_2, and the resultant IOF pattern may be one that is best described as auto-oxidation (Koseluk and others, 1979). An IOF pattern for a kimberlitic clinopyroxene mineral separate taken from an ilmentite-clinopyroxene nodule from the Monastery pipe, South Africa (supplied by J. Gurney) is shown in Figure 9. Note that the original up-temperature IOF slope is near the (WI) buffer, but that a slope change occurs between T* and T**. Subsequent heatings, either up- or down-temperature, after T** has been reached, cause no further slope changes in the IOF signature. As explained in a later section, this re-equilibration may also be related to the oxidation of Ti^{3+} by ferric iron in the sample, but it is likely more related to the 40- to 350-ppmw nonorganic, noncarbonate carbon that we have found by LECO pyrolysis to be present in most ilmenitic samples.[3]

If the emplacement of the xenolith to environments shallow enough to have quenched the xenolith below its solidus is only a matter of a few hours (e.g., see Szekely and Reitan, 1971; McGetchin, 1968; Ozawa, 1984), this quench would have prevented auto-redox re-equilibration in the xenolith. Thus, when autoreduction or auto-oxidation is observed in IOF research, it is inferred to be the result of reheating the sample at 1 atm for tens of hours above the solidus.

This realization leads to the idea that the first heating of a xenolith at least up to T* may represent a 1-bar read-out of the mantle redox memory. That the IOF heating may actually be producing a redox re-equilibration, retrograding because of pressure relaxation from tens of kilobars to 1 bar during IOF testing is examined in the next section.

Retrograde re-equilibrations. The dependence of redox reactions on pressures as high as those in the mantle is only beginning to be understood. The roles of pressure change on redox buffers, on cation redox couples such as Ti^{3+}–Fe^{3+}, on lattice defect concentrations, and on fugacity coefficients of mantle fluid phases are all too imperfectly known to predict whether a mantle sample has survived its ascent intact, as well as a 1-bar

[2]When only 50 ppmw of carbon are present in 25 mg of sample, the sample will control the oxidation products of the carbon. Hence, the gas formed by the oxidation of the carbon in this synthetic lunar basaltic glass [approximately buffered itself at (QFI)] will be rich in CO (Deines and others, 1974).

[3]LECO Corp. (St. Joseph, Michigan) is a commercial laboratory and manufacturer of step-pyrolysis equipment for the determination of carbon by oxidative heating of samples to temperatures as high as 2,000°C, and subsequent measurement of the resultant CO_2.

"recooking" during the IOF research. These problems have been discussed by Ulmer and others (1978, 1980), Duncan (1983), Sato (1984), and Arculus and others (1984).

As shown in Huebner's review (1971), the effects of pressure on the common redox buffers are that the silicate and oxide buffers become more oxidized with increasing pressure. The changes, ranging from 0.005/T to 0.098/T, increase in log fO_2 per bar increase in pressure (temperature in Kelvins). For considerations in shallow crustal plutons there is thus little need for pressure correcting most fO_2 buffers. In contrast to these silicate and oxide buffers with very low oxygen, the effects of pressure on buffers involving large amounts of fluid phase or a carbonate phase are far greater because of the large ΔV of such buffer reactions (Fig. 10). The (EMOG) or (EMOD) buffers involve enstatite, magnesite, olivine, and, respectively, graphite or diamond (Eggler and Baker, 1982); the (CCO) buffer is the pertinent equilibrium in the C-O system (Woermann and others, 1977). The dotted line in each graph is the 1-bar location of the (QFM) buffer; it is clear that, as pressure increases into the mantle range of 30 kb or more, all buffers need pressure correcting. When the pressure is corrected, carbon-containing assemblages become even more oxidized than the pressure-corrected oxide and silicate buffers such as (QFM). When a buffered assemblage that has been poised at mantle pressure is reheated at one bar, a retrograde re-equilibration could occur that would register as an inappropriate IOF value. The auto-oxidation and autoreduction patterns already discussed show the role of carbon (Figs. 8, 9) and indicate that the IOF pattern itself may reveal whether the laboratory heating is causing retrograde re-equilibration. Sato (1984) pointed out that measured, highly reduced values that do not show retrograde patterns are suspect in that they may have retrograded on ascent from the mantle. These would be measured during IOF research as a de facto sample already re-equilibrated in nature (The acronym for this problem is REPROOF, which stands for retrograded pressure recording overprinted oxygen fugacity.)

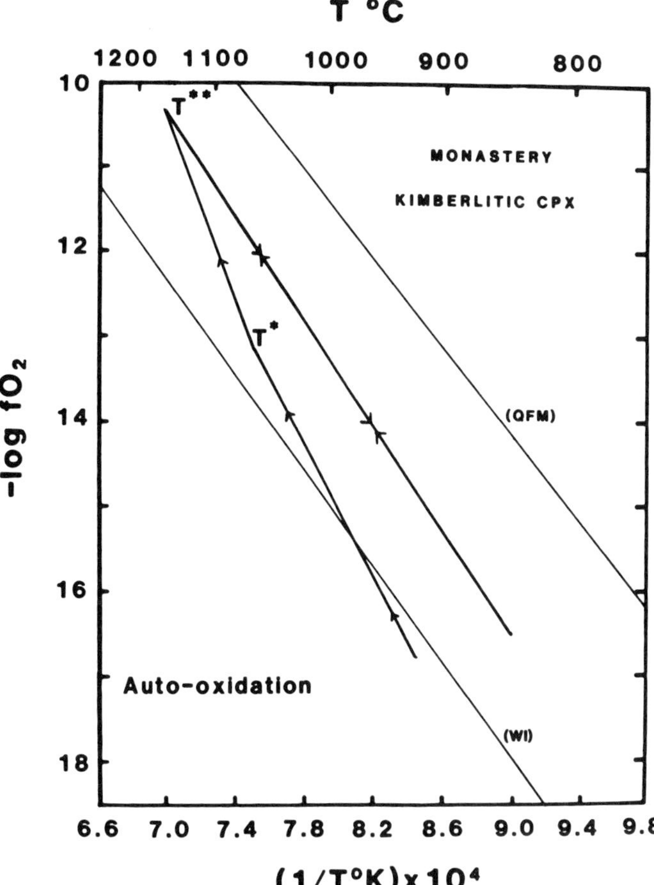

Figure 9. Pattern of suspected auto-oxidation resulting from either oxidation (T*) of carbon and reaction of CO_2 with sample up to T**, and/or auto-oxidation of suspected Ti^{3+} with Fe^{3+}. See text for details.

A second type of retrograde phenomenon has been observed in samples containing high titanium levels. The work of Virgo and others (1976a, b) done in Sato's laboratory on high-titanium garnets from mantle settings indicated that, as their samples were heated, a reaction between Fe^{3+} and Ti^{3+}, stoichiometrically inferred, occurred. This resulted in a decrease in Fe^{2+} (confirmed by Mossbauer data) and the presumable oxidation of the titanium to 4^+. Their study of melanite garnets from several nepheline-bearing localities is exemplified by garnets weathered from a massive nephelnite agglomerate at Rusinga near Lake Victoria, Kenya. In Figure 11, their IOF data for two sample splits, a and b, are plotted on the same axes. For sample split b, Virgo and others (1978a, b) show both $T_{c(b)}$ and a $T_{max(b)}$; in their interpretation these are, respectively, at least similar to our T* and T**. Virgo and others (1967a,b) believed that the garnet is stable in the IOF work until $T_{c(b)}$ is reached, at which point Fe^{3+} and Ti^{3+} begin to react to produce Fe^{2+} and Ti^{4+}. To whatever T_{max} they next heated and then cooled, they retraced, on cooling, a different IOF pattern because of the iron-titanium re-equilibration. As long as they did not exceed their previous T_{max}, the pattern was reproducible, but if a new, higher T_{max} was reached, then a new cooling pattern was the result. Sample split *a* was used to test this interpretation to still higher values of $T_{max(a)}$.

This behavior is very similar to what we have observed in our data on a clinopyroxene-ilmenite nodule from Monastery (Fig. 9; Moats and others, 1985). Unfortunately, neither the garnets nor the clinopyroxene were analyzed by wet chemical techniques because of low availability of sample.

That the Rusinga garnets or ilmenitic assemblages could survive their transport to the surface without having experienced this retrograde re-equilibration would seem to argue that these garnets have some memory of the mantle redox condition. But how can researchers be sure that the data have not been somehow erased, i.e., are not already partially adjusted by retrograde reaction during their ascent? At least the pattern of the IOF data allows consideration of whether this kind of caution is needed in

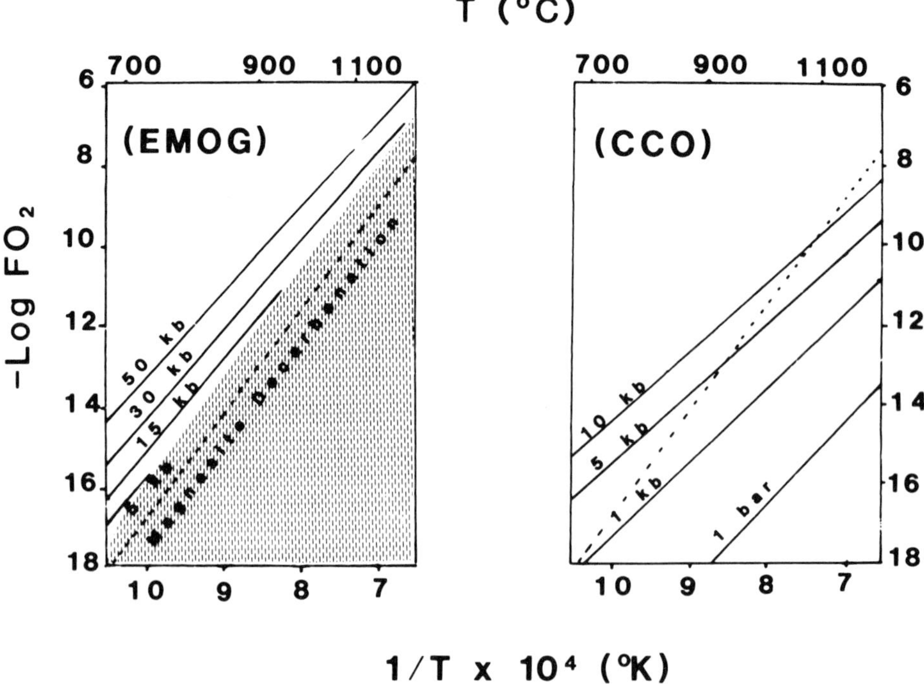

Figure 10. Pressure dependency of assemblage enstatite-magnesite-olivine-graphite (EMOG) for fO_2 buffering (Eggler and Baker, 1982) and pressure dependency on C-O_2 equilibria, i.e., (CCO) buffer (as calculated by Sato and Valenza, 1980). Dotted line marks 1-bar position of (QFM). The shaded area for (EMOG) is disallowed for buffering because of instability of magnesite. The strong pressure effect on both these buffers can cause retrograde reactions, as discussed in text.

interpretation. For example, the gas-mixing redox study of McMahon (1984) on Monastery kimberlite is very careful work, but may be subject to this same question.

The role of defects in possible mantle redox memory and/or retrograde reactions during IOF work has been directly examined, but without conclusive results. $Ag_{60}Pd_{40}$ capsules of both (WI) and (QFM) synthetic buffers were held for 24 hr at 30 kb at 1,100°C in a piston-cylinder apparatus (with the assistance of B. Mysen at the Geophysical Laboratory). Thirty kilobars of pressure would shift the fO_2 of these buffers by 1.20 and 2.01 log units, respectively, in an oxidizing direction (Huebner, 1971). These buffer capsules were run as the "sample," using Pyrex sleeves and graphite furnaces. The quench was standard: power-kill followed by pressure-release, with the piston cylinder cooling-water running. The temperature fell below 700°C within 2 min. These samples were then extracted and run as "unknowns" in the regular IOF procedure. The 30-kb effect on the buffers, if quenched in, should have displaced the IOF readings above the experimental uncertainty of ±0.10 log units in the IOF procedure. Both these buffers read back out within experimental error (±0.10 log units) of their 1-atm values. It could be argued that the starting materials had first been synthesized at 1 bar and thus where not inclined to acquire a different level of defects in a closed capsule, even if taken to high total P and T. It could be further argued that, even if samples became defect at high P and T, the defect concentration was not quenchable. Hence, only conductivity (e.g., Duba, 1979) or theoretical studies (e.g., Duncan, 1983) are available thus far for use in assessing what role defect concentrations play in poising of fO_2 in situ in the mantle.

Multimineral intersects. Sato (1965) originally proposed the separation of a rock into its constituent mineral phases and the acquisition of IOF data on each phase. If any of the fO_2-T curves crossed, the point of intersection could have possible value as petrogenetic fO_2 and/or as geothermometric data. If IOF studies can be compared to other geothermometry data with reasonable agreement, then high confidence can be argued for the IOF data.

The results of such a study on a xenolith are shown in Figure 12. The LBM-11 nodule from the Matsoku kimberlitic diatreme in South Africa contains (in modal percent): 47.5 olivine, 47.0 orthopyroxene, and 5.0 garnet, all of whose compositions are given in Cox and others (1973). These phases were separated (by Ben Harte), and the fO_2 trend for each separate was determined as shown in Figure 12. Olivine and garnet were least-squares fit to a single line (their exact data points are not shown in order to simplify the diagram); the exact data for the orthopyroxene are

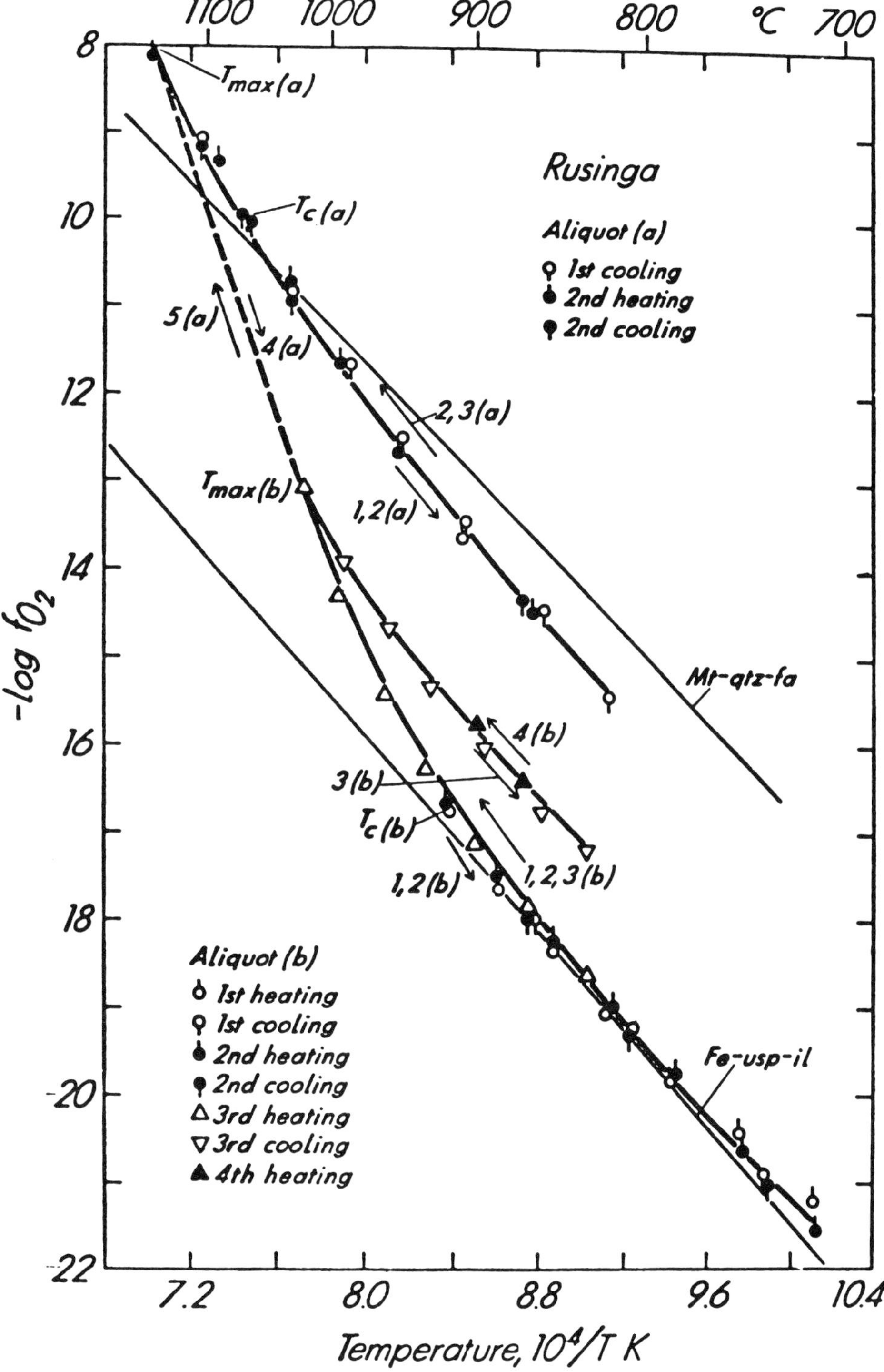

Figure 11. IOF data for Rusinga garnet from sample splits, a and b. (Reproduced from Virgo and others, 1976, a, b.) Patterns that develop at $T_{c(b)}$ and at $T_{max(b)}$ are analogous to pattern between T* and T** in Figure 9. See text for explanation of role of Ti^{3+} in patterns that developed. (Fe-usp-il) is calculated buffer equilibria for assemblage iron-ulvospinel-ilmenite at 1 bar; other buffer line shown is (QFM).

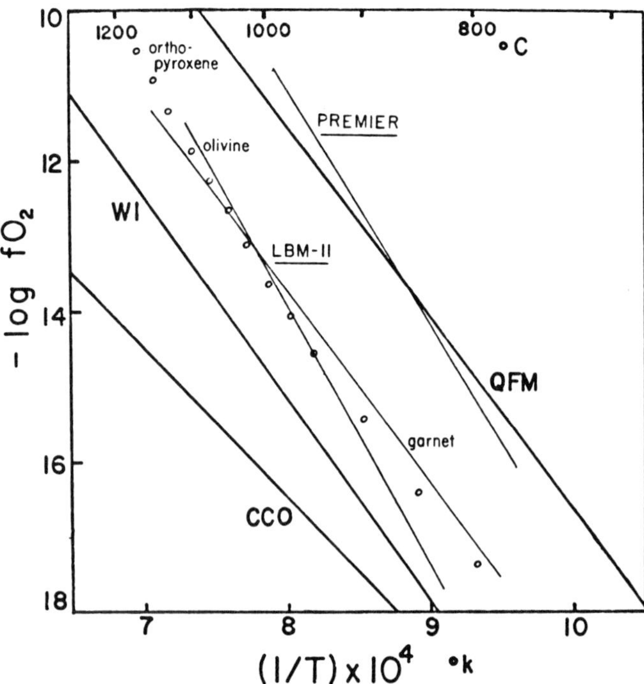

Figure 12. IOF data for three mineral separates from LBM-11 nodules from Matsoku, South Africa. Fourteen data points each for olivine and garnet were least-squares fit to lines shown; data points were left off for clarity. Orthopyroxene data points shown as small open data points. Note the three-mineral intersect is fairly coherent and defines a best temperature of 1,030°C which is in excellent agreement with two geothermometric calculations by F. R. Boyd (personal communication, 1985) for same nodule. Premier data explained in text. These data are called ACME data, as explained in text.

shown without their least-squares fit line. The data for whole-rock gray kimberlite from the Premier Mine (Ulmer and others, 1976) are shown for comparison in that the Matsoku nodule was a xenolith from kimberlite. For the Matsoku nodule, the triple mineral intersect from the three least-squares fit lines define conditions as follows: $-\log fO_2 = 13.0 \pm 0.2$ and 1,030°C ± 5°. This IOF geothermometric temperature is in close agreement with the 1,035°C (and 43.5 kb) determined for the same nodule with the Boyd-Davis geothermometric scheme and also is still reasonably close to the TEMPEST geothermometric calculation (Finnerty and Boyd, 1984) for this nodule, i.e., 1,050°C (and 47.6 kb) (F. R. Boyd, personal communication, 1985). Three totally separate techniques thus converge on the geothermometric value for this nodule, and the IOF fO_2 value thereby gains creditability. (The acronym for these type of data is ACME, which stands for anion-cation measured equivalence.)

Another such study for an extremely interesting eclogite nodule, HRV-247, from the Roberts Victor kimberlite of South Africa, is shown in Figure 13. This nodule contains diopside, garnet, copious graphite flakes, and numerous diamond octahedra. The silicate crystals are weakly zoned with MgO–rich rims for example, and the nodule was discovered to have more graphite at one end and more diamond at the other end (Hatton and Gurney, 1979). A thin slice of the graphite-rich end, sample F-4, was micro-quarried (pressure-flaked) for IOF work; separate garnet and diopside fractions free of visible graphite or diamond were prepared. Utilizing the pyroxene-garnet geothermometer of Raheim and Green (1974), Hatton and Gurney (1979) had concluded that the zoning of the silicates meant that the silicates had "crystallized over a range of temperature and pressure conditions which transects the diamond-graphite equilibrium curve at 1,020 to 1,140° and 42 to 45 kb" (p. 29). From the later Ellis and Green (1979) geothermometer, the silicates crystallized within a temperature-pressure range that transects the diamond-graphite curve at 980 to 1,040°C and 41 to 42 kb. These two calculated sets of values are shown as "ln k_d (H and G) and E and G," respectively, at the top of the figure. The IOF data are least-squares fit for data taken every 30°; they show that the garnet and clinopyroxene have overlapping IOF patterns between 1,010 and 1,120°C and between $-\log fO_2$ values of 12.3 and 14.2, respectively. Errors on the IOF data may be as large as ±5°C and ±0.2 log units fO_2.

The coincidence of the data is good; this is another example of ACME data. The lack of interference by carbon must be attributed to the large and separate nature of the carbon phases. Thus, these types of independently confirmed data lend some increased confidence to the IOF technique.

Gem-quality megacrysts. When microscopic techniques assure that essentially few solid or fluid inclusions exist in megacrysts, it is also possible to obtain IOF patterns on xenolithic materials that are of a high confidence level. (The acronym for such data is NICE, which stands for negligible inclusion components effects.)

Figures 14a,b,c show that the results of such studies of gem-quality megacrysts from the glassy matrix of potassic basalts from Central Mongolia on the Ehangai Highland. The chemical composition of these megacrysts are given in Table 1. Figure 14a gives IOF results for a pyrope with the least-squares fit line of:

$$-\log fO_2 = -3.4306\ (1/T \times 10^4) + 11.3329\ \ (K)$$

with a regression coefficient = 0.9982. Figure 14b gives data for an olivine with the least-squares fit line of:

$$-\log fO_2 = -3.3119\ (1/T \times 10^4) + 11.6088\ \ (K)$$

with a regression coefficient = 0.9979. A third example is given in Figure 14c for a sanidine megacryst that can be expressed by the least-squares fit equation of:

$$-\log fO_2 = -2.25096\ (1/T \times 10^4) + 2.780\ \ (K)$$

with a regression coefficient of 0.9886. The lack of auto-oxidation, autoreduction, T*, T**, or retrograde phenomena in

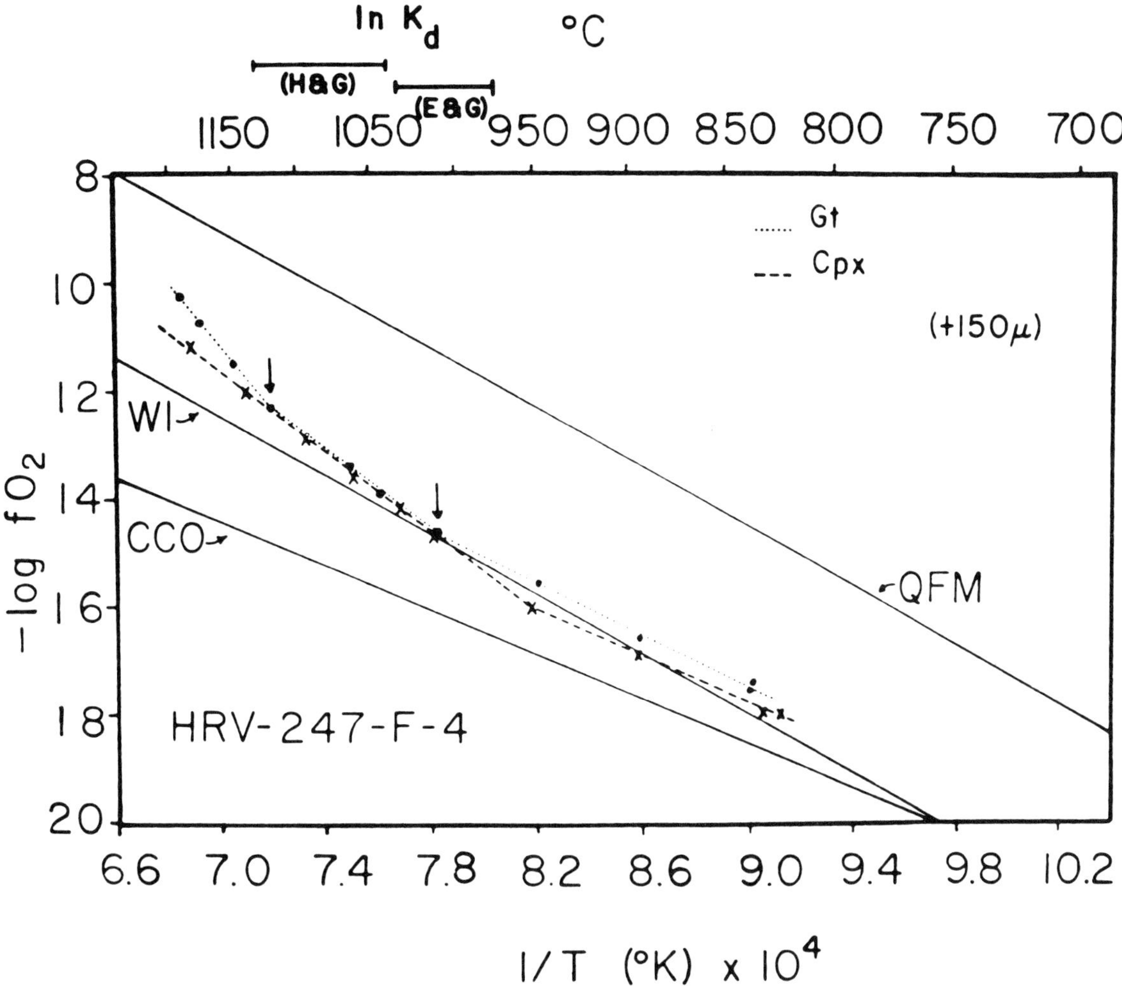

Figure 13. IOF data for garnet eclogite nodule from Roberts Victor Mine in South Africa (HRV-247). Spread in IOF data (the two arrows at 1,010°C and 1,120°C) for garnet-and-clinopyroxene overlapping portion is thought to be result of zonations within silicates. Pyroxene-garnet geothermometer also gives very similar range of temperatures for sample, as shown by two lines indicated "ln K_d." (H & G) represents data calculated by Hatton and Gurney (1979), utilizing Raheim and Green (1974) data for Fe-Mg geothermometry; (E & G) represents our calculation using Ellis and Green (1979) geothermometer, which includes correction for role of Ca on values. These data are also ACME-type data, as explained in text.

these IOF patterns, together with the microscopic evidence that there are few if any fluid inclusions, suggest that all these data may be of high confidence. Furthermore, the regression coefficients for a and b are also very high; even for c, the coefficient is still reasonably high.

Thus, the megacrystic nature of these samples prevents multiple mineral intersect IOF studies, but they have been used to produce some data that are probably accurate enough to elucidate mantle redox conditions. These would be classified as NICE data.

Therefore, it may be concluded that the initial problems in the IOF technique have, in terms of hardware, been overcome. At least for thc ACME and NICE samples, the problems that arise from the very nature of mantle xenoliths may have been sorted out.

Thermodynamic calculation approach

Eggler (1983) has presented a recent improvement and cross-testing scheme for a thermodynamically calculated geothermometer based on the equilibrium: $2\ Fe_2O_3 + 4\ FeSiO_3 = 8\ FeSi_{0.5}O_2 + O_2$. He has shown that even when a "fictive" silicate is proposed to be in equilibrium with an oxide in a xenolith that does not have the needed silicate, the results can be trustworthy

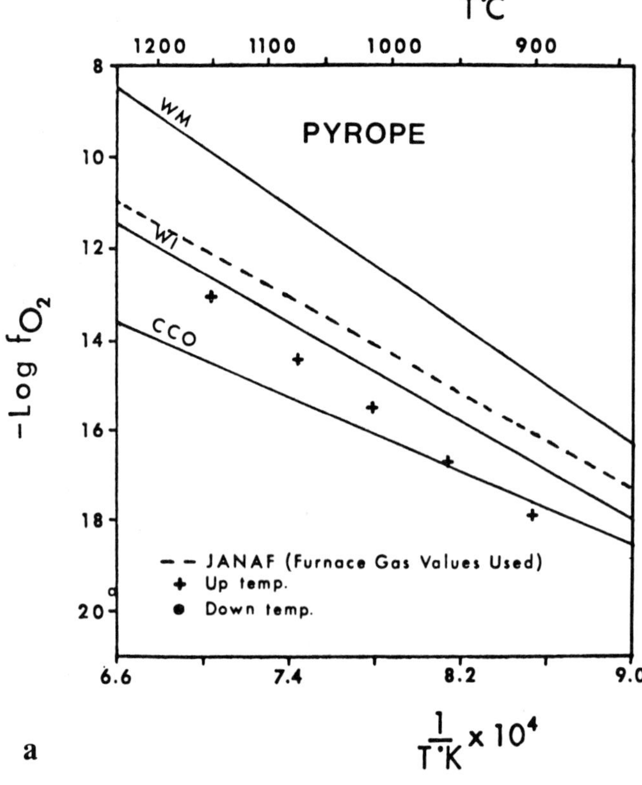

a

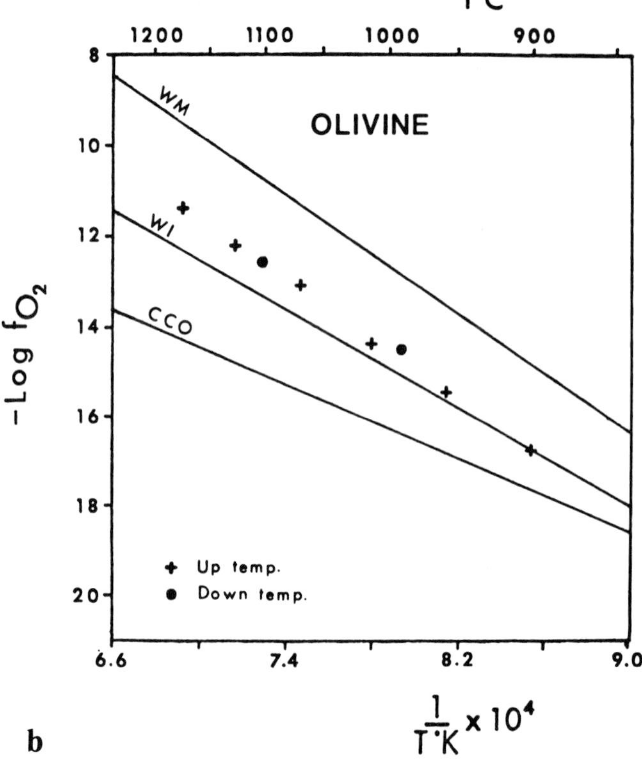

b

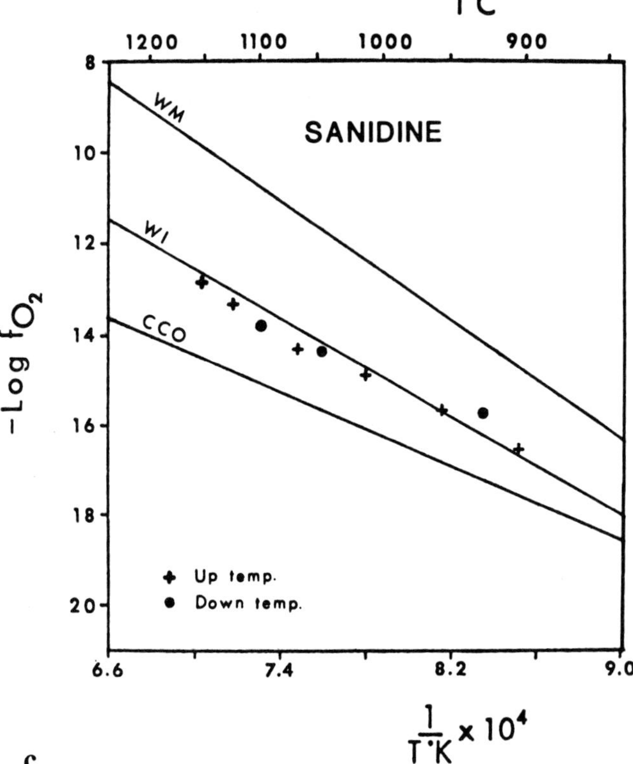

c

Figure 14. IOF data for Mongolian gem-quality megacrysts from potassic glassy basalt. See text for least-squares fit equations for each sample. a. No down-temperature data were available for pyrope due to loss of ZrO_2 cell by cracking during cooling; dashed line traces furnace gas being used so that gas leakage may be ruled out as problem for data. b. Both up- and down-temperature data obtained for this olivine. c. Both up- and down-temperature data obtained for this sanidine. As explained in the text, these data are classified as NICE data.

TABLE 1. WET CHEMICAL ANALYSES OF MONGOLIAN MEGACRYSTS*

Sample	Pyrope (wt. %)	Olivine (wt. %)	Sanidine (wt. %)
SiO_2	39.58	40.81	63.52
TiO_2	0.50	0.02	0.05
Al_2O_3	23.53		19.69
Fe_2O_3	1.91		
FeO	14.28	10.75	0.66
MnO	0.38	0.14	trace
MgO	15.34	48.25	0.15
CaO	5.45	0.03	0.73
Na_2O	0.28		3.26
K_2O	0.22		10.74
p.p.p.**			0.80
Total	101.47	100.00	99.60

*Analyses done at the Vernadsky Institute, Moscow.
**Loss of ignition.

for mantle assemblages involving ilmenite and silicates. The iron-titanium oxide geothermometry also has been recently improved by Lindsley and Spencer (1982). However, some of the assumptions involved in any thermodynamic scheme are still questionable and are deserving of some discussion.

Minor components. Recently Stormer (1983) has shown for the iron-titanium oxide geothermometer that "ignoring, or not analyzing for, minor components can lead to errors in excess of 150°C and 4 units of log fO_2" (p. 586). Since the original Buddington and Lindsley (1964) work on the iron-titanium oxide fO_2-T scheme, many workers, including Lindsley and his co-authors (most recently, Anderson and Lindsley, 1985) have refined the calculations to account for minor components. Stormer pointed out (1983) that among the various more modern approaches he reviewed, differences of not more than 30°C and less than 0.5 units of log fO_2 have been found, and that his computer program gave values in the middle of the range of the variation in modern schemes. Thus, for mantle samples, the clustering of ilmenitic assemblages shown in Figures 2a and b are considered reliable because at least the influence of minor components has been included.

Stoichiometry. The most questionable part of the thermodynamic calculations may well be the stoichiometric assumptions that must be made. These assumptions for mantle xenoliths rule out the existence of defects and of "unusual" valence states such as Ti^{3+}. That these can be very real problems is probably not argumentative, but to evaluate their impact on a calculated value is still impossible at this time.

Whereas Duba and many others have pointed out that the defect concentration is high and real in olivines (for example, see Schock, 1985; specifically, see also Schock and Duba, 1985), and whereas Darken and Gurry (1945) long ago showed that the defect concentration in magnetite affects the Gibbs free energy value, there is still too little quantitative treatment of this subject to allow corrections to the thermodynamic calculations, particularly at mantle pressures. Furthermore, in a pair of papers, Nakamura and Schmalzreid (1983, 1984) studied extensively the dependency of diffusion and point defects upon fO_2 and T in the olivine solid solution Mg_2SiO_4-Fe_2SiO_4. Even though they substantiate the important theoretical work of Kröger and Vink (1956) concerning non-stoichiometry relationships and thermodynamic equilibrium relationships between ΔG, i.e., fO_2 and defect concentrations, the needed data at mantle pressure is lacking. Duncan (1983) has tried to produce corrections, but these also are mainly empirical. Thus, the role of defects on calculated fO_2-T relationships involves errors whose magnitude cannot yet be evaluated.

The existence of Ti^{3+} in terrestrial material has been objected to in the petrologic literature. A search of the crystal-chemical literature (e.g., Burns, 1970) or the materials research literature (e.g., Magneli and others, 1961; Magneli, 1964; Anderssen, 1967a, 1967b), finds, however, that Ti^{3+} is commonly reported and not just from equilibration with metallic phases. Typical Ti^{3+} phases such as Ti_6O_{11} or $Ti_{11}O_{20}$ equilibrate at low fO_2 values. The equilibrium between these two phases is reported (Anderson and Kahn, 1970) at 1,100°C to be fixed at $-\log fO_2 = 17.5$, i.e., lower than the (WI) buffer at this temperature. However, the change in stoichiometry in the rutile phase from TiO_2 to $TiO_{1.975}$ takes place progressively with lowering of the fO_2 at any given temperature and is a complicated process involving repeatable hysteresis in the defect concentration, depending on whether it was obtained from oxidation or reduction of the sample (Merritt and others, 1973). Weighable amounts of Ti^{3+} have been produced at 1,300°C in Temple's thermogravimetric furnace (TGA) under conditions as mildly reducing as pure CO_2 ($-\log fO_2 = 3.42$). Unpublished TGA Temple data at 1,300°C imply that, for pure Ti-O materials, up to 1 wt. percent Ti^{3+} (as Ti_2O_3) can exist at 1,300°C under the redox conditions of the (WM) buffer.

The interpretation of the IOF patterns in Figures 9 and 11 requires the presence of some Ti^{3+} in that Mossbauer data (Virgo and others, 1976a,b) proved that Fe^{3+} was consumed in a closed system, most likely by reaction with Ti^{3+}.

Thus, the influence of the stoichiometric assumption on calculated fO_2 values cannot be properly evaluated at this time.

Disproportionation

The Le Chatelier Principle becomes involved for phases that can disproportionate to denser assemblages under increased pressure. The ideas that either 4 FeO or 2 CO can disproportionate at high pressure to Fe+Fe_3O_4 and C+CO_2, respectively, are not new. Mao (1974) reviewed the stability of wüstite in the lower mantle, and Bassett and Ming (1970) considered the data for Fe_2SiO_4 spinel. In both these studies, it was concluded that disproportionation of either wüstite or fayalitic spinel will not occur until mantle pressures or more than 150 kb were reached. The iron-titanium oxide or ilmenite-olivine geothermometers would thus seem free of disproportionation effects. Eggler and others (1979) have shown that CO and CO_2 behave similarly in terms of the melting relationships of peridotite. However, at room temperature little CO is reported in fluid inclusions (Roedder, 1984). A question arises: Is the lack of CO because of oxidation or is it the result of low-temperature disproportionation, i.e., a change in the CO/CO_2 ratio of fluid inclusions or mantle volatiles by the production of C and CO_2? Mathez and Delaney (1980) and Mathez (1984) proposed that fluids abundant in CO are not possible in the C-O system above moderate pressures because of graphic stability. Thus, the disproportionation of CO may need to be better studied. At the very least, it presents an uncertainty in buffer calculations that involve carbonation reactions of silicates or calculated equilibria in the C-O system.

Thermodynamic data at mantle pressures. The $P\Delta V$ terms of the Gibbs free energy expressions of fO_2-T relationships usually have been determined to only a few kilobars of total pressure. Research (including our own) that considers mantle redox conditions typically has used extrapolations of redox buffers to 50 kbar without hesitation. Consideration of just the

(QFM) buffer shows the fallacies in this practice: both the breakdown of fayalite to ferrosilite and the transformation of quartz to coesite before reaching even 25 kbar underscores the problem. Different polymorphs may have only slightly different free energies and hence only slightly different fO_2-T relations, but data do not exist for correcting the fO_2-T buffers to mantle conditions.

Sato (1984) has, for the first time, treated the simultaneous effects of radial gradients in temperature, pressure, and gravitational potential on the fugacity of elements during planetary evolution. At a depth of 200 km he has calculated the (WI) buffer to be very close to $-\log fO_2 = 9.0$, and similarly, the (WM) buffer to be about 5.0. By comparison, the Eugster and Wones (1962) $-\log fO_2$ values for the (WI) buffer could be extrapolated as 10.0 and for the (WM) buffer as 6.0 at the same temperature and pressure that Sato used. It may be argued that the uncertainties in the temperature at depth are giving rise to more discrepancy than just indicated and therefore the simple $P\Delta V$ upper mantle extrapolation is not altogether wrong. The lack of agreement between Sato's approach and the plain extrapolation may become serious to mantle redox calculations only when considering whether the redox conditions of the core-mantle boundary can be legitimately calculated, and whether these conditions influence the upper mantle.

Thus, some of the uncertainties in the thermodynamic calculations are not fully answerable at this time. The discrepancies between calculated fO_2-T and IOF mantle conditions, as shown in Figure 2a, may exist in part from the pitfalls of each technique. The comparison in Figure 2b, in which both data envelopes are measured by the same IOF technique, thus deserves closer scrutiny.

DISCUSSION AND CONCLUSIONS

Whether examined by IOF alone (Fig. 2b) or by a combination of IOF and thermodynamic calculations (Fig. 2a), the dichotomy between data for ilmenitic mantle assemblages and data for silicate mantle assemblages remains. The fact that the respective envelopes are so similar in these two comparative studies of many xenoliths argues for the existence of a real fO_2 dichotomy.

However, Sato (1984) has argued that IOF data without retrograde patterns may have re-equilibrated in nature before they were subjected to IOF study. As the originator of the IOF technique, Sato obviously believes that the technique can be reliable, although he has written (1987), "How a sample was brought up from the mantle, and how the intrinsic fO_2 measurement was conducted must be carefully examined before these very low [below the 1-bar value of (WI)] values are accepted as reliable" (p. 423). Sato raised this question specifically about some of the data below the (WI) line in Figure 2b, which come from Arculus and Delano (1980, 1981). In our own discussion of Figure 6 herein, we have expressed our concern regarding IOF researchers proving the lack of fluid inclusions in silicate minerals from the mantle. This could also be construed as criticism of the work of Arculus and others (1984) in Figure 2b, since they wrote about their silicates:

> The periodotite samples were sectioned and checked closely for the presence of hydrous phases and glass. We endeavoured to exclude this type of material from the samples chosen for study, but a very minor amount may have been incorporated. . . .We have also studied the intrinsic fO_2 of the bulk peridotite host of one of the spinels studied by Arculus and Delano (1981). . . .In the early part of an experimental run at a specific temperature, fO_2 values coincident with IW were observed but at variable time intervals from minutes to several hours for different charges, an irreversible and progressive oxidation signal was recorded . . . .These results can be interpreted as resulting from the release of CO_2 from the bulk peridotite [they do indeed exclude these data from their confidence]. . . .We are not attempting to ignore the undoubted presence of carbon-bearing species in the samples described here, but find no reason to question the validity of the results given the absence of any complex auto-reduction (or autooxidation) effects (p. 86).

Thus, while these authors saw retrograde reactions, they chose in good faith not to trust any data that show autoreduction or auto-oxidation; instead, they interpreted this lack as proof of there being no problem with fluid inclusions. We, however, have shown that fluid inclusions may in some cases (Fig. 6) totally flood the cell and dominate the IOF reading. Without having a high-enough vapor pressure to be reactive with the solid, the vapor released from fluid inclusions may still be high enough to dominate the argon in the IOF cell, i.e., KRIPPTIC data.

There are three main laboratories doing IOF work: Sato's, Arculus's, and Ulmer's. It might be asked, "If these centers are not in agreement, can any of the IOF data be taken seriously?" In analyzing the body of IOF data we have shown herein that: (1) some of the mantle xenoliths still have coherent multiple mineral intersects that would be impossibly coincidental if retrograding really had occurred; (2) some of the very reduced IOF data are coincident with other geothermometers (ACME data); (3) some of the more oxidized IOF data are coincident with the thermodynamically calculated values for similar xenoliths; and (4) when fluid inclusions are minimized, the fO_2 values may be real and are relatively highly reduced (NICE data). Megacrysts that show no retrograde patterns are suspected by Sato of having retrograded before reaching the laboratory. However, the Mongolian single crystals that we examined were without a priori fluid inclusions, or de facto discolorations, or the IOF offset patterns that occurred in all cases in which we witnessed retrograde re-equilibration.

The scientific community, of course, must judge whether progress has been made on the quantification of the IOF and the thermodynamically calculated redox state of the mantle. We hope that the details of interpretation given in this chapter will serve as a guide to others who have lacked criteria for deciding the validity of the IOF data.

The bulk of the quantitative data we have reviewed for the redox condition of the mantle (Fig. 2a, b) would indicate that there are definite redox inhomogeneities within the mantle. Other major lines of evidence have also led to this conclusion. The large differences in solubilities of H_2O and CO_2 with pressure can have varied effects. The CO_2 solubility in silicate melts increases rapidly with pressure, but as pressures exceed those of relevant carbonation reactions, the partial melts of peridotite may become carbonatitic under high fO_2 conditions. Thus, depending on total pressure, CO_2 may increase or decrease in a residual mantle depleted by water-induced partial melting (e.g., see Eggler, 1976; Mysen, 1977; Wyllie, 1977). The absence of primary carbonate in mantle xenoliths (the only primary occurrences being reported by Deines, 1968; McGetchin and Besancon, 1973; Anderson and others, 1984) has been used to suggest that CO_2 is also inhomogeneously distributed and perhaps present in quantity only in the asthenosphere (for example, Boyd and Nixon, 1973). Many researchers have discussed mantle mineralogy and mantle metasomatism in terms of mantle inhomogeneity (e.g., Haggerty, 1983; Haggerty and Tompkins, 1983; Menzies and others, 1983).

In their paper, Woermann and Rosenhauer (1987) have reviewed the recent literature in order to make extrapolations of mantle conditions based on experimental, petrologic, and theoretical studies. In particular, they have done Schreinemaker's analyses of the systems CaO-MgO-SiO_2-C-H-O and MgO-SiO_2-C-H-O; extensively reviewed the evidences of mantle redox state, including such factors as mantle xenolith fluid inclusion compositions, metasomatism, carbonation equilibria of silicates, mantle mineralogy, IOF measurements, conductivity measurements, the reality or nonreality of indicator minerals such as moissonite (SiC); and the existence of primordial methane. These authors have concluded that there is convincing evidence that supports the existence of a real mantle division into a high fO_2 mantle regime that is (EMOG)-like, and a low fO_2 regime that is (WM)- to (WI)-like. Their study has also concluded that there is little support for intermediate redox conditions between these two regimes.

If this dichotomy is real, then major rethinking of the controlling mechanism for mantle redox is necessary. However, the high-confidence IOF data are far too sparse to support the idea that no intermediate mantle fO_2 values will ever emerge. Thus, from the narrow perspective of IOF data, it is premature to suggest reasons or models for the lack of intermediate values. Stated from another viewpoint, although there seems to be a real dichotomy in the high-confidence IOF mantle data, and whereas this seems to support mantle redox inhomogeneity, there are not enough high-confidence IOF data to decide if there is a systematic change of fO_2 with depth into the mantle. In their paper, Woermann and Rosenhauer (1987) have reviewed, from a wider perspective of metasomatism, magma generation, and phase equilibria, some evidence that supports models that could explain a systematic fO_2 dichtomy with depth. This wider perspective must be the next step in the development of mantle redox models.

REFERENCES CITED

Andersen, D. J., and Lindsley, D. H., 1985, New (and final) models for the Ti-magnetite/ilmenite geothermometer and oxygen barometer: EOS Transactions of the American Geophysical Union, v. 66, p. 416.

Anderson, J. S., and Kahn, A. S., 1970, Equilibria of intermediate oxides in the titanium-oxygen system: Journal of Less Common Metals, v. 22, p. 299–323.

Anderson, T., O'Reilly, S. Y., and Griffin, W. L., 1984, The trapped fluid phase in upper mantle xenoliths from Victoria, Australia; Implications for mantle metasomatism: Contributions to Mineralogy and Petrology, v. 88, p. 72–85.

Andersson, S., 1967a, Aspects on the problems of synthesis and structure of some oxide or oxide-like compounds formed by the transition elements in the Groups 4, 5, and 6 of the periodic table: Ark. Kemi., v. 26, no. 44, p. 521–538.

—— , 1967b, Description of nonstoichiometric transition metal oxides; A logical extension of inorganic crystallography: Bulletin of the Society of French Mineralographers and Crystallographers, v. 90, no. 4, p. 522–527.

Arculus, R. J., and Delano, J. W., 1980, Implications for the primative atmosphere of the oxidation state of earth's upper mantle: Nature, v. 288, no. 5786, p. 72–74.

—— , 1981, Intrinsic oxygen fugacity measurements; Techniques and upper mantle peridotites and megacryst assemblages: Geochimica et Cosmochimica Acta, v. 45, p. 899–913.

Arculus, R. J., Dawson, J. B., Mitchell, R. H., Gust, D. A., and Holmes, R. D., 1984, Oxidation states of the upper mantle recorded by megacryst ilmenite in kimberlite and type A and B spinel lherzolites: Contributions to Mineralogy and Petrology, v. 85, p. 85–94.

Bassett, W. A., and Ming, L. C., 1970, Disproportionation of Fe_2SiO_4 to 2 FeO + SiO_2 at pressures up to 250 kbar and temperatures up to 3000°C: Physics of the Earth and Planetary Interiors, v. 4, p. 154–160.

Boyd, F. R., and Meyer, H.O.A., 1979a, Kimberlites, diatremes, and diamonds; Their geology, petrology, and geochemistry: Proceedings of the Second International Kimberlite Conference: Washington, D.C., American Geophysical Union, v. 1, 400 p.

—— , eds., 1979b, The mantle sample; Inclusions in kimberlites and other volcanics: Proceedings of the Second International Kimberlite Convergence: Washington, D.C., American Geophysical Union, v. 2, 424 p.

Boyd, F. R., and Nixon, D. H., 1973, Origin of the ilmenite-silicate nodules in kimberlites from Lesotho, South Africa, *in* Nixon, D. H., ed., Lesotho kimberlites: Lesotho National Development Corporation, p. 254–268.

Buddington, A. F., and Lindsley, D. H., 1964, Iron-titanium oxide minerals and synthetic equivalents: Journal of Petrology,v. 5, p. 300–357.

Burns, R. G., 1970, Mineralogical applications of crystal field theory: Cambridge, England, Cambridge University Press, 224 p.

Cox, K. G., Gurney, J. J., and Harte, B., 1973, Xenoliths from the Matsoku Pipe, *in* Nixon, P. H., Lesotho kimberlites: Lesotho National Development Corporation, p. 76–100.

Darken, L. S., and Gurry, R. W., 1945, The system iron-oxygen; I. The wüstite field and related equilibria: Journal of the American Ceramic Society, v. 67, p. 1398–1412.

Deines, P., 1968, The carbon and oxygen isotope composition of carbonates from a mica peridotite slice near Dixonville, Pennsylvania: Geochemica et Cosmochimica Acta, v. 32, p. 613–625.

Deines, P., Nafziger, R., Ulmer, G. C., Woermann, E., 1974, Temperature oxygen fugacity tables for selected gas mixtures in the system C-H-O at one atmosphere total pressure: Pennsylvania State University Bulletin of the Earth Mineral Science Experimental Station, v. 88, p. 1–129.

Duba, A., 1979, High electrical conductivity does not an asthenosphere make: EOS Transactions of the American Geophysical Union, v. 60, p. 241–242.

Duba, A., and Shankland, T. J., 1982, Free carbon and electrical conductivity in the earth's mantle: Geophysical Research Letter, v. 9, p. 1271–1274.

Duncan, I. J., 1983, Mantle oxygen fugacities; A re-evaluation of the significance of intrinsic oxygen fugacity data for ultramafic nodules [abs.]: EOS Transactions of the American Geophysical Union, v. 64, p. V51B-09.

Eggler, D. H., 1976, Does CO_2 cause partial melting in the low velocity layer of the mantle?: Geology, v. 2, p. 69–72.

—— , 1983, Upper mantle oxidation state; Evidence from olivine-orthopyroxenes-ilmenite assemblages: Geophysical Research Letters, v. 10, p. 365–368.

Eggler, D. H., and Baker, D. R., 1982, Reduced volatiles in the system C-O-H; Implications to mantle melting, fluid, formation, and diamond genesis, *in* Akimoto, S., and others, eds., High-pressure research in geophysics: Advances in Earth and Planetary Science, v. 12, p. 237–250.

Eggler, D. H., Mysen, B. O., Hoering, T. C., and Holloway, J. R., 1979, The solubility of carbon monoxide in silicate melts at high pressures and its effect on silicate phase relations: Earth and Planetary Science Letters, v. 43, p. 321–330.

Elliott, W. C., Grandstaff, D. E., Ulmer, G. C., Buntin, T., and Gold, D. P., 1982, An intrinsic oxygen fugacity study of platinum-carbon associations in layered intrusions: Economic Geology, v. 77, p. 209–226.

Ellis, D. J., and Green, D. H., 1979, An experimental study of the effect of Ca upon garnet-clinopyroxene Fe-Mg exchange equilibria: Contributions to Mineralogy and Petrology, v. 77, p. 13–22.

Eugster, H. P., and Wones, D. R., 1962, Stability relations of the ferruginous biotite, annite: Journal of Petrology, v. 3, p. 82–125.

Finnerty, A. A., and Boyd, F. R., 1984, Evaluation of thermobarometers for garnet peridotites: Geochimica et Cosmochimica Acta, v. 48, p. 15–28.

French, B. M., and Eugster, H. P., 1965, Experimental control of oxygen fugacities by graphite-gas equilibriums: Journal of Geophysical Research, v. 70, p. 1529–1539.

Freund, F., Leathrein, H., Wengeler, H., Lenobel, R., and Heinen, H. T., 1980, Carbon in solid solution in forsterite; A key to the intractable nature of carbon in terrestrial and conmogenic rocks: Geochimica et Cosmochimica Acta, v. 44, p. 1319–1333.

Frey, F. A., and Prinz, M., 1978, Ultramatic inclusions from San Carlos, Arizona; Petrologic and geochemical data bearing on their petrogenesis: Earth and Planetary Science Letters, v. 38, p. 29–176.

Haggerty, S. E., 1976, Opaque mineral oxides in terrestrial rocks, *in* Rumble, D., Oxide minerals: Washington, D.C., Mineralogical Society of America, 240 p.

—— , 1983, The mineral chemistry of new titanates from the Jagersfontein kimberlite, South Africa; Implications for metasomatism in the upper mantle: Geochimica et Cosmochimica Acta, v. 47, p. 1883–1854.

Haggerty, S. E., and Tompkins, L. A., 1983, Redox state of the earth's upper mantle from kimberlitic ilmenites: Nature, v. 303, p. 295–300.

Hatton, C. J., and Gurney, J. J., 1977, A diamond graphite ecologite from the Roberts Victor mine [extended abs.]: Second International Kimberlite Conference, 3 p.

—— , 1979, A diamond-graphite eclogite from the Roberts Victor mine, *in* Boyd, F. R., and Meyer, H.O.A., eds., The mantle sample; Inclusions in kimberlites and other volcanics: American Geophysical Union, p. 29–36.

Huebner, J. S., 1971, Buffering techniques for hydrostatic systems at elevated pressures, *in* Ulmer, G. C., ed., Research techniques for high pressure and high temperature: New York, Springer-Verlag, p. 123–178.

Koseluk, R. A., Elliott, W. C., and Ulmer, G. C., 1979, Gas inclusions and fO_2-T data for olivines from Can Carlos, Arizona [abs.]: EOS Transactions of the American Geophysical Union, v. 60, p. 419.

Kröger, F. A. and Vink, H. J., 1956, Relations between the concentrations of imperfections in crystalline solids, *in* Seitz, F., and Turnbull, D., eds., Solid State Physics: Advances in Research and Applications, v. 3, p. 307–435.

Lindsley, D. H., and Spencer, K. J., 1982, Fe-Ti oxide geothermometry; Reducing analyses of co-existing Ti-magnetite (Mt) and ilmenite (Ilm) [abs.]: EOS Transactions of the American Geophysical Union, v. 63, p. 471.

Magneli, A., 1964, Crystallographic principles of some nonstoichiometric transition metal oxides: Informal Proceedings of the Buhl International Conference Materials, Pittsburgh, 1963, p. 109–122 (*in* Chemical Abstracts, v. 63, no. 10788, 1965).

Magneli, A., Andersson, S., Asbrink, S., Westman, S., and Homberg, B., 1961, Crystal chemistry of titanium, vanadium, and zirconium oxides at elevated temperatures: U.S. Department of Commerce, Office of Technical Services, P.B. Report 145, no. 923, 82 p. (*in* Chemical Abstracts, v. 57, p. 164, 1962).

Mao, H. K., 1974, A discussion of the iron oxides at high pressure with implications for the chemical and thermal evolution of the earth: Carnegie Institute of Washington Geophysical Laboratory Yearbook, v. 73, p. 510–519.

Mathez, E. A., 1984, Influence of degassing on oxidation states of basaltic magmas: Nature, v. 310, p. 371–375.

Mathez, E. A., and Delaney, J. R., 1980, The nature and distribution of carbon in submarine basalts and peridotite nodules: Earth and Planetary Science Letters, v. 56, p. 217–232.

Mathez, E. A., Dietrich, V. J., and Irving, A. J., 1984, The geochemistry of carbon in mantle peridotites: Geochimica et Cosmochimica Acta, v. 48, p. 1849–1859.

McGetchin, T. R., 1968, The Moses Rock dike; Geology, petrology, and mode of emplacement of a kimberlite-bearing breccia dike, San Juan County, Utah [Ph.D. thesis]: Pasadena, California Institute of Technology, 440 p.

McGetchin, T. R., and Besancon, J. R., 1973, Carbonate inclusions in mantle-derived pyropes: Earth and Planetary Science Letters, v. 18, p. 408–410.

McMahon, B., 1984, Petrologic redox equilibria in the Benfontein sills and in the Allende meteorite and the T-fO_2 stability of kimberlitic ilmenite from the Monastery diatreme [Ph.D. thesis]: Amherst, University of Massachusetts Department of Geology and Geography, 209 p.

Menzies, M. A., Leeman, W. P., Hawkesworth, C. J., 1983, Isotope geochemistry of Cenozoic volcanic rocks reveals mantle heterogeneity below western U.S.A.: Nature, v. 303, p. 205–209.

Merritt, R. R., Hyde, B. G., Bursill, L. A., and Philp, D. K., 1973, The thermodynamics of the titanium oxygen systems; An isothermal gravimetric study of the composition range $T_{13}O_5$ to T_1O_2 at 1304°K: Philosophical Transactions of the Royal Society of London, v. 274, p. 627–661.

Meyer, H.O.A., 1977, A review of minerals in xenoliths from kimberlites: Earth Science Reviews, v. 13, p. 151–181.

Minkin, J. A., Chao, E.C.T., Christian, R. P., Harris, E. E., and Norton, D. R., 1976, Three synthetic lunar glasses: Meteoritics, v. 11, p. 167–171.

Moats, M., Weiss, D., and Ulmer, G., 1985, Evaluating the redox state of ilmenite-bearing xenoliths: EOS Transactions of the American Geophysical Union, v. 66, p. 393.

Mysen, B. O., 1977, The solubility of H_2O and CO_2 under predicted magma genesis conditions and some petrological and geophysical implications: Reviews in Geophysics and Space Physics, v. 15, p. 351–361.

Nakamura, A., and Schmalzreid, H., 1983, On the nonstoichiometry and point defect of olivine. Phys. Chem. Minerals, v. 10, p. 27–37.

—— , 1984, On the Fe^{2+}-Mg^{2+}-interdiffusion in olivine. Ber. Bunsenges. Phys. Chem., v. 88, p. 140–145.

Nixon, P. H., ed., 1973, Lesotho kimberlites: Lesotho National Development Corporation, 350 p.

Osborn, E. F., 1962, Reaction series for subalkaline igneous rocks based on different oxygen pressure conditions: American Mineralogist, v. 47, p. 211–226.

Ozawa, K., 1984, Olivine spinel geospeedometry; Analysis of diffusion-controlled Mg-Fe^{2+} exchange: Geochimica et Cosmochimica Acta, v. 48, p. 2597–2611.

Pasteris, J. D., 1981, Occurrence of graphite in serpentinized olivines in kimberlite: Geology, v. 9, p. 356–359.

—— , 1987, Fluid inclusions in mantle xenoliths, *in* Nixon, P. H., ed., Mantle xenoliths: John Wiley Publications (in press).

Raheim, A., and Green, D. H., 1974, Experimental determination of the temperature and pressure dependence of the Fe-Mg partition coefficient for coexisting clinopyroxene and garnet: Contributions to Mineralology and Petrology, v. 48, p. 179–203.

Richardson, S. H., Gurney, J. J., Erlank, A. J., and Harris, J. W., 1984, Origins of diamonds in old enriched mantle: Nature, v. 310, p. 198–202.

Ringwood, A. E., 1979, Origin of the Earth and Moon: New York, Springer-Verlag, 295 p.

Roedder, E., 1965, Liquid CO_2 in olivine-bearing nodules and phenocrysts in basalts: American Mineralogist, v. 50, p. 1746–1786.

——, 1984, Fluid inclusions: Mineralogical Society of America Reviews in Mineralogy, v. 12, 644 p.

Sato, M., 1965, Electrochemical geothermometry; A possible new method of geothermometry with electroconductive minerals: Economic Geology, v. 60, p. 812–818.

——, 1971, Electrochemical measurements and control of oxygen fugacity and other gaseous fugacities with solid electrolyte sensors, *in* Ulmer, G. C., ed., Research techniques for high pressure and high temperature: New York, Springer-Verlag, p. 43–99.

——, 1977, Oxygen fugacity of the mantle environment, *in* Dick, H., ed., Magma genesis: State of Oregon Department of Geology and Mineral Industries Bulletin 96, 311 p.

——, 1978a, A possible role of carbon in characterizing the oxidation state of a planetary interior and originating a metallic core: IX Lunar and Planetary Science Conference, pt. 2, p. 990–992.

——, 1978b, Oxygen fugacity of basaltic magmas and the role of gas-forming elements: Geophysical Research Letters, v. 5, p. 447–449.

——, 1984, The oxidation state of the supper mantle; Thermochemical modeling and experimental evidence: Proceedings, International Geological Congress, v. 11, Geochemistry and cosmochemistry: Moscow, V.N.U. Science Press, p. 405–433.

Sato, M., and Valenza, M., 1980, Oxygen fugacities of the layered series of the Skaergaard intrusion, east Greenland: American Journal of Science, v. 280A, p. 134–158.

Schock, R. N., 1985, Point defects in minerals; Washington, D.C., American Geophysical Union Monograph 31, Mineral Physics 1.

Schock, R. N., and Duba, A., 1985, Point defects and the mechanisms of electrical conduction in olivine, *in* Schock, R. N., ed., Point defects in Minerals: Washington, D.C., American Geophysical Union Geophysics Monograph 31, p. 88–96.

Stormer, J. C., 1983, The effects of recalculation on estimates of temperature and oxygen fugacity from analyses of multicomponent iron-titanium oxides: American Mineralogist, v. 68, p. 586–594.

Szekely, J., and Reitan, P. H., 1971, Dike filling by magma intrusion and by explosive entrainment of fragments: Journal of Geophysical Research, v. 76, no. 11, p. 2602–2609.

Ulmer, G. C., 1983, Increasing evidence for graphite as a primary phase in many mafic plutons: Geological Society of America Abstracts with Programs, v. 15, p. 709.

——, 1984, ZrO_2 oxygen and hydrogen sensors; A geologic perspective, *in* Science and technology of zirconia II: American Ceramic Society Advances in Ceramics, v. 12, p. 660–671.

Ulmer, G. C., Rosenhauer, M., Woermann, E., Ginder, J., Drory-Wolff, A., and Wasilewski, P., 1976, Applicability of electrochemical oxygen fugacity measurements to geothermometry: American Mineralogist, v. 61, p. 661–670.

Ulmer, G. C., Woerman, E., Knecht, B., and Rosenhauer, M., 1978, The role of pressure on redox buffers: EOS Transactions of the American Geophysical Union, v. 59, p. 398.

Ulmer, G. C., Rosenhauer, M., and Woermann, E., 1980, Glimpses of mantle redox conditions?: EOS Transactions of the American Geophysical Union, v. 61, p. 413.

Ulmer, G., Weiss, D., and Moats, M., 1985a, The mantle redox story: Why the dichotomy?: Geological Society of America Abstracts with Programs, v. 17, p. 67.

Ulmer, G. C., Moats, M. A., and Weiss, D. A., 1985b, Oxygen fugacity, carbon, and the mantle redox state: EOS Transactions of the American Geophysical Union, v. 66, p. 393.

Virgo, D., Rosenhauer, M., and Huggins, F. E., 1976a, Intrinsic oxygen fugacities of some natural melanites and schorlomites and crystal-chemical implications: Carnegie Institution of Washington, Geophysical Laboratory Yearbook, v. 75, p. 720–729.

——, 1976b, Petrologic implications of intrinsic oxygen fugacity measurements on titanium-containing silicate garnets: Carnegie Institution of Washington, Geophysical Laboratory Yearbook, v. 75, p. 730–735.

Weiss, D. S., (1984) Petrography, phase chemistry, volatile inclusions, and intrinsic oxygen fugacity of composite Group II ultramafic nodules from San Carlos, Arizona [M.A. thesis]: Philadelphia, Temple University Geology Department, 133 p.

Weiss, D., Ulmer, G. C., Buntin, T., and Moats, M., 1985, Fluid inclusions and IOF data; Group II composite nodules from San Carlos, Arizona: EOS Transactions of the American Geophysical Union, v. 66, p. 392.

Woermann, E., and Rosenhauer, M., 1987, Fluid phases and the redox state of the Earth's mantle; Extrapolations based on experimental phase theoretical and petrologic data: Fortschritte Mineralogie (in press).

Woermann, E., Knecht, B., Rosenhauer, M., and Ulmer, G. C., 1977, Die stabilitat des graphit in system C-O:Fortschritte Mineralogie, v. 55, p. 155.

Wyllie, P. J., 1977, Mantle fluid compositions buffered by carbonates in peridotite-CO_2-H_2O: Journal of Geology, v. 85, p. 187–207.

MANUSCRIPT ACCEPTED BY THE SOCIETY OCTOBER 14, 1986

Printed in U.S.A.

Geological Society of America
Special Paper 215
1987

Interpretation of the sulfide assemblages in a suite of xenoliths from Kilbourne Hole, New Mexico

Edward L. Dromgoole and Jill D. Pasteris
Department of Earth and Planetary Sciences, McDonnell Center for the Space Sciences, Washington University, Box 1169, St. Louis, Missouri 63130

ABSTRACT

Petrography and electron microprobe analysis were used to investigate the mineralogy, bulk composition, abundance, and distribution of Cu-Fe-Ni sulfide phases in a suite of about 25 alkalic-basalt–hosted xenoliths from Kilbourne Hole, New Mexico. The xenoliths include spinel lherzolites, spinel clinopyroxenites, clinopyroxene megacrysts, mafic granuloblastites, and a sillimanite-quartz granuloblastite. Sulfide abundance is low in all the xenoliths, but shows a maximum (~0.5 vol. %) in pyroxenites and megacrysts. The sulfides occur as interstitial to silicate and oxide grains, fully enclosed and isolated in grains, and along fracture surfaces in grains. The major sulfide phase is monosulfide solid solution (~pyrrhotite), accompanied by variable amounts of pentlandite, and cubanite. Both the mineralogy and bulk sulfide composition are distinguishable among the xenolith types: Ni-rich for the lherzolites, Ni-poor for the crustal xenoliths, and intermediate for the pyroxenites, megacrysts, and composite xenoliths. The sulfides in the Kilbourne Hole spinel lherzolites most likely result from partial retention, in the residual silicate rock, of an immiscible sulfide formed during mantle partial melting.

INTRODUCTION

Xenoliths in alkalic basalts and kimberlites are an important source of information about the composition of the upper mantle and lower crust and about processes that occur in those regions. As such, the mineralogy, petrology, and geochemistry of these xenoliths have been studied extensively during the last decade (e.g., Ahrens and others, 1975; Sobolev, 1977; Boyd and Meyer, 1979; Hawkesworth and Norry, 1983; Nixon, 1987). One result of the studies has been recognition of a sulfide component in a wide variety of xenoliths from many locations. The sulfides have received far less extensive investigation than have other aspects of mantle xenoliths.

The object of the present study is to provide petrographic, mineralogic, and chemical data on the sulfides from a petrographically varied suite of xenoliths from Kilbourne Hole, New Mexico, and to use these data to help interpret upper mantle processes that affected the xenoliths as a whole.

Kilbourne Hole

Located 20 km north of the Mexican border, Kilbourne Hole (Fig. 1) is the largest (~4 km in diameter) of three identified maar volcanoes thought to have formed by phreatomagmatic explosion. The exposed basalt flow is a pre-crater porphyritic olivine basalt associated with the earliest period of lava extrusion in the Aden-Afton region, which has yielded K-Ar whole-rock ages of 141,000 ± 75,000 and 103,000 ± 84,000 yr (Hoffer, 1976).

The mantle and crustal xenoliths generally are encased by a rind of the host basalt as much as several centimeters thick. The basalt (represented by sample KH25 of this study) is vesicular, glass-rich, interstitial to intergranular in texture, and contains olivine and minor pyroxene and plagioclase phenocrysts. Whole-rock analysis (Irving and Frey, 1984) indicates that the basalt is a nepheline-normative alkali olivine basalt (basanite), which is typical of host basalts to spinel lherzolite xenoliths throughout the world (BVSP, 1981a, b).

Previous studies concerning the petrography and petrology of Kilbourne Hole mantle xenoliths include Carter (1965), Reid and Woods (1978), Irving (1979, 1980), BVSP (1981a), Bussod (1981), Bussod and Irving (1981), and Wilshire and others (1985). Trace-element compositions of Kilbourne Hole xenoliths are presented in Jagoutz and others, 1979 (major, minor, trace elements), Morgan and others, 1980 (volatile trace elements), Morgan and others, 1981 (siderophile elements), Morgan and Baedecker, 1983 (siderophile elements in sulfides), Mathez and others, 1984 (carbon), and Irving and Frey, 1984 (trace elements in megacrysts). Other authors have analyzed the crustal and mantle xenoliths from Kilbourne Hole and the surrounding maars to infer subcrustal and upper mantle processes in the region; they include Reid and Woods (1978); Bussod and Irving (1981); Padovani and Hart (1981); and Wasilewski and Padovani (1981).

Background and Goals

The interpretation of small to minute sulfide bodies in mantle xenoliths from Kilbourne Hole, New Mexico, is founded in part on broader scale investigations of sulfide liquid immiscibility in mafic and ultramafic rocks (e.g., Naldrett, 1969, 1981; Skinner and Peck, 1969; Haughton and others, 1974; Buchanan and Nolan, 1979; Pasteris, 1984, 1985), most spectacularly demonstrated in large Cu-Fe-Ni-sulfide ore deposits such as Sudbury, Canada; Norilsk, USSR; and the Duluth Complex, USA. The development of sulfide immiscibility is known to be a complex function of the interactive parameters pressure, temperature, bulk-melt composition, fO_2, and fS_2 (e.g., MacLean, 1969; Shimazaki and Clark, 1973; Haughton and others, 1974; Shima and Naldrett, 1975; Buchanan and Nolan, 1979; Wendlandt, 1982; Boctor, 1982). In spite of the complex interrelationships, there is good experimental and empirical evidence that many natural mafic and ultramafic melts should reach sulfur saturation either before or during crystallization (e.g., Skinner and Peck, 1969; Anderson, 1974). However, the low abundance of probable immiscible sulfides in many mafic and ultramafic rocks seems inconsistent with this hypothesis.

Fleet and others (1981) claimed from evidence of Ni-partitioning between coexisting olivine and sulfides that "an immiscible sulfide liquid cannot be a normal product of upper-mantle magmatic processes" (Fleet and others, 1981, p. 119). Fleet (see, for example, Fleet, 1979; Fleet and others, 1977) repeatedly has offered the hypothesis, not well accepted, that the large sulfide ore deposits in igneous intrusions instead represent hydrothermal introduction of sulfides. Fleet and others (1981) interpreted the sulfide blebs in sea-floor basalts (e.g., Czamanske and Moore, 1977) as an indication that sulfur saturation in this case did not occur until late in the magmatic history (i.e., after crystallization began).

Although sulfide liquid immiscibility probably accounts for the bulk of the sulfides found in mafic and ultramafic rocks, there is indeed evidence for hydrothermal introduction, or at least redistribution, of such sulfides (see Pasteris, 1984, 1985, and references therein). There is also the important question of whether sulfide immiscibility in large intrusions in the crust is a closed-system magmatic process or whether it is usually triggered by assimilation of volatile-rich crustal material (e.g., Mainwaring and Naldrett, 1977; Ripley, 1981). One reason for choosing a suite of upper mantle xenoliths for detailed investigation of its sulfide assemblage, is that these peridotites and pyroxenites are more pristine and less affected by crustal processes than are igneous intrusions such as those at Sudbury, Norilsk, and Duluth.

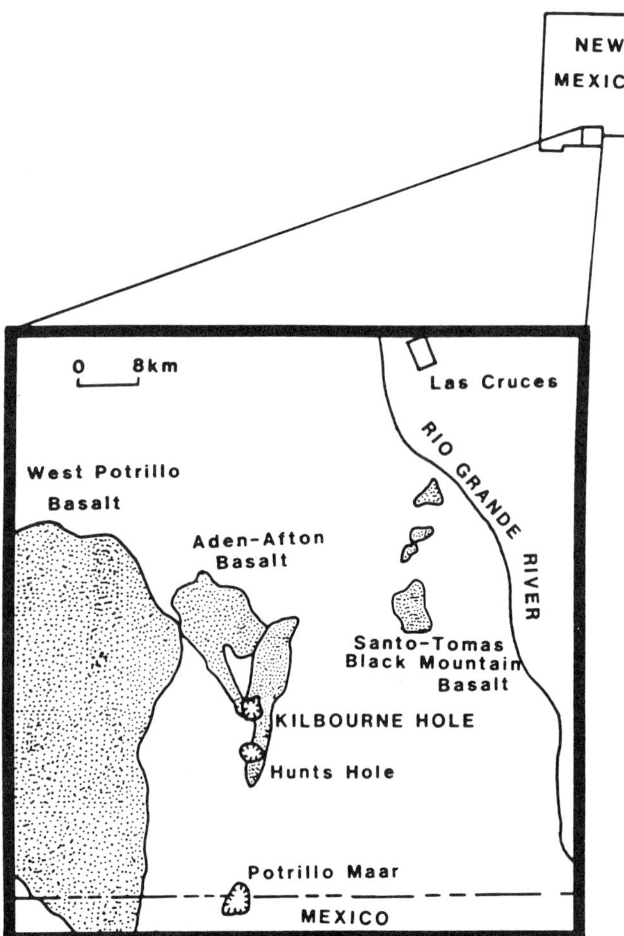

Figure 1. Index map of Potrillo basalt field (from Hoffer, 1976).

The sulfide occurrences in a variety of petrologic types of xenoliths from Kilbourne Hole were studied to determine if the sulfides differed in abundance, mineralogy, bulk composition, and sulfide-silicate association among the types. The aim was to try to determine if the xenolith sulfides were primary (co-crystallized with the host rock) and/or redistributed, and from this information to draw some conclusions about the mechanism of sulfide segregation both in the mantle and in (deep) crustal intrusions.

SAMPLES

The majority of the xenoliths examined in this study were collected in October of 1977, during a field trip in conjunction with the Second Kimberlite Conference. Both crustal (see Padovani and Carter, 1977) and mantle xenoliths were collected, but

TABLE 1. SUMMARY OF TEXTURAL CLASSIFICATION

Porphyroclastic (9 examples)*	
Grain Size:	Two populations, typically porphyroclasts >2 mm, matrix <2 mm; >10% of mineral grains occur as porphyroclasts, usually most abundant modal mineral
Grain shape:	Anhedral, typically equidimentional
Grain boundaries:	Smoothly curving or straight, possibly irregular on porphyroclasts; usually meet at 120° triple junctions
Mosaic-porphyroclastic (3 examples)	
<10% of grains occur as porphyroclasts; otherwise, same as porphyroclastic texture	
Equant granuloblastic (8 examples)	
Grain size:	Typically <2 mm, size range for each mineral may be different; porphyroclasts are rare or absent
Grain shape:	Anhedral; mostly equidimensional and polygonal
Grain boundaries:	Straight or smoothly curving; usually meet at 120° triple junctions
Tabular granuloblastic (5 examples)	
>10% of mineral grains prominently larger in one direction, typically with axis of larger dimension aligned along plane of foliation; otherwise, same as equant granuloblastic texture	
Megacryst (4 examples)	
Grain size:	Dimensions up to 4 cm
Grain shape:	Irregular
Grain boundaries:	Irregular to straight

*The term "examples" is used rather than "samples," because composite xenoliths show more than one textural type.

the latter are the major concern of this study. The most common type of mantle xenolith at Kilbourne Hole is Cr-diopside (type-I) spinel lherzolite (Frey and Prinz, 1978; Irving, 1980; Menzies, 1983), which constitutes the majority of xenoliths in the collection. Less abundant types of mantle xenoliths from Kilbourne Hole included in the collection are Al-augite (type-II) pyroxenites and megacrysts.

Several samples (KH73-1, KH77-7, KH77-16, KH77-20), including thin sections and rock chips from which thin sections were made, were provided by Anthony Irving. These are samples that were part of earlier investigations by Irving (1980; BVSP, 1981a) and Bussod (1981), in which they are described in detail. Mineral chemical data on the silicate and spinel phases in these samples were taken from their earlier works.

The xenoliths included in this study were examined in hand-specimen and in doubly polished thin section with transmitted and reflected light, one thin section per xenolith. They have been separated into groups based on the modal abundance of major phases, textural features (Table 1), and mineral chemistry. Table 2 lists the modes, textures, and classification for each xenolith. Modal abundances were determined by visual estimation. In this work, Harte's (1977) terminology for olivine-bearing xenoliths is applied to all the xenoliths, even those with little or no olivine.

The xenoliths in the present suite include spinel lherzolites, spinel clinopyroxenites, composite xenoliths (composite of the two preceding rock types), clinopyroxene megacrysts, mafic granuloblastites, and a sillimanite-quartz granuloblastite. The lherzolites probably represent a residuum of partial melting in the upper mantle (e.g., Menzies, 1983). The pyroxenite, megacryst, and mafic samples are probably cognate xenoliths that formed (at different depths) by the crystallization of a mantle-derived magma. In some cases, the latter group occurs as veins in the upper mantle country rock, giving rise to composite xenoliths, consisting of two distinct assemblages—spinel lherzolite and spinel clinopyroxenite—in sharp contact (Frey and Prinz, 1978; Irving, 1980; Menzies, 1983). Metamorphism and partial melting of pelitic crustal material probably gave rise to the sillimanite-quartz xenolith that was studied (cf. Padovani, 1977; Padovani and Carter, 1977).

Electron microprobe analyses (Dromgoole, 1984) of olivine, orthopyroxene, clinopyroxene, and spinel further aided classification of the xenoliths. These data also were used in the two-pyroxene geothermometry models of Wells (1977) and Lindsley (1983). There are recognized problems with both temperature estimation procedures. Inconsistent temperature values within xenolith subgroups resulted from using the Lindsley (1983) model, perhaps due to the high proportion of nonquadrilateral components in the Kilbourne Hole pyroxenes. Application of the Wells geothermometer at least allowed comparison with temperature estimates for Kilbourne Hole xenoliths studied in the previous investigations (Padovani, 1977; Bussod, 1981) in which this geothermometer was used. Temperature estimates were made for all xenoliths containing two coexisting pyroxenes (Dromgoole, 1984), and are comparable to those derived by Bussod

TABLE 2. XENOLITH DESCRIPTION SUMMARY*

Sample No.	Modal Proportion of Phases					Texture
	ol	opx	cpx	oxide	amph	
Type I, Spinel Lherzolites						
KH3	15	12	72	1		Mosaic-porphyroclastic
KH33	70	15	12	3		Tabular-granuloblastic
KH34	75	12	10	3		Tabular-granuloblastic
KH77-7	55	20	15	10		Tabular-granuloblastic
KH8	65	20	12	3		Porphyroclastic
KH10	80	10	9	1	<1	Porphyroclastic
KH31	65	25	7	3		Prophyroclastic
KH35	50	30	17	3		Prophyroclastic
KH36	60	20	15	5		Prophyroclastic
KH39	60	23	12	5		Prophyroclastic
KH40	60	20	12	3		Prophyroclastic
KH77-16	60	25	12	3		Prophyroclastic
Type II, Clinopyroxenites						
KH17	5		90	5		
KH19			90	10		Equant-granuloblastic
KH28	10	12	72	3	3	Mosaic-porphyroclastic
Type II, CPX Megacrysts						
KH2	5		92	3		
KH9	1		99			
KH20			100			
KH24	1		99			
Composite						
KH12	55	35	5	5		Mosaic-porphyroclastic
KH16						
(lherz)	60	20	15	5		Porphyroclastic
(pyx)			80	20		Equant-porphyroclastic
KH73-1						
(lherz)	60	20	15	5		Tabular-granuloblastic
(pyx)			85	15		Equant-granuloblastic
KH77-20						
(lherz)	60	15	20	5		Tabular-granuloblastic
(pyx)	5		80	15		Equant-granuloblastiic
Mafic granuloblastites						
	opx	cpx	ox	plag	glass	
KH4	10	8	4	68	7	Equant-granuloblastic
KH5	10	12	7	70	1	Equant-granuloblastic
KH6	15	10	5	70		Equant-granuloblastic
KH7	10	7	3	60	12	Equant-granuloblastic
Sillimanite-quartz granuloblastite						
	qtz	sill	oxide	glass		
KH26	27	30	3	40		Equant-granuloblastic

*Key to abbreviations: ol, olivine; opx, orthopyroxene; ox, oxide; cpx, clinopyroxene; amph, amphibole; plag, plagioclase; qtz, quartz; sill, sillimanite.

(1981) and Padovani (1977) for mantle and crustal xenoliths, respectively, from Kilbourne Hole. The values range from about 900° to 1,070°C, with an estimated uncertainty (Wells, 1977) of 70°C.

The absence of plagioclase and garnet from the spinel lherzolites indicates a pressure range of 9 to 24 kbar. A similar range of pressure can be inferred for some of the pyroxenites and megacrysts based on the coexistence of spinel lherzolite and pyroxenite in composite xenoliths. The presence of plagioclase and the absence of olivine in the mafic granuloblastites indicates a pressure range of 5 to 9 kbar. Since current mineral geobarometers do not seem to be reliable, more accurate pressure estimates cannot be made. No pressure correction was applied, but such a correction would increase the temperature estimates by less than about 50°C at pressures less than about 30 kbar (Wood and Banno, 1973).

SULFIDE PETROGRAPHY

Each xenolith was examined for sulfides in both hand-sample and doubly polished thin section (usually one per thin section). Because the sulfide grains are so small (generally <50 μm), high-magnification, oil-immersion objectives were used for study in reflected light. The relative total sulfide (RTS) content of each thin section was determined, as shown in Table 3.

TABLE 3. ABUNDANCE AND MODAL PROPORTION OF SULFIDES*

Sample No.	Abundance	Average Modal Proportion of Phases** po/mss	cb	pn
Type-I Spinel Lherzolites				
KH3	L	90 (70-100)	2 (0-10)	8 (0-20)
KH8	M	80 (55-100)	3 (0-10)	17 (0-40)
KH10	VL	95 (90-100)	~0 (0- 5)	5 (0-10)
KH31	M	85 (70- 95)	3 (0-10)	12 (5-30)
KH33	M	85 (40-100)	3 (0-10)	12 (0-60)
KH34	L	80 (60- 95)	5 (0-25)	15 (5-35)
KH35	L	60 (30- 90)	5 (0-15)	35 (10-60)
KH36	L	70 (50- 85)	10 (0-20)	20 (10-30)
KH39	VL	90 (80-100)	3 (0-10)	7 (0-15)
KH40	VL	80 (60- 90)	3 (0-10)	17 (10-40)
KH77-7	H	75 (60- 90)	7 (0-25)	18 (10-25)
KH77-6	H	75 (50- 90)	5 (0-20)	20 (10-30)
Type-II Clinopyroxenites				
KH17	M	90 (70-100)	10 (0-30)	
KH19	H	90 (60-100)	10 (0-40)	
KH28	M	80 (55-100)	5 (0-20)	15 (0-40)
Type-II Megacrysts				
KH2	L	75 (60- 90)	5 (0-10)	20 (10-35)
KH9	M	85 (70- 95)	15 (5-30)	
KH20	None			
KH24	VH	100		
Composite				
KH12	None			
KH16	H	90 (60-100)	~0 (0- 5)	10 (0-40)
KH73-1	H	85 (70-100)	12 (0-30)	3 (0-40)
KH77-20	VH	90 (70-100)	10 (0-25)	3 (0-10)
Mafic granuloblastites				
KH4	L	100		
KH5	VL	100		
KH6	M	100		
KH7	L	100		
Sillimanite-quartz granuloblastite				
KH26	VL	100		

*Abbreviations: po, pyrrhoite; mss, monosulfide solid solution; cb, cubanite; pn, pentlandite; VL, very low (<0.02 vol. %); M, moderate; H, high (>0.05 vol. %); VH, very high (~0.5 to 1.0 vol. %).

** Values in parentheses indicate range of phase proportions within individual blebs.

Occurrence of Sulfides in the Xenoliths

Sulfides in the Kilbourne Hole xenoliths (Figs. 2 through 4) have textures similar to the range reported in other xenolith studies (e.g., Frick, 1973; DeWaal and Calk, 1975). The sulfides occur either interstitial to (Fig. 2a, d) or enclosed within (Fig. 2b, c) the major phases. Among other criteria, texture has been used to try to distinguish primary, immiscible sulfide precipitates of the xenolith parent melt from subsequent remobilization of these sulfides or injection of sulfide liquids from another melt unrelated to the xenolith. Interstitial sulfides, i.e., sulfides situated between grains of the dominant phases, account for a major proportion of the total sulfides in all but the megacryst xenoliths. Typically, the interstitial sulfides are small, rounded blebs at triple junctions or blebs and stringers along grain contacts (Fig. 2d), commonly surrounded or accompanied by interstitial brown glassy material. In some cases, particularly in the samples with the highest sulfide content, partially altered interstitial sulfides may be as large as grains of the major phases. When this occurs (see Fig. 3a), the sulfides may exhibit features indicative of textural equilibrium with the major phases, i.e., anhedral, not round, but with curvilinear margins and 120° triple junctions with adjacent phases.

Enclosed sulfides within silicates and spinels are the second major type of sulfide occurrence. They are present in all of the sulfide-bearing xenoliths, but constitute a substantial proportion of the total sulfide only in clinopyroxene-rich xenoliths: pyroxenites, megacrysts, and the mosaic-(clinopyroxene) porphyroclastic lherzolites. Three types of enclosed sulfides (Fig. 3) can be distinguished: (1) sulfides associated with fluid-inclusion arrays (Fig. 3b, c), (2) isolated enclosed sulfides (Fig. 3d), and (3) sulfide inclusions in partial melt zones (Fig. 3e). In the first type, several small sulfide blebs are interspersed with fluid and/or glass inclu-

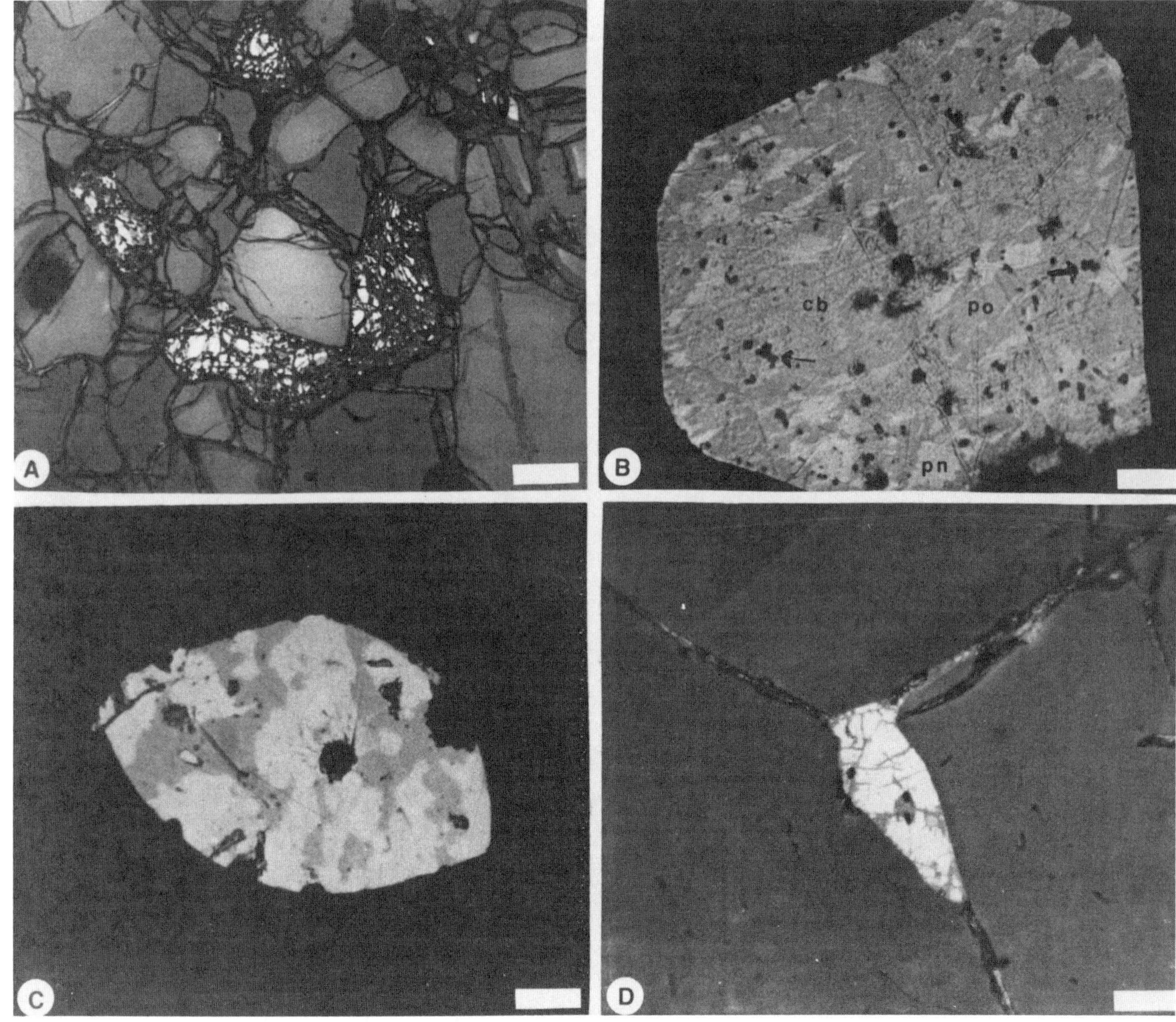

Figure 2. a, Large altered sulfide (white) bleb interstitial to clinopyroxene and spinel (gray) grains. Clinopyroxenite in sample KH77-20. Reflected light, dry objective. Scale bar = 200 μm. b, Sulfide bleb of pyrrhotite/mss (po), pentlandite (pn), cubanite (cb), and magnetite (arrows) isolated in silicate. Composite xenolith KH77-20. Reflected light, oil immersion. Scale bar = 8 μm. c, Sulfide bleb of cubanite (gray) and pentlandite (white) isolated in silicate. Spinel lherzolite KH35. Reflected light, oil immersion. Scale bar = 8 μm. d, Interstitial sulfide (white), somewhat oxidized (light gray intergrowths), at silicate (dark gray) triple junction. Composite xenolith KH77-20. Reflected light, dry objective. Scale bar = 25 μm.

sions within linear or planar arrays. The sulfide blebs are the same size as, and appear to have been introduced with, the fluid/glass inclusions. This type of sulfide inclusion occurs in all of the sulfide-bearing xenoliths and in all phases, but, as with the fluid-inclusion arrays themselves, is more common in clinopyroxene. Although numerically abundant, this type of sulfide occurrence constitutes a volumetrically minor proportion of the total sulfide content. Laser Raman microprobe analyses of many fluid inclusions in the samples show them to be essentially pure CO_2.

The subgroup, isolated enclosed sulfides, consists of single blebs that are much larger than the sulfides in fluid inclusion arrays, but commonly associated with them. Enclosed sulfides occur in all phases but most commonly in clinopyroxene. They compose a major or substantial proportion of the total sulfide content in megacryst xenoliths and clinopyroxenite xenoliths, in which enclosed sulfides in spinel are also common. Generally, enclosed sulfides have round or oval cross sections. However, some inclusions in spinels have square cross sections (Fig. 3f), which may reflect sulfide incorporation during spinel crystal growth or recrystallization, reminiscent of negative-crystal outlines in fluid inclusions. Also, enclosed sulfides in clinopyroxenites may have stellate or decorated borders. It is common to find sulfide-bearing fluid inclusion arrays intersecting or radiating from enclosed sulfides. In rare cases, enclosed sulfides have an "exploded" appearance, which may have resulted from melting and expansion of the sulfide during the ascent and decompression of the xenolith, similar to the decrepitation of fluid inclusions (W. Griffith, personal communication).

Isolated enclosed sulfides are especially abundant and exhibit a unique distribution (Fig. 4a, b) in sample KH24, a clinopyroxene megacryst. Most of the sulfides are relatively large spherical or cylindrical bodies (round blebs or rods with circular or oval cross section in thin section) aligned along parallel planes to which the axes of the cylinders also are parallel. Individual enclosed sulfide bodies within the clusters (and outside them in many cases) are intersected by one or more sulfide-bearing fluid inclusion arrays, typically oriented at high angles to the alignment direction. An x-ray crystallographic study of a clinopyroxene megacryst with oriented sulfide inclusions similar to the ones described above was performed on a xenolith from an alkalic basalt from Nunivak Island, Alaska. The study indicates that the sulfides are aligned along growth planes [(111) and (221)], not cleavage, parting, or common exsolution planes (Peterson and Francis, 1977). This suggests that the sulfides in the Nunivak Island clinopyroxene were occluded during crystal growth in the presence of an immiscible sulfide liquid. Although x-ray examination of the KH24 megacryst was not performed, a similar origin for the enclosed sulfides in this megacryst seems likely. However, it cannot be determined whether the immiscible sulfide liquid formed as a result of a local oversaturation of sulfur in the melt directly adjacent to the crystallizing pyroxene (due to crystal growth rates exceeding sulfur diffusion away from the crystal; Peterson and Francis, 1977) or of sulfur saturation throughout the melt.

Although the fully enclosed, aligned sulfide blebs described above probably represent primary occluded immiscible sulfides from a "megacryst melt," many fully enclosed sulfides are of a more questionable origin. Their association with fractures and with trails of fluid and sulfide inclusions suggest secondary injection or, at least, remobilization and possible compositional change. The marked abundance of sulfides in pyroxene megacrysts may be attributed to the volatile-rich nature of the megacryst parent liquid. However, the abundance of small sulfides enclosed in clinopyroxene grains in lherzolites simply may be due to the brittle deformation features of clinopyroxene, which also accounts for the abundance of fluid inclusion trails in this mineral.

The third subtype of enclosed sulfides is blebs associated with partial melt zones, which are common in clinopyroxene and spinel grains in some xenoliths. The sulfides tend to be concentrated along the margins of the melt zone, but are also common within the melt zone proper, where the irregularly shaped sulfide blebs typically are adjacent to empty or glass-filled cells (Fig. 4c). Textural relationships between interstitial sulfides and melt zones show progressive penetration and eventual segregation of lobes of the interstitial sulfides. Also, in KH24, large sulfide blebs in spongy melt zones around an olivine inclusion are aligned with other sulfides outside of the melt zone. These features suggest that many of the sulfides enclosed in the partial melt zones may be previously existing interstitial and/or enclosed sulfides, which accidentally have been incorporated, and that they are not directly related to or derived from the melting episode. Similar spongy melt halos also surround each individual sulfide bleb in the sulfide-rich clinopyroxene megacrysts from Malaita, Solomon Islands (Pasteris, 1982). It seems reasonable that partial melting would have begun at physical and compositional discontinuities in the rock, which is where the sulfides already would have been concentrated. This hypothesis is also supported by the compositional similarity of melt zone sulfides and interstitial sulfides. Furthermore, sulfide abundances within the melt zones (up to 5 vol. %) are far in excess of what could be expected (<0.02 vol % sulfide, assuming sulfur content of clinopyroxene is less than 100 ppm [Schneider, 1974]) if the sulfides were derived from the clinopyroxene during partial melting.

Textures and Modal Proportions of Sulfide Phases in the Xenoliths

The sulfide blebs in each of the Kilbourne Hole xenoliths have a characteristic sulfide phase assemblage. There is a general relationship between the sulfide assemblage and the type of xenolith. The typical sulfide assemblages are: pyrrhotite (po) only; pyrrhotite $\pm$ cubanite (cb); and pyrrhotite or pyrrhotite-like monosulfide solid solution (mss) $\pm$ pentlandite (pn) $\pm$cubanite. The po assemblage occurs in the mafic and sillimanite-quartz granuloblastite xenoliths and in one megacryst (KH24). The po-cb assemblage occurs in two olivine-poor clinopyroxenites (KH17, KH19) and in one megacryst (KH9). The po-mss-pn-cb assemblage occurs in the spinel lherzolites, one olivine-rich clinopyroxenite (KH28), and one megacryst (KH2) that has relatively abundant olivine and spinel inclusions. Sulfide blebs in the composite xenoliths typically have a po/mss-cb assemblage, but pn is also present in rare cases.

The po-cb assemblage typically consists of a central po mass partially surrounded by a narrow rim of cb. Less commonly, cb also occurs as discrete masses near the edge of the sulfide bleb or as small patches within the po. The typical po-mss-pn-cb assemblage consists of one or more discrete masses of po/mss, pn, and lesser cb (Fig. 2b). The po commonly contains exsolved flames, patches, or lamellae of pentlandite. The discrete pn masses typically lie in the margin of the po, forming partial rims between the po and the enclosing or cross-cutting phases. Cb typically occurs

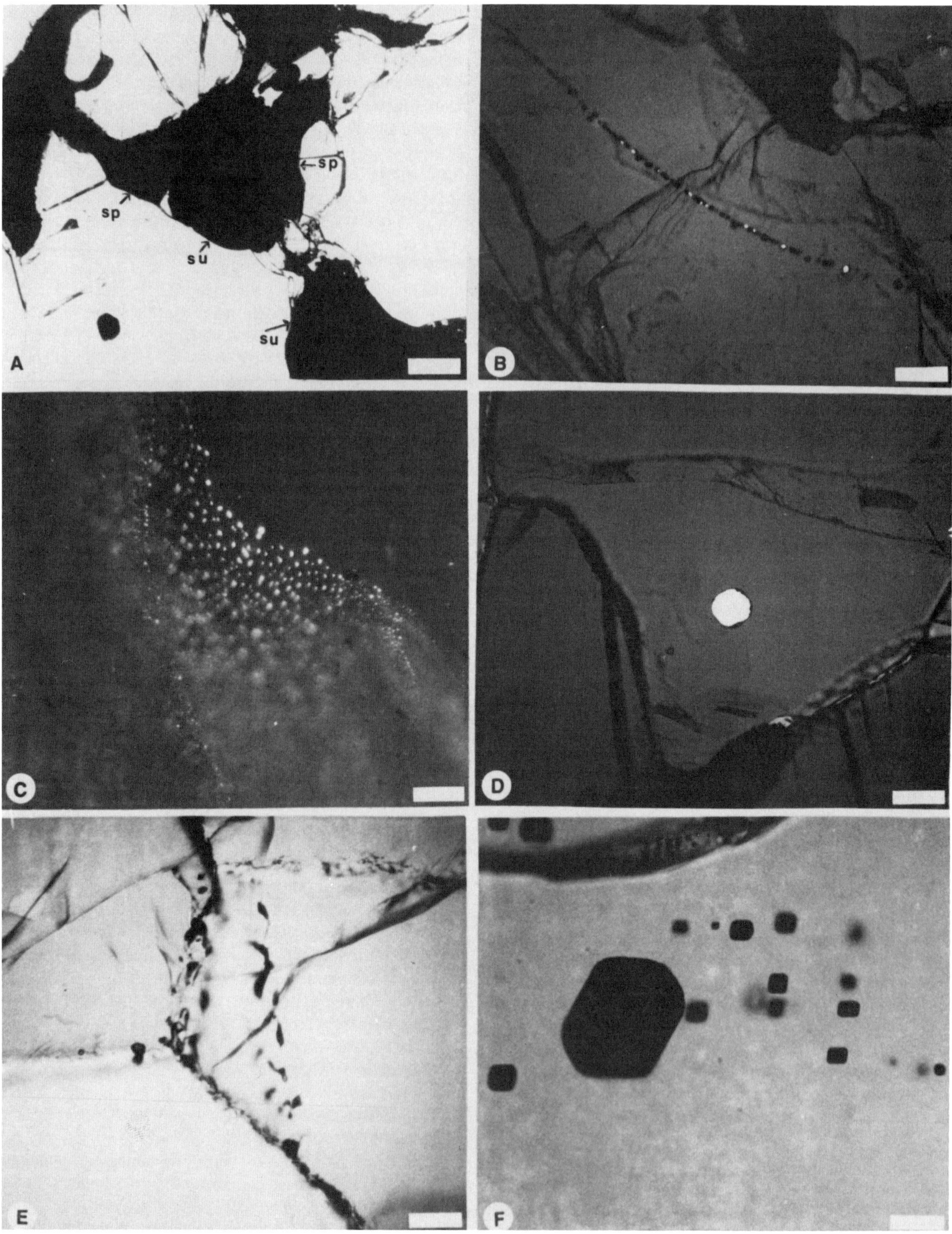

sp
sp
su
su
A
B
C
D
E
F

as smaller discrete masses—for instance, as halos around the outside of the po—in most cases separated from the po by pn. Less commonly, cb can form larger masses adjacent to the po, or exsolved patches in the po. Rarely, cb contains small, round inclusions of pn, possibly exsolved from or nucleated around the cubanite (Fig. 2c).

In addition to sulfide phases, sulfide blebs commonly contain veins or patches of silicate glass and/or iron oxides. In most cases the iron oxides appear to replace the sulfides. The presence of sharp contacts between the sulfides and glass veinlets, the presence of roughly concentric zones of cb, pn, and po/mss on each side of the veinlet, and the presence of some silicate glass inclusions completely enclosed in sulfide suggest that some of the glass was present before low-temperature unmixing of the sulfide and possibly was present during a period when the sulfide was a liquid.

Modal proportions of sulfide phases can vary widely from one sulfide bleb to the next within a single thin section due to nonuniform phase distribution within individual blebs. Some of the variability also may reflect initial chemical inhomogeneity of the sulfide, but sulfide blebs within a given thin section do tend to cluster around an average value. Table 3 lists the range and estimated average modal proportion of the total sulfide phases in each xenolith. Because pn and cb typically are concentrated along the edges of the sulfide blebs, where they are more susceptible to alteration and replacement, the average modes given in Table 3 may underestimate the true contribution of these phases.

Sulfides in Basalt

Sulfides were studied in one host basalt sample (KH25), but the RTS of the sample is very low. The sulfides almost invariably occur as small blebs (typically <10 μm) either enclosed in the matrix glass or adjacent to mineral grains, especially ulvospinel, and are composed of a single-phase, low-Ni pyrrhotite (Table 4).

Rarely, sulfide blebs are present in the basalt rinds on xenoliths, typically located at or near the basalt-xenolith contact. Unlike the sulfides in KH25, these blebs contain pyrrhotite, pentlandite and/or cubanite, similar to the assemblage found in sulfide blebs within the associated xenoliths. Also, semiquantitative and some quantitative analyses indicate that, unlike the sulfides in KH25, the pyrrhotite of those blebs within basalt rinds is Ni-rich, with compositions similar to those of the sulfides in the xenoliths.

Figure 3. a, Subhedral clinopyroxene grains (white) surrounded by large grains of sulfide (su) and spinel (sp), showing textural maturity. Composite xenolith KH77-20. Transmitted light. Scale bar = 200 μm. b, Minute sulfide blebs (white) in trail with fluid inclusions within silicate grain. Clinopyroxenite KH19. Reflected light, oil immersion. Scale bar = 50 μm. c, Curved, dipping fracture plane coated by secondary (postdating silicate crystallization) sulfide inclusions, within silicate grain. Clinopyroxenite KH19. Reflected light, oil immersion. Scale bar = 50 μm. d, Totally isolated sulfide bleb within clinopyroxene (gray) grain. Black grain at bottom is green (in transmitted light) spinel. Lherzolite of sample KH77-20. Reflected light, oil immersion. Scale bar = 50 μm. e, Partial melt zone among three clinopyroxene grains. Sulfides (black) and glass cells (colorless) on edges of and within silicate grains. Clinopyroxenite of KH77-20. Transmitted light, oil immersion, condenser inserted. Scale bar = 50 μm. f, Square (negative crystal) sulfide inclusions (black) in large spinel grain (light gray). Composite xenolith KH16. Transmitted light. Scale bar = 25 μm.

SULFIDE COMPOSITIONAL ANALYSES AND INTERPRETATIONS

Analytic Techniques

Silicate, spinel, and sulfide analyses were performed with a JEOL Superprobe 733 electron microprobe equipped with an energy-dispersive Si(Li) detector and three wavelength spectrometers. Typical operating conditions for sulfide analyses were 15-kV accelerating potential, 30-nanoamp sample current measured on the Faraday cup, 20-second counting times for each element, and generally, a beam diameter of 1 μm.

Natural troilite, millerite, covellite, and synthetic Co_9S_8 were used as standards for sulfide analyses. All elements were calibrated to within 2 standard deviations of the counting statistics on appropriate reference standards. An on-line modified Bence-Albee correction routine (Bence and Albee, 1968; Albee and Ray, 1970) was applied to all analyses.

Analyses of Cu-rich sulfides (cubanite) commonly yielded low totals. Energy dispersive analysis of several cubanites does not indicate sufficient quantities of elements other than those included in the quantitative analytic routine to account for the low totals. In most cases, the low totals are for grains that are quite small, typically less than 5 μm wide. It is believed that matrix excitation of adjacent phases could account for the discrepancies. In any event, molar phase compositions calculated from the cubanite analyses do not appear to be affected significantly.

Sulfide Mineral Chemistry

The sulfide component of the Kilbourne Hole xenoliths is dominated by a pyrrhotite-like monosulfide solid solution phase (mss). The name pyrrhotite (po) is applied to mss containing less than about 5 wt.% Ni. The composition of the po/mss in each individual sulfide bleb is fairly homogeneous, although a slight increase in Ni content commonly occurs toward grain edges in contact with silicates or other sulfide phases. Copper rarely exceeds 0.8 wt.% in the po/mss phases, except in KH28 where the mss phases typically contain about 1 wt.% Cu. (Tables of microprobe analyses of the sulfide phases are available from the authors upon request.)

The nickel content of the po/mss and of the bulk sulfide composition dominated by the po/mss exhibits a general correlation with the sulfide phase assemblage and the host xenolith petrology. In samples in which the only sulfide is po, the Ni

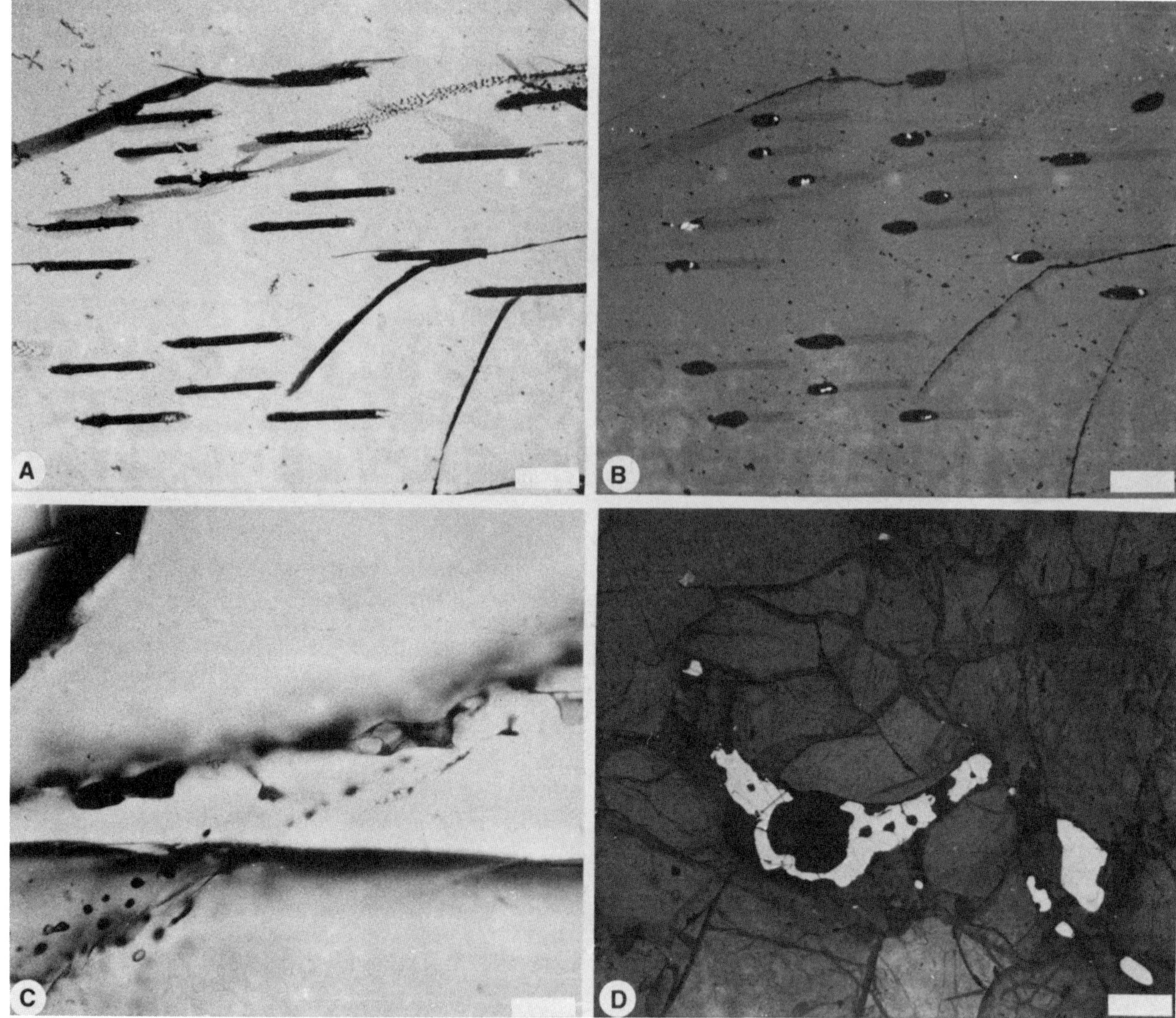

Figures 4. a and b, Oriented sulfide rods in clinopyroxene megacryst KH24. View in transmitted light (a); in reflected light (b). Dry objective. Scale bar = 200 μm. c, Trail of sulfide (black) and glass inclusions within silicate grain. Clinopyroxenite KH19. Transmitted light, oil immersion, condenser inserted. Scale bar = 20 μm. d, Large, abundant, interstitial sulfides (white). Garnet peridotite 70 SAL-2 from Salt Lake Crater, Hawaii. Reflected light, dry objective. Scale bar = 200 μm.

concentrations are low, 0.5 to 3.0 wt.%. Ratios of total-metal/sulfur of the po in these samples typically range from 0.88 to 0.92, except for KH26, in which this ratio is 0.93 to 0.96. In samples containing a po-cb assemblage, Ni can reach slightly higher concentrations in the po, as much as 5.2 wt.% Ni in the po in KH17; the total-metal/sulfur ratios of the po range from 0.88 to 0.95. In samples in which the three-phase assemblage mss-pn-cb occurs, the Ni content of the mss varies widely both within and between samples, with Ni concentrations reaching 30 wt.%. However, the composition of the mss phase within a single sulfide bleb is fairly constant. Within a given xenolith, lower-Ni mss commonly occurs in sulfide blebs containing a somewhat higher proportion of pentlandite or in interstitial sulfide blebs where late oxidation has caused the replacement of some of the sulfide by iron oxides. In two cases (KH31-4, KH34-1, 2), two coexisting mss phases—one Ni-richer (~22 wt.% Ni) than the other (12–13.5 wt.% Ni)—occurred within a single sulfide bleb as grains large enough to analyze separately. In other blebs, mss compositional variations of as much as 2 wt.% Ni result from submicron- to micron-scale, Ni-richer patches within the mss. This may be

TABLE 4. BULK SULFIDE COMPOSITION AND K_d (OLIVINE-SULFIDE)

Sample No.	Fe	Ni	Cu	S	~K_d*
		Element (wt. %)			
Mafic granuloblastites					
KH4	59.15	1.52	0.43	38.44	
KH5	59.66	0.79	0.20	39.69	
KH6A	58.13	2.05	0.43	39.67	
Sillimanite-quartz granuloblastite					
KH26	61.21	0.91	0.32	37.14	
Type-I Spinel Lherzolites					
KH3	35.62	25.32	0.60	38.46	20
KH8	40.32	21.72	0.93	37.03	13
KH10	48.61	12.03	0.50	38.87	9
KH31	39.62	22.06	1.86	36.45	15
KH33	45.43	15.64	1.25	37.68	11
KH34	42.34	17.73	1.73	38.20	13
KH35	37.47	24.55	1.84	36.14	18
KH36	42.75	17.85	3.05	36.36	12
KH39	36.81	24.01	0.96	38.26	17
KH40	45.05	16.62	1.07	37.26	10
KH77-7	48.22	12.60	1.98	37.20	29
KH77-16	36.33	25.65	1.56	36.45	19
Composite xenoliths					
KH16	53.60	8.30	0.88	37.16	10
KH73-1	47.94	10.51	3.31	38.24	9
KH77-20	43.17	8.27	2.78	37.78	
Type-II Clinopyroxenites					
KH17	55.38	3.89	2.26	38.04	13
KH19	57.00	1.48	2.90	38.03	
KH28	52.20	8.41	2.19	37.21	20
Type-II CPX Megacrysts					
KH2	52.34	9.28	2.16	36.22	20
KH9	54.72	2.68	3.46	37.86	24
KH24	58.85	1.55	0.48	38.58	10
Basalt					
KH25	62.40	0.23	0.03	37.01	16

*K_d = $[(X_{Ni}/X_{Fe})_{sulfide}]/[(X_{Ni}/X_{Fe})_{olivine}]$, where X_{Ni} and X_{Fe} are mole fractions of Ni and Fe, respectively.

the result of the exsolution of a Ni-richer mss due to low-temperature miscibility gaps in the monosulfide solid solution (Craig, 1973; Misra and Fleet, 1973) or to incipient pentlandite exsolution. In some cases, this increased Ni content is accompanied by a decrease in total-metal/sulfur ratio, which suggests mss immiscibility rather than pentlandite exsolution.

Pentlandite in the mss-pn-cb assemblage also exhibits a wide range of Ni contents, from 23 to 38 wt.% Ni, and typical Cu concentrations of 1 wt.%. However, pentlandite with less than 31 wt.% Ni is usually found only in samples containing mss relatively low in Ni. No correlation is evident between the Ni contents of pn and the mss or between the proportion of pentlandite and the location of the sulfide bleb. Because of the small size of the pentlandite grains, however, there is a problem of electron beam overlap onto the pyrrhotite matrix.

The only Cu-rich sulfide phase occurring in the Kilbourne Hole xenoliths is similar to cubanite in composition (Fig. 5) and, within the analytic uncertainty, all such analyses lie within the compositional field of quench zone 1 of the intermediate solid solution (iss) at 600°C as defined in the Cu-Fe-S system by Cabri (1973). The iss grains are too small for proper determination of their optical properties. (Distinguishing between cubic iss/cb and orthohombic cb could help establish the temperature of equilibration.) Iss in the Kilbourne Hole xenoliths has Cu:Cu + Fe ratios in the range 0.43 to 0.28 and total-metal/sulfur ratios ranging from 0.97 to 1.03, neither of which appears to be related to the composition of associated mss or pn. Particularly Cu-rich analyses may reflect the presence of chalcopyrite exsolution at a submicron scale, although a discrete chalcopyrite phase was not observed optically or with the SEM. Typically, the cubanite has 1 to 2 wt.% Ni, but up to 4.8 wt.% Ni is present in some analyses. In some cases these higher-Ni cubanites have spherical inclusions of pentlandite; in other cases, the higher-Ni analyses may reflect matrix excitation or beam overlap onto similar, but much

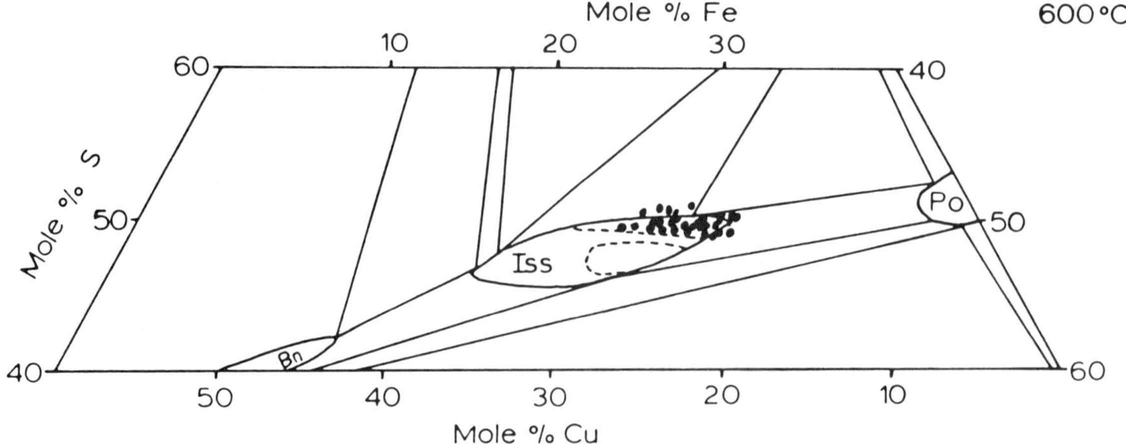

Figure 5. Cu-Fe-S ternary system near iss at 600°C (after Cabri, 1973). Dots represent Cu-rich sulfide analyses from Kilbourne Hole. Key, iss = intermediate solid solution in the Cu-Fe-S system; bn = bornite solid solution field; po = pyrrhotite $(Fe, Ni)_{1-x}S$, with <5 wt. % Ni.

smaller, inclusions. However, no pentlandite inclusions were detected in SEM compositional images (backscattered electron) of these cubanites at 3600× magnification.

Bulk Sulfide Composition

An estimate of the bulk composition of the sulfide component (BSC) of each sample can be made by combining modal proportions of sulfide phases and compositions of the individual phases. Once the bulk sulfide composition is known, experimental studies of phase relationships in metal-sulfur systems can be used to assess, at least qualitatively, the form that the sulfide component would be expected to have in the source region of the Kilbourne Hole xenoliths, as well as the changes, mostly subsolidus, that have occurred to produce the sulfide mineralogy now present.

Calculated bulk sulfide compositions for each of the xenoliths are listed in Table 4. For all calculations the composition of cb, in wt.%, was assumed to be Fe, 37.71; Cu, 25.86; Ni, 1.67; and S, 34.76. This value is more Cu-rich than most of the iss analyses, but it is important to know the maximum likely Cu concentration in the BSC for comparison to the experimental phase diagrams, as done by DeWaal and Calk (1975), Meyer and Boctor (1975), Bishop and others (1975), and Tsai and others (1979). A Cu concentration of 0.6 wt.% was assumed for the po/mss phase, except in KH28 where a value of 1 wt.% Cu was used. Fe, Ni, and S concentrations for the po/mss were selected so that the Ni concentration was near the middle of the range of analyzed Ni concentrations for the po/mss in each xenolith. This seems to be the best way to account for the wide compositional range of po/mss in each sample. For pn, either the composition was assumed to be that of the most Ni-rich pn analysis from each xenolith, or a composition with 34 wt.% Ni was used for samples in which adequate pn analyses were unavailable.

For the sulfide assemblages present in the Kilbourne Hole xenoliths, the two experimentally determined ternary sulfide systems, Cu-Fe-S and Fe-Ni-S systems (see review by Craig and Scott, 1974), are sufficient for consideration. Only the Cu-Fe-S system is necessary for interpretation of the po-cb assemblage, as the calculated Ni concentration in the BSC of these samples is less than 3.9 wt.% and the presence of as much as ~5 wt.% Ni in the bulk composition does not significantly alter the phase relationships in the Cu-Fe-S system (Craig and Kullerud, 1969). For the mss-pn-cb assemblage, the quaternary phase relationships are not substantially different from the relationships defined in the Fe-Ni-S and Cu-Fe-S ternary systems. It should be noted that the phase relationships in these systems are those determined by experiments at 1 atm., and thus only qualitatively indicate the behavior that could occur in the source regions of the xenoliths, where pressure may exceed 20 kbar. Ryzhenko and Kennedy (1973) found an increase of 7°C/kbar in the stability temperature of po. This is probably typical of most sulfides except pn, in which the thermal stability decreases by 5°C/kbar (Bell and others, 1964).

Cu-Fe-S and Fe-Ni-S Systems

Above 600°C, the Cu-Fe-S system (see reviews by Craig and Scott, 1974; Vaughan and Craig, 1978; Barton and Skinner, 1979) is dominated by three extensive solid solution fields and a liquid field, all coexisting with a vapor. The liquid spans the ternary from the Cu-S join to the Fe-S join above 1,100°C. The three solid solution fields that form upon cooling are: the bornite solid solution, appearing below 1,129°C; the intermediate solid solution (iss), appearing below 960°C and including the composition of cubanite; and the pyrrhotite solid solution (po_{ss}). Only the po_{ss} and iss are discussed here. The pyrrhotite solid solution can contain a maximum of about 7 wt.% Cu at 935°C (Kullerud and others, 1969). At lower temperatures the Cu content of the po_{ss}

decreases slowly, still ~4 wt.% Cu at 700°C (Yund and Kullerud, 1966) and tie-lines between po_{ss} and iss remain stable.

The iss can contain 15 wt.% Ni at 700°C (Craig and Kullerud, 1969). Such a Ni-rich iss could exsolve more than 30 vol.% pn at temperatures below ~600°C and still retain 5 wt.% Ni. This could account for the pn inclusions in some of the Ni-rich iss grains present in the Kilbourne Hole xenoliths.

The most Cu-rich calculated BSC of the po-cb assemblage contains 3.5 wt.% Cu. For this composition, and assuming that the temperature estimates from the Wells (1977) two-pyroxene geothermometer (900° to 1,070°C) approximate the ambient temperature for the xenoliths prior to ascent, the sulfide would have existed as a single Cu-rich po_{ss} phase prior to ascent. Exsolution of cb would have occurred only after the xenolith was brought to the surface. (This would also be the case for lower-Cu BSCs.)

Figure 5 shows an enlarged portion of the Cu-Fe-S system around the iss field at 600°C, with the compositions of the Cu-rich sulfides in the Kilbourne Hole xenoliths plotted on it. Cabri (1973) has divided the iss field into three zones, which exhibit different behavior when quenched below 600°C. Allowing for analytic uncertainty and the presence of Ni, all of the Cu-rich sulfide compositions from the Kilbourne Hole xenoliths plot within Cabri's (1973) quench zone 1, where the iss quenches to chalcopyrite and a face-centered cubic phase (fcc) for compositions with $Cu/Cu + Fe \gtrsim 0.36$, or to a single fcc phase for compositions with $Cu/Cu + Fe < 0.36$. Investigations of natural and experimental low-temperature assemblages (e.g., Cabri, 1973) show that the iss is not a stable low-temperature phase. The presence of an iss phase (approximating the composition of cubanite) instead of the stable low-temperature phases, such as talnakhite, mooihoekite, and haycockite in the Kilbourne Hole xenoliths, suggests that subsolidus reequilibration has been blocked at a temperature in excess of 100°C and possibly much higher.

Assuming that the solubility of Cu in Ni-rich mss is about the same as for the po_{ss} in the Cu-Fe-S system, we can effectively evaluate the mss-pn-cb assemblage by looking at only the Fe-Ni-S system. Above about 1,100°C, the sulfide component would consist of a single liquid. Over a small temperature interval somewhere between 1,100° and 1,000°C, a Ni-rich mss would coexist with a Cu-bearing, Ni-richer liquid. Below 1,000°C, in the temperature range indicated by the two-pyroxene geothermometer, only a single solid Ni-rich mss would exist. The cb and pn presently observed would not exsolve until after the xenoliths were brought to the surface and equilibrated to lower temperatures. If the solubility of Cu in Ni-rich mss is not as high as in the po_{ss}, the mss would coexist with a Cu-rich and relatively Ni-rich liquid, until 960°C when iss becomes stable. The presence of such a liquid could produce the commonly observed segregation of cb and pn toward the margins of sulfide blebs and pn exsolution from cb, but the same features could also be produced by exsolution from mss.

Given the temperaturc dependence of assemblages containing two coexisting mss phases (Craig, 1973; Misra and Fleet, 1973) and of the extent of Ni substitution in mssl (decreasing from a limit of ~18 at. % Ni as temperature decreases below 250°C), the minimum blocking temperature or subsolidus reequilibration can be estimated in some of the xenoliths. For most of the xenoliths containing Ni-rich mss, a minimum blocking temperature of about 200°C is indicated, although even lower reequilibration temperatures are permissible for the other xenoliths (Fig. 6).

DISCUSSION

Few studies of sulfides in mantle xenoliths have examined such a wide petrologic variety of samples or have dealt with such great variability in the proposed geneses of these rock types, especially from a single locality. Sulfides may be present throughout the mantle. Their origin is significant to mantle magmatic and metasomatic processes, to the bulk composition of the mantle (for instance, with respect to platinum-group elements), to development of sulfide segregations in large terrestrial magmatic bodies, and to modeling of planetary accretion and evolution (e.g., Mitchell and Keays, 1981; Brett, 1984). However, there is still controversy regarding the origin of the sulfides found in some mantle rocks.

Within the framework of upper mantle silicate petrology, several problems have been addressed on which the sulfide data may shed some light. For instance, the common decoupling in mantle xenoliths of trace-element and major-element signatures has been recognized for many years (e.g., Frey and Green, 1974). This means that a mantle rock may show depletion in major elements (by partial melt extraction), but enrichment in trace elements (by metasomatism) such as light rare-earth elements (LREE) (BVSP, 1981; Menzies, 1983). Furthermore, in type-I xenoliths, it appears that the LREE-enriched component is incorporated in clinopyroxene (Stosch and Seck, 1980). The exact means of this incorporation is unclear; some have inferred the trace elements to be components of a very mobile fluid phase (Stosch, 1982; Zindler and Jagoutz, 1980, 1987). Could the abundance of fluid and sulfide trails in clinopyroxene grains be evidence of this infiltration? Because sulfides form a visible precipitate, unlike rare-earth elements, they represent a mobile component whose movements can be traced to some extent.

Over the years many petrologists have characterized mantle rocks as parental (primary, "pristine"), residual (typical characterization of type I), or cumulate/precipitate (typical characterization of type II; see Menzies, 1983, for a review). Because of the chemical overprinting effects mentioned above, such categorization is not always reliable. One thought has been to see if the sulfide component in a suite of variably depleted xenoliths confirms the classification ascribed to the silicates (DeWaal and Calk, 1975).

Sulfides previously have been examined in several types of mantle rocks, including (1) kimberlite-hosted eclogite, garnet peridotite, and spinel lherzolite xenoliths; garnet and diamond

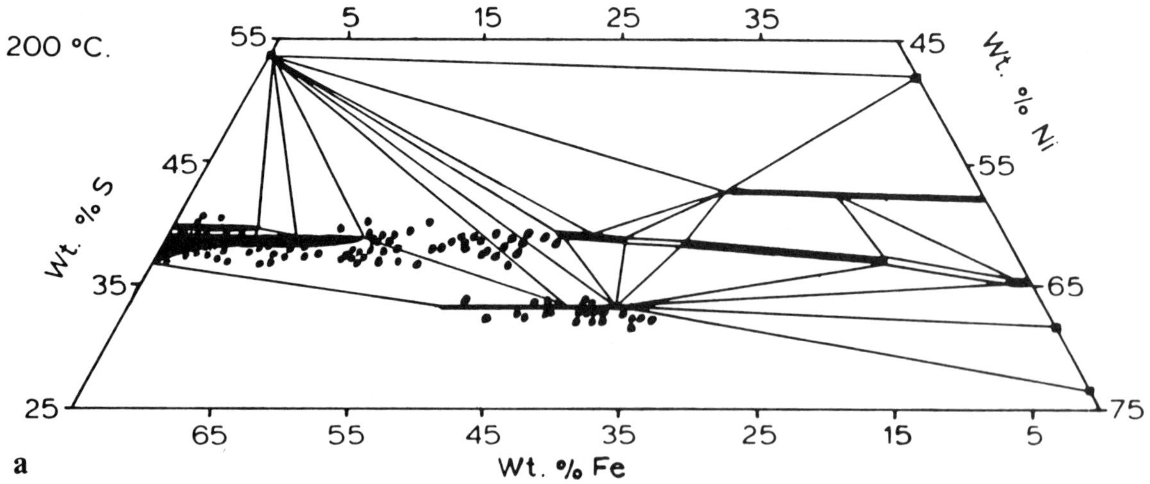

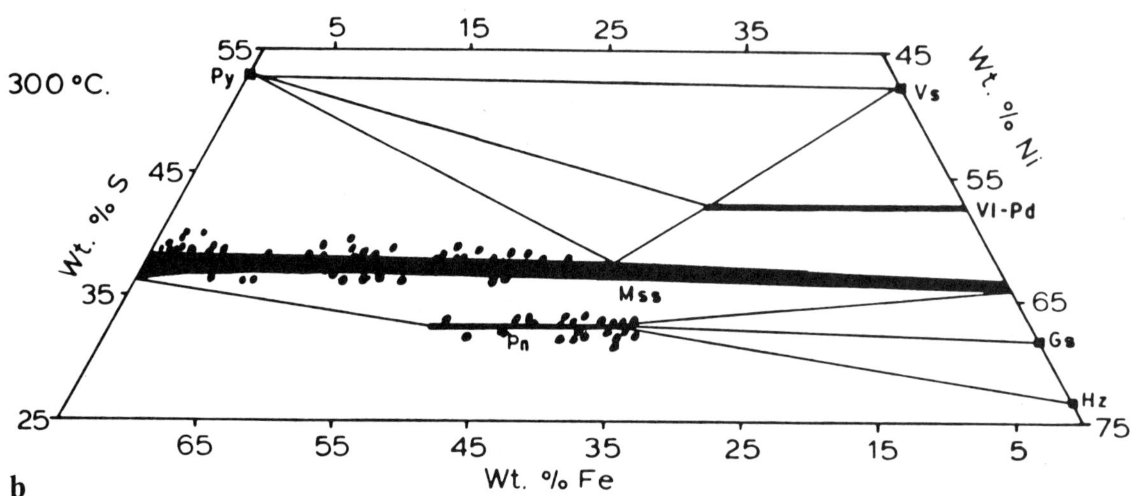

Figure 6. Kilbourne Hole mss and pn analyses (dots) plotted on Fe-Ni-S ternary (after Craig, 1973): a, at 200°C; b, 300°C. Key, mss = monosulfide solid solution = $(Fe, Ni)_{1-x}S$; pn = pentlandite = $(Fe, Ni)_9S_8$.

xenocrysts (e.g., Tsai and others, 1979 and references therein); (2) alkalic-basalt–hosted xenoliths of garnet clinopyroxenite and websterite (DeWaal and Calk, 1975) and of spinel lherzolite (MacRae, 1979; Lorand and Conquere, 1983); and (3) spinel lherzolite and clinopyroxenite in ultramafic complexes presumed to be tectonically emplaced mantle fragments (e.g., Garuti and others, 1984).

The sulfide textures, mineralogy, and, to a lesser extent, abundances observed at Kilbourne Hole are similar to the range reported for other mantle xenolith suites (e.g., eclogites: Frick, 1973, Meyer and Boctor, 1975, Tsai and others, 1979; garnet pyroxenites: DeWaal and Calk, 1975; pyroxene megacrysts: Pasteris, 1982). One striking exception is that chalcopyrite (Meyer and Boctor, 1975; DeWaal and Calk, 1975; Tsai and others, 1979) usually is reported as the Cu-Fe-sulfide, rather than cubanite. (Meyer and Boctor, 1975, reported identifiable cubanite exsolution bodies in chalcopyrite in eclogite from the Stockdale kimberlite, Kansas). The existence of more than one sulfide assemblage in the same xenolith or xenolith suite has been reported previously (in a very detailed study by DeWaal and Calk, 1975). The eclogites studied by Frick (1973) and Tsai and others (1979) have assemblages of Ni-po alone, as well as po + pn ± cp. Some diamonds contain po-cp assemblages with lesser pn (Tsai and others, 1979), similar to the po-cb assemblages at Kilbourne Hole. The latter, however, does not have the Ni-pyrite that is quite abundant in the eclogites. The Kilbourne Hole xenoliths also do not contain the unusual sulfide assemblage reported by Desborough and Czamanske (1973) for a South African eclogite: low-Ni $(Fe, Ni)_9S_{11}$, Ni-pyrite, and a violarite-like phase, $(Fe, Ni)_9S_{11}$.

Many of the sulfide bodies in other xenolith suites are on average much larger than those at Kilbourne Hole. Blebs in the garnet pyroxenite of Salt Lake Crater, Hawaii, may exceed 300 μm in diameter and may contain large inclusions of glass (DeWaal and Calk, 1975; J. D. Pasteris, unpublished data), whereas the sulfide blebs from Stockdale can exceed 1 mm (Meyer and Boctor, 1975).

Sulfide abundances also vary greatly among mantle suites. One observation that has been made repeatedly, however, for suites of varied petrologic types, is that there is a greater abundance of sulfides in pyroxenites (including pyroxene megacrysts) than in the other ultramafic rock types (Frick, 1973; DeWaal and Calk, 1975; Irving, 1980; Pasteris, 1982; Garuti and others, 1984). Among the *spinel lherzolite* suites studied by Irving (1980; BVSP, 1981a) from alkalic basalt localities around the world, samples from Kilbourne Hole were the highest two in sulfide content. He estimated as much as 0.5 modal % (based on bulk-rock sulfur analyses). In particular, Irving (1980; BVSP, 1981a) noted that Kilbourne Hole samples with a fine-grained, tabular texture have a higher abundance of spinel and sulfide than do the more common coarse-granular xenoliths. The garnet *clinopyroxenites* from Salt Lake Crater, however, have much more abundant sulfides *on average* (0.06 to 0.26 vol. %; DeWaal and Calk, 1975) than at Kilbourne Hole (Fig. 5d).

One indisputable observation at Kilbourne Hole is that the sulfide mineralogies and bulk compositions differ among the xenolith types. This suggests that the sulfides had *distinctive* origins. However, one cannot rule out *homogeneous* contamination by a sulfide liquid followed by sulfide-silicate re-equilibration (unlikely). In their suite of a dozen eclogites from various localities, Tsai and others (1979) also noticed distinct compositional and textural differences among the sulfide assemblages. As the authors point out, some of the differences are due to cooling rates. They also could have added that some differences are due to secondary replacement. The mineralogic and bulk sulfide composition differences at Kilbourne Hole, however, appear to be original.

Sulfide liquid immiscibility somewhere in the mantle system probably is the ultimate source of sulfides in all the mantle xenoliths, but the mechanism of sulfide segregation remains unclear. The factors controlling the solubility of sulfur and the development of an immiscible sulfide liquid have been investigated experimentally (mostly at 1 atm., with a few high-pressure studies) in various simple systems analogous to natural melts (e.g., Shima and Naldrett, 1975; Buchanan and Nolan, 1979; Mysen and Popp, 1980; Wendlandt, 1982). The effect of pressure on the development of sulfide immiscibility is disputed (Mysen and Popp, 1980; Wendlandt, 1982). It is difficult to separate changing fO_2, fS_2, and bulk composition (e.g., Fe-free versus Fe-bearing melt) from changing pressure effects (Mysen and Popp, 1980; Wendlandt, 1982). The complex interplay of factors affecting sulfur solubility suggests that substantial differences may exist between natural systems and the simple experimental systems that have been examined as potential analogs for the behavior of sulfur-bearing systems during magmatic processes (e.g., Thompson and others, 1984; Barnes and Naldrett, 1985). For instance, the experiments of Naldrett (1969) and the empirical analysis of Skinner and Peck (1969) suggest that magnetite should be a common phase in immiscible sulfide blebs. Almost no primary magnetite was found associated with the Kilbourne Hole sulfides, which in part may be due to a pressure effect (Wendlandt, 1982) or to fO_2 (Boctor and Yoder, 1983; N. Boctor, personal communication, 1985). It is also a reminder that the processes of sulfide immiscibility may differ significantly between crustal and upper mantle environments due to differences in pressure and to crustal or other contamination. In a word, the triggering mechanism for sulfide immiscibility in mantle melts remains unknown. However, other aspects of the origin of sulfides in the Kilbourne Hole xenoliths are clearer.

Origin of Sulfides in Clinopyroxenites and Megacrysts

The sulfides in the megacrysts and clinopyroxenites, including the clinopyroxenites in the composite xenoliths, present the least arguable interpretation. Many features, including the abundance of fully enclosed sulfide blebs, point to an immiscible sulfide phase formed in a magma at depth. The clinopyroxenites and megacrysts probably represent phases that crystallized at high pressure from an alkalic basalt similar to their host. Immiscible sulfides have been observed in basalts from many locations (e.g., Desborough and others, 1968; Moore, 1973; Mathez, 1976; Czamanske and Moore, 1977; MacLean, 1977; Pedersen, 1979) and are present in the Kilbourne Hole basalts.

MacLean (1969) suggested that for extents of partial melting reasonably expected in the mantle, an immiscible liquid may always be present. If these melts do contain a sulfide liquid at the time of their extraction, it is quite possible that some of the sulfide could remain as a residual phase (Helz, 1977). Following extraction of a partial melt, sulfide immiscibility could be initiated or maintained due to decreasing temperature, increasing fO_2, or changes in the composition of the magma due to reaction with wallrock or to crystallization, especially if an Fe-Ti-oxide (such as the Fe-rich spinels of the pyroxenite, megacryst, and mafic xenoliths) is part of the solid assemblage (Haughton and others, 1974; Buchanan, 1976).

If the sulfides in the xenoliths formed as an immiscible sulfide liquid in equilibrium with the silicate melt that crystallized the pyroxenite or megacryst, the sulfide liquid would be expected to have compositions that reflect the bulk composition of the coexisting silicate melt through sulfide-melt/silicate-melt distribution coefficients (e.g., MacLean and Shimazaki, 1976; Rajamani and Naldrett, 1978). In lieu of the bulk composition of the silicate melt, the composition of a phase that crystallized in equilibrium with both melts can be used, if the distribution coefficients are known. Olivine is the only suitable silicate phase for which distribution coefficients with sulfide liquid have been determined experimentally. Unfortunately, the results of these studies, all performed under significantly different conditions, are difficult to interpret and somewhat inconsistent (Clark and Naldrett, 1972; Mysen and Kushiro, 1976; Fleet and others, 1977; Boctor, 1982; Fleet and MacRae, 1983). Whereas the effects of pressure and sulfur fugacity may be negligible, other factors, such as temperature and oxygen fugacity, may have a significant effect on the distribution coefficients (e.g., Boctor, 1982). Reported values of the distribution coefficient for the Ni-Fe exchange between olivine and sulfide, K_d = (Ni/Fe)sulf/(Ni/Fe)ol, range from 4 to 33. Expected distribution coefficients for olivine-sulfide pairs equilibrated in natural, mafic bodies under normal magmatic conditions range from 5 to 20 (Thompson and others, 1984; Barnes and Naldrett, 1985). The calculated values for this ratio (Table 4) using the bulk sulfide composition and average olivine compositions determined for Kilbourne Hole mantle xenoliths (Dromgoole, 1984) range from 9 to 29. Considering the uncertainties, these values cannot be used to prove that an immiscible sulfide liquid coexisted with the crystallizing silicate liquid. The data, however, certainly are not inconsistent with this hypothesis.

Further evidence for an immiscible sulfide liquid as the origin of the sulfides in this group of xenoliths is the oriented sulfide blebs in megacryst KH24, suggesting that sulfide, probably still liquid, was trapped during crystal growth (cf. Peterson and Francis, 1977; Pasteris, 1982; Haggerty and Tompkins, 1982). There also are apparent equilibrium textures between some larger sulfide blebs and primary silicates and spinel. In addition, sulfide inclusions occur isolated within silicates; the latter two features suggest that sulfides probably were present prior to deformation and recrystallization.

Origin of Sulfides in Mafic Granuloblastites

The mafic granuloblastites at Kilbourne Hole are believed to be the metamorphosed product of a mantle-derived melt (James and others, 1980). The presence of fully enclosed sulfide blebs in these rocks suggests that sulfides were present prior to incorporation of the xenoliths into the host basalt, probably as an immiscible sulfide liquid within the magma that crystallized to form the xenoliths. In general, the Ni and Ni/Cu values in the sulfides tend to decrease with decreasing Mg/Mg + Fe ratios in pyroxene and with decreasing modal proportions of ferromagnesian minerals, evidence supporting silicate melt–sulfide melt equilibration (MacLean and Shimazaki, 1976; Rajamani and Naldrett, 1978; Boctor and Yoder, 1983). Alternatively, if these xenoliths represent metamorphosed crustal rocks, the sulfides could represent metamorphosed primary sulfides. This mechanism is discussed below in regard to the sillimanite-quartz granuloblastite.

Origin of Sulfides in the Sillimanite Granuloblastite

Sulfides are rare in the one xenolith of this type, and their origin is not readily apparent. They probably reflect primary sulfides present in the crustal rocks prior to metamorphism, perhaps resulting from desulfidation of pyrite during metamorphism. At 1 atm., pyrite is unstable above 742°C, at which temperature it converts to pyrrhotite and sulfur vapor. Alternatively, reduction of sulfate during metamorphism could result in sulfide formation. The small Ni and Cu content of the sulfides in this sample could have been present in original (sedimentary) pyrite or acquired through equilibration with other phases in the rock. However, the possibilities of an immiscible sulfide associated with the extensive partial melting that has occurred in this sample, or of externally derived sulfides, cannot be ruled out. Sulfur isotopic studies might distinguish between these sources, depending on the effects and extent of metamorphism and late-stage partial melting (cf. Ripley, 1981).

Origin of Sulfides in the Spinel Lherzolites

A controversy exists concerning the origin of sulfides in "residual" mantle material, i.e., not pyroxenites or other rocks that probably represent crystallization of a mantle-derived mafic melt. The question is whether the sulfides represent a cogenetic immiscible sulfide liquid, or whether they were introduced during a relatively late "contamination" event.

MacRae (1979) investigated the distribution of sulfides in spinel lherzolite xenoliths in the alkalic Newer Basalts of Victoria, Australia. He proposed that the sulfide originated as an immiscible sulfide liquid separated from internally derived silicate partial melts formed during the ascent of the xenoliths. He further proposed that these melts became sulfur-saturated due to the influx of a sulfur-bearing, CO_2-rich vapor from the host basalt, which triggered partial melting of the silicates. MacRae's main reason for this proposition is his observation that most of the sulfides are associated with partial melt glasses. In addition, based on his petrologic studies, MacRae concluded that the Ni content of the bulk sulfide of the Newer Basalt xenoliths was not high enough to have equilibrated with primary mantle olivine.

Contrary to MacRae's hypothesis, an external magmatic source of sulfur seems unnecessary to account for the sulfides in the Kilbourne Hole spinel lherzolite xenoliths (and possibly for the sulfide in the Newer Basalt xenoliths as well), and is inconsistent with some features. First, the fact that the bulk sulfide composition of the Kilbourne Hole xenoliths is Ni-rich suggests equilibrium with a relatively primitive melt. The calculated partition coefficients for olivine-sulfide Ni exchange in the Kilbourne Hole xenoliths, although difficult to evaluate, are within the range of possible values indicated by experiment. (Although experimental studies [e.g., Clark and Naldrett, 1972] also have shown that at temperatures exceeding only 900°C, there can be rapid reequilibration of the Ni partitioning between adjacent fayalite and sulfide grains, such reequilibration with Mg-rich olivine is much less likely [N. Boctor, personal communication, 1985]). Second, the mechanism proposed by MacRae (1979) would be incapable of producing the large enclosed sulfide blebs (especially those without fractures connecting them to the grain edge) that apparently were present before crystallization/recrystallization of the primary phases. The latter blebs, in turn, are inferred from other textural analysis to have existed prior to incorporation of the xenoliths into the host basalt (Bussod, 1981). Also, the chemical similarity between the enclosed and the interstitial sulfides argues against a different origin for the two types of sulfides. Third, the presence of fluid inclusions in the primary minerals of the xenoliths does not prove that the fluid or sulfide in them was derived from the host basalt, as assumed by MacRae. They could have been present even before incorporation of the xenolith, but simply redistributed (e.g., Pasteris, 1987). In the latter case, the composition of the sulfide-associated glass inclusions and the interstitial glasses probably would be noticeably different, a test that could be made in a future study. Finally, the hypothesis of an external magmatic source for sulfur would not account for the presence of relatively high abundances of sulfides in xenoliths, such as KH79, where partial melt textures and interstitial glasses are virtually absent.

A second "contamination" hypothesis is that some sulfides in mantle xenoliths, in particular K-Fe-Ni-sulfides, have been introduced or remobilized by metasomatic fluids in the mantle or by the transporting magma (Gurney and others, 1975; Clarke and others, 1977; Tsai and others, 1979; Irving, 1979; BVSP, 1981a; Pasteris, 1982). The LIL-rich component proposed by Frey and Prinz (1978) to account for the trace-element pattern of some spinel lherzolite xenoliths might be such a fluid. In particular, Irving (1979, 1980) has suggested S-bearing, CO_2- and/or H_2O-rich fluids that were derived from basanitic magma within the upper mantle, as the source for some of the sulfides in Kilbourne Hole xenoliths, especially some of the composite xenoliths included in this study (BVSP, 1981a). Sufficient data are not available on the trace element composition of the xenoliths in this study to evaluate the extent of metasomatism, but the absence of hydrous phases in most of the spinel lherzolites and the general pattern of LREE depletion in Kilbourne Hole xenoliths (BVSP, 1981a) suggest that minimal metasomatism has occurred. No obviously metasomatic sulfides (i.e., K-Fe-Ni sulfides) were detected in the Kilbourne Hole xenoliths, nor have they been identified in spinel lherzolites from any other location. The presence in the Kilbourne Hole xenoliths of small sulfide blebs associated with fluid/glass inclusions along fractures in primary minerals might be the result of remobilization of sulfides in the presence of a (metasomatic) fluid. However, if these sulfides were remobilized, it could be expected that they would be Cu and Ni rich, which is not the case for the few analyses obtained for them. The abundance of platinum-group elements in the sulfides from Kilbourne Hole also led Mitchell and Keays (1981) to conclude that a metasomatic origin for the sulfides is unlikely.

The third "contamination" hypothesis is that an immiscible sulfide liquid separated from a mantle-derived melt not cogenetic with the spinel lherzolites, and that part of this sulfide liquid infiltrated the mantle "country rock." Such a mechanism would be most likely to have affected the composite xenoliths. The sulfides in the lherzolite (i.e., wallrock) portion of the composite xenoliths examined in this study are chemically indistinguishable from the sulfides in the pyroxenite (i.e., dike or vein) portion, which could be interpreted as the result of essentially isochemical infiltration (unlikely) of a sulfide liquid from the pyroxenite-forming mafic magma into the lherzolite. However, considering the substantial chemical modification and reequilibration exhibited by the silicates and spinels of the lherzolite in response to the introduced pyroxenite (Irving, 1980; Dromgoole, 1984), it is likely that sulfides in the lherzolite would have equilibrated to compositions similar to those in the pyroxenite. Unfortunately, the bulk composition of the sulfides in the composite xenoliths cannot be used to associate them clearly with either the spinel lherzolites or the pyroxenites and megacrysts in the rest of the suite. The calculated compositions are similar to the compositions of the Ni-richest pyroxenites, but not drastically different from the Ni-poorest lherzolite compositions. If sulfides were introduced during the plating out of the pyroxenite dikes (Irving, 1980), it seems likely that the abundance of sulfides would be higher than actually observed for lherzolites in composite xenoliths. The sulfide abundance would be expected to increase toward the pyroxenite contact, but this has not been noted. Although recrystallization or other factors might homogenize the sulfide distribution, the above observations do suggest that the sulfides in the lherzolite portion of these xenoliths were modified by, but present prior to, the introduction of the apparently sulfide-rich pyroxenite. Contamination from mafic magmas becomes even more dubious for the simple spinel lherzolite xenoliths, which may have never been in contact with a noncogenetic melt other than the host basalt. Different bulk sulfide compositions in lherzolite and pyroxenite lithologies in an ultramafic complex, and the absence of increased sulfide abundance toward pyroxenite dikes are criteria also used by Garuti and others (1984) to reject the possibility of sulfide infiltration from external sources to account for sulfides in the lherzolite.

The hypothesis favored by most investigators of sulfides in mantle rocks and the one that seems most likely to account for the sulfides present in the spinel lherzolite xenoliths in this study, is that an immiscible sulfide liquid formed during partial melting of mantle material and that a part of this liquid was trapped in the solid-liquid residuum represented by the spinel lherzolite (Frick, 1973; DeWaal and Calk, 1975; Tsai and others, 1979 and references therein; Mitchell and Keays, 1981; Lorand and Conquere, 1983; Garuti and others, 1984). Supersaturation of sulfur in mantle-derived melts is indicated by the presence of sulfides in the pyroxenites and megacrysts derived from such melts and the common occurrence of sulfides in extruded, mantle-derived basalts (e.g., Czamanske and Moore, 1977). Experimentally determined metal partitioning between sulfide liquids and mafic/ultramafic melts indicates that the sulfide liquid would be Ni-rich, with high Ni/Cu ratios, as are the sulfides in the spinel lherzolites from Kilbourne Hole and elsewhere. In addition, calculated olivine-sulfide Ni partition coefficients are similar to experimentally determined equilibrium coefficients (MacLean, 1977; Rajamani and Naldrett, 1978; Fleet and MacRae, 1983, 1986; Lorand and Conquere, 1983; Garuti and others, 1984).

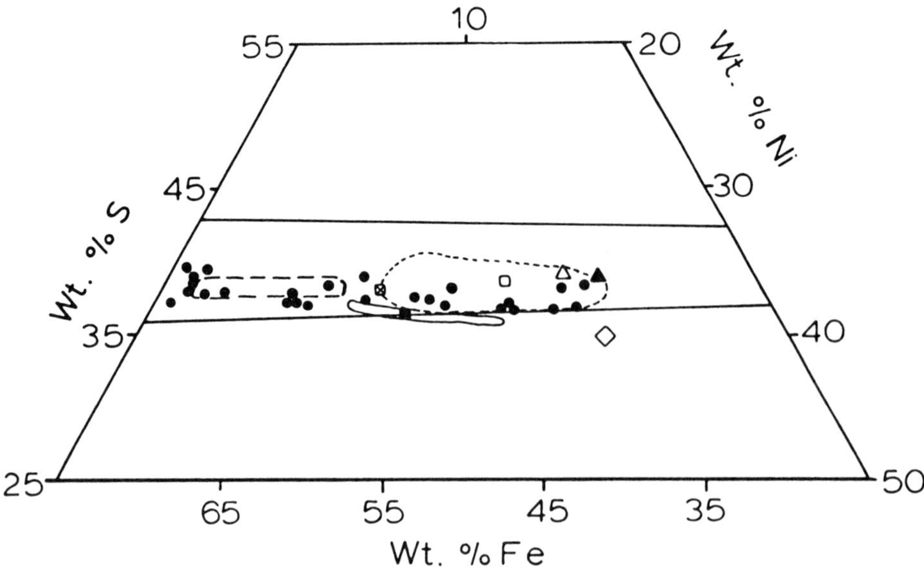

Figure 7. Portion of Fe-Ni-S ternary system at 1,000°C, showing approximate compositional limits of mss and projections of sulfide compositions in several suites of mantle rocks, including Kilbourne Hole. Key to symbols: black circle = bulk sulfide composition for Kilbourne Hole xenoliths; open square = average bulk sulfide composition of French xenoliths; open square with cross-marks = average bulk sulfide composition of interstitial sulfides in Montferrier xenoliths; black square = average bulk sulfide composition of enclosed sulfides in Montferrier xenoliths; open triangle = average bulk sulfide composition of enclosed sulfides in Bestiac ultramafic body; black triangle = average bulk sulfide composition of interstitial sulfides in Bestiac ultramafic body; open diamond = average bulk sulfide composition of enclosed sulfides in Beni Bousera ultramafic body; long dashes = range of bulk sulfide compositions for individual xenoliths from Salt Lake Crater, Hawaii; short dashes = range of bulk sulfide composition of enclosed sulfides for individual French xenoliths; solid line = range of bulk sulfide compositions of interstitial sulfides for individual xenolith from Montferrier. Salt Lake Crater data from DeWaal and Calk, 1975; French, Montferrier, Bestiac, and Beni Bousera data from Lorand and Conquere, 1983.

An alternate hypothesis that cannot be eliminated based on the observations made here, and also involving a cogenetic origin, is that the sulfides are primordial mantle material, present since the formation of the mantle. It seems doubtful, however, that a primordial sulfide phase, with its relatively high reactivity and low melting temperature, could survive unaltered through partial melting and other processes that have affected mantle lherzolites.

COMPARISON WITH PREVIOUS INVESTIGATIONS OF MANTLE SULFIDES

Only a few other studies addressed the sulfides in xenolith suites of a petrologic variability comparable to the Kilbourne Hole suite of the present investigation. The results (see Fig. 7) and conclusions of these studies in several cases differ strongly from those presented above, emphasizing the importance of studying sulfides to help infer differences in the processes affecting different mantle suites.

DeWaal and Calk (1975) investigated sulfides (see Fig. 7) in garnet pyroxenite xenoliths from Salt Lake Crater, Hawaii. They determined that the xenoliths define two groups based on their bulk sulfide composition, and that these subdivisions also correlate with the bulk composition of the entire xenolith. This led DeWaal and Calk (1975) to believe that the silicate and sulfide compositions confirmed the inferred petrologic range of the xenoliths from primitive, through derivative, to residual mantle material. They also found that, in the Ni-Cu–rich samples, the interstitial sulfides are substantially lower in Ni and Cu content than the enclosed sulfides, unlike the Kilbourne Hole xenoliths. DeWaal and Calk interpreted this difference as due to reequilibration of the interstitial sulfides with the host magma, a type of "contamination" hypothesis.

Lorand and Conquere (1983) investigated sulfides in spinel lherzolites from several localities in the Tertiary basalt province of France (see Fig. 7). They typically found the assemblage mss/po ± pn ± cb/cp, with modal proportions similar to those at Kilbourne Hole. There are some distinct differences, however, between the New Mexico and the French sulfides, mostly with regard to the interstitial sulfides. The relative proportion of interstitial sulfides at Kilbourne Hole far exceeds that in many of the French samples. Furthermore, in one of the French suites (Montferrier) that does contain abundant interstitial sulfides, the compositions (<1 wt.% Ni in po cf. Ni-rich mss) and modes (pn = po >>cb) of the interstitial sulfides differ from those of the enclosed sulfides in the same suite (Fig. 7). Lorand and Conquere (1983) interpreted the difference as due to desulfidation of the interstitial sulfides, a process that they believed occurred in the mantle. Since all the xenoliths they examined are reportedly rich in amphibole, suggestive of mantle metasomatism, the postulated desulfidation event might have coincided with the metasomatism. The virtual absence of amphibole in the Kilbourne Hole xenoliths suggests that a similar metasomatic event has not affected them, which may account for their lack of compositional difference between enclosed and interstitial sulfides.

Another difference between the French and the Kilbourne Hole xenoliths is in the apparent behavior of the sulfides in response to partial melting of the host rock. Lorand and Conquere (1983) documented a bimodal distribution of the modes of the sulfide phases in xenoliths that had undergone extensive partial melting. One group consistently has dominant pn with lesser cb and very minor mss (cf. Garuti and others, 1984), and the other has predominantly mss with lesser pn and cb. Lorand and Conquere interpreted this as the result of the formation and mobilization of a Ni-(Cu)–rich liquid separated from a previously solid mss in response to heating of the xenolith during incorporation by the host basalt (similar to Meyer and Boctor's, 1975, idea of why some sulfide globules have an mss rim around their po-pn-cp cores). Such compositional distinctions are not found in the Kilbourne Hole suite, perhaps because its partial melting was predominantly a decompression effect, as suggested by Kleeman and others (1969) and Irving (1980).

A more extensive treatment of ultramafic mantle rocks with a pn-rich assemblage is presented by Garuti and others (1984), who examined peridotites of the Ivrea-Verbano zone in the Italian Alps, i.e., ultramafic complexes. They studied rocks exhibiting three different degrees of partial-melt depletion and concluded that partial melting during the ascent of the peridotite extracted Fe (and presumably sulfur) from a solid mss phase that had formed during an earlier partial melting event. Garuti and others thus concluded that po-rich sulfide assemblages, characteristic of volcanic xenoliths, are present only in relatively primitive mantle lherzolites.

SUMMARY AND CONCLUSIONS

The abundance of sulfides is low in all of the Kilbourne Hole xenoliths examined. The highest abundances (<0.5 vol. %) generally are found in pyroxenites and megacrysts, in keeping with the results of other studies that show pyroxene-rich mantle rocks to have high sulfide concentrations (e.g., Frick, 1973; DeWaal and Calk, 1975; Irving, 1980; Pasteris, 1982 and unpublished data; Garuti and others, 1984). The sulfides at Kilbourne Hole occur as irregular blebs interstitial to and enclosed in silicates, and as inclusions in partial melt zones and along fracture planes, where they may be associated with fluid ± glass inclusions. Mss is the predominant sulfide phase in all of the xenoliths examined. In the crustal xenoliths and one megacryst, mss is the only sulfide phase, but in the other xenoliths it is associated with pn and/or cb. The sulfide assemblages indicate that subsolidus equilibration was blocked at temperatures as high as 200°C.

Estimates of the bulk composition of the total sulfides in each xenolith were calculated from mineral chemistry and modal abundance of sulfide phases. In support of the belief that this petrologically varied suite of xenoliths was not infiltrated homogeneously by some late-stage sulfide contaminant, the petrologic types have distinct sulfide chemistries. In addition, no significant difference was observed between the mineral chemistry or modal abundance in the interstitial compared to the enclosed sulfides. This is in contrast to the findings of DeWaal and Calk (1975) for garnet pyroxenites and Lorand and Conquere (1983) for spinel lherzolites. This may be due to a lack of postsulfide metasomatism or partial melt extraction at Kilbourne Hole.

The bulk sulfide compositions at Kilbourne Hole are Ni-rich for the lherzolite xenoliths (12 to 27 wt.%), Ni-poor for the crustal xenoliths (<2 wt. %), and intermediate for the pyroxenites, megacrysts, and composite xenoliths (1.5 to 10.5 wt.%). For all the xenoliths, Cu is less than 3.5 wt. % of the BSC. The bulk sulfide composition in all cases is within the mss one-phase field of the Fe-Ni-Cu-S system at the temperature of last equilibration of the silicates as indicated by the Wells two-pyroxene geothermometer. Thus, the present mineralogy is the result of subsolidus reequilibration.

A number of mechanisms have been proposed to account for the presence of sulfides in mantle spinel lherzolites ("residua"), but for the Kilbourne Hole xenoliths the one that seems most likely is partial retention, in the residual rock, of an immiscible sulfide formed during mantle partial melting. For xenoliths ascribed to formation by crystallization from mantle-derived magmas, the sulfides are interpreted as having originated as an immiscible sulfide liquid that developed within these magmas. This is suggested by the occurrence of sulfides in the Kilbourne Hole basalt and other mantle-derived melts, experimental studies indicating the likelihood of sulfur saturation of mantle melts, the consistence of observed and experimental olivine-sulfide Ni partition coefficients, and apparent equilibrium textures between larger sulfide blebs and surrounding silicates. As with the lherzolite xenoliths, the host basalt can be eliminated as a possible source for the sulfides in these xenoliths.

Although studies like the present one involve very painstaking analysis, it appears worthwhile to continue such work on petrologically well-documented suites. The very preservation of sulfides in xenoliths reveals something about their introduction. The specific mineralogy (e.g., pn-dominant versus po-pn-cb/cp) of the sulfides, and whether it differs among textural sites within a xenolith or among members of a petrologically related suite, reveals information about postsulfide processes such as partial melting. More suites need to be analyzed in order to make generalizations. Another line of evidence to be followed is the analysis of the different types of glass (interstitial, spongy partial melt halos on silicates, in fluid inclusion trails) that are found alone or associated with sulfides. Glass compositions may help reveal whether most of the frozen liquid was injected from outside the xenolith or instead represents in situ decompression melting. Sulfur isotope analysis on the sulfide assemblages could confirm or refute the assumption of a dominant mantle sulfur component.

ACKNOWLEDGMENTS

We thank Anthony Irving for the loan of several Kilbourne Hole samples and Lewis Calk for providing unpublished sulfide analyses for Salt Lake Crater xenoliths. The loan of a suite of Salt Lake Crater thin sections to J.P.D. from the Smithsonian Institution is appreciated. We also appreciate the comments and suggestions by Anthony Irving and Nabil Boctor on an earlier draft of the manuscript, but we accept full responsibility for all data and interpretations expressed in the paper. This project was funded in part by grant EAR 8025255 from the National Science Foundation.

REFERENCES CITED

Ahrens, L. H., Dawson, J. B., Duncan, A. R., and Erlank, A. J., eds., 1975, Physics and chemistry of the earth, v. 9: Oxford, Pergamon Press.

Albee, A. L., and Ray, L., 1970, Correction factors for electron probe microanalysis of silicates, oxides, carbonates, phosphates, and sulfates: Analytical Chemistry, v. 42, p. 1408–1414.

Anderson, A. T., Jr., 1974, Chlorine, sulfur, and water in magmas and oceans: Geological Society of America Bulletin, v. 85, p. 1485–1492.

BVSP (Basalt Volcanism Study Project), 1981a, Ultramafic xenoliths in terrestrial volcanics and mantle magmatic processes, *in* Basaltic volcanism on the terrestrial planets: New York, Pergamon Press, p. 282–310.

—— , 1981b, Experimental petrology of basalts and their source rocks, *in* Basaltic volcanism on the terrestrial planets: New York, Pergamon Press, p. 493–631.

Barnes, S. J., and Naldrett, A. J., 1985, Geochemistry of J-M (Howland) Reef of the Stillwater Complex, Minneapolis Adit area. I. Sulfide chemistry and sulfide-olivine equilibrium: Economic Geology, v. 80, p. 627–645.

Barton, P. B., and Skinner, B. J., 1979, Sulfide mineral stabilities, *in* Barnes, H. L., ed., Geochemistry of hydothermal ore deposits: New York, John Wiley & Sons, p. 278–403.

Bell, P. M., England, S. L., and Kullerud, G., 1964, Pentlandite; Pressure effects on breakdown: Carnegie Institution of Washington Yearbook, v. 63, p. 206–207.

Bence, A. E., and Albee, A. L., 1968, Empirical correction factors for the electron microanalysis of silicates and oxides: Journal of Geology, v. 76, p. 382–403.

Bishop, F. C., Smith, J. V., and Dawson, J. B., 1975, Pentlandite-magnetite intergrowth in DeBeers spinel lherzolite; Review of sulphides in nodules, *in* Ahrens, L. H., and others, eds., Physics and chemistry of the earth, v. 9: Oxford, Pergamon Press, p. 323–337.

Boctor, N. Z., 1982, The effect of fO_2, fS_2, and temperature on Ni partitioning between olivine and iron sulfide: Carnegie Institution of Washington Yearbook, v. 81, p. 366–368.

Boctor, N. Z., and Yoder, H. S., 1983, Partitioning of nickel between silicate and iron sulfide melts: Carnegie Institution of Washington Yearbook, v. 82, p. 275–277.

Boyd, F. R., and Meyer, H.O.A., eds., 1979, The mantle sample: Washington, D.C., American Geophysical Union, 424 p.

Brett, R., 1984, Chemical equilibration of the Earth's core and upper mantle: Geochimica et Cosmochimica Acta, v. 48, p. 1183–1188.

Buchanan, D. L., 1976, The sulfide and oxide assemblages in the Bushveld Complex rocks of the Bethal area: Transactions of the Geological Society of South Africa, v. 79, p. 76–80.

Buchanan, D. L., and Nolan, J.,1979, Solubility of sulfur and sulfide immiscibility in synthetic tholeiitic melts and their relevance to Bushveld-Complex rocks: Canadian Mineralogist, v. 17, p. 483–494.

Bussod, G.Y.A., 1981, Thermal and kinematic history of mantle xenoliths from Kilbourne Hole, New Mexico [M.S. thesis]: Seattle, University of Washington, 72 p.

Bussod, G.Y.A., and Irving, A. J., 1981, Thermal and rheological history of the upper mantle beneath the southern Rio Grande Rift; Evidence from Kilbourne Hole xenoliths, *in* Papers presented to the conference on the processes of planetary rifting: Lunar and Planetary Institute, Contribution No. 457, p. 145–148.

Cabri, L. J., 1973, New data on phase relations in the Cu-Fe-S system: Economic Geology, v. 68, p. 443–454.

Carter, J. L., 1965, The origin of olivine bombs and related inclusions in basalts [Ph.D. thesis]: Houston, Rice University, 275 p.

Clark, T., and Naldrett, A. J., 1972, The distribution of Fe and Ni between synthetic olivine and sulfide at 900°C: Economic Geology, v. 67, p. 939–952.

Clarke, D. B., Pe, G. G., MacKay, R. M., Gill, K. R., O'Hara, M. J. and Gard, J. A., 1977, A new potassium-nickel sulfide from a nodule in kimberlite: Earth and Planetary Science Letters, v. 35, p. 421–428.

Craig, J. R., 1973, Pyrite-pentlandite assemblages and other low temperature relations in the Fe-Ni-S system: American Journal of Science, v. 273-A, p. 496–510.

Craig, J. R., and Kullerud, G., 1969, Phase relations in the Cu-Fe-Ni-S system and their application to magmatic ore deposits, *in* Wilson, H. D., ed., Magmatic ore deposits: Economic Geology Monograph, v. 4, p. 344–358.

Craig, J. R., and Scott, S. D., 1974, Sulfide phase equilibria, *in* Ribbe, P. H., ed., Sulfide minerals: Washington, D.C., Mineralogical Society of America, Reviews in Mineralogy, v. 1, p. CS1–CS110.

Czamanske, G. K., and Moore, J. G., 1977, Composition and phase chemistry of sulfide globules in basalt from the Mid-Atlantic Ridge rift valley near 37°N. lat.: Geological Society of America Bulletin, v. 88, p. 587–599.

Desborough, G. A., and Czamanske, G. K., 1973, Sulfides in eclogite nodules from a kimberlite pipe, South Africa, with comments on violarite stoichiometry: American Mineralogist, v. 58, p. 195–202.

Desborough, G. A., Anderson, A. T., and Wright, T. L., 1968, Mineralogy of sulfides from certain Hawaiian basalts: Economic Geology, v. 63, p. 636–644.

DeWaal, S. A., and Calk, L. C., 1975, The sulfides in the garnet pyroxenite xenoliths from Salt Lake Crater, Oahu: Journal of Petrology, v. 16, p. 134–153.

Dromgoole, E. L., 1984, Petrology and chemistry of sulfides in xenoliths from Kilbourne Hole, New Mexico [M.A. thesis]: St. Louis, Washington University, 199 p.

Fleet, M. E., 1979, Partitioning of Fe, Co, Ni, and Cu between sulfide liquid and basaltic melts and the composition of Ni-Cu sulfide deposits; A discussion: Economic Geology, v. 74, p. 1517–1519.

Fleet, M.E., and MacRae, N. D., 1983, Partition of Ni between olivine and sulfide and its application to Ni-Cu sulfide deposits: Contributions to Mineralogy and Petrolgoy, v. 83, p. 75–81.

Fleet, M. E., and MacRae, N. D., 1986, Partition of Ni between olivine and sulfide; The effect of temperature, fO_2, and fS_2 [abs.]: Geological Society of America Abstracts with Programs, v. 18, p. 602.

Fleet, M. E., MacRae, N. D., and Herzberg, C. T., 1977, Partition of nickel between olivine and sulfide; A test for immiscible sulfide liquids: Contribu-

tions to Mineralogy and Petrology, v. 65, p. 191–197.
Fleet, M. E., MacRae, N. D., and Osborne, M. D., 1981, The partition of nickel between olivine, magma and immiscible sulfide liquid: Chemical Geology, v. 32, p. 119–127.
Frey, F. A., and Green, D. H., 1974, The mineralogy, geochemistry, and origin of lherzolite inclusions in Victorian basanites: Geochimica et Cosmochimica Acta, v. 38, p. 1023–1059.
Frey, F. A., and Prinz, M., 1978, Ultramafic inclusions from San Carlos, Arizona; Petrologic and geochemical data bearing on their petrogenesis: Earth and Planetary Science Letters, v. 38, p. 129–176.
Frick, C., 1973, The sulphides in griquaite and garnet-peridotite xenoliths in kimberlite: Contributions to Mineralogy and Petrology, v. 39, p. 1–16.
Garuti, G., Gorgoni, C., and Sighinolfi, G. P., 1984, Sulfide mineralogy and chalcophile and siderophile element abundances in the Ivrea-Verbano mantle peridotites (Western Italian Alps): Earth and Planetary Science Letters, v. 70, p. 69–87.
Gurney, J. J., Harte, B., and Cox, K. G., 1975, Mantle xenoliths in Matsoku kimberlite pipe, *in* Ahrens, L. H., and others, eds., Physics and chemistry of the earth, v. 9: Oxford, Pergamon Press, p. 507–524.
Haggerty, S. E., and Tompkins, L. A., 1982, Sulfur solubilities in mantle derived nodules from kimberlites [abs.]: EOS Transactions, American Geophysical Union, v. 63, p. 463.
Harte, B., 1977, Rock nomenclature with particular relation to deformation and recrystallization textures in olivine-bearing xenoliths: Journal of Geology, v. 85, p. 279–288.
Haughton, D. R., Roeder, P. L., and Skinner, B. J., 1974, Solubility of sulfur in mafic magmas: Economic Geology, v. 69, p. 451–467.
Hawkesworth, C. J., and Norry, M. J., eds., 1983, Continental basalts and mantle xenoliths: Nantwich, United Kingdom, Shiva Publishing, 272 p.
Helz, R. T., 1977, Determination of the P-T dependence of the first appearance of FeS-rich liquid in natural basalts to 20 kb [abs.]: EOS Transactions, American Geophysical Union, v. 58, p. 533.
Hoffer, J. M., 1976, The Potrillo basalt field, south-central New Mexico, *in* Elston, W., and Northrop, S. A., eds., Cenozoic volcanism in southwestern New Mexico: New Mexico Geological Society Special Publication 5, p. 89–92.
Irving, A. J., 1979, Kilbourne Hole spinel lherzolites; Samples of multiply depleted, enriched and deformed mantle [abs.]: EOS Transactions, American Geophysical Union, v. 60, p. 418.
——, 1980, Petrology and geochemistry of composite ultramafic xenoliths in alkali basalts and implications for magmatic processes in the mantle: American Journal of Science, v. 280-A, p. 389–426.
Irving, A. J., and Frey, F. A., 1984, Trace element abundances in megacrysts and their host basalts; Constraints on partition coefficients and megacryst genesis: Geochimica et Cosmochimica Acta, v. 48, p. 1201–1221.
Jagoutz, E., Palme, H., Baddenhausen, H., and others, 1979, The abundances of major, minor, and trace elements in the earth's mantle as derived from primitive ultramafic nodules: Proceedings of the 10th Lunar and Planetary Science Conference, p. 2031–2050.
James, D. E., Padovani, E. R., and Hart, S. R., 1980, Oxygen isotopic composition of lower crustal xenoliths, Kilbourne Hole, New Mexico: Carnegie Institution of Washington Yearbook, v. 79, p. 455–458.
Kleeman, J. D., Green, D. H., and Lovering, J. F., 1969, Uranium distribution in ultramafic inclusions from Victorian basalts: Earth and Planetary Science Letters, v. 5, p. 449–458.
Kullerud, G., Yund, R. A., and Moh, G. H., 1969, Phase relations in the Cu-Fe-S, Cu-Ni-S, Fe-Ni-S systems, *in* Wilson, H. D., ed., Magmatic ore deposits: Economic Geology Monograph, v. 4, p. 323–343.
Lindsley, D. H., 1983, Pyroxene thermometry: American Mineralogist, v. 68, p. 477–493.
Lorand, J., and Conquere, F., 1983, Contribution a l'etude des sulfures dans les enclaves de lherzolite a spinelle des basaltes alcalins (Massif Central et Languedoc, France): Bulletin de Mineralogie, v. 106, p. 585–605.
MacLean, W. H., 1969, Liquidus phase relations in the FeS-FeO-Fe_3O_4-SiO_2 system, and their application in geology: Economic Geology, v. 64, p. 865–884.
——, 1977, Sulfides in Leg 37 drill core from the Mid-Atlantic Ridge: Canadian Journal of Earth Sciences, v. 14, p. 674–683.
MacLean, W. H., and Shimazaki, H., 1976, The partition of Co, Ni, Cu, and Zn between sulfide and silicate liquids: Economic Geology, v. 71, p. 1049–1057.
MacRae, N. D., 1979, Silicate glass and sulfides in ultramafic xenoliths, Newer Basalts, Victoria, Australia: Contributions to Mineralogy and Petrology, v. 68, p. 275–280.
Mainwaring, P. R., and Naldrett, A. J., 1977, Country-rock assimilation and genesis of Cu-Ni sulfides in the Water Hen intrusion, Duluth Complex, Minnesota: Economic Geology, v. 72, p. 1269–1284.
Mathez, E. A., 1976, Sulfur solubility and magmatic sulfides in submarine basalt glass: Journal of Geophysical Research, v. 81, p. 4269–4276.
Mathez, E. A., Dietrich, V. J., and Irving, A. J., 1984, The geochemistry of carbon in mantle peridotites: Geochemica et Cosmochimica Acta, v. 48, p. 1849–1859.
Menzies, M., 1983, Mantle ultramafic xenoliths in alkaline magmas; Evidence for mantle heterogeneity modified by magmatic activity, *in* Hawkesworth, C. J., and Norry, M. J., eds., Continental basalts and mantle xenoliths: Nantwich, United Kingdom, Shiva Publishing, p. 92–110.
Meyer, H.O.A., and Boctor, N. Z., 1975, Sulfide-oxide minerals in eclogite from Stockdale kimberlite, Kansas: Contributions to Mineralogy and Petrology, v. 52, p. 57–68.
Misra, K. C., and Fleet, M. E., 1973, The chemical composition of synthetic and natural pentlandite assemblages: Economic Geology, v. 68, p. 518–539.
Mitchell, R. H., and Keays, R. R., 1981, Abundance and distribution of gold, palladium, and iridium in some spinel and garnet lherzolites; Implications for the nature and origin of precious metal-rich intergranular components in the upper mantle: Geochimica et Cosmochimica Acta, v. 45, p. 2425–2442.
Moore, J. G., 1973, Vesicles, water, and sulfur in Reykjanes Ridge basalts: Contributions to Mineralogy and Petrology, v. 41, p. 105–118.
Morgan, J. W., and Baedecker, P. A., 1983, Elemental composition of sulfide particles from an ultramafic xenolith and the siderophile element content of the upper mantle, *in* Lunar and Planetary Science XIV: Houston, Lunar and Planetary Institute, p. 513–514.
Morgan, J. W., Wandless, G. A., Petrie, R. K., and Irving, A. J., 1980, Composition of the earth's upper mantle; II, Volatile trace elements in ultramafic xenoliths: Proceedings of the Lunar and Planetary Science Conference 11th, p. 213–233.
Morgan, J. W., Wandless, G. A., Petrie, R. K., and Irving, A. J., 1981, Compositions of the earth's upper mantle; I, Siderophile trace elements in ultramafic nodules: Tectonophysics, v. 75, p. 47–67.
Mysen, B. O., and Kushiro, I., 1976, Partitioning of iron, nickel, and magnesium between metal, oxide, and silicate in Allende meteorite as a function of fO_2: Carnegie Institute of Washington Yearbook, v. 75, p. 678–684.
Mysen, B. O., and Popp, R. K., 1980, Solubility of sulfur in $CaMgSi_2O_6$ and $NaAlSi_3O_6$ melts at high pressure and temperature with controlled fO_2 and fS_2: American Journal of Science, v. 280, p. 78–92.
Naldrett, A. J., 1969, A portion of the system Fe–S–O between 900 and 1080°C and its application to sulfide ore magmas: Journal of Petrology, v. 10, p. 171–201.
——, 1981, Nickel sulfide deposits; Classification, composition, and genesis, *in* Skinner, B. J., ed., Economic Geology; Seventy-fifth anniversary volume: Economic Geology, p. 628–685.
Nixon, P. H., ed., 1987, Mantle xenoliths: New York, John Wiley.
Padovani, E. R., 1977, Granulite facies xenoliths from Kilbourne Hole Maar, New Mexico and their bearing on deep crustal evolution [Ph.D. thesis]: Dallas, University of Texas, 158 p.
Padovani, E. R., and Carter, J. L., 1977, Aspects of the deep crustal evolution beneath south-central New Mexico, *in* Heacock, J. G., ed., The earth's crust; Its nature and physical properties: American Geophysical Union, Geophysical Monograph, 20, p. 19–55.

Padovani, E. R., and Hart, S. R., 1981, Geochemical constraints on the evolution of the lower crust beneath the Rio Grande rift, *in* Papers presented to the conference on the processes of planetary rifting: Lunar and Planetary Institute Contribution No. 457, p. 149–152.

Pasteris, J. D., 1982, Evidence of potassium metasomatism in mantle xenoliths [abs.]: EOS Transactions, American Geophysical Union, v. 63, p. 462.

——, 1984, Further interpretation of the Cu-Fe-Ni sulfide mineralization in the Duluth Complex, northeastern Minnesota: Canadian Mineralogist, v. 22, p. 39–53.

——, 1985, Relationships between temperature and oxygen fugacity among Fe–Ti oxides in two regions of the Duluth Complex: Canadian Mineralogist, v. 23, p. 111–127.

——, 1987, Fluid inclusions in mantle xenoliths, *in* Nixon, P. H., ed., Mantle xenoliths: New York, John Wiley (in press).

Pedersen, A. K., 1979, Basaltic glass with high-temperature equilibrated immiscible sulphide bodies with native iron from Disko, central west Greenland: Contributions to Mineralogy and Petrology, v. 69, p. 397–407.

Peterson, R., and Francis, D., 1977, The origin of sulfide inclusions in pyroxene megacrysts: American Mineralogist, v. 62, p. 1049–1051.

Rajamani, V., and Naldrett, A. J., 1978, Partitioning of Fe, Co, Ni, and Cu between sulfide liquid and basaltic melts and the composition of Ni-Cu sulfide deposits: Economic Geology, v. 73, p. 82–93.

Reid, J. B., and Woods, G. A., 1978, Oceanic mantle beneath the southern Rio Grande Rift: Earth and Planetary Science Letters, v. 41, p. 303–316.

Ripley, E. M., 1981, Sulfur isotopic studies of the Dunka Road Cu-Ni deposit, Duluth Complex, Minnesota: Economic Geology, v. 76, p. 610–620.

Ryzhenko, B. N., and Kennedy, G. C., 1973, The effect of pressure on the eutectic in the system Fe-FeS: American Journal of Science, v. 273, p. 803–810.

Schneider, A., 1974, Sulfur-abundance in rock-forming minerals, phase equilibria, sulfur minerals, *in* Wedepohl, H., ed., Handbook of geochemistry: Berlin, Springer–Verlag, p. 16D-19.

Shima, H., and Naldrett, A. J., 1975, Solubility of sulfur in an ultramafic melt and the relevance of the system Fe-S-O: Economic Geology, v. 70, p. 960–967.

Shimazaki, H., and Clark, L. A., 1973, Liquidus relations in the FeS-FeO-SiO_2-Na_2O system and geological implications: Economic Geology, v. 68, p. 79–96.

Skinner, B. J., and Peck, D. L., 1969, An immiscible sulfide melt from Hawaii, *in* Wilson, H. D., ed., Magmatic ore deposits: Economic Geology Monograph, v. 4, p. 310–322.

Sobolev, N. V., 1977, Deep-seated inclusions in kimberlites and the problem of the composition of the upper mantle (trans. by D. A. Brown): Washington, D.C., American Geophysical Union, 279 p.

Stosch, H.-G., 1982, Rare earth element partitioning between minerals from anhydrous spinel peridotite xenoliths: Geochimica et Cosmochimica Acta, v. 46, p. 793–811.

Stosch, H.-G., and Seck, H. A., 1980, Geochemistry and mineralogy of two spinel peridotite suites from Dreiser Weiher, West Germany: Geochimica et Cosmochimica Acta, v. 44, p. 457–470.

Thompson, J.F.H., Barnes, S. J., and Duke, J. M., 1984, The distribution of nickel and iron between olivine and magmatic sulfides in some natural assemblages: Canadian Mineralogist, v. 22, p. 55–66.

Tsai, H., Shieh, Y., and Meyer, H.O.A., 1979, Mineralogy and $^{34}S/^{32}S$ ratios of sulfides associated with kimberlite, xenoliths, and diamonds, *in* Boyd, F. R., and Meyer, H.O.A., eds., The mantle sample: Washington, D.C., American Geophysical Union, p. 87–103.

Vaughan, D. J., and Craig, J. R., 1978, Mineral chemistry of metal sulfides: Cambridge University Press, 493 p.

Wasilewski, P. J., and Padovani, E. R., 1981, Crustal magnetization beneath the Rio Grande Rift based on xenoliths from Kilbourne Hole and Potrillo Maar, *in* Papers presented to the conference on the processes of planetary rifting: Lunar and Planetary Institute Contribution No. 457, p. 153–155.

Wells, P.R.A., 1977, Pyroxene thermometry in simple and complex systems: Contributions to Mineralogy and Petrology, v. 62, p. 129–139.

Wendlandt, R. F., 1982, Sulfide saturation of basalt and andesite melts at high pressures and temperatures: American Mineralogist, v. 67, p. 877–885.

Wilshire, H. G., and Shervais, J. W., 1975, Al-augite and Cr-diopside ultramafic xenoliths in basaltic rocks from western United States, *in* Ahrens, L. H., and others, eds., Physics and chemistry of the earth, v. 9: Oxford, Pergamon Press, p. 257–272.

Wilshire, H. G., Meyer, C. E., Nakata, J. K., and others, 1985, Mafic and ultramafic xenoliths from volcanic rocks of the western United States: U.S. Geological Survey Open-File report 85-139.

Wood, B., and Banno, S., 1973, Garnet-orthopyroxene and orthopyroxene-clinopyroxene relationships in simple and complex systems: Contributions to Mineralogy and Petrology, v. 42, p. 109–124.

Yund, R. A., and Kullerud, G., 1966, Thermal stability of assemblages in the Cu-Fe-S system: Journal of Petrology, v. 7, p. 454–488.

Zindler, A., and Jagoutz, E., 1980, Isotope and trace element systematics in mantle-derived peridotite nodules from San Carlos [abs.]: EOS Transactions, American Geophysical Union, v. 61, p. 374.

——, 1987, Mantle cryptology: Geochimica et Cosmochimica Acta (in press).

Manuscript Accepted by the Society October 14, 1986

Printed in U.S.A.

Geological Society of America
Special Paper 215
1987

A model of mantle metasomatism

H. G. Wilshire
U.S. Geological Survey, 345 Middlefield Road, Menlo Park, California 94025

ABSTRACT

Structural, lithologic, and geochemical relationships are integrated to provide a basis for modeling upper mantle metasomatism, the evidence of which is observed in mafic and ultramafic rocks of peridotite massifs and xenoliths from basalts and kimberlites. Mafic rock types associated with peridotite occur as dikes, and most metasomatic phenomena observed in peridotites appear to be directly related to the magmatic rocks. The chronology of mafic dike emplacement, established by crosscutting relationships, is the same in peridotite massifs and xenoliths in basalts, and progresses toward lower pressure assemblages. Metasomatic alteration of peridotite wallrock in the mantle occurs locally adjacent to the dikes, which are crystallized initial magmas or magmatic differentiates separated from their parent dikes. The alteration is caused both by diffusion and infiltration processes. Additional metasomatism ("cryptic") can be caused by a CO_2-rich, light rare earth element (LREE)-enriched gas phase evolved from the mafic dikes and distributed through hydro-fractures in peridotite. Crosscutting relationships of mafic dikes indicate that local metasomatic events have occurred serially in the melting history of peridotite, so that superposition of different types of metasomatic effects is to be expected. The sequential melting and metasomatic events are interpreted as consequences of diapiric rise of the host peridotite. The type and number of melting-metasomatic events evident in xenoliths is thought to be limited by the level of diapiric rise at which eruption occurs.

INTRODUCTION

A number of models of mantle metasomatism have been advanced since the pioneering work of Bailey (1970, 1972), itself drawn from important preceding (Oxburgh, 1964) and contemporary (Varne, 1970) observations. Bailey's concepts (1970, 1972) were based on theoretical-experimental considerations of melting behavior under the influence of introduced volatiles and other mobile components. Lloyd and Bailey (1975) applied these ideas to East African and German alkaline basalt provinces based on field observations of the relative abundance of pyroxenitic xenoliths, commonly containing mica and/or amphibole, that evidently comprised a significant part of the mantle source regions of those basalts. The pyroxenites were viewed as products of metasomatic alteration of the peridotite from which the host basalts were subsequently derived. The mantle source regions of other alkaline basalt provinces were similarly interpreted by Menzies (1983) and Menzies and Wass (1983), but the concept of wholesale conversion of previously depleted peridotite to pyroxenites and various assemblages enriched in amphibole, mica, and apatite has not been widely endorsed. The influx of metasomatizing components is considered by Bailey (1970, 1972) and Dick and others (1984) to be the first stage of alkaline volcanism and the ultimate cause of it, and by other proponents of this idea, to represent a necessary precursor of alkaline volcanism. Bailey (1984) extended the concept to explain the origin of kimberlites as products of ultrametasomatism.

A second concept of mantle metasomatism is that previously depleted peridotite is locally enriched in Fe, Ti, and other components by interaction with mafic magmas, now represented by pyroxenites with or without hydrous phases (Wilshire and Shervais, 1975; Wilshire and Jackson, 1975; Harte and others, 1977). This concept is based on field observations of the analogous distribution and structural relationships of mafic and ultramafic lithologies in peridotite massifs and in xenoliths, particularly those contained in alkali basalts. Detailed electron-probe studies (Wilshire and Shervais, 1975; Wilshire and Jackson, 1975; Pike and others, 1980; Irving, 1980; Kempton and others, 1984) of major element compositions of minerals in composite peridotite-pyroxenite xenoliths have documented metasomatic conversion of magnesian peridotite to more Fe-rich peridotite adjacent to the pyroxenite intrusions. This idea has been extended as a comprehensive explanation of relatively Fe-, Ti-rich peridotite xenoliths in kimberlite and minette by Gurney and Harte (1980), Ehrenberg (1979, 1982), and Harte (1983)—an alternative to derivation of Fe-rich peridotites from undepleted mantle sources (Nixon and Boyd, 1973; Nixon and others, 1981). Observations such as those made on composite peridotite/pyroxenite inclusions have led to the conclusion that magnesian peridotite is altered to more Fe, Ti-, and alkali-rich compositions adjacent to hydrous mineral veins by infiltration of the vein-forming fluids (Wilshire and Trask, 1971; Francis, 1976; Stewart and Boettcher, 1977; Wass, 1979; Boettcher and others, 1979; Boettcher and O'Neil, 1980; Stosch and Seck, 1980; Wilshire and others, 1980; Kempton and others, 1984; Roden and others, 1984a; Menzies and others, 1985).

A third concept of mantle metasomatism is based on trace element and isotopic compositions of xenoliths. The observation that otherwise refractory peridotite xenoliths commonly are enriched in LREE (Nagasawa and others, 1969; Frey and Green, 1974; Frey and Prinz, 1978; Menzies, 1983; Kempton and others, 1984; Menzies and others, 1985) is explained by introduction of metasomatic fluids from a source lower in the mantle. The introduced incompatible elements are contained in clinopyroxene or hydrous minerals or both. Additional material that is rich in incompatible elements occurs as acid-leachable interstitial deposits and in fluid inclusions in the same xenoliths. Many authors consider relative enrichment of incompatible elements to be a necessary precursor to alkaline volcanism because the depleted peridotites are not a viable source of alkali basalts (Menzies and Murthy, 1980a, b; Wass and Rogers, 1980; Boettcher and O'Neil, 1980; Wass and others, 1980).

The third model has spawned subdivisions labeled "cryptic," where peridotites are relatively enriched in incompatible elements without other obvious changes; and "patent," where the enrichment is accompanied by introduction of hydrous minerals (Dawson, 1982, 1984; Kramers and others, 1983; Menzies, 1983; Kempton and others, 1984; Menzies and others, 1985; Roden and Murthy, 1985). Some of these authors consider "cryptic" metasomatism to be a universal process and "patent" metasomatism to be a local phenomenon related to igneous events in the mantle (Dawson, 1984; Kramers and others, 1983; Harte, 1983). It is important to recognize, however, that whereas secondary mineralization of depleted peridotites is *prima facie* evidence of metasomatism, trace element patterns identified as "cryptic" metasomatism are not. In fact, some authors consider relative LREE enrichment of clinopyroxenes in anhydrous depleted peridotite to be a residual effect of partial melting, not metasomatic enrichment (Ottonello, 1980; P. D. Kempton, written communication, 1985).

The general concepts of metasomatism cited above are clearly not mutually exclusive. To some extent they reflect the special focus of their authors—a focus on xenoliths from kimberlites, or alkali basalts, or on massif peridotites. In this paper I attempt to integrate lithologic, structural, textural, and geochemical evidence from the three principal modes of occurrence of mantle samples; with this as a basis, I then present a model of metasomatism that may explain some of the similarities and difference among them.

TERMINOLOGY

Metasomatism describes a process or processes whereby the mineralogy and/or chemical composition of a solid rock is altered by the introduction of chemical components from an external source; the alteration process is commonly accompanied by loss of other components from the altered rock (modified from *AGI Glossary,* 1980). Where the components are introduced without the aid of intergranular fluids, the alteration process is called diffusion-metasomatism. Where it is accomplished by intergranular fluids (including liquids), it is called infiltration metasomatism (Korshinsky, 1970). Commonly, "metasomatism" is used synonymously with "enrichment," which includes introduction of materials that have crystallized in fractures in the preexisting rock. Although such processes may have important effects on the compositions of melts later derived from the aggregate of fracture fillings and wallrocks (see, for example, Wilshire and Pike, 1975; Hanson, 1977; Wood, 1979; Wass and Rogers, 1980; Eggler and others, this volume), they do not constitute metasomatism in the strictest sense.

PRINCIPAL LITHOLOGIC VARIANTS

The basis for classifying rock types in basalt- and kimberlite-xenolith asemblages and in peridotite massifs is not uniform. For example, spinel- and garnet-bearing peridotites are classified together for kimberlite occurrences and separately for basalt occurrences. Significant differences also result from differing emphasis on metasomatic features. Pyroxenites in massifs are commonly classified as low- or high-alumina types whatever the mineralogy may be, whereas equivalent lithologies in xenoliths are classified according to their mineralogy. Table 1 lists the common rock types in a way that emphasizes the overlap among the three modes of occurrence; that is, by omitting special place names not in common use and not conveying the mineralogy of the rocks, and by omitting names that are acronyms. Group names are those used by Wilshire and others (1985), with equivalent designations of Frey and Prinz (1978) indicated.

Common rock types in xenoliths from basalts are the same as those in massifs (Table 1; Wilshire and Pike, 1975; Conquere, 1977; Shervais, 1979; Sinigoi and others, 1980; Wilshire, 1984). Many authors classify pyroxenites in massifs as low- or high- (ariegite) alumina types irrespective of the aluminous phase, which makes it difficult to compare them with xenolith lithologies. Fortunately, many authors provide straightforward descriptions that allow correlation of minor lithologies with those occurring in xenoliths (for example, Dickey, 1970; Etienne, 1971; Jackson, 1979; Quick, 1981; Jackson and Ohnenstetter, 1981).

Xenoliths in kimberlites have been subdivided by Dawson (1980) into groups composed of (1) peridotite and pyroxenite;

TABLE 1. COMPARISON OF ROCK TYPES IN XENOLITH AND MASSIF OCCURRENCES*

Group	Rock Types			
	Ultramafic	Anhydrous Mafic	Hydrous	Megacrysts
Xenoliths in Basalts and Other Volcanic Rocks				
Garnetiferous ultramafic	Garnet lherzolite Garnet-spinel lherzolite	Garnet clinopyroxenite Garnet websterite	Phlogopite lherzolite Phlogopite pyroxenite Mg-Cr glimmerite	OLIVINE Clinopyroxene
Cr-Diopside (Type I of Frey and Prinz, 1978)	SPINEL LHERZOLITE Spinel harzburgite Spinel dunite	Spinel clinopyroxenite Spinel websterite Spinel orthopyroxenite Garnet-spinel pyroxenite		
Al-Augite (Type II of Frey and Prinz, 1978)	Spinel lherzolite Spinel wehrlite	Spinel clinopyroxenite Spinel websterite	Phlogopite pyroxenite Phlogopite lherzolite Kaersutite pyroxenite Kaersutite lherzolite Fe-Ti glimmerite Kaersutite hornblendite	CLINOPYROXENE KAERSUTITE Olivine Phlogopite Spinel
Feldspathic ultramafic	Feldspathic lherzolite	Feldspathic garnet clinopyroxenite Feldspathic clino-pyroxenite Feldspathic websterite		Plagioclase Alkali feldspar
Gabbro/metagabbro		Gabbro Metagabbro	Kaersutite gabbro Kaersutite metagabbro	
Xenoliths in Kimberlites				
Garnetiferous ultramafic	GARNET HARZBURGITE Garnet-spinel lherzolite Garnet-spinel harzburgite	Garnet clinopyroxenite	Phlogopite Mg-Cr glimmerite Richterite peridotite Richterite hornblendite	PHLOGOPITE GARNET Ilmenite Olivine Orthopyroxene Clinopyroxene
Cr-Diopside	Spinel lherzolite			
Feldspathic ultramafic		Feldspathic garnet clinopyroxenite		
Gabbro/metagabbro		Metagabbro		
Massifs				
Garnetiferous ultramafic	Garnet lherzolite	Garnet websterite	Phlogopite peridotite Phlogopite pyroxenite	
Cr-Diopside	SPINEL LHERZOLITE SPINEL HARZBURGITE Dunite	Spinel clinopyroxenite Spinel websterite Spinel orthopyroxenite Garnet-spinel clinopyroxenite		
Al-Augite	Spinel lherzolite	Spinel clinopyroxenite Spinel websterite	Kaersutite peridotite Kaersutite pyroxenite Kaersutite hornblendite Fe-Ti glimmerite	
Feldspathic ultramafic	Feldspathic lherzolite	Feldspathic pyroxenite		
Gabbro/metagabbro		Gabbro Metagabbro	Amphibole gabbro Amphibole metagabbro	

*Overall relative abundances of rock types: all capital letters = abundant; lower case letters = less abundant.

(2) eclogite, which includes feldspathic types that grade into "granulite"; (3) metasomatized peridotites of the "patent" variety (Dawson, 1982, 1984); and (4) glimmerite and MARID-suite. The peridotite-pyroxenite group includes both spinel and garnet types, but spinel peridotite is not commonly present in African kimberlites, even though it is thought to comprise the upper 30 km or so of the mantle there (Carswell and others, 1984). More complete sampling of garnet and spinel lithologies as well as lower crustal rocks is found in some U.S. kimberlites (e.g., McCallum and Eggler, 1976; Hearn and McGee, 1983; Eggler and others, this volume). Harte's (1983) classification of peridotites based on texture and presence or absence of patent metasomatism (further subdivided into Mg or Fe and hot or cold types) is apparently not universally applicable (see Eggler and others, this volume).

Megacrysts associated with the main xenolith groups are listed in Table 1. Megacrysts comprise single crystals or crystal fragments interpreted as pieces of pegmatitic members of the main lithologies. Such pegmatites occur in massif peridotites.

STRUCTURAL RELATIONSHIPS OF THE VARIOUS ROCK TYPES

Peridotite Massifs

Structural relationships of the minor rock types observed in peridotite massifs are much less equivocal than those in xenoliths because of visible structural continuity over much greater distances than can be observed in xenoliths. Pyroxenitic and/or gabbroic layers typically make up about 1 to 3 percent of the massifs (Kornprobst, 1969; Boudier, 1978; Quick, 1981), but they may locally make up as much as 35 percent of substantial outcrops (Shervais, 1979). There is generally a large modal gap between pyroxenites and even the most pyroxene-rich peridotites. Layers of all main groups may crosscut foliation in the peridotite, and inclusions of peridotite are fairly common in all types of layers. Dilational offset of layers crosscut by younger layers of the same or different groups is common. Internal phase layering is uncommon but has been observed, especially in gabbro layers (e.g., Jackson and Ohnenstetter, 1981) and hornblende pyroxenites. Dickey (1970) and Kornprobst (1969) reported common gneissose texture with compositional banding in mafic dikes of the Ronda, Lherz, and Beni Bouchera massifs. Grain-size layering occurs in both pyroxenites (Dick and Sinton, 1979) and gabbros (Jackson and Ohnenstetter, 1981); pegmatitic pyroxenites, plagioclasites, and hornblendites occur as segregations in dikes.

In some massifs, some or all of the lithologies—Cr-diopside spinel websterite, garnet pyroxenite, Al-augite pyroxenite, and feldspathic pyroxenite—form parallel layers (Beni Bouchera: Kornprobst, 1969; Lanzo: Boudier, 1978; Bogota: Prinzhofer and Nicolas, 1980), but in other massifs, layers of the same and different types may crosscut one another or form branching networks (Cypress Island: Raleigh, 1965; Red Hill: Walcott, 1969; Ronda: Dickey, 1970; Baldissero: Etienne, 1971, Jackson, 1979, Sinigoi and others, 1980, Nicolas, 1984; Balmuccia: Lensch, 1976, Shervais, 1979, Jackson, 1979, Sinigoi and others, 1983; Lherz: Ave Lallemant, 1967, Wilshire and Pike, 1975, Conquere, 1977; Monte Maggiore: Jackson and Ohnenstetter, 1981; Antalya: Juteau and others, 1977; Del Puerto: Evarts, 1978; and Trinity: Quick, 1981).

Hornblendite and glimmerite layers occur in Al-augite pyroxenites and peridotite. They commonly occupy fractures that appear planar on a scale of xenoliths (1 to 25 cm), but may be irregular on a scale of meters, or form net-vein systems. Inclusions of peridotite are common in hydrous mineral layers, but rocks that would be described as breccias with hydrous mineral matrices are rare.

Xenoliths in Basalts

Composite xenoliths containing more than one rock type are widespread but not abundant. Minor rock types generally form thin, sharply defined layers in peridotite host rock. Mafic layers of all the main groups, including Cr-diopside pyroxenites, may crosscut foliation where present in peridotite. Mafic layers of all the groups may form irregular, anastomosing net-vein systems, but this is more common in the Al-augite and feldspathic ultramafic groups than in other groups. Mafic layers of the same or different groups may crosscut one another; in places there is dilational offset of one layer by another.

Angular to subround inclusions or lenticular screens of the host peridotite occur in all types of mafic layers (Irving, 1974; Wilshire and Pike, 1975; Griffin and others, 1984). Such inclusions are considered by Lloyd and Bailey (1975) to be relics of peridotite replaced by pyroxenite. Mineral debris derived by disaggregation of peridotite host rock is locally abundant in pyroxenites and hornblendites of the Al-augite group. Relic disaggregated debris in pyroxenites of the Cr-diopside group would probably not be detectable because of extensive recrystallization. Lithic and mineral inclusions in Al-augite layers may either be little-modified Cr-diopside peridotite or Al-augite peridotite. Al-augite peridotite commonly forms thin septa that separate Al-augite mafic layers from Cr-diopside peridotite (Wilshire and Shervais, 1975; Wilshire and Jackson, 1975).

Internal grain size and/or phase layering of the minor lithologies is not common but is locally prominent in garnet pyroxenites (Arculus and Smith, 1979; Schulze and Helmstaedt, 1979) and in gabbros. Pegmatitic pyroxenites (Irving, 1974), gabbros, and hornblendites occur rarely. Mafic and ultramafic rocks with porphyroclastic or tabular textures (Pike and Schwarzman, 1977) and prominent foliation are locally common (e.g., Trask, 1969; Bergman, 1982; Bussod, 1983).

Peridotite xenoliths, and in some localities xenoliths of other lithologies, commonly are bounded by planar surfaces. Less commonly, peridotite xenoliths have internal unhealed planar fractures (Wilshire and Trask, 1971). The planar bounding surfaces of xenoliths are fractures along which the xenoliths were excavated by the host magma, or along which the xenoliths broke in transit to the surface. The planar facets formed a complex system of intersecting fractures with spacings of 1 to 10 cm or more in the mantle. Hydrous mineral veins of the Al-augite group commonly occupy some of the fractures. Xenoliths of peridotite breccia with a kaersutite-rich matrix occur rarely.

Xenoliths in Kimberlite

Composite xenoliths in kimberlites are widespread but uncommon, and there is much less variety than in xenoliths in basalts. Pyroxene-rich layers are found in peridotite host rocks (Harte and others, 1977), and anastomosing pyroxenite layers in peridotite occur rarely. Crosscutting relationships between different mafic layers have not been reported. Composite peridotite-"eclogite" xenoliths are quite rare, only two having been reported (Gurney and Harte, 1980). "Eclogites" may show both grain size and phase layering (Gurney and others, 1969; Mathias and others, 1970; Lappin and Dawson, 1975), and phase layering of phlogopite clinopyroxenites ("MARID" suite) has been described (Dawson and Smith, 1977; Dawson, 1980). Composite xenoliths of peridotite and hydrous mineral layers are also widespread (Gurney and Harte, 1980; Harte, 1983), but are only locally relatively common (Erlank and Rickard, 1977). As in xenoliths from basalt, phlogopite and/or amphibole layers commonly occupy planar fractures in the peridotite (Dawson, 1980, 1981, 1984). Breccias of peridotite and "eclogite" clasts in a phlogopite-rich matrix occur in at least three kimberlites (Lawless and others, 1979). Peridotite with sheared texture is widespread (Boyd, 1973; Nixon and Boyd, 1973; Gurney and Harte, 1980; Dawson, 1980; Harte, 1983; McGee and Hearn, 1984); these peridotites commonly are more Fe- and Ti-rich than unsheared peridotites, but not always (Harte and others, 1975; Dawson and others, 1975; Boyd and Nixon, 1978).

ORIGIN OF THE MINOR ROCK TYPES

Physical Conditions of Emplacement and Differentiation

It is evident from the foregoing account that the minor lithologies in all three principal modes of occurrence of mantle peridotite share many characteristics. In all, the common mafic lithologies form thin, generally sharply defined layers in peridotite host rock, and commonly contain inclusions of the peridotite. The mafic layers may crosscut foliation of the peridotite, form anastomosing vein systems, and crosscut one another, commonly with dilational offset of earlier by later layers. The structural relations clearly indicate that the mafic layers were emplaced as fluids (including liquids) in solid Cr-diopside spinel peridotite or garnet peridotite. Although there is much evidence of wallrock-fluid interaction, the structural evidence is decidedly in favor of emplacement of the fluids in dilational fractures and not in favor of a replacement origin such as proposed by Lloyd and Bailey (1975), Lloyd (1981), Menzies and Wass (1983), and Menzies (1983). The physical conditions attending the opening of the fractures into which the fluids were injected were quite variable. In places—especially prominent in xenoliths from kimberlite, moderately common in xenoliths from basalts, and less common in massif peridotite—the wallrocks were sheared and subsequently recrystallized to fine-grained mosaic or tabular textures. In other places, common in xenoliths from basalts and massif peridotites, there was substantial disaggregation of the wallrock and impregnation by the melts; disaggregation resulted in incorporation of variable, locally large, amounts of lithic and crystal debris in both pyroxenite and hornblendite (see Irving, 1974; Griffin and others, 1984; Wilshire and others, 1985).

The postemplacement deformational history of dike rocks in xenoliths and massif peridotites is similar. Older dikes of the Cr-diopside group contain fewer relics of igneous features and more commonly form plane-parallel layers or isoclinal folds in the foliation plane of the peridotite than younger dikes. Analysis of dike orientations in massifs (e.g., Jackson, 1979; Jackson and Ohnenstetter, 1981; Nicolas and Jackson, 1982) indicates that dikes emplaced before or early in the history of plastic deformation are deformed and tectonically rotated toward the foliation and lineation directions. Younger dikes postdating plastic deformation, more commonly contain relics of igneous textures and may show crosscutting and anastomosing relationships.

In general, pyroxenites of the Cr-diopside and Al-augite groups from basalts and peridotite massifs, and "eclogites" and pyroxenites from kimberlites lack an array of minor and incompatible elements that should be present if these pyroxenitic rocks represented unmodified liquid compositions (e.g., Frey and Prinz, 1978; Frey, 1980, 1984). The lack of these elements has led to the supposition that the pyroxenites are differentiates representing early crystal fractionates from a mafic liquid. Liquid fractions of the differentiation are commonly thought to be no longer present in or near the pyroxenites. The best evidence that at least some of the liquid residues remain trapped in the immediate vicinity of their parent intrusions is found in massif peridotites where composite dikes of Al-augite pyroxenite, feldspathic pyroxenite, and gabbro record the fractionation processes that yielded isolated dikes of each of the lithologies (Shervais, 1979; Sinigoi and others, 1983). Further differentiation of gabbroic liquids is similarly recorded by mineralogic and chemical trends in different gabbro phases in composite dikes of Monte Maggori (Jackson and Ohnenstetter, 1981) and other massifs. Direct field evidence of derivation of hornblendites from composite hornblende pyroxenite-hornblendite dikes is seen in the Lherz massif (Wilshire and others, 1980) where late liquids of hornblendite composition have been segregated within and squeezed from pyroxenite dikes.

All of the types in composite pyroxenite-gabbro and gabbro-gabbro dikes in massifs also occur in xenolith assemblages in basalts, indicating a common genetic relationship. Direct evidence of derivation of hornblendites from hornblende pyroxenites is also seen in xenoliths from basalts (Roden and others, 1984a; Wilshire and others, 1985). Because of the lithologic and structural similarity of hydrous mineral occurrences in kimberlite xenoliths, a case also may be made for derivation of glimmerite

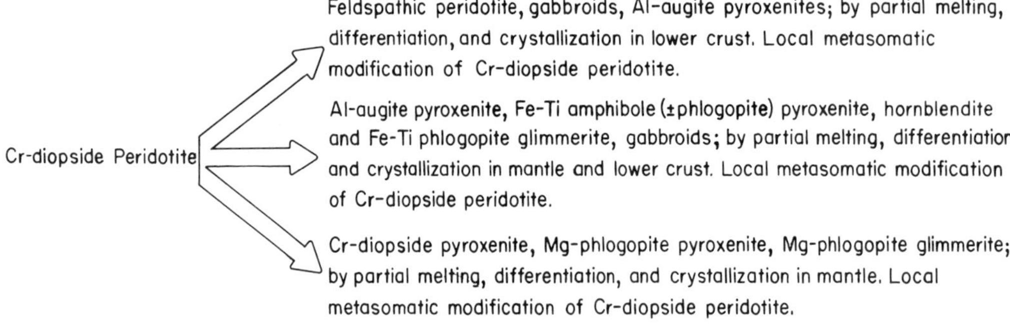

Feldspathic peridotite, gabbroids, Al-augite pyroxenites; by partial melting, differentiation, and crystallization in lower crust. Local metasomatic modification of Cr-diopside peridotite.
Al-augite pyroxenite, Fe-Ti amphibole (±phlogopite) pyroxenite, hornblendite and Fe-Ti phlogopite glimmerite, gabbroids; by partial melting, differentiation, and crystallization in mantle and lower crust. Local metasomatic modification of Cr-diopside peridotite.
Cr-diopside Peridotite
Cr-diopside pyroxenite, Mg-phlogopite pyroxenite, Mg-phlogopite glimmerite; by partial melting, differentiation, and crystallization in mantle. Local metasomatic modification of Cr-diopside peridotite.

Figure 1. Schematic representation of source rock and processes involved in evolution of rocks in Cr-diopside, Al-augite, and feldspathic groups.

and richterite hornblendites as liquid fractionates of hydrous pyroxenites of the MARID suite in kimberlite xenoliths. A genetic relation between MARID pyroxenites and anhydrous pyroxenites is possible.

Proposed differentiation mechanisms include cumulus processes, crystal plating (Dickey and others, 1977; Irving, 1978, 1980; Suen and Frey, 1978; Sinigoi and others, 1983), filter pressing or dilation in a structurally dynamic environment (e.g., Kornprobst, 1969; Moores, 1969; Kornprobst and Conquere, 1972; Dick, 1977; Evarts, 1978; Wilshire and others, 1980), wallrock reaction (Wilshire and Pike, 1975; Quick, 1981), and partial refusion (Wilshire and Pike, 1975; Wilshire and others, 1980; Loubet and Allegre, 1982; Wilshire, 1984). The clear record of deformation of some dikes and their wallrocks in massifs, and similar deformation textures in xenoliths indicate emplacement of dikes in a structurally dynamic system. Filter pressing is therefore likely to be a significant process in separating liquid residues from their parent dikes, but any or all of the other processes also may have played a role in determining the present compositions of the minor rock types. The inferred source and the general character of processes responsible for minor rock types in the Cr-diopside, Al-augite, and feldspathic ultramafic groups are shown in Figure 1.

Chronology of Dike Emplacement

Many detailed structural studies of peridotite massifs have elucidated the chronology of dike emplacement on the basis of cross-cutting relationships. Virtually all show a consistent pattern of dike emplacement: in order, Cr-diopside pyroxenite, Al-augite pyroxenite, and gabbro (e.g., Etienne, 1971; Lensch, 1976; Shervais, 1979; Jackson, 1979; Jackson and Ohnenstetter, 1981; Nicolas and Jackson, 1982; Sinigoi and others, 1983). In a very few massifs, the chronology of other minor lithologies is also exposed. At Lherz (Conquere, 1977; Wilshire and others, 1985), garnet pyroxenite, hornblendite, and glimmerite dikes crosscut Cr-diopside pyroxenite dikes; hornblende pyroxenites and their derivative hornblendites are younger than Al-augite pyroxenites. At Baldissero, Jackson (1979) recognized a pyroxenite intermediate between Cr-diopside and Al-augite pyroxenite, a lithology also recognized in xenoliths from basalts (Wilshire and others, 1985). In the Bogota segment of the ophiolitic Massif du Sud (Prinzhofer and Nicolas, 1980), feldspathic and hornblende pyroxenites postdate Al-augite(?) pyroxenites.

Although there is much less documentation, the same chronologic relationships have been observed in xenoliths from basalts (Wilshire and Pike, 1975; Wilshire, 1984; Wilshire and others, 1985). In addition, mineralogic relationships suggest that uncommon Mg-phlogopite + Cr-diopside veins in xenoliths are derivatives of Cr-diopside pyroxenite, a relationship that also may hold for kimberlite xenoliths.

The documented and inferred chronology of dike emplacement is shown in Figure 2. Considering the fact that the chronology is established on cross-cutting relationships, it is evident that older, higher pressure dikes (to about 20 kb) had to be transported, along with their host peridotite, to a lower pressure environment (less than about 9 kb) for crystallization of gabbros. The most obvious mechanism to achieve this is diapirism. Inasmuch as the chronology illustrated in Figure 2 is reconstructed from many occurrences of mantle peridotite, no implication is intended that all samples were derived from diapirs that passed through all P regimes from spinel or garnet facies to crustal conditions. Rather, it is inferred that, at any stage, volcanism can terminate the diapir-related processes that affected the mantle ultramafic and mafic rocks sampled by volcanism (Wilshire and Pike, 1975). Massifs, of course, require tectonic transport into the crust to be seen.

Wallrock Reactions and Metasomatism

Emplacement of mafic partial melts and of fractionated liquids derived from them into spinel or garnet peridotite host rock generally resulted in significant local chemical interchanges between the liquids and wallrocks. In xenoliths in basalts, the nature

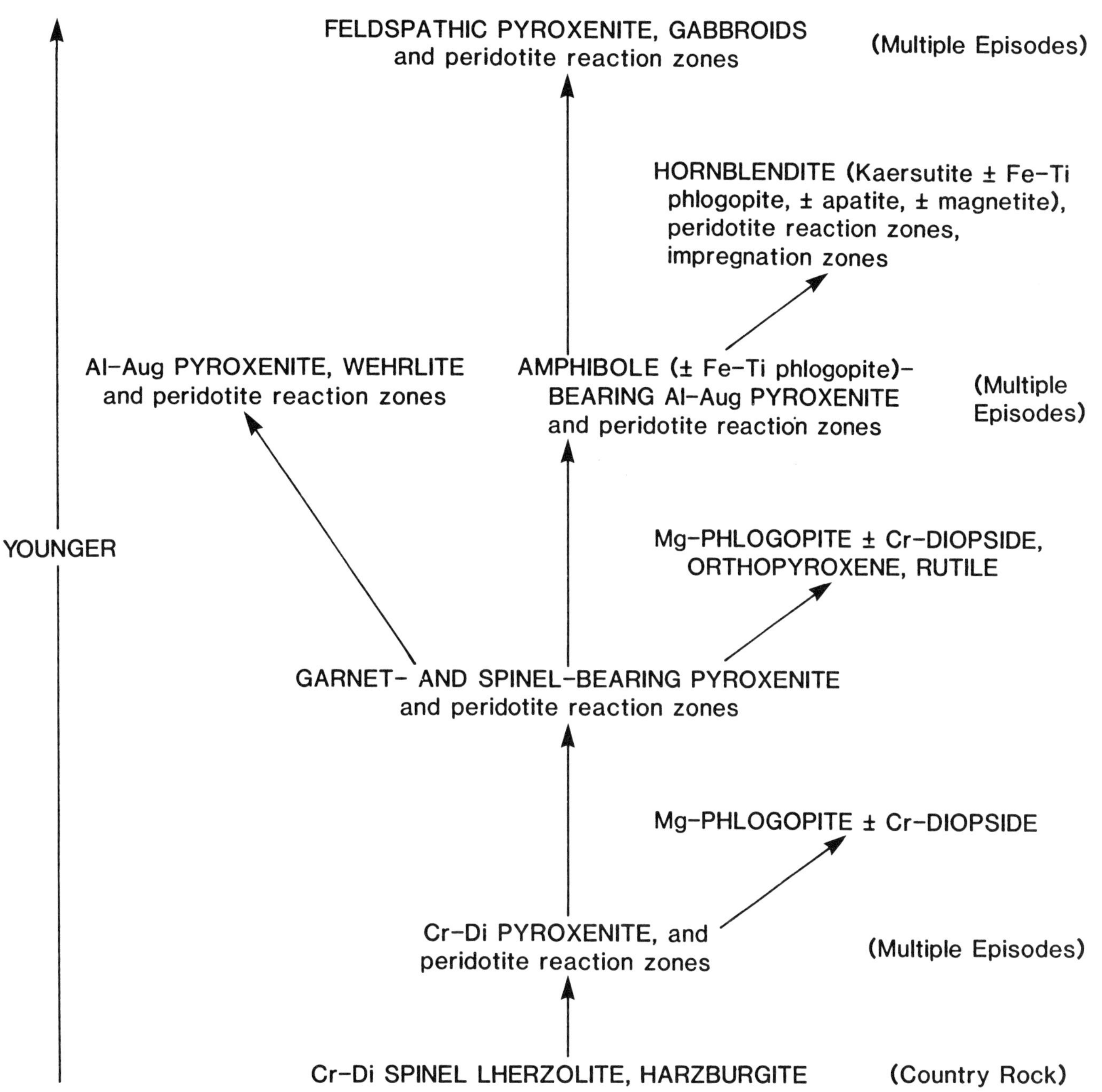
FELDSPATHIC PYROXENITE, GABBROIDS
and peridotite reaction zones
(Multiple Episodes)
HORNBLENDITE (Kaersutite ± Fe-Ti
phlogopite, ± apatite, ± magnetite),
peridotite reaction zones,
impregnation zones
Al-Aug PYROXENITE, WEHRLITE
and peridotite reaction zones
AMPHIBOLE (± Fe-Ti phlogopite)-
BEARING Al-Aug PYROXENITE
and peridotite reaction zones
(Multiple
Episodes)
YOUNGER
Mg-PHLOGOPITE ± Cr-DIOPSIDE,
ORTHOPYROXENE, RUTILE
GARNET- AND SPINEL-BEARING PYROXENITE
and peridotite reaction zones
Mg-PHLOGOPITE ± Cr-DIOPSIDE
Cr-Di PYROXENITE, and
peridotite reaction zones
(Multiple Episodes)
Cr-Di SPINEL LHERZOLITE, HARZBURGITE
(Country Rock)

Figure 2. Chronology of dike emplacement.

TABLE 2. COMPARISON OF BULK CHEMICAL DIFFERENCES BETWEEN Cr-DIOPSIDE PERIDOTITE/Al-AUGITE PERIDOTITE AND GARNET PERIDOTITE/SHEARED OR FERTILE PERIDOTITE*

	Cr-Diopside Lherzolite Avg. (24)	Cr-Diopside Harzburgite Avg. (13)	Cr-Diopside Lherzolite + Harzburgite Avg. (37)	Al-Augite Peridotite Avg. (11)	Coarse Garnet Lherzolite[1] Avg. (16)	Sheared Garnet Lherzolite[1] Avg. (11)
SiO_2	44.3	43.8	44.1	42.9	43.0	43.1
Al_2O_3	2.71	1.30	2.22	2.43	1.41	1.60
Fe_2O_3	1.02	0.61	0.88	1.01		
FeO	7.34	7.69	7.47	10.79	8.05	9.48
MgO	41.22	44.87	42.51	38.11	43.6	41.6
CaO	2.31	0.82	1.79	2.63	1.32	1.54
Na_2O	0.19	0.06	0.14	0.13	0.08	0.14
K_2O	0.04	0.06	0.05	0.03	0.01	0.03
TiO_2	0.09	0.04	0.07	0.20	0.11	0.26

	Granular Garnet Lherzolite[2] Avg. (4)	Sheared Garnet Lherzolite[2] Avg. (6)	Garnet Lherzolite + Harzburgite[3] Avg. (11)	Fertile Lherzolite[3] Avg. (6)
SiO_2	44.8	42.9	43.9	41.7
Al_2O_3	0.49	1.81	1.03	2.56
Fe_2O_3	1.84	1.48	2.02	2.67
FeO	4.43	7.06	5.27	7.83
MgO	45.04	41.99	42.92	39.83
CaO	0.80	1.82	0.73	1.98
Na_2O	0.06	0.21	0.07	0.22
K_2O	0.03	0.04	0.12	0.06
TiO_2	0.03	0.16	0.06	0.19

*Number in parentheses in each column is number of samples.
[1]Ehrenberg (1982b).
[2]Nixon and Boyd (1973).
[3]Nixon and others (1981).

of major element effects of reaction between Cr-diopside peridotite and mafic intrusions has been known for some time (e.g., Wilshire and Shervais, 1973, 1975), and was subsequently reported for garnetiferous ultramafic assemblages (Harte and others, 1977; Gurney and Harte, 1980; Ehrenberg, 1982). These local effects are much less documented in massifs (exceptions are the Trinity massif [Quick, 1981] and the Zambales ophiolite [Evans, 1985]). The principal effects of the least differentiated intrusions (those leaving pyroxenite residues) are enrichment of the wallrock in Fe, Ti (Wilshire and Shervais, 1975; Wilshire and Jackson, 1975; Harte and others, 1977; Irving, 1980; Ehrenberg, 1982), and for some, LREE (Irving, 1980; Ehrenberg, 1982; Roden and Murthy, 1985) and presumably other incompatible elements. As shown in Table 2, the nature and magnitude of major-element compositional changes in lherzolites of the Cr-diopside group resulting from reaction with igneous intrusions are the same as those for granular and sheared garnet lherzolite xenoliths in minette (Ehrenberg, 1982) and kimberlite (Nixon and Boyd, 1973; Nixon and others, 1981). The main difference in xenoliths from basalts is that the altered wallrock is less commonly sheared than it is in kimberlite samples, which may reflect differences in mechanical behavior of fractures at different pressures.

Other differentiates of the mafic magmas are volatile-rich liquids that crystallize to hornblendite and glimmerite, and less volatile-rich liquids that crystallize to anhydrous gabbroic assemblages. These liquids were physically separated from the parent intrusion and injected into surrounding unaltered peridotite with which they are commonly not in equilibrium. Disaggregation of the wallrock allowed extensive local infiltration of the liquids, similar to infiltration adjacent to pyroxenites. Unlike the pyroxenite reactions, however, the more volatile fluids that crystallized hydrous minerals effectively penetrated the wallrock along grain boundaries and through microfractures for distances of a few to 15 cm or more. Feldspathic melts also penetrated the wallrock peridotite, forming interstitial deposits and microscopic veinlet networks (see Evans, 1985; Dick and Friesz, 1984). Although amphibole and plagioclase reaction rims on spinel are common, and replacement of clinopyroxene by amphibole occurs in places, much of the introduced amphibole and plagioclase is in apparent "textural equilibrium" with the peridotite, so that distinctions between "primary" and "secondary" minerals are not always simple. Use of internally contradictory terms such as "primary metasomatic" (Harte and others, 1975; Gurney and Harte, 1980), however, is unnecessarily misleading.

Many systematic studies have been made of major element variations in minerals in the hydrous vein–wallrock systems in xenoliths from basalts (Francis, 1976; Stewart and Boettcher, 1977; Boettcher and others, 1979; Boettcher and O'Neil, 1980; Irving, 1980; Wilshire and others, 1980), but far fewer data on bulk compositional effects are available than for reaction zones adjacent to Al-augite clinopyroxenites. Limited bulk compositional data (Wilshire and others, 1985) indicate enrichment of the wallrock in Fe, Ti, Al, Ca, and alkalis, and reduced Si, Mg, Cr, and Ni. The chemical effects of fluid-wallrock reactions are such that vein minerals are substantially richer in Fe, Ti, and other mobile or incompatible components than interstitial hydrous minerals in adjacent wallrock. The same differences are found between micas of the MARID suite and interstitial micas in garnet lherzolites (Jones and others, 1982). Although the detailed work required to document this in massifs has not been done, the compositional trends in amphiboles from dikes into the wallrock in the St. Paul's Rocks massif (Melson and others, 1972) appears to be like that found in xenoliths.

The relatively high REE abundances and LREE-enriched patterns of kaersutitic amphibole, and especially of the commonly associated apatite, (Wass and others, 1980; Irving and Frey, 1984; Menzies and others, 1985) indicate that a net result of fluid infiltration is bulk enrichment of peridotite in REE and other incompatible elements. It would be surprising if substantial quantities of introduced incompatible elements did not remain in the rocks as leachable intergranular deposits, as suggested by Basu and Murthy (1977) and Nielson and Noller (this volume). Net enrichment of the REE is much less if only micas crystallize, although the micas also have LREE-enriched patterns (Wass and others, 1980; Irving and Frey, 1984; Menzies and others, 1985).

Reaction between the introduced fluids and wallrocks may also have significant effects on isotopic compositions (Roden and others, 1984b; Menzies and others, 1985). The effect of isotopic equilibration appears to be to shift the composition of the wallrock within the "mantle array" toward the bulk earth composition; reaction presumably has the opposite isotopic effect on the intrusions. Varying degrees of reaction may alleviate the necessity of having a separate fluid of distinctive isotopic composition for every observed reaction product (Menzies and others, 1985).

The relationship between hydrous mineral veins and an intricate system of planar fractures in peridotites that are unoccupied by veins is quite clear at many xenolith localities. Similar relationships are found in garnet lherzolite/harzburgite inclusions in kimberlite (Dawson, 1980, 1981, 1984). Wilshire and others (1985) postulated that propagation of a fracture system by hydraulic fracturing (Nicolas and Jackson, 1982) beyond the limits reached by hydrous liquids was caused by a gas phase derived from the hornblendite/glimmerite liquids and their parent mafic liquids. Evolution of a gas phase from these sources is attested by the abundance of macroscopic vesicles in amphibole and pyroxene megacrysts derived from vein systems, and in some wallrocks (Griffin and others, 1983; Wilshire and others, 1985; Nielson and Noller, this volume). The gas phase is considered to be CO_2-rich from the abundance of CO_2 fluid inclusions in peridotites and associated minor lithologies (e.g., Bergman and others, 1981; Menzies and Wass, 1983). Menzies and others (1985) also postulated that evolution of a CO_2-rich gas phase from hydrous silicate liquids extends the zone of infiltration around veins.

The significance of distributing a gas phase beyond the limits reached by silicate liquids lies in fractionation of LREE into the gas phase (Wendlandt and Harrison, 1979; Harrison, 1981; Mysen, 1983). Such a process may allow LREE—enrichment of peridotite and its constituent clinopyroxene by infiltration and reaction, with much less tangible influence on major element compositions than occurs when silicate liquids are involved. The process will also affect the REE radiogenic system by decreasing the Sm/Nd ratio; potential major effects on the Rb/Sr system are indicated by the presence of Rb and radiogenic Sr on grain boundaries in xenoliths (e.g., Basu, 1979; Jagoutz and others, 1980).

The short life-expectancy of fine-grained sheared textures under mantle conditions (Goetze, 1975; Mercier, 1979) may also apply to the planar fracture systems. If so, observed fracture systems were probably generated in the last stages of deformation before incorporation of xenoliths in the host magma. These fracture systems have not been identified in peridotite massifs, but probably have been destroyed by deformation accompanying or following massif emplacement.

A MODEL OF MANTLE METASOMATISM

The documented local metasomatic interactions between peridotite wallrock and intrusions of mafic melts may be combined with the effects of a gas phase derived from the melts to establish a model of metasomatism capable of yielding both cryptic and patent metasomites. The ultimate metasomatic products are results of an interrelated sequence of events: partial melting, injection of melts, reaction with and local infiltration of wallrock; differentiation of the intrusions, separation and injection of the liquid fractionates, which themselves can cause a sequence of reaction and infiltration; and, finally, separation and distribution of a gas phase through a fracture system propagated in front of the liquid intrusion zone, at depths appropriate to existence of a gas phase (Schneider and Eggler, 1984). The gas phase also infiltrates the wallrock and reacts with it. This sequence is illustrated in Figure 3, which shows three relationships between dikes and wallrocks and between different members of composite dikes, and the cryptic effects of a gas phase evolved from the liquids: (1) in the zone of origin of the melts, wallrock interactions with melts may be on a scale that is large compared to the size of xenoliths, so that chemical disequilibrium is detectable only in massif occurrences (A, Fig. 3); (2) the most common composite relationship seen in xenoliths, and in dikes and their wallrocks in massifs, results from injection of melts and liquid fractionates of these melts into peridotite with which they are not in equilibrium (B, Fig. 3); (3) residual liquids are segregated but not removed from their parent dikes (C, Fig. 3); here there may be no chemical variations with proximity to lithologic contacts as is the rule in the point B (Fig. 3) situation; (4) residual liquids are separated from their parent dike and react with peridotite wallrock (D, Fig. 3); (5) a gas phase is distributed through a fracture system propagated beyond the zone of hydrous liquid injection (E, Fig. 3).

DISCUSSION

Crosscutting relationships among the various intrusive veins and dikes, and geochemical evidence that all of the varieties of intrusions have reacted with their wallrocks, clearly indicate that local patent metasomatic events have occurred serially in the igneous history of massifs, and in the source regions of xenoliths

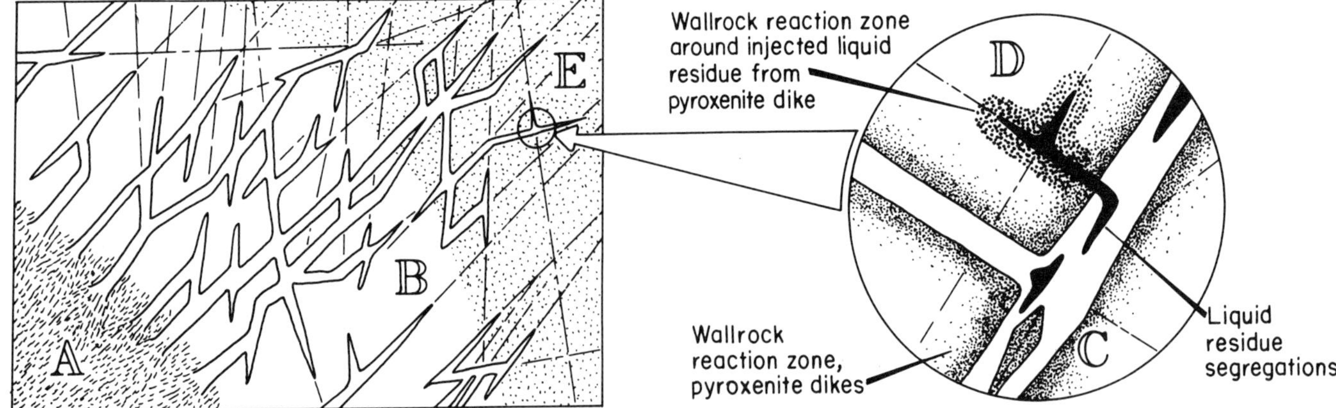

Figure 3. Model of igneous-metasomatic events. A, Zone of partial melting. B, Zone of dike injection; magmas not in equilibrium with wallrock. C, Segregation of residual liquids within differentiating dike. D, Separation of residual liquids from parent dike, injection of those liquids, and wallrock reaction. E, Zone in which gas phase evolved from dikes is distributed through hydrofractures.

before eruption of the host basalts. If cryptic metasomatism is a consequence of evolution of gas from these melts, then cryptic metasomatic events may have also occurred serially in the history of melting events. Because the various minor lithologies can occur in close proximity to one another in the host peridotite, superposition of patent and cryptic events is likely (Nielson and Noller, this volume).

The correlation between Fe-Ti metasomatism and fine-grained (sheared) textures in kimberlite xenoliths (and in some xenoliths in basalt) suggests a short time between the metasomatism and eruption of kimberlite because such textures are unstable under mantle conditions. Similarly, the spatial relationship between hydrous mineral veins and unoccupied planar fractures in peridotite also suggests contemporanity with hydrous mineral emplacement shortly before eruption, again because such structures may be unstable at mantle T and P. Smith and Ehrenberg (1985) suggested that Fe-Mg zoning of garnets in metasomatized peridotite inclusions in minette could not have survived in the mantle longer than about a thousand years, indicating that metasomatism is essentially contemporaneous with eruption of the host minette. Boyd and others (1984; see also Pike and others, 1980) noted the presence of zoned olivine in Fe-rich olivine neoblasts in dunites with sheared textures from Kimberley area kimberlite; they concluded that the metasomatic zoning was caused by kimberlitic fluids before explosive eruption.

The timing of metasomatic events as assessed from isotopic compositions of metasomatized rocks has been reviewed recently (Roden and Murthy, 1985). These authors discussed the difficulties in using Rb-Sr and U-Pb systems because of variability in fractionation of parent and daughter elements in metasomatic processes. The Sm-Nd system is believed preferable because of the chemical similarity of Sm and Nd, consistent enrichment of Nd relative to Sm in metasomatized rocks, and the resistance of Sm and Nd to remobilization. However, the behavior of these elements and other parent-daughter pairs in a gas phase (probably rich in CO_2) under mantle conditions is poorly known (Wendlandt and Harrison, 1979; Mysen, 1983). Evidence from inclusions in basalts indicates that metasomatism commonly precedes basalt eruption by a time period shorter than the resolution of the Sm-Nd system (±200–500 m.y.; Menzies and Murthy, 1980a; Roden and others, 1984a; Kempton and others, 1984; Roden and Murthy, 1985). This is also consistent with common Sr isotopic similarity of host basalts and the younger vein assemblages (Al-augite hornblendite, pyroxenite) in mantle xenoliths.

In contrast, metasomatism in African xenoliths, mostly from kimberlites, is commonly inferred to have preceded the host kimberlite by about 60 m.y. (Erlank and others, 1980; Hawkesworth and others, 1983) to as much as 2 b.y. (Menzies and Murthy, 1890b; Hawkesworth and others, 1983; Cohen and others, 1984). The "ages" of metasomatism of these rocks are inferred from high $^{87}Sr/^{86}Sr$, low $^{143}Nd/^{144}Nd$ ratios (Menzies and Murthy, 1980b), apparent isochronal relations of a number of xenoliths and megacrysts (Erlank and others, 1980; Hawkesworth and others, 1983), and a Nd model age of clinopyroxene from a micaceous garnet lherzolite xenolith (Cohen and others, 1984). The pitfalls of constructing "isochrons" from the type of data presented by Erlank and others (1980) and Hawkesworth and others (1983) are lucidly described by O'Reilly and Griffin (1984); this, combined with the very great scatter of points used to define the isochrons, suggests that these data do not constrain the age of metasomatism or initial isotopic compositions of the metasomatizing fluids.

In addition, the planar fracture systems, some of which were healed by hydrous mineral veins, in African xenoliths (Dawson, 1980, 1981, 1984) would have to survive at mantle T and P for about 60 m.y. before kimberlite eruption according to the present interpretation. The xenoliths described by Menzies and Murthy (1980b) and Cohen and others (1984) have clinopyroxenes that extend the "mantle array" well into the enriched quadrant of the Sr/Nd diagram (high $^{87}Sr/^{86}Sr$, low $^{143}Nd/^{144}Nd$), and for this reason are inferred to have undergone ancient metasomatic events. All of these rocks, however, have evidence of significant patent metasomatism in the form of introduced phlogopite. Isotopic compositions of two micas (Menzies and Murthy, 1980b) straddle and plot well off the "mantle array." Menzies and Murthy considered them (1980b) to be secondary and unrelated to the process of enrichment of the clinopyroxene. All mineral components of the xenolith studied by Cohen and others (1984) have extremely radiogenic Sr and Nd isotopic compositions, and the clinopyroxene is used to compute a model Nd age. A puzzling feature of this rock is that both garnet and clinopyroxene, with much lower Rb contents than the phlogopite, have higher $^{87}Sr/^{86}Sr$ ratios than the phlogopite so that an internal Sr isochron has a negative slope. The isotopic compositions and LREE contents of clinopyroxenes are considered to be "intrinsic features" (Roden and Murthy, 1985) or "most representative of mantle conditions" (Hawkesworth and others, 1983; Menzies and Murthy, 1980b) on which ages can be modeled, but such assumptions are doubtful when leachable interstitial components of possibly multiple origins or the products of patent metasomatism are present and have modified clinopyroxene compositions to unknown extents (see Nielson and Noller, this volume).

A small amount of data available on patently metasomatized xenoliths (Roden and others, 1984b; Menzies and others, 1985) indicates that the isotopic compositions of clinopyroxenes in the peridotite wallrock are shifted toward older apparent ages by a young metasomatic event. That is, the resultant isotopic composition of the wallrock reflects the isotopic composition of the metasomatizing fluid, not the time of metasomatism. Although these data indicate that some caution is in order in interpreting the age of metasomatism, the extreme compositions of introduced phlogopite recorded by Menzies and Murthy (1980b) and Cohen and others (1984) cannot be obviously reconciled with fluids related to the host kimberlite or ankaramite.

Setting aside for the present whether large time gaps are demonstrable between metasomatism and eruption of host magmas—which would preclude any genetic relation between them—the record from all independent lines of evidence for xenoliths in alkali basalts is clearer. Here, the time elapsed between metasomatism identified with the Al-augite and younger groups and the time of eruption of host basalts can be short, ranging from less than a thousand years to unknown longer time spans, depending on the sensitivity of the dating technique. A close time relation between metasomatism and eruption is consistent with both of the principal contending hypotheses: (1) that the metasomatism is not only a necessary precursor but is also a causal factor in alkaline magmatism (Bailey, 1972, 1982, 1984); (2) that the metasomatism is a consequence of alkaline magmatism, as is asserted here. In both hypotheses, the ultimate sources of elements whose relative abundance characterizes alkali basaltic rocks are in the unknown depths of the mantle. Both hypotheses have sought, as their strongest basis, the integration of field and geochemical data. Clearly, much has yet to be learned from all disciplines of earth science to resolve these differences, or to find new paths.

SUMMARY

The similarity of lithologic, petrologic, and structural features in mafic-ultramafic assemblages occurring as xenoliths and massifs indicates a common history. Sequential melting events have resulted in emplacement of dikes of varying composition in peridotite. The dikes have commonly undergone differentiation, physical separation of residual liquids, which themselves form dikes, and recrystallization and remelting. Diffusive reaction between the various generations of dikes and their peridotite wallrock has caused local Fe-Ti metasomatism of the peridotite; infiltration of peridotite wallrock by liquids from which pyroxene and hydrous minerals crystallize also caused local metasomatic alteration of the peridotite. Separation of a CO_2-rich gas phase enriched in incompatible elements from the dikes is postulated to cause further metasomatic changes beyond the limits of liquid injection. The dikes were emplaced serially in the peridotite during diapiric rise, so that different diking and associated metasomatic events may be superimposed. These events are believed to represent a broadly continuous sequence that led, in xenolith occurrences, to eruption. The special abundance of metasomatized rocks in xenolith occurrences is ascribed to selective sampling of the volcanic conduit system. Thus, metasomatized rocks are not representative of the mantle.

ACKNOWLEDGMENTS

I am grateful to Frank Dodge, Carter Hearn, Jr., James Meen, and Martin Prinz for their helpful criticisms of the manuscript.

REFERENCES CITED

American Geological Institute, 1980, Bates, R. L., and Jackson, J. A., eds., Glossary of geology: Falls Church, Virginia, American Geological Institute, 749 p.

Arculus, R. J., and Smith, D., 1979, Eclogite, pyroxenite, and amphibolite inclusions in the Sullivan Buttes latite, Chino Valley, Yavapai County, Arizona, *in* Boyd, F. R., and Meyer, H.O.A., eds., The mantle sample; Inclusions in kimberlites and other volcanics; Proceedings of the Second International Kimberlite Conference: Washington, D.C., American Geophysical Union, v. 2, p. 309–317.

Ave Lallemant, H. G., 1967, Structural and petrofabric analysis of an "alpine-type" peridotite; The lherzolite of the French Pyrenees: Leidsche Geologische Mededlingen, v. 42, p. 1–57.

Bailey, D. K., 1970, Volatile flux, heat focusing, and generation of magma: Geological Journal, Special Issue No. 2, p. 177–186.

——, 1972, Uplift, rifting, and magmatism in continental plates: Leeds Journal of Earth Sciences, v. 8, p. 225–239.

——, 1982, Mantle metasomatism—Continuing chemical change within the Earth: Nature, v. 296, p. 525–530.

——, 1984, Kimberlite; "The mantle sample" formed by ultrametasomatism, *in* Kornprobst, J., ed., Kimberlites. I, Kimberlites and related rocks: Amsterdam, Elsevier, p. 323–333.

Basu, A. R., 1979, Geochemistry of ultramafic xenoliths from San Quintin, Baja California, *in* Boyd, F. R., and Meyer, H.O.A., eds., The mantle sample; Inclusions in kimberlites and other volcanics; Proceedings of the Second

International Kimberlite Conference: Washington, D.C., American Geophysical Union, v. 2, p. 391–399.

Basu, A. R., and Murthy, V. R., 1977, Ancient lithospheric lherzolite xenolith in alkali basalt from Baja California: Earth and Planetary Science Letters, v. 35, p. 239–246.

Bergman, S. C., 1982, Petrogenetic aspects of the alkali basaltic layers and included megacrysts and nodules from the Lunar Crater volcanic field, Nevada, U.S.A. [Ph.D. thesis]: Princeton, New Jersey, Princeton University, 432 p.

Bergman, S. C., Foland, K. A., and Spera, F. J., 1981, On the origin of an amphibole-rich vein in a peridotite inclusion from the Lunar Crater volcanic field, Nevada, U.S.A.: Earth and Planetary Science Letters, Pt. 2, p. 594–621.

Boettcher, A. L., and O'Neil, J. R., 1980, Stable isotope, chemical, and petrographic studies of high-pressure amphiboles and micas; Evidence for metasomatism in the mantle source regions of alkali basalts and kimberlites: American Journal of Science, v. 280-A, The Jackson Volume, Pt. 2, p. 594–621.

Boettcher, A. L., O'Neil, J. R., Windom, K. E., Stewart, D. C., and Wilshire, H. G., 1979, Metasomatism of the upper mantle and the genesis of kimberlites and alkali basalts, *in* Boyd, F. R., and Meyer, H.O.A., eds., The mantle sample; Inclusions in kimberlites and other volcanics; Proceedings of the Second International Kimberlite Conference: American Geophysical Union, v. 2, p. 173–182.

Boudier, F., 1978, Structure and petrology of the Lanzo peridotite massif (Piedmont Alps): Geological Society of America Bulletin, v. 89, p. 1574–1591.

Boyd, F. R., 1973, A pyroxene geotherm: Geochimica et Cosmochimica Acta, v. 37, p. 2533–2546.

Boyd, F. R., and Nixon, P. H., 1978, Ultramafic nodules from the Kimberley pipes, South Africa: Geochimica et Cosmochimica Acta, v. 42, p. 1367–1382.

Boyd, F. R., and Jones, R. A., and Nixon, P. H., 1984, Mantle metasomatism; The Kimberley dunites: Geological Society of America Abstracts with Programs, v. 16, p. 453.

Bussod, G.Y.A., 1983, Thermal and kinematic history of mantle xenoliths from Kilbourne Hole, New Mexico: Los Alamos, New Mexico, U.S. Scientific Laboratory, Publication LA-9616-T, 74 p.

Carswell, D. A., Griffin, W. L., and Kresten, P., 1984, Peridotite nodules from the Ngopetsoeu and Lipelaneng kimberlites, Lesotho; A crustal or mantle origin, *in* Kornprobst, J., ed., Kimberlites. II, The mantle and crust-mantle relationships: Amsterdam, Elsevier, p. 229–243.

Cohen, R. S., O'Nions, R. K., and Dawson, J., 1984, Isotope geochemistry of xenoliths from East Africa; Implications for development of mantle reservoirs and their interaction: Earth and Planetary Science Letters, v. 68, p. 209–220.

Conquere, F., 1977, Petrologie des pyroxenites litees dans les complexes ultramafiques de l'Ariege (France) et autres gisements de lherzolite a spinelle; I. Compositions mineralogiques et chimiques, evolution des conditions d'equilibre des pyroxenites: Bulletin de la Societe Francaise de Mineralogie et de Cristallographie, v. 100, p. 42–80.

Dawson, J. B., 1980, Kimberlites and their xenoliths: New York, Springer-Verlag, 252 p.

—— , 1981, The nature of the upper mantle: Mineralogical Magazine, v. 44, p. 1–18.

—— , 1982, Contrasting types of mantle metasomatism: Terra Cognita, v. 2, p. 232.

—— , 1984, Contrasting types of upper-mantle metasomatism? *in* Kornprobst, J., ed., Kimberlites. II, The mantle and crust-mantle relationships: Amsterdam, Elsevier, p. 289–294.

Dawson, J. B., and Smith, J. V., 1977, The MARID (mica-amphibole-rutile-ilmenite-diopside) suite of xenoliths in kimberlite: Geochimica et Cosmochimica Acta, v. 41, p. 309–323.

Dawson, J. B., Gurney, J. J., and Lawless, P. J., 1975, Paleogeothermal gradients derived from xenoliths in kimberlite: Nature, v. 257, p. 299–300.

Dick, H.J.B., 1977, Evidence of partial melting in the Josephine peridotite, *in* magma genesis: Oregon Department of Geology and Mineral Industries Bulletin 96, p. 59–62.

Dick, H.J.B., and Friesz, B. L., 1984, Abyssal plagioclase peridotites and melt storage in the mantle beneath mid-ocean ridges: Geological Society of America Abstracts with Programs, v. 16, p. 487.

Dick, H.J.B., and Sinton, J. M., 1979, Compositional layering in alpine peridotites; Evidence for pressure solution creep in the mantle: Geology, v. 87, p. 403–416.

Dick, H.J.B., Fisher, R. L., and Bryan, W. B., 1984, Mineralogic variability of the uppermost mantle along mid-ocean ridges: Earth and Planetary Science Letters, v. 69, p. 88–106.

Dickey, J. S., Jr., 1970, Partial fusion products in alpine-type peridotites; Serrania de la Ronda and other examples: Mineralogical Society of America Special Publication 3, p. 33–49.

Dickey, J. G., Jr., Obata, M., Suen, C. J., Frey, F. A., and Lundeen, M., 1977, The Ronda ultramafic complex, southern Spain: Geological Society of America Abstracts with Programs, v. 9, p. 949–950.

Ehrenburg, S. N., 1979, Garnetiferous ultramafic inclusions in minette from the Navajo volcanic field, *in* Boyd, F. R., and Meyer, H.O.A., eds., The mantle sample; Inclusions in kimberlites and other volcanics; Proceedings of the Second International Kimberlite Conference: Washington, D.C., American Geophysical Union, v. 2, p. 330–344.

—— , 1982, Petrogenesis of garnet lherzolite and megacrystalline nodules from The Thumb, Navajo volcanic field: Journal of Petrology, v. 23, p. 507–547.

Erlank, A. J., and Rickard, R. S., 1977, Potassic richterite bearing peridotites from kimberlite and the evidence they provide for upper mantle metasomatism: Second International Kimberlite Conference, Extended Abstracts, not paginated.

Erlank, A. J., Allsop, H. L., Duncan, A. R., and Bristow, J. W., 1980, Mantle heterogeneity beneath southern Africa; Evidence from the volcanic record: Philosophical Transactions of the Royal Society of London, Series A, v. 297, p. 295–307.

Etienne, F., 1971, La lherzolite rubanee de Baldissero Canavese [Ph.D. thesis]: Nancy, France, Universite de Nancy, 157 p.

Evans, C. A., 1985, Magmatic "metasomatism" in peridotites from the Zambales ophiolite: Geology, v. 13, p. 166–169.

Evarts, R. C., 1978, The Del Puerto ophiolite complex, California; A structural and petrologic investigation [Ph.D. thesis]: Stanford, California, Stanford University, 409 p.

Francis, D. M., 1976, The origin of amphibole in lherzolite xenoliths from Nunivak Island, Alaska: Journal of Petrology, v. 17, p. 357–378.

Frey, F. A., 1980, The origin of pyroxenites and garnet pyroxenites from Salt Lake Crater, Oahu, Hawaii; Trace element evidence: American Journal of Science, v. 280-A, The Jackson Volume, p. 427–449.

—— , 1984, Rare earth element abundances in upper mantle rocks, *in* Henderson P., ed., Rare earth geochemistry: Amsterdam, Elsevier, p. 153–204.

Frey, F. A., and Green, D. H., 1974, The mineralogy, geochemistry, and origin of lherzolite inclusions in Victoria basanites: Geochimica et Cosmochimica Acta, v. 38, p. 1023–1059.

Frey, F. A., and Prinz, M., 1978, Ultramafic inclusions from San Carlos, Arizona: Petrologic and geochemical data bearing on their petrogenesis: Earth and Planetary Science Letters, v. 34, p. 129–176.

Goetze, C., 1975, Sheared lherzolites; From the point of view of rock mechanics: Geology, v. 3, p. 172–173.

Griffin, W. L., Wass, S. Y., and Hollis, J. D., 1984, Ultramafic xenoliths from Bullenmerri and Gnotuk maars, Victoria, Australia; Petrology of a subcontinental crust-mantle transition: Journal of Petrology, v. 25, p. 53–87.

Gurney, J. J., and Harte, B., 1980, Chemical variations in upper mantle nodules from southern African kimberlites: Philosophical Transactions of the Royal Society of London, v. 297, p. 273–293.

Gurney, J. J., Siebert, J. C., and Whitfield, G. G., 1969, A diamondiferous eclogite from Roberts Victor Mine: Geological Society of South Africa, Special Publication 2, p. 351–357.

Hanson, G. N., 1977, Evolution of the suboceanic mantle: Journal of Geological Sciences, v. 134, pt. 2, p. 235–254.

Harrison, W. J., 1981, Partitioning of REE between minerals and coexisting melts during partial melting of a garnet lherzolite: American Mineralogist, v. 66, p. 242–252.

Harte, B., 1983, Mantle peridotites and processes; The kimberlite sample, *in* Hawkesworth, C. J., and Norry, M. J., eds., Continental basalts and mantle xenoliths: Norwich, United Kingdom, Shiva Publishing, p. 46–91.

Harte, B., Cox, K. G., and Gurney, J. J., 1975, Petrography and geological history of upper mantle xenoliths from the Matsoku kimberlite pipe: Physics and Chemistry of the Earth, v. 9, p. 477–506.

Harte, B., Gurney, J. J., and Cox, K. G., 1977, Clinopyroxene-rich sheets in garnet-peridotite; Xenolith specimens from the Matsoku kimberlite pipe, Lesotho: Second International Kimberlite Conference, Extended Abstracts, not paginated.

Hawkesworth, C. J., Erlank, A. J., Marsh, J. S., Menzies, M. A., and van Calsteren, P., 1983, Evolution of the continental lithosphere; Evidence from volcanics and xenoliths in southern Africa, *in* Hawkesworth, C. J., and Norry, M. J., eds., Continental basalts and mantle xenoliths: Norwich, United Kingdom, Shiva Publishing, p. 111–138.

Hearn, B. C., Jr., and McGee, E. S., 1983, Garnet peridotites from Williams kimberlites, north-central Montana, USA: U.S. Geological Survey, Open-File Report 83-172, 26 p.

Irving, A. J., 1974, Megacrysts from the newer basalts and other basaltic rocks of southeastern Australia: Geological Society of America Bulletin, v. 85, p. 1503–1514.

——, 1978, Flow crystallization; A mechanism for fractionation of primary magmas at mantle pressures: EOS American Geophysical Union Transactions, v. 59, p. 1214.

——, 1980, Petrology and geochemistry of composite ultramafic xenoliths in alkali basalts and implications for magmatic processes within the mantle: American Journal of Science, v. 280-A, The Jackson Volume, p. 389–426.

Irving, A. J., and Frey, F. A., 1984, Trace element abundances in megacrysts and their host basalts; Constraints on partition coefficients and megacryst genesis: Geochimica et Cosmochimica Acta, v. 48, p. 1201–1221.

Jackson, M. D., 1979, Structures des filons dans les massifs de peridotites; Mechanismes d'injection et relation avec la deformation plastique (These 3eme cycle): Nantes, France, Universite de Nantes, 147 p.

Jackson, M. D., and Ohnenstetter, M., 1981, Peridotite and gabbroic structures in the Monte Maggiore massif, alpine Corsica: Journal of Geology, v. 89, p. 703–719.

Jagoutz, E., Carlson, R. W., and Lugmair, G. W., 1980, Equilibrated Nd-unequilibrated Sr isotopes in mantle xenoliths: Nature, v. 286, p. 708–710.

Jones, A. P., Smith, J. V., and Dawson, J. B., 1982, Mantle metasomatism in 14 veined peridotites from Bultfontein Mine, South Africa: Journal of Geology, v. 90, p. 435–453.

Juteau, T., Nicolas, A., Dubessy, J., Fruchard, J. C., and Bouchez, J. L., 1977, Structural relationships in the Antalya ophiolite complex, Turkey; Possible model for an oceanic ridge: Geological Society of America Bulletin, v. 88, p. 1740–1748.

Kempton, P. D., Menzies, M. A., and Dungan, M. A., 1984, Petrography, petrology, and geochemistry of xenoliths and megacrysts from the Geronimo volcanic field, southeastern Arizona, *in* Kornprobst, J., ed., Kimberlites. II, The mantle and crust-mantle relationships: Amsterdam, Elsevier, p. 71–83.

Kornprobst, J., 1969, Le massif ultrabasique des Beni Bouchera (Rif Interne, Maroc); Etude des peridotites de haute temperature et de haute pression, et des pyroxenolites a grenat ou sans grenat, qui leur sont associees: Contributions to Mineralogy and Petrology, v. 23, p. 283–322.

Kornprobst, J., and Conquere, F., 1972, Les pyroxenolites a grenat du massif de lherzolite de Moncaup (Haut Garonne–France); Caracteres communs avec certaines enclaves des basaltes alcalins: Earth and Planetary Science Letters, v. 16, p. 1–14.

Korshinsky, D. S., 1970, The theory of metasomatic zoning (translated by J. Agrell): Oxford, Clarendon Press, 162 p.

Kramers, J. D., Roddick, J.C.M., and Dawson, J. B., 1983, trace element and isotope studies on veined, metasomatic and "MARID" xenoliths from Bultfontein, South Africa: Earth and Planetary Science Letters, v. 65, p. 90–106.

Lappin, M. A., and Dawson, J. B., 1975, Two Roberts Victor cumulate eclogites and their re-equilibration: Physics and Chemistry of the Earth, v. 9, p. 351–366.

Lawless, P. J. Gurney, J. J., and Dawson, J. B., 1979, Polymict peridotites from the Bultfontein and De Beers Mines, Kimberley, South Africa, *in* Boyd, F. R., and Meyer, H.O.A., eds., The mantle sample; Inclusions in kimberlites and other volcanics; Proceedings of the Second International Kimberlite Conference: Washington, D.C., American Geophysical Union, p. 145–155.

Lensch, G., 1976, Ariegite und websterite im lherzolite von Balmuccia (Val Sesia/Zone von Ivrea): Neues Jahrbuch fur Mineralogie Abhandlungen, v. 128, p. 189–208.

Lloyd, F. E., 1981, Upper-mantle metasomatism beneath a continental rift; Clinopyroxenes in alkali mafic lavas and nodules from southwest Uganda: Mineralogical Magazine, v. 44, p. 315–323.

Lloyd, F. E., and Bailey, D. K., 1975, Light element metasomatism of the continental mantle; The evidence and the consequences: Physics and Chemistry of the Earth, v. 9, p. 389–416.

Loubet, M., and Allegre, C. J., 1982, Trace elements in orogenic lherzolites reveal the complex history of the upper mantle: Nature, v. 198, p. 809–814.

Mathias, M., Siebert, J. C., and Rickwood, P. C., 1970, Some aspects of the mineralogy and petrology of ultramafic xenoliths in kimberlites: Contributions to Mineralogy and Petrology, v. 26, p. 75–123.

McCallum, M. E., and Eggler, D. H., 1976, Diamonds in an upper mantle peridotite nodule from kimberlite in southern Wyoming: Science, v. 192, p. 253–256.

Melson, W. G., Hart, S. R., and Thompson, G., 1972, St. Paul's Rocks, Equatorial Atlantic; Petrogenesis, radiometric ages, and implications on sea-floor spreading: Geological Society of Ameria Memoir 132, p. 241–272.

Menzies, M. A., 1983, Mantle ultramafic xenoliths in alkaline magmas; Evidence for mantle heterogeneity modified by magmatic activity, *in* Hawkesworth, C. J., and Norry, M. J., eds., Continental basalts and mantle xenoliths: Nantwich, United Kingdom, Shiva Publishing, p. 92–110.

Menzies, M. A., and Murthy, V. R., 1980a, Mantle metasomatism as a precursor to the genesis of alkaline magmas; Isotopic evidence: American Journal of Science, v. 280-A, The Jackson Volume, p. 622–638.

——, 1980b, Nd and Sr isotopes in diopsides from kimberlite nodules: Nature, v. 283, p. 634–636.

Menzies, M. A., and Wass, S. Y., 1983, CO_2- and LREE-rich mantle below eastern Australia; A REE and isotopic study of alkaline magmas and apatite-rich mantle xenoliths from the southern highlands province, Australia: Earth and Planetary Science Letters, v. 65, p. 287–302.

Menzies, M. A., Kempton, P. D., and Dungan, M. A., 1985, Interaction of continental lithosphere and asthenospheric melts below the Geronimo volcanic field, Arizona, U.S.A.: Journal of Petrology, v. 26, p. 663–693.

Mercier, J. C., 1979, Peridotite xenoliths and the dynamics of kimberlite intrusion, *in* Boyd, F. R., and Meyer, H.O.A., eds., The mantle sample; Inclusions in kimberlites and other volcanics; Proceedings of the Second International Kimberlite Conference: Washington, D.C., American Geophysical Union, p. 197–212.

Moores, E. M., 1969, Petrology and structure of the Vourinos ophiolitic complex, northern Greece: Geological Society of America Special Paper 118, 74 p.

Mysen, B. O., 1983, Rare earth element partitioning between (H_2O + CO_2) vapor and upper mantle minerals; Experimental data bearing on the conditions of formation of alkali basalt and kimberlite: Neues Jahrbuch fur Mineralogie Abhandlungen, v. 146, p. 41–65.

Nagasawa, H., Wakita, H., Higuchi, H., and Onuma, N., 1969, Rare earths in peridotite nodules; An explanation of the genetic relationships between basalt and peridotite nodules: Earth and Planetary Science Letters, v. 5, p. 377–381.

Nicolas, A., 1984, Lherzolites of the western Alps; A structural review, *in* Kornprobst, J., ed., Kimberlites II; The mantle and crust-mantle relationships:

Amsterdam, Elsevier, p. 333–345.

Nicolas, A., and Jackson, M., 1982, High temperature dikes in peridotites; Origin by hydraulic fracturing: Journal of Petrology, v. 23, p. 568–582.

Nixon, P. H., and Boyd, F. R., 1973, Petrogenesis of the granular and sheared ultrabasic nodule suite in kimberlites, *in* Nixon, P. H., ed., Lesotho Kimberlites: Cape Town, Cape and Transvaal Printers, p. 48–56.

Nixon, P. H., Rogers, N. W., Gibson, I. L., and Grey, A., 1981, Depleted and fertile mantle xenoliths from southern African kimberlites: Annual Review of Earth and Planetary Sciences, v. 9, p. 285–309.

O'Reilly, S. Y., and Griffin, W. L., 1984, Sr isotopic heterogeneity in primitive basaltic rock, southeastern Australia; Correlation with mantle metasomatism: Contributions to Mineralogy and Petrology, v. 87, p. 220–230.

Otonello, G., 1980, Rare earth abundances and distribution in some spinel peridotite xenoliths from Assab (Ethiopia): Geochimica et Cosmochimica Acta, v. 44, p. 1885–1901.

Oxburgh, E. R., 1964, Petrological evidence for the presence of amphibole in the upper mantle and its petrogenetic and geophysical implications: Geological Magazine, v. 101, p. 1–19.

Pike, J.E.N., and Schwarzman, E. C., 1977, Classification of textures in ultramafic xenoliths: Journal of Geology, v. 85, p. 49–61.

Pike, J.E.N., Meyer, C. E., and Wilshire, H. G., 1980, Petrography and chemical composition of a suite of ultramafic xenoliths from Lashaine, Tanzania: Journal of Geology, v. 88, p. 343–352.

Prinzhofer, A., and Nicolas, A., 1980, The Bogota Peninsula; A possible oceanic transform fault: Journal of Geology, v. 88, p. 387–398.

Quick, J. E., 1981, Petrology and petrogenesis of the Trinity peridotite, an upper mantle diapir in the eastern Klamath Mountains, northern California: Journal of Geophysical Research, v. 86, p. 11837–11863.

Raleigh, C. B., 1965, Structure and petrology of an alpine peridotite on Cypress Island, Washington, U.S.A.: Contributions to Mineralogy and Petrology, v. 11, p. 719–741.

Roden, M. F., and Murthy, V. R., 1985, Mantle metasomatism: Annual Review of Earth and Planetary Sciences, v. 13, p. 269–296.

Roden, M. F., Frey, F. A., and Francis, D. M., 1984a, An example of consequent mantle metasomatism in peridotite inclusions from Nunivak Island, Alaska: Journal of Petrology, v. 25, p. 546–577.

Roden, M. F., Murthy, V. R., and Irving, A. J., 1984b, Isotopic heterogeneity and evolution of uppermost mantle, Kilbourne Hole, New Mexico: EOS American Geophysical Union Transactions, v. 65, p. 306.

Schneider, M. E., and Eggler, D. H., 1984, Compositions of fluids in equilibrium with peridotite; Implications for alkaline magmatism-metasomatism, *in* Kornprobst, J., ed., Kimberlites. I, Kimberlites and related rocks: Amsterdam, Elsevier, p. 383–394.

Schulze, D. J., and Helmstaedt, H., 1979, Garnet pyroxenite and eclogite xenoliths from the Sullivan Buttes latite, Chino Valley, Arizona; *in* Boyd, F. R., and Meyer, H.O.A., eds., The mantle sample; Inclusions in kimberlites and other volcanics; Proceedings of the Second International Kimberlite Conference: Washington, D.C., American Geophysical Union, v. 2, p. 318–329.

Shervais, J. W., 1979, Ultramafic and mafic layers in the alpine type lherzolite massif at Balmuccia (Italy): Memoire di Scienze Geologiche, Padova, v. 33, p. 135–145.

Sinigoi, S., Comin-Chiaramonti, P., and Alberti, A., 1980, Phase relations in the partial melting of the Baldissero spinel-lherzolite (Ivrea-Verbano zone), western Alps, Italy: Contributions to Mineralogy and Petrology, v. 75, p. 111–121.

Sinigoi, S., Comin-Chiaramonti, P., Demarchi, G., and Siena, F., 1983, Differentiation of partial melts in the mantle; Evidence from the Balmuccia peridotite, Italy: Contributions to Mineralogy and Petrology, v. 82, p. 351–359.

Smith, D., and Ehrenberg, S. N., 1985, Zoned minerals in garnet peridotite nodules from the Colorado Plateau; Implications for mantle metasomatism and kinetics: Contributions to Mineralogy and Petrology, in press.

Stewart, D. C., and Boettcher, A. L., 1977, Chemical gradients in mantle xenoliths: Geological Society of America Abstracts with Programs, v. 9, p. 1191–1192.

Stosch, H.-G., and Seck, H. A., 1980, Geochemistry and mineralogy of two spinel peridotite suites from Dreiser Weiher, West Germany: Geochimica et Cosmochimica Acta, v. 44, p. 457–470.

Suen, C. J., and Frey, F. A., 1978, Origin of mafic layers in alpine peridotite bodies as indicated by the geochemistry of the Ronda massif, Spain: EOS American Geophysical Union Transactions, v. 59, p. 401.

Trask, N. J., 1969, Ultramafic xenoliths in basalt, Nye County, Nevada: U.S. Geological Survey Professional Paper 650-D, p. D43–D48.

Varne, R., 1970, Hornblende lherzolite and the upper mantle: Contributions to Mineralogy and Petrology, v. 27, p. 45–51.

Walcott, R. I., 1969, Geology of the Red Hill Complex, Nelson, New Zealand: Transactions of the Royal Society of New Zealand, Earth Science, v. 7, p. 57–88.

Wass, S. Y., 1979, Fractional crystallization in the mantle of late-stage kimberlitic liquids; evidence in xenoliths from the Kiama area, N.S.W., Australia, *in* Boyd, F. R., and Meyer, H.O.A., eds., The mantle sample; Inclusions in kimberlites and other volcanics: Washington, D.C., American Geophysical Union, p. 366–373.

Wass, S. Y., and Rogers, N. W., 1980, Mantle metasomatism; Precursor to continental alkaline volcanism: Geochimica et Cosmochimica Acta, v. 44, p. 1811–1823.

Wass, S. Y., Henderson, P., and Elliott, C. J., 1980, Chemical heterogeneity and metasomatism in the upper mantle; Evidence from rare earth and other elements in apatite-rich xenoliths in basaltic rocks from eastern Australia: Philosophical Transactions of the Royal Society of London, v. 297-A, p. 333–346.

Wendlandt, R. F., and Harrison, W. J., 1979, Rare earth partitioning between immiscible carbonate and silicate liquids and CO_2 vapor; Results and implications for the formation of light rare earth-enriched rocks: Contributions to Mineralogy and Petrology, v. 69, p. 409–419.

Wilshire, H. G., 1984, Mantle metasomatism: The REE story; Geology, v. 12, p. 395–398.

Wilshire, H. G., and Jackson, E. D., 1975, Problems in determining mantle geotherms from pyroxene compositions of ultramafic rocks: Journal of Geology, v. 83, p. 313–329.

Wilshire, H. G., and Pike, J.E.N., 1975, Upper-mantle diapirism; Evidence from analogous features in alpine peridotite and ultramafic inclusions in basalt: Geology, v. 3, p. 467–470.

Wilshire, H. G., and Shervais, J. W., 1973, Al-augite and Cr-diopside ultramafic xenoliths in basaltic rocks from western United States; Structural and textural relationships: International Conference on Kimberlites, South Africa, Extended Abstracts, p. 321–324.

——, 1975, Al-augite and Cr-diopside ultramafic xenoliths in basaltic rocks from western United States; Structural and textural relationships: Physics and Chemistry of the Earth, v. 9, p. 257–272.

Wilshire, H. G., and Trask, N. J., 1971, Structural and textural relationships of amphibole and phlogopite in peridotite inclusions, Dish Hill, California: American Mineralogist, v. 56, p. 240–255.

Wilshire, H. G., Pike, J.E.N., Meyer, C. E., and Schwarzman, E. C., 1980, Amphibole-rich veins in lherzolite xenoliths, Dish Hill and Deadman Lake, California: American Journal of Science, v. 280-A, The Jackson Volume, Pt. 2, p. 576–593.

Wilshire, H. G., Meyer, C. E., Nakata, J. K., Calk, L. C., Shervais, J. W., Nielson, J. E., and Schwarzman, E. C., 1985, Mafic and ultramafic xenoliths from volcanic rocks of the western United States: U.S. Geological Survey Open-File Report 85-139, 505 p.

Wood, D. A., 1979, A variably veined suboceanic upper mantle; Genetic significance for mid-ocean ridge basalts from geochemical evidence: Geology, v. 7, p. 499–503.

Manuscript Accepted by the Society October 14, 1986

Printed in U.S.A.

Geological Society of America
Special Paper 215
1987

Processes of mantle metasomatism; Constraints from observations of composite peridotite xenoliths

Jane E. Nielson and Jay S. Noller
U.S. Geological Survey, 345 Middlefield Road, Menlo Park, California 94025

ABSTRACT

Available major, trace element, and isotopic data from peridotite xenoliths show that the chemical signature of "cryptic" metasomatism (characterized by enrichment of light rare-earth elements, LREE, relative to heavy rare-earth elements, HREE, in bulk rock and clinopyroxenes) also is seen in rocks affected by "patent" metasomatism (characterized by the presence of secondary hydrous minerals). Detailed major element, REE, and isotopic analyses of minerals in two similar composite xenoliths from Dish Hill, California, demonstrate that patent and cryptic metasomatism, as currently defined, can be caused by the intrusion of a mafic melt. Clinopyroxenes may become enriched in LREE at the same time that hydrous minerals are deposited and minerals of the peridotite wallrock are enriched in Fe, Ti, and Al by reaction with volatile-rich fluid from a dike (Fe-Ti metasomatism).

Microscopic solid and fluid inclusions have been identified as sources of incompatible elements in peridotite xenoliths. Distributions of such inclusions in the composite xenoliths show that they can be added by mantle fluids. Limited data suggest that bulk rock enrichment of LREE/HREE in clinopyroxene-poor, low-CaO metasomatized anhydrous peridotite xenoliths can be controlled by olivine rather than by pyroxene. This control may be due to LREE-enriched inclusions in olivine and other minerals that do not accommodate REE in structural sites. If the chemical signature of cryptic metasomatism is caused by microscopic additions, the distinction between "patent" and "cryptic" categories becomes arbitrary.

The evidence suggests that mantle metasomatism results from reaction between fluid differentiates of mafic melts and peridotite wallrocks. Variations of initial melt compositions, degrees of melt differentiation, and ratios of peridotite to melt all influence the LREE and isotopic compositions of metasomatized mantle rock.

INTRODUCTION

Metasomatic processes change the composition of a rock by the action of pore fluids or gases that deposit new minerals or alter the compositions of old minerals, either by nearly simultaneous dissolution and deposition or by addition of new chemical components from external sources (*AGI Glossary,* 1980). Metasomatic fluids or vapors invade rock volumes by infiltration along cracks and grain boundaries or by diffusion across grain boundaries (Korzhinsky, 1970). These processes are well demonstrated in some types of ore deposits (Lindgren, 1928), where evidence suggests that metasomatizing agents were introduced by intrusions.

Mantle rocks appear to be dominantly spinel- or garnet-bearing Mg- and Cr-rich peridotites. Nearly everywhere, these peridotites are found in association with mafic rock types of lesser abundance, many of which can be identified as dikes (Wilshire and Pike, 1975; Nicolas and Jackson, 1982). Metasomatism of mantle regions is indicated by several lines of evidence, derived dominantly from peridotite xenoliths. This evidence includes: (1) the presence of secondary, commonly hydrous, minerals in peridotite and pyroxenite xenoliths (for example, Lloyd and Bailey, 1975; Varne and Graham, 1971); (2) identification of major element reactions between peridotite and dike rocks in composite xenoliths (for example, Wilshire and Shervais, 1975; Irving, 1980; Kempton and others, 1984); and (3) incompatible element enrichments both in mantle-derived lavas (Boettcher, 1984) and in bulk rock and mineral compositions of otherwise refractory peridotite xenoliths (for example, Frey, 1984; Dawson, 1984; Kempton and others, 1984).

Two apparently contrasting styles of metasomatism identified by Dawson (1984) are called patent and cryptic metasomatism. Patent metasomatism (also called modal metasomatism by Harte, 1983) is indicated by the presence of clearly secondary alkali-bearing hydrous minerals such as phlogopite and amphibole, accompanied by apatite, Ti-magnetite, and sphene, in peridotite xenoliths from worldwide localities (Lloyd and Bailey, 1975; Varne, 1970; Wilshire and Trask, 1971; Best, 1974; Francis, 1976). Similar rocks occur in massifs (Green, 1964; Melson and others, 1972; Bonatti and others, 1981). The introduced minerals formed by infiltration of metasomatizing H_2O-CO_2 fluids (Lloyd and Bailey, 1975). Cryptic metasomatism is identified by enrichment of the bulk rock and clinopyroxene compositions of otherwise refractory anhydrous peridotite xenoliths in light rare-earth elements (LREE) relative to heavy rare-earth elements (HREE). These chemical relations occur in the absence of textural or modal evidence of metasomatism.

Harte (1983) defined a third category of metasomatism (or "fertilization") that is here called Fe-Ti metasomatism. Fe, Al, and Ti are added to the primary minerals of refractory peridotite wallrocks adjacent to hydrous and anhydrous mafic dikes in composite xenoliths. Such relations are known in xenoliths from the western United States (Wilshire and Shervais, 1975; Wilshire and others, 1980; Irving, 1980; Stewart and Boettcher, 1977; Kempton and others, 1984) and from Victoria, Australia (Irving, 1980). According to Harte (1983), composite xenoliths affected by this form of metasomatism have no textural or modal evidence of modification. Harte also assumed that this type of metasomatism would change any preexisting relative LREE-enrichments to "primordial" abundances.

The three types of metasomatism thus are identified by different descriptive criteria: patent metasomatism by introduced minerals, cryptic metasomatism by enriched incompatible trace elements, and Fe-Ti metasomatism by increased Fe, Ti, and Al. Patent metasomatism is ascribed to infiltration of fluids, whereas both cryptic and Fe-Ti metasomatism are thought to occur by mass diffusion across grain boundaries (Korzhinsky, 1970; Gurney and Harte, 1980). These apparent distinctions have led some authors (Kramers and others, 1983; Harte, 1983; Dawson, 1984; Menzies and Murthy, 1980) to conclude that cryptic metasomatism is a separate process unrelated to patent metasomatism. One difficulty in applying these ideas arises for rocks that show both cryptic and patent metasomatism. Kramers and others (1983) ascribed relative LREE-enrichment in clinopyroxene to an event separate from the one that deposited secondary hydrous and associated anhydrous minerals within the same rock, but observed that the chemical signatures of cryptic and patent metasomatism are indistinguishable.

Not all authors subscribe to the idea that the processes differ. For example, Frey and Green (1974), Frey and Prinz (1978), and Frey (1984) have identified mantle metasomatism by distinctive bulk major and trace element relations in both hydrous and anhydrous peridotite xenoliths, and have ascribed them both to an event that mixed "depleted" and "enriched" components. According to Frey (1984), the enriched component is likely to be a melt.

Identification of mantle metasomatism based on contrasting criteria raises substantive questions, such as, "Is patent metasomatism chemically distinct from cryptic metasomatism?" Is relative LREE/HREE enrichment incompatible with enrichment of Fe and Al in clinopyroxene? Are the processes of diffusion and infiltration metasomatism restricted to separate metasomatic events? In this investigation we have sought a context for the interpretation of metasomatic processes and the events that cause them. In so doing, we first reviewed the critical bulk and mineral major and trace element compositions of mantle rocks available in the literature. Data that approach a statistically reasonable size are available predominantly from studies of peridotite xenoliths from basalts. We conclude that the data reviewed indicate that some geochemical correlations may be artifacts of mode-mineral composition variations.

To put the geochemical data into perspective, we have examined the processes of metasomatism as indicated by detailed studies of two composite hornblendite-peridotite xenoliths from Dish Hill, California. The Dish Hill samples provide a case history of dike intrusion, accompanied by Fe-Ti and patent metasomatism, in which relative LREE-enrichments and isotopic compositions of clinopyroxenes (cryptic metasomatism) can be related directly to the same fluids. We conclude that the contrasting criteria of metasomatism may result from a single intrusive event.

As defined above, the chemical evidence of metasomatism is based on relative enrichment of LREE over HREE in bulk rock peridotite and constituent minerals, and to a lesser extent on enrichment of other incompatible elements in mantle rocks and minerals (Frey, 1984). Rare-earth element chondrite-normalized (REEcn) values were calculated from tabulated data in the literature, using the chondrite abundances of Evensen and others (1978). We used only data that fit our criterion of matching major element and REE analyses of rock and/or mineral samples of dominant spinel- and garnet-bearing peridotite (Table 1). We assume that amphibole- or mica-bearing peridotites have been patently metasomatized. As far as can be determined, none of the plotted anhydrous samples contains secondary metasomatic minerals. In all diagrams hydrous samples are distinguished from anhydrous (unmetasomatized or cryptically metasomatized) samples.

MODAL AND COMPOSITIONAL SIGNATURES OF PATENT AND CRYPTIC MANTLE METASOMATISM: A REVIEW

Mantle Rock Compositions

Most xenoliths from basaltic lavas have been assigned to one

TABLE 1. MINERAL AND BULK COMPOSITIONS OF TYPE I PERIDOTITE XENOLITHS: SOURCES AND KINDS OF DATA

Locality	References	Bulk-Rock Analyses*			Mineral Analyses**								
		Kind of Data	Number of Samples Plotted		Kind of Data						Number of Samples Plotted		
			Hydrous	Anhydrous	Ol	Opx	Cpx	Sp	Gt	Other	Cpx	Ol	Other
Africa: Central Sahara	Miller and Richter, 1982	MR	-	2	--	--	--	--	--	--	-	-	-
Australia: Victoria	Frey and Green, 1974	MR	4	2	M	MR	MR	M	--	MR,am	6	-	-
										M,ph,ap	-	-	-
	Irving, 1980	MR	-	1	--	--	--	--	--	--	-	-	-
Austria: Kapfenstein	Jagoutz and others, 1979	--	-	-	--	--	MR	--	--	--	2	-	-
	Kurat and others, 1980	MR	1	2	MR	MR	MR	MR	--	MR,am	3	3	-
Ethiopia: Assab	Piccardo and Ottonello, 1978	--	-	-	M	M	M	M	--	--	9	-	-
	Ottonello, 1980	MR	-	9	MR	MR	MR	MR	--	--	-	9	-
France: Massif Central	Jagoutz and others, 1979	--	-	-	--	--	MR	--	--	--	7	-	-
Germany:													
Dreiser Weiher	Jagoutz and others, 1979	--	-	-	--	--	MR	--	--	--	10	-	-
	Stosch and Seck, 1980	MR	8	8	M	MR	MR	MR	--	--	2	2	-
Vogelsberg	Jagoutz and others, 1979	--	-	-	--	--	MR	--	--	--	4	-	-
Japan: Itinome-gata	Kuno and Aoki, 1970	M	-	-	--	--	--	--	--	--	-	-	-
	Aoki and Shiba, 1973	M	-	-	M	M	--	--	--	--	-	-	-
	Tanaka and Aoki, 1981	R	1	2	R	R	R	--	--	--	2	2	-
Lesotho:													
Matsoku	BVSP, 1980	MR	-	1***	--	--	--	--	--	--	-	-	-
Thaba Putsoa, Mothae	Nixon and Boyd, 1973a	--	-	-	--	--	M	--	M	--	-	-	-
	Shimizu, 1975	--	-	-	--	--	R	--	R	--	4	-	-
Liqhobong	Nixon and Boyd, 1973b	--	-	-	M	M	M	--	M	--	-	-	-
Mexico: San Quintin	BVSP, 1980	MR	-	1	--	--	--	--	--	--	-	-	-
Tanzania: Lashaine	Reid and others, 1975	--	-	-	M	M	M	M	M	--	-	-	-
	Ridley and Dawson, 1975	MR	1	3	--	--	M	--	R	R,ph	1	-	-
	BVSP, 1980	MR	-	1	M	M	--	M	M	--	-	-	-
United States:													
Alaska: Nunivak	Francis, 1974	--	-	-	M	M	M	M	--	M,am	-	-	-
	Roden and others, 1984a	R	-	-	--	R	--	--	--	R,am	2	-	-
	BVSP, 1980	MR	1	-	--	M	M	--	--	M,am	-	-	-
Arizona:													
Cochise Crater	BVSP, 1980	MR	-	1	--	--	--	--	--	--	-	-	-
Geronimo	Kempton, 1984	M	-	-	M	M	MR	M	--	MR,am	9	-	-
San Carlos	Frey and Prinz, 1978	MR	1	1	--	M	M	MR	M	--	1	-	-
	Irving, 1980	MR	-	2	--	--	--	--	--	--	-	-	-
The Thumb	Ehrenberg, 1979	--	-	-	--	M	M	--	M	M,ph,il	-	-	-
	Ehrenberg, 1982	--	-	-	--	--	R	--	R	--	3	-	-
California: Dish Hill	Wilshire and others, 1980	--	-	-	M	M	M	M	--	M,am	-	-	-
	Irving and Frey, 1984	--	-	-	--	--	--	--	--	R,am	-	-	2,am
	Menzies and others, 1985	--	-	-	--	--	R	--	--	R,am	2	-	2,am
Hawaii: Salt Lake Crater	BVSP, 1980	MR	-	1	--	--	--	--	--	--	-	-	-
New Mexico: Kilbourne Hole	Irving, 1980 (BVSP, 1980)	MR	-	8	M	M	M	M	--	--	-	-	-
Yemen: Ataq	Varne, 1970	--	-	-	--	--	--	--	--	M,am	-	-	-
	Varne and Graham, 1971	MR	1	-	--	--	MR	--	--	R,am	1	-	-

* M = majors; R = rare-earth elements. Data set in plots are rocks and minerals analyzed for both M and R.

** Ol = olivine; Opx = orthopyroxene; Cpx = clinopyroxene; Sp = spinel; Gt = garnet. Other: am = amphibole; ap = apatite; ph = phlogopite; il = ilmenite.

*** Data for Matsoku not plotted because Yb abundance not determined.

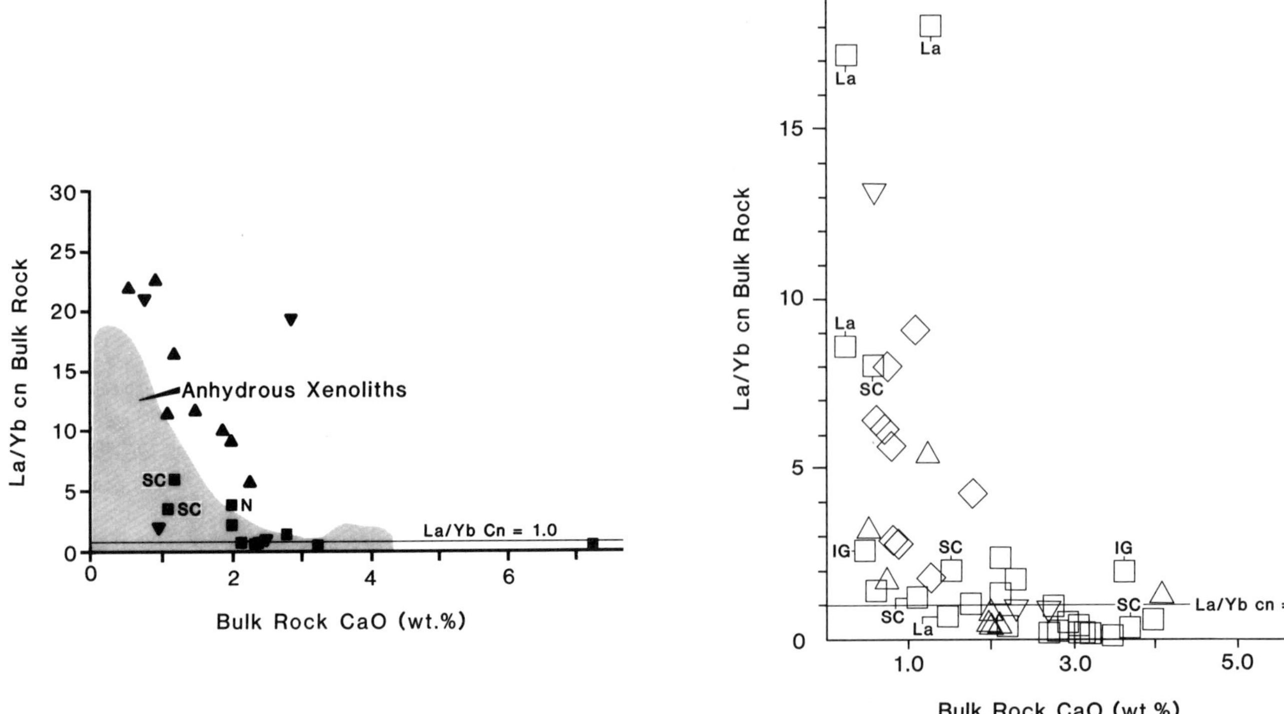

Figure 1. Bulk-rock CaO (wt. %) versus chondrite-normalized La/Yb (La/Ybcn) for type I spinel and garnet peridotite: a, Hydrous xenoliths. Patterned area is the field of anhydrous peridotite compositions. b, Anhydrous xenoliths, plotted at expanded scale. Filled symbols indicate hydrous peridotite; open, anhydrous peridotite; inverted triangles, from Victoria, Australia; upright triangles, from Dreiser Weiher, West Germany; diamonds, from Assab, Ethiopia; SC, from San Carlos, Arizona; IG, from Itinome-gata, Japan; N, from Nunivak Island, Alaska; La, from Lashaine, Tanzania. Squares without letter identification from Kilbourne Hole, New Mexico; Kapfenstein, Austria; Cochise Crater, Arizona; Hoggar, central Sahara; Salt Lake Crater, Hawaii.

of two groups: Cr-rich rocks, which may contain Cr-diopside and which have bulk Mg-ratio [Mg/(Mg + ΣFe)] greater than or equal to 0.85; and rocks that commonly contain Al-augite, with bulk Mg-ratios less than 0.85, low Cr, and higher Al and Ti contents. These lithologic groups have been called, respectively, Cr-diopside and Al-augite groups by Wilshire and Shervais (1975), but they also are referred to as types I and II (nomenclature of Frey and Prinz, 1978). Type I (Cr-diopside group) spinel- or garnet-bearing peridotite (lherzolite and/or harzburgite) is the dominant lithology of massifs and most xenolith suites. By extension, type I peridotites probably represent the dominant lithology of the upper mantle. Minor lithologies include type II peridotites and related mafic rocks that may be common locally. (See Wilshire and others, 1985, for a review of the chemical and structural relations between minor rock types and dominant peridotite.)

Major and trace element compositions of type I lherzolites from massifs are consistent with an origin of the lherzolites as residues from various small percentages of partial melting (Frey, 1984). Theoretical models indicate that low-Ca lherzolite and harzburgite—which have lower modal clinopyroxene, Ca, Al, and incompatible trace element contents than lherzolite (Dick, 1977)—may be simple residues of larger melt volumes than those that produce lherzolite. Thus, lherzolite compositions may change to harzburgite when a large percentage of melt removes a substantial fraction of clinopyroxene from mantle protolith (Frey, 1984). Major element compositions of harzburgites from peridotite massifs, ophiolites, and ocean basins all support this hypothesis (Dick and others, 1984). However, trace element studies of harzburgites are few and do not demonstrate a simple relation between lherzolite and harzburgite residues for most samples (Frey, 1984).

Bulk Compositions of Type I Peridotite Xenoliths

Frey (1984) demonstrated significant correlations between bulk rock major and trace element compositions of lherzolite and harzburgite xenoliths, which are not known from massifs. Specifically, Frey (1984) pointed out that type I harzburgite and low-Ca lherzolite xenoliths from many localities are preferentially enriched in LREE relative to HREE (Fig. 1a,b). If low-Ca peri-

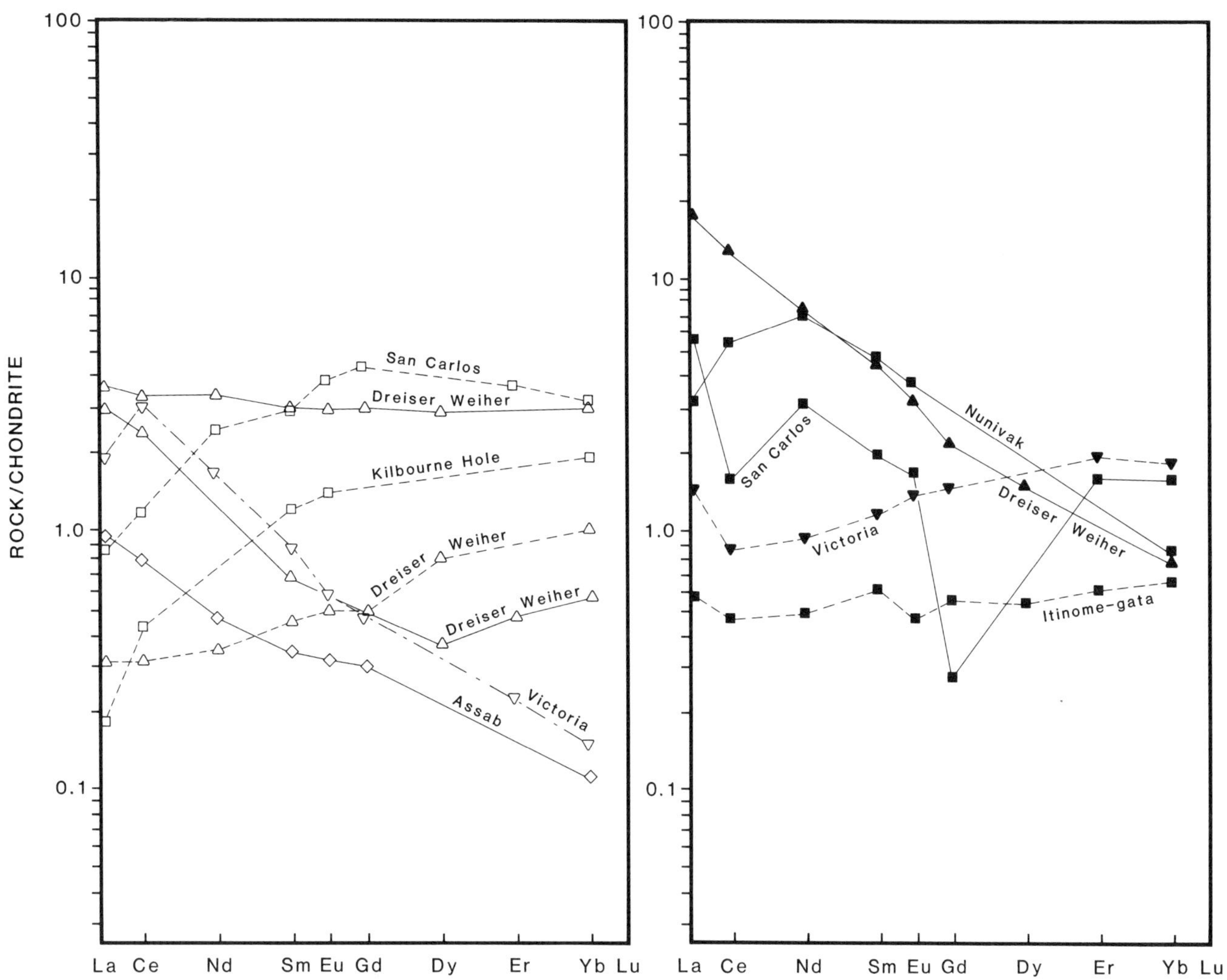

Figure 2. Chondrite-normalized bulk-rock REE (REEcn) patterns of type I peridotites: a, Range of variation for anhydrous lherzolite and harzburgite xenoliths. b, Range of variation for hydrous lherzolite and harzburgite xenoliths. Symbols as in Figure 1.

dotites are simple residues of larger percentage partial melt events than are lherzolites, this negative correlation of incompatible major and trace elements is paradoxical. Frey and Green (1974) suggested that these chemical signatures arise in two stages. The first stage is a partial melt event that leaves residues depleted in incompatible elements. The relative LREE enrichments were ascribed by those authors to a later event that may be metasomatic. An alternate suggestion by Ottonello (1980) is that LREE partition coefficients change during the late stages of equilibrium partial melting, so that LREE cease to behave as incompatible elements. Both hypotheses pose problems. Frey's model suggests no reason for metasomatism to have preferentially affected the most refractory rocks, and Ottonello's idea implies that a negative correlation between CaO and LREE/HREE in anhydrous harzburgites should be observed in rocks of massifs and ophiolites, as well as in xenolith suites.

REEcn Patterns

The REE abundances of type I spinel peridotite xenoliths are widely variable (for a review, see Frey, 1984). Chondrite-normalized LREE (LREEcn) abundances range from <1.0 (depleted) to >1.0 (enriched). Figures 2a,b, show the range of REE chondrite-normalized patterns in the literature for type I hydrous and anhydrous lherzolite and harzburgite xenoliths. In most cases the relative enrichment of LREEcn over HREEcn corresponds to LREEcn abundances >1.0, but Figure 2 shows that there are exceptions among both anhydrous and hydrous samples. For example, a sample from Assab, Ethiopia (Ottonello, 1980), with high LREE/HREE, has REEcn abundances of <1.0 (Fig. 2a). A sample from The Anakies, Victoria (Frey and Green, 1974), has Lacn and Cecn values that are only slightly >1.0 but has very high LREE/HREE because of extremely low HREEcn abundances. Thus, some relative LREE enrichments cannot be explained by the simple addition of LREE. In these cases the relation of LREE to HREE may represent removal of HREE.

There are relatively few analyses of hydrous peridotite xenoliths available in the literature. We find that most have REEcn patterns with negative slopes that are markedly steeper than the slopes of relatively LREE-enriched anhydrous peridotite xenoliths, although a few hydrous samples have bulk rock LREE/HREE <1.0 (for example, Itinome-gata, Japan: Table 1, Fig. 2b). These compositional differences reflect in part the variations of mode and mineral compositions between samples. In addition, there may be contamination by the host lavas, crustal contamination, and weathering, all of which tend to increase LREE/HREE. The low LREE/HREE values found in some hydrous peridotite xenoliths may reflect the composition of metasomatic fluid (melt).

Major and Trace Element Correlations

In Figure 1, CaO (wt. %) is plotted against bulk rock La/Ybcn values for hydrous (symbols) and anhydrous (patterned area) type I peridotite xenolith compositions currently available in the literature (1985, Table 1). Most are spinel peridotites, but samples from Lesotho, Lashaine, Tanzania, and The Thumb, Arizona (Table 1) contain garnet. Compositions of anhydrous xenoliths are plotted at expanded scale in Figure 1b. By the criteria of Dawson (1984), all the hydrous samples may be identified as patently metasomatized. By the criteria of Frey (1984), all the samples in Figures 1a,b with La/Ybcn <1.0 are unmetasomatized, and those with La/Ybcn >1.0 may be called cryptically metasomatized. Dawson's (1984) criterion for cryptic metasomatism requires relative LREE enrichment of clinopyroxene. Whether or not this entails enrichment of the bulk rock is unclear, but most authors have assumed that LREE/HREE values of the bulk rock are controlled by clinopyroxene.

Comparison of the fields of hydrous and anhydrous samples in Figure 1a shows that anhydrous peridotite xenoliths that have marked enrichments of LREE/HREE are more restricted in CaO contents than those with low relative LREE contents. The published data also indicate that anhydrous peridotites have less marked relative enrichments of LREE over HREE than do hydrous types. Many hydrous samples with high LREE/HREE are lherzolites with CaO contents >1 wt. % in contrast to more harzburgitic compositions of the relatively enriched anhydrous peridotites (Fig. 1b). However, the most CaO-poor hydrous peridotite xenoliths all are relatively enriched in LREE/HREE. Limited data on the absolute abundance of REE for these rocks suggest that the CaO-poor hydrous peridotite xenoliths also have higher total REE than CaO-rich hydrous samples. No such correlation is evident for the anhydrous samples.

Both hydrous and anhydrous depleted and enriched peridotites occur together in some suites. Comparison of Figures 1a and 1b shows that the hydrous Dreiser Weiher samples have a strong linear trend of increasing La/Ybcn with decreasing CaO, in contrast to the scatter of the anhydrous xenoliths. The opposite relation occurs for localities in Victoria, Australia, which have CaO and La/Ybcn values for hydrous xenoliths that are widely scattered, in contrast to a linear trend for anhydrous samples. The scatter of CaO contents at high relative LREE/HREE values for hydrous xenoliths can be at least partly explained by metasomatic additions of LREE and CaO in secondary pargasite, apatite, and associated minerals (Frey and Green, 1974; Francis, 1976; Wass and others, 1980; Menzies and Wass, 1983; Roden and others, 1984a).

Anhydrous peridotite xenoliths from all occurrences predominantly have bulk CaO contents that range from 0.3 to 4 wt. % and widely variable LREE/HREE. Samples with CaO contents >2 wt. % range from relatively LREE-depleted (La/Ybcn, <1.0) to slightly enriched (La/Ybcn, 1.0 to 2.5). Samples with CaO of 2 wt. % or less range in LREE/HREE from slightly to highly enriched (La/Ybcn, 1.2 to 13.4), as noted by Frey (1984). The samples in Figure 1b with high LREE relative to HREE are from Assab, Ethiopia, Victoria, Australia, and San Carlos, Arizona (Table 1). Three samples from Australia (inverted triangles) and four from San Carlos (SC) show a linear increase of LREE/HREE with decrease in CaO (Fig. 1b), but samples from Dreiser Weiher (upright triangles), Assab, Ethiopia (diamonds), and Itinome-gata, Japan (IG) show only moderate relative LREE/HREE enrichment and considerable scatter at low CaO contents.

If one exceptional hydrous peridotite sample (Mount Noorat, Victoria, Australia, Frey and Green, 1974) is ignored, the trend of high bulk-rock LREE/HREE at low CaO content shown in Figure 1a appears to be the same for hydrous and anhydrous type I peridotite xenoliths. Thus, a negative correlation between these parameters may be characteristic of both patently and cryptically metasomatized rocks. The Mt. Noorat sample indicates that patent metasomatism does affect mantle rocks that have not undergone high degrees of partial melting. Some hydrous peridotites have LREE/HREE values of <1.0, which indicates that this chemical criterion cannot distinguish patently metasomatized from unmetasomatized peridotites. Thus, although these bulk compositions are the most abundant chemical data available for mantle peridotite, they cannot be interpreted unambiguously.

Mineral Composition, Bulk Composition, and Mode

Bulk REE contents of spinel peridotite xenoliths are generally higher than those calculated from the compositions of either water-washed or acid-leached mineral separates (Frey and Green, 1974; Frey and Prinz, 1978; Ottonello, 1980; Jagoutz and others, 1980; Ehrenberg, 1982; Feigenson and Stolz, 1984; Zindler and Jagoutz, in press). This discrepancy is most commonly attributed to intergranular contamination by LREE- and alkali-rich host lavas during transit (Jagoutz and others, 1980; Ottonello, 1980; Stosch and Seck, 1980). Kurat and others (1980) discovered that olivines with inclusions ("black dots"), found within grains and at

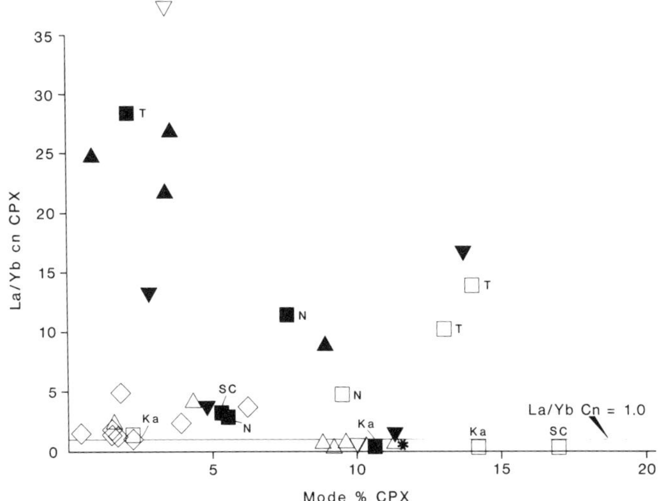

Figure 3. Modal content of clinopyroxene (cpx) from peridotite xenoliths versus La/Ybcn of clinopyroxene. Filled symbols indicate cpx in hydrous xenoliths; open, cpx from anhydrous xenoliths; T, from The Thumb, Arizona; Ka, from Kapfenstein, Austria; asterisk, sample from Dreiser Weiher with reported range of 1.6 to 21.1 percent modal cpx, plotted at average value. Other symbols as in Figure 1.

grain margins, are richer in some incompatible trace elements than olivines lacking black dots. Also, Zindler and Jagoutz (in press) found that inclusions, coatings, and fillings of probable mantle origin contribute a significant proportion of the incompatible elements reported for olivine and spinel. Zindler and Jagoutz (in press) did not find substantial amounts of LREE (Nd) in these inclusions, but Jagoutz and others (1980), Kurat and others (1980), and Stosch (1982) found that washing and acid leaching of orthopyroxenes and olivines may cause losses of REE in the residue, or may fractionate La relative to Lu (Stosch and Seck, 1980).

Mineral Compositions

Clinopyroxene is widely viewed as the most sensitive primary mantle mineral for monitoring depletion and enrichment events (for example, Jagoutz and others, 1979; Frey, 1984; Kempton and others, 1984; Roden and Murthy, 1985). Clinopyroxenes clearly contain higher abundances of all REE than do other peridotite minerals, and may dominate the total REE content of the rock. Empirical and theoretical studies of REE partitioning (Nagasawa and others, 1969; Ottonello, 1980; Stosch, 1982) have shown that clinopyroxenes accommodate LREE, particularly Nd, in structural sites (Ottonello, 1980). Thus it is widely held that the high bulk-rock LREE/HREE values of metasomatized anhydrous peridotite xenoliths reflect the abundance of LREE in constituent clinopyroxenes because there are no secondary mineral sources of LREE such as occur in hydrous xenoliths (Frey, 1984). If this is correct, clinopyroxene compositions must account for both the CaO contents and La/Ybcn values of anhydrous peridotite xenoliths.

Figure 3 shows that there is no correlation between estimated clinopyroxene abundance and the La/Ybcn of constituent clinopyroxene in anhydrous peridodite xenoliths. Thus, while the CaO content of the bulk rock is provided by clinopyroxene, there are other constituents that make important contributions to the balance of light to heavy rare-earth elements. With the exception of clinopyroxene from The Anakies (Victoria) harzburgite, the clinopyroxenes in Figure 3 that have marked relative LREE enrichments are dominantly from hydrous xenoliths. The LREE/HREE enrichments in clinopyroxenes from these rocks show a rough correlation with decrease in modal clinopyroxene, although there is considerable scatter. However, much of the LREE in these rocks occurs in secondary minerals, such as hornblende and apatite, which also contain calcium.

A significant test of the dominance by clinopyroxene of the LREE/HREE value in peridotite xenoliths is shown in Figures 4a,b, plots of La/Ybcn in bulk rock versus La/Ybcn in constituent minerals. Concordant LREE/HREE values of rock and mineral analyses are expected to lie close to a line with slope of unity. Figure 4a shows that values for clinopyroxene plot close to the line only for unmetasomatized samples and three cryptically metasomatized samples with moderate relative LREE/HREE enrichments. Significantly, clinopyroxenes from most anhydrous metasomatized peridotite xenoliths have lower LREE/HREE values than the bulk rock. This is particularly true of most harzburgites and low-Ca lherzolites from Assab (diamonds in Figs. 4a,b; Ottonello, 1980), which were not acid-leached before analysis. Most clinopyroxenes from the hydrous samples generally have much higher LREE/HREE values than do the corresponding bulk rocks, but the sample population is too limited for this difference to be significant. Figure 4b is a plot of the more limited olivine data versus bulk-rock LREE/HREE for many of the same rocks (Table 1). The figure shows that La/Ybcn of acid-leached olivine samples from anhydrous metasomatized rocks is close to that of the corresponding bulk rock more consistently than is La/Ybcn of clinopyroxene. In general, Figure 4b suggests that metasomatic enrichment of LREE/HREE in olivine can have a greater influence than clinopyroxene on bulk LREE/HREE, especially in olivine-rich (CaO-poor) peridotite. However, two samples from Assab (not leached) and one from Dreiser Weiher (leached) have higher bulk La/Ybcn than can be accounted for by either olivine or clinopyroxene (Fig. 4). An important question is whether these compositions are created by mantle processes or are due to later contamination.

Figures 3 and 4 suggest several significant points about the chemical signature of metasomatism. First, relative enrichment of nonstructural LREE in olivine and other peridotite minerals may be as important an effect of metasomatism as is LREE/HREE enrichment in clinopyroxene, although the effect is not easily observed except in harzburgite. Second, clinopyroxenes in rocks that have undergone an early depletion event may have

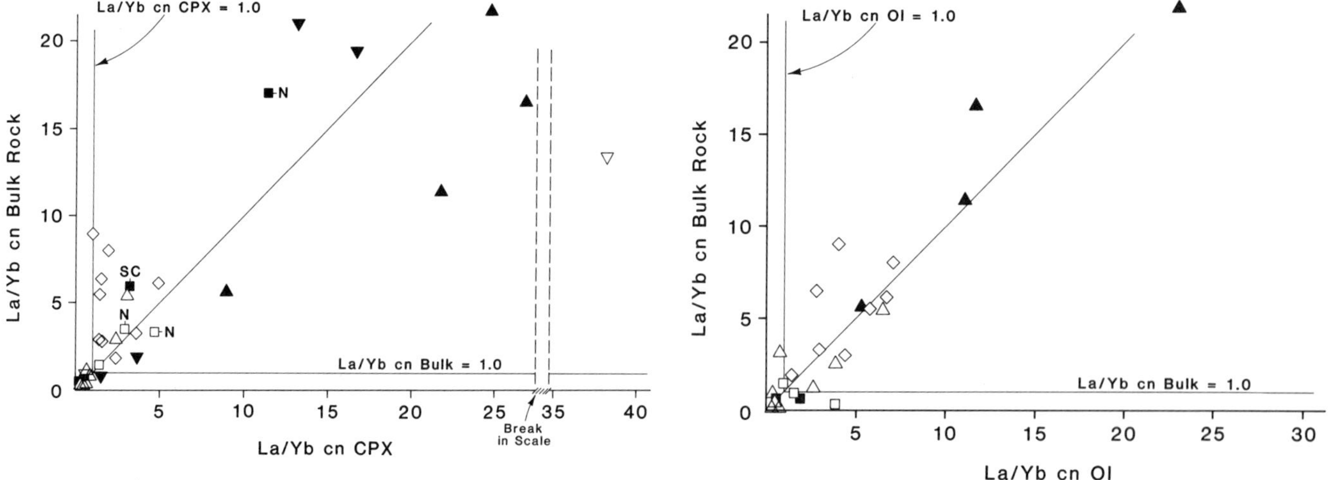

Figure 4. La/Ybcn of peridotite bulk rocks and constituent minerals. a, La/Ybcn of cpx versus bulk rock. b, La/Ybcn of olivine versus bulk rock. Symbols as in Figures 1 through 3.

La/Yb<<1. In a later event, significant LREE may be added to the clinopyroxenes by metasomatic fluids but the final La/Yb value will be the result of both events, thus masking the magnitude of the metasomatic enrichment. Third, enrichment of LREE/HREE in clinopyroxenes correlates strongly with the presence of hydrous minerals in peridotite. However, not all hydrous peridotites contain clinopyroxene with high LREE/HREE values (for example, Fig. 3). The most common explanation of this observation is that patent metasomatism does not cause enrichment of LREE/HREE and the patently metasomatized rocks that have high LREE/HREE in clinopyroxenes were also metasomatized by a separate "cryptic" event. However, the data allow the possibility that patent metasomatism can cause relative enrichment of LREE/HREE in clinopyroxene (Menzies and others, 1985).

Other Indices of Cryptic Enrichment in Clinopyroxenes

Kempton and others (1984) distinguished cryptic enrichment of depleted type I nodules using Ce/Smcn or Ce/Ybcn values of constituent clinopyroxene (see also Menzies and others, 1985). These authors have shown that high relative LREE/HREE enrichments correlate with high Cr/Al in clinopyroxenes of xenoliths from the Geronimo Volcanic Field, Arizona. Shimizu (1975) and Kempton and others (1984) have noted the "conflict" or "decoupling" of incompatible trace and major element contents in these minerals, which is like the relation noted by Frey and Green (1974) and Frey and Prinz (1978) for bulk rock compositions.

All available Ce/Smcn and Cr/Al data for clinopyroxene from hydrous and anhydrous type I peridotite xenoliths (Table 1) are plotted in Figure 5. The compilation shows that the maximum variation of LREE/HREE values (<1.0 to <6.0) correlates with low Cr/Al values (<0.30). Compositions of clinopyroxenes from hydrous and anhydrous metasomatized rocks are virtually indistinguishable at most Cr/Al and Ce/Smcn values, although the highest relative LREE/HREE enrichments are in clinopyroxenes from patently metasomatized rocks. Samples from Assab (all anhydrous; Ottonello, 1980) show a strong positive correlation between Cr/Al and Ce/Smcn, like the trend identified for Geronimo (Kempton, 1984), but clinopyroxenes of other suites, such as Dreiser Weiher, do not show a consistent trend.

Some samples represented in Figure 5 vary substantially in Cr/Al within single xenoliths. For example, a hydrous lherzolite with LREE-enriched clinopyroxene from Nunivak (sample 10002; Francis, 1974; Roden and others, 1984a) is zoned in Cr, and clinopyroxenes in hydrous Dish Hill xenolith samples Ba-1-72 and Ba-2-1 have a range of Cr/Al within each rock (Wilshire and others, 1985). The plotted Ce/Smcn values are averages derived from clinopyroxene separates (Roden and others, 1984a; Menzies and others, 1985). It is possible that REE contents of clinopyroxenes vary along with Cr/Al and other major elements in these samples.

METASOMATIC EVENTS AND PROCESSES OBSERVED IN COMPOSITE XENOLITHS

Evidence of Fe-Ti mantle metasomatism comes from studies of composite xenoliths. Wilshire and Shervais (1975), Wilshire and Jackson (1975), Harte and others (1977), Irving (1980), and Roden and others (1984a) showed that primary minerals of type I peridotite become enriched in Fe, Al, and Ti, with increasing proximity to a contact between the peridotite and type I or II mafic dikes. The magnitude of the enrichment is much greater for a type I peridotite in contact with type II pyroxenite or hornblendite than for type I dikes (Wilshire and others, 1980;

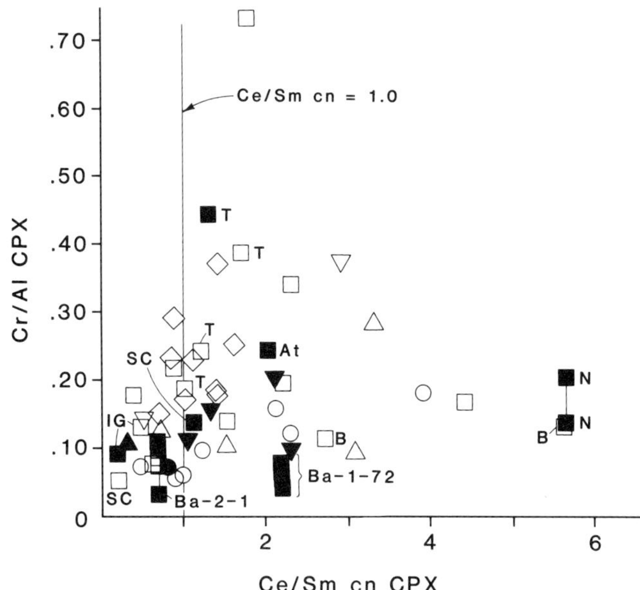

Figure 5. Cr/Al versus Ce/Smcn in clinopyroxene from xenoliths. Circle symbols indicate cpx from Geronimo Volcanic Field, Arizona; At, from Ataq, Yemen; B, from Beyzac, Massif Central, France; Ba-2-1 and Ba-1-72, from Dish Hill, California. Localities without identification from Itinome-gata, Japan; Kapfenstein, Austria; Vogelsburg, West Germany, and other Massif Central localities. Other symbols, same as Figures 1 through 4.

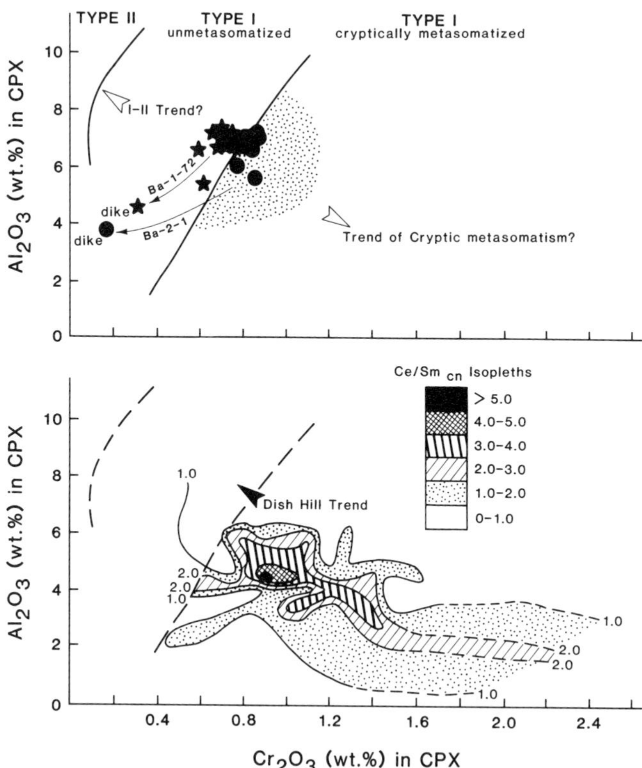

Figure 6. Al_2O_3-Cr_2O_3 compositions of clinopyroxene from type I peridotite xenoliths. a, Areas and trends identified by Kempton (1985), compared to compositions of cpx from Dish Hill composite xenoliths. Type II, cpx compositions from type II peridotite and pyroxenite; type I cryptically metasomatized, compositions of peridotite cpx with LREE/HREE values >1.0; Type I unmetasomatized, compositions of peridotite cpx with LREE/HREE <1.0. Solid lines separate compositional regions; patterned area shows area of overlap of type I metasomatized and unmetasomatized cpx. Star symbols indicate cpx from Dish Hill composite peridotite Ba-1-72; circles, cpx from Ba-2-1. Arrows show trend of Al_2O_3-Cr_2O_3 in cpx with proximity to kaersutite dikes. Open darts indicate trends identified by Kempton (1985) from cpx compositions of worldwide localities (Table 1). b, LREE/HREE enrichment factor superimposed on Figure 6b. Contours are isopleths of Ce/Smcn values derived from analyses of noncomposite type I peridotite xenoliths (Table 1). Filled dart indicates trend of Al_2O_3-enrichment in cpx of Ba-1-72 compared to Ba-2-1. Long dashes indicate same region boundaries shown as solid lines in Figure 6a.

Kempton and others, 1984). Also, Wilshire and others (1985) and Stewart (1980) showed that type I peridotite xenoliths from Dish Hill, California, contain interstitial pargasite formed by reaction between the peridotite and fluids that emanated from kaersutite-rich dikes.

The magnitude of major element interchanges between dikes and their wallrocks suggests that trace element compositions of peridotite also may be affected by the reactions. For example, Roden and others (1984b) have shown that clinopyroxenes in peridotite wallrocks of type II pyroxenite dikes are in or close to isotopic equilibrium with the dikes. In contrast, Kempton and others (1984) and Kempton (1985) concluded that the major element trends of clinopyroxenes induced by Fe-Ti and patent metasomatism are opposite those seen in cryptic metasomatism. Evidence that bears on these conflicting expectations is described below.

Case History: Metasomatism of Composite Xenoliths from Dish Hill, California

Two composite xenoliths from Dish Hill, California, are hydrous type I peridotites with interstitial pargasite and selvages of kaersutite-rich (type II) dikes. The selvages are thin (<1 to 7 mm) remnants of dike margins (Wilshire and Trask, 1971; Stewart and Boettcher, 1977; Wilshire and others, 1980; Stewart, 1980) similar to kaersutite dikes observed in massifs (Wilshire, 1984). Sample Ba-2-1 is large (17 × 16 × 9.5 cm) with a single dike selvage composed of kaersutite, apatite, and phlogopite. Ba-1-72 is smaller (12.5 × 7.5 × 7.5 cm), and has two subparallel kaersutite + apatite selvages about 6.4 cm apart. Peridotite in both xenoliths contains about 15 percent modal clinopyroxene. Only mineral major and trace element analyses are available for these samples (Table 1, Figs. 6, 7).

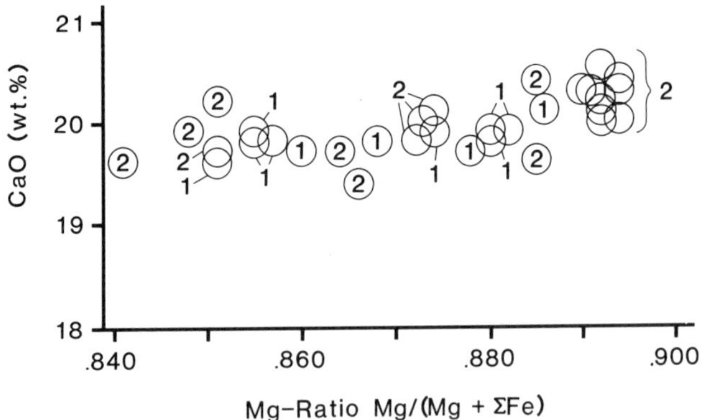

Figure 7. Mg ratios versus CaO (wt. %) of clinopyroxene in composite xenoliths from Dish Hill, California, determined by electron microprobe analysis. Symbol 1 indicates cpx from Ba-1-72; 2, from Ba-2-1.

Mineral Compositions

All minerals in both Dish Hill xenoliths vary significantly in major element composition with respect to the dike-wallrock contacts (Wilshire and others, 1980, 1985). The relation between dike injection and metasomatism is indicated in both xenoliths by continuous variation of amphibole compositions from the dike into the wallrock, where transition from kaersutite to pargasite occurs within 1 cm of the contact. Clinopyroxene, orthopyroxene, olivine, and spinel are Fe-rich at the contact. The Fe-enrichment of anhydrous minerals, especially spinel, is due to reactions between peridotite wallrock and volatile-rich fluid from the dike. By these reactions the fluid became progressively enriched in Mg and formed pargasite in the wallrock (Wilshire and others, 1980).

The variations in Mg ratios for clinopyroxenes of Ba-2-1 and Ba-1-72 (Fig. 7) contrast with nearly constant and virtually identical CaO contents. We have concluded that clinopyroxene CaO was relatively unaffected by reaction with fluids from the dike. Slightly higher CaO and Mg ratios in Ba-2-1, a much larger xenolith, occur in clinopyroxenes that are farther from the influence of a dike than any analyzed spot in Ba-1-72. These compositional similarities suggest that the peridotite of Ba-2-1 and Ba-1-72 are from the same mantle level. Figure 6a shows that clinopyroxene Al_2O_3 and Cr_2O_3 compositions are closely similar for the two peridotites, but Ba-1-72 clinopyroxene compositions (stars) are shifted to Al_2O_3 contents that are slightly higher than those of Ba-2-1 separates (circles). This is the trend expected of Fe-Ti metasomatism (Kempton, 1985). As noted by Kempton (1985) for Geronimo hydrous composites xenoliths, the trend of clinopyroxene compositions within both Dish Hill samples is toward lowered Al_2O_3 and Cr_2O_3, with proximity to the dike (arrows).

Figure 5 shows that clinopyroxenes from the two Dish Hill composite xenoliths have overlapping Cr/Al values, but Ba-1-72 clinopyroxenes are relatively enriched in LREE/HREE compared to Ba-2-1. REEcn for amphibole dikes from the two samples (Menzies and others, 1985; Irving and Frey, 1984; Fig. 8a), and Nd and Sr isotopic ratios (Menzies and others, 1985; Fig. 8b) are closely similar; thus the dikes in both samples crystallized from fluids of similar composition. Clinopyroxenes of Ba-2-1 are relatively depleted in LREE/HREE and also are isotopically more depleted (lower $^{87}Sr/^{86}Sr$, higher $^{143}Nd/^{144}Nd$) than MORBS, whereas Ba-1-72 clinopyroxenes are relatively enriched in LREE/HREE and isotopically are more enriched than MORBS. Ba-1-72 wallrock clinopyroxenes are isotopically near equilibrium with the kaersutite dikes, whereas Ba-2-1 clinopyroxenes remain isotopically distinct from the dikes.

Inclusions in the Dish Hill Xenoliths

Types and distribution of solid and fluid inclusions (inter- and intragranular particles, coatings, and fillings) were determined for xenoliths Ba-2-1 and Ba-1-72. The distribution of inclusions in previously analyzed and oriented thin sections (Wilshire and others, 1980) is displayed as contoured diagrams in Figure 9.

Three stages of formation of inclusions can be distinguished from cross-cutting relations. From earliest to latest, these are: (1) parallel trains of euhedral dark minerals (to 20 μm) aligned along grain boundaries and intragrain fractures; (2) clear glass and trapped vapor along grain boundaries and intragrain fractures in and near reaction rims around amphiboles; and (3) dark basaltic glass along grain boundaries, intragrain fractures, and crystallographic discontinuities. Figure 9a shows that dominantly intergranular stage 1 inclusions decrease in abundance with distance from the kaersutite selvage of sample Ba-2-1, whereas stage 2 and 3 features have no apparent relation to the dike contact. Stage 2 apparently represents decompression melting, and stage 3 invasion of the host lava. The distribution of stage 1 inclusions in Ba-1-72 is more complex because of the presence of two dikes.

The stage 1 inclusions clearly are related to injection of the kaersutite dike in xenolith Ba-2-1, and are of mantle origin. Thus, stage 1 inclusions in these rocks are modal evidence of mantle metasomatism. We do not yet know the inclusion compositions, but their abundance in olivines and along grain boundaries supports the findings of Zindler and Jagoutz (in press) that microscopic inclusions in peridotite minerals may be of mantle origin. The concentration of dike-related inclusions at grain boundaries in Ba-2-1 (Fig. 9b) indicates that acid leaching of mineral separates may remove mantle-derived trace elements along with contaminants from the host basalts and other sources.

DISCUSSION

Interpretation of Metasomatic Processes and Events Observed in Composite Xenoliths

Detailed study of mineral major and trace elements from composite xenoliths Ba-2-1 and Ba-1-72 clarifies the relation

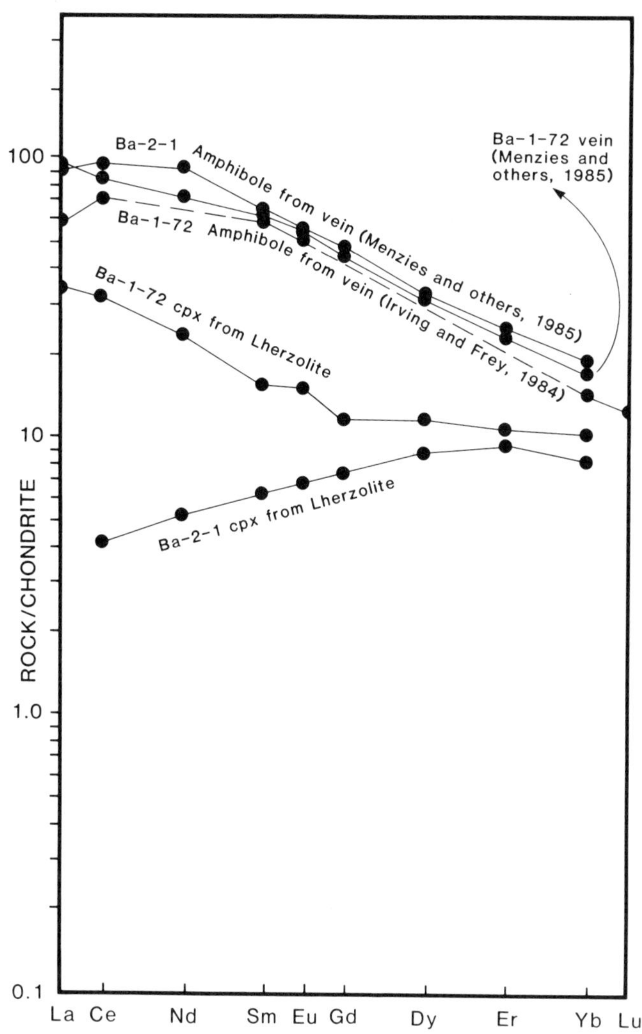

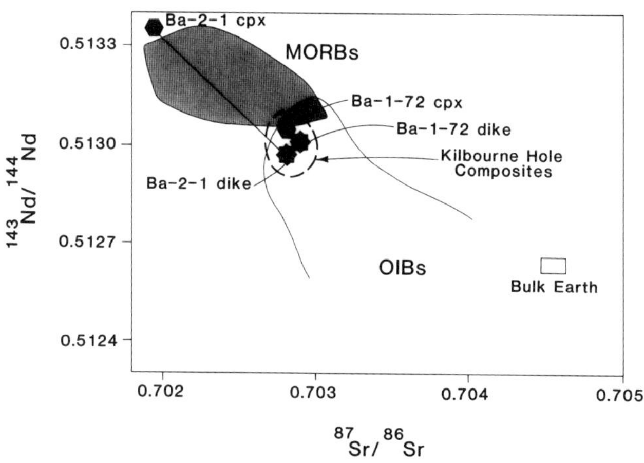

Figure 8. REE and isotopic compositions of mineral separates from Dish Hill, California, composite xenoliths. a, REEcn patterns of cpx separates. b, Isotopic ratios of Nd and Sr in cpx and kaersutite dikes. Dark stippled area indicates Mid-Ocean Ridge Basalts (MORBs); light stippled area, high $^{143}Nd/^{144}Nd$, low $^{87}Sr/^{86}Sr$ end of compositions of ocean island basalts (OIBs); dashed line, field of Kilbourne Hole composite type II/II peridotite and pyroxenite bulk-rock compositions; hexagons, cpx separates from peridotite; stars, kaersutite from dikes, Dish Hill, California.

among apparently contrasting types of metasomatism. Both these samples have undergone patent and Fe-Ti metasomatism. Similarities of mineral compositions, modes, and textures suggest that both represent the same protolith. Patent mantle metasomatism added both hydrous minerals and microscopic inclusions to the peridotite of both xenoliths. Similar compositional trends within each sample demonstrate that the same kind of reactions occurred between wallrock protolith and metasomatic fluids.

The principal chemical differences between Ba-1-72 and Ba-2-1 are that clinopyroxenes of Ba-1-72 are relatively enriched in LREE/HREE, whereas the clinopyroxenes of Ba-2-1 are not enriched. Also, the REEcn patterns of clinopyroxenes in the two samples are different (Fig. 8a). Thus, Ba-1-72 has the chemical signature of cryptic metasomatism and Ba-2-1 does not. One possible explanation of these differences (Kramers and others, 1983) is that Ba-1-72 underwent a separate cryptic metasomatic event unrelated to the patent and Fe-Ti metasomatism. However, if these samples came from the same mantle region, as seems likely, such a cryptic event could not have been pervasive. This interpretation would also require that the Nd isotopic similarity between the dikes of both samples and the clinopyroxenes of Ba-1-72 is coincidental.

A more plausible interpretation is that the LREE/HREE enrichment of Ba-1-72 clinopyroxenes was caused by emplacement of the kaersutite dike (Menzies and others, 1985). The same event affected the two samples differently because of the different sizes of the two xenoliths, and the fact that the smaller xenolith (Ba-1-72) comprises peridotite sandwiched between two dikes. We speculate that the amount of metasomatizing fluid per unit volume of peridotite was much greater in Ba-1-72 than in Ba-2-1. We suggest further that the REE and isotopic differences between Ba-1-72 and Ba-2-1 may result from varying degrees of soaking of the wallrock by fluids from the dikes. However, the compositional trends of major elements in the narrow zone adjacent to the dikes are identical in the two xenoliths and both may show similar LREE enrichments within that zone. This hypothesis can be tested by analyzing serial samples from Ba-2-1 with proximity to the dike, which we currently are doing.

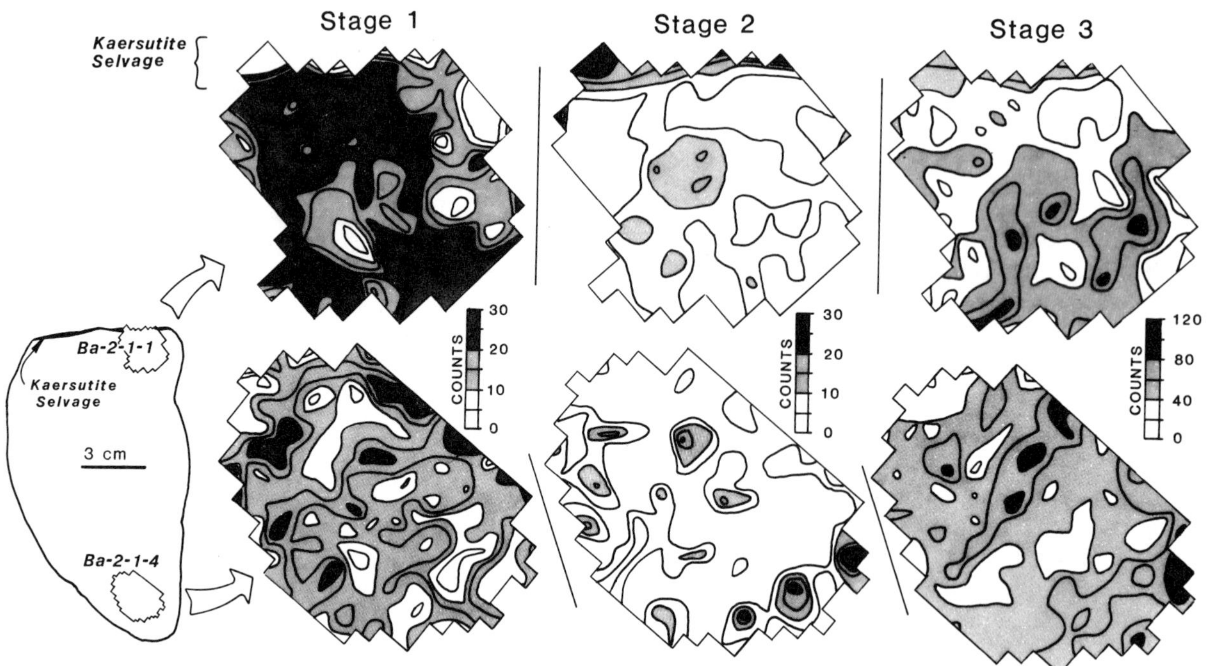

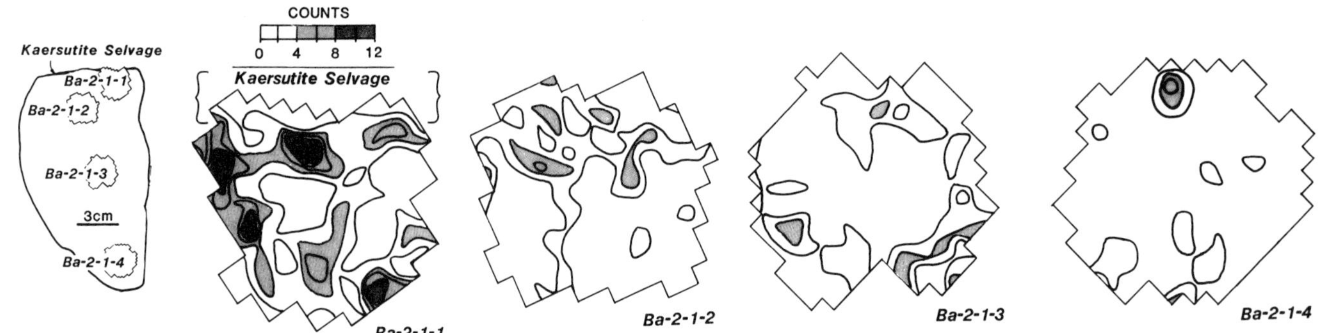

Figure 9. Contoured maps of composite hydrous xenolith Ba-2-1 from Dish Hill, California, showing distribution of inter- and intragranular fluid, and solid inclusions, coatings, and fillings. a, Distribution of inter- and intragranular features: upper diagrams represent areas adjacent to dike selvage, and lower diagrams are farthest from dike. Stage 1 features are concentrated adjacent to hornblendite dike selvage. Stage 2 features are randomly distributed. Stage 3 features permeate friable rock. b, Distribution of solely intragranular inclusions. Abundance of stage 1 features is greatest adjacent to dike, but is considerably reduced compared to Figure 9a.

"Contrasting" Styles of Mantle Metasomatism

Bulk and mineral compositions of peridotite xenoliths indicate that the definitions of cryptic and patent metasomatism probably describe different aspects of the same process. The moderately abundant data on noncomposite xenoliths show that metasomatized peridotite xenoliths have similar bulk rock and clinopyroxene major and trace element signatures, whether the rocks are hydrous or anhydrous. The pattern of increasing bulk LREE/HREE enrichment with decreasing CaO common to both kinds of metasomatism is more marked in patently than in cryptically metasomatized peridotite. However, the correlation is strongly dependent on modal abundances of olivine and clinopyroxene, and may not be a reliable index of metasomatism, particularly for clinopyroxene-rich peridotite.

Clinopyroxenes of patently metasomatized peridotite range from being relatively depleted in LREE/HREE to being more highly enriched than clinopyroxenes from cryptically metasomatized samples. The evidence from composite samples Ba-2-1 and Ba-1-72 suggests that the degree of REE enrichment of clinopyroxenes in patently metasomatized rocks may depend on a number of factors, including pressure and temperature, clinopyroxene abundance, fluid volume, compositions of mineral and fluid, and available structural sites. Except for pressure, the influence of these factors is likely to be inhomogeneously distributed, even over the dimensions of a single xenolith.

Although the considerations discussed above indicate that clinopyroxene LREE/HREE values likely are equivocal indices of cryptic metasomatism, Menzies and others (1985) and Kempton (1985) proposed chemical criteria that they believe distinguish cryptic from patent metasomatism. Kempton (1985) asserted that compositional trends of patent metasomatism cannot yield the chemical signature of cryptic metasomatism (Fig. 6). Clinopyroxenes in cryptically metasomatized type I peridotite (LREE/HREE, >1.0) generally have lower Al_2O_3/Cr_2O_3 than those from apparently unmetasomatized type I peridotite, although there is considerable overlap (Fig. 6a, patterned area). Type II rocks have even higher Al_2O_3/Cr_2O_3 than unmetasomatized type I xenoliths. Kempton (1985) concluded that the major element trend of patent and Fe-Ti metasomatism shown as an increase of Al_2O_3 in clinopyroxene (Fig. 6a), defines a trend toward type II compositions that is opposite to the inferred trend of relative LREE enrichment in clinopyroxenes of cryptically metasomatized samples. The enrichment of LREE/HREE in clinopyroxenes is thought by Kempton (1985) to be inconsistent with Al_2O_3/Cr_2O_3 compositional variations in clinopyroxenes with proximity to kaersutite dikes, like those shown for the Dish Hill composites in Figure 6a (arrows). Thus, it is Kempton's belief that cryptic metasomatism cannot be caused by fluids that also deposit hydrous minerals.

In Figure 6b we show that simple plots of clinopyroxene Al_2O_3 versus Cr_2O_3 are not adequate to deduce the direction of LREE-enrichment trends. For example, clinopyroxenes of Ba-1-72 are relatively enriched in LREE/HREE and are also displaced to higher Al_2O_3/Cr_2O_3, compared to clinopyroxenes of Ba-2-1 (Fig. 6a). The direction of relative LREE enrichment in clinopyroxene (Fig. 6b, black dart) between these two samples is opposite to the trend inferred by Kempton (1985, Fig. 6a). To test whether or not the cryptic metasomatic trend is opposite that of Fe-Ti metasomatism, isopleths of Ce/Smcn for clinopyroxenes of noncomposite xenoliths (Table 1) were superimposed on Figure 6b. The pattern exhibited depends on the parameter chosen, since some samples with La/Ybcn >1.0 may have Ce/Smcn or Ce/Ybcn <1.0, and vice versa. Figure 6b shows that clinopyroxenes with highest Ce/Smcn enrichments plot at the Al_2O_3-enriched end of Kempton's field of cryptically metasomatized samples. The pattern displayed in Figure 6b is consistent with a trend of LREE-enrichment that correlates positively with enrichment of Al_2O_3 in clinopyroxene, as suggested by the Dish Hill samples.

Our evidence indicates that both hydrous and anhydrous metasomatized peridotites may contain features that constitute "patent" metasomatism. Anhydrous samples may contain obvious anhydrous secondary minerals (such as apatite, sulfides, or oxides), but the anhydrous metasomatized rocks in our sources (Table 1) apparently do not. However, we anticipate that these rocks may contain microscopic to submicroscopic features, such as solid and fluid inclusions, coatings, and fillings, deposited by metasomatic fluids in the mantle. The presence of such inclusions in olivines of cryptically and patently metasomatized peridotite appears to explain the lack of correlation between LREE/HREE enrichments in metasomatized bulk rock and those in constituent clinopyroxenes. These minute features probably are present also in orthopyroxene, spinel, and clinopyroxene rims.

Metasomatic Events and Processes in the Mantle

Patent metasomatism and Fe-Ti metasomatism both can result from intrusion of mafic dikes that may differentiate to produce volatile-rich fluids, whether or not the fluids are hydrous. Hydrous melts may fractionate to produce amphibole- and/or mica-rich dikes or veins, which evidently are relatively enriched in LREE/HREE. Anhydrous melts fractionate to produce clinopyroxenites that have variously depleted to slightly enriched LREE/HREE values (Irving, 1980). Relative enrichment of LREE/HREE may occur in wallrocks without introduction of hydrous minerals (Roden and others, 1984b), probably from differentiates that are dominantly CO_2-bearing fluids and vapor (Menzies and Wass, 1983; Bergman and Dubessy, 1984; Griffin and others, 1984; Menzies and others, 1985). Fluids of dominantly CO_2 composition may occur in mantle regions at depths around 70 km (Schneider and Eggler, 1984). The isotopic compositions of hydrous and anhydrous melts and fluid differentiates depend on the mineralogy and isotopic composition of their source (Menzies and Wass, 1983). The processes considered here include infiltration along cracks and grain boundaries in wallrock adjacent to conduits, accompanied by diffusion of Fe, Al, and Ti across grain boundaries.

SUMMARY

The accumulated evidence from discrete and composite xenoliths of spinel- and garnet-bearing peridotite suites indicates that all identified metasomatic processes—Fe-Ti, cryptic, and patent—can result from the same igneous intrusive event. The magnitude of diffusive effects is determined at least in part by compositions and relative volumes of both wallrock and fluids. Fluid compositions are buffered by reaction with peridotite minerals. Thus, the chemical signatures produced by metasomatic processes are influenced by modal mineral abundances, and the signatures left by previous depletion or enrichment events. As a result, none of the indices of metasomatism based on relative LREE/HREE enrichment in bulk rock or constituent clinopyroxenes is reliable. Similarly, "trends" deduced from scatter diagram plots of bulk and mineral compositions from noncomposite xenoliths may be misleading.

ACKNOWLEDGMENTS

Probe analysis of samples Ba-1-72 and Ba-2-1 were made by Charles Meyer, and many of the petrographic data were collected by Beth Schwarzman. Elaine McGee, Dave Clague, Art Boettcher, Mike Roden, and Fred Frey carefully reviewed the manuscript. They made many helpful suggestions that aided us to better focus and reorganize early versions of the manuscript. Howard Wilshire helped us to interpret several literature sources. Lisa Wells patiently allowed long computer sessions in her home; and Paul, Ben, and Owen gave needed sympathetic support and remained patient when this work interfered with more important matters.

REFERENCES CITED

American Geological Institute, 1980, Glossary of geology, Bates, R. L., and Jackson, J. A., eds., Falls Church, Virginia, American Geological Institute, 749 p.

Aoki, K.-I., and Shiba, I., 1973, Pyroxenes from lherzolite inclusions of Itinomegata, Japan: Lithos, v. 6, p. 41–51.

Bergman, S. C., and Dubessy, J., 1984, CO_2-CO fluid inclusions in a composite peridotite xenolith; Implications for upper mantle fugacity: Contributions to Mineralogy and Petrology, v. 85, p. 1–13.

Best, M. G., 1974, Mantle-derived amphibole within inclusions in alkalic-basaltic lavas: Journal of Geophysical Research, v. 27, p. 25–44.

Boettcher, A. L., 1984, The source regions of alkaline volcanoes, *in* Explosive volcanism; Inception, evolution, and hazards: Washington, D.C., National Academy of Sciences, p. 13–22.

Bonatti, E., Hamlyn, P., and Ottonello, G., 1981, Upper mantle beneath a young oceanic rift; Peridotites from the island of Zabargad (Red Sea): Geology, v. 9, p. 474–479.

BVSP (Basaltic Volcanism Study Project), 1980, Ultramafic xenoliths in terrestrial volcanics and mantle magmatic processes; Basaltic volcanism on the terrestrial planets: New York, Pergamon Press, p. 282–310.

Dawson, J. B., 1984, Contrasting types of upper-mantle metasomatism? *in* Kornprobst, J., ed., Kimberlites II; The mantle and crust-mantle relationships: Amsterdam, Elsevier, p. 289–294.

Dick, H.J.B., 1977, Partial melting in the Josephine peridotite, I; The effect of mineral composition and its consequence for geobarometry and geothermometry: American Journal of Science, v. 277, p. 801–830.

Dick, H.J.B., Fisher, R. L., and Bryan, W. B., 1984, Mineralogic variability of the uppermost upper mantle along mid-ocean ridges: Earth and Planetary Science Letters, v. 69, p. 88–106.

Ehrenberg, S. N., 1979, Garnetiferous ultramafic inclusions in minette from the Navajo volcanic field, *in* Boyd, F. R., and Meyer, H.O.A. eds., The mantle sample; Inclusions in kimberlites and other volcanics: Washington, D.C., American Geophysical Union, Proceedings of the Second International Kimberlite Conference, v. 2, p. 330–344.

——, 1982, Petrogenesis of garnet lherzolite and megacrystalline nodules from The Thumb, Navajo volcanic field: Journal of Petrology, v. 23, p. 507–547.

Evensen, N. M., Hamilton, P. J., and O'Nions, R. K., 1978, Rare-earth abundances in chondritic meteorites: Geochimica et Cosmochimica Acta, v. 42, p. 1199–1213.

Feigenson, M. D., and Stolz, R., 1984, Geochemistry of a Kilbourne Hole ultramafic nodule: Geological Society of America Abstracts with Programs, v. 16, p. 507.

Francis, D. M., 1974, Xenoliths and the nature of the upper mantle and lower crust, Nunivak Island, Alaska: [Ph.D. thesis]: Cambridge, Massachusetts Institute of Technology, 237 p.

——, 1976, The origin of amphibole in lherzolite xenoliths from Nunivak Island, Alaska: Journal of Petrology, v. 17, p. 357–378.

Frey, F. A., 1984, Rare earth element abundances in upper mantle rocks, *in* Henderson, P., ed., Rare earth geochemistry: Amsterdam, Elsevier, p. 153–203.

Frey, F. A., and Green, D. H., 1974, The mineralogy, geochemistry, and origin of lherzolite inclusions in Victoria basanites: Geochimica et Cosmochimica Acta, v. 38, p. 1023–1059.

Frey, F. A., and Prinz, M., 1978, Ultramafic inclusions from San Carlos, Arizona; Petrologic and geochemical data bearing on their petrogenesis: Earth and Planetary Science Letters, v. 34, p. 129–176.

Green, D. H., 1964, The petrogenesis of the high-temperature peridotite intrusion in the Lizard area, Cornwall: Journal of Petrology, v. 5, p. 134–188.

Griffin, W. L., Wass, S. Y., and Hollis, J. D., 1984, Ultramafic xenoliths from Bullenmerri and Gnotuk maars, Victoria, Australia; Petrology of a subcontinental crust-mantle transition: Journal of Petrology, v. 25, p. 53–87.

Gurney, J., and Harte, B., 1980, Chemical variations in upper mantle nodules from southern African kimberlites: Philosophical Transactions of the Royal Society of London, v. A297, p. 273–293.

Harte, B., 1983, Mantle peridotites and processes; The kimberlite sample, *in* Hawkesworth, C. J., and Norry, M. J., eds., Continental basalts and mantle xenoliths: Nantwich, United Kingdom, Shiva Publishing Ltd., p. 46–91.

Harte, B., Gurney, J. J., and Cox, K. G., 1977, Clinopyroxene-rich sheets in garnet-peridotite; Xenoliths specimens from the Matsoku kimberlite pipe, Lesotho: Second International Kimberlite Conference, Extended Abstracts (no pagination).

Irving, A. J., 1980, Petrology and geochemistry of composite ultramafic xenoliths in alkali basalts and implications for magmatic processes within the mantle: American Journal of Science, v. 280-A, The Jackson Volume, p. 389–426.

Irving, A. J., and Frey, F. A., 1984, Trace element abundances in megacrysts and their host basalts; Constraints on partition coefficients and megacryst genesis: Geochimica et Cosmochimica Acta, v. 48, p. 1201–1221.

Jagoutz, E., Lorenz, V., and Wanke, H., 1979, Major trace elements of Al-augites and Cr-diopsides from ultramafic nodules in European alkali basalts, *in* Boyd, F. R., and Meyer, H.O.A., eds., The mantle sample; Inclusions in kimberlites and other volcanics: Washington, D.C., American Geophysical Union, Proceedings of the Second International Kimberlite Conference, v. 2, p. 382–390.

Jagoutz, E., Carlson, R. W., and Lugmair, G. W., 1980, Equilibrated Nd-unequilibrated Sr isotopes in mantle xenoliths: Nature, v. 286, p. 708–710.

Kempton, P. D., 1984, Petrography, petrology and geochemistry of xenoliths and megacrysts from the Geronimo Volcanic Field, southeastern Arizona: [Ph.D. thesis]: Dallas, Texas, Southern Methodist University, p. 92–217.

—— , 1985, Styles of metasomatism in the mantle; Implications for alkali basalt genesis: Geological Society of America Abstracts with Programs, v. 17, p. 163.

Kempton, P. D., Menzies, M. A., and Dungan, M. A., 1984, Petrography, petrology, and geochemistry of xenoliths and megacrysts from the Geronimo volcanic field, southeastern Arizona, *in* Kornprobst, J., ed., Kimberlites II; The mantle and crust-mantle relationships: Amsterdam, Elsevier, p. 71–83.

Korzhinsky, D. S., 1970, Theory of metasomatic zoning (trans. by A. Augrell): Oxford, Clarendon Press, 162 pp.

Kramers, J. D., Roddick, J.C.M., and Dawson, J. B., 1983, Trace element and isotope studies on veined, metasomatic, and "MARID" xenoliths from Bulfontein, South Africa: Earth and Planetary Science Letters, v. 65, p. 90–106.

Kuno, H., and Aoki, K.-I., 1970, Chemistry of ultramafic nodules and their bearing on the origin of basaltic magmas: Physics of Earth and Planetary Interiors, v. 3, p. 273–301.

Kurat, G., Palme, H., Spettel, B., Baddenhausen, H., Hoffmeister, H., Palme, C., and Wanke, H., 1980, Geochemistry of ultramafic xenoliths from Kapfenstein, Austria; Evidence for a variety of upper mantle processes: Geochimica et Cosmochimica Acta, v. 44, p. 45–60.

Lindgren, W., 1928, Mineral deposits, 3rd ed.: New York, McGraw-Hill, 1049 pp.

Lloyd, F. E., and Bailey, D. K., 1975, Light element metasomatism of the continental mantle; The evidence and the consequences: Physics and Chemistry of the Earth, v. 9, p. 389–416.

Melson, W. G., Hart, S. R., and Thompson, G., 1972, St. Paul's Rocks, equatorial Atlantic; Petrogenesis, radiometric ages, and implications on sea-floor spreading: Geological Society of America Memoir 132, p. 241–272.

Menzies, M. A., and Murthy, V. R., 1980, Mantle metasomatism as a precursor to the genesis of alkaline magmas; Isotopic evidence: American Journal of Science, v. 280-A, The Jackson Volume, p. 622–638.

Menzies, M. A., and Wass, S. Y., 1983, CO2- and LREE-rich mantle below eastern Australia; A REE and isotopic study of alkaline magmas and apatite-rich mantle xenoliths from the southern highlands province, Australia: Earth and Planetary Science Letters, v. 65, p. 287–302.

Menzies, M. A., Kempton, P. D., and Dungan, M. A., 1985, Interaction of continental lithosphere and asthenospheric melts below the Geronimo volcanic field, Arizona, U.S.A.: Journal of Petrology, v. 26, p. 663–693.

Miller, C., and Richter, W., 1982, Solid and fluid phases in the lherzolite and pyroxenite inclusions from the Hoggar, central Sahara: Geochemical Journal, v. 16, p. 263–277.

Nagasawa, H., Wakita, H., Higuchi, H., and Onuma, N., 1969, Rare earths in peridotite nodules; An explanation of the genetic relationships between basalt and peridotite nodules: Earth and Planetary Science Letters, v. 5, p. 377–381.

Nicolas, A., and Jackson, M. D., 1982, High temperature dikes in peridotites; Origin by hydraulic fracturing: Journal of Petrology, v. 23, p. 568–582.

Nixon, P. H., and Boyd, F. R., 1973a, Petrogenesis of the granular and sheared ultrabasic nodule suite in kimberlites, *in* Nixon, P. H., ed., Lesotho kimberlites: Maseru, Lesotho National Development Corporation, p. 48–56.

—— , 1973b, The Liquobong intrusions and kimberlites olivine composition, *in* Nixon, P. H., ed., Lesotho kimberlites: Maseru, Lesotho National Development Corporation, p. 141–148.

Ottonello, G., 1980, Rare earth abundances and distribution in some spinel peridotite xenoliths from Assab (Ethiopia): Geochimica et Cosmochimica Acta, v. 44, p. 1885–1901.

Piccardo, G. B., and Ottonello, G., 1978, Partial melting effects on coexisting minerals compositions in upper mantle xenoliths from Assab (Ethiopia): Rendiconti Societa di Mineralogia e Petrologia, v. 34, p. 499–526.

Reid, A. M., Donaldson, C. H., Brown, R. W., Ridley, W. I., and Dawson, J. B., 1975, Mineral chemistry of peridotite xenoliths from the Lashaine volcano, Tanzania: Physics and Chemistry of the Earth, v. 9, p. 525–544.

Ridley, W. I., and Dawson, J. B., 1975, Lithophile trace element data bearing on the origin of peridotite xenoliths, anakaramite, and carbonatite from Lashaine volcano, Tanzania: Physics and Chemistry of the Earth, v. 9, p. 559–570.

Roden, M. F., and Murthy, V. R., 1985, Mantle metasomatism: Annual Reviews of Earth and Planetary Sciences, v. 13, p. 269–296.

Roden, M. F., Frey, F. A., and Francis, D. M., 1984a, An example of consequent mantle metasomatism in peridotite inclusions from Nunivak Island, Alaska: Journal of Petrology, v. 25, p. 546–577.

Roden, M. F., Murthy, V. R., and Irving, A. J., 1984b, Isotopic heterogeneity and evolution of uppermost mantle, Kilbourne Hole, New Mexico: EOS American Geophysical Union Transactions, v. 65, p. 306.

Schneider, M. E., and Eggler, D. H., 1984, Compositions of fluids in equilibrium with peridotite; Implications for alkaline magmatism-metasomatism, *in* Kornprobst, J., ed., Kimberlites II; The mantle and crust-mantle relationships: Amsterdam, Elsevier, p. 333–394.

Shimizu, N., 1975, Rare earth elements in garnets and clinopyroxenes from garnet lherzolite nodules in kimberlites: Earth and Planetary Science Letters, v. 25, p. 26–32.

Stewart, D. C., 1980, A petrographic, chemical, and experimental study of kaersutite occurrences at Dish Hill, California, with implications for volatiles in the upper mantle: [Ph.D. thesis]: University Park, Pennsylvania State University, 88 p.

Stewart, D. C., and Boettcher, A. L., 1977, Chemical gradients in mantle xenoliths: Geological Society of America Abstracts with Programs, v. 9, p. 1191–1192.

Stosch, H.-G., 1982, Rare earth element partitioning between minerals from anhydrous spinel peridotite xenoliths: Geochimica et Cosmochimica Acta, v. 46, p. 793–812.

Stosch, H.-G., and Seck, H. A., 1980, Geochemistry and mineralogy of two spinel peridotite suites from Dreiser Weiher, West Germany: Geochimica et Cosmochimica Acta, v. 44, p. 457–470.

Tanaka, T., and Aoki, K.-I., 1981, Petrogenic implications of REE and Ba on mafic and ultramafic inclusions from Itinome-gata, Japan: Journal of Petrology, v. 89, p. 369–390.

Varne, R., 1970, Hornblende lherzolite and the upper mantle: Contributions to Mineralogy and Petrology, v. 27, p. 45–51.

Varne, R., and Graham, A. L., 1971, Rare earth abundances in hornblende and clinopyroxene of a hornblende lherzolite xenolith; Implications for upper mantle fractionation processes: Earth and Planetary Science Letters, v. 13, p. 11–18.

Wass, S. Y., Henderson, P., and Elliott, C. J., 1980, Chemical heterogeneity and metasomatism in the upper mantle; Evidence from rare earth and other elements in apatite-rich xenoliths in basaltic rocks from eastern Australia: Philosophical Transactions of the Royal Society of London, v. A297, p. 333–346.

Wilshire, H. G., 1984, Mantle metasomatism: The REE story: Geology, v. 12, p. 395–398.

Wilshire, H. G., and Jackson, E. D., 1975, Problems in determining mantle geotherms from pyroxene compositions of ultramafic rocks: Journal of Geology, v. 83, p. 313–329.

Wilshire, H. G., and Pike, J.E.N., 1975, Upper-mantle diapirism; Evidence from analogous features in alpine peridotite and ultramafic inclusions in basalt: Geology, v. 3, p. 467–470.

Wilshire, H. G., and Shervais, J. W., 1975, Al-augite and Cr-diopside ultramafic xenoliths in basaltic rocks from western United States; Structural and textural relationships: Physics and Chemistry of the Earth, v. 9, p. 257–272.

Wilshire, H. G., and Trask, N. G., 1971, Structural and textural relationships of amphibole and phlogopite in peridotite inclusions, Dish Hill, California: American Mineralogist, v. 56, p. 240–255.

Wilshire, H. G., Pike, J.E.N., Meyer, C. E. and Schwarzman, E. L., 1980, Amphibole-rich veins in lherzolite xenoliths, Dish Hill and Deadman Lake, California: American Journal of Science, v. 280-A, The Jackson Volume, Part 2, p. 576–593.

Wilshire, H. G., Meyer, C. E., Nakata, J. K., Calk, L. C., Shervais, J. W., Nielson, J. E., and Schwarzman, E. C., 1985, Mafic and ultramafic xenoliths from volcanic rocks of the western United States: U.S. Geological Survey Open-File Report 85-139, 505 p.

Zindler, A., and Jagoutz, E., 1987, Trace element and Sr isotope systematics of peridotite nodules from Peridot Mesa, San Carlos, Arizona: Geochimica et Cosmochimica Acta, in press.

MANUSCRIPT ACCEPTED BY THE SOCIETY OCTOBER 14, 1986

Printed in U.S.A.

Geological Society of America
Special Paper 215
1987

Kimberlite-transported nodules from Colorado-Wyoming; A record of enrichment of shallow portions of an infertile lithosphere

David H. Eggler
Department of Geosciences, The Pennsylvania State University, University Park, Pennsylvania 16802
M. E. McCallum
Department of Earth Resources, Colorado State University, Fort Collins, Colorado 80521
M. B. Kirkley
Department of Geochemistry, Cape Town University, Rondebosch 7700, Cape Town, Republic of South Africa

ABSTRACT

Nodules of Devonian age in Colorado-Wyoming kimberlites include infertile, enriched, and ilmenite-bearing peridotites; garnet websterites and associated garnet pyroxenites; garnet-spinel websterites; plus megacrysts, eclogites, and granulites. Both enriched and infertile peridotites equilibrated in the spinel, spinel-garnet, and garnet facies, but enriched garnet peridotites were sampled only at depths of less than 100 km, whereas infertile garnet peridotites extend to depths of 200 km. Occurrence of infertile rocks and cool paleogeothermal gradient define the sampled mantle as lithosphere.

Infertile peridotites are interpreted to be residua from a Precambrian melting event involving the entire lithosphere; spinel and spinel-garnet peridotites are generally more infertile than garnet peridotites. The infertile lithosphere was subsequently enriched at shallow depths by igneous processes in two events: one produced Fe- and Ti-enriched ilmenite peridotites; the other, more widespread event was associated with websteritic rocks. A few composite nodules as well as geothermobarometry suggest that websterites formed a network of dikes, veins, and layers cutting spinel and spinel-garnet infertile peridotite, and that these were bordered by zones of enriched peridotite. Websterites are Na-, Ca-, and slightly Ti-enriched, but are magnesian, which requires that the protoliths were crystalline assemblages, not melts.

The ratio of collected infertile spinel and spinel-garnet peridotites to enriched peridotites and websterites indicates that enriched rocks occupied 80 percent of the lithosphere sampled between the crust and a depth of 100 km. This very high percentage, unusual among xenolith populations, suggests that ancient enriched rocks occurred along conduits subsequently exploited by kimberlite eruptions.

INTRODUCTION

In several studies of Laramide igneous rocks in the Rocky Mountains, it has been found that strontium isotopic signatures are approximately those of bulk earth, whereas neodymium signatures are markedly less radiogenic than bulk earth (e.g., Vollmer and others, 1984; Meen, 1983; Dudas and others, 1987). These signatures imply prolonged light rare-earth element (LREE) enrichment of the source relative to an asthenospheric reservoir. Arrays of Pb isotopes, interpreted as ages, are 1.5 to 3.8 b.y. (Carlson and others, 1985). Both isotopic systems show that the magmas, many of which are alkalic, were derived from continental lithosphere that was old, in the sense of time since stabilization from asthenosphere.

There are few occurrences of nodules of deep continental lithosphere in the Rocky Mountains. One is in Montana (Hearn

TABLE 1. LITHOSPHERIC NODULES FROM COLORADO-WYOMING KIMBERLITE PIPES

Group	No.	Comments	Reference
Granulites	20	Two-px granulites, two-px gar granulites, cpx garnet granulites, and gabbro-norites. Interpreted to represent lower crust.	Bradley and McCallum (1984)
Eclogites	158	Cpx-gar eclogites with acc. rutile, sanidine, quartz, and sphene; kyanite eclogites with acc. corundum, rutile, and sanidine. Paleotemps. 700-1300°C. Interpreted to represent subducted oceanic crust (Archean?) now at 80-200 km.	Ater, Eggler, and McCallum (1984)
Infertile peridotites		A suite dominated by harzburgites occurring in 3 facies. Interpreted to represent infertile residua from extensive Archean(?) melting. Some garnet-facies rocks contain chromite. Textures in all facies coarse equant to porphyroclastic.	This paper
a. spinel	36		
b. spinel-garnet	5		
c. garnet	74		
Enriched peridotites		A suite dominated by lherzolites occurring in 3 facies. The few nodules in garnet facies are shallow. In spinel-garnet facies, rimming of spinels by garnets and exsolution in pyroxenes suggest high-temperature protoliths. Textures dominantly coarse equant.	Kirkley and others (1984) and this paper
a. spinel	56		
b. spinel-garnet	19		
c. garnet	23		
Ilmenite-garnet peridotites	7	Ilmenite-bearing enriched perids. More Fe- and Ti-rich than other enriched peridotites. Equant to porphyroclastic textures.	This paper
Spinel-garnet peridotites	11	Olivine-free spinel or spinel-gar pyroxenites and websterites. Includes "green spinel" grp of Kirkley and others(1984). Textures coarse equant, but dominated by exsolution. Low-P rocks.	Kirkley and others (1984)
Garnet pyroxenites and websterites	52	Olivine-free pyroxenites and websterites, related compositionally to enriched perids. Acc. rutile, sulfides, and phlogopite. Can occur as layers in perids. Coarse equant or porphyroblastic, few trans. porphyroblastic.	This paper
Megacrysts	353	Two groups, Cr-rich and Cr-poor; neither identical in mineral chemistry to any peridotites. Interpreted as phenocrysts in alkaline magmas.	Eggler and others (1979)

and McGee, 1984); another is the Colorado-Wyoming kimberlite province. That province consists of several districts: (1) the Colorado-Wyoming State Line district, approximately 16 km wide by 24 km long, which is located astride the state line in the Precambrian Front Range and contains more than 40 occurrences of hypabyssal to diatreme-facies kimberlite (McCallum and others, 1975; Smith and others, 1979); the kimberlite is Devonian in age (Chronic and others, 1969; Naeser and McCallum, 1977; Smith, 1979); (2) the Iron Mountain district in the southern Laramie Range of Wyoming, which consists of more than 57 occurrences of kimberlite in diatremes, dikes, and a few sills (Smith, 1977); and (3) isolated occurrences, which include a dike in Estes Park, Colorado, and a diatreme west of Boulder, Colorado.

Nodules have been collected from these kimberlite occurrences since 1964. A new classification of peridotite and pyroxenite nodules (excluding eclogites) is proposed, based on the new synthesis (Table 1) in this paper.

FIELD RELATIONS

Nodules from the Sloan, Nix, and Moen groups of pipes range in size from 0.5 to 10 cm in their largest dimension. Xenoliths from the Schaffer 3 diatreme are the largest, ranging in size from 3 to 15 cm. Unbroken nodules in all pipes are subspherical to ovoid in shape.

Previously, we (Eggler and others, 1979) showed that differences in chemistries of megacrysts exist between groups of kimberlite pipes. Subtle differences are present in nodule suites as well. They are not emphasized in this paper, in part because of space constraints and in part because they are obscured by an alteration problem: the Sloan 2 pipe, and, to a lesser extent, the Nix 2 and 4 pipes, are the only pipes in which most nodules (>90 percent) occur without extensive serpentinization, silicification, or carbonatization, and in which unaltered olivine or orthopyroxene grains occur.

METHODS

Electron probe microanalyses of minerals were performed on the MAC400 probe at the Geophysical Laboratory of the Carnegie Institution of Washington and on the ETEC probe at the Pennsylvania State University, using a single set of mineral and glass standards. Ferric iron contents of spinel and ilmenite were calculated by assuming cation/anion stoichiometry.

Whole-rock compositions were calculated by converting the volume modes (Kirkley, 1980) to weight modes through an algorithm that computes densities of minerals from constituent end members, assuming additivity of molar volumes.

Textural terms are from Harte (1977), except for two new terms—porphyroblastic-coarse and porphyroblastic-granuloblastic—added by Kirkley (1980) to describe textures of some pyroxenitic nodules. The first term refers to the occurrence of garnet porphyroblasts about 1 cm diameter in a matrix of garnet and pyroxene that is equant to tabular. The other refers to porphyroblasts of garnet or pyroxene in a granuloblastic matrix.

SAMPLE HOMOGENEITY

Compositions of minerals in nodules with paleotemperatures in excess of about 1,000°C were found to be homogeneous. Homogeneity is taken to represent standard deviations for analyses, within and between grains, that do not exceed standard deviations expected from microprobe counting statistics.

Compositions of minerals in lower temperature nodules, especially in nodules with exsolution features, are slightly inhomogeneous. An exsolved spinel lherzolite nodule was studied in considerable detail by Kirkley and others (1983). Standard deviations for Cr/(Cr + Al) are 0.004 in garnet and 0.02 in diopside; for Mg/(Mg + Fe) they are 0.006 in garnet and 0.005 in diopside. The standard deviation of pressure, calculated from Al_2O_3 in orthopyroxene, is 2.0 kb.

ROCK CLASSIFICATION

Olivine-bearing rocks occur in two chemical groupings (Table 1): infertile and enriched. Infertile and enriched groups in turn occur in three facies (Table 1): spinel, spinel-garnet, and garnet. Parameters that differentiate infertile from enriched groups include molar Mg/(Mg + Fe), hereafter called mg; molar Cr/(Cr + Al), hereafter called cr; and bulk rock contents of Na_2O and CaO. Enriched groups have lower mg, lower cr, and higher whole-rock Na_2O and CaO. They also contain clinopyroxenes with greater jadeite component. Rocks were classified into groups using the full array of parameters. For a two-parameter plot, we chose one parameter to be a whole-rock measure of fertility. We chose the other to be a mineralogic parameter in order to classify nodules for which modal analyses are not available or to classify small nodules that cannot be accurately point-counted. For the whole-rock parameter, either Na_2O or CaO (both of which are magmaphilic oxides), could serve; in the enriched groups their correlation coefficients (r) exceed 0.94. For a mineralogic parameter, cr is discriminating. The cr value in either diopside, garnet, or spinel can be used, because correlation coefficients for cr between those minerals exceed 0.8; in most cases they exceed 0.9.

In the resulting plot (Fig. 1), infertile and enriched rocks are separated, except for a few rocks with low Na_2O and low cr. Such rocks have "enriched" mineral signatures but do not have enriched modal signatures, possibly because they occurred near margins of metasomatic zones. Mineral cr values are consistently related to mg (Fig. 2) and to mg and cr of coexisting minerals (Fig. 2).

Diopsides of enriched rocks are generally more sodic (Fig. 3; Table 2). Enriched rocks, like infertile rocks, contain diopsides with essentially no ferric iron (Fig. 3). Contents of CaTschermaks molecule (Fig. 3) denote facies (garnet or spinel) more than fertility.

Modally (Fig. 4), enriched peridotites are dominated by lherzolites with few harzburgites or olivine websterites. The infertile group is dominated by harzburgites with few lherzolites and no olivine websterites.

Ilmenite-bearing peridotites are garnet peridotites that contain ilmenite as interstitial masses that are subspherical to weblike, which generally range in size from about 0.2 to 1 mm, although some exceed 1 cm. Mineral analyses (Fig. 2; Table 2) indicate that ilmenite peridotites are an enriched group.

Olivine-free rocks are classified modally (Fig. 4) in two groups, garnet websterites with associated garnet pyroxenites and garnet-bearing spinel websterites. Garnet websterites and pyroxenites contain diopsides and garnets that in part overlap in composition with minerals of enriched garnet-spinel and garnet peridotites in mg and cr (Fig. 2), but that are more iron-rich as well. Diopsides, however, are somewhat less sodic than diopsides of garnet peridotites (Fig. 3). Websterites also typically contain accessory rutile and sulfides and may contain coarse-grained primary phlogopite. In previous publications (e.g., Eggler and McCallum, 1974), this group was combined with the enriched peridotites as a "websterite group."

The other group of orthopyroxene-poor websterites (Fig. 4) is spinel-bearing and may contain garnet. Minerals are dominantly low-Cr (Fig. 2), and hence spinels are typically green, a reflection of the high Mg-Al spinel component.

WHOLE-ROCK CHEMISTRY

By definition, enriched peridotites are higher in whole-rock Na_2O and lower in mineral cr and mg (Figs. 1, 2) than infertile peridotites. Enriched peridotites are also lower, on average, in whole-rock mg and cr than infertile peridotites and higher in TiO_2, Al_2O_3, and CaO (Fig. 5; Table 3). In both enriched and infertile groups, there is correspondence of whole-rock chemistries across facies, from spinel to spinel-garnet to garnet, although some systematic differences exist (Fig. 5).

Garnet websterites and spinel-garnet websterites have mg values ranging downward from those in the peridotites (Fig. 5). Whole-rock contents of Na_2O, as well as TiO_2 and CaO, are elevated relative to peridotites (Fig. 5; Table 3).

GEOTHERMOBAROMETRY

Thermobarometric Methods

In most cases, geothermometry (with the BASIC program TEEPEE) was based on the Ca/(Ca + Mg) of clinopyroxene

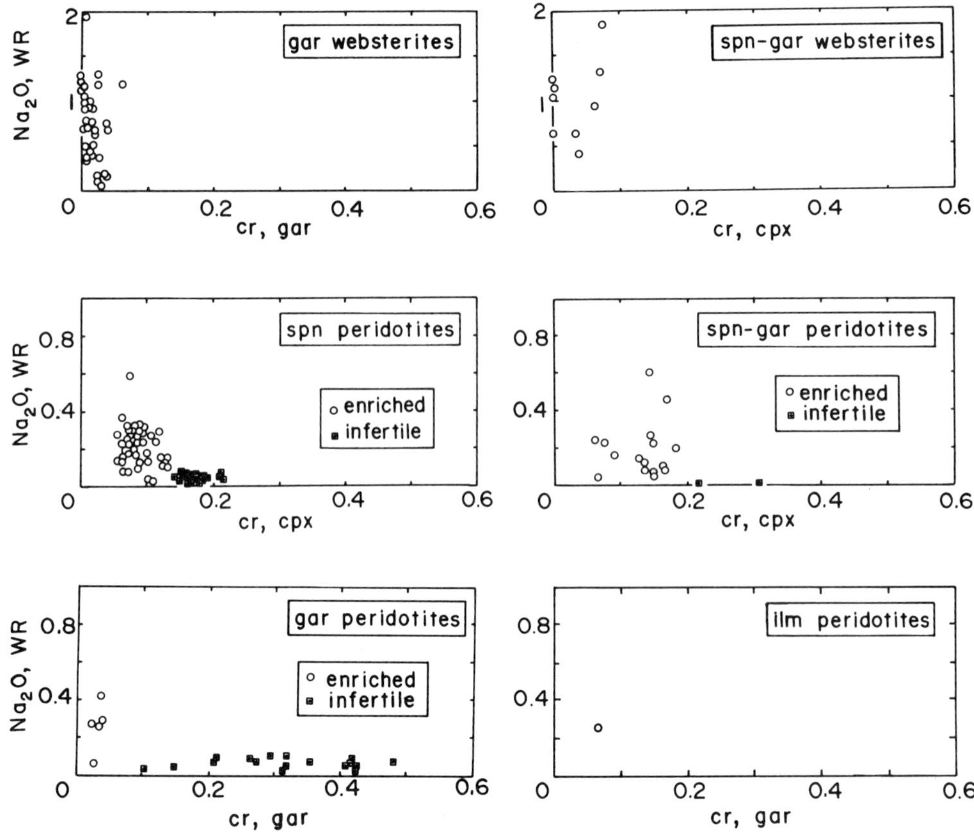

Figure 1. Definitions of groups of websterites and peridotites: Na_2O in whole rocks versus Cr/(Cr + Al) (cr) in garnet or clinopyroxene.

(LD20: the 20-kbar pseudosolvus of Lindsley and Dixon, 1976); the barometry was based on the Al_2O_3 content of orthopyroxene (M74: MacGregor, 1974), as recommended by Finnerty and Boyd (1984). For lower temperature rocks, pyroxene thermometry yields scattered results because of the steepness of the diopside solvus. Both a thermometer based on alumina content of orthopyroxene in the spinel facies (GNWW: Gasparik and Newton, 1984, as modified by Webb and Wood, 1986) and the olivine-garnet Mg-Fe partitioning thermometer (OW79: O'Neill and Wood, 1979) were tested, especially on the spinel-garnet peridotite rocks. They yielded temperatures comparable to pyroxene thermometry but with much less scatter. The olivine-garnet thermometer was also applied to high-temperature garnet peridotites lacking clinopyroxene.

Those peridotite nodules containing both spinel and garnet can be fixed in *P-T* space on the spinel-garnet peridotite transition, provided that the effects of minor elements are accounted for. Here, pressures along the reaction curve in the CMAS system (Wood and Holloway, 1984) were corrected for effects of Fe and Cr with an equation incorporating various expressions in Webb and Wood (1986). The resulting garnet-spinel barometer is termed GASP.

Geothermobarometry Results

Pressures and temperatures of nodules (Fig. 6) were calculated by LD20/M74 except for enriched spinel-garnet peridotites, enriched garnet peridotites, and those infertile garnet peridotites lacking modal diopside. Calculations for those exceptions were by OW79/M74, either because clinopyroxene was lacking or because of the low-temperature insensitivity of LD20. Three infertile spinel-garnet peridotites shown were calculated by LD20/GASP.

Spinel-garnet transition zones shown (Fig. 6) were calculated by GASP, over a range of temperatures, for actual mineral compositions in Colorado-Wyoming nodules. Because a transition is shifted to higher pressure by higher cr, each transition has finite width, and the transition for infertile nodules lies at higher pressures than the transition for enriched nodules. Comparison of zones to independent thermobarometry is encouraging: *P-T* coordinates for enriched spinel-garnet peridotites lie within or at slightly higher pressures than the calculated zone (Fig. 6), and coordinates for enriched garnet peridotites properly lie at higher pressures than the zone.

Spinel-garnet websterites lack mineral assemblages for

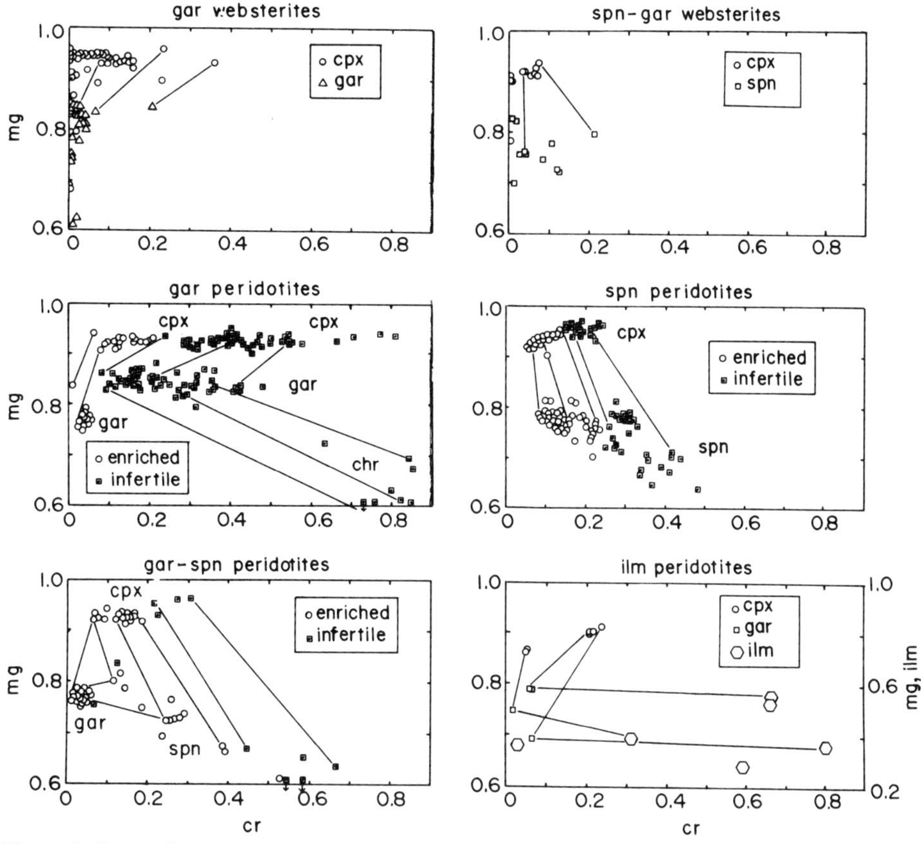

Figure 2. Compositions of minerals (clinopyroxene, garnet, spinel, ilmenite) in websterites and peridotites: Mg/(Mg + Fe) (mg) versus Cr/(Cr + Al) (cr). Representative tielines are shown.

thermometry by OW79 or GNWW, and Ca/(Ca + Mg) values are typically too high for thermometry by LD20. Maximum *P-T* limits (Fig. 6) were estimated, in the case of temperature, by highest LD20 values calculated, and, in the case of pressure, by GASP, on the realization that the mineral assemblages cannot occur at pressures within or above the spinel-garnet peridotite transition zone (at such pressures, nodules would contain olivine).

Temperature limits on the *P-T* fields of spinel peridotites (Fig. 6) represent ranges of LD20 values. (GNWW thermometry, which is generally consistent with LD20, yields high temperatures of 800°–950°C for spinel peridotites.) Maximum pressure limits represent low-pressure limits of the spinel-garnet peridotite transition zones, calculated by GASP for actual spinel peridotite mineral compositions.

TEXTURES

The most common texture of all rock groups, except infertile garnet peridotites, is coarse equant (Fig. 7; Table 1). Porphyroclastic (deformed) textures are more prevalent in deeper and hotter infertile garnet peridotites (Fig. 7), but they also occur in shallow-seated groups, particularly in spinel peridotites (Fig. 7).

Porphyroblastic textures occur chiefly in the group of garnet pyroxenites and websterites (seven nodules with porphyroblastic coarse texture and five with porphyroblastic-granuloblastic texture). Nodules whose textures are dominated by exsolution of pyroxene and garnet are common in the spinel-garnet websterite group (Kirkley and others, 1984), but exsolution features are also found in two garnet websterite nodules and in most nodules of the enriched spinel-garnet peridotite group.

COMPOSITE NODULES

In a few composite nodules, garnet websterites or pyroxenites can be observed to occur as dikes, veins, and layers in enriched peridotites. Such nodules have been found particularly in Schaffer pipes. A few enriched spinel peridotites, as well, contain clinopyroxene- and spinel-rich layers. From these observations we surmise that the general occurrence of pyroxenites and websterites in lithosphere was as dikes and layers cutting infertile but locally enriched peridotite.

OTHER LITHOSPHERIC ROCKS

In addition to rocks discussed above, kimberlites contain

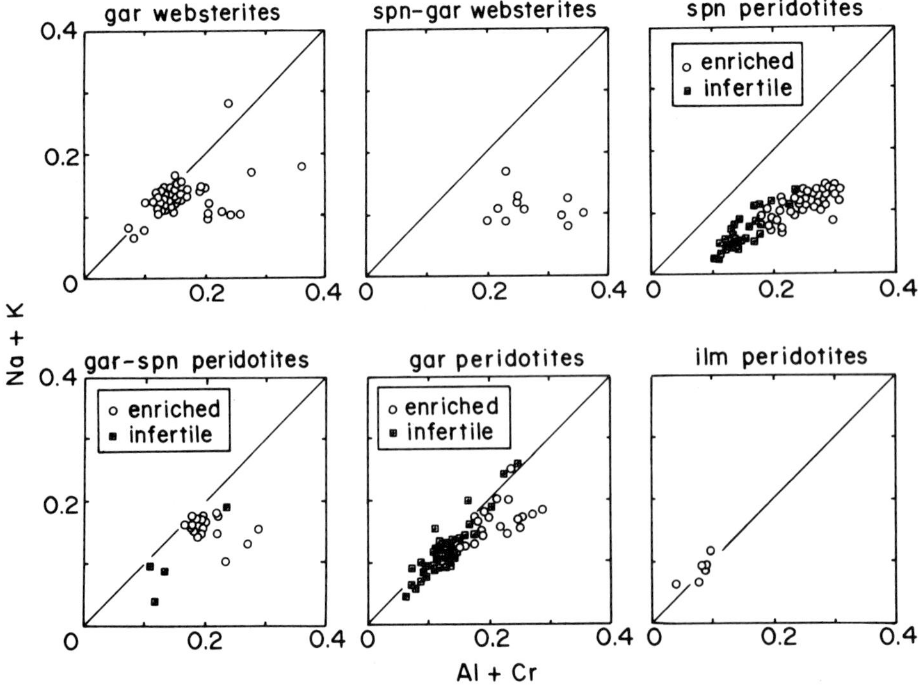

Figure 3. Composition of diopsides: Na + K versus Al + Cr. Pyroxene stoichiometry requires that diopsides plotting above 1:1 line contain ferric iron, and that diopsides plotting below line contain CaTschermaks molecule.

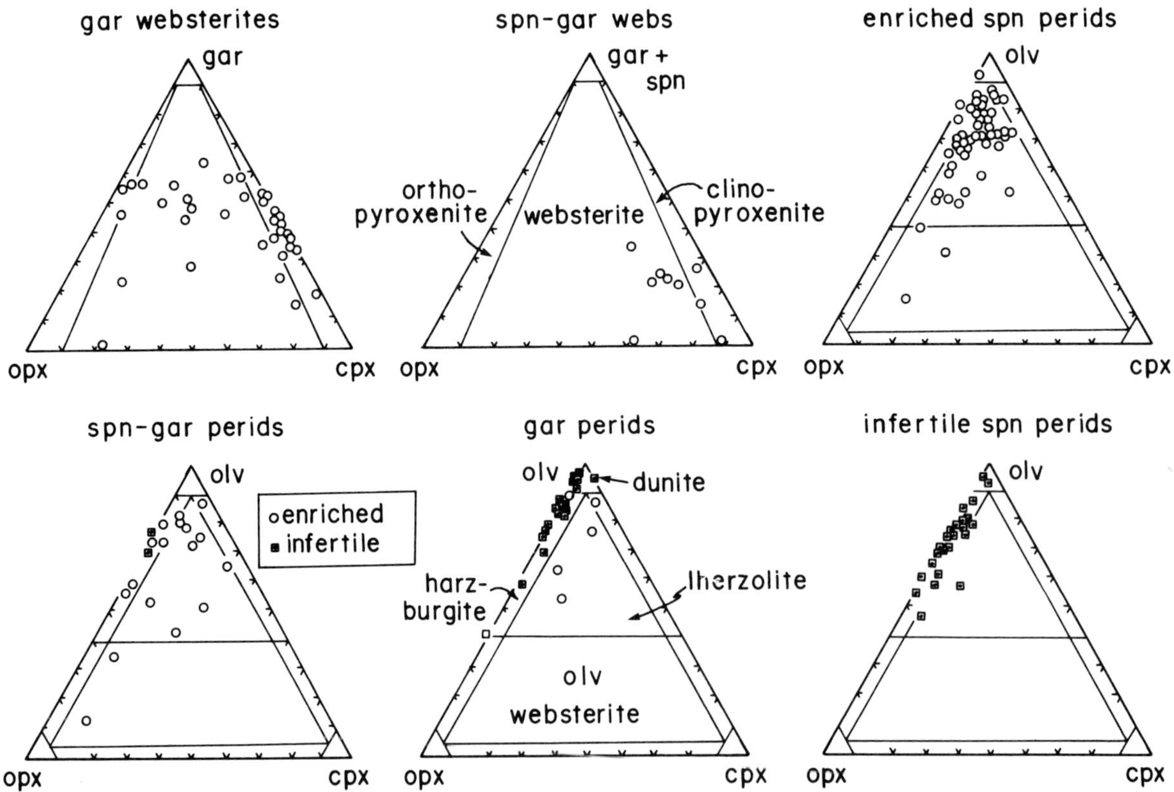

Figure 4. Modal compositions, in volume percent, of peridotites and olivine-free websterites.

TABLE 2. REPRESENTATIVE ANALYSES OF COLORADO-WYOMING NODULES

SD2-L70 enriched spn peridotite

	cpx	opx	olv	spn	WR
SiO_2	52.7	56.1	39.9	0.05	44.2
TiO_2	0.29	0.06	0.01	0.07	0.06
Al_2O_3	6.0	3.59	0.00	56.6	1.82
Cr_2O_3	0.80	0.27	0.00	11.0	0.23
Fe_2O_3	0.00	0.00	0.00	1.84	.01
FeO	1.88	6.1	9.7	9.6	7.9
MnO	0.11	0.18	0.16	0.14	0.16
NiO	0.03	0.04	0.44	0.28	0.31
MgO	15.1	34.1	49.9	20.3	41.8
CaO	21.3	0.32	0.01	0.01	3.26
Na_2O	1.57	0.05	0.02	0.00	0.26
K_2O	0.00	0.00	0.00	0.00	0.00
total	99.7	100.8	100.2	99.9	100.1

SD2-L186 enriched spn-gar peridotite

	cpx	opx	gar	olv	spn	WR
SiO_2	53.4	57.9	42.1	40.9	0.03	42.7
TiO_2	0.55	0.12	0.06	0.00	0.13	0.06
Al_2O_3	4.63	0.95	22.9	0.00	43.8	3.71
Cr_2O_3	0.70	0.05	0.86	0.00	22.4	.62
Fe_2O_3	0.00	0.00	0.00	0.00	4.22	.09
FeO	2.30	5.8	10.1	9.1	12.0	8.5
MnO	0.07	0.11	0.54	0.17	0.60	0.21
NiO	0.00	0.07	0.03	0.35	0.11	0.26
MgO	15.1	35.1	19.2	50.0	17.3	42.3
CaO	20.5	0.23	4.62	0.00	0.00	1.75
Na_2O	2.55	0.05	0.00	0.00	0.02	0.16
K_2O	0.01	0.00	0.00	0.00	0.00	.00
total	99.8	100.4	100.4	100.4	100.5	100.4

SD2-L66 enriched gar peridotite

	cpx	opx	gar	olv	WR
SiO_2	54.9	56.4	41.5	40.8	43.9
TiO_2	0.24	0.07	0.05	0.00	0.04
Al_2O_3	3.30	1.22	21.8	0.00	2.01
Cr_2O_3	0.79	0.23	1.18	0.00	0.21
Fe_2O_3	0.00	0.00	0.00	0.00	0.00
FeO	2.12	6.1	11.3	9.2	8.1
MnO	0.05	0.13	0.55	0.09	0.12
NiO	0.03	0.05	0.00	0.32	0.23
MgO	15.6	35.3	19.0	49.6	41.6
CaO	21.4	0.20	4.85	0.01	3.22
Na_2O	1.80	0.02	0.05	0.00	0.25
K_2O	0.00	0.00	0.00	0.00	0.00
total	100.2	99.7	100.3	100.0	99.6

SD2-L6 infertile spn peridotite

	cpx	opx	olv	spn	WR
SiO_2	54.0	55.1	40.7	0.09	45.0
TiO_2	0.06	0.00	0.00	0.01	0.00
Al_2O_3	2.58	2.58	0.00	40.9	0.78
Cr_2O_3	0.75	0.47	0.00	27.4	0.15
Fe_2O_3	0.00	0.00	0.00	2.64	0.00
FeO	1.50	4.80	7.2	9.1	6.4
MnO	0.09	0.21	0.10	0.23	0.13
NiO	0.16	0.12	0.44	0.15	0.34
MgO	17.6	35.1	51.6	18.8	46.2
CaO	23.4	0.40	0.01	0.01	0.70
Na_2O	0.72	0.04	0.00	0.02	0.03
K_2O	0.00	0.00	0.00	0.00	0.00
total	100.9	98.7	100.1	99.4	99.7

SD2-L90 infertile spn-gar peridotite*

	cpx	opx	olv	spn	WR
SiO_2	54.2	56.0	41.0	0.07	44.6
TiO_2	0.00	0.00	0.00	0.08	0.00
Al_2O_3	2.21	2.23	0.00	31.0	0.77
Cr_2O_3	0.92	0.47	0.00	37.4	0.37
Fe_2O_3	0.00	0.00	0.00	3.61	0.02
FeO	1.65	5.1	7.9	13.3	7.2
MnO	0.06	0.16	0.13	1.12	0.14
NiO	0.03	0.04	0.34	0.28	0.26
MgO	17.3	35.0	51.0	14.9	46.4
CaO	23.5	0.52	0.00	0.04	0.43
Na_2O	0.56	0.00	0.00	0.02	0.01
K_2O	0.02	0.00	0.00	0.00	0.00
total	100.4	99.5	100.4	101.8	100.2

SD2-L102 infertile gar peridotite

	cpx	opx	gar	olv	chr	WR
SiO_2	54.7	57.4	41.5	41.0	0.14	42.4
TiO_2	0.07	0.04	0.16	0.00	0.75	0.02
Al_2O_3	1.91	0.72	19.5	0.02	11.8	2.14
Cr_2O_3	1.96	0.30	5.1	0.00	55.3	0.59
Fe_2O_3	0.00	0.00	0.00	0.00	3.07	0.00
FeO	2.32	4.79	7.2	8.1	14.7	7.7
MnO	0.12	0.13	0.36	0.12	0.31	0.15
NiO	0.05	0.06	0.00	0.35	0.00	0.29
MgO	17.3	35.8	20.7	50.7	12.6	46.1
CaO	19.7	0.53	5.2	0.04	0.00	0.99
Na_2O	1.71	0.07	0.05	0.00	0.01	0.04
K_2O	0.00	0.00	0.00	0.00	0.00	0.00
total	99.8	99.8	99.6	100.3	98.6	100.4

NX4-L7 spinel-garnet websterite

	cpx	opx	gar	spn	WR
SiO_2	52.5	56.0	41.4	0.07	42.5
TiO_2	0.87	0.04	0.12	0.01	0.38
Al_2O_3	5.2	2.08	24.1	65.4	18.8
Cr_2O_3	0.00	0.00	0.00	0.83	0.11
Fe_2O_3	0.00	0.00	0.00	5.00	0.67
FeO	3.01	7.2	11.2	8.6	7.0
MnO	0.07	0.16	0.45	0.11	0.21
NiO	0.02	0.01	0.04	0.35	0.07
MgO	15.5	34.8	18.4	22.3	20.5
CaO	22.5	0.23	5.5	0.00	10.5
Na_2O	1.56	0.05	0.05	0.03	0.63
K_2O	0.00	0.01	0.00	0.00	0.00
total	101.2	100.6	101.3	102.6	101.3

SD2-L80 garnet websterite†

	cpx	opx	gar	WR
SiO_2	54.8	57.8	42.7	48.7
TiO_2	0.40	0.09	0.21	1.00
Al_2O_3	3.24	0.94	22.9	12.4
Cr_2O_3	0.43	0.09	0.73	0.52
Fe_2O_3	0.00	0.00	0.00	0.00
FeO	1.73	4.16	7.5	4.86
MnO	0.01	0.09	0.29	0.16
NiO	0.06	0.13	0.03	0.05
MgO	16.3	35.9	21.0	21.2
CaO	21.6	0.25	4.59	10.1
Na_2O	1.80	0.05	0.03	0.67
K_2O	0.00	0.00	0.00	0.00
total	100.5	99.5	100.0	99.6

SD2-L52 gar cpxite§

	cpx	gar	WR
SiO_2	54.3	42.2	47.5
TiO_2	0.65	0.13	0.65
Al_2O_3	3.07	23.9	12.6
Cr_2O_3	0.00	0.00	0.00
Fe_2O_3	0.00	0.00	0.00
FeO	2.07	7.9	4.69
MnO	0.10	0.35	0.21
NiO	0.07	0.04	0.05
MgO	17.7	21.2	18.9
CaO	20.5	4.68	12.7
Na_2O	1.76	0.04	0.92
K_2O	0.02	0.01	0.09
total	100.2	100.4	98.3

SD2-L40 ilmenite peridotite

	cpx	opx	olv	gar	ilm	WR
SiO_2	53.8	57.5	40.1	41.1	0.13	44.2
TiO_2	0.08	0.05	0.00	0.15	42.0	0.16
Al_2O_3	1.42	0.70	0.00	20.9	0.80	1.54
Cr_2O_3	0.66	0.15	0.00	2.28	4.87	0.33
Fe_2O_3	0.00	0.00	0.00	0.00	20.4	0.06
FeO	2.88	8.2	13.0	13.1	24.0	10.1
MnO	0.11	0.19	0.16	0.73	0.23	0.18
NiO	0.02	0.05	0.21	0.00	0.31	0.14
MgO	16.7	34.7	47.6	16.3	7.5	36.5
CaO	22.9	0.28	0.01	5.9	0.05	6.7
Na_2O	0.91	0.02	0.00	0.01	0.00	0.26
K_2O	0.00	0.00	0.00	0.00	0.00	0.00
total	98.5	101.9	101.1	100.5	100.4	100.6

*Contains 0.2% garnet that could not be analyzed.

†Also contains 0.6% rutile and 0.4% sulfide.

§Also contains 0.9% phlogopite, 0.4% sulfide, and 0.2% rutile.

Abbreviations:
- cpx = clinopyroxene
- opx = orthopyroxene
- gar = garnet
- spn = spinel
- olv = olivine
- chr = chromite
- WR = whole rock

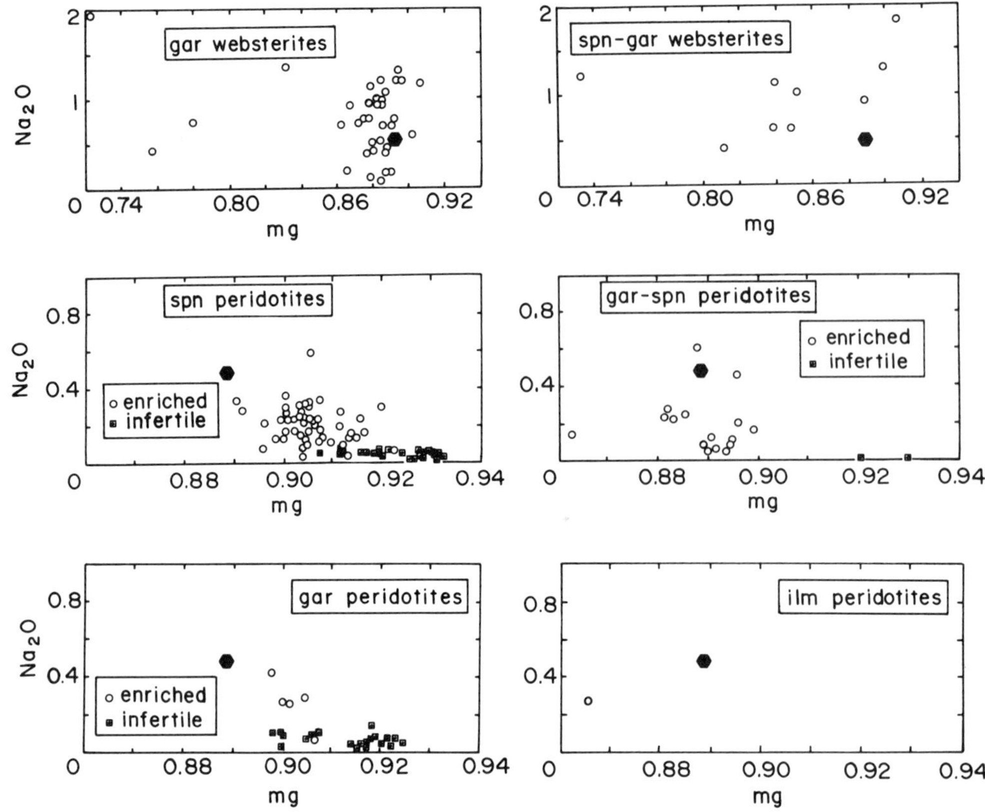

Figure 5. Bulk-rock compositions of peridotites and websterites: Na_2O (wt.%) versus Mg/(Mg + Fe) (mg). Octagon denotes Pyrolite II model peridotite composition.

nodules of lower crustal granulites, eclogites, and megacrysts (Table 1). Eclogites, with equilibration temperatures of 700°–1,300°C (Ater and others, 1984), probably occurred over the entire mantle lithosphere. Their large numbers (Table 1) may be overrepresented because of their resistance to mechanical degradation and to weathering. Megacrysts, interpreted as phenocrysts in alkalic magmas, occurred in about the same depth range as infertile garnet peridotites—160 to 200 km.

THERMAL AND CHEMICAL HISTORY OF LITHOSPHERE

An Ancient Depletion

Highly infertile peridotites in the garnet facies were sampled by kimberlite eruptions between 150 and 200 km (Fig. 6; note one point at 250 km). If Schaffer-group peridotites, with paleotemperatures down to 900°C, were placed along the Sloan 2 geotherm, the depth of sampling of infertile garnet peridotites would be 120–200 km. Infertile spinel-garnet peridotites were sampled at about 75 km (Fig. 6), and infertile spinel peridotites were sampled at depths less than about 75 km. Apparently, then, the entire mantle to a depth of at least 200 km contains infertile peridotite. It is in fact petrologically reasonable that a melting event or events of sufficient extent to produce highly infertile harzburgites did indeed affect that entire depth range.

Principal occurrences of highly infertile rocks and cool geothermal gradient define mantle, to a depth of at least 200 km, as lithosphere.

Depletion probably was not a simple, single event, inasmuch as there is no correlation between Na_2O and mg (Fig. 5) or cr (Fig. 1) within facies. Moreover, spinel and spinel-garnet peridotites appear to be more infertile (Fig. 5 and Table 3: lower Na_2O and TiO_2, higher mg) than garnet peridotites.

No isotopic data constrain the age or ages of depletion. Depletion certainly preceded kimberlite emplacement by at least hundreds of million years, because the lithosphere had cooled from the high temperatures associated with extensive melting to temperatures close to a geotherm of 40 mW/m^2 characteristic of shields (Fig. 6).

Geologic evidence also suggests that the depletion, which presumably would have resulted in major crustal basaltic activity, was Precambrian, because no igneous activity apart from kimberlites occurred in the Front Range in the Paleozoic. The oldest events in the Front Range, at 1.65–1.7 b.y. (Peterman and others, 1968), are granitic plutonism and metamorphism of 1.8-b.y.-old

TABLE 3. AVERAGE CALCULATED WHOLE-ROCK COMPOSITIONS OF WEBSTERITES AND PERIDOTITES

	garnet pyroxenites-websterites n=36	spinel-garnet websterites n=9	enriched spinel peridotites n=51	enriched spn-garnet peridotites n=16	enriched garnet peridotites n=5	infertile spinel peridotites n=26	infertile spn-garnet peridotites n=2	infertile garnet peridotites n=20
SiO_2	49.2 (2.0)	45.0 (7.5)	43.8 (2.0)	43.4 (1.7)	44.5 (2.2)	44.5 (2.2)	45.0	43.4 (2.0)
TiO_2	0.46 (0.50)	0.38 (0.12)	0.08 (0.04)	0.09 (0.04)	0.10 (0.06)	0.02 (0.02)	0.02	0.07 (0.05)
Al_2O_3	11.9 (3.0)	14.0 (8.1)	2.45 (1.8)	6.91 (3.2)	1.86 (0.8)	1.31 (0.8)	0.53	0.85 (1.1)
Cr_2O_3	0.40 (0.34)	0.60 (0.60)	0.42 (0.42)	1.00 (0.71)	0.18 (0.09)	0.49 (0.44)	0.35	0.48 (0.48)
Fe_2O_3	0.00	0.11 (0.22)	0.05 (0.05)	0.15 (0.21)	0.00	0.04 (0.05)	0.02	0.00
FeO	5.5 (1.8)	6.0 (2.5)	7.9 (0.7)	8.2 (0.6)	8.2 (0.6)	6.9 (0.7)	6.9	7.8 (0.8)
MnO	0.19 (0.04)	0.16 (0.05)	0.16 (0.03)	0.22 (0.05)	0.14 (0.03)	0.13 (0.03)	0.14	0.14 (0.02)
NiO	0.04 (0.02)	0.06 (0.06)	0.29 (0.06)	0.18 (0.08)	0.30 (0.08)	0.31 (0.04)	0.29	0.31 (0.07)
MgO	21.0 (4.0)	18.1 (2.2)	42.6 (3.5)	36.8 (4.1)	42.5 (3.9)	46.1 (2.3)	47.6	46.5 (3.1)
CaO	10.8 (4.6)	13.6 (3.7)	2.35 (1.2)	2.56 (1.3)	2.43 (1.3)	0.71 (0.63)	0.26	0.61 (0.51)
Na_2O	0.76 (0.43)	1.01 (0.42)	0.20 (0.10)	0.19 (0.15)	0.25 (0.13)	0.04 (0.02)	0.01	0.07 (0.03)
K_2O	0.01 (0.02)	0.00	0.00	0.00	0.00	0.00	0.00	0.00
	-----	-----	-----	-----	-----	-----	-----	-----
total	100.0	99.1	100.3	99.7	100.5	100.5	101.1	100.2
cr	0.022	0.028	0.103	0.089	0.061	0.201	0.307	0.275
mg	0.872	0.843	0.906	0.889	0.902	0.923	0.925	0.914

Note: Figures in parentheses are one s.d. of the mean.

protoliths derived from asthenosphere (DePaolo, 1981). Younger granitic batholiths, dated at 1.4 and 1.0 b.y. (Peterman and others, 1968), largely represent mobilization of crust, with minor addition of mantle-derived magma (DePaolo, 1981). Plausible mantle-derived rocks now outcrop as a few basic intrusions (Eggler, 1968) and basaltic dike swarms. Lack of huge volumes of basaltic material in Proterozoic crust might suggest that the depletion event was Archean. In fact the pipes lie just south of a major suture separating Proterozoic crust from Archean crust of the Wyoming province. Karlstrom and Houston (1984) have suggested that development of that suture was predated by an Archean subduction event.

A Less Ancient Websterite-Associated Enrichment

As discussed, kimberlite pipes contain a group of garnet websterites and pyroxenites and a group of spinel-garnet websterites, which are interpreted as having been dikes and layers. They also contain enriched peridotites in the spinel, spinel-garnet, and garnet peridotite facies, which are interpreted as having been metasomatic zones. Very few nodules occur that are modally intermediate between pyroxenites-websterites and peridotites, namely, olivine websterites (Fig. 4). Geochemical similarities between minerals of all the groups suggest that they are related.

Geothermometry (Fig. 6) is consistent with a view that garnet pyroxenite and websterite dikes cut those infertile peridotites occupying lithosphere above 100 km, namely, spinel, spinel-garnet, and uppermost garnet-facies peridotites. Enriched spinel-garnet peridotites could very well represent reaction zones between garnet pyroxenite or websterite and infertile spinel peridotite. Spinel-garnet websterites occurred, probably as dikes, within spinel peridotite in the mantle (Fig. 6) but may have populated the lower crust as well.

The relative abundances of collected nodules of websterite, enriched peridotite, and infertile spinel and spinel-garnet peridotite (Table 1) suggest that enriched rocks (websterites and peridotites) occupied about 80 percent of the lithosphere between the crust and a depth of 100 km. This enormously high percentage indicates considerable density of dikes and layers, at least in the lithosphere sampled by kimberlites. It is possible, however, that the population of websterites and enriched peridotites has been oversampled by kimberlite eruption due to breakage of wallrocks preferentially near the contacts of dikes and layers. Websterites may also be overrepresented as nodules, relative to peridotites, because of their greater resistance to mechanical degradation and to weathering.

Comparison of enriched peridotites to infertile peridotites

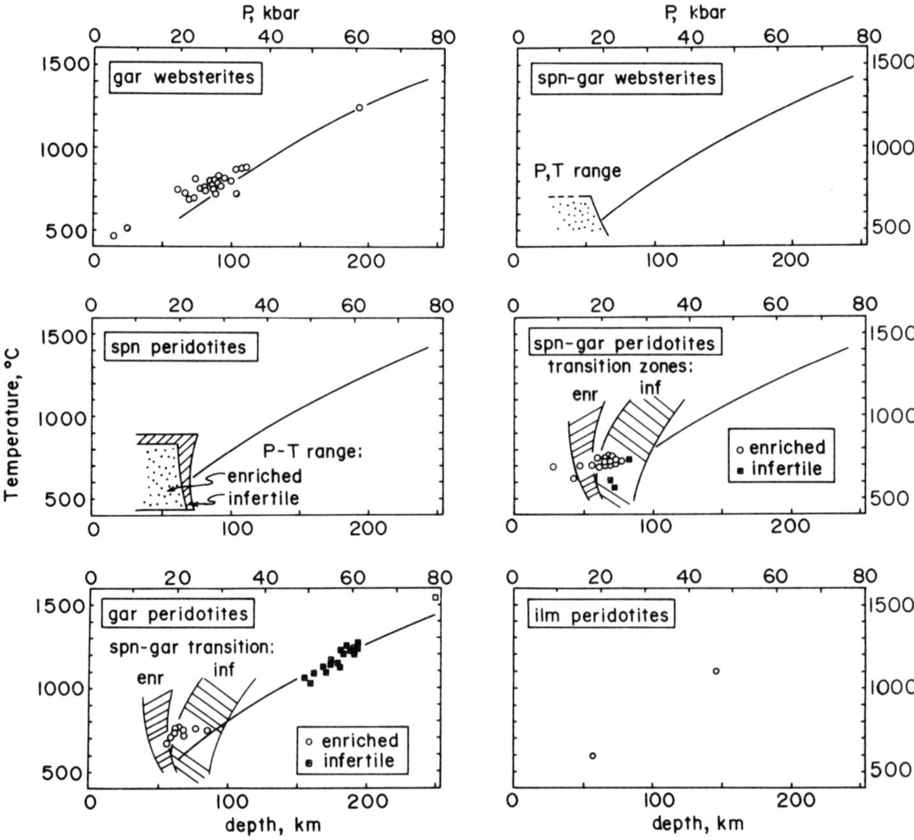

Figure 6. Geothermobarometry of websterites and peridotites; methods discussed in text. For spinel-garnet websterites and spinel peridotites, maximum temperatures and pressures are shown (see text). Spinel-garnet peridotite transition zones were calculated for actual mineral composition ranges in nodules. Geotherm shown has been calculated for continental shield with heat flow of 40 mW/m^2 (Pollack and Chapman, 1977).

(Fig. 5; Table 3) indicates that contents of Fe, Ca, and Na are elevated. Slight elevation has also occurred in Ti, particularly in the spinel facies (Table 3). Although locally containing phlogopite, the websterites and pyroxenites themselves contain too little K_2O and Na_2O and are too magnesian (Fig. 5) to represent crystallized melts. Protoliths probably were cumulate assemblages, perhaps plated out along dike walls (although present mineral compositions reflect subsequent metamorphism). We surmise that protoliths of enriched rocks once existed at higher temperatures; this is because garnets rim spinels in garnet-spinel peridotites (Kirkley and others, 1984), representing cooling from the spinel into the garnet-spinel peridotite facies, and because coarse exsolution textures are present in garnet pyroxenites-websterites, enriched spinel-garnet peridotites, and spinel-garnet websterites (Kirkley and others, 1984; Table 1).

Some petrologists (e.g., Wilshire and Pike, 1975) have ascribed melting and metasomatic events to the diapiric rise of peridotite that terminates in the eruption of magma containing nodules bearing evidence of the metasomatism. We do not consider the websterite-associated enrichment to be a consequence of kimberlite magmatism. Our primary evidence is the cold paleotemperatures of both the websterites and the enriched peridotites. The temperatures, all below 900°C (Fig. 6), are subsolidus to any peridotite melting curve. Not only are the paleotemperatures cold, but they are not perturbed from the 40 mW/m^2 geotherm typical of continental shields. Both thermal characteristics seem incompatible with a model in which hot asthenospheric diapirs have risen to depths as shallow as 50 km at a time essentially contemporaneous with kimberlite eruption.

It is also questionable whether diapirs could have penetrated a cold, thick (at least 200 km) lithosphere. Of course, lithosphere at depths as shallow as 50 km could simply have been intruded by protokimberlite magmas, undergone metasomatism, and then cooled substantially before its incorporation in erupting kimberlite magma. Such a model does not explain the lack of enrichment in the deeper lithosphere, where protokimberlite magma might be more abundant. A near-total lack of phlogopite in enriched rocks also appears inconsistent with kimberlitic metasoma-

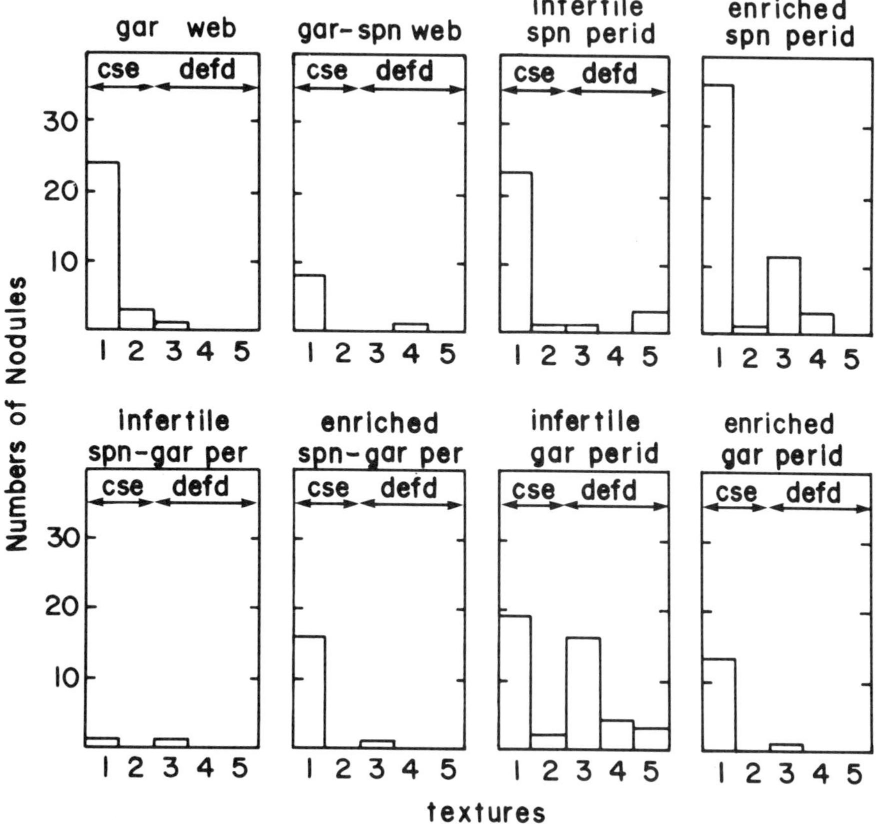

Figure 7. Textures of websterites and peridotites. Textures: 1, coarse equant; 2, tabular; 3, transitional porphyroclastic; 4, porphyroclastic; 5, mosaic porphyroclastic.

tism. Until isotopic data are available, we conclude that lithosphere throughout its sampled depth was cold, and therefore old.

The age of enrichment, if not Devonian, was probably at least 1.65–1.7 b.y. ago, following arguments above for infertile peridotites. If the depletion event was Archean in age, then it is possible that the enrichment was Archean in age as well.

An Ilmenite Peridotite Enrichment Event

Minerals in ilmenite peridotites are enriched in Fe relative to minerals in other peridotites (Fig. 2); the one bulk rock calculated has distinctively low mg (Fig. 5) and somewhat elevated TiO_2 (Table 2). The enrichment event is thus distinctive from the websterite-associated enrichment. Analogous rocks are recognized from other kimberlites (e.g., Cox and others, 1973; Harte and Gurney, 1975) and are commonly attributed to interaction with magma, quite possibly in a precursory kimberlitic event that produces Fe- and Ti-enriched peridotites (Ehrenburg, 1982). In Colorado-Wyoming, ilmenite peridotites are relatively cold and in one case quite shallow (Fig. 6; the total range of LD20 paleotemperatures is 580° to 1,100°C). The arguments noted above for websterites apply here: we would interpret the enrichment not to be contemporaneous with kimberlite magmatism but to be Precambrian.

COMPARISON TO OTHER PERIDOTITE SUITES

Suites in Kimberlite and Related Igneous Rocks

Harte (1983) has summarized characteristics of suites of peridotites from southern African kimberlites in terms of Mg- or Fe-rich (olivine mg more or less than 0.91), hot or cold (temperature more or less than 1,100°C), and coarse or deformed. Cold rocks are commonly Mg-rich and coarse, and hot rocks are commonly Fe-rich and deformed, the classic pattern established by Boyd and Nixon (1975) for Lesotho nodules. The common pattern is not seen in Colorado-Wyoming rocks, where the hottest rocks (infertile garnet peridotites) are not exceptionally hot (temperature, less than 1,200°C), do not define an inflected geotherm (Fig. 6), and are not Fe-rich. In those respects the nodules resemble Kimberley nodules (Boyd and Nixon, 1978). The more Fe-rich rocks are, in fact, cold and coarse. There is a correlation between depth and deformation, however, because half the deep, infertile garnet peridotites have deformed textures, whereas the other, shallower suites are dominated by coarse textures (Fig. 7).

Chemically, the infertile peridotites in Colorado-Wyoming pipes are not unlike the common southern African infertile peridotites (Harte, 1983), except that they are not quite as magnesian and are even lower in Na_2O.

Colorado-Wyoming pipes contain many spinel (cr, <0.6) and garnet-spinel (cr, <0.7) peridotites (Table 1). Spinel in kimberlite nodules typically is accessory chromite (cr, >0.7) in garnet-facies peridotite (e.g., Boyd and Nixon, 1978). Carswell and others (1984), however, on the basis of a limited number of infertile spinel (0.25 < cr < 0.65) peridotites at Ngopetsoeu and other southern African localities, surmised a general occurrence of spinel peridotites in uppermost African lithosphere. They discussed a few nodules with more aluminous spinel as well, but assigned them to a crustal source.

Finally, Colorado-Wyoming kimberlite suites are distinguished by the abundant occurrence of olivine-free pyroxenites and websterites. Olivine-free eclogites and MARID-suite clinopyroxenites (Dawson and Smith, 1977) are common in kimberlites. Cr-diopside pyroxenites or websterites are uncommon, however (Harte, 1983), a principal occurrence being Fe-rich peridotites associated with pyroxenite layers in nodules in the Matsoku pipe (Harte and others, 1977).

Suites in Alkali Basalts and Peridotite Massifs

Wilshire and Shervais (1975) noted the common occurrence in alkali basalts of two groups of nodules, the Cr-diopside group of spinel lherzolite, websterite, and clinopyroxenite, and the Al-augite group of spinel clinopyroxenite, wehrlite, lherzolite, kaersutite pyroxenite, and hornblendite. Less abundant are groups of garnetiferous clinopyroxenites and lherzolites and Mg-Cr phlogopite pyroxenite and glimmerite, feldspathic rocks identical to the Cr-diopside group save for the presence of plagioclase, and a group of gabbros and metagabbros (Wilshire, 1986). Mafic layers of any group may crosscut foliation in Cr-diopside peridotites (Wilshire, this volume), but the common mafic layers are of the Al-augite group.

In reviewing massifs, Wilshire (this volume) concludes that the principal lithologies are essentially identical to principal groups in alkali basalt xenoliths. Pyroxenitic or gabbroic layers typically comprise 1 to 3 percent of massifs and range in form from bands to parallel layers to crosscutting layers. Dike emplacement follows the order Cr-diopside pyroxenite; Al-augite pyroxenite; gabbro (Wilshire, this volume).

Comparison of Colorado-Wyoming rocks to alkali basalt nodules or to peridotite massifs is hindered by the relative lack of composite nodules in the Colorado-Wyoming suites. From the data available, it appears that analogies certainly exist between observed peridotites crosscut by ultramafic to mafic dikes in nodules and massifs and our interpreted peridotites crosscut by ultramafic dikes in Colorado-Wyoming kimberlite pipes. In detail, however, the situations are not comparable, either in lithospheric structures or in lithologies.

Because massifs have undergone complicated emplacement histories, lithospheric structure is best referenced to alkali basalt nodules. In the United States alkali basalt nodules occur principally in the tectonically young Basin-Range and Rio Grande Rift provinces (Wilshire and others, 1985). The Nd and Sr isotopic signatures of Al-augite group rocks (Menzies and others, 1985) are essentially the same as those of alkali basalt host rocks and lie within the "mantle array" of asthenospheric signatures. These signatures indicate the contemporaneity of basalts and enriched rocks and the derivation of both from reservoirs that were asthenospheric or else were separated from the asthenosphere for only a short period of time. All these characteristics mark lithosphere represented in alkali basalt suites as young, warm, and thin, in contrast to lithosphere in kimberlite suites that is old, cold, and thick.

Lithologic differences between Colorado-Wyoming rocks and alkali basalt suites are highlighted by the absence of Al-augite pyroxenites and peridotites or gabbros in the former, although the Cr-diopside pyroxenites and websterites of massifs have analogues in Colorado-Wyoming. The inferred density of dikes and enriched rocks in the Colorado-Wyoming subcrustal mantle (80 percent) is atypical of alkali basalt nodules or of massifs. The dominant lithology in alkali basalt nodules and in massifs—spinel peridotite—is more fertile than the infertile spinel peridotites of Colorado-Wyoming (e.g., the average Cr-diopside lherzolite of alkali basalt nodules has Na_2O = 0.17 and mg = 0.900; Wilshire and others, 1985).

Hawkesworth and others (1984) and Menzies and others (1985) identified two types of modal metasomatism in nodule suites in basalts and in kimberlites: one representing passage of Fe- and Ti-rich silicate melts and one passage of K-rich "fluids." The dominant websterite-associated metasomatism in Colorado-Wyoming nodules is yet another type, characterized by negligible to slight enrichment in K and only mild enrichment in Fe and Ti (Table 3).

KIMBERLITE GENERATION AND EMPLACEMENT

We have reasoned that metasomatic events represented in nodules are unrelated chemically or temporally to kimberlite magmatism. The Nd isotopic signature of Colorado-Wyoming kimberlite is only slightly depleted relative to bulk earth (ϵ_{Nd} = +1: Basu and Tatsumoto, 1980), implying an asthenospheric source with isotopic characteristics akin to those of oceanic island basalts (Smith, 1983).

We surmise that kimberlite was emplaced from a depth of at least 200 km in a tectonic regime favorable for fracturing or refracturing of thick lithosphere. The very high ratio of enriched to infertile rocks in shallow Colorado-Wyoming lithosphere (see above) implies that ancient magmatic conduits lined with websterites and enriched peridotites were preferentially exploited by younger kimberlite eruptions.

SUMMARY

Ultramafic rocks sampled by Colorado-Wyoming kimberlites bear faint resemblance, in tectonic setting or in rock types, to nodules from alkali basalts or to peridotite massifs. The kimberlite nodule suite has been interpreted, however, to contain diked and layered rocks that are evident in basalt nodules and in massifs. Either an abundance of diked or layered rocks or an abundance of websteritic rocks, interpreted to have been dikes, is uncharacteristic of kimberlite nodule suites. In that sense, Colorado-Wyoming rocks provide links between kimberlite suites and peridotite massifs.

Websterites and chemically related enriched peridotites represent enrichment of an infertile lithosphere that is still preserved as nodules derived from spinel, spinel-garnet, and garnet peridotite facies. Such enrichment occurred only in shallow lithosphere. The igneous event could be related to kimberlite magmatism, but lack of enrichment in deep lithosphere and cold paleotemperatures of enriched rocks suggest that the enrichment event was old and therefore Precambrian in age.

ACKNOWLEDGMENTS

This research was supported by several grants from the Earth Sciences Section of the National Science Foundation, most recently grant EAR-8308292 (D.H.E.). Previous support by the Geophysical Laboratory, Carnegie Institution of Washington, was essential. Manuscript reviews by L. A. Taylor, and especially by J. K. Meen and H. G. Wilshire, whose comments materially improved the manuscript, are appreciated.

REFERENCES CITED

Ater, P. C., Eggler, D. H., and McCallum, M. E., 1984, Petrology and geochemistry of mantle eclogite xenoliths from Colorado-Wyoming kimberlites; Recycled ocean crust?, *in* Kornprobst, J., ed., Kimberlites. II, The mantle and crust–mantle relationships: Amsterdam, Elsevier, p. 309–318.

Basu, A. R., and Tasumoto, M., 1980, Nd-Isotopes in selected mantle-derived rocks and minerals and their implications for mantle evolution: Contributions to Mineralogy and Petrology, v. 75, p. 43–54.

Boyd, F. R., and Nixon, P. E., 1975, Origins of the ultramafic nodules from some kimberlites of northern Lesotho and the Monastery Mine, South Africa: Physics and Chemistry of the Earth, v. 9, p. 431–454.

——, 1978, Ultramafic nodules from the Kimberley pipes, South Africa: Geochimica et Cosmochimica Acta, v. 42, p. 1267–1382.

Bradley, S. D., and McCallum, M. E., 1984, Granulite facies and related xenoliths from Colorado–Wyoming kimberlite, *in* Kornprobst, J., ed., Kimberlites. II, The mantle and crust–mantle relationships: Amsterdam, Elsevier, p. 205–218.

Carlson, R. W., Dudas, F. O., Meen, J. K., and Eggler, D. H., 1985, Formation and evolution of the Archean subcontinental mantle beneath the northwestern U.S. [abs.]: EOS Transactions of the American Geophysical Union, v. 66, p. 1109.

Carswell, D. A., Griffin, W. L., and Kresten, P., 1984, Peridotite nodules from the Ngopetsoeu and Lipelaneng kimberlites, Lesotho; A crustal or mantle origin, *in* Kornprobst, J., ed., Kimberlites. II, The mantle and crust-mantle relationships: Amsterdam, Elsevier, p. 229–243.

Chronic, J., McCallum, M. E., Ferris, C. S., Jr., and Eggler, D. H., 1969, Lower Paleozoic rocks in diatremes, southern Wyoming and northern Colorado: Geological Society of America Bulletin, v. 80, p. 149–156.

Cox, K. G., Gurney, J. J., and Harte, B., 1973, Xenoliths from the Matsoku Pipe, *in* Nixon, P. H., ed., Lesotho kimberlites: Maseru, Lesotho National Development Corporation, p. 76–100.

Dawson, J. B., and Smith, J. V., 1977, The MARID (mica-amphibole-rutile-ilmenite-diopside) suite of xenoliths in kimberlite: Geochimica et Cosmochimica Acta, v. 41, p. 309–323.

DePaolo, D. J., 1981, Neodymium isotopes in the Colorado Front Range and crust-mantle evolution in the Proterozoic: Nature, v. 291, p. 193–196.

Dick, H.J.B., and Fisher, R. L., 1984, Mineralogic studies of the residues of mantle melting; Abyssal and alpine-type peridotites, *in* Kornprobst, J., ed., Kimberlites. II, The mantle and crust-mantle relationships: Amsterdam, Elsevier, p. 295–308.

Dudas, F. O., Carlson, R. W., and Eggler, D. H., 1987, Regional middle Proterozoic enrichment of the subcontinental mantle source of igneous rocks from central Montana: Geology (in press).

Eggler, D. H., 1968, Virginia Dale ring-dike complex, Colorado-Wyoming: Geological Society of America Bulletin, v. 79, p. 1545–1564.

Eggler, D. H., and McCallum, M. E., 1974, Preliminary upper mantle–lower crust model for the Colorado-Wyoming Front Range: Carnegie Institution of Washington Yearbook 73, p. 295–300.

Eggler, D. H., McCallum, M. E., and Smith, C. B., 1979, Megacryst assemblages in kimberlites from northern Colorado and southern Wyoming; Petrology, geothermometry-barometry, and areal distribution, *in* Boyd, F. R., and Meyer, H.O.A., eds., The mantle sample: Washington, D.C., American Geophysical Union, p. 213–226.

Ehrenburg, S. N., 1982, Petrogenesis of garnet lherzolite and megacrystalline nodules from The Thumb, Navajo Volcanic Field: Journal of Petrology, v. 23, p. 507–547.

Finnerty, A. A., and Boyd, F. R., Jr., 1984, Evaluation of thermobarometers for garnet peridotites: Geochimica et Cosmochimica Acta, v. 48, p. 15–27.

Gasparik, T., and Newton, R. C., 1984, The reversed alumina contents of orthopyroxene in equilibrium with spinel and forsterite in the system MgO-Al_2O_3-SiO_2: Contributions to Mineralogy and Petrology, v. 85, p. 186–196.

Harte, B., 1977, Rock nomenclature with particular relation to deformation and recrystallization textures in olivine-bearing xenoliths: Journal of Geology, v. 85, p. 279–288.

——, 1983, Mantle peridotites and processes; The kimberlite sample, *in* Hawkesworth, C. J., and Norry, M. J., eds., Continental basalts and mantle xenoliths: Nantwich, United Kingdom, Shiva Publishing, p. 46–91.

Harte, B., and Gurney, J. J., 1975, Ore mineral and phlogopite mineralization within ultramafic nodules from the Matsoku kimberlite pipe, Lesotho: Carnegie Institute of Washington Yearbook 74, p. 528–536.

Harte, B., Gurney, J. J., and Cox, K. G., 1977, Clinopyroxene-rich sheets in garnet peridotite; Xenolith specimens from the Matsoku kimberlite pipe, Lesotho: Santa Fe, New Mexico, Second International Kimberlite Conference, Extended Abstracts (no pagination).

Hawkesworth, C. J., Rogers, N. W., Van Calsteren, P.W.C., and Menzies, M. A., 1984, Mantle enrichment processes: Nature, v. 311, p. 331–335.

Hearn, B. C., Jr., and McGee, E. S., 1984, Garnet peridotites from Williams kimberlites, north-central Montana, U.S.A., *in* Kornprobst, J., ed., Kimberlites. II, The mantle and crust-mantle relationships: Amsterdam, Elsevier, p. 57–70.

Karlstrom, K. E., and Houston, R. B., 1984, The Cheyenne Belt; Analysis of a Proterozoic suture in southern Wyoming: Precambrian Research, v. 25, p. 415–446.

Kirkley, M. B., 1980, Peridotite xenoliths in Colorado-Wyoming kimberlites [M.S. thesis]: Fort Collins, Colorado State University, 289 p.

Kirkley, M. B., McCallum, M. E., and Eggler, D. H., 1983, Coexisting garnet and spinel in upper mantle xenoliths from Colorado-Wyoming kimberlites; Appendix, *in* Kornprobst, J., ed., Kimberlites. III, Documents: Clermont-Ferrand, France, Annales Scientifique de L'Universite de Clermont-Ferrand II, p. 149–156.

——, 1984, Coexisting garnet and spinel in upper mantle xenoliths from Colorado-Wyoming kimberlites, *in* Kornprobst, J., ed., Kimberlites. II, The mantle and crust-mantle relationships: Amsterdam, Elsevier, p. 85–96.

Lindsley, D. H., and Dixon, S. A., 1976, Diopside-enstatite equilibria at 850 to 1,400°C, 5 to 35 kb: American Journal of Science, v. 276, p. 1285–1301.

MacGregor, I. D., 1974, The system $MgO-Al_2O_3-SiO_2$; Solubility of Al_2O_3 in enstatite for spinel and garnet peridotite compositions: American Mineralogist, v. 59, p. 110–119.

McCallum, M. E., Eggler, D. H., and Burns, L. K., 1975, Kimberlite diatremes in northern Colorado and southern Wyoming: Physics and Chemistry of the Earth, v. 9, p. 149–161.

Meen, J. K., 1983, Isotopic composition of some Laramide volcanics, Absaroka Mountains, Montana: Carnegie Institution of Washington Yearbook 82, p. 481–486.

Menzies, M. A., 1983, Mantle ultramafic xenoliths in alkaline magmas: Evidence for mantle heterogeneity modified by magmatic activity, *in* Hawkesworth, C. J., and Norry, M. J., eds., Continental basalts and mantle xenoliths: Nantwich, United Kingdom, Shiva Publishing, p. 92–110.

Menzies, M. A., Kempton, P., and Dungan, M., 1985, Interaction of continental lithosphere and asthenospheric melts below the Geronimo Volcanic Field, Arizona: Journal of Petrology, v. 26, p. 663–693.

Naeser, C. W., and McCallum, M. E., 1977, Fission-track dating of kimberlitic zircons: Santa Fe, Second International Kimberlite Conference, Extended Abstracts (no pagination).

O'Neill, H.S.C., and Wood, B.J., 1979, An experimental study of Fe-Mg partitioning between garnet and olivine and its calibration as a geothermometer: Contributions to Mineralogy and Petrology, v. 70, p. 59–70.

Peterman, Z. E., Hedge, C. E., and Braddock, W. A., 1968, Age of Precambrian events in the northeast Front Range, Colorado: Journal of Geophysical Research, v. 73, p. 2277–2296.

Pollack, H. N., and Chapman, D. S., 1977, On the regional variation of heat flow, geotherms, and lithospheric thickness: Tectonophysics, v. 38, p. 279–296.

Smith, C. B., 1977, Kimberlite and mantle derived xenoliths at Iron Mountain, Wyoming [M.S. thesis]: Fort Collins, Colorado State University, 218 p.

——, 1979, Rb-Sr mica ages of various kimberlites: Cambridge, Cambridge Kimberlite Symposium II, p. 61–66.

——, 1983, Pb, Sr, and Nd isotopic evidence for sources of southern African Cretaceous kimberlites: Nature, v. 304, p. 51–54.

Smith, C. B., McCallum, M. E., Coopersmith, H. G., and Eggler, D. H., 1979, Petrochemistry and structure of kimberlites in the Front Range and Laramie Range, Colorado-Wyoming, *in* Boyd, F. R., and Meyer, H.O.A., eds., Kimberlites, diatremes, and diamonds; Their geology, petrology, and geochemistry: Washington, D.C., American Geophysical Union, p. 178–189.

Vollmer, R., Ogden, P., Schilling, J.-G., Kingsley, R. H., and Waggoner, 1984, Nd and Sr isotopes in ultrapotassic volcanic rocks from the Leucite Hills, Wyoming: Contributions to Mineralogy and Petrology, v. 87, p. 359–368.

Webb, S.A.C., and Wood, B.J., 1986, Spinel-pyroxene-garnet relationships and their dependence on Cr/Al ratio: Contributions to Mineralogy and Petrology, v. 92, p. 471–480.

Wilshire, H. G., and Pike, J.E.N., 1975, Upper-mantle diapirism; Evidence from analogous features in alpine peridotite and ultramafic inclusions in basalt: Geology, v. 3, p. 467–470.

Wilshire, H. G., and Shervais, J. W., 1975, Al-augite and Cr-diopside ultramafic xenoliths in basaltic rocks from western United States; Structural and textural relationships: Physics and Chemistry of the Earth, v. 9, p. 257–272.

Wilshire, H. G., Meyer, C. E., Nakata, J. K., Calk, L. C., Shervais, J. W., Nielson, J. E., and Schwarzman, E. C., 1985, Mafic and ultramafic xenoliths from volcanic rocks of the western United States: U.S. Geological Survey Open-File Report 85-139, 49 p.

Wood, B. J., and Holloway, J. R., 1984, A thermodynamic model for subsolidus equilibria in the system $CaO-MgO-Al_2O_3-SiO_2$: Geochimica et Cosmochimica Acta, v. 48, p. 159–176.

MANUSCRIPT ACCEPTED BY THE SOCIETY OCTOBER 14, 1986

Printed in U.S.A.

Geological Society of America
Special Paper 215
1987

Mantle metasomatism and carbonatites; An experimental study of a complex relationship

*James K. Meen**
Department of Geosciences, The Pennsylvania State University, University Park, Pennsylvania 16802

ABSTRACT

The solidus of peridotite in the presence of water and carbon dioxide displays two cusps. Consequently, water- and carbon dioxide–bearing magmas rising from depths in excess of 70 km will cross the peridotite solidus if they are at temperatures of less than 1,100°C. The melt may react with the peridotitic wallrock to produce a variety of mineral assemblages that will remain as veins in the mantle after the altered magma has passed on. The main effect of high-temperature interactions is shown here to be replacement of orthopyroxene in the wallrocks with clinopyroxenes. Greater degrees of reaction (at lower temperatures) produce more exotic mineral assemblages. At pressures above 17 kbar, the magma may react with the peridotite to completion, producing a carbonated and amphibolitized mantle. At lower pressures, at which carbonate is not stable, reaction of peridotite and magma may produce an exotic mineral assemblage (including feldspathoids and sanidine with or without amphibole and, possibly, apatite, sulfides, and oxides), while enriching the melt in alkalis and CO_2. The composition of the melt thus changes from ijolitic to carbonatitic, and may ultimately produce a CO_2 fluid.

Long-term storage of the alkaline magma in the mantle at $17 < P < 22$ kbar will result in the production of low Nd isotopic ratios, moderate Sr ratios, and high Pb ratios. Melting of this altered mantle at a later time will produce carbonated hyperalkaline magmas that also possess these isotopic characteristics. Several occurrences of such rocks with the suggested isotopic compositions are noted.

INTRODUCTION

An extensive body of data has appeared during the last decade to suggest that the source regions of some magmas underwent enrichment prior to melting. The difference in time between enrichment and melting may have been too small to be resolved by isotopic systematics (e.g., Nunivak, Menzies and Murthy, 1980), in which case the enrichment is shown only by trace element patterns. On the other hand, enrichment may have occurred hundreds of millions of years (e.g., Barreiro, 1983; Barreiro and Cooper, this volume) or even billions of years (Meen, 1985) before melting, in which case the disturbed trace element patterns could have produced unusual isotopic compositions with the passage of time. Mantle-derived igneous rocks and xenoliths from a number of regions (St. Helena: Cohen and O'Nions, 1982; Tubuai: Vidal and others, 1984; southeast Australia: Menzies and Wass, 1983; New Zealand: Barreiro, 1983, Barreiro and Cooper, this volume; the western United States: Vollmer and others, 1981; Meen, 1983, 1985, Dudas and others, 1985) have lower values of $(^{143}Nd/^{144}Nd)_i$ at a given value of $(^{87}Sr/^{86}Sr)_i$ than does the mantle array (they fall to the "left" of the array). These Sr-Nd systematics may be accompanied by Pb isotopic signatures that are relatively radiogenic (Barreiro, 1983; Barreiro and Cooper, this volume) or nonradiogenic (Meen, 1985). Menzies and Wass (1983) have suggested that the relatively low values of Rb:Sr and Sm:Nd in the source regions of some alkaline magmas from Australia are the result of carbonic metasomatism, because CO_2-rich fluids are presumed to have low values of Rb:Sr and Sm:Nd. However, as emphasized by Eggler (1986), CO_2-rich fluids cannot exit in equilibrium with peridotite at $P > 17$ kbar, as carbonate is a stable mineral. Furthermore, carbonated magmas cannot be produced by melting of peridotite at $P < 17$ kbar, so carbonic

*Present address: Department of Geology, University of North Carolina at Chapel Hill, Chapel Hill, N.C. 27514

metasomatism cannot be responsible for enrichment of the source regions of such magmas.

An alternate model of metasomatism involving interaction of carbonated hyperalkaline magma with peridotite is discussed herein. This model differs from models involving metasomatism by fluids in that a magmatic agent may move very large quantities of major and trace elements compared to a fluid agent (Fregeau, 1985) and in that the metasomatic effects occur only in a restricted pressure range. Further, it is shown that this mantle enrichment process does not require the fortuitous freezing of magma, but is a necessary result of the phase relations of peridotite-H_2O-CO_2. Interaction of such magma with peridotite at $P<17$ kbar results in evolution to strongly carbonated alkaline magma and ultimately in the creation of carbonatite melt and/or carbonic fluid. The association of carbonated hyperalkaline silicic and carbonatitic magmas is well known (see, for example, Le Bas, 1977).

PHASE RELATIONS OF HYDRATED CARBONIC PERIDOTITE

Figure 1 shows the schematic phase relations of hydrated and carbonated peridotite from Olafsson and Eggler (1983). At pressures above 17 kbar, carbonate is stable on the peridotite solidus; below 22 kbar, amphibole is stable. Consequently the subsolidus mineral assemblage in this system may be divided into three general types: carbonate-peridotite (P >22 kbar); carbonate-amphibole-peridotite ($17 < P < 22$ kbar); amphibole-peridotite ($P < 17$ kbar). The carbonate-in and amphibole-in curves are associated with displacements of the solidus to higher temperatures at lower pressures.

A carbonated alkaline magma, formed by low degrees of partial melting of volatile-bearing peridotite at $P>22$ kbar, will, upon rising, cross the H_2O-CO_2-peridotite solidus. The magma will be above its own solidus and so will not freeze. The margins of the magma will be in intimate contact with the host peridotite, however, so melt-solid interactions will occur to armor the conduit. The extent of the reaction will depend on the temperature of the system and on the amount of wallrock with which a given mass of magma interacts.

Two possible scenarios are feasible. In the first, a magma formed at point V (27 kbar, 1,070°C) on Figure 1 rises to cross the H_2O-CO_2-peridotite solidus at W (22 kbar, 1,040°C), i.e., within the carbonate and amphibole stability fields. The extent to which the melt and peridotite interact is controlled primarily by the temperature of the system and the effective rock-melt ratio. If the temperature of the system is well above the solidus (≃1,080°C), then only slight modifications of mantle and magma will occur. Furthermore, if the magmatic conduit is wide, so that only a small portion of the melt is in contact with the wallrock, effects on magma and peridotite will be limited. In such a case, a thin, highly modified border zone might be expected to form. Intimate contact of melt and host-rock at relatively low temperatures will allow formation of carbonates, amphiboles, and other

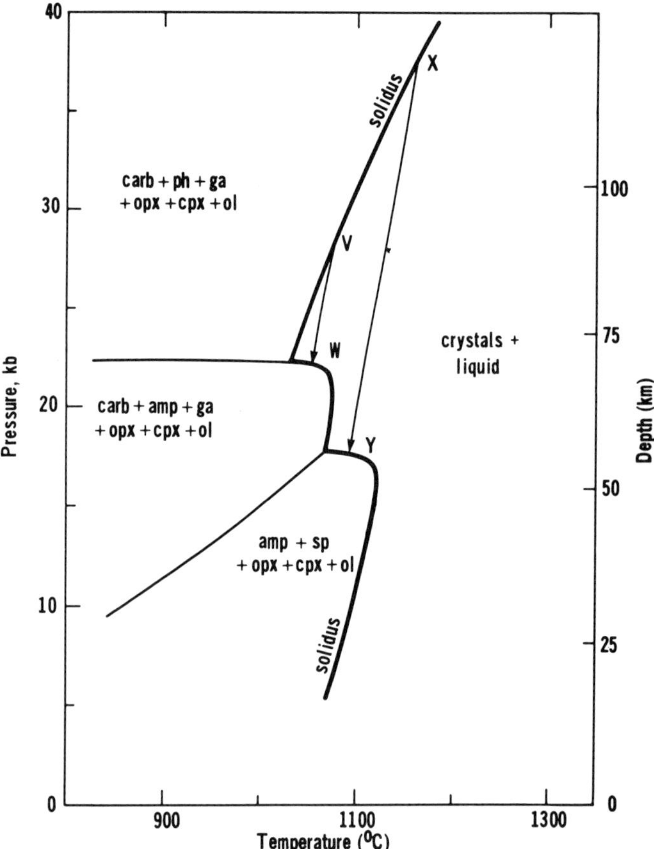

Figure 1. Pressure-temperature (P-T) projection of phase relations of peridotite in presence of small amounts of H_2O and CO_2; after Olafsson and Eggler (1983). Solidus is applicable to volatile composition (molar) $CO_2/(CO_2+H_2O)$ of 0.1 to 0.8, and to volatile contents less than about 0.4 wt %. A low-temperature partial melt of peridotite at V will, on rising, intersect the peridotite-H_2O-CO_2 solidus at W, whereas a partial melt at X will intersect the solidus at Y. Abbreviations used: ol, olivine; opx, orthopyroxene; cpx, Ca-rich clinopyroxene; ga, garnet; ph, phlogopite; amp, amphibole; sp, spinel; carb, carbonate.

volatile-bearing phases. Complete reaction may occur to produce a carbonated and amphibolitized peridotite, and to completely consume any melt or fluid phase.

In the second scenario, a magma created at point X (38 kbar, 1,180°C) rises to cross the solidus at a point (Y: 17 kbar, 1,090°C) at which carbonate is not a stable phase. Interactions at relatively high temperatures and rock-melt ratios presumably will be similar to those at higher pressures. More extensive reactions will lead to enrichment of the melt in CO_2. The products of complete reaction will thus be a highly modified but carbonate-free peridotite or pyroxenite, and a CO_2-rich fluid phase. Schneider and Eggler (1984) have shown that such a fluid can dissolve only a small amount of major elements.

These hypothetical relations have been tested by phase equilibrium studies on the ijolite-harzburgite join, and are discussed below. Rocks of the ijolite series are composed almost entirely of nepheline and mafic minerals (Le Bas, 1977). Rocks with 0 to 30 percent of mafic minerals are termed urtite, and those with 30 to 70 percent are ijolites. More mafic rocks are melteigites and pyroxenites (>90% pyroxene). The rock chosen as a starting material here is an ijolite from the Oka Complex, dominated by nepheline and sodian augite (in the ratio 35:65). Although this rock has an Mg number (atomic Mg/[Mg + Fe^{2+}]) too low to be a mantle melt, it is considered a good model for a magma that has suffered some olivine fractionation from a small degree partial melt of a carbonated and alkali-bearing mantle (cf. Wendlandt and Eggler, 1980).

PHASE STUDIES ON THE IJOLITE-HARZBURGITE JOIN

Experimental Techniques

Starting materials. The ijolite chosen for this study was P-061 of Deines (1967) from the Oka Complex, Quebec. It was ground to <10 μm (determined by microscopic examination), and dried at 700°C for 1 hr. Analysis of the powder after drying (Table 1, column 1) showed it to be nominally free of H_2O and CO_2. The harzburgite used was artificially prepared from MgO, Fe_2O_3, and SiO_2, equivalent to a ratio of 3:1 olivine:orthopyroxene with ≈90 percent of the Mg end member (Table 1, column 5). The chemicals were ground together under acetone, and fired at 1 atm, 1,400°C, fO_2 = 10^{-12} bars, reground, refired, ground again, and fired for a final time.

Four starting compositions were employed in this study (Table 1). Besides ijolite, the other three were made by grinding together accurately known weights of ijolite and harzburgite (in weight ratios 1:1, 1:2, and 1:3) under acetone, and were mixed with a mass of silver oxalate equivalent to CO_2, amounting to 5 wt.% of the ijolite in the charge, and a small amount of graphite.

Experimental method. All experiments were accomplished in solid-media, high-pressure apparatus similar to that of Boyd and England (1960), using the piston-out technique. Talc-Pyrex assemblies with a 25.4-mm diameter were used. When these were calibrated against the decarbonation reaction enstatite + magnesite = forsterite + CO_2 (Newton and Sharp, 1975), it was found that a −11% pressure correction was needed (D. R. Baker, personal communication). Pressures are believed to be accurate to ±0.5 kbar. Temperatures were measured with Pt-$Pt_{90}Rh_{10}$ thermocouples and automatically controlled. No correction for the pressure effect on the emf output of the thermocouple was made. Temperatures are believed to be accurate to ±10°C.

Olafsson (1980) noted that the fO_2 of unbuffered hydrous samples in the equipment used was above the NiNiO-buffer and below the hematite-magnetite buffer. The retention of graphite in run products (see below) suggests that the fO_2 of the samples was controlled by graphite-CO_2 equilibrium. Such a condition is consistent with those suggested as applicable to the upper mantle by Eggler (1983).

TABLE 1. COMPOSITIONS OF MATERIALS EMPLOYED IN THIS STUDY*

	1†	2	3	4	5
SiO_2	41.2	43.3	43.9	44.3	45.3
TiO_2	0.39	0.20	0.13	0.10	0.00
Al_2O_3	15.1	7.55	5.03	3.78	0.00
FeO*§	7.56	7.88	7.99	8.04	8.20
MnO	1.16	0.58	0.39	0.29	0.00
MgO	6.20	26.4	33.1	36.4	46.5
CaO	16.7	8.35	5.57	4.18	0.00
Na_2O	6.75	3.38	2.25	1.69	0.00
K_2O	3.18	1.59	1.06	0.80	0.00
P_2O_5	0.47	0.24	0.16	0.12	0.00
S	1.20	0.60	0.40	0.30	0.00
Totals	99.9	100.0	100.0	100.0	100.0
CIPW Norms					
Or	---	---	1.00	4.73	---
Ab	---	---	---	2.22	---
An	1.51	0.73	0.49	0.36	---
Lc	14.7	7.37	4.13	---	---
Ne	30.9	15.5	10.3	6.55	---
Di	13.0	19.4	20.7	15.5	---
Hy	---	---	---	---	24.8
Ol	16.1	50.6	62.2	69.7	75.2
La	19.4	4.56	---	---	---
Mt	0.50	0.18	0.14	0.12	---
Il	0.74	0.38	0.25	0.19	---
Ap	1.03	0.52	0.35	0.26	---
Py	2.02	0.75	0.46	0.33	---
Totals	99.9	100.0	100.0	100.0	100.0

*Normative composition calculated with FeO:Fe_2O_3 calculated by equations of Sack and others (1980), using T = 1200°C for f_{O_2} = 10^{-12}.

†Analyst: H. Gong.

§FeO* represents total iron expressed as FeO.

Key to column heads:
1 - ijolite
2 - 50% ijolite, 50% harzburgite
3 - 33% ijolite, 67 % harzburgite
4 - 25% ijolite, 75% harzburgite
5 - harzburgite

The experimental charges were loaded in iron-presaturated Pt capsules. Saturation with Fe was accomplished by loading the capsules with wustite and running them in a 1-atm. furnace at 1,150°C and fO_2 = 10^{-12} bar for 8 hr. The wustite was removed and the capsules cleaned ultrasonically. The capsules were sealed for experimental runs and checked for weight loss during loading and welding. Run durations varied from 3 to 20 hr according to the temperature. Run data are presented in Table 2. The subsolidus runs (T = 1,050°C) were held at 1,300°C for 3 hr to induce melting, and then cooled to the run temperature at a rate of 50°C/hr.

Material characterization. Experimental charges were examined in polished sections by petrographic microscope and electron microprobe. An accelerating potential of 15 kV, approx-

TABLE 2. RESULTS OF EXPERIMENTAL RUNS

Solid Composition Harzburgite:Ijolite (wt)	CO2 (wt.%)	Temperature (°C)	Time (hr)	Stable Phase Assemblage
0	5.0	1400	3.0	ol,L
1	2.5	1400	3.0	ol,L
2	1.67	1400	3.0	ol,L
3	1.25	1400	3.0	ol,L
0	5.0	1300	5.0	ol, cpx, L
1	2.5	1300	5.0	ol, cpx, L
2	1.67	1300	5.0	ol, opx, L
3	1.25	1300	5.0	ol, opx, L
0	5.0	1200	9.0	ol, cpx, L
1	2.5	1200	9.0	ol, opx, cpx, L
2	1.67	1200	9.0	ol, opx, cpx, L
3	1.25	1200	9.0	ol, opx, cpx, L
0	5.0	1100	12.0	ol, opx, cpx, sp, L
1	2.5	1100	12.0	ol, opx, cpx, sp, L
2	1.67	1100	12.0	ol, opx, cpx, L
3	1.25	1100	12.0	ol, opx, cpx, L
0	5.0	1050	20.0*	ol, opx, cpx, sp, ne, ks, sa†
1	2.5	1050	20.0*	ol, opx, cpx, sp, ne, ks, sa†
2	1.67	1050	20.0*	ol, opx, cpx, sp, ne, ks, sa†
3	1.25	1050	20.0*	ol, opx, cpx, sp, ne§

*Includes 3 hr at 1300°C and 5 hr decreasing temperature to run conditions at 50°C/hr.

†Determined by x-ray diffrection. Assemblage may also include apatite and sulfide.

§Determined by x-ray diffrection. Assemblage may also include sanidine or kalsilite, apatite, and a sulfide.

Abbreviations:
ol - olivine
opx - orthopyroxene
cpx - Ca-rich clinopyroxene
sp - spinel
ne - nepheline
ks - kalsilite
sa - sanidine
L - melt phase

imately 1.2-μamp specimen current, and a 2-μm beam diameter were used when analyzing crystalline materials. Quenched liquids were analyzed with a defocused beam (10 μm in diameter) to minimize alkali devolatilization, and as far from any mineral grain as possible. Alkali counts were monitored for 200 sec of impingement of the electron beam at 20-sec intervals; no change in the number of counts was observed. Matrix corrections of Bence and Albee (1968) and Albee and Ray (1970) were applied to all microprobe data. Phase compositions used are averages of at least four different analyses (involving at least three grains).

The quenched melt in two samples (1,100°C—harzburgite:ijolite = 0 and 1) contained too much quench carbonate for accurate microprobe analysis. The finer-grained samples (the runs at 1,050° and 1,100°C) were examined by x-ray diffractometer. The 1,050°C runs contained extremely fine-grained (<1 μm) aggregates, dominated by feldspathoids, which are interpreted as representing pools of melt that formed at 1,300°C, and crystallized on cooling to 1,050°C.

Experimental Results

Part of the ijolite-harzburgite join was determined at 14 kbar; the phase relations are shown in Figure 2. Note that the composition coordinate in this diagram is the harzburgite-ijolite weight ratio, not weight fraction of the end members.

Olivine is the liquidus phase at all the compositions studied. The second phase to appear with continued crystallization is a pyroxene. The more harzburgite-rich compositions crystallize as hypersthene, whereas more alkalic compositions have clinopyroxene as a near-liquidus phase. At lower temperatures, the solid assemblage is composed of olivine and two pyroxenes. Spinel occurs at supersolidus conditions at more alkalic compositions but was not observed in the more "harzburgitic" supersolidus runs, although x-ray diffraction suggests its presence at all compositions at subsolidus conditions. The subsolidus assemblage includes, in addition to the spinel lherzolite assemblage, nepheline, kalsilite, and sanidine. The relatively high P_2O_5 and S contents of the ijolite suggest that apatite and pyrrhotite will crystallize near the solidus, although these were not observed in the run products, presumably because of their low modal amounts.

Table 3 gives the melt and solid compositions in each of the 14 runs in which analysis of the glass was possible. The compositions were determined by microprobe analysis, and melt analyses were normalized to 100%. Modal proportions of phases were determined by mass balance of all analyzed phases, using all

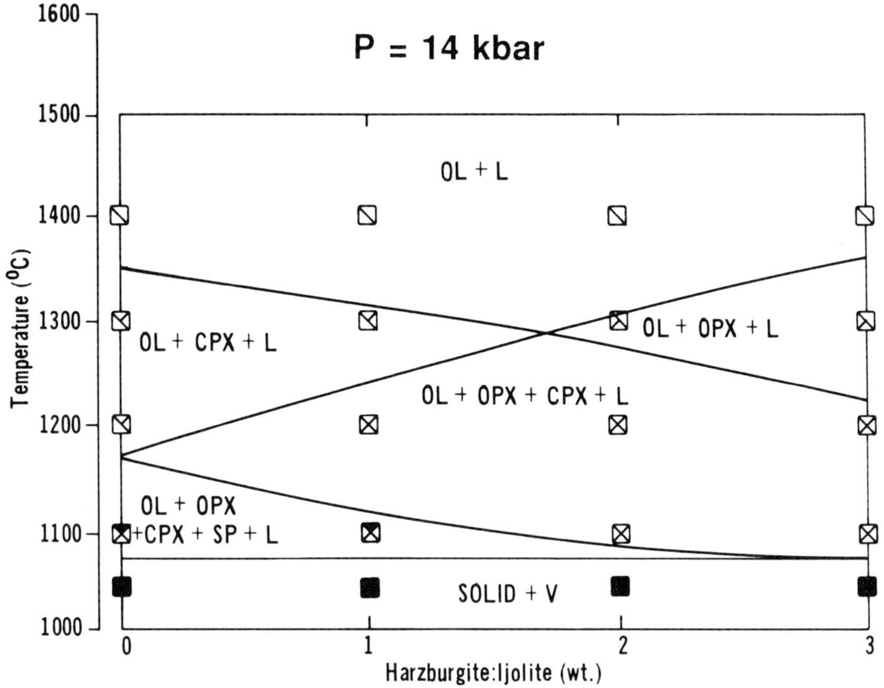

Figure 2. Phase relations of part of ijolite-harzburgite join at 14 kbar. Note that abscissa is weight ratio of harzburgite and ijolite.

elements determined. A simple petrologic mixing calculation was employed to achieve mass balance using a least-squares approach. The CO_2 content of the melt quoted assumes that all CO_2 in the system was dissolved in the melt (an untested presumption).

Calculated distribution coefficients of Fe/Mg between olivine and melt are close to 0.33 (Roeder and Emslie, 1970). Clinopyroxenes have 0.8–1.4% Na_2O and 4–7% Al_2O_3; orthopyroxenes contain 0.1–0.2% and 3–4% of these oxides. Consequently, the values of Na/K and Al/(Na + K) of the melts are lower at lower temperatures. Analyzed melts at 1,100°C (fraction of melt = 0.1–0.14) are thus potassic and strongly peralkaline. Graphite was present in all run products as determined either optically or by x-ray diffraction.

TABLE 3. COMPOSITIONS (IN WT%) OF MELTS AND CRYSTAL ASSEMBLAGES OBTAINED ON THE IJOLITE-HARZBURGITE JOIN (P = 14 kbar)*

Run Composition	Run Characteristics							
(Harzburgite: Ijolite-wt.)	0	1	2	3	0	1	2	3
T (oC)	1400	1400	1400	1400	1300	1300	1300	1300
Fraction Melt	0.957	0.588	0.403	0.352	0.772	0.405	0.348	0.307
	Composition of Melts							
% CO_2 (melt)	5.22	4.25	4.14	3.55	6.48	6.17	4.79	4.07
SiO_2	42.3	43.8	49.8	52.9	41.2	46.0	40.8	44.5
TiO_2	0.50	0.45	0.42	0.22	0.38	0.45	0.29	0.22
Al_2O_3	17.0	14.1	12.2	9.82	19.3	15.8	13.2	11.1
FeO*	6.68	7.38	7.10	9.21	5.92	6.12	6.01	8.79
MgO	5.75	10.4	9.49	10.6	2.55	10.2	10.2	15.2
CaO	18.5	14.8	12.5	10.2	17.4	9.93	15.2	12.1
Na_2O	6.83	6.13	5.54	4.52	8.71	7.58	6.57	5.16
K_2O	2.44	2.94	2.81	2.54	4.48	3.88	3.03	2.92
CIPW Norm								
Or	---	---	9.05	15.0	---	---	---	---
Ab	---	---	---	14.2	---	---	---	---
An	8.52	2.27	0.12	---	0.33	---	---	---
Kp	---	---	---	---	1.75	---	---	---
Lc	11.3	13.6	5.92	---	18.3	18.0	14.0	13.5
Ne	31.3	28.1	25.4	12.0	39.9	32.3	27.6	22.1
Ac	---	---	---	0.30	---	0.35	0.39	0.41
Ns	---	---	---	0.36	---	0.95	0.96	0.55
Di	12.7	19.2	50.1	41.2	31.9	24.3	9.70	21.8
Ol	14.1	21.1	8.31	16.5	---	17.3	22.4	30.9
Cs	21.0	14.7	---	---	6.79	5.92	19.6	10.2
Mt	0.15	0.13	0.13	---	0.29	---	---	---
Il	0.95	0.85	0.80	0.42	0.72	0.85	0.55	0.42
	Composition of Crystal Assemblages							
Olivine	100.0	100.0	100.0	100.0	22.4	70.3	80.5	77.5
Orthopyroxene	---	---	---	---	---	---	19.5	22.5
Clinopyroxene	---	---	---	---	77.6	29.7	---	---

*Melt analyses normalized to 100%. Percentage of CO_2 in melt calculated by mass balance. Normative compositions calculated with FeO:Fe_2O_3 calculated by equations of Sack and others (1980), using run temperature and $\underline{f}_{O_2}$ of graphite-CO_2 buffer.

RESULTS OF IJOLITE-PERIDOTITE INTERACTIONS

Compositions of Metasomatized Mantle

If the ijolite, which was employed for the experiments discussed above, is a reasonable model for a liquid formed by low degrees of partial melting at pressures in excess of 22 kbar, then the phase relations can be used to model the effects of melt-peridotite interaction at $P < 17$ kbar. The products of reaction will depend on the temperature of reaction and the effective rock-melt ratio. The effects of the last two parameters on the composition of the modified peridotite are shown in Figure 3 for a pressure of less than 17 kbar (i.e., carbonate not stable). The abscissa in Figure 3 is the ratio of harzburgite to ijolite (i.e., the effective peridotite-melt ratio). Interaction at higher temperatures (for example, along the margins of major conduits) will result in some modal metasomatism (Dawson, 1984) to produce a mineral assemblage enriched in clinopyroxene. More extensive reaction (at lower temperatures) will create an even more altered mineralogy, including carbonates (if $P > 17$ kbar).

No amphibole was recognized in the nominally anhydrous run products. Olafsson (1980) found in melting experiments in the system peridotite-H_2O-CO_2 at 13.5 kbar that amphibole appeared only when $X_{fluid}^{CO_2}$(mole) <0.85. It melts over a very narrow temperature range unless $X_{fluid}^{CO_2} < 0.10$. A relatively small amount of H_2O would stabilize amphibole in place of the feldspathoids, but would otherwise leave the phase relations in Figure 2 unaltered. Amphiboles would fix the halides of the melt as well as the water, and might have relatively high values of (F +

TABLE 3 (CONTINUED)*

Run Composition	Run Characteristics					
(Harzburgite: Ijolite-wt.)	0	1	2	3	2	3
T (°C)	1200	1200	1200	1200	1100	1100
Fraction Melt	0.751	0.265	0.228	0.133	0.136	0.104
	Composition of Melts					
% CO_2 (melt)	6.66	9.43	7.31	9.40	12.3	12.0
SiO_2	41.9	42.2	44.2	44.7	45.2	44.7
TiO_2	0.40	0.52	0.42	0.53	0.33	0.42
Al_2O_3	19.2	21.2	18.2	22.5	25.2	24.7
FeO*	5.61	4.23	4.42	2.76	1.72	1.25
MgO	2.81	6.62	9.88	5.11	2.99	3.32
CaO	16.6	6.70	9.10	6.99	2.12	6.70
Na_2O	8.99	12.1	9.32	11.3	14.2	12.4
K_2O	4.49	6.35	4.51	6.15	8.26	8.82
CIPW Norm						
Or	---	---	---	---	---	---
Ab	---	---	---	---	---	---
An	---	---	---	---	---	---
Kp	0.39	11.6	---	---	8.72	13.9
Lc	20.3	13.4	20.9	28.5	26.2	21.7
Ne	40.0	39.9	37.1	44.1	45.3	42.2
Ac	0.94	0.56	0.47	0.37	0.50	0.38
Ns	0.29	6.53	2.29	3.19	8.37	6.18
Di	23.8	19.8	6.49	0.41	6.15	12.5
Ol	---	7.36	11.5	11.8	4.13	---
Cs	13.5	---	20.6	10.6	---	2.29
Mt	---	---	---	---	---	---
Il	0.76	0.99	0.80	1.01	0.63	0.80
	Composition of Crystal Assemblages					
Olivine	13.5	54.9	64.6	67.5	64.1	66.2
Orthopyroxene	---	1.53	15.8	16.1	6.70	17.2
Clinopyroxene	86.5	43.6	19.6	16.4	29.2	16.6

*Melt analyses normalized to 100%. Percentage of CO_2 in melt calculated by mass balance. Normative compositions calculated with $FeO:Fe_2O_3$ calculated by equations of Sack and others (1980), using run temperature and f_{O_2} of graphite-CO_2 buffer.

Cl)/H_2O. Such amphiboles may contain CO_2-rich fluid inclusions and show other evidence of having crystallized in the presence of a CO_2-rich fluid. Such amphiboles have been described from mantle xenoliths by Andersen and others (1984).

The phase relations on the ijolite-harzburgite join at pressures in the region of 17 to 22 kbar were not determined in this study. It is considered, however, that they would differ only in a few respects from those obtained at $P < 17$ kbar. The most important difference is the stability of a carbonate phase at the higher pressures. Eggler and Wendlandt (1979) determined the melting relations of a kimberlite at pressures of 30 and 55 kbar as a function of the volatile constituents ($CO_2 = 5.2$ wt.%; H_2O varied from 0 to 10 wt.%). These runs showed that carbonate was a supersolidus mineral at all except the most hydrous conditions, and that, if $X^{CO_2}_{system} > \simeq 0.5$, carbonate persisted to temperatures more than 100° above the solidus. Olafsson and Eggler (1983) found that carbonate persisted above the solidus of Figure 1 ($P > 17$ kbar). The presence of carbonate fixes CO_2 in the solid assemblage, thus removing it from the melt. The melt is thus relatively enriched in H_2O, and hydrous minerals (phlogopite in kimberlite of Eggler and Wendlandt, 1979; amphibole in peridotite of Olafsson and Eggler, 1983) may crystallize near the solidus even with $X^{CO_2}_{system}$ near 1. Consequently, the entire hyperalkaline magma, including the H_2O and CO_2, may be fixed in the mantle at pressures of 17 to 22 kbar.

Chemical Compositions of Fugitive Phases Following Metasomatism

The chemical composition of the melt or fluid in the system ijolite-harzburgite must vary antithetically with that of the solid assemblage (Table 3). If the ijolite-harzburgite system were at a high temperature and if the rock-melt ratio were also high, modification of the melt would occur primarily by dissolution of orthopyroxene of the wallrock, resulting in decreases in the degree of silica undersaturation and in the alkali content of the melt. At lower temperatures (if $P<17$ kbar), the magma becomes increasingly alkalic and carbonated due to extraction of augite from the melt. At temperatures immediately above the solidus, the melt is rich in the feldspathoid components (an urtite, according to the definitions of Le Bas, 1977). In the system studied, the final melts may have had >10 wt.% CO_2. This amount is merely a reflection of the initial starting conditions, particularly of the value of the ratio of (Na_2O + K_2O) to CO_2.

Koster van Groos (1975) suggested that the miscibility gap between silicate and carbonatitic fluids, which was well known in simple systems at low pressures (Koster van Groos and Wyllie, 1966, 1968, 1973), persisted into more complex systems at 10 kbar with bulk CO_2 >25%. Liquid immiscibility may occur near the solidus in the join ijolite-harzburgite if the CO_2 of the melt is sufficiently elevated. Presumably, the final liquid will, in all cases, unmix. However, if the CO_2 content of the initial melt were low, then the amount of carbonatite melt formed would be extremely small. In this case, the total reaction of urtitic melt and peridotite is to produce only feldspathoidal (or amphibole-) peridotite and CO_2-fluid. If the initial magma were sufficiently carbonated, the amount of carbonatite would be significant, and this phase concentrates $CaCO_3$ and Na_2CO_3, whereas the silicate phase retains the MgO, SiO_2, Fe_2O_3, K_2O, and Al_2O_3 (more syenitic in composition). The carbonatitic melt will be buoyant and at a temperature significantly above its solidus, so it will probably rise out of the system. In this case, the remaining urtitic melt may react with the solid assemblage to produce an alkali-enriched mantle, although this mantle will have a considerably higher value of K_2O:Na_2O than did the original ijolite.

The CO_2-rich $X^{CO_2}_{fluid}$ >0.85 at 13.5 kbar) will rise to shallower depths, and cannot be fixed in the solid assemblages until it reaches the crust. It thus has potential to cause some metasomatic alteration at higher levels of the mantle. Schneider and Eggler (1984) have shown, however, that carbonic fluids can carry only small amounts of the major elements, although they may be capable of transporting large amounts of some trace elements (e.g., light rare-earth elements: Mysen, 1983, Wendlandt and Harrison, 1979). Consequently, carbonic metasomatism will affect only trace element concentrations in peridotite, leaving the major element and mineralogic constitutions of the latter essentially unaltered (cryptic metasomatism in the terminology of Dawson, 1984).

Products of Melting of Metasomatized Mantle

It has been shown that the reaction of ijolite and peridotite may produce a wide array of mantle mineralogies. The products of high-temperature metasomatism will be veins of varying width that are enriched in clinopyroxene over the surrounding depleted mantle. Reheating of such veined mantle will produce magmas that would be somewhat enriched in CaO, Na_2O, and certain trace elements over a "normal" mantle melt. Such melting will, in the absence of outside influences, be at the volatile-free solidus (as it will be of olivine, pyroxenes, and an aluminous phase only), and the required temperatures (1,400° to 1,500°C in the pertinent pressure range; Kushiro and others, 1968) may not be commonly realized in stable lithosphere.

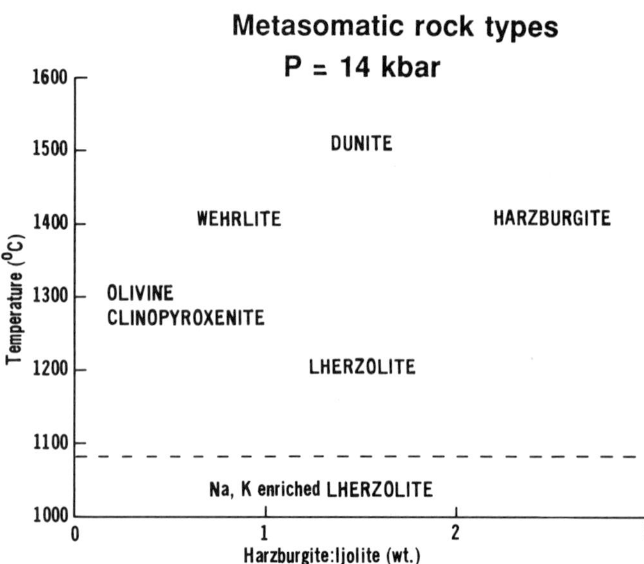

Figure 3. Compositions of solid assemblages coexisting with fugitive phase on join ijolite-harzburgite at 14 kbar (see Fig. 2). Supersolidus assemblages named according to Streckeisen (1980). Dashed line is trace of solidus from Figure 2.

If reaction occurs at $17 < P < 22$ kbar and is extensive enough, carbonate-lherzolite (with or without amphibole) is formed. Later heating of these metasomatized regions of the mantle will produce carbonated alkalic magmas (carbonatitic or nephelinitic). Essentially the carbonated alkaline magma has been stored in the mantle for a certain length of time, and may be released by heating. At $P<17$ kbar, metasomatism by strongly carbonated magma produces an assemblage enriched in augite and alkali-bearing phases. This assemblage may melt at volatile-free conditions to an alkalic magma. Hydrous carbonic magma can, however, produce amphibole, at least upon complete crystallization of the melt. Such a mantle may melt at ≃1,100°C to a basanitic magma (Olafsson, 1980, p. 37).

Mantles metasomatized by CO_2 fluids ($P<17$ kbar) will not change their mineralogy, and so will melt at a similar temperature and to the same products after metasomatism as they would have done before it occurred.

The trace element compositions of all except one of the mantles described above are controlled by the partitioning of each element between the mantle minerals and the melt or fluid (the values for such partitioning are not known). The one exception to this is in the case of complete melt-peridotite reaction at $17 < P < 22$ kbar, in which the mineral assemblage formed will possess enrichments in the incompatible trace elements similar to those in the metasomatizing magma. (The initial peridotite would have been strongly depleted in such elements.) Carbonated alkaline magmas show extreme enrichment in the light rare-earth elements over heavy rare-earth elements (Loubet and others, 1972, measured La-Yb ratios at 60 to 740 for various carbonatites), and in U and Th over Pb, but the extreme enrichment in the alkaline earth elements means that Rb-Sr ratios are not elevated greatly over that of bulk mantle. The metasomatized mantle will thus develop, over time, relatively low $^{143}Nd/^{144}Nd$, moderate $^{87}Sr/^{86}Sr$, and high Pb isotopic ratios. Such compositions have been described by Barreiro (1983; Barreiro and Cooper, this volume) for Tertiary lamprophyres and carbonatites from New Zealand, and she suggests, on the basis of Pb isotopic compositions, that the U enrichment occurred within the last few hundred million years. Bell and others (1982) have shown that a number of alkaline and carbonatitic bodies from the Canadian Shield define an $^{87}Sr/^{86}Sr$ growth curve that suggests a lower Rb-Sr ratio for the source mantle than "bulk" mantle for the last 2.8 b.y. These two groups of magmas could have been formed from sources 55 to 70 km deep that had been extensively metasomatized by carbonated nephelinitic magma in the Paleozoic and Archean, respectively. (Note that carbonated magmas formed by melting at $17 < P < 22$ kbar may, on rising, interact with the mantle at $P < 17$ kbar in exactly the same manner as outlined above and undergo evolution from ijolite through urtite to carbonatite.)

SUMMARY

The shape of the peridotite-H_2O-CO_2 solidus suggests certain models for the interaction of hydrous and carbonated alkaline magmas and peridotite. Some magmas, on rising from depths greater than 70 km, may intersect the solidus at $T = 1{,}000°$ to 1,050°C, $P = 22$ kbar (amphibole and carbonate stable), or at T = 1,050° to 1,100°C, $P = 17$ kbar (amphibole stable, carbonate not stable). Melt-rock reactions are similar at both *P-T* regimes, if the effective volume of peridotite that reacts with the magma is small and the temperature is high, with clinopyroxene formed at the expense of orthopyroxene and olivine. At lower values of melt/rock and of temperature, the products of reaction differ according to the pressure. At $P > 17$ kbar, carbonate joins the mineral assemblage and the magma may be ultimately consumed to produce carbonated garnet lherzolite containing amphibole, apatite, sulfides, oxides, and, in some cases, phlogopite or sanidine. The isotopic composition of this mantle will be displaced to relatively low $^{143}Nd/^{144}Nd$, moderate $^{87}Sr/^{86}Sr$, and high Pb isotopic ratios. The mantle described may later melt to give a carbonated nephelinite or basanite of similar isotopic characteristics. At least some of these characteristics have been reported for alkaline rocks from New Zealand (Barreiro, 1983; Barreiro and Cooper, this volume), Australia (Menzies and Wass, 1983), the western United States (Vollmer and others, 1981; Dudas and others, 1985), and the Canadian Shield (Bell and others, 1982).

At $P < 17$ kbar, extensive melt-rock reaction produces enrichment of the alkali elements and CO_2 in the magma. The carbonated feldspathoidal magma produced at temperatures only a little above the solidus will unmix into a silicate melt and a carbonatitic melt if it is sufficiently carbonated. In this way, a complete transition from ijolite to carbonatite may be produced. If the CO_2 content of the magma is too low for liquid immiscibility to occur, extensive reaction of melt and peridotite produces an enriched mantle (including amphibole if $X_{system}^{CO_2} < 0.85$), and CO_2-fluid. The peridotite may later melt to an alkalic (low in CO_2) magma. The carbonic fluid may result in cryptic metasomatism at shallow levels.

The rise of H_2O-CO_2-bearing alkaline magmas is clearly a powerful method of creating a number of enriched mantle sources. The metasomatism required to create these assemblages leads to increased carbonation of the alkaline magma (particularly at $P < 17$ kbar), and, at $17 < P < 22$ kbar, it can produce a potential carbonatite source rock. The production of carbonatites and the form of metasomatism described above occur in a parallel manner over a wide pressure range but are both products of a single, although complex, process—interaction of hyperalkaline magma and peridotite—and neither is related in a simple manner to the other.

ACKNOWLEDGMENTS

Much of the development of this model was conducted while I was a predoctoral fellow at the Department of Terrestrial Magnetism, Carnegie Institution of Washington, in discussions with B. A. Barreiro and R. W. Carlson. I thank G. Wetherill and the staff of DTM for the opportunity to work there. The experimental work was aided by E. J. Fregeau. Various versions of this manuscript were reviewed by D. H. Eggler, B. A. Barreiro, D. B. Joyce, R. W. Carlson, A. H. Treiman, and V. R. Murthy, whose comments are appreciated. I am grateful to P. Deines for supplying the ijolite starting mateirals. The study was supported by grant EAR 8206769 from the National Science Foundation to D. Eggler.

REFERENCES CITED

Albee, A. L., and Ray, L., 1970, Correction factors for electron probe microanalysis of silicates, oxides, carbonates, phosphates, and sulfates: Analytical Chemistry, v. 42, p. 1408–1414.

Andersen, T., O'Reilly, S. Y., and Griffin, W. L., 1984, The trapped fluid phase in upper mantle xenoliths from Victoria, Australia; Implications for mantle metasomatism: Contributions to Mineralogy Petrology, v. 88, p. 72–85.

Barreiro, B. A., 1983, An isotopic study of Westland dike swarm, South Island, New Zealand: Carnegie Institution Washington Yearbook, v. 82, p. 471–475.

Bell, K., Blenkinsop, J., Cole, T.J.S., and Menagh, D. P., 1982, Evidence from Sr isotopes for long-lived heterogeneities in the upper mantle: Nature, v. 298, p. 251–253.

Bence, A. E., and Albee, A. L., 1968, Empirical correction factors for the electron microanalysis of silicates and oxides: Journal of Geology, v. 76, p. 382–403.

Boyd, F. R., and England, J. L., 1960, Apparatus for phase equilibrium measurements at pressures up to 50 kilobar and temperatures up to 1750°C: Journal of Geophysical Research, v. 65, p. 741–748.

Cohen, R. S., and O'Nions, R. K., 1982, Identification of recycled continental material in the mantle from Sr, Nd, and Pb isotope investigations: Earth and Planetary Science Letters, v. 61, p. 73–84.

Dawson, J. B., 1984, Contrasting types of upper-mantle metasomatism?, *in* Kornprobst, J., ed., Kimberlites II; The mantle and crust–mantle relationships: Amsterdam, Elsevier, p. 289–294.

Deines, P., 1967, Stable carbon and oxygen isotopes of carbonatite carbonates and their interpretation [Ph.D. thesis]: University Park, Pennsylvania State University, 230 p.

Dudas, F. O., Carlson, R. W., and Eggler, D. H., 1985, Ancient enriched mantle sources for magmatism in the Crazy Mountains, Montana: EOS American Geophysical Union Transactions, v. 66, p. 414.

Eggler, D. H., 1983, Upper mantle oxidation state; Evidence from olivine-orthopyroxene-ilmenite assemblages: Geophysical Research Letters, v. 10, p. 365–368.

——, 1986, Solubility of major and trace elements in mantle metasomatic fluids; Experimental constraints, *in* Menzies, M. A., and Hawkesworth, C. J., eds., Mantle metasomatism: London, Academic Press (in press).

Eggler, D. H., and Wendlandt, R. F., 1979, Experimental studies on the relationship btween kimberlite magmas and partial melting of peridotite, *in* Boyd, F. R., and Meyer, H.O.A., eds., Kimberlites, diatremes, and diamonds: Washington, D.C., American Geophysical Union, p. 330–338.

Fregeau, E. J., 1985, Trace element partitioning between a silicate melt and a super-critical hydrous fluid; Implications to mantle metasomatism [M.S. paper]: University Park, Pennsylvanian State University, 71 p.

Koster van Groos, A. F., 1975, The effect of high CO_2 pressures on alkalic rocks and its bearing on the formation of alkalic ultrabasic rocks and the associated carbonatites: American Journal of Science, v. 275, p. 163–185.

Koster van Groos, A. F., and Wyllie, P. J., 1966, Liquid immiscibility in the system Na_2O-Al_2O_3-SiO_2-CO_2 at pressures to 1 kilobar: American Journal of Science, v. 264, p. 234–255.

——, 1968, Liquid immiscibility in the join $NaAlSi_3O_8$-Na_2CO_3-H_2O and its bearing on the genesis of carbonatites: American Journal of Science, v. 266, p. 932–967.

——, 1973, Liquid immiscibility in the join $NaAlSi_3O_8$-$CaAl_2Si_2O_8$-Na_2CO_3-H_2O: American Journal of Science, v. 273, p. 465–487.

Kushiro, I., Syono, Y., and Akimoto, S., 1968, Melting of peridotite nodule at high pressure and high water pressure: Journal of Geophysical Research, v. 73, p. 6023–6029.

Le Bas, M. J., 1977, Carbonatite-nephelinite volcanism; An African case history: New York, John Wiley & Sons, 347 p.

Loubet, M., Bernat, M., Javoy, M., and Allegre, C. J., 1972, Rare earth contents in carbonatites: Earth and Planetary Science Letters, v. 14, p. 226–232.

Meen, J. K., 1983, Isotopic composition of some Laramide volcanics, Absaroka Mountains, Montana: Carnegie Institution Washington Yearbook, v. 82, p. 481–486.

——, 1985, The origin and evolution of a continental volcano; Independence, Montana [Ph.D. thesis]: University Park, Pennsylvania State University, 917 p.

Menzies, M. A., and Murthy, V. R., 1980, Nd and Sr isotope geochemistry of hydrous mantle nodules and their host alkali basalts; Implications for local heterogeneities in metasomatically veined mantle: Earth and Planetary Science Letters, v. 46, p. 323–334.

Menzies, M., and Wass, S. Y., 1983, CO_2- and LREE-rich mantle below eastern Australia; A REE and isotopic study of alkaline magmas and apatite-rich mantle xenoliths from the Southern Highlands Province, Australia: Earth and Planetary Science Letters, v. 65, p. 287–302.

Mysen, B. O., 1983, Rare earth element partitioning between ($H_2O + CO_2$) vapor and upper mantle minerals; Experimental data bearing on the conditions of formation of alkali basalts and kimberlite: Neues Jahrbuch für Mineralogie Abhandlungen, v. 146, p. 41–65.

Newton, R. C., and Sharp, W. E., 1975, Stability of forsterite + CO_2 and its bearing on the role of CO_2 in the mantle: Earth and Planetary Science Letters, v. 26, p. 239–244.

Olafsson, M., 1980, Partial melting of peridotite in the presence of small amounts of volatiles, with special reference to the low-velocity zone [M.S. thesis]: University Park, Pennsylvania State University, 59 p.

Olafsson, M., and Eggler, D. H., 1983, Phase relations of amphibole, amphibole-carbonate, and phlogopite-carbonate peridotite; Petrologic constraints on the asthenosphere: Earth and Planetary Science Letters, v. 64, p. 305–315.

Roeder, P. L., and Emslie, R. F., 1970, Olivine-liquid equilibrium: Contributions to Mineralogy and Petrology, v. 29, p. 275–289.

Sack, R. O., Carmichael, I.S.E., Rivers, M., and Ghiorso, M. S., 1980, Ferric-ferrous equilibria in natural silicate liquids at 1 bar: Contributions to Mineralogy and Petrology, v. 75, p. 369–376.

Schneider, M. E., and Eggler, D. H., 1984, Compositions of fluids in equilibrium with peridotite; Implications for alkaline magmatism-metasomatism, *in* Kornprobst, J., ed., Kimberlites. I; Kimberlites and related rocks: Amsterdam, Elsevier, p. 383–394.

Streckeisen, A., 1980, Classification and nomenclature of volcanic rocks, lamprophyres, carbonatites, and melilitic rocks; IUGS subcommission on the systematics of igneous rocks: Geologische Rundschau, v. 69, p. 194–207.

Vidal, P., Chauvel, C., and Brousse, R., 1984, Large mantle heterogeneity beneath French Polynesia: Nature, v. 307, p. 536–538.

Vollmer, R., Ogden, P., and Schilling, J. G., 1981, Nd isotope geochemistry of Leucite Hills ultra-potassic volcanism: EOS American Geophysical Union Transactions, v. 62, p. 431.

Wendlandt, R. F., and Eggler, D. H., 1980, The origins of potassic magmas; 1. Melting relations in the systems $KAlSiO_4$-Mg_2SiO_4-SiO_2 and $KAlSiO_4$-MgO-SiO_2-CO_2 to 30 kilobars: American Journal of Science, v. 280, p. 385–420.

Wendlandt, R. F., and Harrison, W. J., 1979, Rare earth partitioning between immiscible carbonate and silicate liquids and CO_2 vapor; Results and implications for the formation of light rare earth-enriched rocks: Contributions to Mineralogy and Petrology, v. 69, p. 409–419.

MANUSCRIPT ACCEPTED BY THE SOCIETY OCTOBER 14, 1986

Printed in U.S.A.

Geological Society of America
Special Paper 215
1987

Magma mixing and kimberlite genesis; Mineralogic, petrologic, and trace element evidence from eastern U.S.A. kimberlites

John W. Shervais
Department of Geological Sciences, University of Tennessee, Knoxville, Tennessee 37996, and Department of Geology, University of South Carolina, Columbia, South Carolina 29208
Lawrence A. Taylor
Department of Geological Sciences, University of Tennessee, Knoxville, Tennessee 37996
J. C. Laul
Department of Radiological Sciences, Battelle Northwest Laboratories, Richland, Washington 99253

ABSTRACT

Kimberlites in the eastern United States contain two suites of megacrysts/inclusions that are mineralogically similar but compositionally distinct. One suite (olivine, garnet, diopside, Cr-spinel) has higher Cr and Mg than the other (olivine, garnet, diopside, picroilmenite). Based on detailed petrologic studies of megacrysts from the Fayette County, Pennsylvania, kimberlite, Hunter and Taylor (1984) suggested that these two suites represent the crystallization products of separate magmas that mixed in the low-velocity zone (LVZ) to form kimberlite magma.

Major and trace element abundances of individual garnet megacrysts from eastern U.S. kimberlites (i.e., from Kentucky, New York, Pennsylvania, and Tennessee) support the magma-mixing hypothesis but also indicate additional complications. Eclogite garnets have Cr_2O_3 <0.3 wt.%, CaO >7 wt. %, and chondrite-normalized Lu/Hf <<1. Peridotite garnets have Cr_2O_3 >2 wt.%, MG# >83, and chondrite-normalized Lu/Hf <1. Garnet megacrysts from Kentucky and Pennsylvania form two groups, one with TiO_2 <0.5 wt.%, and one with TiO_2 >0.5 wt.%. Both groups span a similar range in Cr_2O_3 (≅1.0 to 9.0 wt.% Cr_2O_3), but the high-Ti garnets may have Cr_2O_3 as low as 0.1 wt.%. The low-Ti garnets have chondrite-normalized Lu/Hf <1 and are probably derived by the disaggregation of peridotite xenoliths and wall rock. The high-Ti garnet megacrysts have chondrite-normalized Lu/Hf ⩾1 and are interpreted here as cognate "phenocrysts" that crystallized in a kimberlite or proto-kimberlite magma. Two suites of high-Ti garnet megacrysts are recognized: a low-Cr to very low-Cr suite (Cr_2O_3 <4 wt.%) with flat to slightly positive heavy rare-earth element (HREE) slopes, and a high-Cr suite with steeply negative HREE slopes. These suites correspond to the "Cr-poor" and "Cr-rich" suites, respectively, defined by Hunter and Taylor (1984) for the Pennsylvania kimberlite. These data are consistent with the mixing of two magma batches to form kimberlite, as proposed by Hunter and Taylor (1984). Mixing prpobably occurred in the LVZ prior to eruption of the hybrid kimberlite magma.

INTRODUCTION

Kimberlite is a volatile-rich, potassic, ultrabasic igneous rock that occurs most commonly in continental areas underlain by ancient cratons. Although relatively common in some cratons (e.g., South Africa, Siberia), kimberlites are rare in the United States. Diatremes and dikes from many locations that were originally classified as kimberlites or "kimberlitic" are now considered to represent varieties of lamproites (e.g., Prairie Creek, Arkansas: Mitchell and Lewis, 1983; Scott-Smith and Skinner, 1984a, 1984b) or lamprophyres (various diatremes in the Four Corners region: Ehrenberg and others, 1984; Smith and Levy, 1976). Kimberlite crops out in at least four locations in the Cumberland Plateau region of the eastern United States (Fig. 1), forming a chain parallel to, and west of, the axis of the Appalachian orogen (Meyer, 1975, 1976; Taylor, 1984). From north to south these are: Ithaca, New York (Jackson and others, 1982; Kay and others, 1983), Fayette County, Pennsylvania (Hunter and Taylor, 1982, 1984; Hunter and others, 1984), Elliott County, Kentucky (Garrison and Taylor, 1980; Agee and others, 1982; Schulze, 1984; Shervais and Taylor, 1984), and Norris, Tennessee (Taylor and others, 1983). These occurrences provide a unique opportunity to study mantle-derived igneous rocks and the subcontinental mantle of the eastern United States.

Kimberlite at these locations occurs as both dikes (New York, Pennsylvania) and diatremes (Kentucky, Tennessee). The previous investigations cited above and our own unpublished data show that, like kimberlites elsewhere, they contain a wide variety of both crustal and mantle xenoliths, as well as megacrysts and macrocrysts. Crustal xenoliths include altered shale, limestone, and sandstone. The most common mantle-derived xenoliths are garnet peridotite and eclogite; spinel lherzolite is also found.

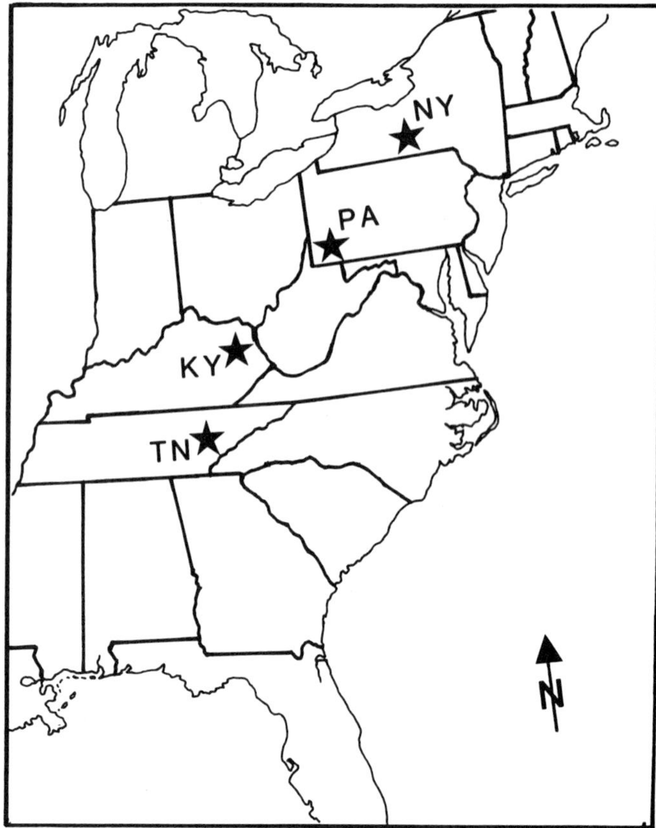

Figure 1. Map showing location of eastern U.S. kimberlites studied here. NY refers to Ithaca site; PA, to Fayette County site; KY, to Elliott County site; TN, to Norris site.

Megacrysts and macrocrysts of olivine, pyroxene, garnet, phlogopite, and picroilmenite are a characteristic feature of kimberlites worldwide and are also found in the kimberlites of the eastern United States. Because of the (auto?)metasomatic alteration that is ubiquitous during and after kimberlite emplacement, olivine megacrysts are commonly altered to serpentine; in highly altered kimberlite, garnet and ilmenite may be the only megacrysts preserved.

Megacrysts are generally defined as anhedral or rounded single crystals larger than 1 to 2 cm in diameter (Nixon and Boyd, 1973; Dawson, 1980). Ilmenite-silicate intergrowths are included in this term (e.g., Nixon and Boyd, 1973). Megacrysts may be as large as 15 cm in their maximum dimension, but most are between 0.5 to 3 cm (Dawson, 1980). "Macrocryst" is a relatively new term applied to anhedral insets of these same minerals as large as 0.5 cm in diameter (Clement and others, 1984). Although these terms are specified to be nongenetic, "macrocrysts" are commonly construed to represent disaggregated mafic and ultramafic xenoliths (Clement and others, 1984). Megacrysts from kimberlites in the eastern U.S. are typically 0.5 to 1.5 cm in size, but are rarely as large as 3 cm (Garrison and Taylor, 1980; Hunter and Taylor, 1984). While these are near the lower size limit of typical megacrysts from other kimberlites, the eastern U.S. megacrysts are larger than typical "macrocrysts" or the constituent grains of coexisting xenoliths, and many are compositionally similar to megacrysts from other kimberlites. The garnet analyses reported in Table 2 include 2 garnets from eclogites, 1 from a garnet peridotite, and 17 "megacrysts." We use "megacryst" here in its nongenetic sense, which includes both large cognate crystals and possible xenocrysts.

The origin of these megacrysts has been the subject of considerable debate: are they phenocrysts that crystallized from a kimberlitic or proto-kimberlitic magma (e.g., Gurney and others, 1979; Eggler and others, 1979; Boyd and Nixon, 1973; Nixon and Boyd, 1973; Harte and Gurney, 1981; Garrison and Taylor, 1980; Hunter and Taylor, 1984)? Or do they represent disaggregated mantle wallrock from the kimberlite magma conduit (Meyer and McCallister, 1984; Kempton and others, 1984)?

In this chapter, we review briefly the petrology and mineral chemistry of these four kimberlites, and then present new data on the major and trace element geochemistry of garnet megacrysts and the kimberlite matrix. We interpret these data to show that,

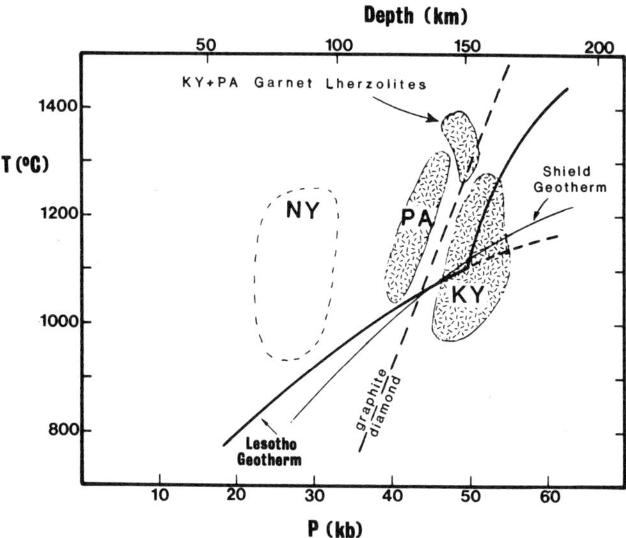

Figure 2. Calculated *P-T* regimes of eastern U.S. kimberlites studied here. Data from Garrison and Taylor, 1980; Jackson and others, 1982; Hunter and Taylor, 1984.

in addition to xenocrysts formed by disaggregation of both eclogite and garnet lherzolite, there are two suites of megacrysts that formed as phenocrysts in separate proto-kimberlite magmas that mixed prior to intrusion into the crust. The major and trace element systematics of the garnet megacrysts suggest that these garnets correspond to the Cr-poor and Cr-rich megacryst suites defined by Hunter and Taylor (1984) for the Fayette County, Pennsylvania, kimberlite. This study supports the conclusions of Hunter and Taylor (1984) concerning magma mixing in the low-velocity zone (LVZ), and extends them to embrace the data presented here.

MINERALOGY AND PETROGRAPHY

Kimberlite is defined petrographically as an inequigranular micaceous peridotite consisting of megacrysts, macrocrysts, phenocrysts, and xenocrysts of olivine, phlogopite, picroilmenite, chromian spinel, pyropic garnet, diopside, and/or enstatite set in a finer grained matrix. Common matrix phases include olivine, phlogopite, carbonate (usually calcite), serpentine, pyroxene, monticellite, apatite, spinels, perovskite, and ilmenite (Skinner and Clement, 1979; Dawson, 1980; Clement and others, 1984). Similar rocks that contain melilite are classified as alnöite, and may be gradational into kimberlite (e.g., Hearn, 1985). Three of the locations reviewed here consist of kimberlite *sensu stricto*; the fourth (Ithaca, New York) includes kimberlite, alnöite, and related rock types.

In this section we review briefly the mineralogy, petrology, and occurrence of our sample locales, and summarize the variety of mantle xenoliths and megacrysts found at each site.

Ithaca, New York. More than 80 occurrences of igneous rock have been described from the Finger Lake region near Ithaca, New York (Foster, 1970). These rocks occur primarily as small vertical dikes a few centimeters to meters thick that parallel a north-south joint set in early Paleozoic shales and limestones (Foster, 1970; Jackson and others, 1982; Kay and others, 1983). The dikes range in composition from mica peridotite to alnöite; however, the rocks from Taughannock Creek (our sample location) contain megacrysts of pyrope, Cr-diopside, and picroilmenite, and are considered true kimberlite (Jackson and others, 1982; Kay and others, 1983; Taylor, 1984). Other megacrysts found in the Taughannock Creek kimberlites include olivine (→ serpentine), enstatite, Cr-spinel, and pargasite. The groundmass contains these same phases, plus calcite, phlogopite, perovskite, sphene, and opaque oxides (Jackson and others, 1982; Kay and others, 1983).

In addition to abundant supracrustal xenoliths, the New York kimberlites contain rare polymineralic fragments of mantle origin, and a few lower crustal granulites (Kay and others, 1983). We have separated garnet from two such fragments: a garnet peridotite and an eclogite. Paleothermobarometry on these and other mantle-derived xenoliths indicates a relatively shallow origin for the Ithaca kimberlite (≅30 ± 2 kbar), but temperature estimates vary from ≅800 to 1,000°C (Jackson and others, 1982; Kay and others, 1983). Temperature and pressure estimates (Fig. 2) are hampered by the lack of complete mineral assemblages in single xenoliths. Nonetheless, the anomalously shallow origin for these kimberlites distinguishes them from other eastern U.S. kimberlites (Taylor, 1984).

Fayette County, Pennsylvania. The Fayette County kimberlite crops out as a dike 1 to 2 m thick, trending northwest-southeast, that follows a preexisting fault zone for some 4 km along strike (Roen, 1968). Although alteration is locally extensive, most of the dike is remarkably fresh. Two facies are exposed: porphyritic and autolithic (Hunter and others, 1984). The porphyritic facies comprises abundant phenocrysts of olivine, phlogopite, and spinel in a fine-grained groundmass of carbonate, phlogopite, and serpentine. The autolithic facies contains abundant megacrysts (1 to 3 cm), and rare garnet peridotite xenoliths, with autolithic rims of groundmass phlogopite as large as 5 mm thick (Hunter and others, 1984).

Two megacryst suites have been identified (Hunter and Taylor, 1984): a Cr-rich suite of olivine, garnet, pyroxenes, Cr-spinel, and sulfides, and a Cr-poor suite of olivine, garnet, picroilmenite, and pyroxene. The Cr-poor suite is more Fe-rich than the Cr-rich suite, but olivines in both suites have identical rim compositions. Similarly, picroilmenite megacrysts of the Cr-poor suite have Mg-enriched rims (Hunter and Taylor, 1984). Hunter and Taylor (1984) attributed these zoning patterns to mixing of two proto-kimberlite magmas in the LVZ. We have analyzed two garnet megacrysts from the autolithic facies; however, one of these is a xenocryst.

Thermobarometry of the Fayette County kimberlite based on garnet peridotite xenoliths suggests an origin at about 1,300°C and 50 kbar, just above the graphite diamond phase boundary

(Hunter and Taylor, 1982; Fig. 2). Enstatite inclusions in garnet show that the Cr-rich megacrysts equilibrated over a range of temperatures (1,310° to 1,055°C) and pressures (48 to 39 kbar). These data define an inflected geotherm limb, parallel to the Lethoso geotherm of Boyd and Nixon (1975) but at shallower depths (Hunter and Taylor, 1982, 1984; Taylor, 1984).

Elliott County, Kentucky. Two kimberlite pipes crop out in the vicinity of Ison Creek in Elliott County, Kentucky (Bolivar, 1972; Meyer, 1976; Garrison and Taylor, 1980; Agee and others, 1982; Schulze, 1984; Shervais and Taylor, 1984). The kimberlite is extremely fresh in both pipes. The western pipe, called "Ison Creek Pipe" by Schulze (1984) and "Pipe 2" by Garrison and Taylor (1980), straddles Ison Creek and is nearly devoid of xenoliths. The eastern pipe, called "Pipe 1" by Garrison and Taylor (1980) and "Hamilton Branch pipe" by Schulze (1984), lies 0.5 km south of Ison Creek. This pipe contains abundant xenoliths of metasomatized limestone, siltstone, and shale, and less common mantle-derived xenoliths. The latter include spinel lherzolite, garnet lherzolite, harzburgite, spinel websterite, and dunite (Garrison and Taylor, 1980).

Megacrysts are abundant, and include olivine, garnet, picroilmenite, phlogopite, diopside (both Cr-rich and Cr-poor), enstatite, and, rarely, clinopyroxene-ilmenite intergrowths (Bolivar, 1972; Garrison and Taylor, 1980; Schulze, 1984). We have sampled 14 garnet megacrysts for major and trace element analysis. Eight of these are probably derived from disaggregated peridotite; the rest may be magmatic in origin. The origin of these suites is discussed in detail below.

Thermobarometry of the Elliott County kimberlite megacrysts and xenoliths yields the following temperature-pressure estimates (Fig. 2): 1,240° to 1,360°C at 47 to 49 kbar (garnet peridotites), 1,165° to 1,255°C at 51 to 53 kbar (high-Ca enstatite megacrysts), and 970° to 1,020°C at 4,656 kbar (low-Ca enstatite megacrysts). These estimates are consistent with an origin on an infected geotherm and within the diamond stability field (Garrison and Taylor, 1980; Taylor, 1984).

Norris, Tennessee. The Norris kimberlite crops out as two plugs or pipes of deeply weathered mica peridotite that are largely covered by Norris Lake, a Tennessee Valley Authority (TVA) reservoir about 50 km north of Knoxville, Tennessee (Hall and Amick, 1944; Johnson, 1961). It consists almost entirely of serpentine, carbonate, phlogopite (altered to vermiculite), magnetite, and rutile (Meyer, 1975; Hsu and Taylor, 1978). Garnet and picroilmenite are the only megacrysts that are found unaltered. These phases have compositions that are typical of kimberlite: the garnet is pyrope-rich, and the ilmenite contains 8 to 15 percent MgO (Meyer, 1975; Hsu and Taylor, 1978) and up to 30 percent MnO (Hsu and Taylor, 1978).

Mantle-derived xenoliths are rare. One eclogite xenolith has been found that consists of Ca-rich garnet and serpentine after pyroxene. We have separated the garnet from this eclogite for major and trace element analysis, along with one garnet megacryst. Because garnet is the only unaltered primary silicate, it is not possible to estimate *P-T* conditions for this body.

MAJOR AND TRACE ELEMENT GEOCHEMISTRY

Methods

Seventeen garnet megacrysts, ranging from about 0.5 to 1.5 cm in diameter, were separated manually from the matrix of their host kimberlite and abraded, using hand tools to remove surface contaminants. In addition, garnet was separated from two eclogite xenoliths and one garnet peridotite xenolith. All of the garnets were examined optically and determined to be free of fractures, alteration, or large inclusions. However, no acid leaches were used. Thus, we cannot discount the possible occurrence of small mineral or matrix inclusions in some of the separated garnets, which could account for minor irregularities observed in the light rare-earth element (LREE) concentrations of a few samples. However, our conclusions are based almost entirely on the middle and heavy rare-earth elements (MREE, HREE), and on other elements that are not affected significantly by the presence of trace inclusions. The geochemical regularity displayed by these elements in our analyses supports this contention.

Kimberlite matrix was separated from two locations in which unweathered kimberlite is found—Fayette County, Pennsylvania, and Elliott County, Kentucky. Two samples of matrix were obtained from each location by slabbing the freshest material available and chipping to remove megacrysts and xenoliths. Smaller macrocrysts and phenocrysts are included in the "matrix."

The garnet separates were analyzed for 23 major and trace elements by instrumental neutron activation analysis (INAA) using a high-efficiency Ge(Li) detector. Matrix samples were analyzed by INAA for the same elements as the garnet separates, plus Ce, Nd, Ta, and Th. The matrix analyses are reported in Table 1, and the analyses of garnet separates are reported in Table 2. Relative uncertainties estimated from counting statistics are ±0.5 to 3 percent for TiO_2, Al_2O_3, FeO, MnO, Na_2O, Cr_2O_3, Sc, Co, La, Sm, Eu, Yb, and Lu; ±5 percent for MgO, CaO, K_2O, V, Tb, Dy, Hf, and Ta; ±5 to 10 percent for Ce, Ho, Nd, Tm, Th, and Ni. Details of the INAA procedure were described by Laul (1979).

Major element data for additional garnet megacrysts were obtained by electron microprobe analyses using the automated MAC 400 S instrument at the University of Tennessee, and from our previously published results (e.g., Garrison and Taylor, 1980; Jackson and Taylor, 1982; Hunter and Taylor, 1982, 1984). The microprobe data are *not* reported here in Table 2, which includes only INAA data. In the figures that follow, major element data from microprobe analyses are shown by open symbols; major and trace element data by INAA are shown by solid symbols or circled dots. An exception to this system are the chondrite normalized REE plots, in which both open and solid symbols are used to identify specific garnet analyses.

Kimberlite Matrix

The "matrix" analyses of kimberlite reported here (Table 1)

TABLE 1. MAJOR AND TRACE ELEMENT ANALYSES OF KIMBERLITE MATRIX*

Element	Kimberlite**			
	KMX-1	KMX-2	PMX-1	PMX-2
Oxide Element (wt. %)				
TiO_2	3.0	2.1	2.7	2.4
Al_2O_3	2.0	3.0	2.9	2.2
FeO*	8.80	8.00	8.90	8.45
MnO	0.16	0.17	0.17	0.16
MgO	33.8	31.8	21.2	17.3
CaO	6.7	8.7	12.5	17.2
Na_2O	0.058	0.290	0.150	0.16
K_2O	1.2	1.15	2.00	1.70
Cr_2O_3	0.26	0.27	0.20	0.20
MG#	87.3	87.6	80.9	78.5
Trace Element (ppm)				
SC	14.0	15.0	15.0	12.5
V	150	130	170	200
Co	105	105	76	76
Ni	1400	1150	700	700
La	86	89	145	140
Ce	170	170	280	260
Nd	58	50	90	80
Sm	7.2	7.5	13.8	12.9
Eu	1.9	2.0	3.3	3.2
Tb	0.62	0.70	0.90	0.90
Dy	2.6	2.3	3.6	3.2
Ho	<0.4	<0.4	<0.4	<0.4
Tm	<0.4	<0.4	<0.4	<0.4
Yb	0.72	0.85	1.0	1.1
Lu	0.10	0.12	0.15	0.13
Hf	3.4	3.4	7.6	6.0
Ta	12.0	11.5	12.5	12.0
Th	11.3	11.4	18.0	16.0

*Analyses of samples made by instrument neutron activation.
**KMX-1 and KMX-2 kimberlites from Elliott County, Kentucky; PMX-1 and PMX-2 kimberlites from Fayette County, Pennsylvania.

are equivalent to kimberlite whole-rock analyses, minus megacrysts and xenoliths. The Kentucky and Pennsylvania kimberlites are similar in both major and trace element compositions to kimberlites from South Africa (Fesq and others, 1975) and Lesotho (Gurney and Ebrahim, 1973). They differ from one another largely in calcite and olivine content, as shown by their concentrations of CaO and MgO (Table 1). The Kentucky kimberlite is rich in olivine and is geochemically similar to "hardebank" from the Lemphane kimberlite, Lesotho (Gurney and Ebrahim, 1973). The Kentucky kimberlite has high concentrations of MgO and Ni, but low CaO (Table 1). In contrast, the Pennsylvania kimberlite is richer in the calcite component, with high CaO and lower MgO (Table 1). Differences between the separate aliquots of each sample reflect modal inhomogeneity of the samples analyzed. These differences are generally small; however, sample KMX-1 has about 1 percent more TiO_2 than KMX-2, suggesting a higher content of ilmenite macrocrysts.

The trace element concentrations of both the Kentucky and Pennsylvania kimberlites are within the same range as kimberlites from South Africa and elsewhere (Fesq and others, 1975; Dawson, 1980). In particular, the LREE are enriched relative to the HREE (La/Yb = 105 to 145) and chondrite-normalized La concentrations range from 250 to 425 × chondrite (Fig. 3).

Garnet Megacrysts

Major Elements. Major element variations in garnets from the four kimberlites studied here are summarized in Figure 4. The open symbols represent new and previously published microprobe analyses; the solid symbols represent INAA data from Table 2. Most are garnet megacrysts, but garnets from eclogite and peridotite xenoliths are also shown. These garnets form two distinct groups on a plot of CaO versus Cr_2O_3 (wt.% oxide): a very low-Cr group with $Cr_2O_3 < 0.3$ wt.% and CaO = 4.5 to 9.0, and a group with higher Cr concentrations (Cr_2O_3 = 1 to 9 wt.%) and CaO = 4 to 7 wt.% (Fig. 4a). The very low-Cr garnets display low MG#s (=100*Mg/(Mg + Fe)) that range from 80 to 67, whereas garnets with higher Cr concentrations display a restricted range of higher MG#s, mostly between MG# = 80-85 (Fig. 4b). The very low-Cr group includes eclogite garnets with

TABLE 2. MAJOR AND TRACE ELEMENT ANALYSES OF GARNETS FROM EASTERN U.S. KIMBERLITES

	Tennessee		New York		Pennsylvania								Kentucky*							
	TN-1	TN-2	NY-1	NY-2	PA-1	PA-2	KY-1	KY-2	KY-3	KY-4	KY-5	KY-6	KY-7	KY-8	KY-9	KY-10	KY-11	KY-12	KY-13	KY-14
Oxide (wt. %)																				
TiO_2	0.2	0.6	0.2	0.1	0.5	0.8	0.2	0.85	1.1	0.75	0.3	0.1	0.65	0.3	0.3	0.2	0.2	0.3	0.9	1.2
Al_2O_3	20.9	20.6	22.4	22.1	20.7	20.2	20.3	20.2	18.3	19.7	21.3	19.5	21.5	21.1	21.1	20.7	20.2	21.9	20.1	38.2
FeO*	10.1	8.4	9.0	8.5	7.4	9.2	8.9	10.2	6.7	7.75	7.75	7.10	6.90	8.00	8.20	7.31	8.40	8.0	8.20	8.30
MnO	0.25	0.23	0.244	0.395	0.27	0.32	0.365	0.320	0.250	0.290	0.290	0.375	0.260	0.330	0.380	0.370	0.390	0.380	0.260	0.310
MgO	18.0	22.0	19.5	21.0	21.8	20.0	20.0	20.0	21.7	20.5	20.2	19.5	22.6	21.1	19.2	19.5	20.0	20.0	21.7	19.5
CaO	7.9	4.4	7.3	4.9	7.1	5.0	4.9	4.7	5.3	5.7	5.0	5.7	5.2	5.4	4.3	5.0	4.6	4.8	5.6	6.7
Na_2O	0.036	0.078	0.050	0.059	0.073	0.085	0.066	0.087	0.070	0.076	0.057	0.053	0.066	0.070	0.065	0.063	0.027	0.067	0.080	0.086
K_2O	0.001	0.001	0.10	0.008	0.0013	0.025	0.020	0.026	0.006	0.013	0.035	0.052	0.025	0.025	0.035	0.010	0.002	0.030	0.050	0.051
Cr_2O_3	0.087	1.30	0.300	1.90	1.60	2.10	2.5	0.11	4.6	3.35	2.77	4.92	2.27	2.10	2.24	3.18	3.85	1.80	2.89	1.85
MG#	76.1	82.3	79.5	81.5	84.0	79.4	80.0	77.7	85.2	82.5	82.2	83.0	85.3	82.5	80.7	82.6	81.0	81.7	82.6	80.8
Trace Element (ppm)																				
Sc	74	75	58	75	70	95	90	110	110	100	90	140	91	87	83	91	125	81	110	100
V	140	210	100	130	210	360	170	200	340	360	220	220	250	200	180	180	290	190	320	430
Co	60.0	47.6	55.0	70.0	42.6	43.8	43.6	45.2	40.8	44.0	44.0	45.0	44.0	42.0	42.0	42.0	41	43	43	45
Ni	30	140	50	30.0	80	60	40	30	110	60	40	90	110	60	50	40	40	20	100	60
La	0.45	0.34	0.36	0.60	0.090	0.25	0.25	0.58	0.23	0.25	0.32	0.55	0.25	0.40	0.19	0.33	0.12	0.29	0.55	0.33
Sm	0.39	0.87	0.64	0.52	0.56	1.0	0.57	1.1	1.1	0.93	0.47	1.3	1.0	0.58	0.46	0.85	0.49	0.43	1.1	1.0
Eu	0.25	0.46	0.35	0.46	0.31	0.49	0.31	0.50	0.54	0.46	0.29	0.57	0.47	0.35	0.27	0.42	0.19	0.32	0.50	0.60
Tb	0.33	0.49	0.32	0.40	0.42	0.41	0.54	0.65	0.60	0.48	0.32	0.50	0.46	0.40	0.36	0.38	0.1	0.38	0.56	0.69
Dy	2.6	3.7	2.0	3.0	3.5	3.1	5.0	4.6	4.4	3.4	2.4	3.0	3.0	3.2	2.8	2.4	3.0	3.2	4.1	5.0
Ho	0.69	0.90	0.51	0.68	0.94	0.70	1.0	1.6	0.90	0.80	0.66	0.70	0.80	0.80	0.68	0.65	0.3	0.7	1.0	1.2
Tm	0.32	0.37	0.29	0.32	0.38	0.35	0.44	0.68	0.39	0.33	0.32	0.24	0.35	0.40	0.35	0.30	0.1	0.35	0.45	0.52
Yb	2.3	2.6	2.0	2.4	2.6	2.2	2.6	4.5	2.1	2.5	2.1	1.5	2.3	2.4	2.4	1.9	0.99	2.4	3.4	3.9
Lu	0.35	0.40	0.33	0.37	0.42	0.33	0.41	0.68	0.32	0.39	0.40	0.23	0.36	0.41	0.37	0.32	0.19	0.43	0.52	0.64
Hf	0.10	1.3	0.40	0.60	0.80	2.0	0.60	4.4	1.9	2.2	0.80	0.75	1.7	0.55	0.40	1.0	0.1	0.55	2.5	3.4
ΣREE	7.68	10.13	6.8	8.75	9.22	8.33	11.12	14.89	10.58	9.54	5.28	8.59	8.99	8.94	7.88	7.55	3.48	8.5	12.18	13.88

*Analyses of samples made by instrument neutron activation.

Key to kimberlite locations: TN-1, garnet from eclogite nodule, Norris, Tennessee; TN-2, garnet megacryst from altered kimberlite, Norris, Tennessee; NY-1, garnet from ecologite nodule, Taughannock Creek kimberlite, Ithaca, New York; NY-2, garnet from periodotite nodule, Ithaca, New York; PA-1 and PA-2, garnet megacrysts from autolithic kimberlite, Fayette County kimberlite, Pennsylvania; KY-1 through KY-14, garnet megacrysts from Elliott County kimberlite, Kentucky.

high CaO contents that are similar to garnets from eclogites associated with basal gneiss complexes (Dobretsov and Sobolev, 1970). Garnet megacrysts with similar high-Ca and very low-Cr contents probably formed by disaggregation of eclogite nodules. Very low-Cr garnet megacrysts that are also low in CaO may form by eclogite disaggregation, or they may be more evolved members of the megacryst group with higher Cr contents.

Garnets from peridotite xenoliths have high Cr concentrations ($Cr_2O_3 \cong 2$ to 8 percent), which span about the same range as the associated megacrysts. Schulze (1984) found that peridotite garnets from Kentucky have Cr_2O_3 = 2.1 to 7.4 wt.% and MG#s of 81 to 86, while megacrysts that he interpreted as cognate phenocrysts have Cr_2O_3 = 0.13 to 3.6 wt.% and MG#s of 76 to 80. Garrison and Taylor (1980) argued that all of the megacrysts in their study are cognate; however, the results of Schulze (1984) and this study suggest a xenocrystic origin for some garnet megacrysts from this kimberlite.

Taylor and coworkers observed a gap in the Cr_2O_3 contents of garnet megacrysts from Kentucky and Pennsylvania, between 0.3 and 1 percent Cr_2O_3 (Garrison and Taylor, 1980; Hunter and Taylor, 1984). They referred to megacrysts with $Cr_2O_3 <0.3$ wt.% as "Cr-poor" and megacrysts with $Cr_2O_3 >1$ wt.% as "Cr-rich;" they observed no gap in Cr_2O_3 contents between 1 and 12 percent Cr_2O_3. The terms "Cr-rich" and "Cr-poor" were originally used by Eggler and others (1979) to describe megacryst suites from the Colorado-Wyoming State Line kimberlites. These suites are defined on the basis of their color and composition, and by correlation with other members of the megacryst suite (D. H. Eggler, personal communication, 1985; Eggler and others, 1979). Garnet megacrysts described by Eggler and others (1979) as "Cr-rich" contain $Cr_2O_3 >6$ wt.%, whereas garnet megacrysts described as "Cr-poor" contain $Cr_2O_3 <5$ wt.%. Because many of the garnets studied here do not correspond to either the "Cr-rich" or "Cr-poor" megacryst suites as defined by Eggler and others (1979), we have revised our terminology to emphasize these differences and to avoid a direct correlation with the terminology of Eggler and others (1979). We refer to garnet megacrysts with $Cr_2O_3 >4$ wt.% as "high-Cr" garnets; these may correspond to the "Cr-rich" group of Eggler and others (1979), but they are also similar to garnet from peridotite xenoliths. Garnet megacrysts with $Cr_2O_3 <4$ wt.% are referred to as either "low-Cr" (Cr_2O_3 = 1 to 4 wt.%) or "very low-Cr" ($Cr_2O_3 <0.3$ wt.%) to recognize the gap observed in our data. As discussed below, some low-Cr and very low-Cr garnets from this study may correspond to garnets from the Cr-poor megacryst suite of Eggler and others (1979).

Garnets with $Cr_2O_3 >1$ percent exhibit a wide range of TiO_2 concentrations that do not correlate with Cr concentrations (Fig. 4c). Most of our trace element data are from the diatremes of Elliott County, Kentucky. These megacrysts define two distinct groups: one with high TiO_2 (>0.5 %), one with low TiO_2 (<0.5 %). Megacrysts from the Fayette County, Pennsylvania, kimberlite span the gap between these two groups, and include both high-Ti and low-Ti garnets. Peridotite garnets have either

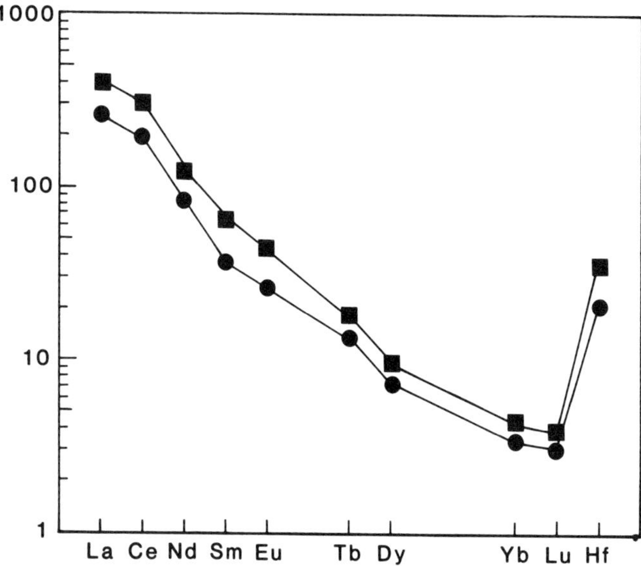

Figure 3. Chondrite-normalized REE and Hf in Kimberlites from Elliott County, Kentucky, and Fayette County, Pennsylvania. Squares refer to Pennsylvania data; dots, to Kentucky data.

low or intermediate TiO_2 concentrations. Eclogite garnets are low in TiO_2, but other "very low-Cr" garnets have TiO_2 concentrations as high as 1 wt.% (Fig. 4c).

In summary, the major element data on garnet megacrysts and garnets from xenoliths show that several groups may be defined on the basis of CaO, Cr_2O_3, and TiO_2 variations. The high-Cr garnets are similar to Cr-rich garnet megacrysts from the Colorado-Wyoming kimberlites and to garnets from garnet peridotite xenoliths. Low-Cr garnets that are high in TiO_2 are similar to Cr-poor garnet megacrysts from Colorado (Eggler and others, 1979). Very low-Cr garnets are found in eclogites and in some Cr-poor megacryst suites (e.g., Monastary kimberlite: Gurney and others, 1979; Iron Mountain kimberlite: Eggler and others, 1979). The low-Cr, low-Ti garnets do not correspond to any previously described megacryst suite but are similar to some peridotite garnets. These low-Cr, low-Ti garnets may represent either a new megacryst suite or disaggregated xenolithic material. It is necessary, however, to use trace element data to further unravel the significance of these groups.

Trace Elements. The distinction between high-Ti and low-Ti garnets persists for trace elements. Garnets with high TiO_2 (>0.5 %) are enriched in other incompatible elements, such as Hf (Fig. 5). The coherent behavior of these trace elements in the high-Ti garnets is demonstrated by the positive correlation seen between TiO_2 and Hf (Fig. 5). In contrast, the low-Ti garnets show no correlation between TiO_2 and Hf, aside from low concentrations of each element. The low-Ti group includes both eclogite garnets and the peridotite garnet (Fig. 5).

The trace element data for garnets from each of the four

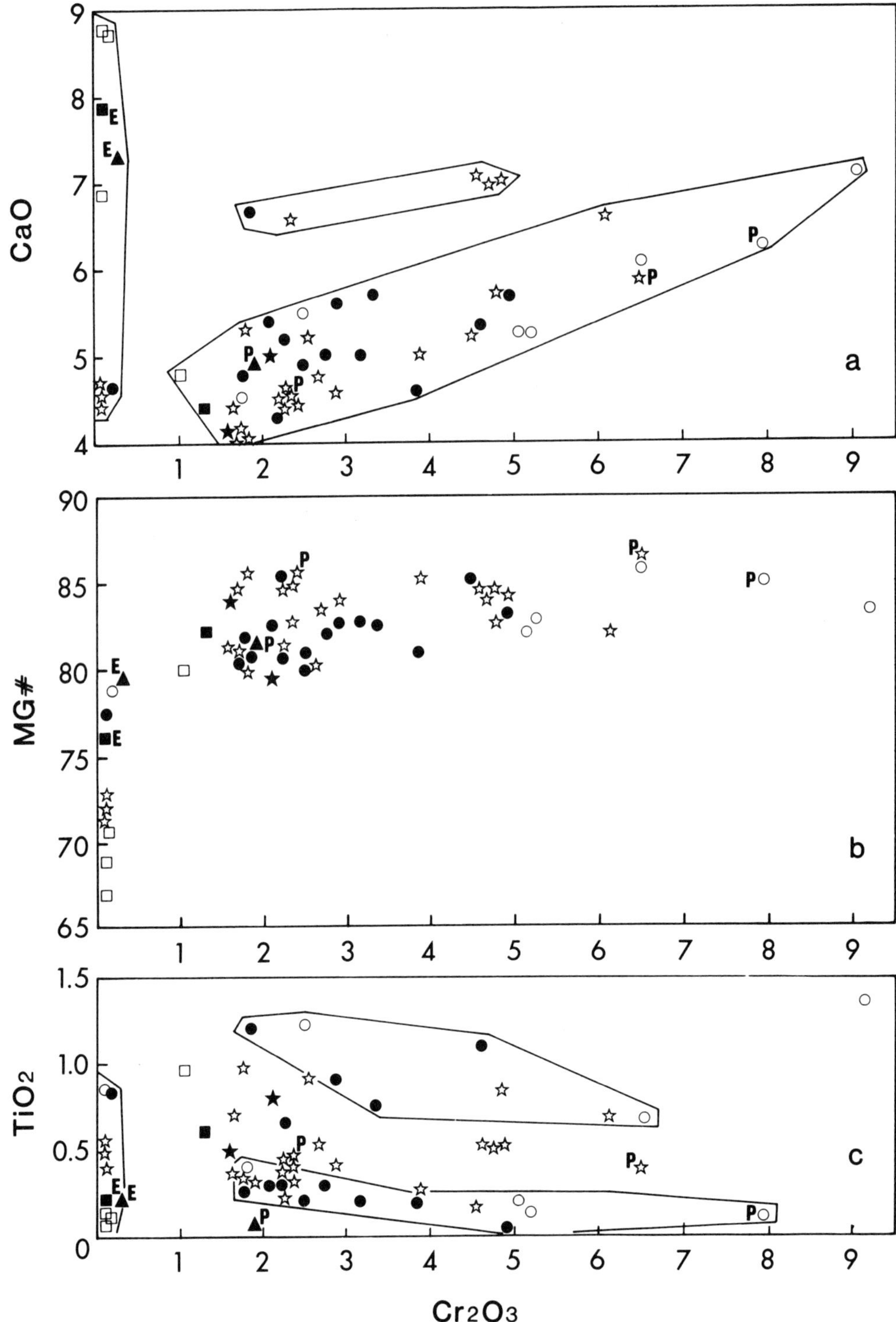

Figure 4. Cr_2O_3 variation diagrams for garnet megacrysts, garnets from periodotite (P), and garnets from eclogite (E). Triangles refer to New York data; stars, to Pennsylvania data; circles, to Kentucky data; squares, to Tennessee data. Solid symbols are garnets analyzed by INAA; open symbols, garnets analyzed by electron microprobe. a, Cr_2O_3 versus CaO wt. %. Garnets form two series: peridotite garnets and megacrysts trend toward increasing uvarovite solid solution, garnets from eclogites trend toward increasing grossular solid solution. b, Cr_2O_3 versus MG# (=100 × Mg/(Mg + Fe)). Garnets with high MG#s vary from 1 to 9 wt. % in Cr_2O_3; garnets that are low in Cr_2O_3 have low MG#s as well (67 to 80). c, Cr_{203} versus TiO_2. Garnets form two series: a high-Ti group with TiO_2 >0.5 wt. %, and a low-Ti group with TiO_2 <0.5 wt. %. Both span similar range in Cr_{203}.

locations studied are compared using chondrite-normalized concentrations of REE and Hf in Figures 6 and 7. Garnets from eclogite and peridotite xenoliths in the kimberlite from Ithaca, New York, have similar chondrite normalized concentrations and both are depleted in Hf relative to Lu (Fig. 6). The eclogite garnet and garnet megacryst from Norris, Tennessee, also have similar REE concentrations, but the eclogite garnet is more severely depleted in Hf than the megacryst (Fig. 6). The garnet megacrysts from the kimberlite dike in Fayette County, Pennsylvania, have crossing REE patterns and contrasting Hf/Lu ratios (Fig. 6). The low-Ti garnet megacryst has chondrite-normalized Hf/Lu <1, and lower LREE/HREE ratio than the high-Ti garnet. The high-Ti garnet megacryst has Hf/Lu >1 and is less depleted in the LREE (Fig. 6).

The contrast between the high-Ti and low-Ti megacryst suites is best displayed by the garnets from Elliott County, Kentucky. The high-Ti garnets have higher overall concentrations of the REE, a rough positive correlation between La and the HREE, and chondrite-normalized Hf concentrations that are greater than or equal to Lu. In general, Hf/Lu ratios increase with increasing HREE (Fig. 7). In contrast, the low-Ti garnet group has lower overall REE concentrations, a vague negative correlation between La and the HREE, and chondrite-normalized Hf concentrations that are depleted relative Lu. There is also a tendency for Hf concentrations to *decrease* as HREE concentrations increase (Fig. 7).

The Kentucky garnet megacrysts are almost all low-Cr or very low-Cr garnets. However, two of these megacrysts, one with high-Ti and one with low-Ti, are high-Cr garnets that are also rich in Ni and have high MG#s. Unlike the low-Cr garnet megacrysts, which have flat or slightly positive HREE slopes, the high-Cr garnets have convex REE patterns, with the MREE >HREE, and negative HREE slopes (Fig. 7). Similar convex REE patterns have been observed in a Cr-rich garnet megacryst from Colorado (Eggler and others, 1979) and in garnets from granular peridotite xenoliths from South African kimberlites (Shimizu, 1975).

DISCUSSION

The major and trace element data discussed above show that there are two suites of garnet megacrysts present in the kimberlites from Pennsylvania and Kentucky. Each suite has about the same range in Cr_2O_3 contents, but they contrast significantly and systematically in their minor element (TiO_2) and trace element (REE, Hf) chemsitry.

High-Ti garnets that are also low in Cr_2O_3 ("low-Cr" and "very low-Cr" using the terminology defined earlier) correspond to the "Cr-poor" megacryst suite of Eggler and others, (1979). Eggler and others (1979) considered the "Cr-poor" megacryst suite to represent phenocrysts that grew in a fractionating protokimberlite magma that was saturated with olivine, orthopyroxene, clinopyroxene, garnet, and, at lower temperatures, ilmenite. Similar conclusions have been proposed for the origin of Cr-poor megacryst suites from kimberlites in South Africa (Gurney and others, 1979), Lesotho (Boyd and Nixon, 1973; Nixon and Boyd, 1973), and the eastern U.S. (Garrison and Taylor, 1980; Hunter and Taylor, 1984; Schulze, 1984). The low-Cr, high-Ti garnet megacrysts from Kentucky have coherent fractionation trends in which TiO_2, Hf, and the HREE increase simultaneously, consistent with an origin as cognate phenocrysts in a magma undergoing fractional crystallization. The decrease in TiO_2 in the high-Ti garnet with the lowest Cr content (Table 2, KY-2) probably signals the onset of ilmenite fractionation (e.g., Gurney and others, 1979). The increase in HREE with increasing TiO_2 and Hf implies multiphase fractionation dominated by phases other than garnet. These phases must have partition coefficients for the HREE that are less than one.

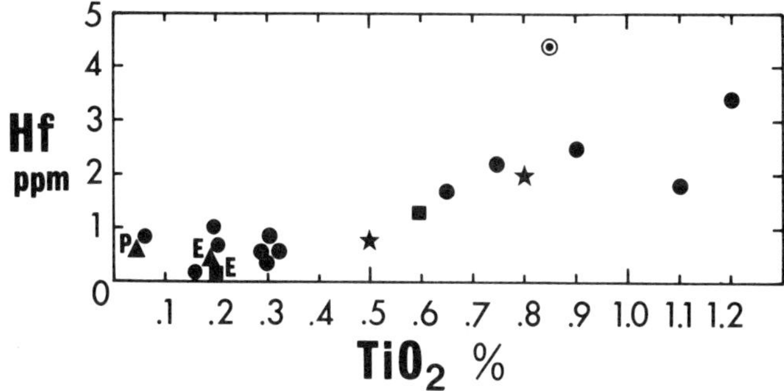

Figure 5. Hf versus TiO_2 for garnets analyzed by INAA. High-Ti garnets have HF >1 ppm; low-Ti garnets have Hf <1 ppm. Symbols as in Figure 4, except circle with dot = Cr, Ca-rich garnet that plots above main Cr-rich group in Figure 4.

One of the high-Ti garnets for which we have trace element data is also high in Cr (Table 2, KY-3). This garnet has a convex REE pattern with a negative HREE slope (Fig. 7a). Shimizu (1975) observed similar convex REE patterns in garnets from granular peridotite xenoliths from South African kimberlites. Negative HREE slopes are characteristic of the LREE-enriched clinopyroxenes that occur as megacrysts in kimberlite and in eclogite and peridotite xenoliths (Shimizu, 1975; Ehrenberg, 1982; Shervais and others, 1985). Shimizu (1975) has suggested that garnets with negative HREE slopes formed by recrystallization of a clinopyroxene-rich, garnet-free protolith. This implies that these garnets must represent preexisting peridotite wall rock, and are not cognate cumulates. However, similar convex REE patterns have been observed in a Cr-rich garnet megacryst that Eggler and others (1979) interpreted as igneous. We interpret garnet KY-3 as an igneous megacryst similar to the Cr-rich garnets from Colorado (Eggler and others, 1979). This conclusion is based on the chondrite-normalized Hf/Lu ratio of this garnet (Hf/Lu >1), which is similar to the ratios observed in the high-Ti, low-Cr garnet megacrysts (Fig. 7).

The low-Ti garnets for which we have trace element data do not correspond to garnets in any previously described megacryst

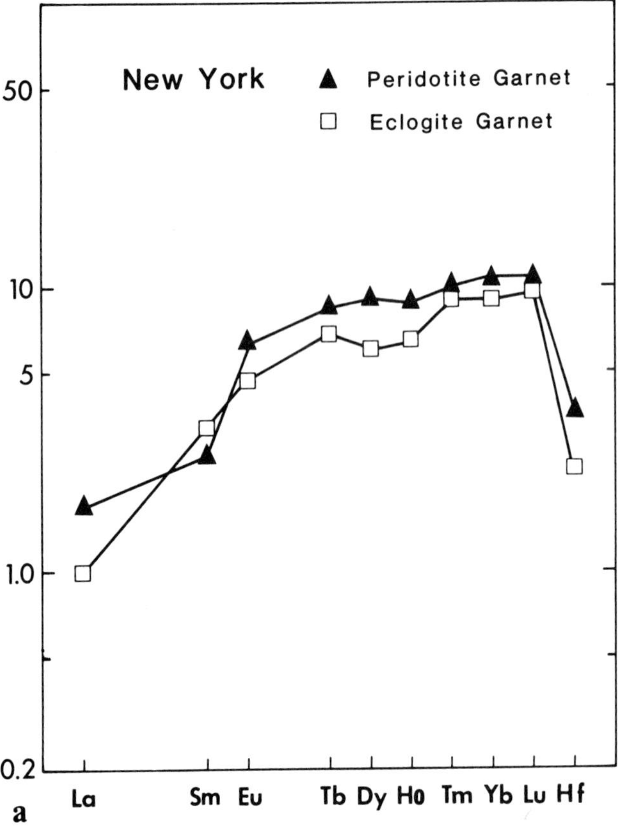

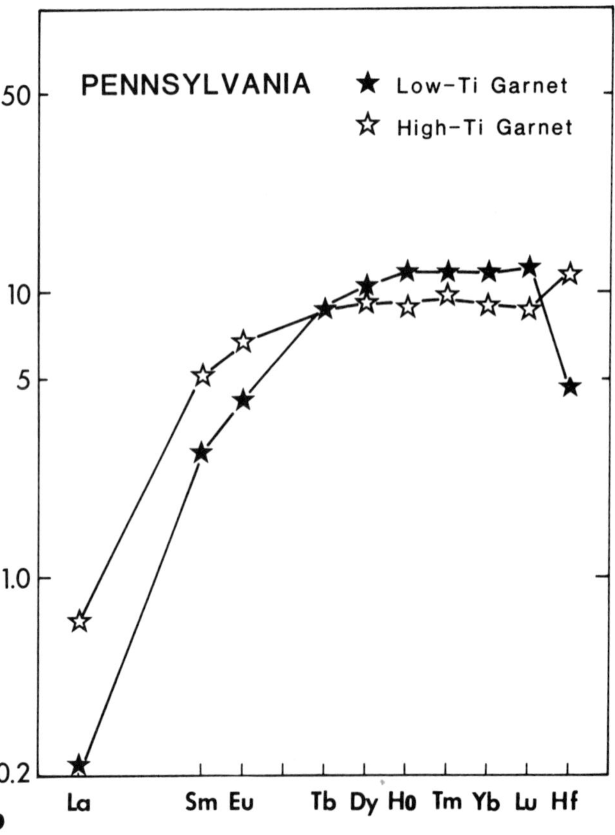

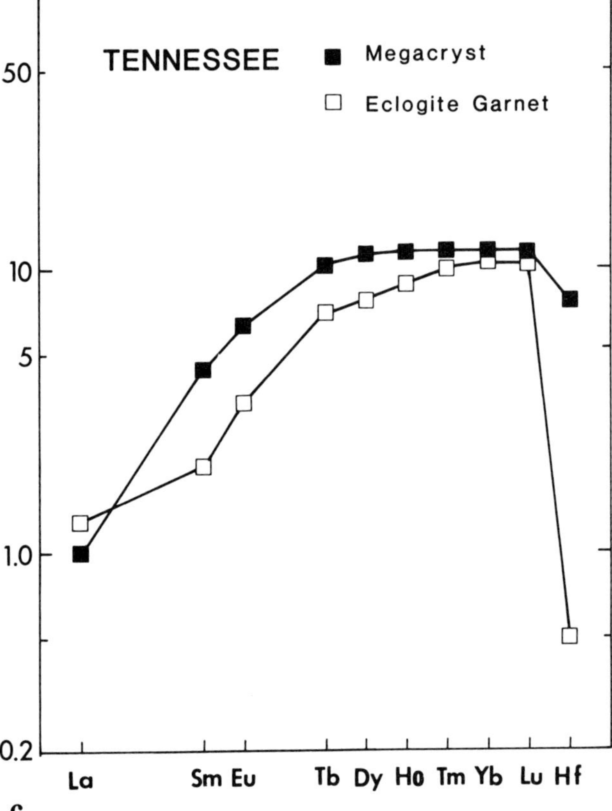

Figure 6. a, Chondrite-normalized REE + Hf concentrations for garnets from Ithaca, New York, kimberlites. Note depleted Hf/Lu ratios. b, Chondrite-normalized REE + Hf concentrations for garnets from Norris, Tennessee, kimberlite. c, Chondrite-normalized REE + Hf concentrations for garnets from Fayette County, Pennsylvania, kimberlites. Low-Ti garnet has Lu/Hf $<<1$; high-Ti garnet, Lu/Hf $\cong 1$.

suite. Two central characteristics of these garnets, besides their low TiO_2 contents, are their low Hf concentrations and low chondrite–normalized Hf/Lu ratios (<1). In contrast to the high-Ti garnet suite, Hf/Lu ratios in the low-Ti garnets generally *decrease* with increasing HREE (Fig. 7b). If these trends were caused by fractional crystallization, then phases with high crystal/liquid partition coefficients for Hf, such as ilmenite and/or zircon, must have composed a major part of the fractionating assemblage. However, although ilmenite and zircon are commonly associated in kimberlite megacryst suites (e.g., Kresten and others, 1975; Raber and Haggerty, 1979), most investigators have concluded that ilmenite does not join the silicate fractionation assemblage until relatively low temperatures are reached (e.g., Boyd and Nixon, 1973; Eggler and others, 1979; Gurney and others, 1979; Pasteris and others, 1979). Because of the wide range of Cr_2O_3 concentrations over which this Hf/Lu depletion is observed in the low-Ti garnets, fractional crystallization of ilmenite and/or zircon cannot be responsible.

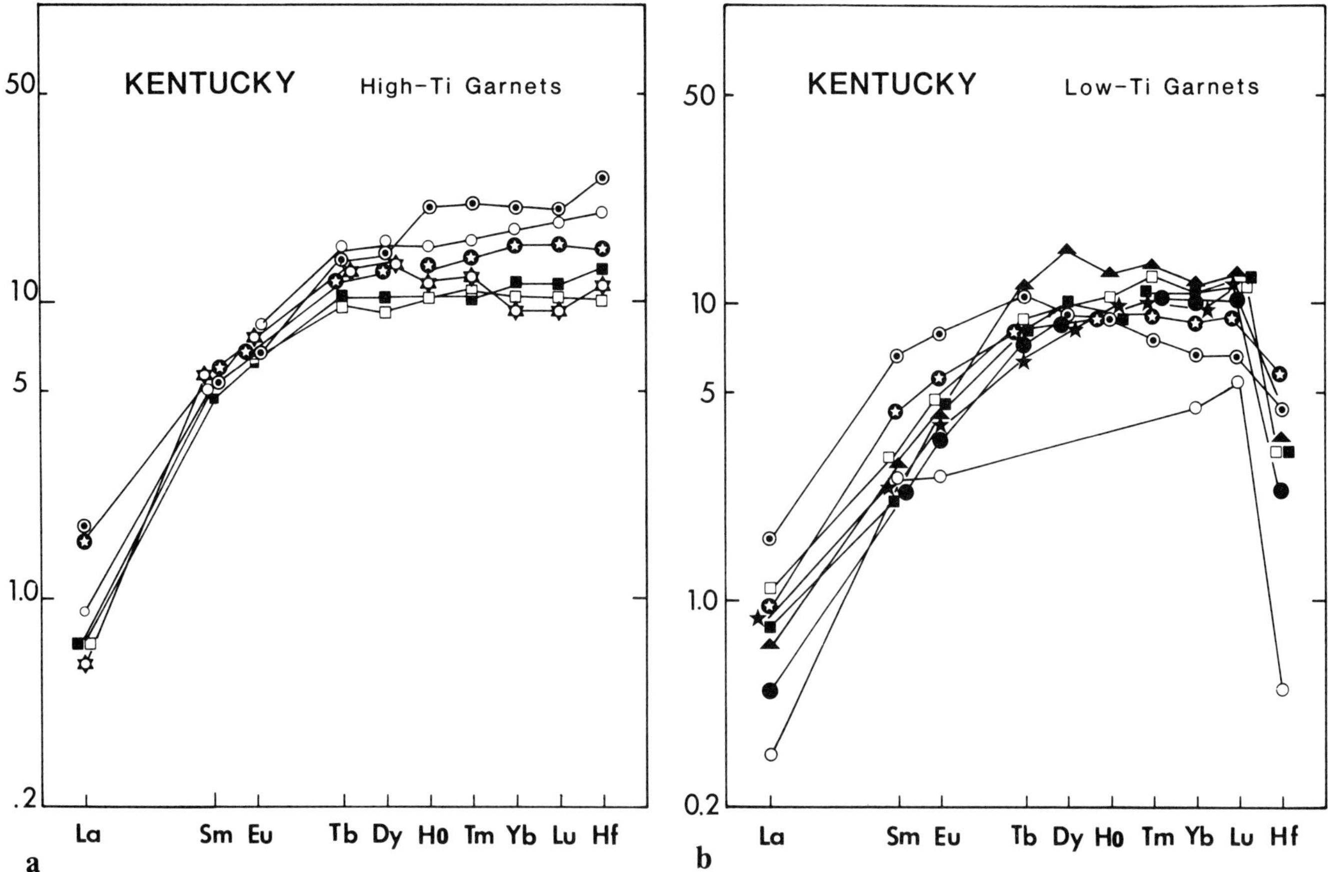

Figure 7. Chondrite-normalized REE + Hf concentrations for garnets from Elliott County, Kentucky, kimberlite. a, High-Ti garnet megacrysts with chondrite-normalized Lu/Hf ≅1.0. These garnets have trace element systematics consistent with fractional crystallization and may be cognate phenocrysts. Six-point stars refer to KY-3; open squares, to KY-7; solid squares, to KY-4; circles with stars, KY-13; open circles, KY-14; circles with dots, KY-2. b, Low-Ti garnet megacrysts with chondrite-normalized Lu/Hf<1.0. These garnets probably derive from disaggregated peridotite. Open circles refer to KY-11; solid circles, to KY-9; triangles, KY-1; solid squares, KY-12; stars, KY-5; circles with stars, KY-10; open squares, KY-8; circles with dots, KY-6.

The depletion of Hf/Lu in the low-Ti garnet megacrysts is also observed in garnets from eclogite and peridotite (Fig. 6). Eclogite garnets are low in TiO_2, but have Cr_2O_3 <0.3 wt.%, much lower than the low-Ti garnet megacrysts (Table 2). Peridotite garnets have Cr_2O_3 and TiO_2 in the same range as the low-Ti garnet megacrysts, suggesting that the low-Ti garnet megacrysts are derived from disaggregated peridotite xenoliths and wallrock. This interpretation is supported by the trace element data, which are consistent with a partial melting origin for the Hf/Lu variations (Fig. 7). Hf is incompatible with garnet and will be partitioned into the liquid during progressive partial melting, decreasing its concentration in the refractory garnet. The HREE, however, are partitioned preferentially into refractory garnet during partial melting because garnet/liquid partition coefficients for the HREE are between 5 and 10 (Shimizu, 1975; Harrison, 1981; Fujimak and others, 1983). The combined effect will be an increase in the HREE and a decrease in Hf in refractory garnet during partial melting. This is the general trend observed in the low-Ti garnets (Fig. 7).

The high-Cr, low-Ti garnet megacryst in this group (Table 2, KY-6) has a convex REE pattern similar to the high-Ti, high-Cr megacryst KY-3. Unlike KY-3, however, KY-6 displays the same Hf/Lu depletion characteristic of all the low-Ti garnets studied here. This garnet is probably refractory in origin, but the negative HREE slope suggests it may have formed originally by recrystallization of a clinopyroxene-rich protolith (e.g., Shimizu, 1975). Alternatively, the negative HREE slope in both of the high-Cr garnets from Kentucky, and in the Cr-rich garnet megacryst from Colorado (Eggler and others, 1979) may reflect a change in the partition coefficients of the HREE that is coupled to increasing Cr_2O_3 in garnet.

Hunter and Taylor (1984) found two suites of megacrysts in

the Fayette County, Pennsylvania, kimberlite: a "Cr-rich" suite consisting of olivine, garnet, Cr-diopside, Cr-spinel, and an immiscible sulfide melt; and a "Cr-poor" suite consisting of olivine, garnet, diopside, and picroilmenite. Both of these suites contain garnet, but the suites are defined by the complete assemblage of phenocrysts in each. Hunter and Taylor (1984) proposed that the occurrence of these two suites in a single kimberlite dike was due to the mixing of two proto-kimberlite magmas in the low-velocity zone (LVZ) prior to intrusion of the crust. Boyd and Nixon (1973) proposed a similar hypothesis to explain the origin of ilmenite-silicate nodules in kimberlites from South Africa and Lesotho. The fundamental conclusion of both investigations is that kimberlite magmas are hybrid in origin and result from the mixing of magmas derived from different levels of the upper mantle.

This conclusion is supported by the trace element data presented here for garnet megacrysts from the Kentucky kimberlite. These data show that, in addition to garnet derived from peridotite, two megacryst suites are represented by garnets that are probably magmatic in origin: a high-Cr, high-Ti suite, and a low-Cr, high-Ti suite. Previous investigations of the Kentucky megacrysts have found only a "Cr-poor" suite (Garrison and Taylor, 1980; Schulze, 1984), although Garrison and Taylor (1980) considered the high-Cr garnets megacrysts also. Our data show that it may not be possible to distinguish high-Cr garnet megacrysts from perioditite garnet.

CONCLUSIONS

Kimberlites from the eastern United States provide a rare opportunity to study the origin and evolution of this complex rock type and its subcontinental source region.

The Kentucky and Pennsylvania kimberlites have major and trace element compositions similar to kimberlites from South Africa and Lesotho, and mineral assemblages that are characteristic of kimberlite. The Norris, Tennessee, kimberlite is highly weathered, but this pipe also contains mineral assemblages and mineral compositions that are characteristic of kimberlite. Some of the Ithaca, New York, dikes contain melilite and are thus alnoïite, but others may be true kimberlite.

Garnet megacrysts may be divided into two groups based on their TiO_2 content and trace element characteristics. The low-Ti garnets are depleted in Hf relative to Lu, as are garnets in peridotite and eclogite xenoliths. The low-Ti garnets probably represent dissaggregated peridotite xenoliths whose Hf/Lu variations were caused by partial melting. The high-Ti garnets form two "suites": one low in Cr that corresponds to the "Cr-poor" suite of megacrysts from the Sloan-Nix pipes in Colorado, and one with high-Cr that corresponds to garnets of the "Cr-rich" suite from the Colorado kimberlites (Eggler and others, 1979; D. H. Eggler, personal communication, 1985).

The occurrence of two suites of high-Ti garnet megacrysts, one with high Cr and one with low Cr, in the Fayette County, Pennsylvania, and Elliott County, Kentucky, kimberlites probably reflects the mixing of two proto-kimberlite magmas in the low-velocity zone, as originally proposed by Hunter and Taylor (1984) for the Fayette County kimberlite. Hunter and Taylor (1984) showed that one of these proto-kimberlite magmas was saturated with olivine, garnet, diopside, Cr-spinel, and an immiscible sulfide melt (the "Cr-rich" suite); the other was saturated with olivine, garnet, diopside, and picroilmenite (the "Cr-poor" suite).

Smith (1983) has shown recently that South African kimberlites can be subdivided into two groups, based on differences in age, mineral chemistry, whole-rock composition, and isotope geochemistry. Group I kimberlites from South Africa are Cretaceous in age and are associated with melilites and alnoïites (Smith, 1983; J. B. Dawson, personal communication, 1985). They are characterized by scarce matrix phlogopite and diopside, negative ϵSr, and positive ϵNd, similar to ocean island basalts (Smith, 1983). Group II kimberlites from South Africa are older (Jurassic to early Cretaceous) and are associated with ultrapotassic lamprophyres and olivine lamprophyres. The Group II kimberlites are characterized by abundant matrix phlogopite and diopside, positive ϵSr, and negative ϵNd, similar to the ultrapotassic diamond-bearing olivine lamproites of Western Australia (Smith, 1983; McCulloch and others, 1983). Dawson (1986, and personal communication, 1985) has noted another important difference between these two kimberlite groups: matrix ilmenite is abundant in the group I kimberlites but scarce in the group II kimberlites. The ilmenite-saturated group I kimberlites were derived from a "depleted" mantle reservoir (the LVZ?); the ilmenite-undersaturated group II kimberlites from an "enriched" mantle reservoir (Smith, 1983). Thus, these two kimberlite groups *may* correspond to the two "proto-kimberlite" magmas that we infer from our trace element data. Alternatively, these two kimberlite groups may be themselves mixtures of more extreme "proto-kimberlite" compositions, as suggested by McCulloch and others (1983). In either case, a mixing origin for kimberlite is supported by the isotope data as well as by trace element data.

The major and trace element data for garnet megacrysts presented in this chapter enable us to refine the magma-mixing model for kimberlite origin of Hunter and Taylor (1984), but leave it essentially intact. The origin of eruptive kimberlite by mixing of proto-kimberlite magmas that form at different horizons in the earth's mantle implies that *all* kimberlites are hybrid in nature, as first suggested by Boyd and Nixon (1973). Variations in the proportions of the two proto-kimberlite magmas may account for much of the variability observed between different kimberlite pipes.

ACKNOWLEDGMENTS

We thank our many colleagues for helpful discussions about this paper. Thorough reviews by D. G. Brookins, B. H. Scott-Smith, and especially D. H. Eggler improved the manuscript substantially. This research was supported by National Aeronautics and Space Administration grant NAG 9-73 (to L.A.T.).

REFERENCES CITED

Agee, J. J., Garrison, J. R., Jr., and Taylor, L. A., 1982, Petrogenesis of oxide minerals in kimberlite, Elliott County, Kentucky: American Mineralogist, v. 67, p. 28–42.

Bolivar, S. L., 1972, Kimberlite of Elliott County, Kentucky [M.S. thesis]: Richmond, Eastern Kentucky University, 61 p.

Boyd, F. R., and Nixon, P. H., 1973, Origin of the ilmenite-silicate nodules in kimberlites from Lesotho and South Africa, *in* Nixon, P. H., ed., Lesotho kimberlites: Lesotho National Development Company, 254–268.

—— , 1975, Origins of the ultramafic nodules from some kimberlites of northern Lesotho and the Monastery Mine, South Africa: Physics and Chemistry of the Earth, v. 9, p. 431–454.

Clement, C. R., Skinner, E.M.W., and Scott-Smith, B. H., 1984, Kimberlite redefined: Journal of Geology, v. 92, p. 223–228.

Dawson, J. B., 1980, Kimberlites and their xenoliths: Berlin, Springer-Verlag, 252 p.

—— , 1986, The kimberlite clan; Relationship to olivine and leucite lamproites, and inferences for upper mantle metasomatism, *in* Fitton, C., and Upton, B.J.C., eds., Alkaline rocks: Geological Society of London Special Publications (in press).

Dobretsov, N. L., and Sobolev, N. V., 1970, Eclogites from metamorphic complexes of the U.S.S.R.: Physics of the Earth and Planetary Interiors, v. 3, p. 462–470.

Eggler, D. H., McCallum, M. E., and Smith, C. B., 1979, Megacryst assemblages in kimberlites from northern Colorado and southern Wyoming; Petrology, geothermometry-barometry, and areal distribution, *in* Boyd, F. R., and Meyer, H.O.A., eds., The mantle sample; Inclusions in kimberlites and other volcanics: Washington, D.C., American Geophysical Union, 330 p.

Ehrenberg, S. N., 1982, Rare earth element geochemistry of garnet lherzolite and megacrystalline nodules from minette of the Colorado Plateau province: Earth and Planetary Science Letters, v. 57, p. 191–210.

Fesq, H. W., Kable, E.J.D., and Gurney, J. J., 1975, Aspects of the geochemistry of kimberlites from the Premier Mine and other selected South African occurrences with particular reference to the rare earth elements: Physics and Chemistry of the Earth, v. 9, p. 687–707.

Foster, B. P., 1970, A study of the kimberlite-alnoite dikes in central New York [M.S. thesis]: State University of New York at Buffalo, 55 p.

Fujimaki, H., Tatsumoto, M., and Aoki, K., 1983, Partition coefficients of Hf, Zr, and REE between phenocrysts and groundmass; Proceedings of the Lunar and Planetary Science Conference 14: Journal of Geophysical Research suppl., v. 89.

Garrison, J. R., and Taylor, L. A., 1980, Megacrysts and xenoliths in kimberlite, Elliott County, Kentucky: Contributions to Mineralogy and Petrology, v. 75, p. 27–42.

Gurney, J. J., and Ebrahim, S., 1973, Chemical composition of Lesotho kimberlites, *in* Nixon, P. H., ed., Lesotho kimberlites: Lesotho National Development Company, p. 280–284.

Gurney, J. J., Jakob, W.R.O., and Dawson, J. B., 1979, Megacrysts from the Monastery kimberlite pipe, South Africa, *in* Boyd, F. R., and Meyer, H.O.A., eds., The mantle sample; Inclusions in kimberlites and other volcanics: Washington, D.C., American Geophysical Union, p. 227–243.

Hall, G. M., and Amick, H. C., 1944, Igneous rock regions in the Norris region, Tennessee: Journal of Geology, v. 52, p. 424–430.

Harrison, W. J., 1981, Partition coefficients for REE between garnets and liquids; Implications for non-Henry's Law behaviour for models of basalt origin and evolution: Geochimica et Cosmochimica Acta, v. 45, p. 1529–1544.

Harte, B., and Gurney, J. J., 1981, The model for formation of chromium-poor megacryst suites from kimberlites: Journal of Geology, v. 89, p. 749–753.

Hearn, B. C., 1985, Alkalic ultramafic magmas, Missouri Breaks, Montana; The kimberlite-alnoite continuum: Geological Society of America Abstracts with Programs, v. 17, p. 161.

Hsu, J. A., and Taylor, L. A., 1978, An unusual mineralogy in the Norris kimberlite of Tennessee: Geological Society of America Abstracts with Programs, v. 10, p. 172.

Hunter, R. H., and Taylor, L. A., 1982, Instability of garnet from the mantle glass as evidence of metasomatic melting: Geology, v. 10, p. 617–620.

—— , 1984, Magma-mixing in the low velocity zone; Kimberlitic megacrysts from Fayette County, Pennsylvania: American Mineralogist, v. 69, p. 16–29.

Hunter, R. H., Kissling, R. D., and Taylor, L. A., 1984, Mid- to late-stage kimberlitic melt evolution; Phlogopites and oxides from the Fayette County kimberlite, Pennsylvania: American Mineralogist, v. 69, p. 30–40.

Jackson, D. E., Hunter, R. H., and Taylor, L. A., 1982, Shallow-level kimberlite from the northeastern United States; An unusual sample: EOS American Geophysical Union Transactions, v. 63, p. 463–464.

Johnson, R. W., Jr., 1961, Dimensions and attitude of the peridotite in Clark Hollow, Union County, Tennessee; An aeromagnetic study: Southeastern Geology, v. 2, p. 137–154.

Kay, S. M., Snedden, W. T., Foster, B. P., and Kay, R. W., 1983, Upper mantle and crustal fragments in the Ithaca kimberlites: Journal of Geology, v. 91, p. 277–290.

Kempton, P. D., Menzies, M. A., and Dungan, M. A., 1984, Petrography, petrology, and geochemistry of xenoliths and megacrysts from the Geronimo Volcanic field, southeastern Arizona, *in* Kornprobst, J., ed., Kimberlites II; The mantle and crust-mantle relationships: Amsterdam, Elsevier, p. 97–108.

Kresten, P., Fels, P., and Berggren, G., 1975, Kimberlitic zircons; A possible aid to prospecting for kimberlites: Mineralium Deposita, v. 10, p. 47–56.

Laul, J. C., 1979, Neutron activation analysis of geological materials: Atomic Energy Review, v. 17, no. 3, p. 603–607.

McCulloch, M. T., Jaques, A. L., Nelson, D. R., and Lewis, J. D., 1983, Nd and Sr isotopes in kimberlites and lamproites from western Australia; An enriched mantle origin: Nature, v. 302, p. 400–403.

Meyer, H.O.A., 1975, Kimberlite from Norris Lake, eastern Tennessee; Mineralogy and petrology: Journal of Geology, v. 83, p. 518–526.

—— , 1976, Kimberlites of the Continental United States; A review: Journal of Geology, v. 84, p. 377–403.

Meyer, H.O.A., and McCallister, R. H., 1984, Two-pyroxene megacrysts from South African kimberlites, *in* Kornprobst, J., ed., Kimberlites II; The mantle and crust-mantle relationships: Amsterdam, Elsevier, p. 97–108.

Mitchell, R. H., and Lewis, R. D., 1983, Priderite-bearing xenoliths from the Prairie Creek mica peridotite, Arkansas: Canadian Mineralogist, v. 21, p. 59–64.

Nixon, P. H., and Boyd, F. R., 1973, The discrete nodule (megacryst) association in kimberlites from Norther Lesotho, *in* Nixon, P. H., ed., Lesotho kimberlites: Lesotho National Development Company, p. 67–75.

Pasteris, J. D., Boyd, F. R., and Nixon, P. H., 1979, The ilmenite association at the Frank Smith Mine, R.S.A., *in* Boyd, F. R., and Meyer, H.O.A., eds., The mantle sample; Inclusions in kimberlites and other volcanics: Washington, D.C., American Geophysical Union, p. 265–278.

Raber, E., and Haggerty, S. E., 1979, Zircon-oxide reactions in diamond-bearing kimberlites, *in* Boyd, F. R., and Meyer, H.O.A., eds., Kimberlites, diatremes, and diamonds: Washington, D.C., American Geophysical Union, p. 229–240.

Roen, J. B., 1968, A transcurrent structure in Fayette and Green Counties, Pennsylvania: U.S. Geological Survey Professional Paper 600-C, p. 149–152.

Schulze, D. J., 1984, Cr-poor megacrysts from the Hamilton Branch kimberlite, Elliott County, Kentucky, *in* Kornprobst, J., ed., Kimberlites II; The mantle and crust-mantle relationships: Amsterdam, Elsevier, p. 97–108.

Scott-Smith, B. H., and Skinner, E.M.W., 1984a, A new look at Prairie Creek, Arkansas, *in* Kornprobst, J., ed., Kimberlites I; Kimberlites and related rocks: Amsterdam, Elsevier, p. 255–284.

—— , 1984b, Diamondiferous lamproites: Journal of Geology, v. 92, p. 433–438.

Skinner, E.M.W., and Clement, C. R., 1979, Mineralogical classification of southern African kimberlites, *in* Boyd, F. R., and Meyer, H.A.O., eds.,

Kimberlites, diatremes, and diamonds: Washington, D.C., American Geophysical Union, p. 129–130.

Smith, C. B., 1983, Pb, Sr, and Nd isotopic evidence for sources of southern African Cretaceous kimberlites: Nature, v. 307, p. 51–54.

Smith, D., and Levy, S., 1976, Petrology of the Green Knobs diatreme and implications for the upper mantle below the Colorado Plateau: Earth and Planetary Science Letters, v. 29, p. 107–125.

Shervais, J. W., and Taylor, L. A., 1984, Mineralogy, petrology, and geochemistry of a kimberlite from Elliott County, northeast Kentucky: Geological society of America Abstracts with Programs, v. 16, p. 195.

Shervais, J. W., Taylor, L. A., and Korotev, R. L., 1985, Petrology and mineral chemistry of some African eclogites and the evolution of subcontinental mantle and continental crust, Lunar and Planetary Science 16: Houston, Texas, Lunar and Planetary Institute, p. 769–770.

Shimizu, N., 1975, Rare earth elements in garnets and clinopyroxenes from garnet lherzolite nodules in kimberlites: Earth and Planetary Science Letters, v. 25, p. 26–32.

Taylor, L. A., 1984, Kimberlitic magmatism in the eastern U.S.; Relationships to mid-Atlantic tectonism, *in* Kornprobst, J., ed., Kimberlites I: Kimberlites and related rocks: Amsterdam, Elsevier, p. 417–424.

Taylor, L. A., Shervais, J. W. Hunter, R. H., and Laul, J. C., 1983, Major and trace element geochemistry of garnets and ilmenites from eastern U.S.A. kimberlites: Geological Society of America Abstracts with Programs, v. 15, p. 704.

Manuscript Accepted by the Society October 14, 1986

Printed in U.S.A.

Geological Society of America
Special Paper 215
1987

A Sr, Nd, and Pb isotope study of alkaline lamprophyres and related rocks from Westland and Otago, South Island, New Zealand

Barbara A. Barreiro
Department of Earth Sciences, Dartmouth College, Hanover, New Hampshire 03755, and Department of Terrestrial Magnetism, Carnegie Institution of Washington, 5241 Broad Branch Road NW, Washington, D.C. 22015
Alan F. Cooper
Department of Geology, University of Otago, Dunedin, New Zealand

ABSTRACT

The Westland dike swarm of South Island, New Zealand, consists of a variety of ultrabasic, alkaline rock types including camptonite lamprophyres, ouachitite peridotites, and carbonatites. At least part of the swarm is as young as Oligocene-Miocene age. The majority of Nd and Sr isotope analyses for 16 dikes ϵ_{Nd} = +3.5 to +5.2, $^{87}Sr/^{86}Sr_i$ = 0.7028 to 0.7035) fall to the left of the mantle array in Nd-Sr isotope correlation diagram. The data are similar to the ocean islands St. Helena and Tubuai and to some continental volcanics and mantle nodules (Menzies and Wass, 1983). The dikes contain radiogenic Pb ($^{206}Pb/^{204}Pb$ = 19.19 to 20.59, $^{207}Pb/^{204}Pb$ = 15.64 to 15.71, $^{208}Pb/^{204}Pb$ = 39.0 to 40.2), which also is similar to the most radiogenic Pb in ocean islands.

The trace element and isotope data are consistent with an origin involving a Paleozoic enrichment event, which added CO_2, Sr, light rare-earth elements, and U to a depleted mantle. We propose that this first event was the manifestation of a deep-seated intrusion of carbonated hyperalkaline magma. A second event triggered the melting of this newly metasomatized mantle and the emplacement of the Westland dikes.

INTRODUCTION

There is a growing consensus among geochemists that the mantle sources of many within-plate basalts experienced some degree of trace element enrichment; this is inferred from concentrations and ratios of incompatible elements and radiogenic isotopes (see Hawkesworth and others, 1984, for a brief review). The effects of these processes are observed not only in volcanic rocks and mantle nodules that originate within the subcontinental lithosphere, but from the ocean basins as well.

Several mechanisms for the enrichment have been proposed, including trapping of silicate melts, which themselves originate by small degrees of partial melting of spinel or garnet peridotite (Frey and Green, 1974); incorporation of a continental crustal component through subduction (Hoffman and White, 1982) or exchange with the lower crust (Semkin and DePaolo, 1983); and migration of trace elements in fluids of either hydrous or carbonic composition (Menzies and Murthy, 1980).

One result of an uncommon style of trace element enrichment has been the creation of regions of so-called "anomalous mantle" (Menzies and Murthy, 1980). This particular process is characterized by enrichment in many incompatible elements, particularly in the light rare-earth elements (LREE), without an accompanying increase in Rb or Rb/Sr, leading with time to Nd and Sr isotopic compositions that plot to the left of the mantle array. We discuss Nd, Sr, and Pb isotope data on a suite of alkaline ultrabasic rocks from a young continental fragment—South Island, New Zealand. These rocks exhibit the trace element and isotopic signatures of anomalous mantle; we present a model that accounts for their origin.

GEOLOGIC SETTING

The Westland and Otago Dike swarm has been described previously in detail (Cooper, 1971, 1979, 1986); the following summary represents only a brief synopsis of those reports. The dikes intrude the Upper Paleozoic–Mesozoic Haast Schist of South Island in an elongate north–northeast-trending zone more than 100 km long and 24 to 30 km wide (Fig. 1). The dikes intrude the country rock in bodies from a few centimeters to 85 m in thickness, which in many cases cross-cut the latest foliation and thus postdate the late metamorphic and early postmetamorphic Rangitata deformation of the Haast Schist. In addition to dikes and more concordant sills, diatreme intrusions are known from a small number of locations (Wallace, 1975; Cooper, 1986).

Potassium-argon mineral and whole-rock dates for the southern part of the swarm cluster about 16 to 28 Ma (Adams, 1980). The northern part of the swarm produces very discordant K-Ar results, and some degree of argon inheritance is inferred from the old dates obtained. New U-Pb and Rb-Sr dates of ~24 Ma on petrologically evolved members from the Haast River area, a tinguaite and a microsyenite (D. Kimbrough, personal communication, and unpublished data of B.B.) are within the range of dates from the southern part of the swarm, and suggest that the entire swarm may be of latest Oligocene–early Miocene age.

In the Burke River and Haast River areas, the dike swarm attains both its greatest intensity and its greatest lithologic diversity. There the rocks consist of peridotite (ouachitite, Streckeisen, 1980), volumetrically dominant camptonitic lamprophyre, sodalite tinguaite, trachyte, and carbonatite. The trachytes exhibit complex chemical and isotopic compositions that imply some degree of country rock interaction, and are not discussed further.

The ouachitites are rather coarse-grained and contain olivine, kaersutite, Ti-rich clinopyroxene, and phlogopite, an Fe-Ti oxide, apatite, and occasional perovskite, sphene, carbonate, feldspar, or analcime. The lamprophyres contain phenocrysts of olivine and complexly zoned kaersutite, titansalite, and titanbiotite. Characteristic leucocratic ocelli are set in groundmass amphibole, pyroxene, mica, feldspar, titanomagnetite, apatite, carbonate, and analcime. The major element compositions of the ouachitites and lamprophyres (Table 1) correspond to water- and CO_2-rich nephelinites and basanites. The ouachitites are all very undersaturated in silica, containing up to 15.5 percent normative nepheline and 1.6 percent normative Cs, although modal nepheline and melilite have not been seen. The lamprophyres range from highly undersaturated to moderately saturated in silica. Their Mg/(Mg + Fe) ratios (0.52 to 0.71) and relatively high Cr (120 to 960 ppm) and Ni (203 to 308 ppm; Cooper, 1986) concentrations indicate that some of the rocks are close to primitive compositions. These ultrabasic rocks have high concentrations of the incompatible elements: the large ion lithophiles (K, Rb, Ba, Sr, and Th) and the high field strength elements (Ti, P, Zr, Nb, and Y). They have TiO_2/K_2O and Na/K >1 and an average Rb/Sr = 0.045 (Tables 1 and 2, and Cooper, in press).

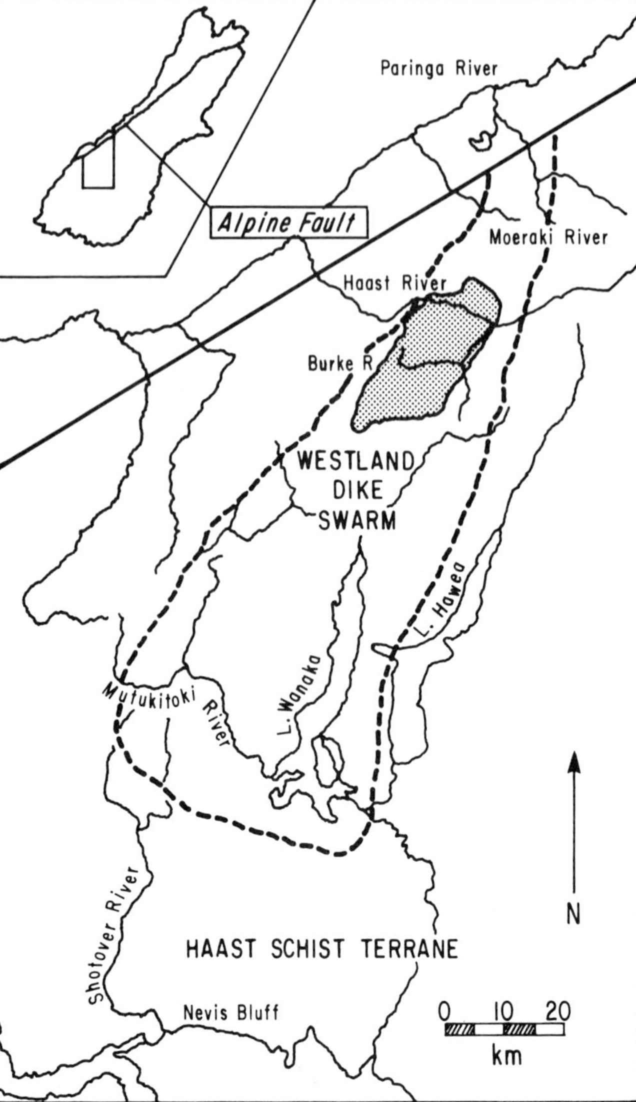

Figure 1. Regional extent of lamprophyre dike swarm in Westland and Otago, South Island, New Zealand. Shaded area represents region of most intense dike intrusion (after Cooper, 1979).

Relative abundance diagrams (Fig. 2) show the range of trace element concentrations normalized to midocean ridge basalts. The ouachitites and lamprophyres exhibit trace element patterns typical of within-plate basalts and are very similar to eastern Australia alkaline volcanic rocks (Wass, 1979).

We believe that the other dike rock types—the tinguaites, trachytes and carbonatites—evolved from a parental lamprophyric magma. Interstitial glasses of phonolitic composition found in the groundmass and ocelli of camptonitic lamprophyres are similar in composition to the sodic nepheline-sodalite-potassium feldspar-pyroxene-bearing tinguaites (Cooper, 1979), suggesting that they evolved via efficient fractional crystallization

TABLE 1. MAJOR ELEMENT DATA FOR WESTLAND DIKE SWARM

	Peridotites Sample Number				Lamprophyres Sample Number						Carbonatites Sample Number			Tinguaites Sample Number		
	59*	33	181	208	1236*	1163*	52*	CB 56	CB 82	CB107	1207*†	CB 14	CB 26	1229	CB 88	CB 99
SiO_2	38.40	35.0	33.6	37.7	43.69	44.28	45.65	38.5	40.9	38.8	8.60	2.77	7.42	52.64	48.7	51.8
TiO_2	3.65	3.30	3.69	2.36	3.00	2.64	2.48	3.14	2.76	3.11	0.39	0.32	0.66	0.75	1.17	0.61
Al_2O_3	9.67	7.50	8.54	9.20	12.07	14.04	14.90	9.47	9.81	11.30	3.90	0.78	2.20	20.02	19.6	19.1
Fe_2O_3	6.66	8.12	9.40	6.24	3.05	5.13	3.65	3.72	4.04	4.45	1.41	0.73	9.62	1.99	4.24	3.60
FeO	7.39	6.50	7.40	6.00	8.93	6.43	6.83	8.90	6.90	10.50	8.84	14.2	1.60	1.24	1.40	1.10
MnO	0.21	0.17	0.22	0.27	0.18	0.20	0.20	0.16	0.17	0.21	0.99	2.07	1.02	0.14	0.17	0.17
MgO	12.79	18.4	11.9	12.6	8.71	7.25	5.98	13.5	10.2	9.54	8.29	3.26	2.75	1.14	0.94	1.11
CaO	12.99	10.7	13.4	12.3	9.48	9.21	8.05	9.95	10.5	10.4	3.71	31.2	36.4	2.67	3.95	1.68
Na_2O	1.62	2.00	2.34	4.02	3.14	4.89	6.03	1.06	3.52	3.50	1.80	0.99	2.20	9.66	9.46	10.8
K_2O	2.06	1.86	2.64	2.01	1.16	2.06	2.58	1.65	1.45	2.48	0.01	0.25	0.14	4.59	4.35	4.54
P_2O_5	1.02	0.64	1.73	0.95	0.53	0.77	0.69	0.53	0.56	0.82	1.37	4.99	2.46	0.20	0.29	0.11
H_2O^+	1.71	3.37	2.39	2.27	2.02	1.52	1.52	5.01	1.86	1.90	0.16			2.06	0.61	3.40
H_2O^-	0.10				0.30	0.21	0.21				0.03			0.03		
CO_2	1.06	2.2	1.9	4.1	3.27	0.46	1.16	3.6	6.3	1.2	21.33	34.4	32.8	0.99	4.4	0.8
S	0.24				0.15						3.98			0.06		
Total	99.57	99.76	99.15	100.02	99.68	99.09	99.93	99.19	98.97	98.21	64.81	95.96	99.27	98.18	99.28	98.82
Mg/Mg+Fe	0.63	0.70	0.57	0.66	0.57	0.54	0.51	0.66	0.63	0.54	0.59	0.28	0.32	0.40	0.24	0.31
CIPW Norm																
Cc	2.47	5.19	4.45	9.54	7.65	1.08	2.69	8.69	14.75	2.83				2.35		1.91
Ap	2.48	1.57	4.24	2.3	1.29	1.87	1.66	1.33	1.37	2.02				0.49		0.27
Il	7.11	6.5	7.24	4.59	5.86	5.15	4.8	6.33	5.4	6.13				1.48		1.21
Mt	9.9	12.21	14.09	9.27	4.55	7.64	5.39	5.73	6.03	6.7				1.02		
Or	9.33	3.17		12.15	7.05	12.5	15.26	10.35	8.82	13.18				28.25		28.12
Ab				6.21	27.33	16.49	19.02	9.52	30.67					28.4		21.89
An	13.36	6.22	5.17	1.15	15.86	10.55	6.08	17.21	6.88	8.1						
Aug	30.61	23.43	26.63	22.3	6.21	22.39	18.12	5.86	1.57	25.87				4.72		2.17
Hyp					12.24			17.9	9.34							
Olv	14.71	25.78	12.91	17.09	12	8.28	8.91	17.1	15.2	16.96				1.06		2.34
Ne	7.62	9.51	11.09	15.49		14.09	17.84			16.66				28.29		29.57
Lc	2.47	6.46	12.64							1.59						
Cs			1.62													
Ac														3.96		10.92
Ns																1.61

*Analyses from Cooper, 1986. All other analyses by x-ray fluorescence, X-Ray Assay Laboratories.
†Includes BaO, 25.69%; ZnO, 5.95%; PbO, 1.21%; ReO, 1.92%; SrO, 2.39%.

TABLE 2. TRACE ELEMENT DATA FOR WESTLAND DIKE SWARM

Sample	Sr* (ppm)	Rb* (ppm)	$^{87}Rb/^{86}Sr$	Zr† (ppm)	Nb† (ppm)	Y† (ppm)	Cr† (ppm)	Nd§ (ppm)	Sm§ (ppm)	$^{147}Sm/^{144}Nd$
CB 59	1,155	33.6	0.084	348	142	27	278	77.7	13.2	0.1028
CB 33	950	44.1	0.134	250	90	30	960	58.2	10	0.1039
181	1,478	68.3	0.134	450	110	20	120	101	17.5	0.1048
208	1,456	70.6	0.140	420	120	20	270	85.2	13.5	0.0959
1236	726	35.0	0.139	247	59	24	226	38.5	7.56	0.1189
1163	1,236	55.5	0.165	351		28				
52-A	1,491	53.5	0.123	360				47	10	0.1287
CB 56	642	94.8	0.426	210	70	10	370	42	7.7	0.1109
CB 82	1,028	50.3	0.141	230	90	30	380			
CB 107	1,141	64.9	0.165	320	120	20	230			
1207	14,940	0	0.000		111	60	71	3,541	264	0.0451
CB 14	12,630	3.14	0.001	410	140	230	10	2,160	567	0.1589
CB 26	18,400	2.47	0.000	760	300	240	200	704	94.1	0.0809
1229	1,121	131	0.318	552	164	20	20			
CB 88	1,464	126	0.249	840	230	10	10	55.3	8.56	0.0936
CB 99	376	176	1.460	1,150	260	20	30	25.8	4.51	0.1059

*Isotope dilution.
†X-ray fluorescence, X-Ray Assay Laboratories and Cooper, 1986.
§Instrumental neutron activation analysis, N. W. Rogers, analyst; and Cooper, 1986.

of amphibole, titanbiotite, and apatite. As a result, the tinguaites and microsyenites are tremendously enriched in Na, K, Rb, Zr, and Nb, but depleted in P and Ti, relative to the lamprophyres.

Much attention has been focused on the carbonatite dikes (Cooper, 1971; Blattner and Cooper, 1974). They exhibit a diverse mineralogy that includes calcite, dolomite, ankerite, and siderite, as well as the more unusual norsethite ($BaMg[CO_3]_2$), strontianite, and daqingshanite ($3\,[SrBa]CO_3REE[PO_4]$). Among the accessory minerals are Na-silicates albite and acmite, sulfides, rutile, apatite, and monazite. Cooper (1979) has proposed that the carbonatites originated as immiscible liquids in equilibrium with silicate magmas of phonolitic composition (the tinguaites). They show typical extreme concentrations in Sr, Nb, Ba, and other trace elements. The three carbonatites analyzed in this study show an extreme diversity in chemical composition. One evolved norsethite-bearing dike (1207) is extremely rich in Ba; two others are more Ca-Fe rich.

The more mafic members of the dike swarm occasionally contain ultramafic inclusions and xenocrysts interpreted to be of mantle origin. The Moeraki diatreme, for instance, contains spinel lherzolite, dunite, and harzburgite nodules (Wallace, 1975), and similar Cr-diopside–suite nodules are also found in ouachitites and lamprophyres south of the Burke River (Brodie, 1985). In addition to these rather typical inclusions, some ultramafic dikes contain green clinopyroxene-kaersutite-apatite-mica–bearing inclusions and amphibole, clinopyroxene, and titanbiotite megacrysts. A similar inclusion and megacryst assembalge has previously been described from alkali basalts in New South Wales (Wass, 1979; Menzies and Wass, 1983), and mineralogic, trace element, and isotopic studies have concluded that they represent crystallized magmas or fluids rich in incompatible elements.

The presence of kaersutite, phlogopite, titanbiotite, and carbonate in the Westland rocks is a clear sign that there has been a volatile component involved in their genesis and evolution. To investigate some of the short-term and long-term effects of this relationship, 16 members of the suite were selected for Nd, Sr, and Pb isotopic analysis. The samples include four ouachititic peridotites (all megacryst or inclusion bearing), six lamprophyres, three carbonatites, and three of the more evolved sodic tinguaites or microsyenites. Table 3 contains the Sr, Nd, and Pb isotopic data; analytical procedures are outlined in the Appendix.

ISOTOPIC RESULTS

The Sr isotopic compositions measured in the 16 samples range from $^{87}Sr/^{86}Sr = 0.70285$-$0.70476$. Lamprophyre CB 56 is distinct in having far more radiogenic Sr than any of the other members of the suite. Among the 10 ultrabasic rocks analyzed, it has a lower Sr concentration ($\sim$600 ppm versus $>$1,000), a higher Rb/Sr ratio, and lower Zr and Nb contents. It seems possible, therefore, that this one dike has experienced some degree of chemical interaction with the country rocks. With the exception of that sample, the range in initial $^{87}Sr/^{86}Sr$ (calculated for an age of 25 m.y.) is between 0.70279 and 0.70348, and no relationship is observed with respect to rock type or Rb/Sr.

Present-day $^{143}Nd/^{144}Nd$ shows a narrow range in values between 0.5182 and 0.51291, which corresponds to a present-

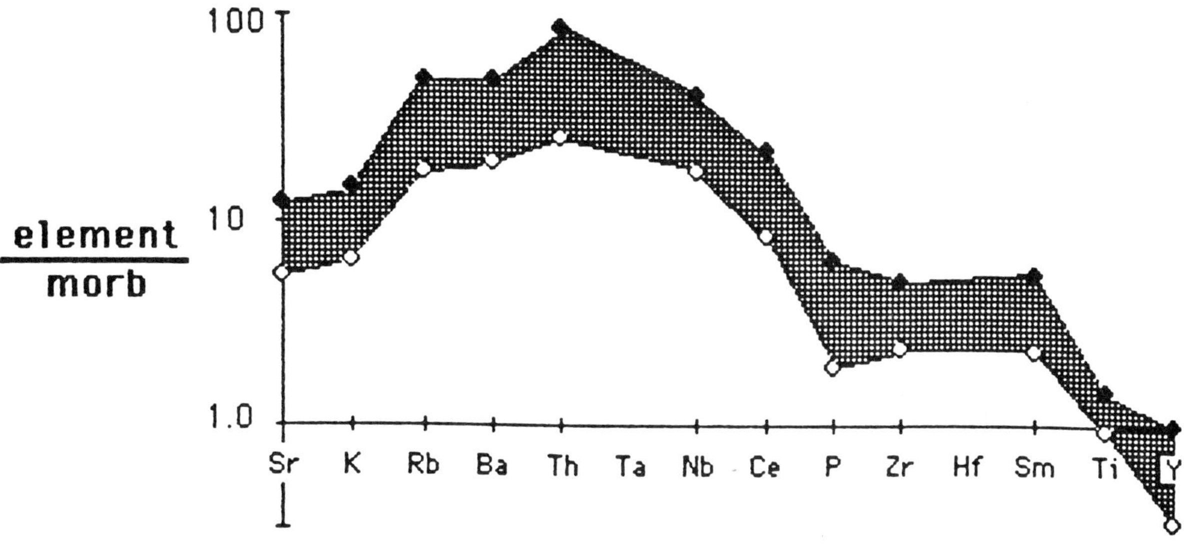

Figure 2. Range of mid-ocean ridge basalt (MORB) normalized trace element variations in peridotite and camptonite dikes. Ba, Th, Ce, and Sm data from Cooper, 1986; other elements from Tables 1 and 2. MORB-normalizing factors are Sr, 120; K, 0.15 percent; Rb, 2; Ba, 20; Th, 0.2; Ta, 0.18; Nb, 3.5; Ce, 10; P, 0.12 percent; Zr, 90; Hf, 2.4; Sm, 3.3; Ti, 1.5 percent; Y, 30 (from Pearce, 1983).

day ϵ_{Nd} = +3.5 to +5.2. If an age of 25 m.y. is assumed for the entire suite, the $\epsilon_{Nd(25\ Ma)}$ ranges from +3.6 to +5.5 (Table 3).

On a $^{143}Nd/^{144}Nd$-$^{87}Sr/^{86}Sr$ diagram (Fig. 3), 15 rocks of the Westland dike swarm fall to the left of the mantle array defined by midocean ridge basalts and most oceanic islands. The suite as a whole has $^{87}Sr/^{86}Sr$ ratios similar to those of the more radiogenic midocean ridge basalt, but lower $^{143}Nd/^{144}Nd$ ratios. Although the rocks are presently enriched in the LREE (Cooper, 1986), their Nd isotopic compositions show that their mantle source experienced a long-term depletion in the light rare-earth elements. In addition, a long-term depletion in Rb/Sr relative to bulk earth, similar in magnitude to that experienced by the source of the midocean ridge basalts, is suggested by the Sr isotope data, despite an average Rb/Sr value of 0.045 for the nine most primitive samples.

The dikes contain very radiogenic Pb, with values of $^{206}Pb/^{204}Pb$ = 19.19 to 20.59, $^{207}Pb/^{204}Pb$ = 15.64 to 15.71, and $^{208}Pb/^{204}Pb$ = 39.03 to 40.22 (Table 3, Fig. 4). Although these values do not account for the addition of radiogenic Pb since the rocks were intruded, the spread of values cannot have been generated by an isotopically homogeneous suite in 25 m.y. without impossibly high values of $^{238}U/^{204}Pb$ = 360. Thus, the source of the rocks was already heterogeneous with respect to Pb isotopic composition prior to magma generation. These values fall within the range of isotopic compositions previously observed for ocean island basalts (OIB). However, they seem to be characterized by high $^{207}Pb/^{204}Pb$ relative to many OIB. The rocks appear to be examples of the "Dupal" anomaly (Hart, 1984), an isotopic anomaly that may affect parts of the mantle in the Southern Hemisphere.

The combination of highly radiogenic Pb and Nd-Sr values to the left of the mantle array is characteristic of the isotopic compositions of the oceanic islands of St. Helena (O'Nions and others, 1980) and Tubuai (Vidal and others, 1984). In this respect the New Zealand rocks differ from suites of continental basalts whose Nd and Sr isotopic compositions fall to the left of the mantle array but whose Pb is decidedly nonradiogenic, such as the Skye basalts (Thirlwall and Jones, 1983).

DISCUSSION

Models that involve very small degrees of partial melting of a normal or midocean ridge basalt-type source to account for the abundances of incompatible elements in within-plate magmas (e.g., Kay and Gast, 1973) fail to satisfy other major and trace element criteria. In addition, the discrepancy between long-term temporal integration of Sm/Nd and Rb/Sr ratios, evidenced by Sr and Nd isotopic compositions and the short-term concentration measurements, has been noted for many suites of alkali basalts from both continental and oceanic regimes (see, for instance, Clague and Frey, 1982) and has led to the proposal that these rocks are generated subsequent to a recent episode of trace element enrichment that added LREE and other trace elements to the source of the magma.

The major element and trace element chemistry suggests that the Westland lamprophyres cannot be generated by even

TABLE 3. ISOTOPIC DATA FOR WESTLAND DIKE SWARM

Sample	$^{87}Sr/^{86}Sr(m)$*	$^{87}Sr/^{86}Sr(i)$†	$^{143}Nd/^{144}Nd$	εNd(O)§	εNd(T)†§	$T_{(DM)}$**	$^{206}Pb/^{204}Pb$	$^{207}Pb/^{204}Pb$	$^{208}Pb/^{204}Pb$
CB 59	0.70299±0.00005	0.70296	0.512849±0.000026	4.1	4.4	415			
CB 33	0.70325±0.00004	0.70320	0.512857±0.000028	4.3	4.6	408			
181	0.70285±0.00004	0.70281	0.512896±0.000028	5.0	5.3	357	20.59	15.71	40.22
208	0.70305±0.00004	0.70300	0.512856±0.000027	4.3	4.6	381	19.19	15.64	39.03
1236	0.70328±0.00005	0.70323	0.512906±0.000024	5.2	5.5	393	19.65	15.69	39.42
1163	0.70308±0.00004	0.70303	0.512897±0.000032	5.1			19.99	15.71	39.61
52-A	0.70307±0.00005	0.70302	0.512877±0.000028	4.7	4.9	491	20.52	15.65	40.22
CB 56	0.70476±0.00006	0.70461	0.512867±0.000021	4.5	4.7	421	19.49	15.65	39.25
CB 82	0.70309±0.00008	0.70304	0.512890±0.000031	4.9			19.45	15.67	39.34
CB 107	0.70285±0.00005	0.70279	0.512861±0.000025	4.4			20.06	15.70	39.70
1207	0.70348±0.00004	0.70348	0.512863±0.000027	4.4	4.9	260	19.92	15.68	39.50
CB 14	0.70346±0.00004	0.70346	0.512817±0.000024	3.5	3.6	425	19.90	15.65	39.99
CB 26	0.70299±0.00007	0.70299	0.512825±0.000028	3.6	4.0	374	20.31	15.68	40.42
1229	0.70298±0.00008	0.70287	0.512835±0.000031	3.8	n.d.		20.37	15.68	40.04
CB 88	0.70311±0.00006	0.70302	0.512857±0.000029	4.3	4.6	373	20.39	15.64	39.96
CB 99	0.70351±0.00008	0.70299	0.512853±0.000027	4.2	4.5	359	20.31	15.67	39.96

*2-sigma error.

†Calculated using values in Table 2, an age of 25 m.y., and $\lambda^{87}Rb=1.42 \times 10^{-11}$ and $\lambda^{147}Sm=6.54 \times 10^{-12}$.

§$\epsilon Nd = (((({}^{143}Nd/{}^{144}Nd)_{m,T}) / (({}^{143}Nd/{}^{144}Nd)_{CHUR,T}))-1) \times 10^4$

**$T_{(DM)} = (1/\lambda)\ln[\, 1+(0.51315-{}^{143}Nd/{}^{144}Nd) / (0.21362-{}^{147}Sm/{}^{144}Nd)]$.

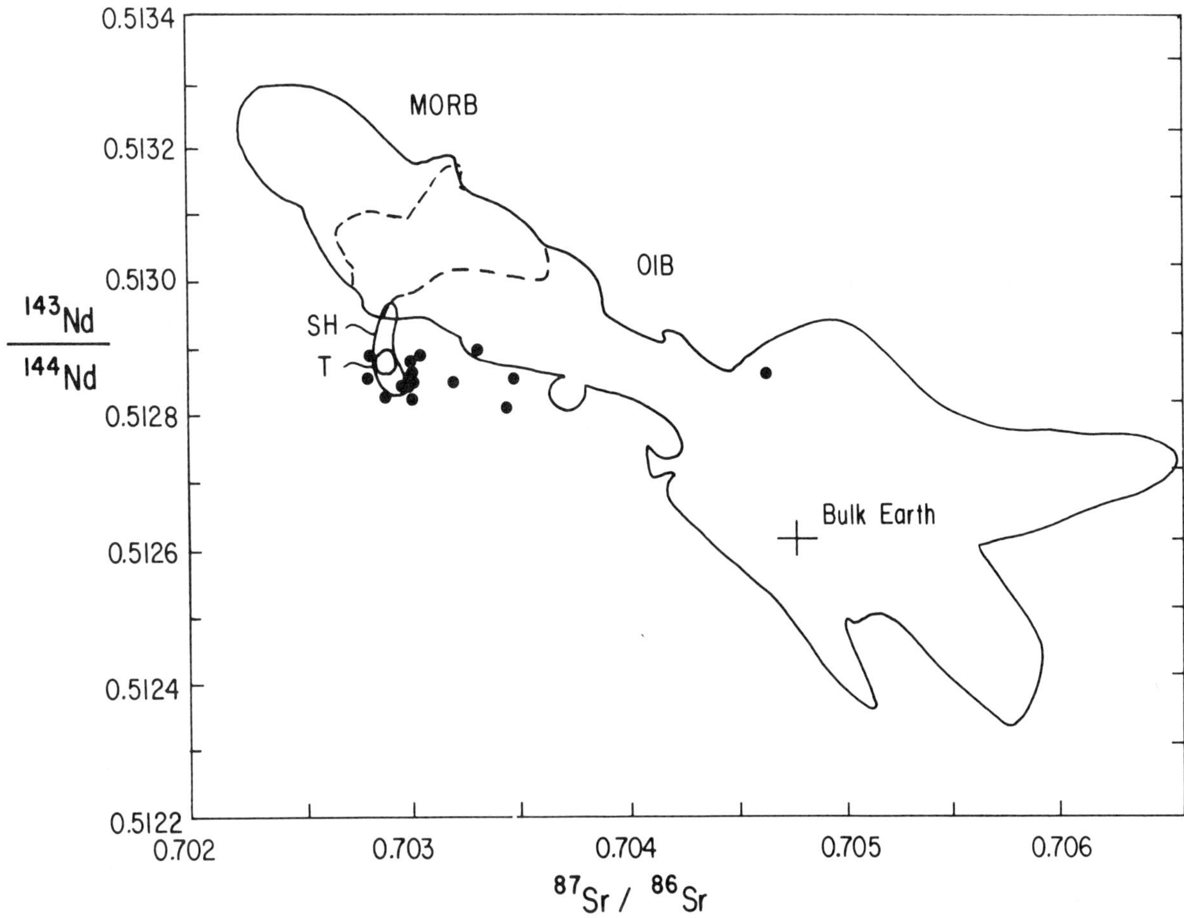

Figure 3. Nd and initial Sr isotopic compositions of Westland dikes. MORB and Ocean Island Basalt data from literature. SH, St. Helena (Cohen and O'Nions, 1982); T, Tubuai (Vidal and others, 1984).

small degrees of partial melting of a garnet peridotite with trace element abundances similar to either chondrites or to the source of midocean ridge basalt. The nonradiogenic Sr and positive ϵ_{Nd} show that the source of the Westland rocks was in fact depleted in Rb and the light rare-earth elements relative to bulk earth over most of the earth's history, and that it had at least a Rb/Sr and Sm/Nd very similar to the depleted midocean ridge basalt source. The data presented here, however, indicate that the mantle source of the Westland rocks, like that of many other alkalic rocks, is enriched in many trace elements, particularly in the high field strength elements (Ti, P, Nb), the large ion lithophile-incompatible elements (Sr, K, Rb, U, and Th), and LREE compared to average "primordial" mantle. Thus this source has experienced a recent episode of trace element addition to a previously depleted peridotite.

If the extent of partial melting of mantle peridotite to produce the primitive Westland dikes was large enough to preclude significant fractionation of Sm from Nd, the measured parent/daughter ratios and Nd isotopic compositions can be used to calculate model ages dating the trace element enrichment of a depleted mantle that has present-day values of $^{147}Sm/^{144}Nd$ = 0.21362 and $^{143}Nd/^{144}Nd$ = 0.51315 (Miller and O'Nions, 1984). These calculations yield Nd model ages (T_{DM}) of between 260 and 491 m.y., with a mean model age of 409 ±42 m.y. years for the peridotites and lamprophyres (Table 3). If some degree of Nd-Sm separation did occur during partial melting, and the source of the lamprophyres had a $^{147}Sm/^{144}Nd$ ratio as high as 0.15 (still representative of a slightly LREE-enriched source), the Sm/Nd model age would increase to 660 m.y. However, additional model age constraints are provided by the Pb isotope data, which scatter around a zero-age line. The small spread of $^{207}Pb/^{204}Pb$ prohibits any U enrichment of a single source for all the rocks older than approximately 200 m.y. Together, then, the two isotope systems imply a Mid- to Late Paleozoic, but no older, model age for the trace element enrichment. It is unclear what relation this event may have had with regard to the late Paleozoic–Mesozoic deposition and crustal stabilization of the Haast Schist Terrain. Coombs and others (1986) in fact suggested that

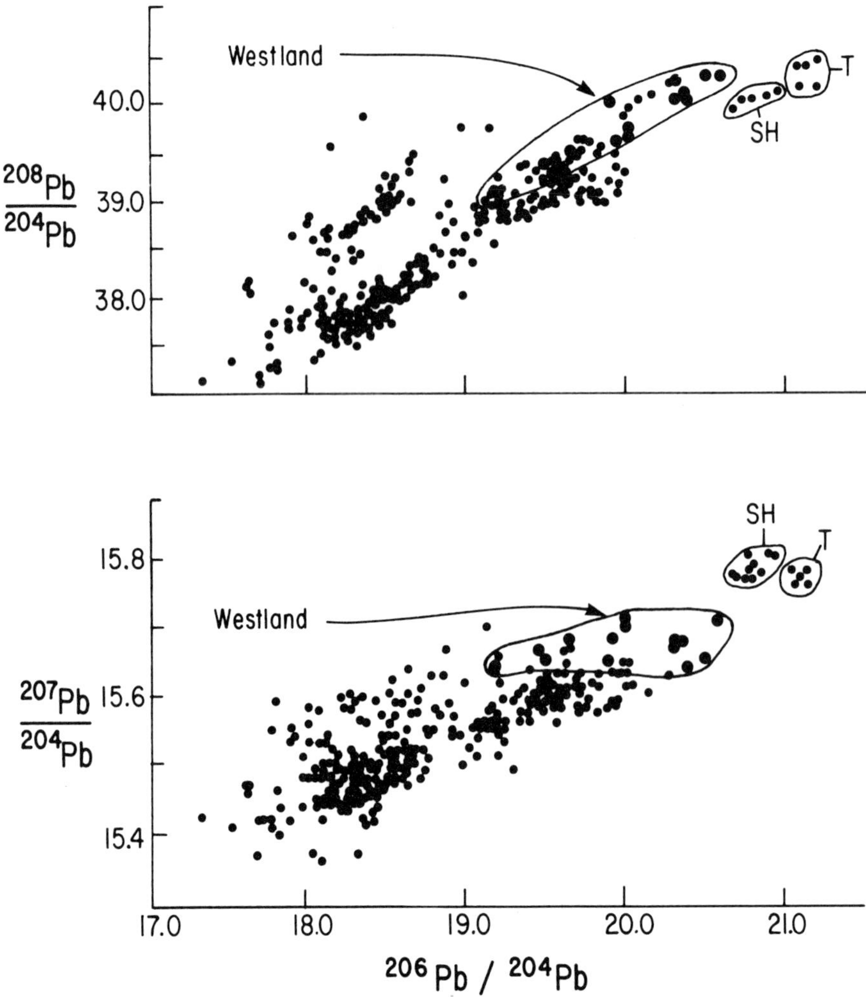

Figure 4. Pb isotopic compositions of Westland dikes compared to data from MORB and OIB from literature. SH, St. Helena, T, Tubuai. Modified from Barreiro (1983), which contains individual references.

this trace element enrichment of depleted mantle peridotite was more widespread, and affected the lithosphere beneath what is now Eastern Australia, New Zealand, and Antarctica. Whatever its geographic extent, it is clear from the model ages that the enrichment preceded the intrusion of the dike swarm by a few hundred million years.

Many workers consider that mantle metasomatism is responsible for the production of at least some of the "anomalous" mantle compositions suggested by trace element and isotopic studies of within-plate basalts and mantle nodules. A number of possible mechanisms have been suggested, most requiring that the original mantle suffered influx of either a magma (i.e., a melt possessing a high content of silicate or carbonate material) or a supercritical fluid (i.e., a volatile-dominated phase with little dissolved silicate component). For example, hydrous fluids rich in the large ion lithophile elements but not in the high field-strength elements (such as Ti, P, Zr, Nb, and Ta) may be implicated in the enrichment of the sources of island-arc magmas and in certain suites of continental basalts and mantle nodules from South Africa, particularly the suites characterized by high K_2O/TiO_2 and Rb/Sr ratios (Hawkesworth and others, 1984). In contrast, the strongly silica-undersaturated, alkali-rich nature of many igneous rock suites considered to have been derived from metasomatized mantle has led several authors to suggest that their source regions experienced influx of CO_2-rich fluids, which carry appreciable amounts of Sr and LREE (Wendlandt and Harrison, 1979), but perhaps not much Rb (see Anderson and others, 1984, for a review). The Westland and Otago lamprophyre suite might be considered an appropriate candidate for the second type of fluid metasomatism described in light of their CO_2-rich nature.

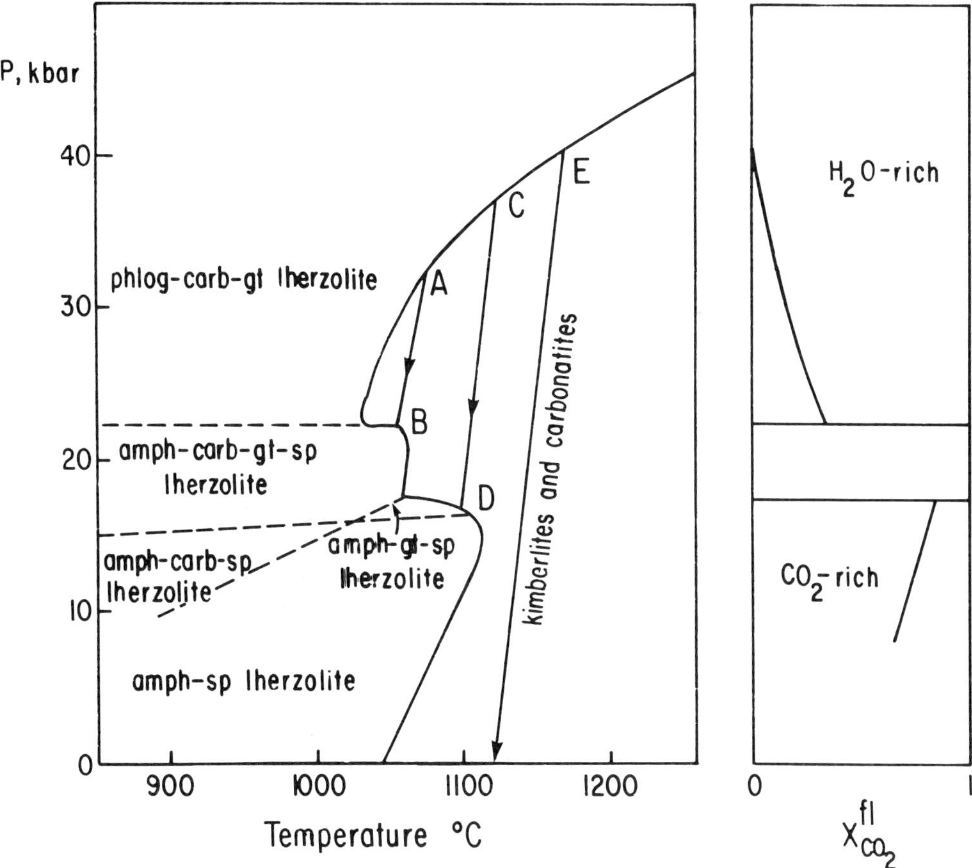

Figure 5. Phase relations of carbonated, hydrated peridotite; after Olafsson and Eggler (1983). Partial melting and migration along path A->B discussed in text. Partial melting at much higher pressures and depths produces melts that do not intercept solidus and reach surface relatively unmodified as kimberlites or carbonatites.

However, experimental studies of mantle materials show not only that the source region of the Westland dikes cannot have suffered the influx of carbonic fluids, but also offer alternate models for the generation of the lamprophyre suite.

Partial melting experiments conducted on carbonated, hydrated peridotites (Fig. 5) show that ne-normative melts are produced by low degrees of melting at 13 kbar under H_2O-CO_2–buffered conditions (Olafsson and Eggler, 1983). Lamprophyric and carbonatitic magmas require greater depths (>17.5 kbar or 58 km) of origin, where carbonate is stable. However, there is a region at depths between 58 and 72 km that can support a free fluid phase only if the mantle is richer in volatiles than the 0.3 percent H_2O and 5 percent CO_2 buffered by the presence of amphibole and carbonate. Even if oversaturation occurs locally, migration of any fluid into regions of unsaturated mantle will result in the consumption of the fluid in this depth interval. At pressures greater than 22 kbar, amphibole breaks down and a fluid is again stable; yet the stability of carbonate means that this fluid is necessarily hydrous rather than carbonic (Fig. 5). Thus, although CO_2-rich fluids might explain the trace element enrichments observed in the Westland dikes, carbonic fluids are not stable at the depths at which the generation of these magmas is believed to have occurred. However, examination of Figure 5 suggests an alternate explanation for the origin of the Westland dikes.

Because the peridotite solidus passes through two cusps or abrupt changes in slope, a carbonated hyperalkaline magma formed at point A by a small degree of partial melting of carbonate-bearing peridotite with depleted trace element characteristics, will, upon rising, define a *P-T* path that intersects the solidus again at point B. At this lower pressure, consumption of the melt by the wallrock peridotite will create an exotic mineral assemblage containing such species as carbonate, amphibole, and apatite. We believe that fragments of this assemblage are preserved in the Westland lamprophyres as the aforementioned apatite–amphibole–green clinopyroxene nodules. Since the characteristic signatures of carbonated hyperalkaline magmas include enrichments in LREE, alkaline earth metals, Nb, Ta, P, and other high field strength elements, as well as U and Th, the newly metasomatized peridotite was presumably enriched in Nd relative to Sm, U and Th relative to Pb, and in some cases Sr relative to Rb. Therefore, over time, the enriched mantle developed radiogenic Pb but nonradiogenic Sr and Nd isotopic compositions in response to the new U-Th/Pb, Sm/Nd, and Rb/Sr ratios. The Paleozoic Sm/Nd model ages of the Westland dike swarm give an estimate of the timing of this interaction between the carbonated hyperalkaline magma and the normal "depleted" peridotite.

After some hundreds of million years of isotopic evolution, a second thermal or chemical event triggered partial melting of the metasomatized peridotite, at which time the more exotic or volatile and trace element–rich phases were preferentially incorporated into the melt. The Westland dikes represent early, small percentage partial melts. Eocene-Oligocene volcanics, in part alkaline, and widespread in the northern and eastern region of South Island, may represent greater melt fractions. Miocene alkaline volcanism in East Otago is in part synchronous with the youngest Westland dike intrusions (Coombs and others, 1986) and is likely another manifestation of the same triggering event.

SUMMARY AND CONCLUSIONS

In a recent paper, Hawkesworth and others (1984) compared trace element enrichment patterns of basalts from oceanic and continental environments, and used a TiO_2 versus K_2O plot to distinguish two classes of basalts. Basalts with high Ti/K ratios (like the lamprophyres described in this study), do not seem to be related to water-rich subduction fluids. These authors proposed that this style of mantle enrichment is due to the addition of small quantities of partial melts of basanite composition that were themselves derived from miniscule degrees of partial melting of "normal" depleted mantle.

The model presented here is similar in that magma, rather than fluid, is invoked as the agent of trace element enrichment; however, we envisage interaction with a magma that is more CO_2-rich than basanite. Our mechanism does not require the fortuitous freezing of rising magmas; peridotite-magma interaction is a natural consequence of the shape of the H_2CO_2-peridotite solidus and of the path followed by rising carbonated hyperalkaline magmas derived from fairly shallow depths within the garnet lherzolite stability field. This process may be locally important within the old continental lithosphere, beneath young continental fragments such as New Zealand, and within the oceanic lithosphere.

Consideration of the evolution of magmas formed by small degrees of partial melting at even greater depths, resulting in either consumption of the magma at lower *P-T* conditions within the field of CO_2-rich fluid compositions (path C-D) has been discussed in part by Anderson and others (1984) and is beyond the scope of this paper (but is addressed in part by Meen, this volume).

ACKNOWLEDGMENTS

This work was done while one of us (B.B.) was a postdoctoral fellow at the Department of Terrestrial Magnetism, Carnegie Institution of Washington. Discussions with Rick Carlson, Jim Meen, and Bill Hart contributed to the work. Comments from Kiyoto Futa and an anonymous reviewer also improved the manuscript. The Faculty Research Fund of Dartmouth College partially supported the research.

APPENDIX

ANALYTICAL PROCEDURES

All isotopic analyses were performed at the Department of Terrestrial Magnetism, Carnegie Institution of Washington. Powdered samples were dissolved in closed Teflon PFA screw-top beakers using HF and HNO_3 acids. The samples were spiked with ^{85}Rb and ^{84}Sr prior to dissolution for simultaneous isotope dilution and composition measurements. Conventional cation-exchange chromatography using 2N and 4N HCl as eluants separated Rb, Sr, and the REE. A second ion-exchange column with a 2-methyllactic acid eluant extracted Nd from the other REE.

Sr isotopic compositions were measured on a 6-in. mass spectrometer with ion currents of 1 to 2×10^{-11} A. Results are fractionation corrected to $^{86}Sr/^{86}Sr = 0.1194$ and reported relative to $^{87}Sr/^{86}Sr = 0.70800$ for the E&A Sr standard.

Nd was loaded onto a single Re filament and isotopic compositions were measured on a 15-in. mass spectrometer as NdO+ with currents of 3 to 6×10^{-12} A. Ratios are fractionation corrected to $^{146}NdO/^{144}NdO = 0.722251$ ($^{146}Nd/^{144}Nd = 0.7219$) and then for oxygen using the isotopic compositions of Nier (1950). Interference by Sm was monitored by comparing the sample's $^{142}Nd/^{144}Nd$ or $^{148}Nd/^{142}Nd$ to those measured in the La Jolla Nd standard. Small corrections of less than 1 part in 10,000 were made for about one-third of the samples. No other interferences were observed. The data in Table 3 are reported relative to $^{143}Nd/^{144}Nd = 0.511680$ for the La Jolla Nd standard.

A separate dissolution was used for Pb isotopic composition measurement. Extraction of Pb was achieved using two HCl-HBr anion exchange columns. Pb was loaded onto a Re filament with H_3PO_4 and silica gel and measured on the 15-in. mass spectrometer with typical ion currents of 5×10^{-12} to 5×10^{-11} A. The ratios are corrected for fractionation based on repeated analyses of NBS 981 and 982. Within-run errors on Pb isotope ratios did not exceed 0.07 percent for $^{204}Pb/^{206}Pb$, and the reported ratios have an average estimated error of 0.1 percent.

Sr, Nd, and Pb blanks are 0.4, 0.2, and 0.7 to 1.6 ng, respectively; no blank corrections were made to the data.

Major and trace element data for samples 52-A, 1163, 1236, 1229, and 1207 are from Cooper (1986). The rest were measured by X-Ray Assay Laboratories Limited of Ontario, using x-ray fluorescence and wet chemical methods. They cite uncertainties of 0.01 percent for major elements and 10 ppm for trace elements.

REFERENCES CITED

Adams, C. J., 1980, New K-Ar age data for South Island lamprophyre dyke swarms in the Buller-S. Westland and Haast-Wanaka areas: Geological Society of New Zealand Conference, Christchurch, Abstracts, p. 14.

Anderson, T., O'Reilly, S. Y., and Griffin, W. L., 1984, The trapped fluid phase in upper mantle xenoliths from Victoria, Australia: Implications for mantle metasomatism: Contributions to Mineralogy and Petrology, v. 88, p. 72–85.

Barreiro, B., 1983, An isotopic study of Westland dike swarm, South Island, New Zealand: Carnegie Institution of Washington Year Book 82, p. 471–475.

Blattner, P., and Cooper, A. F., 1974, Carbon and oxygen isotopic composition of carbonatite dikes and metamorphic country rock of the Haast Schist Terrain, New Zealand: Contributions to Mineralogy and Petrology, v. 44, p. 17–27.

Brodie, C. G., 1985, Geology of the Blue River–Burke River area and a study of the nodule-bearing lamprophyres [M.Sc thesis]: University of Otago.

Clague, D. A., and Frey, F. A., 1982, Petrology and trace element geochemistry of the Honolulu Volcanics, Oahu; Implications for the oceanic mantle below Hawaii: Journal of Petrology, v. 23, p. 447–504.

Cohen, R. S., and O'Nions, R. K., 1982, Identification of recycled continental material in the mantle from Sr, Nd, and Pb isotope investigations: Earth and Planetary Science Letters, v. 61, p. 73–84.

Coombs, D. S., Cas, R., Kawachi, Y., Landis, C. A., McDonough, W. F., and Reay, A., 1986, Cenozoic volcanism in north and east Otago, *in* Smith, I.E.M., ed., Late Cenozoic volcanism in New Zealand: Publications of the Royal Society of New Zealand (in press).

Cooper, A. F., 1971, Carbonatites and fenitization associated with a lamprophyric dike-swarm intrusive into schists of the New Zealand geosyncline: Geological Society of America Bulletin, v. 82, p. 1327–1340.

—— , 1979, Petrology of ocellar lamprophyres from western Otago, New Zealand: Journal of Petrology, v. 20, p. 139–163.

—— , 1986, A carbonatitic lamprophyre dyke swarm from the Southern Alps, Otago, and Westland, New Zealand: International Volcanological Association meeting (in press).

Frey, F. A., and Green, D. H., 1974, The mineralogy, geochemistry, and origin of lherzolite inclusions in Victorian basanites: Geochimica et Cosmochimica Acta, v. 38, p. 1023–1059.

Hart, S. R., 1984, A large-scale isotope anomaly in the Southern Hemisphere mantle: Nature, v. 309, p. 753–757.

Hawkesworth, C. J., Rogers, N. W., van Calsteren, P.W.C., and Menzies, M., 1984, Mantle enrichment processes: Nature, v. 311, p. 331–335.

Hoffman, A. W., and White, W. M., 1982, Mantle plumes from ancient oceanic crust: Earth and Planetary Science Letters, v. 57, p. 421–436.

Kay, R. W., and Gast, P. W., 1973, The rare earth content and origin of alkali-rich basalts: Journal of Geology, v. 81, p. 653–682.

Menzies, M. A., and Murthy, V. R., 1980, Nd and Sr isotope geochemistry of hydrous mantle nodules and their host alkali basalts; Implications for local heterogeneities in metasomatically veined mantle: Earth and Planetary Science Letters, v. 46, p. 323–334.

Menzies, M. A., and Wass, S. Y., 1983, CO_2- and LREE-rich mantle below eastern Australia; A REE and isotopic study of alkaline magmas and apatite-rich mantle xenoliths from the Southern Highlands Province, Australia: Earth and Planetary Science Letters, v. 65, p. 287–302.

Miller, R. G., and O'Nions, R. K., 1984, The provenance and crustal residence ages of British sediments in relation to paleogeographic reconstructions: Earth and Planetary Science Letters, v. 68, p. 459–470.

Nier, A. O., 1950, A redetermination of the relative abundances of the isotopes of carbon, nitrogen, oxygen, argon, and potassium: Physics Review, v. 77, p. 789–793.

Olafsson, M., and Eggler, D. H., 1983, Phase relations of amphibole, amphibole-carbonate, and phlogopite-carbonate peridotite; Petrologic constraints on the asthenosphere: Earth and Planetary Science Letters, v. 64, p. 305–315.

Pearce, J. A., 1983, Role of the sub-continental lithosphere in magma genesis at active continental margins, *in* Hawkesworth, C. J., and Norry, M. J., eds., Continental basalts and mantle xenoliths: Nantwich, United Kingdom, Shiva Publishing, p. 230–249.

Semken, S. C., and DePaolo, D. J., 1983, Sm-Nd and Rb-Sr constraints on Cenozoic basalt petrogenesis and mantle heterogeneity in the SW Great Basin [abs.]: EOS American Geophysical Union Transactions, v. 64, p. 338.

Streckeisen, A., 1980, Classification and nomenclature of volcanic rocks, lamprophyres, carbonatites, and melilitic rocks IUGS Subcommission on the systematics of igneous rocks: Geologiches Rundschau, v. 69, p. 194–207.

Thirlwall, M. F., and Jones, N. W., 1983, Isotope geochemistry and contamination mechanics of Tertiary lavas from Skye, northwest Scotland, *in* Hawkesworth, C. J., and Norry, M. J., eds., Continental basalts and mantle nodules: Nantwich, United Kingdom, Shiva Publishing, p. 186–208.

Vidal, P., Chauvel, C., and Brousse, R., 1984, Large mantle heterogeneity beneath French Polynesia: Nature, v. 307, p. 536–538.

Wallace, R. C., 1975, Mineralogy and petrology of xenoliths in a diatreme from south Westland, New Zealand: Contributions to Mineralogy and Petrology, v. 49, p. 191–199.

Wass, S. Y., 1979, Fractional crystallisation in the mantle of late-stage kimberlitic liquids; Evidence in xenoliths from the Kiama area, N.S.W., Australia, *in* Boyd, F. R., and Meyer, H.O.A., eds., The mantle sample; Inclusions in kimberlites and other volcanics: Washington, D.C., American Geophysical Union, p. 366–373.

Wendlandt, R. F., and Harrison, W. J., 1979, Rare earth partitioning between immiscible carbonate and silicate liquids and CO_2 vapour; Results and implications for the formation of light rare earth-enriched rocks: Contributions to Mineralogy and Petrology, v. 69, p. 409–419.

MANUSCRIPT ACCEPTED BY THE SOCIETY OCTOBER 15, 1986

Printed in U.S.A.

Geological Society of America
Special Paper 215
1987

Rb/Sr and Sm/Nd ratios of metasomatized mantle; Implications for the role of metasomatized mantle in the petrogenesis of Na_2O-rich alkaline basalts

Michael F. Roden
Department of Geology, University of Georgia, Athens, Georgia 30602

ABSTRACT

Relatively low Rb/Sr and Sm/Nd ratios are typical in peridotites from at least three examples (Nunivak, Kiama, St. Paul's Rocks) postulated to be metasomatized mantle. With time, such metasomatized mantle will develop isotopic characteristics unlike most Na_2O-rich alkaline basalts (but similar to some K_2O-rich alkaline basalts). Consequently, it is unlikely that old metasomatized mantle is the source of Na_2O-rich alkaline basalts. The relative abundance of metasomatized spinel peridotites indicates that such geochemically anomalous mantle may be common at relatively shallow depths. Mantle metasomatism largely may occur at the Moho, where the density contrast between crustal rocks and mantle peridotite leads to the trapping and crystallization of alkaline basalts. Residual melts and fluids derived from these alkaline magmas are the agents of metasomatism.

It is also emphasized that most metasomatized mantle peridotites, even those with high La/Ce ratios, crystallized in an open system; i.e., some of the metasomatic fluid escaped during or following crystallization. Thus, it is difficult to infer the specific fluid compositions from bulk rock compositions without considering elemental fractionation between solids and fluids.

INTRODUCTION

Mantle metasomatism is a process in which the bulk composition and, commonly, the phase assemblages of mantle peridotites are modified by reaction between peridotite and melt or fluid (see Roden and Murthy, 1985, and references therein). Two of the hallmarks of this process are the presence of hydrous and other incompatible element–rich phases that texturally appear to replace the primary phases, and the marked relative enrichment in the light rare earth elements (LREE) and other incompatible trace elements. Many peridotite xenoliths in basalts exhibit these metasomatic hallmarks. The metasomatic phases are stable at high temperatures and pressures and frequently exhibit disequilibrium textures where in contact with the host basalt. Consequently, it is unlikely that these phases formed from the near-surface contamination by host basalt or aqueous fluids. In some cases, isotopic data are consistent with a genetic relationship existing between host or related basalt and xenoliths (e.g., Menzies and Murthy, 1980); in other cases, isotopic data preclude such a relationship (e.g., Stosch and others, 1980). Often, parent/daughter element ratios of the xenolith and/or constituent minerals are inconsistent with their present-day isotopic composition (especially in the case of the ^{147}Sm-^{143}Nd decay system). This inconsistency requires that the metasomatism be relatively recent (e.g., Menzies and Murthy, 1980; Zindler and Jagoutz, 1987). Metasomatized regions appear to be common in the uppermost mantle beneath regions of alkaline volcanism (e.g., Frey and Prinz, 1978; Stosch and Seck, 1980; Menzies and Murthy, 1980); moreover, relatively large bodies of metasomatized peridotites also occur in ultramafic massifs such as St. Paul's Rocks, which were tectonically emplaced into oceanic crust (M. K. Roden and others, 1984). Still uncertain, however, is the role of metasomatism in the source region of alkaline basalts, especially the more abundant Na_2O-rich basalts (for the purposes of this paper, these latter basalts are defined as having $K_2O/(K_2O + Na_2O < 0.5$ [wt. %]). It is this last problem that this paper addresses.

It is my purpose to reconsider the significance of parent/

daughter element ratios in two typical examples of mantle metasomatism: Nunivak Island, Alaska, where metasomatized peridotite xenoliths have been brought to the surface by basanites (Francis, 1976), and St. Paul's Rocks, equatorial Atlantic Ocean, where an ultramafic body composed of hornblendites and amphibole-peridotites has been tectonically emplaced into oceanic crust (Melson and others, 1972). Peridotites from both localities are characterized by relative LREE–enrichment (i.e., relatively low Sm/Nd ratios) and $^{143}Nd/^{144}Nd$ ratios greater than chondrites (Frey, 1970; Roden and others, 1984a; M. K. Roden and others, 1984). These isotopic systematics require recent (<350 Ma) metasomatic events that resulted in increased relative abundances of the LREE and presumably other incompatible elements. It has been suggested that both of these examples are compositionally similar to the source material for alkaline basalts; indeed, there is no question that the geochemistry of these peridotites resembles the inferred source of alkaline basalts (Menzies and Murthy, 1980; M. K. Roden and others, 1984). However, if the Rb/Sr and Sm/Nd ratios of the peridotites from these two localities are typical of metasomatized mantle, then these metasomatized regions will evolve with time to become highly heterogeneous material largely characterized by low $^{87}Sr/^{86}Sr$ ratios and low $^{143}Nd/^{144}Nd$ ratios. Such "aged" metasomatic mantle is isotopically unlike the source regions for typical Na_2O-rich alkaline basalts, which are characterized by relatively high $^{143}Nd/^{144}Nd$ ratios (Fig. 1). These arguments do not eliminate the possibility of recent metasomatism in the source regions of these basalts, but with the advent of models for the separation of small amounts of melt from peridotite, it may no longer be necessary to postulate enriched sources for incompatible element-rich basalts (e.g., McKenzie, 1984). In contrast to the Na_2O-rich alkaline basalts, the isotopic characteristics of some K_2O-rich alkaline basalts (Fig. 1) are consistent with derivation from such heterogeneous, metasomatized regions (Menzies and Wass, 1983).

I also postulate that most metasomatized mantle is localized near the crust-mantle boundary, that it results from the trapping and differentiation of alkaline basalts near this density discontinuity, and that the localization of metasomatized mantle at shallow levels in the upper mantle effectively prevents this material from contributing to basaltic magmas (see below).

GEOLOGIC BACKGROUND AND DATA SOURCES

Nunivak Island

Nunivak Island (166°W, 60°N; diameter, 75 km) is a basaltic island located on the Bering Sea shelf 50 km west of the Alaska coast and 600 km north of the Aleutian Islands. Thin olivine tholeiite flows are capped by xenolith-bearing alkaline basalt flows with associated cinder cones and maars. All the volcanism is younger than 6.1 Ma. The metasomatized peridotites are coarse-grained Group I peridotites (nomenclature of Frey and Prinz, 1978) and are associated with fine-grained unmetasomatized lherzolites, pyroxene granulites, and Group II amphibole pyroxenites. Further details about field relations and petrology can be found in Francis (1976, 1978) and the references therein. The data for this paper are from Roden and others (1984a).

St. Paul's Rocks

St. Paul's Rocks (0°56′N, 29°22′W) is an amphibole-rich, mylonitized peridotite massif (total volume estimated to be ~400 km^3) emplaced into the oceanic crust of the equatorial Atlantic. The age of emplacement is uncertain but must be between mid-Miocene and Quaternary. Lithologies include spinel peridotite, amphibole peridotite, and hornblendites; these rocks are remarkably enriched in incompatible elements relative to most peridotites from ultramafic massifs. Further details on the geology, petrology, and geochemistry of St. Paul's Rocks may be found in the primary data sources for the present paper: Frey (1970), Melson and others (1972), and M. K. Roden and others (1984).

DISCUSSION

Relationship of Measured Bulk Rock Parent/Daughter Element Ratios to Mantle Values

As stated in the introduction, the Sm-Nd isotopic characteristics of peridotites from St. Paul's Rocks and Nunivak Island require relatively recent, metasomatic events. Trace element abundances, normalized to primitive mantle abundances, of metasomatized Group I peridotites from Nunivak Island and metasomatized peridotites and hornblendites from St. Paul's Rocks are shown in Figure 2. Both rock suites show the same general pattern of increasing absolute abundance of elements with increasing incompatibility, although the range for St. Paul's rocks extends to higher absolute abundances. Even though a few samples have lower abundances of some incompatible elements than primitive mantle, most samples have higher abundances of these elements than primitive mantle, and typically have La/Yb ratios greater than chondrites or primitive mantle (compare data in Tables 1 and 2 to primitive mantle abundances in Table 3). Both suites show a similar relative abundance pattern that is characterized by a comparative flattening in the region bounded by Nd and Rb. As described below, I believe that this flattening reflects open-system crystallization of metasomatic phases.

The following discussion is predicated on the assumption that the measured bulk rock Rb/Sr and Sm/Nd ratios of the Nunivak xenoliths and the peridotites and hornblendites from St. Paul's Rocks approximate mantle values. Because of the possibility that these ratios have changed—either by interaction with basaltic magma (in the case of the Nunivak inclusions) or with sea water (in the case of peridotites from St. Paul's Rocks)—this assumption must be critically examined. In principle, if it can be shown that the measured bulk rock ratios are consistent with mass balance calculations based on mineral analyses and modal

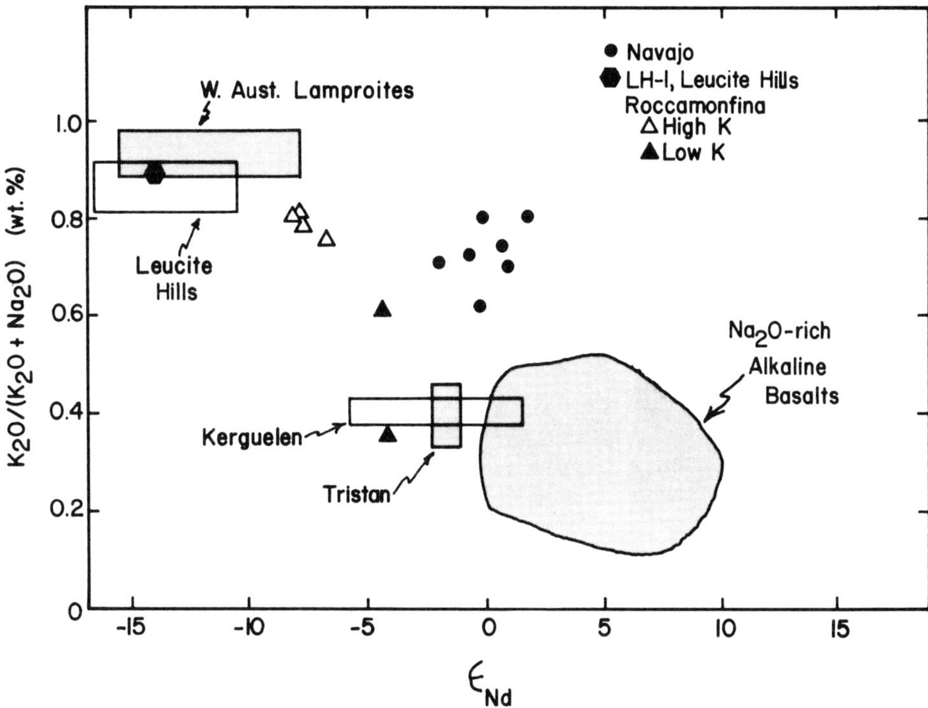

Figure 1. $K_2O/(K_2O + Na_2O)$ versus ϵ_{Nd} for various alkaline suites. Individual symbols indicate samples for which both major element and initial Nd isotopic compositions known. Rectangles indicate approximate plotting position of given suites, based on isotopic and major element compositions determined on separate samples (western Australian lamproites: Jaques and others, 1984, McCulloch and others, 1983; Leucite Hills: Carmichael, 1967, Vollmer and others, 1984; Tristan da Cunha: Baker and others, 1964, O'Nions and others, 1977, White and Hofmann, 1982; Kerguelen: Watkins and others, 1974, Dosso and Murthy, 1980). Field for Na_2O-rich alkaline basalts defined by data for basalts from Grenada (Arculus, 1976; Hawkesworth and others, 1979a), Honolulu Volcanic Series (Clague and Frey, 1982; Roden and others, 1984b), Azores (White and others, 1979; White and Hofmann, 1982), Nunivak Island (D. M. Francis, unpublished data; Roden, 1982), St. Paul's Rocks (Melson and others, 1972; M. K. Roden and others, 1984), Nyiragongo (Bell and Powell, 1969; Vollmer and Norry, 1983a), Massif Central (Chauvel and Jahn, 1984), Samoa (Hedge and others, 1972; White and Hofmann, 1982), Guadalupe Island (Engel and others, 1965; White and Hofmann, 1982), Iceland (O'Nions and others, 1976, 1977), St. Helena (Daly, 1927; White and Hofmann, 1982), West Antarctica (Futa and Le Masurier, 1983), and the approximate plotting position of the Kenya Rift lavas (Norry and others, 1980). Data for Roccamonfina and Navajo lavas from Cox and others (1976), Hawkesworth and Vollmer (1979), and Williams (1936), Roden and Smith (1979), Ehrenberg (1978), M. F. Roden and V. Murthy (in preparation), respectively.

abundances, then the measured elemental ratios approximate mantle values. Implicit in this contention is another assumption, namely, that the minerals last equilibrated in the mantle. The characteristic perservation of relatively high-pressure mineral assemblages (e.g., spinel + clinopyroxene + orthopyroxene + olivine) in xenoliths supports this latter assumption.

These calculations have been carried out for Nunivak but not for St. Paul's Rocks. Nonetheless, reasonable estimates (discussed below) of the effect of sea-water interaction on the elemental ratios of the St. Paul's samples can be made based on studies of sea-water–ocean crust equilibria.

Nunivak

Roden and others (1984a) reported mass balance calculations for two metasomatized peridotites: one chosen to be representative of the amphibole- or glass-bearing, coarse-grained suite, and one chosen to be representative of the amphibole-free, coarse-grained suite. These calculations were based on clinopyroxene and glass analyses and the assumption of equilibrium partitioning between clinopyroxene and the other phases. For both inclusions the calculated Sm/Nd ratio is within 20 percent of the measured bulk rock ratio; moreover, for two clino-

TABLE 1. ELEMENTAL ABUNDANCES IN SELECTED SAMPLES FROM ST. PAUL'S ROCKS*

Sample	Element (ppm)						
	K	Rb	Sr	La	Ce	Nd	Sm
Spinel peridotites							
H-2	99.3	0.25	9.06	0.58	0.6	0.31	0.095
7-327		0.801	42.8	4.2	7.3	1.48	0.24
SE-22		0.437	37.7		7.3	2.17	0.39
Amphibole peridotites							
7-479	760.0	3.36	40.5	2.0	2.8	2.12	0.49
NE-4	310.0	0.558	25.9	8.1	6.0	3.92	0.30
PH-i	1,284.0	1.70	40.7	4.5	9.3	6.2	1.11
18-900	971.0	2.65	16.7	5.9	18	6.92	1.68
PI-i	1,447.0	1.46	90.4	4.6	18.7	8.0	1.7
Hornblendites							
SE-30	9,730	10.82	1,030			24.03	6.01
SE-31		8.77	1,029	10.4	40	27.08	6.78
31-219	8,620	11.2	1,500			28.69	6.49
7-320	13,000	34.6	2,960	39.0	84	47.3	9.5
SE-13	6,720	4.69	1,293	59.3	102	56.0	12.6
Alkali basalt							
43-49	13,500	36.7	726	42.0	97	42.0	7.7

*Data from Frey (1970), Melson and others (1972, and M. K. Roden and others (1984).

pyroxene-bulk rock pairs, the chondrite-normalized REE abundance pattern of the bulk rock closely mimics the REE pattern of clinopyroxene separated from the bulk rock.

This observation is consistent with the preferential concentration of the REE in clinopyroxene during equilibrium between this phase and olivine, orthopyroxene, and spinel (Stosch, 1982), and indicates that the measured bulk rock Sm/Nd ratio is a mantle value. Furthermore, several other studies of xenoliths have shown that bulk rock Sm/Nd ratios commonly reflect ratios of the constituent minerals, and so have not been significantly perturbed by basalt-xenolith interaction (Frey and Prinz, 1978; Stosch and Seck, 1980; Zindler and Jagoutz, 1987).

The petrogenetic significance of the measured Rb/Sr ratios is not so clear. In the amphibole-bearing inclusions, a significant amount of the Rb is contained in the amphibole or glass (Menzies and Murthy, 1980; Roden and others, 1984a), and for the peridotite for which mass balance calculations were performed, the calculated bulk rock Rb/Sr ratio is within 20 percent of the measured bulk rock ratio. Hence, the bulk rock ratio may closely approach the mantle value. However, the anhydrous metasomatized peridotite for which calculations were carried out, has a calculated Rb/Sr ratio that is almost an order of magnitude lower than the measured bulk rock Rb/Sr ratio (Roden and others, 1984a). The problem lies in the location of Rb; the bulk rock Sr is essentially accounted for by the Sr content of the anhydrous phases (especially clinopyroxene). The problem of "occult" Rb has been observed in other xenolith suites and the missing Rb has been thought to reside in minor, mantle-derived phases, to have been added during low-temperature, ground-water–xenolith equilibria, or by contamination by host basalt (e.g., Zindler and Jagoutz, 1987). The detailed leaching study of Zindler and Jagoutz (1987) showed that a labile component attributed mainly

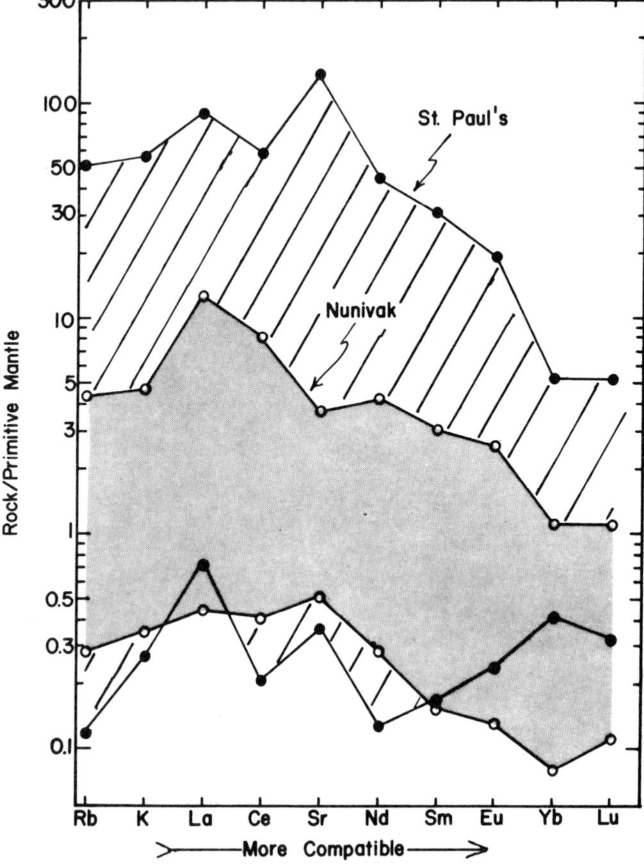

Figure 2. Ranges of abundances of Rb, K, Sr, and some REE in metasomatized Group I peridotite xenoliths from Nunivak Island (shaded area), and peridotites and hornblendites from St. Paul's Rocks (lined area) normalized to primitive mantle abundances (Sun, 1982).

TABLE 2. ELEMENTAL ABUNDANCES IN SELECTED WHOLE-ROCK SAMPLES FROM NUNIVAK ISLAND, ALASKA*

Sample	Element (ppm)						
	K	Rb	Sr	La	Ce	Nd	Sm
Group I Peridotites							
Coarse-grained, anhydrous							
10007	118	0.058	12.2	0.29	0.70		0.061
10004	172	0.042	31.1	0.91	2.5	1.1	0.180
10008	80.4	0.187	11.0	0.88	1.6	0.68	0.099
Coarse-grained, hydrous							
10006	195	0.215	53.5	8.3	14.0	4.6	0.54
10002	206	0.226	82.2	3.65	6.5	2.1	0.25
10055	682	0.323	50.0	1.68	4.9	3.3	0.77
10051	1,065	2.83	42.4	1.48	4.6	3.3	0.82
10075	968	1.9	67.1	3.59	8.5	5.3	1.22
Fine-grained, anhydrous							
10026	102	0.090	12.0	0.20	0.83	0.84	0.28
10068	16.5	0.107	12.9	0.26	1.1	0.844	0.334
10070	9.01	0.011	15.8	0.27	0.99	0.959	0.374
Group II Amphibole Pyroxenites							
13004		0.348	54.8	1.83	6.1	7.1	2.98
13007		0.265	50.1	1.76	6.4	7.1	3.05
13002		8.58	207	6.89	16.9	3.40	1.19
13005		2.38	129	4.33	11.1	2.55	0.90
Nunivak Basalts							
Olivine tholeiites							
B-1	6,990	10.9	469	15.3	33.5	16.8	4.27
B-5	8,640	14.2	524	17.4	38.0	19.6	4.86
B-12	7,600	12.3	526	13.1	30.1	16.5	4.21
Basanites							
B-6	14,850	27.9	785	29.2	63.7	27.5	6.34
B-7	12,990	26.7	700	32.9	62.8	27.1	6.11
B-9	20,990	45.7	1,046	55.6	101	38.8	7.40
B-10	18,790	37.5	897	43.3	90.0	34.1	7.23
B-13	11,400	21.1	639	27.1	58.0	23.8	5.30

*Data for coarse-grained peridotites and pyroxenites from Roden and others (1984a). Other analyses followed methods described in Roden and other (1984a): K, Rb, Sr, by isotope dilution; peridotite REE by radiochemical neutron activation analysis; except Sm and Nd in 10070 and 10068 analyzed by isotope dilution at University of Minnesota (see Roden and others, 1985); basalt REE by instrumental neutron activation analysis (BHVO-1 analyzed concomitantly with basalts yielded La = 15.3 ppm; Ce = 40.3; Nd = 22.9; Sm = 5.59).

TABLE 3. SOME ESTIMATES OF ELEMENTAL ABUNDANCES IN PRIMITIVE AND DEPLETED MANTLE

Mantle/Reference	Element (ppm)						
	K	Rb	Sr	La	Ce	Nd	Sm
Primitive Mantle							
Sun (1982)	230	0.66	22	0.658	1.73	1.26	0.406
Taylor and McLennan (1981)	180	0.48	15.5	0.50	1.30	0.967	0.314
Wood and others (1979)	252	0.86	23	0.71	1.90	1.29	0.385
Depleted Mantle							
Wood and others (1979)	106	0.1	13.2	0.31	0.95	0.86	0.32
O'Nions and others (1979)	50	0.11	16.3			0.79	0.27
Range, fertile lherzolite xenoliths; Jagoutz and others (1979	8.1-138	0.02-0.19	5.88-28.4	0.051-0.51	0.21-1.70		0.20-0.54

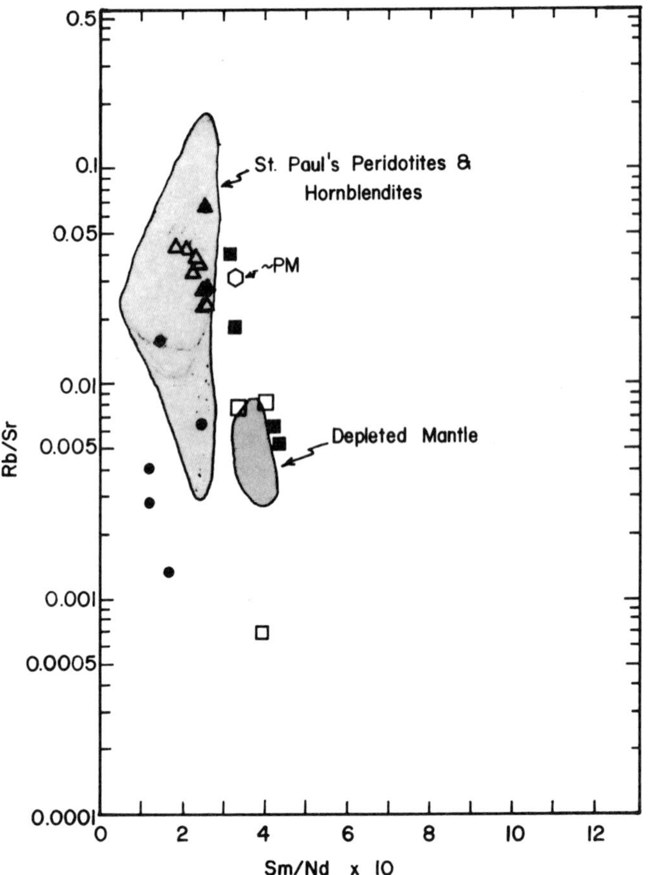

Figure 3. Rb/Sr and Sm/Nd ratios for various materials discussed in the text. Approximate primitive mantle value indicated by open hexagon. All other symbols refer to Nunivak samples: open triangles, Nunivak basalts; open squares, granuloblastic equant lherzolites (clinopyroxene-rich, LREE-depleted); solid squares, Group II pyroxenites; solid circles, coarse-grained peridotites with no phlogopite (in one case a trace is present); solid triangles, coarse-grained peridotites with significant modal phlogopite (Roden, 1982; Roden and others, 1984a). Also shown are ranges of Rb/Sr and Sm/Nd ratios in peridotites and hornblendites from St. Paul's Rocks (Frey, 1970; Melson and others, 1972; M. K. Roden and others, 1984), and estimated range in depleted mantle (O'Nions and others, 1979; Wood and others, 1979; Jacobsen and Wasserburg, 1979).

to a ground-water source had a higher Rb/Sr ratio than the bulk rock.

Without further detailed study, it is uncertain whether the occult Rb in the *anhydrous* Nunivak inclusions can also be attributed in part to ground-water–xenolith or sea-water–xenolith interaction. However, the bulk rock Rb/Sr ratios of the three anhydrous, metasomatized xenoliths are similar to, or less than, the amphibole- and phlogopite-bearing peridotites (Fig. 3). Thus, there is no obvious *elevation* in Rb/Sr ratios that can be attributed to secondary effects. Although in the following section I treat the Rb/Sr ratio of the anhydrous metasomatized lherzolites as primary, the conclusions are not dependent on this assumption, but rather on the overall low Rb/Sr ratio of all the Nunivak xenoliths (Fig. 3). This low Rb/Sr ratio appears to be consistent with the modal mineralogy of the amphibole-bearing inclusions as described above.

St Paul's Rocks

Insufficient mineral data are available to evaluate whether the bulk rock abundance ratios are consistent with modal data. However, the following factors are all consistent with the measured Sm/Nd ratios reflecting mantle (albeit heterogeneous mantle) values: the similarity between Sm/Nd ratios of the samples from St. Paul's Rocks and the peridotites from Nunivak (Fig. 3), the inference that the Sm/Nd ratios of the Nunivak samples reflect mantle Sm/Nd ratios, and the presence in the samples from St. Paul's Rocks of primary phases (clinopyroxene, amphibole; Melson and others, 1972) that are mineral hosts for the REE.

The case for the significance of the measured Rb/Sr ratios is, as for Nunivak, more equivocal. The Rb/Sr ratios of the samples from St. Paul's rocks overlaps that of the Nunivak samples, but extends to somewhat higher Rb/Sr ratios (Fig. 3). Moreover, a number of samples from St. Paul's Rocks have Rb/Sr ratios similar to that of primitive mantle or to estimates of depleted mantle (Fig. 3). However, the relationships of the bulk rock Rb/Sr ratio to a primary Rb/Sr ratio is complicated by the mobility of Rb during sea-water–oceanic crust equilibria (e.g., Hart and Staudigel, 1982).

The presence of Cl-bearing minerals in some of the St. Paul's samples (Melson and others, 1972), as well as stable isotope data, suggests that some reaction with seawater occurred: hydrogen isotope ratios show evidence for exchange with sea water, although oxygen isotope ratios do not (Sheppard and Epstein, 1970). Sr isotopic ratios show no apparent sea-water effect (M. K. Roden and others, 1984). Thus, the relation of the bulk rock Rb/Sr ratios to mantle values remains problematic. This uncertainty only affects the following discussion if Rb has been lost and the Rb/Sr ratio decreased as a result of sea-water–rock equilibria. The similarity in Rb/Sr ratios between metasomatized peridotites from Nunivak, models for primitive and depleted mantle, and many of the samples from St. Paul's Rocks is circumstantial evidence that the Rb/Sr ratios of the latter samples have not *decreased* significantly as a consequence of secondary effects.

The implication of Figure 3 is that, although mantle metasomatism is conceptually thought of as an influx of incompatible elements into relatively depleted peridotite, with a concomitant increase in both incompatible element abundances and the ratios of highly incompatible elements to moderately incompatible elements, many of the samples from Nunivak and St. Paul's Rocks have relatively low Rb/Sr ratios. In contrast, the low Sm/Nd ratios of these metasomatized peridotites are more consistent with expectations.

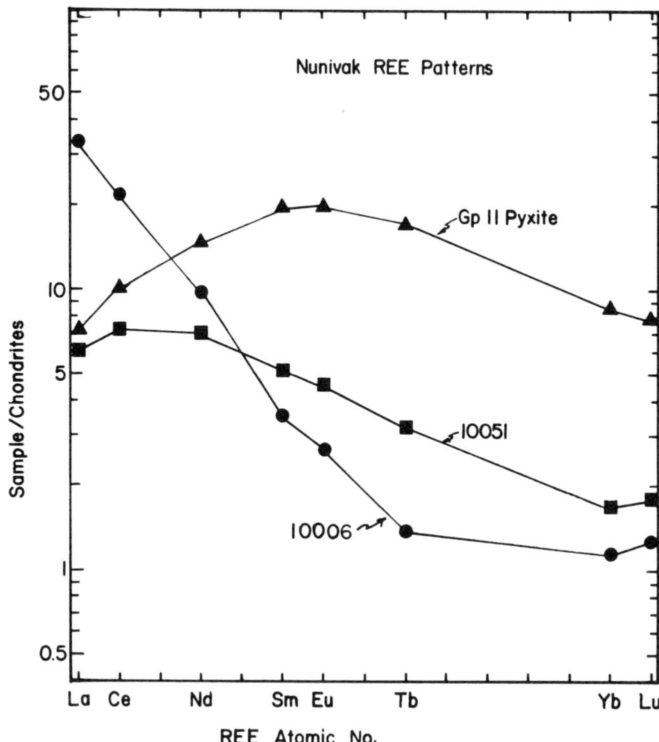

Figure 4. Typical chondrite-normalized REE patterns for Group II pyroxenites (Gp II Pyxite) and Group I peridotite xenoliths (10006, 10051) from Nunivak Island (chondrite values: Evensen and others, 1978; REE data: Roden and others, 1984a).

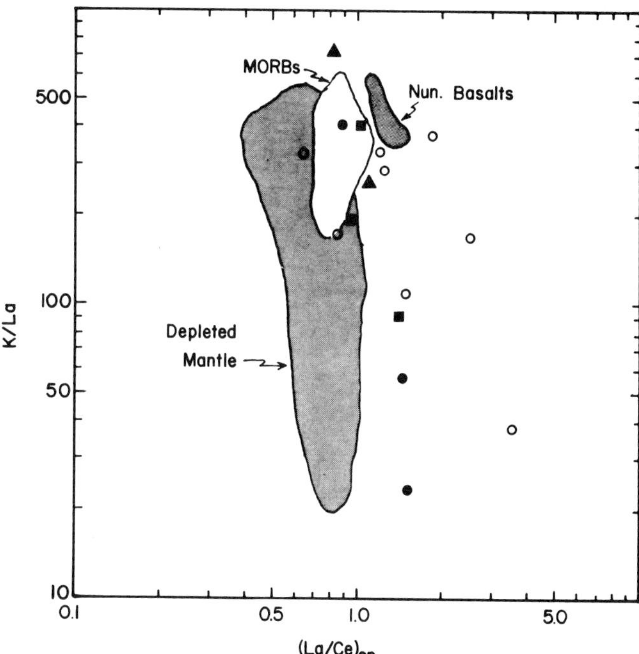

Figure 5. K/La and chondrite-normalized La/Ce ratios for various materials discussed in text. Solid symbols represent Nunivak samples: solid squares, coarse-grained peridotites without amphibole or phlogopite; solid circles, coarse-grained peridotites with amphibole but no phlogopite; solid triangles, coarse-grained peridotites with phlogopite. Open symbols represent peridotites and hornblendites from St. Paul's Rocks (data sources as in Fig. 3). MORB field defined by isotope dilution data from Kay and others (1970). Depleted mantle field defined by RNAA or ID data for LREE-depleted peridotite xenoliths (Table 3; Jagoutz and others, 1979), as well as MORB-type mantle of Wood and others (1979).

Relation of Bulk Rock Chemistry to Metasomatic Fluid Composition

An important question in attempts to estimate the composition of the metasomatic fluid is whether the bulk chemistry of metasomatized peridotites accurately reflects the chemistry of the fluid. Stosch and Seck (1980) pointed out that peridotites with convex-upward REE patterns, and especially with La/Ce ratios less than chondrites, probably did not crystallize as closed peridodite + fluid systems. This interpretation follows from the relative incompatibilities of La versus Ce and the light–REE versus the middle–REE, and the inference that the metasomatic fluid would have high La/Ce ratios (Frey and Prinz, 1978). For example, if an LREE–depleted protolith similar to the primitive lherzolites (Jagoutz and others, 1979) equilibrated with a fluid having REE abundances typical of alkaline magmas, then mass balance calculations show that this metasomatized peridotite should have a chondrite-normalized La/Ce ratio of approximately 1.2, if it had crystallized as a closed peridotite plus fluid system. This calculation is based on Ce abundance in Nunivak sample 10051 (Fig. 4) and component compositions based on the upper limit to REE abundances in basanites (chondrite-normalized La/Ce = 1.4, Ce = 100 ppm) and typical REE abundances of granuloblastic equant peridotites (chondrite-normalized La/Ce = 0.65, Ce = 1 ppm) from Nunivak (Table 3). However, 10051 has a chondrite-normalized La/Ce ratio of 0.89 (Fig. 4).

Although model-dependent, these calculations support the contention that some of the fluid escaped during or following metasomatism and carried with it a significant amount of the highly incompatible elements such as La. Consequently, trace element abundances and ratios of peridodites with convex-upward REE patterns (e.g., 10051, Fig. 4), cannot be used to estimate fluid compositions without careful modeling to account for mineral-fluid fractionation. Lack of partitioning data precludes detailed modeling at present.

There are other samples in the various metasomatized suites (e.g., sample 10006, Fig. 4) that have relatively high La/Ce ratios and thus may reflect closed system reaction of peridotite and metasomatic fluid. In order to examine this possibility, K/La ratios have been plotted versus chondrite-normalized La/Ce ratios for the Nunivak and St. Paul's samples, as well as depleted mantle, Nunivak basalts, and midocean ridge basalts (MORB; Fig. 5). Of particular interest is the observation that several of the Nunivak and St. Paul's samples with high La/Ce ratios have very low K/La ratios. Insofar as K is a highly incompatible element in typical mantle assemblages, it is unlikely that the relatively low K/La ratios in the peridotites with high La/Ce ratios reflect fluid composition. This contention is supported by the mineralogic control over K/La ratios in the Nunivak suite: two phlogopite-rich peridotites have relatively high K/La ratios (Fig. 5). Curiously, sample 10051, which has a La/Ce ratio less than chondrites, also has the highest K/La ratio of any of the Nunivak samples, including the alkaline basalts from Nunivak (Fig. 5). The samples with the lowest K/La ratios from Nunivak contain amphibole and apatite with only traces of phlogopite. The low K/La ratios can be understood in terms of the very high preference of apatite for the REE relative to K (Watson and Green, 1981; Deer and others, 1962).

This mineralogic control of elemental ratios is similar to that of cumulate rocks where fractionation during crystal-melt equilibria accounts for elemental ratios in the cumulates that are very different from that of the parent melt (e.g., Haskin, 1984). In the present case, the fractionation is believed to have occurred during solid-fluid equilibration and must be followed by the egress of fluid from the system. Consequently, elemental ratios of bulk metasomatized peridotites from Nunivak are not reliable guides to fluid composition unless fluid-mineral fractionation is considered. In that the St. Paul's samples also exhibit variable K/La and La/Ce ratios (Fig. 5), fluid-mineral fractionation effects may be common during mantle metasomatism. Consequently, bulk rock elemental ratios can be used only with extreme caution when inferring fluid composition.

Rb/Sr and Sm/Nd Ratios: Implications for the Role of Metasomatized Sources for Alkaline Basalts

In Figure 6, initial $^{143}Nd/^{144}Nd$ and $^{87}Sr/^{86}Sr$ ratios are plotted for representative Na_2O-rich alkaline suites (nephelinites, basanites, etc.) and representative K_2O-rich alkaline suites (minettes, lamproites, etc.) Also shown are ϵ_{Nd} values for "aged" Nunivak and St. Paul's mantles. These ϵ values are model present-day isotopic compositions calculated for primitive mantle that was metasomatized 1 b.y. ago. Following metasomatism, small closed systems within this metasomatized mantle were assumed to be characterized by the present-day Rb/Sr and Sm/Nd ratios of the various Nunivak and St. Paul's samples (the approximate 1.5 atom percent and <1 atom percent decay of ^{87}Rb and ^{147}Sm over the last 1 b.y. was ignored). The present-day isotopic compositions of these closed systems were then calculated using the standard equation relating present-day Sr isotopic composition to initial $^{87}Sr/^{86}Sr$ ratio, present-day $^{87}Rb/^{86}Sr$ ratio, and time of closed-system behavior. The calculated present-day $^{87}Sr/^{86}Sr$ ratios are plotted on Figure 6. Several intriguing observations can be made by comparing the isotopic composition and isotopic heterogeneity of the various rock suites to the model systems.

1. Most Na_2O-rich alkalic suites (Nyiragongo nephelinites, St. Helena basalts, Grenada basalts, Azores basalts, Honolulu Volcanic Series, Nunivak basalts, Ahoggar basalts, Kenya Rift basalts; Fig. 6) tend to have lower $^{87}Sr/^{86}Sr$ and higher $^{143}Nd/^{144}Nd$ ratios than the bulk earth, and to plot within the mantle array defined by MORB and ocean island basalts. Moreover, a number of these suites (e.g., Honolulu Volcanic Series, Fig. 6) are remarkably homogeneous with respect to isotopic composition. Na_2O-rich alkaline basalts from Kerguelen and Tristan da Cunha are notable exceptions to these generalizations because of their relatively low initial $^{143}Nd/^{144}Nd$ ratios (Fig. 6). Nonetheless, basalts from these islands are not as extreme in isotopic composition as the K_2O-rich basalts (Fig. 1); moreover, basalts from these islands are comparatively K_2O-rich relative to Na_2O-rich basalts in general (Fig. 1).

2. In contrast, the K_2O-rich suites (Navajo, Leucite Hills, Roccamonfina lavas, Virunga, western Australian lamproites) display marked isotopic heterogeneity both within suites and between suites, and commonly do not plot within the mantle array (Fig. 6). This heterogeneity could reflect crustal contamination, but the most extreme members of these suites (e.g., Leucite Hills, Australian lamproites) are isotopically the most anomalous and have major and trace element characteristics (especially high Mg, Ni, *and* Sr and Nd contents) inconsistent with contamination (McCulloch and others, 1983; Vollmer and others, 1984). Thus, although crustal contamination may cause some intra-suite variation, it is most likely not the explanation for the unusual isotopic composition of the potassium-rich magmas.

3. The aged metasomatized mantles display significant internal isotopic heterogeneity, but in both cases the majority of the small, closed systems have lower $^{143}Nd/^{144}Nd$ ratios *and* lower $^{87}Sr/^{86}Sr$ ratios than the bulk earth. Menzies and Wass (1983) predicted that based on their studies of xenoliths from Kiama, Australia, that some metasomatized mantle would age in just this manner. They attributed this behavior to metasomatism by a low Rb/Sr, CO_2rich fluid. Present calculations bear out their predictions as to the isotopic characteristics of some aged metasomatic mantle. However, mineralogic evidence for metasomatism by a CO_2-rich fluid is lacking or equivocal at Nunivak and St. Paul's Rocks: the assemblage at Nunivak is characterized by relatively abundant hydrous phases and lesser amounts of CO_2-rich(?) fluid inclusions (Francis, 1976; Roden and others, 1984a). The peridotites and hornblendites from St. Paul's Rocks are also characterized by relatively abundant hydrous phases and only rare carbonates that appear to have crystallized relatively late in the history of the rocks (Melson and others, 1972).

The peculiar isotopic composition of the aged metasomatic mantle is a time-integrated response to the relatively low Rb/Sr and Sm/Nd ratios of this material (e.g., Menzies and Wass, 1983); these characteristics could reflect fluid composition (the low Rb/Sr, CO_2-rich fluid of Menzies and Wass, 1983) or elemental fractionation during metasomatism and the subsequent migration of residual fluid out of the immediate vicinity. It has been maintained in a previous section that the unsympathetic variation of K/La ratios (and by inference, the Rb/La ratios) with La/Ce ratios of the Nunivak samples reflects fluid-solid

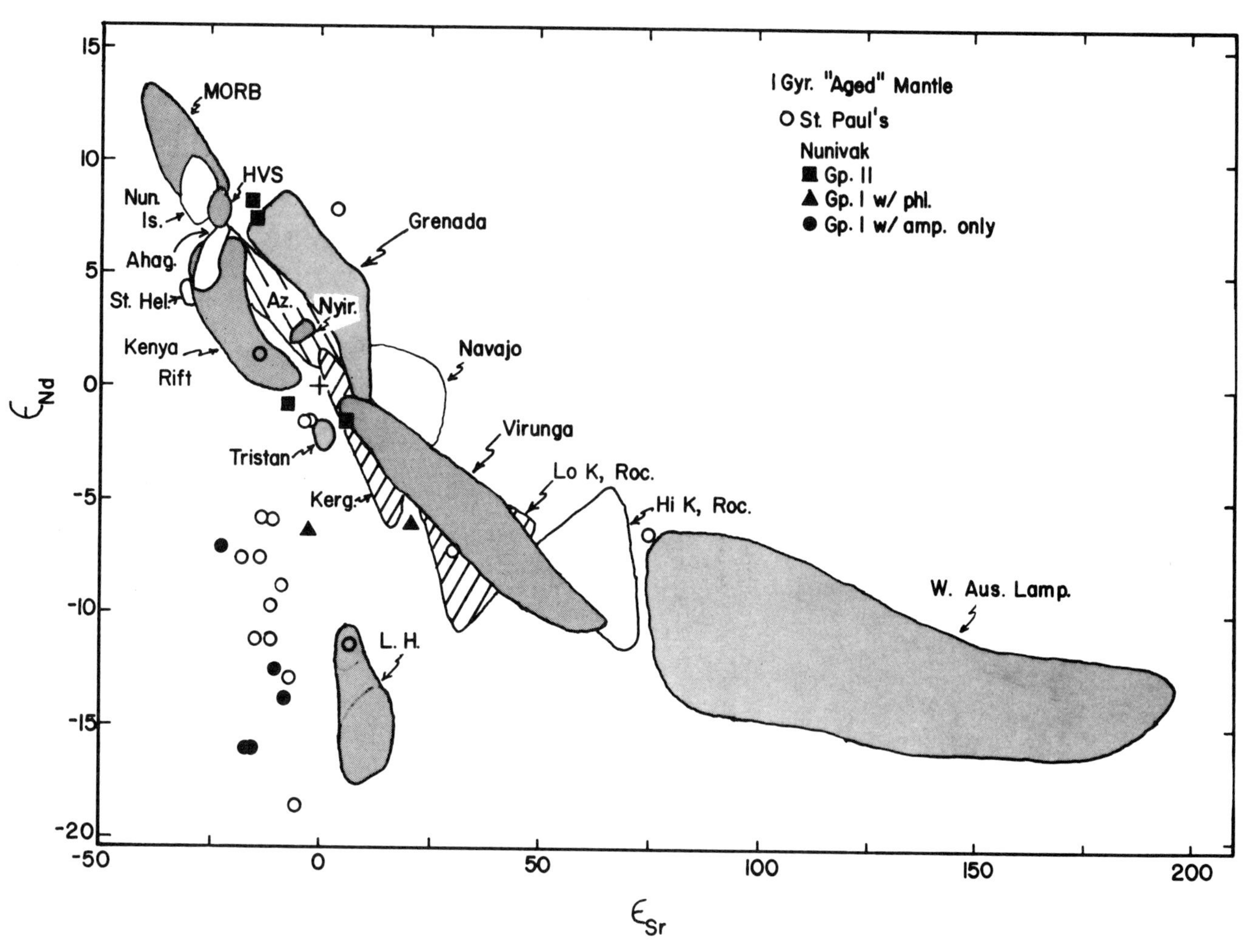

Figure 6. Sr-Nd isotopic correlation diagram (epsilon values calculated after DePaolo and Wasserburg, 1976; UR and CHUR taken to be 0.705 and 0.51264; all data relative to E&A standard = 0.70800 and U.S.G.S. standard rock BCR-1 = 0.51264, where possible). Solid symbols represent "aged" (see text) Nunivak xenoliths, conventions and data sources as in Figure 3; open circles represent aged peridotites and pyroxenites from St. Paul's rocks (data sources as in Fig. 2). Fields for MORB (data from literature) and various continental alkaline suites are shown: Nunivak Island (Nun. Is.: Roden, 1982); Honolulu Volcanic Series (HVS: Ahaggar basalts (Ahag., Allegre and others, 1981); Kenya Rift lavas (Kenya Rift, Norry and others, 1980); Nyiragongo nephelinites (Nyir., Vollmer and Norry, 1983a); Navajo minettes (Navajo, Roden and Murthy, 1984); Virunga K20-rich lavas (Virunga, Vollmer and Norry, 1983b); Leucite Hills (L. H., Vollmer and others, 1984); high (Hi K) and low (Lo K) K O series from Roccamonfina (Roc., Carter and others, 1978; Hawkesworth and Vollmer, 1979); Australian lamproites (W. Aus. Lamp., McCulloch and others, 1983). Also shown are fields for selected oceanic alkaline suites: Honolulu Volcanic Series (HVS, Stille and others, 1983; Roden and others, 1984b); Grenada (Grenada, Hawkesworth and others, 1979a); Azores (Az., Hawkesworth and others, 1979b; White and Hofmann, 1982); St. Helena (St. He.: White and Hofmann, 1982); Tristan da Cunha (Tirstan: O'Nions and others, 1977, White and Hofmann, 1982; alkaline Kerguelen rocks (Kerg: Dosso and Murthy, 1980).

fractionation. In short, the low K/La ratios of some inclusions reflect the precipitation of REE-rich and K-poor phases from the fluid and are not related necessarily to an intrinsically low K/La fluid. Because Sr is concentrated in the same phases (amphibole, clinopyroxene) as the REE (Irving and Frey, 1984; Zindler and Jagoutz, 1987), the low Rb/Sr ratios of the various metasomatized peridotites can be explained simply as a result of fluid-solid fractionation and do not necessarily reflect precipitation from a low Rb/Sr, CO_2-rich fluid.

A second inference derived from consideration of the data in Figure 6 is that the isotopic data appear to exclude an important role for old, metasomatized mantle in the petrogenesis of most sodium-rich alkalic basalts. The rarer potassic lavas (compare field for Leucite Hills lavas to present-day composition of aged Nunivak and St. Paul's mantles; Fig. 6) may form by melting of old metasomatized mantle, as supported by intrasuite isotopic heterogeneity and their lack of coherence with the mantle array (Menzies and Wass, 1983; Vollmer and others, 1984). Most Na_2O-rich alkalic basalts, however, plot within the mantle array; they commonly have initial $^{143}Nd/^{144}Nd$ ratios greater than the bulk earth, and many of these suites exhibit remarkable isotopic homogeneity (e.g., Honolulu Volcanic Series; Figs. 1, 6).

These data preclude derivation of these alkalic lavas from old, metasomatized mantle but do not eliminate a role for metasomatism, provided it was very recent (e.g., less than 200 m.y. ago). One of the reasons for postulating a metasomatized source for these lavas was to avoid the problem of extracting a very small percentage of melt from a peridotite reservoir; this very small percentage of melt (commonly less than 3 percent) was required to explain the very high incompatible element abundances in the melt if primitive or depleted mantle was the source material (e.g., Kay and Gast, 1973). However, recent models suggest that low percentages of melt may separate efficiently from the residue in the mantle (McKenzie, 1984). Consequently, the main underpinning of the metasomatized source model has been questioned. Moreover, in view of several other models (e.g., Navon and Stolper, 1984) that can explain the derivation of incompatible element-rich melts from depleted mantle and the consistency of these models with the relatively high initial $^{143}Nd/^{144}Nd$ ratios of these basalts, it no longer seems necessary to invoke a metasomatized source for Na_2O-rich alkaline basalts.

Nonetheless, sizeable volumes of the mantle (e.g., Nunivak, St. Paul's, Kiama) appear to be metasomatized. As these regions age, they will develop anomalous isotopic compositions with respect to the mantle array. There is no *a priori* reason to expect that this process has occurred only in the recent past; indeed, the isotopic composition of peridotite xenolith BD738 from Lashaine volcano (Cohen and others, 1984), as well as the isotopic composition of some K_2O-rich volcanics (Figs. 1, 6), suggests that the process has been active for more than 1 b.y. If that is the case, then the existence of the mantle array and the rarity of the isotopically anomalous basalts (such as the highly potassic basalts) is evidence that much of metasomatized mantle is located in a region that does not commonly participate in basalt genesis, provided that the peridotites from St. Paul's Rocks and Nunivak Island are typical of metasomatized mantle.

The common occurrence of metasomatized spinel-bearing peridotites (e.g., Stosch and Seck, 1980; Frey and Prinz, 1978) indicates that metasomatized peridotites occur in the uppermost mantle and perhaps may be located in the vicinity of the Moho in continental regions. The development of metasomatized mantle may result from the density contrasts between crust, mantle, and alkaline basalt. Basaltic melts may be trapped at the base of the crust and differentiate (Cox, 1980); this contention is strengthened by experimental evidence of the relatively high density of some alkaline magmas as well as the occurrence of spinel lherzolite xenoliths in felsic rocks (Stolper and Walker, 1980; Irving and Price, 1981). Complete crystallization of an alkaline basalt will lead to the evolution of H_2O- and CO_2-rich fluid, which will be rich in incompatible elements and which will react with peridotite wallrock to form metasomatized mantle (Mysen, 1983).

It is suggested that metasomatized mantle commonly exists immediately below the Moho in continental regions affected by alkaline volcanism, and that much of this mantle is isotopically anomalous. Because it is relatively shallow, this material is generally not involved in the genesis of basalts. The localization of metasomatized material in the uppermost mantle is a consequence of the density contrast between crust and mantle and the crystallization of alkaline magmas near or at this density discontinuity.

SUMMARY

Based on the consideration of the present-day Rb/Sr and Sm/Nd ratios in metasomatized mantle from Nunivak Island, Alaska, and St. Paul's Rocks, it appears unlikely that metasomatized mantle plays an important part in the source region of Na_2O-rich alkaline basalts because peridotites from Nunivak Island and St. Paul's Rocks are characterized by relatively low Rb/Sr and Sm/Nd ratios. Insofar as these examples are typical samples of metasomatized mantle, then old metasomatized mantle will plot off the mantle array and will be isotopically heterogeneous because of the large range in Rb/Sr and Sm/Nd ratios in the metasomatized peridotites. These features are in conflict with the source regions of many nephelinites and basanites, which are characterized by high $^{143}Nd/^{144}Nd$ ratios, plot within the mantle array, and isotopically are comparatively homogeneous. Thus it is considered unlikely that old metasomatized mantle can be the source of the Na_2O-rich alkaline basalts. The metasomatic source model for alkaline basalts seems viable only if one invokes a fortuitous, recent metasomatic event. In contrast to the Na_2O-rich alkaline basalts, some K_2O-rich alkaline basalts have isotopic characteristics consistent with derivation from a metasomatized source region.

With the exception of the ultrapotassic lineage, metasomatized mantle does not appear to be involved in basalt petrogenesis. This observation, combined with the common occurrence of metasomatized spinel lherzolites, indicates that regions of metasomatism may be localized at shallow depths in the mantle and away from regions of partial melting. Metasomatized peridotites may be concentrated near the crust-mantle boundary because of the reaction between peridotite and fluids or residual melts derived from crystallizing alkaline magmas. These magmas may be trapped by the density discontinuity at the Moho.

ACKNOWLEDGMENTS

Analytic data were obtained in the laboratories of F. A. Frey and S. R. Hart, and were supported by NSF grants EAR 7823423 and EAR 8218982 to Frey. Critical reviews by J. W. Hawkins and D. O. Nelson were of assistance in revising the manuscript. Thanks are also extended to P. Gary for help in preparing the manuscript.

REFERENCES CITED

Allegre, C. J., Dupre, B., Lambret, B., and Richard, P., 1981, The subcontinental versus suboceanic debate, I. Lead-neodymium-strontium isotopes in primary alkali basalts from a shield area; The Ahaggar volcanic suite: Earth and Planetary Science Letters, v. 52, p. 85–96.

Arculus, R. J., 1976, Geology and geochemistry of the alkali basalt–andesite association of Grenada, Lesser Antilles island arc: Geological Society of America Bulletin, v. 87, p. 612–624.

Baker, P. E., Gass, I. G., Harris, P. G., and Le Maitre, R. W., 1964, The volcanological report of the Royal Society Expedition to Tristan da Cunha 1962: Philosophical Transactions of the Royal Society of London, v. A256, p. 439–575.

Bell, K., and Powell, J. L., 1969, Strontium isotopic studies of alkalic rocks; The potassium-rich lavas of the Birunga and Toro-Ankole Regions, East and Central Equatorial Africa: Journal of Petrology, v. 10, p. 536–572.

Carmichael, I.S.E., 1967, The mineralogy and petrology of the volcanic rocks from the Leucite Hills, Wyoming: Contributions to Mineralogy and Petrology, v. 15, p. 24–66.

Carter, S. R., Evensen, N. M., Hamilton, P. J., and O'Nions, R. K., 1978, Continental volcanics derived from enriched and depleted source regions; Nd- and Sr-isotope evidence: Earth and Planetary Science Letters, v. 37, p. 401–408.

Chauvel, C., and Jahn, B-M., 1984, Nd-Sr isotope and REE geochemistry of alkali basalts from the Massif Central, France: Geochimica Cosmochimica Acta, v. 48, p. 93–110.

Clague, D. A., and Frey, F. A., 1982, Petrology and trace element geochemistry of the Honolulu Volcanics, Oahu; Implications for the oceanic mantle beneath Hawaii: Journal of Petrology, v. 23, p. 447–504.

Cohen, R. S., O'Nions, R. K., and Dawson, J. B., 1984, Isotope geochemistry of xenoliths from East Africa; Implications for development of mantle reservoirs and their interaction: Earth and Planetary Science Letters, v. 68, p. 209–220.

Cox, K. G., 1980, A model for flood basalt vulcanism: Journal of Petrology, v. 21, p. 629–650.

Cox, K. G., Hawkesworth, C. J., O'Nions, R. K., and Appleton, J. D., 1976, Isotopic evidence for the derivation of some Roman Region Volcanics from anomalously enriched mantle: Contributions to Mineralogy and Petrology, v. 56, p. 173–180.

Daly, R. A., 1927, The geology of St. Helena Island: Proceedings of the American Academy of Arts and Sciences, v. 62, p. 31–92.

Deer, W. A., Howie, R. A., and Zussman, J., 1962, Rock-forming minerals. Vol. 5, Non-silicates: London, Longmans, Green, & Co., 371 p.

DePaolo, D. J., and Wasserburg, G. J., 1976, Nd isotopic variations and petrogenetic models: Geophysical Research Letters, v. 3, p. 249–252.

Dosso, L., and Murthy, V. R., 1980, A Nd isotopic study of the Kerguelen Islands; Inferences on enriched oceanic mantle sources: Earth and Planetary Science Letters, v. 48, p. 268–276.

Enrenberg, S. N., 1978, Petrology of potassic volcanic rocks and ultramafic xenoliths from the Navajo volcanic field, New Mexico and Arizona [Ph.D. thesis]: Los Angeles, University of California, 259 p.

Engel, A.E.G., Engel, C. G., and Havens, R. G., 1965, Chemical characteristics of oceanic basalts and the upper mantle: Geological Society of America Bulletin, v. 76, p. 719–734.

Evensen, N. M., Hamilton, P. J., and O'Nions, R. K., 1978, Rare-earth abundances in chondritic meterorites: Geochimica Cosmochimica Acta, v. 42, p. 1199–1213.

Francis, D. M., 1976, The origin of amphibole in lherzolite xenoliths from Nunivak Island, Alaska: Journal of Petrology, v. 17, p. 357–378.

——, 1978, The implications of the compositional dependence of texture in spinel lherzolite xenoliths: Journal of Geology, v. 86, p. 473–485.

Frey, F. A., 1970, Rare earth and potassium abundances in St. Paul's Rocks: Earth and Planetary Science Letters, v. 7, p. 351–360.

Frey, F. A., and Prinz, M., 1978, Ultramafic inclusions from San Carlos, Arizona; Petrologic and geochemical data bearing on their petrogenesis: Earth and Planetary Science Letters, v. 38, p. 129–176.

Futa, K., and Le Masurier, W. E., 1983, Nd and Sr isotopic studies on Cenozoic mafic lavas from West Antarctica; Another source for continental alkali basalts: Contributions to Mineralogy and Petrology, v. 83, p. 38–44.

Hart, S. R., and Staudigel, H., 1982, The control of alkalies and uranium by ocean crust alteration: Earth and Planetary Science Letters, v. 58, p. 202–212.

Haskin, L. A., 1984, Petrogenetic modelling; Use of rare earth elements, *in* Henderson, P., ed., Rare earth element geochemistry: Amsterdam, Elsevier, p. 115–152.

Hawkesworth, C. J., and Vollmer, R., 1979, Crustal contamination versus enriched mantle; $^{143}Nd/^{144}Nd$ and $^{87}Sr/^{86}Sr$ evidence from the Italian Volcanics: Contributions to Mineralogy and Petrology, v. 69, p. 151–165.

Hawkesworth, C. J., O'Nions, R. K., and Arculus, R. J., 1979a, Nd and Sr isotope geochemistry of island arc volcanics, Grenada, Lesser Antilles: Earth and Planetary Science Letters, v. 45, p. 237–248.

Hawkesworth, C. J., Norry, M. J., Roddick, J. C., and Vollmer, R., 1979b, $^{143}Nd/^{144}Nd$ and $^{87}Sr/^{86}Sr$ ratios from the Azores and their significance in LIL-element enriched mantle: Nature, v. 280, p. 28–30.

Hedge, C. E., Peterman, Z. E., and Dickinson, W. R., 1972, Petrogenesis of lavas from western Samoa: Geological Society of America Bulletin, v. 83, p. 2709–2714.

Irving, A. J., and Frey, F. A., 1984, Trace element abundances in megacrysts and their host basalts; Constraints on partition coefficients and megacryst genesis: Geochimica Cosmochimica Acta, v. 48, p. 1201–1221.

Irving, A. J., and Price, R. C., 1981, Geochemistry and evolution of lherzolite-bearing phonolitic lavas from Nigeria, Australia, East Germany, and New Zealand: Geochimica Cosmochimica Acta, v. 45, p. 1309–1320.

Jacobsen, S. B., and Wasserburg, G. J., 1979, The mean age of mantle and crustal reservoirs; Journal of Geophysical Research, v. 84, p. 7411–7427.

Jagoutz, E., Palme, H., Baddenhausen, H., Blum, K., Cendales, M., Dreibus, G., Spettel, B., Lorenz, V., and Wanke, H., 1979, The abundances of major, minor, and trace elements in the earth's mantle as derived from primitive ultramafic nodules: Proceedings of the Lunar and Planetary Science Conference 10, p. 2031–2050.

Jaques, A. L., Lewis, J. D., Smith, C. B., Gregory, G. P., Ferguson, J., Chappell, B. W., and McCulloch, M. T., 1984, The diamond-bearing ultrapotassic (lamproitic) rocks of the West Kimberley region, Western Australia, *in* Kornprobst, J., ed., Kimberlites I; Kimberlites and related rocks: Amsterdam, Elsevier, p. 225–254.

Kay, R. W., and Gast, P. W., 1973, The rare earth content and origin of alkali-rich basalts: Journal of Geology, v. 81, p. 653–682.

Kay, R. W., Hubbard, N. J., and Gast, P. W., 1970, Chemical characteristics and origin of oceanic ridge volcanic rocks: Journal of Geophysical Research, v. 75, p. 1585–1613.

McCulloch, M. T., Jaques, A. L., Nelson, D. R., and Lewis, J. D., 1983, Nd and Sr isotopes in kimberlites and lamproites from Western Australia; An enriched mantle origin: Nature, v. 302, p. 400–403.

McKenzie, D., 1984, The generation and compaction of partially molten rock: Journal of Petrology, v. 25, p. 713–765.

Melson, W. G., Hart, S. R., and Thompson, G., 1972, St. Paul's Rocks, equatorial Atlantic; Petrogenesis, radiometric ages, and implications on sea-floor spreading: Geological Society of America Memoir 132, p. 241–272.

Menzies, M. A., and Murthy, V. R., 1980, Mantle metasomatism as a precursor to the genesis of alkaline magmas; Isotopic evidence: American Journal of Science, v. 280-A, p. 622–638.

Menzies, M. A., and Wass, S. Y., 1983, CO_2- and LREE-rich mantle below eastern Australia; A REE and isotopic study of alkaline magmas and apatite-rich mantle xenoliths from Southern Highlands Province, Australia: Earth and Planetary Science Letters, v. 65, p. 287–302.

Mysen, B. O., 1983, Rare earth element partitioning between ($H_2O + CO_2$) vapor and upper mantle minerals; Experimental data bearing on the conditions of formation of alkali basalt and kimberlite: Neues Jahrbuch Mineralogie Abhandlungen, v. 146, p. 41–65.

Navon, O., and Stolper, E., 1984, The upper mantle as an ion exchange column [abs.]: Geological Society of America Abstracts with Programs, v. 16, p. 608.

Norry, M. J., Truckle, P. H., Lippard, S. J., Hawkesworth, C. J., Weaver, S. D., and Marriner, G. F., 1980, Isotopic and trace element evidence from lavas bearing on mantle heterogeneity beneath Kenya: Philosophical Transactions of the Royal Society of London, v. A297, p. 259–271.

O'Nions, R. K., Pankhurst, R. J., and Gronwold, K., 1976, Nature and development of basalt magma sources beneath Iceland and Reykjanes Ridge: Journal of Petrology, v. 3, p. 315–338.

O'Nions, R. K., Hamilton, P. J., and Evensen, N. M., 1977, Variations in $^{143}Nd/^{144}Nd$ and $^{87}Sr/^{86}Sr$ ratios in oceanic basalts: Earth and Planetary Science Letters, v. 34, p. 13–22.

O'Nions, R. K., Evensen, N. M., and Hamilton, P. J., 1979, Geochemical modelling of mantle differentiation and crustal growth: Journal of Geophysical Research, v. 84, p. 6091–6101.

Roden, M. F., 1982, Geochemistry of the earth's mantle, Nunivak Island, Alaska, and other areas [Ph.D. thesis]: Cambridge, Massachusetts Institute of Technology, 412 p.

Roden, M. F., and Murthy, V. R., 1984, Isotopic (Sr, Nd) composition of the source for the Navajo minettes, Colorado Plateau [abs.]: Geological Society of America Abstracts with Programs, v. 16, p. 637.

——, 1985, Mantle metasomatism: Annual Reviews of Earth and Planetary Science, v. 13, p. 269–296.

Roden, M. F., and Smith, D., 1979, Field geology, chemistry, and petrology of Buell Park minette diatrem, Apache County, Arizona, *in* Boyd, F. R., and Meyer, H.O.A., eds., Kimberlites, diatremes, and diamonds; Their geology, petrology, and chemistry: Washington, D.C., American Geophysical Union, p. 364–381.

Roden, M. F., Frey, F. A., and Francis, D. M., 1984a, An example of consequent mantle metasomatism in peridotite inclusions from Nunivak Island, Alaska: Journal of Petrology, v. 25, p. 546–577.

Roden, M. F., Frey, F. A., and Clague, D. A., 1984b, Geochemistry of tholeiitic and alkalic lavas from the Koolau Range, Oahu, Hawaii; Implications for Hawaiian volcanism: Earth and Planetary Science Letters, v. 69, p. 141–158.

Roden, M. F., Murthy, V. Rama, and Gaspar, J. C., 1985, Sr and Nd isotopic composition of the Jacupiranga carbonatite: Journal of Geology, v. 93, p. 212–220.

Roden, M. K., Hart, S. R., Frey, F. A., and Melson, W. G., 1984, Sr, Nd, and Pb isotopic and REE geochemistry of St. Paul's Rocks; The metamorphic and metasomatic development of an alkali basalt mantle source: Contributions to Mineralogy and Petrology, v. 85, p. 376–390.

Sheppard, S.M.F., and Epstein, S., 1970, D/H and $^{16}O/^{18}O$ ratios in minerals of possible mantle or lower crustal origin: Earth and Planetary Science Letters, v. 9, p. 232–239.

Stille, P., Unruh, D. M., and Tatsumoto, M., 1983, Pb, Sr, Nd, and Hf isotopic evidence of multiple sources for Oahu, Hawaii basalts: Nature, v. 304, p. 25–31.

Stolper, E., and Walker, D., 1980, Melt density and the average composition of basalt: Contributions to Mineralogy and Petrology, v. 74, p. 7–12.

Stosch, H., 1982, Rare earth element partitioning between minerals from anhydrous spinel peridotite xenoliths: Geochimica and Cosmochimica Acta, v. 46, p. 793–811.

Stosch, H., and Seck, H. A., 1980, Geochemistry and mineralogy of two spinel peridotite suites from Dreiser Weiher, West Germany: Geochimica and Cosmochimica Acta, v. 44, p. 457–470.

Stosch, H., Carlson, R. W., and Lugmair, G. W., 1980, Episodic mantle differentiation; Nd and Sr isotopic evidence: Earth and Planetary Science Letters, v. 47, p. 263–271.

Sun, S.-S., 1982, Chemical composition and origin of the earth's mantle: Geochimica and Cosmochimica Acta, v. 46, p. 179–193.

Taylor, S. R., and McLennan, S. M., 1981, The composition and evolution of the continental crust; Rare earth element evidence from sedimentary rocks: Philosophical Transactions of the Royal Society of London, v. A301, p. 381–399.

Vollmer, R., and Norry, M. J., 1983a, Unusual isotopic variations in Nyiragongo nephelinites: Nature, v. 301, p. 141–143.

——, 1983b, Possible origin of K-rich volcanic rocks from Virunga, East Africa, by metasomatism of continental crustal material; Pb, Nd, and Sr isotopic evidence: Earth and Planetary Science Letters, v. 64, p. 374–386.

Vollmer, R., Ogden, P., Schilling, J.-G., Kingsley, R. H., and Waggoner, D. G., 1984, Nd and Sr isotopes in ultrapotassic volcanic rocks from the Leucite Hills, Wyoming: Contributions to Mineralogy and Petrology, v. 87, p. 359–368.

Watkins, N. D., Gunn, B. M., Nougier, J., and Baksi, A. K., 1974, Kerguelen; Continental fragment or oceanic island?: Geological Society of America Bulletin, v. 85, p. 201–212.

Watson, E. B., and Green, T. H., 1981, Apatite/liquid partition coefficients for the rare earth element and strontium: Earth and Planetary Science Letters, v. 56, p. 405–421.

White, W. M., and Hofmann, A. W., 1982, Sr and Nd isotope geochemistry of oceanic basalts and mantle evolution: Nature, v. 296, p. 821–825.

White, W. M., Tapia, M.D.M., and Schilling, J-G., 1979, The petrology and geochemistry of the Azores Islands: Contributions to Mineralogy and Petrology, v. 69, p. 201–213.

Williams, H., 1936, Pliocene volcanoes of the Navajo-Hopi country: Geological Society of America, v. 47, p. 111–172.

Wood, D. A., Joron, J.-L., Treuil, M., Norry, M., and Tarney, J., 1979, Elemental and Sr isotope variations in basic lavas from Iceland and the surrounding ocean floor: Contributions to Mineralogy and Petrology, v. 70, p. 319–339.

Zindler, A., and Jagoutz, E., 1987, Trace element and Nd and Sr isotope systematics of peridotite nodules Peridot Mesa, San Carolos, Arizona: Geochimica and Cosmochimica Acta (in press).

Manuscript Accepted by the Society October 15, 1986

Printed in U.S.A.

Geological Society of America
Special Paper 215
1987

Oceanic magmas with alkalic characteristics; Evidence from basal cumulate rocks in the Zambales ophiolite, Luzon, Philippine Islands

*Cynthia A. Evans**
Department of Geology, Colgate University, Hamilton, New York 13346

ABSTRACT

This paper examines the compositions of early cumulus minerals, particularly chromite, from two different sections (Acoje and Coto) in the Zambales Range ophiolite, Luzon, Philippines, and then infers compositional characteristics of the magmas from which they were derived.

The cumulus chromite compositions are compared with compositions of phenocrysts that occur in different types of basaltic glasses. Alkali and transitional alkalic basalts, midocean-ridge basalts, and primitive island arc basalts (boninites) are considered. Chromite, one of the earliest minerals to crystallize from basaltic magma and common in the basalt cumulates of ophiolites, is found to be the most useful mineral to compare cumulate occurrences with phenocryst occurrences. Data from basalt-chromite pairs suggest that the TiO_2 contents of chromites are strongly correlated with TiO_2 and other incompatible element abundances in basalts.

Large differences in TiO_2 (0.05 to 1.2 percent) occur in the basal cumulate chromites from the different sections in the Zambales ophiolite. Ti-rich chromites occur at the base of the Coto section. Their compositions are similar to compositions of chromite phenocrysts in alkalic and transitional alkalic basalts. These data suggest that the early magma from which the Coto cumulates were derived had alkalic characteristics. In contrast, chromites from the Acoje section of the Zambales ophiolite are depleted in TiO_2 and appear to have crystallized from a magma depleted in magmaphilic elements. Clinopyroxene compositions from the respective sections can be similarly interpreted. Thus, compositional variability of oceanic basalts might be identified in ophiolite sections by careful examination of cumulate mineral compositions and sequences.

INTRODUCTION

Basaltic rocks recovered from the oceans have, on one level, remarkably uniform compositions (e.g., Engel and others, 1965). Recent investigations, however, which benefit from larger data bases and greater sampling density, indicate that substantial compositional variability does exist within oceanic crust (e.g., Morel and Hekinian, 1980; Sigurdsson, 1981; Bryan and Dick, 1982).

For consideration of first-order compositional heterogeneity in oceanic crust, the most primitive basaltic rocks from the ocean floor must be examined. However, most basaltic rocks recovered by dredging or drilling from the ocean floor show evolved (nonprimitive) compositions. Clearly, these rocks have been modfiied by crystal fractionation or other processes prior to their eruption on the sea floor. Furthermore, if relatively primitive ocean basalts do erupt, they are rarely recovered.

Assuming that most oceanic magmas undergo extensive

*Present address: Lamont-Doherty Geological Observatory of Columbia University, Palisades, New York 10964.

crystal fractionation in a crustal magma chamber, the early fractionation products would reflect the early magma composition. The earliest fractional crystallization products are likely, but not necessarily, to be preserved in the cumulate layers at or near the base of the oceanic crust. Unfortunately, basal cumulates are seldom exposed on the ocean crust (except in fracture zones and oceanic trenches) and are therefore seldom sampled. Thus, rigorous characterization of oceanic crust is limited by sampling capabilities.

Examination of the cumulate sequences of ophiolites (exposed sections of oceanic crust) is one way to examine oceanic magma composition and evolution. Ophiolite sections can have well-exposed, stratigraphically intact cumulate sections. The basal cumulates of many ophiolites are dominated by olivine and chromite. Both of these mineral phases crystallize early from basaltic magmas. The mineral assemblages and compositions (e.g., high Mg, Ni, Cr) suggest that the basal sections of the cumulates represent the accumulation of the early fractionation products of oceanic basaltic magmas. Thus, the mineral associations and compositions of basal cumulates from ophiolites preserve a record of the early magma composition and the subsequent magma evolution, and can be used to characterize the magmas from which they crystallized.

For this study, data from the Zambales ophiolite, Luzon, Philippine Islands, is used to demonstrate that information regarding magma composition and evolution can be extracted from the mineral chemistry of the cumulate sequence. In particular, it is shown that mineral compositions, especially chromite, in the basal cumulates from different sections of the Zambales suggest derivations from an alkalic basalt and a basalt depleted in magmaphilic elements, respectively.

Because the cumulates are the products of complicated physical and chemical processes, and because mineral crystallization is likely to be more complicated as differentiation proceeds and more phases are involved, the focus of this paper is on the basal cumulates. It is assumed, based on mineralogy and composition, that these rocks represent accumulations from early stages of crystal fractionation in the crustal magma chamber. Supporting data are derived from the phenocryst compositions in a variety of oceanic basalts and from other ophiolites.

DESCRIPTIONS AND GEOCHEMICAL DATA FROM ZAMBALES OPHIOLITE

The Zambales Ophiolite in western Luzon, Philippine Islands, preserves at least two ophiolite sections (Fig. 1). The two sections, with dissimilar basalt compositions and plutonic units, have been interpreted by Hawkins and Evans (1983) and Evans (1983) as preserved oceanic crustal sections of a young island arc (Acoje) and back-arc basin (Coto). The existence of two different ophiolite sections provides a good opportunity to examine and contrast the oceanic crustal sections of two different tectonic environments.

Although the plutonic sections (excluding the tectonized peridotites) of both ophiolites are composed primarily of cumulus minerals, they have different cumulate sequences, mineral abundances, and mineral compositions. The differences in the basal cumulate sections, described below, are later used to document differences in the basaltic magma compositions from which they were respectively derived.

Cumulates at the base of both sections are dominated by olivine, and have variable abundances of chromite, clinopyroxene, and some postcumulus plagioclase and/or clinopyroxene. Photomicrographs of typical basal cumulates from both sections are given in Figure 2. The basal cumulates in the Acoje (island arc) section comprise thick (about 1,000 m) accumulations of dunite, wehrlite, and pyroxenite, with some harzburgite and gabbronorite (Evans, 1983; Abrajano, 1984). Although chromite is always present as an accessory phase in dunites, thin layers of chromitite are restricted to the lower few hundred meters of the cumulate section (Evans, 1983). The upper cumulates include gabbronorites and gabbros, with some dunite, wehrlite, and pyroxenite interlayers.

The accumulation sequence for the base of the Acoje section is olivine plus chromite, clinopyroxene and/or orthopyroxene, and plagioclase. Compositional and textural data indicate that this is also the crystallization sequence for the base of the cumulates (Evans, 1983; Hawkins and Evans, 1983). These rocks have very Mg-rich olivine ($Fo_{92\text{-}89}$) and orthopyroxene ($En_{91\text{-}89}$), Cr-rich chromite with low abundances of TiO_2 ($Cr/(Cr + Al) = 0.7$; TiO_2 = 0.3 to 0.3 percent), diopsidic clinopyroxene, and calcic plagioclase ($An_{90\text{-}95}$). Representative mineral compositions are given in Tables 1 and 2.

The basal cumulates from the Coto section are significantly different from those of the Acoje section. The thickness of ultramafic cumulates (dunite and troctolite) is only about 50 m. Chromite is an important accessory phase in these rocks, but is less abundant than in the Acoje section; it does not occur in chromitite layers within the cumulate pile.

The accumulation sequence in the Coto section is olivine plus chromite, plagioclase, and clinopyroxene. Compositional and textural data suggest that this is also the crystallization sequence (Evans, 1983; Hawkins and Evans, 1983). The upper cumulates are gabbros and olivine gabbros and appear to be typical of fractionation products from normal midocean-ridge basalt (MORB).

The basal cumulus minerals at Coto have compositions different from those at Acoje, which are also given in Table 1. The olivine is less magnesian ($Fo_{89\text{-}85}$), the plagioclase is less calcic ($An_{80\text{-}85}$), and the chromite is less Cr-rich but has a much higher abundance of TiO_2 ($Cr/(Cr + Al) = 0.5$; TiO_2 = 0.4 to 1.2 percent). The clinopyroxene has the composition of diopside, but has slightly higher abundances of Na and Ti than Acoje diopside.

CHROMITE COMPOSITIONS FROM OCEANIC BASALTS AND OTHER OPHIOLITES

Unlike many silicate minerals that occur as phenocrysts in basalts, chromites have highly variable compositions; these com-

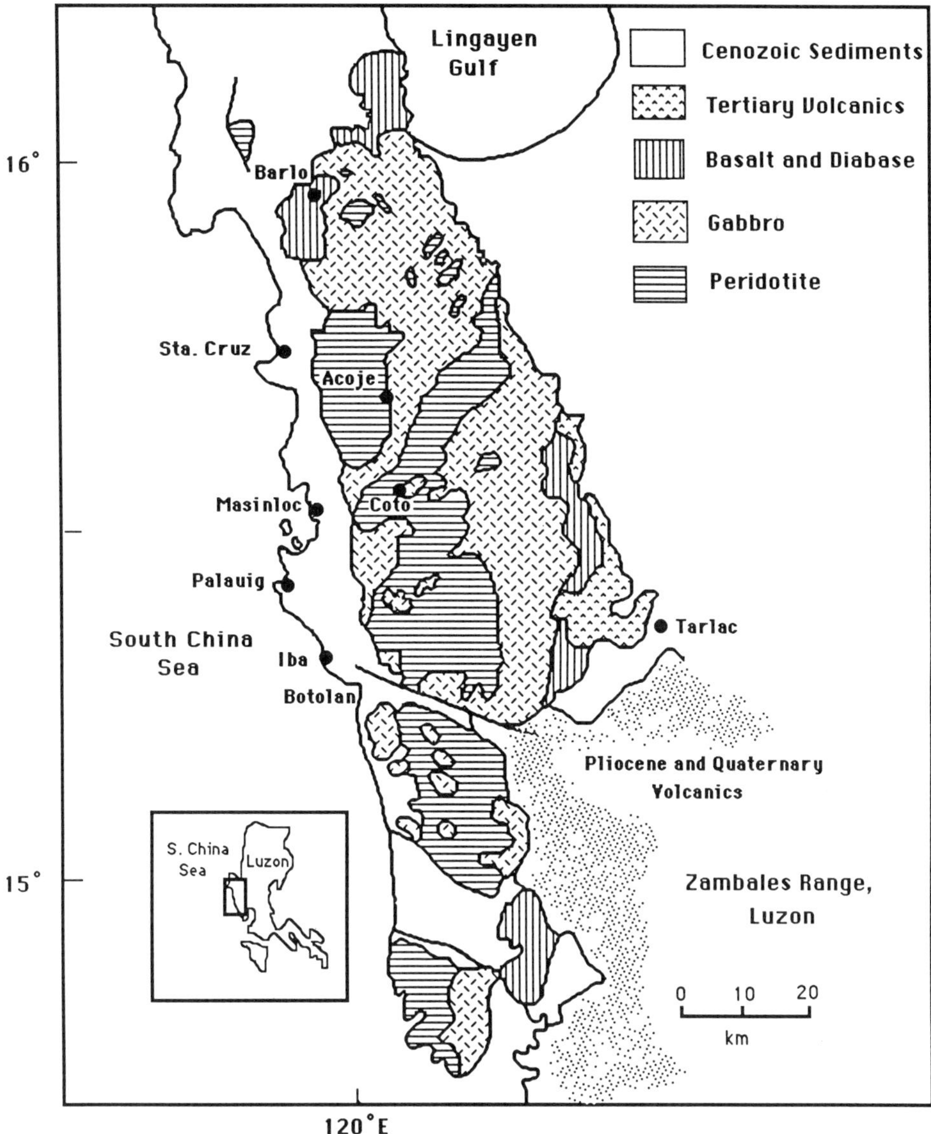

Figure 1. Location map of Zambales ophiolite, western Luzon, Philippine Islands. Acoje and Coto are locations of basal cumulates discussed in text.

positional variabilities give indications of chemical and physical conditions of early crystallization. Experimental crystallization of chromite from basaltic liquids by Fisk and Bence (1980) and Hill and Roeder (1974) indicates that magma composition, pressure (*P*), temperature (*T*), and fugacity (fO_2), are all important factors influencing chromite composition.

Compositional data from a variety of oceanic basalts with chromite phenocrysts are presented to demonstrate the relationship between basalt composition and chromite composition. In particular, a strong positive correlation between TiO_2 content in basalts and coexisting chromite phenocryst is displayed (Fig. 3). The relationship between chromite composition and fO_2 and temperature is also discussed. To demonstrate that the high-Ti chromites from the Coto cumulates are similar in composition to chromites derived from alkalic basalts and that the Acoje cumulate chromites are similar to chromites from alkali-poor basaltic magmas, chromite phenocryst data are compared with data from the Zambales ophiolite. These include compositions of chromite phenocryst in alkalic basalts from Kilauea (Evans and Wright, 1971), Loihi (Hawkins and Melchior, 1983), normal and alkalic basalts from the Mid-Atlantic Ridge (Sigurdsson and Schilling, 1976), alkalic basalts from Rhum (Ridley, 1977), basalts from the Mariana Trough (Hawkins and Melchior, 1985) and the Costa Rica Rift (Natland and others, 1983), and boninites from

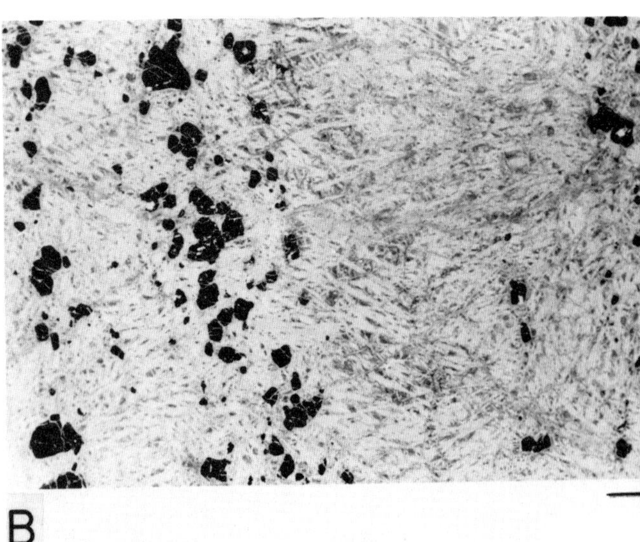

Figure 2. Photomicrographs of basal cumulate assemblages. Bars = 1 mm. a, Cumulus olivine and chromite assemblage from Coto section, with postcumulus clinopyroxene and plagioclase. b, Dunite (heavily serpentinized) with chromite from base of Acoje layered cumulate series.

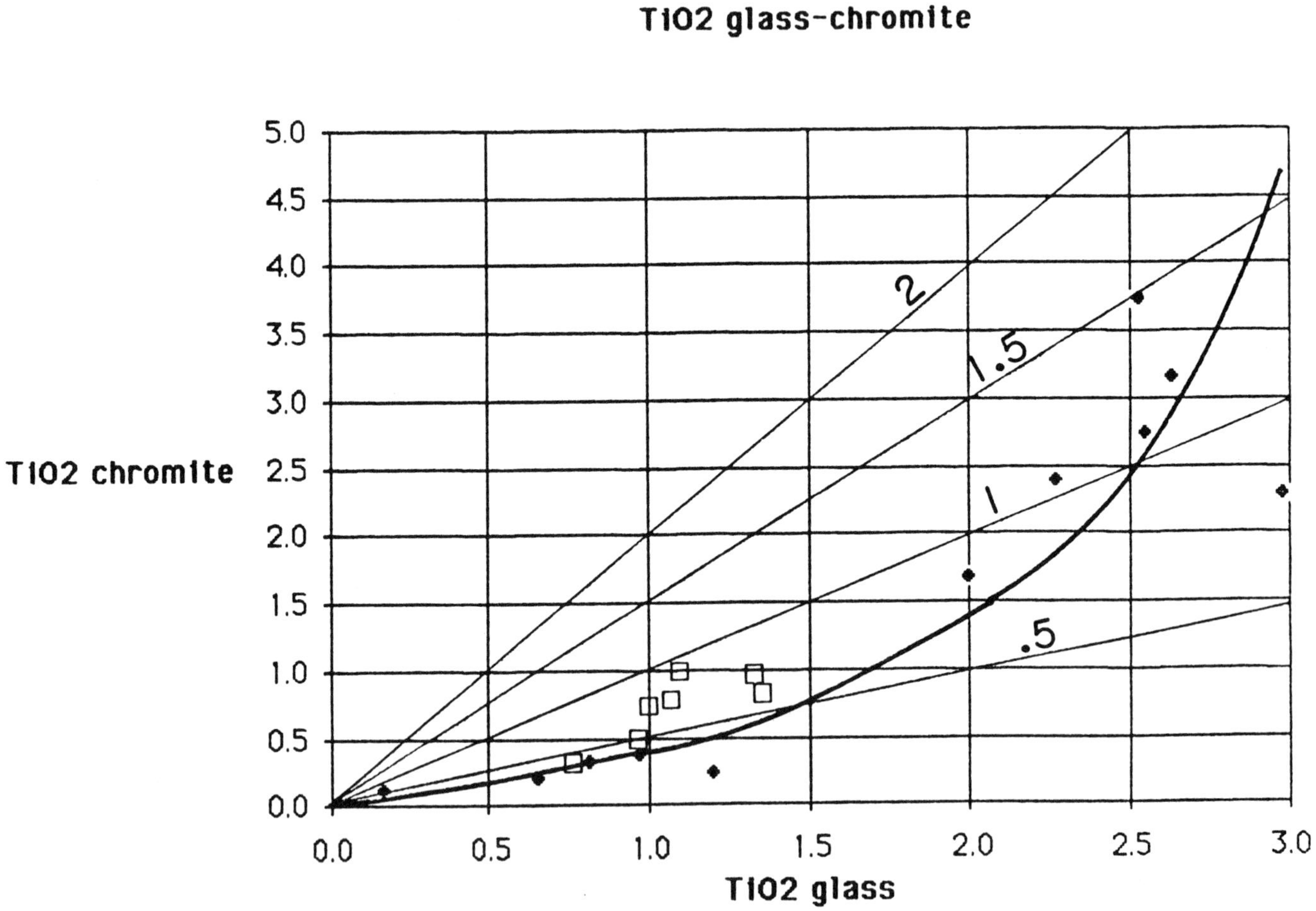

TiO2 glass-chromite

TiO2 chromite

TiO2 glass

Figure 3. Distribution of TiO_2 between coexisting chromite-basaltic glass pairs. Open squares represent Mariana Trough; solid symbols, MOR basalts. Curved line is exponential curve fit to data; radiating lines represent chromite-liquid distribution for TiO_2 of 2, 1.5, 1.0, 0.5, respectively.

TABLE 1. REPRESENTATIVE ANALYSES OF ZAMBALES RANGE CHROMITE, COTO AND ACOJE BASAL CUMULATES

Sample	Coto			Acoje		
	397E	210A	454E	176D	181-G	276-ZS
Oxide (wt. %)						
TiO_2	1.18	0.94	0.63	0.31	0.18	0.09
Al_2O_3	23.10	20.85	23.96	15.83	11.98	37.93
Cr_2O_3	40.27	39.66	39.56	46.29	53.87	29.97
FeO*	24.27	29.44	25.34	26.14	23.46	15.97
MnO	0.29	0.45	0.43	0.36	0.42	0.24
MgO	10.4	8.46	11.21	9.26	9.58	15.71
Total	99.51	99.80	101.13	98.18	99.49	99.91
Cations/32 oxygen						
Ti	0.221	0.176	0.118	0.058	0.034	0.017
Al	6.791	6.130	7.044	4.651	3.522	11.150
Cr	7.974	7.853	7.833	9.165	10.666	5.934
Fe^{+3}	0.667	1.566	1.392	1.356	0.955	1.682
Fe^{+2}	3.883	4.572	3.892	4.094	3.936	3.497
Mn	0.061	0.095	0.091	0.076	0.089	0.051
Mg	3.869	3.147	4.17	3.445	3.564	5.844
$Cr/(Cr+Al+Fe^{+3})$	0.517	0.505	0.482	0.604	0.706	0.337
$Fe/(Cr+Al+Fe^{+3})$	0.043	0.100	0.085	0.088	0.060	0.028
Mg no.	0.50	0.408	0.517	0.455	0.471	0.672

TABLE 2. REPRESENTATIVE ANALYSES OF CLINOPYROXENE, BASAL CUMULATE GABBROS

Sample	Coto			Acoje	
	397F	210A	213C	276ZN	276ZS
Oxide (wt. %)					
SiO_2	51.59	51.92	50.91	51.96	51.80
TiO_2	0.57	0.39	0.60	0.16	0.12
Al_2O_3	3.40	2.50	2.62	2.51	3.42
FeO*	3.42	3.27	4.12	1.98	2.55
MnO	0.16	0.28	0.18	0.06	0.15
MgO	17.64	17.10	16.72	16.71	18.06
CaO	20.28	22.99	22.41	24.65	22.01
Na_2O	0.39	0.37	0.19	0.18	0.09
Cr_2O_3	1.25	0.91	0.44	1.04	1.02
Total	98.70	99.73	98.19	99.24	99.22
Wo	42.7	46.6	45.9	49.9	44.8
En	51.7	48.2	47.6	47.0	51.1
Fs	5.6	5.2	6.6	3.1	4.0

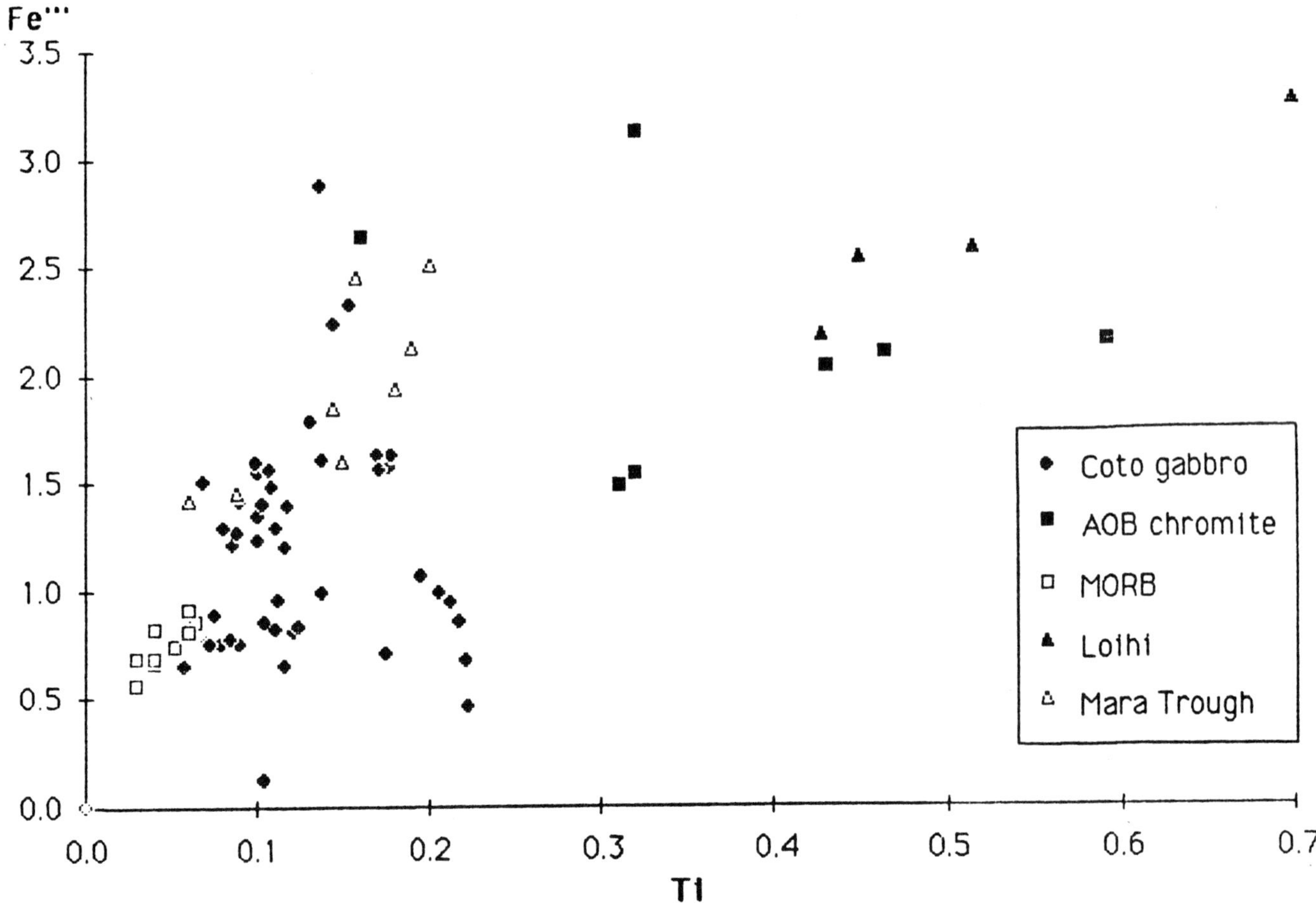

Figure 4. Cation abundances of Ti and Fe^{3+} (based on 32 oxygen) for Coto chromite and variety of chromite phenocrysts. AOB represents chromite in alkali-olivine basalt; MORB, in midocean-ridge tholeiites; and Mara Trough in Mariana Trough basalts. References cited in text. Coto chromite show variable abundances of Fe^{3+} and Ti abundances transitional between MORB and AOB chromite.

the Mariana Trench (Bloomer, 1982). Comparative chromite data from the basal cumulate sections or other ophiolites are also presented.

Because chromite is a very early liquidus mineral, it is assumed that the basalts with chromite phenocrysts are relatively primitive basalts. This assumption is corroborated by the occurrence of magnesian olivine phenocrysts in these basalts and the high-Mg number of the basaltic groundmass (Sigurdsson and Schilling, 1976; Natland and others, 1983; Hawkins and Melchior, 1985). Hence, comparisons between the chromite phenocrysts in basalts and the early cumulate chromites from ophiolites are valid because both are presumably the fractionation products from early, primitive basaltic magmas.

Alkali Olivine Basalts

Alkali olivine basalts are characterized by high contents of alkali elements. They also have high concentrations of TiO_2 relative to MORB. Basalts from Hawaii (Kilaeau and Loihi) have 2 to 3 percent TiO_2 (Evans and Wright, 1972; Hawkins and Melchior, 1983). These basalts have chromite phenocrysts with correspondingly high concentrations of TiO_2 (2 to 3.2 percent). These data are plotted in Figure 3. Compositional data from chromite phenocrysts are displayed in Figs. 4 through 7.

Mid-ocean Ridge Basalts

Sigurdsson and Schilling (1976) correlate the composition of chromite occurring in basalts from the Mid-Atlantic Ridge with the basalt composition. They identified three groups of spinels: (1) Titaniferous magnesiochromites occur in alkali-olivine basalts from certain segments of the mid-Atlantic ridge; the basalts are relatively enriched in the incompatible elements and have approximately 2 percent TiO_2. (2) Magnesiochromites with Cr number = 0.4 to 0.5 are typical of olivine tholeiites. These chromites have relatively low TiO_2 abundances (0.4 wt. %). Data

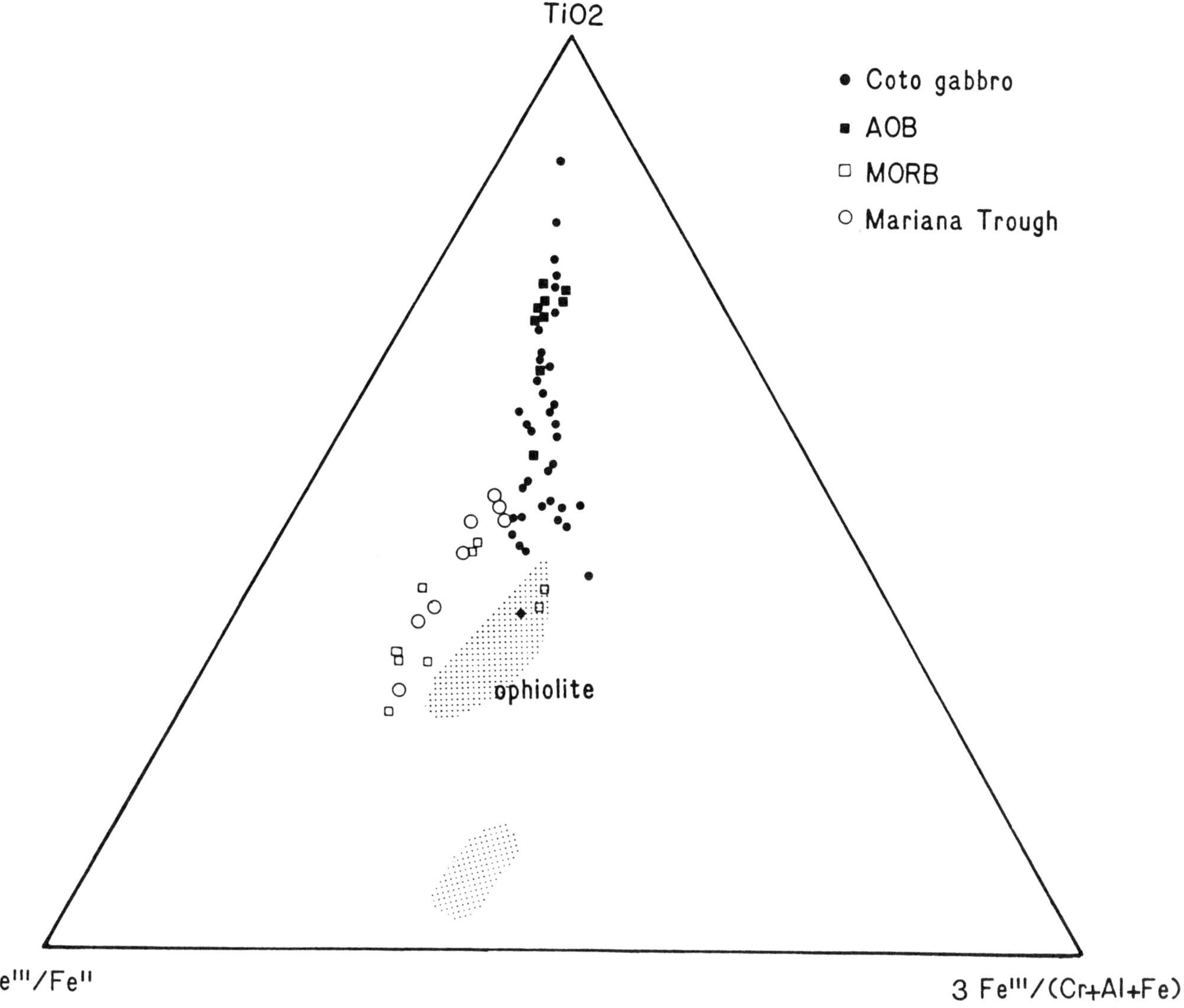

Figure 5. Triangular plot of Fe^{3+}/Fe^{2+} versus TiO_2 versus 3× $Fe^{3+}/(Cr + Al + Fe)$ in chromite. Chromite in Mariana Trough and midocean-ridge basalts plot in same field. Chromite from Coto cumulates plot in same field as chromite from AOB. Chromite in cumulates from Acoje plots with other ophiolites, close to MORB field. This plot indicates that high TiO_2 in Coto and AOB chromite is not due to high abundance of Fe^{3+}.

from FAMOUS basalts (Fisk and Bence, 1980) and the Costa Rica Rift (Natland and others, 1983) corroborate this observation. (3) Chrome spinels with Cr number = 0.23 occur in high-Al picrites.

The titaniferous chromite in the alkali basalts from the Mid-Atlantic Ridge (Sigurdsson and Schilling, 1976, group 1 spinels) have compositions that are similar to chromite compositions in Kilauea and Loihi basalts (Evans and Wright, 1972; Hawkins and Melchior, 1984); see data in Figures 4 through 7. Sigurdsson and Schilling (1976) conclude that the different compositions of chromite crystals in the basalts from the Mid-Atlantic Ridge were controlled by the basalt compositions. In particular, the titaniferous chromites reflect the high-TiO_2 content of the alkali basalts in which they occur.

Mariana Trough Basalts

Back-arc basin basalts have compositions similar to normal midocean-ridge basalts but with greater concentrations of volatiles and some incompatible elements, higher Fe^{3+}/Fe^{2+} and variable isotopic ratios (Hawkins, 1980; Hawkins and Melchior, 1985). In particular, the Mariana Trough basalts (MTB) are enriched in the alkali metals, alkaline earth elements, light rare-earth elements (LREE), and H_2O relative to MORB, in addition

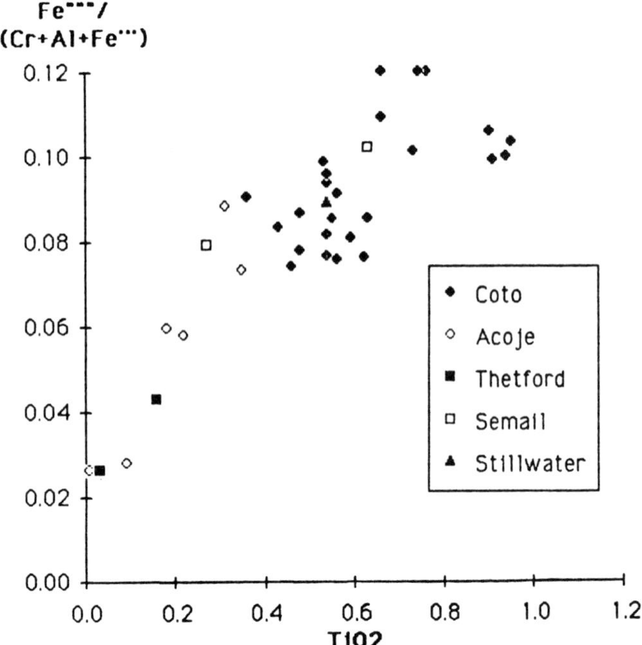

Figure 6. TiO_2 versus $Fe^{3+}/(Cr + Al + Fe^{3+})$ in chromite from basal cumulates from Zambales, other ophiolites, and Stillwater Complex. Data from references in text.

to having a high ferric-ferrous iron ratio (Hawkins and Melchior, 1985). Despite the low Cr contents of the Mariana Trough glasses, the basalts have high abundances of chromite, and chromite phenocrysts occur in all magma types, not just the most primitive basalt compositions. Chromites from Mariana Trough basalts display TiO_2 concentrations similar to chromites from alkali basalts (1–2% TiO_2; Table 3). However, the high TiO_2 concentrations in the Mariana Trough chromites may not reflect the TiO_2 concentrations in the basaltic glasses. The high ferric iron (Fe^{3+}/Fe^{2+} = 0.47 to 0.63) and TiO_2 contents in the Mariana Trough chromites (Figs. 4 through 7) are interpreted as the result of chromite crystallization at relatively low temperatures and high fO_2 (Hawkins and Melchior, 1985). This interpretation is supported by observations from the experimental crystallization of chromite from basaltic magmas under changing conditions of temperature and fO_2 (Hill and Roeder, 1974; Fisk and Bence, 1980).

Boninites

Boninitic rocks from the Mariana Trench are extremely depleted in TiO_2 (0.15–0.25%; Bloomer, 1982). These rocks, thought to be early island arc magmas, are also depleted in other magmaphilic elements. Chromite phenocrysts in these rocks are Cr-rich (Cr/(Cr + Al) = 0.8) but have very low TiO_2 concentrations (0.11%).

Ophiolites

Many ophiolites have chromite-bearing rocks (generally dunites) at the base of their cumulate sections. A rough correlation between cumulus chromite composition and composition of the associated basaltic rocks within an ophiolite section is noted. For example, chromites from the layered ultramafic cumulates at Thetford Mines ophiolite have less than 0.1 percent TiO_2 (Table 2, Figs. 5, 6, 7b). The Thetford cumulates are associated with basaltic rocks with extremely low (0.1–0.3%) TiO_2 concentrations (C. Evans, unpublished data; Sequin and Laurent, 1975; Church, 1977). Chromites from basal dunites in the Samail ophiolite have slightly higher concentrations of TiO_2 (0.3–0.63%), and the associated basaltic rocks have TiO_2 concentrations typical of MORB (Pallister and Hopson, 1981). Similarly, the ultramafic cumulates from the Bay of Islands have chromite with 0.2 to 0.4 percent TiO_2 (Komor and others, 1985). These data are presented in Figures 5 through 7.

DISCUSSION

Chromite is an early (perhaps the earliest) phase to crystallize from a basaltic magma; it is generally associated with only the most primitive basalts (Sigurdsson and Schilling, 1976; Fisk and Bence, 1980; Natland and others, 1983). It has been established that chromite composition is a sensitive indicator of prevailing petrogenetic conditions like magma composition, temperature, pressure, and oxygen fugacity (Irvine, 1967; Hill and Roeder, 1974; Fisk and Bence, 1980).

The chromites in both sections of the Zambales ophiolite are restricted in occurrence to the basal cumulates, or, in Coto, to a few gabbroic dikes in the tectonized peridotite immediately below the basal cumulates. They coexist with magnesian olivine (Table 1), which suggests that the basaltic magmas from which these minerals were derived were relatively primitive basaltic magmas. Thus, the chromite compositions from the Zambales ophiolite may be useful indicators of early and relatively primitive basaltic compositions in both the Coto and Acoje sections.

The data from chromite phenocrysts in different types of basalts indicate that, except under conditions of high fO_2, there is a strong correlation between TiO_2 concentration in chromite and the associated basaltic magma. Titaniferous chromites occur in basaltic rocks with high-Ti and alkali contents, whereas low-Ti chromites occur in basalts with low concentrations of Ti. Data from basaltic glass–chromite pairs can be plotted to calculate distribution coefficients for TiO_2 in chromite (Fig. 3).

The data in Figure 3 indicate that the partitioning coefficient of TiO_2 into the chromite lattice changes with the composition of the melt. An exponential curve is fit to the data. Using this curve, chromite with 0.75 to 1.2 percent TiO_2 (compositions typical of Coto chromite) would coexist with liquids with 1.5 to 1.85 percent TiO_2, a relatively high concentration for a primitive MORB. Chromites with compositions similar to the chromite from Acoje (0.2–0.3% TiO_2) would coexist with liquids with only 0.4 to 0.7

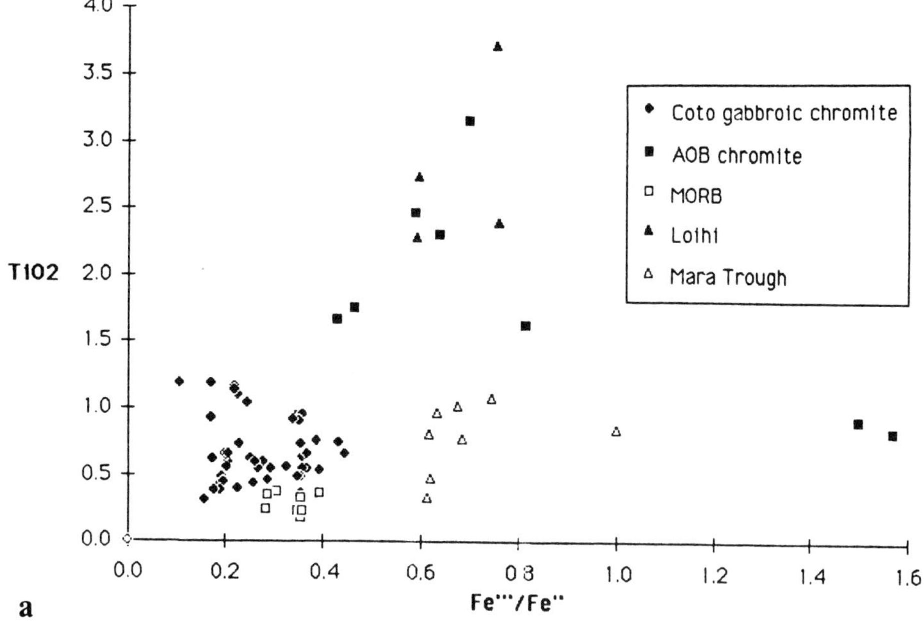

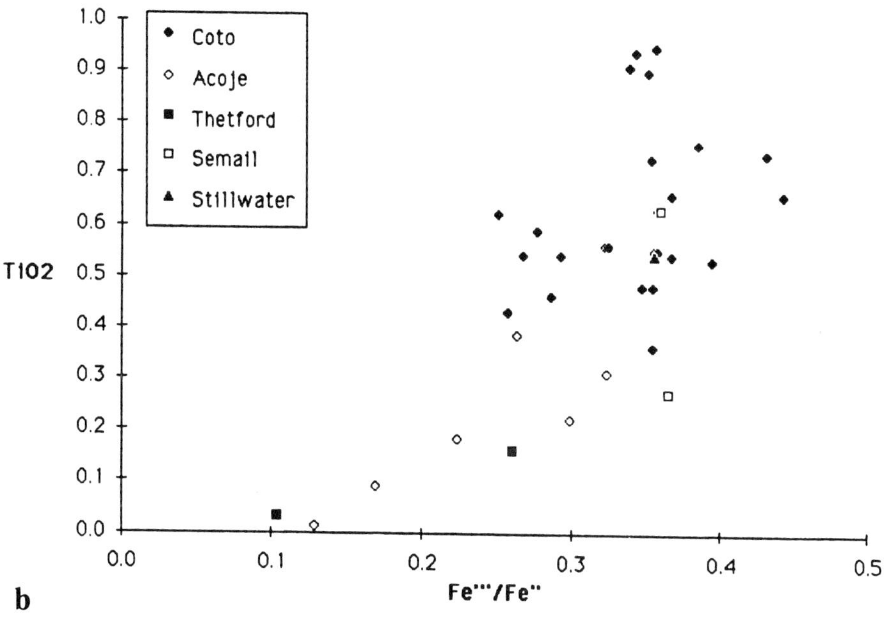

Figure 7. a, TiO_2 versus Fe^{3+}/Fe^{2+} for Coto chromite and chromite from variety of oceanic basalts. MORB chromites have uniformly low abundances of TiO_2 and Fe^{3+}/Fe^{2+}. Coto chromites have values intermediate between MORB and AOB chromites. Mariana Trough samples are distinguished by high Fe^{3+}/Fe^{2+}. b, TiO_2 versus Fe^{3+}/Fe^{2+} for comparative data from basal cumulates of other ophiolites, and chromite from Stillwater Complex, Montana.

TABLE 3. COMPARATIVE CHROMITE COMPOSITIONS

Sample	Mid-Atlantic Ridge*	Alkali Basalt, MAR*	Loihi**	Semail Ophiolite#	Thetford Mines Ophiolite
Oxide (wt. %)					
TiO_2	0.24	1.67	2.37	0.63	0.33
Al_2O_3	26.6	21.3	15.08	20.26	11.60
Cr_2O_3	42.1	38.6	39.79	40.45	57.41
Fe_2O_3	3.7	7.9	13.49	8.48	---
FeO	11.3	16.6	17.88	19.20	19.87
MnO	0.19	0.21	0.25	0.32	0.43
MgO	16.4	12.9	12.04	9.92	10.00
Total	100.53	98.90	100.92	99.59	99.34
$Cr/(Cr+Al+Fe^{+3})$	0.493	0.496	0.529	0.514	0.749
$Fe^{+3}/(Cr+Al+Fe^{+3}$	0.041	0.096	0.170	0.102	0.026
Mg no.	0.722	0.582	0.545	0.480	0.498

*Sigurdsson and Schilling (1976).
**Hawkins and Melchior (1983).
#Pallister and Hopson (1981).

percent TiO_2. Note that the Mariana Trough chromites, which formed at high fO_2, plot slightly above the curve.

It can further be assumed that the high TiO_2 contents of the Coto chromites were probably not the result of chromite crystallization under conditions of high fO_2. The data displayed in Figures 4 through 7 indicate that, in addition to high TiO_2 contents, the Coto chromites have variable contents of ferric iron and Fe^{3+}/Fe^{2+}. Both of these compositional parameters appear to be independent of Cr/Al ratios. The Coto chromites are distinct from MORB chromites in these respects. The variable Fe^{3+}/Fe^{2+} in the Coto chromites suggest that, while they may be crystallized under variable oxidizing conditions, the TiO_2 concentrations in the chromites were not affected.

Figure 5 summarizes compositional differences between chromite in alkali-olivine basalts (AOB) and MORB. In this diagram, Mariana Trough basalts and MORB plot in the same field, and the Coto chromites plot in a field which includes the AOB and is separate from the MORB field. Although chromite from other ophiolites are dominated by their low abundances of TiO_2, they appear to fall close to the MORB field.

Using a published distribution coefficient of 0.32 for TiO_2 between clinopyroxene and liquid (Duke, 1976; Irving, 1978), the relative abundances of TiO_2 in the presumed parent magmas for both Coto and Acoje sections can also be calculated from clinopyroxene compositions. A basaltic liquid coexisting with Coto clinopyroxene with 0.5 to 0.7 percent TiO_2 would have 1.56 to 2.18 percent TiO_2. This calculation for the Acoje clinopyroxenes (0.2–0.3 percent TiO_2) indicates they would coexist with a liquid with 0.62 to 0.94 percent TiO_2. These calculations are in agreement with those made using data from chromites. It should be noted that these calculations do not take into consideration the *P-T* regime during crystallization; both isothermal (at about 1,200°C) and isobaric conditions are assumed.

The basal cumulates of the different sections of the Zambales ophiolite can be interpreted, using the data presented from different basalt-phenocryst pairs, as having crystallized from different types of basaltic magmas. The uniformly low-TiO_2 concentrations of Acoje chromites are typical of phenocryst compositions from primitive basalts that are depleted in alkali and (HFS) elements. For example, the calculated concentration of TiO_2 in the parent magmas is less than 1 percent. Indeed the volcanic rocks associated with the Acoje section have low abundances of TiO_2 (0.2–0.45%) and other HFS and alkali elements—compositions typical of island arc tholeiites (Hawkins and Evans, 1983).

Conversely, the compositions of the basal cumulate minerals at Coto suggest that the early basaltic magma had transitional alkalic characteristics, with relatively high concentrations of TiO_2 (1.5–2.2 %). In addition, it appears that the alkalic signature in the Coto cumulates changes upsection. The chromite-bearing basal cumulates are restricted in occurrence and they are overlain by a large thickness of cumulate gabbros and olivine gabbros that are similar to oceanic gabbros (Hawkins and Evans, 1983; Evans, 1983). There is nothing to suggest that these overlying gabbros may have been derived from a magma with alkalic characteristics. The coexisting basaltic rocks in the Coto section have compositions that are typical of MORB and back-arc basin basalts (Hawkins and Evans, 1983). It appears that the occurrence of the more alkalic magma that produced the basal cumulates was restricted to early stages of crust formation; it was later overwhelmed by larger volumes of basaltic magma that was more typical of MORB. Although the reason for the apparent compositional change is unknown, I suggest that the initial magmas producing the Coto section may have been produced by small degrees of partial melting, yielding a basaltic magma with alkalic characteristics. This was followed by more extensive melting of the mantle source, producing a magma of primitive MORB composition.

Compositional variations in MORB and back-arc basin basalts are well documented (e.g., Kay and Gast, 1973; Hawkins, 1980; Morel and Hekinian, 1980; Sigurdsson, 1981; Langmuir and Bender, 1984; Batiza and Vanko, 1984; Hawkins and Melchior, 1985). In fact, compositional variations, which range from alkalic to normal MORB compositions, occur along individual ridge segments. Such compositional differences should be reflected in the phenocryst and cumulate compositions. The compositional differences noted in the basal cumulates of the Zambales ophiolite are consistent with such differences in the basalt compositions.

SUMMARY

Many ophiolites are floored by cumulates dominated by olivine, but also having variable amounts of chromite, pyroxene, and plagioclase. The chromite compositions are sensitive to magma compositions and the conditions of crystallization. In particular, the TiO_2 abundance in the chromite (except under conditions of very high oxygen fugacity) is indicative of the TiO_2 concentration in the parent magma. Thus, cumulus chromite may be an important mineral to trace the composition and evolution of early oceanic magmas, or the succession of magma compositions that were involved in the formation of oceanic crust.

The compositions of the basal chromites of the Coto section of the Zambales Range ophiolite have compositions that approach the compositions of chromites in alkalic basalts (Kilauea, Loihi, alkalic basalts erupted along the Mid-Atlantic Ridge and are different from chromite phenocrysts in normal MORB. This suggests that the early basaltic magmas associated with initial stages of crustal formation had compositions that were transitional alkalic or alkalic basalt compositions. The early magmas were perhaps composed of initial melts that were generated by a small degree of partial melting of the mantle source. The overlying cumulate assemblage in the Coto comprises olivine gabbros and gabbros with compositions consistent with their derivation from a magma similar to normal MORB. This suggests that, with time, the Coto magma chamber evolved into a more steady state and received large influxes of magma with compositions similar to normal ocean tholeiites, derived by greater degrees of partial melting. This mechanism is consistent with the interpretation that the Coto section represents oceanic crust generated in a back-arc basin setting.

The Acoje ophiolite section has chromite-rich horizons as well. The olivine-dominated basal cumulate section is much thicker than the Coto section and is associated with basalts with low TiO_2. The chromite in the basal cumulate section has compositions that are uniformly low in TiO_2 (0.2–0.3%) and were apparently derived from magmas that were depleted in TiO_2.

Because detailed compositional data from the basal cumulates of other ophiolite sections is not readily available in the literature, it is difficult to compare the observations presented in this paper with data from other ophiolites and to assess whether or not the Coto section preserves a unique record of early magmatic evolution. Examination of the basal cumulate minerals in ophiolite sections may be the only way to examine primitive magma evolution in an oceanic environment.

ACKNOWLEDGMENTS

This work was part of my Ph.D. dissertation, and I gratefully acknowledge the help and efforts of J. Hawkins. Discussions and editing by J. Powell are also appreciated. Thoughtful comments from J. Pasteris, J. Abrajano, and an anonymous reviewer improved the manuscript. Some of the analytic work was supported by the Colgate University Research Council.

REFERENCES CITED

Abrajano, T. A., Jr., 1984, The petrology and low-temperature geochemistry of the sulfide-bearing and mafic-ultramafic units of the Acoje Massif, Zambales ophiolite, Philipines; Characterization of mineral and fluid equilibria [Ph.D. thesis]: St. Louis, Missouri, Washington University, 473 p.

Batiza, R., and Vanko, D., 1984, Petrology of young Pacific seamounts: Journal of Geophysical Research, v. 89, p. 11235–11260.

Bloomer, S. H., 1982, Mariana Trench; Petrologic and geochemical studies; Implications for the structure and evolution of the inner slope [Ph.D. thesis]: University of California at San Diego, 273 p.

Bryan, W. B., and Dick, H.J.B., 1982, Contrasted abyssal basalt liquidus trends; Evidence for mantle major element heterogeneity: Earth and Planetary Science Letters, v. 58, p. 15–26.

Church, W. R., 1977, The ophiolites of southern Quebec; Oceanic crust of Betts Cove type: Canadian Journal of Earth Sciences, v. 14, p. 1668–1673.

Duke, J. M., 1976, Distribution of the period four transition elements among olivine, calcic clinopyroxene and mafic silicate liquid; Experimental results: Journal of Petrology, v. 17, p. 499–521.

Engle, A.E.J., Engle, C. G., and Havens, R. G., 1965, Chemical characteristics of oceanic basalts and the upper mantle: Geological Society of America Bulletin, v. 76, p. 719–734.

Evans, B. W., and Wright, T. L., 1972, Composition of liquidus chromite from the 1959 (Kilauea Iki) and 1965 (Makaopuhi) eruptions of Kilauea volcanoe, Hawaii: American Mineralogist, v. 57, p. 217–230.

Evans, C. A., 1983, Petrology and geochemistry of the transition from mantle to crust beneath an island arc–backarc pair; Implications from the Zambales Range ophiolite, Luzon, Philippines [Ph.D. thesis]: San Diego, University of California, 299 p.

Fisk, M., and Bence, A. E., 1980, Experimental crystallization of chrome-spinel in FAMOUS basalt 527-1-1: Earth and Planetary Science Letters, v. 48, p. 111–123.

Hawkins, J. W., 1980, Petrology of back-arc basins and island arcs; Their possible role in the origin of ophiolites, *in* Panayiotou, A., ed., Ophiolites; Proceedings, International Ophiolite Symposium, Cyprus, 1979: Cyprus Geological Survey Department, p. 244–253.

Hawkins, J. W., and Evans, C. A., 1983, Geology of the Zambales Range, Luzon, Philippine Islands; Ophiolite derived from an island arc–backarc basin pair, *in* Hayes, D., ed., The tectonic and geologic evolution of southeast Asian Seas and islands, Pt. 2: Washington, D.C., American Geophysical Union Monograph 27, p. 95–123.

Hawkins, J. W., and Melchoir, J., 1983, Petrology of basalts from Loihi Seamont,

Hawaii: Earth and Planetary Science Letters, v. 66, p. 3356–3368.
—— , 1985, Petrology of Mariana Trough and Lau Basin basalts: Journal of Geophysical Research, v. 90, p. 11431–11468.

Hill, R., and Roeder, P., 1974, The crystallization of spinel from basaltic liquid as a function of oxygen fugacity: Journal of Geology, v. 82, p. 709–729.

Irvine, T. N., 1967, Chromian spinel as a petrogenetic indicator, Pt. 2; Petrologic applications: Canadian Journal of Earth Sciences, v. 4, p. 71–103.

Irving, A. J., 1978, A review of experimental studies of crystal/liquid trace element partitioning: Geochimica et Cosmochimica Acta, v. 42, p. 743–770.

Kay, R. W., and Gast, P. W., 1973, The rare earth content and the origin of alkali-rich basalts: Journal of Geology, v. 81, p. 653–682.

Komor, S., Elthon, D., and Casey, J. F., 1985, Mineralogic variation in a layered ultramafic cumulate sequence at the North Arm Mountain Massif, Bay of Islands ophiolite, Newfoundland: Journal of Geophysical Research, v. 90, p. 7705–7736.

Langmuir, C. H., and Bender, J. F., 1984, The geochemistry of oceanic basalts in the vicinity of transform faults; Observations and implications: Earth and Planetary Science Letters, v. 69, p. 107–127.

Morel, J. M., and Hekinian, R., 1980, Compositional variations of volcanoes along segments of recent spreading ridges: Contributions to Mineralogy and Petrology, v. 72, p. 425–436.

Natland, J. H., Adamson, A. C., Laverne, C., Melson, W. G., O'Hearn, T., 1983, A compositionally nearly steady state magma chamber at the Costa Rica Rift; Evidence from basalt glass and mineral data, Deep Sea Drilling Project sites 501, 504, and 505, *in* Cann, J. R., and others, eds., Initial reports of the Deep Sea Drilling Project: Washington, D.C., U.S. Government Printing Office, p. 811–858.

Pallister, J. S., and Hopson, C. A., 1981, Samail ophiolite plutonic suite; Field relations, phase variations, cryptic variation and layering, and a model of a spreading ridge magma chamber: Journal of Geophysical Research, v. 86, p. 2593–2644.

Sequin, M. K., and Laurent, R., 1975, Petrological features and magnetic properties of pillow lavas from the Thetford Mines ophiolite (Quebec): Canadian Journal of Earth Sciences, v. 12, p. 1406–1420.

Sigurdsson, H., 1981, First-order major element variation in basalt glasses from the Mid-Atlantic Ridge; 29°N to 73°N: Journal of Geophysical Research, v. 86, p. 9483–9502.

Sigurdsson, H., and Schilling, J. G., 1976, Spinels in Mid-Atlantic Ridge basalts; Chemistry and occurrence: Earth and Planetary Science Letter, v. 29, p. 7–20.

Thompson, R. W., 1973, Titanian chromite and chromian titanomagnetite from a Snake River Plain basalt; A terrestrial analogue to lunar spinels: American Mineralogy, v. 58, p. 826–830.

MANSCRIPT ACCEPTED BY THE SOCIETY OCTOBER 15, 1986

Printed in U.S.A.

Geological Society of America
Special Paper 215
1987

Geochemical characteristics of boninite- and tholeiite-series volcanic rocks from the Mariana forearc and the role of an incompatible element–enriched fluid in arc petrogenesis

Sherman H. Bloomer
Department of Geology, Duke University, Box 6729, College Station, Durham, North Carolina 27708

ABSTRACT

Samples dredged from the forearc west of the Mariana Trench include boninite and island-arc tholeiite series volcanic rocks. These are part of a late Eocene–early Oligocene arc complex that forms most of the forearc basement; the complex has been exposed by tectonic erosion. The boninites are depleted in TiO_2 (0.20%), Y (5–9 ppm), and heavy rare-earth elements (Yb_N = 2.4), and have low Ti/Zr and Y/Zr ratios. They are variously enriched in alkali metals, alkaline earths, and light rare-earth elements. These boninitic samples, in common with other such suites, appear to be hydrous melts of a once-melted peridotitic mantle that was enriched prior to the boninite-melting event by a light rare-earth element Ba-, Sr-, and Zr-rich fluid or melt. The tholeiitic series samples, which are spatially associated with the boninites, include basalts, andesites, and dacites. The least fractionated basalts have higher concentrations of TiO_2 (0.4–0.5%), Y (10–15 ppm), and heavy rare-earth elements (Yb_N = 6) than the boninites, but similar concentrations of Ba, Sr, Zr, and Ce. The least fractionated samples from both series have similar ratios of moderately incompatible elements (Ti/Y, Ti/Yb) and highly incompatible elements (Ce/Zr, Ba/Zr); values of Ti/Zr and Ce/Sm vary antithetically. The melts parental to the boninite and arc-tholeiite series cannot be derived by different degrees of melting of a single, homogeneous source; their distinct chemical compositions reflect initially different sources. The similarities of some element ratios and the continuous variation of others, between the two suites, suggest that there is some similarity or gradation between the boninite source and that of the arc tholeiites.

Calculations indicate that many of the characteristics of the two suites can be produced by melting mixtures of variously depleted mantle residual from the production of ocean-ridge basalts, and an incompatible element–enriched fluid. The postulated fluid has high Ce/Sm; is enriched in Ba, Sr, and Zr; and is nearly devoid of Ti, Y, V, and the heavy rare-earth elements. The more depleted portions of the mantle yield boninitic melts; the less depleted portions, tholeiitic melts. Concentrations of high field-strength cations are controlled principally by the percentage of melting of the mixed source and by the initial degree of depletion of the ORB-type mantle. Light rare-earth enrichments, Ti/Zr, and Ba and Sr abundances are controlled primarily by the amount of fluid added. The postulated incompatible element–enriched fluid may be derived from a less depleted mantle, from the subducted slab, or from both; this fluid is not by itself adequate to produce the Ba enrichments in typical arc tholeiites. An additional enrichment process may be required.

INTRODUCTION

Tholeiite-series volcanic rocks are an important volcanic product in intraoceanic island arcs. They are chemically distinct from most ocean-ridge basalts (ORB) in their lower concentrations of Ti, Zr, Y, and heavy rare-earth elements (HREE); their relative enrichments of Ba, Sr, K, and Rb; their high Ba/La ratios; and their commonly higher $^{87}Sr/^{86}Sr$ and lower $^{143}Nd/^{144}Nd$ values (Jakes and Gill, 1970; DePaolo and Wasserburg, 1977; Perfit and others, 1980; Kay, 1980; Stern, 1981; McCulloch and Perfit, 1981; Morris and Hart, 1983; Kay, 1984). The subducted slab, a depleted ORB–source mantle, a less depleted, ocean-island–basalt-source mantle, and subducted sediments are all possible sources of intraoceanic-arc magmas. These sources and the resulting melts may be modified by metasomatic fluids from subducted sediments or crust, or from another part of the mantle; residual accessory phases present during melting may also affect magma compositions (e.g., Perfit and others, 1980; Kay, 1980; McCulloch and Perfit, 1981; Green, 1981; Morris and Hart, 1983). The suggestion that an ocean-island–type mantle source may contribute to arc volcanism is particularly important as the only mantle generally considered to contribute to the production of arc magmas is that above the subduction zone. If those magmas are in part produced by a less depleted mantle, the distribution of that less depleted source in the upper mantle can be constrained.

Chemical data for two spatially and temporally related island-arc volcanic rock series from the forearc of the Mariana Trench are discussed here. One series is a typical tholeiitic suite, the other is boninitic (an unusual silica-saturated, high-Mg series; Meijer, 1980). Both suites include samples that are less fractionated than those typically found in arc environments. These samples are of interest because they include both tholeiites, whose origin is intensely debated (e.g., Perfit and others, 1980; Kay, 1980; Morris and Hart, 1983), and boninites, whose petrogenesis is better understood (e.g., Hickey and Frey, 1982; Cameron and others, 1983). It is difficult to understand the petrogenesis of arc-tholeiite magmas because of their complex geochemistry and the extensive fractionation common in arc suites. In contrast, boninites have distinct compositions that require very specific source compositions and melting conditions (Hickey and Frey, 1982; Cameron and others, 1983). The spatial association of boninites and primitive arc tholeiites in the Mariana forearc and chemical similarities between the two series suggest that the tholeiites may be generated by processes similar to those postulated for the boninites. Results from some preliminary melt models that address this idea are presented here.

The samples are from five dredge hauls along the landward slope of the Mariana Trench (Fig. 1). Boninite (samples D28, D50, and D51) and tholeiite (samples D28, D36, and D52) series rocks are the principal volcanic rock type in all five collections; other rock types recovered include serpentinites, gabbros, and volcaniclastic siltstones (Bloomer, 1983). Boninites are olivine, orthopyroxene, clinopyroxene, and plagioclase phyric, with a clear glassy matrix. Tholeiites are clinopyroxene-plagioclase phyric, with rare orthopyroxene or olivine. Smectites, zeolites, and calcite fill veins and vesicles in some samples from both suites. Several tholeiitic samples have been intensely altered; some are recrystalilized to greenschist facies assemblages.

Similar boninitic and tholeiitic rocks have been dredged south of Guam (Beccaluva and others, 1980; Sharaskin and others, 1983) and drilled at 18°N in the forearc at DSDP sites 458 (boninites and tholeiites) and 459 (arc tholeiites) (Meijer and others, 1981; Hussong and Uyeda, 1981; Natland and Tarney, 1981). At site 458, tholeiites and boninites are interlayered (Wood and others, 1981). Boninitic and tholeiite series volcanics also are found in the oldest sections on Guam (Reagan and Meijer, 1984). Boninites form most of the Eocene volcanic section on Guam; they are locally overlain by arc tholeiites. The Oligocene formations include arc tholeiites, boninites, and calcalkaline basalts (Reagan and Meijer, 1984). Boninites have not been recovered to the west or southwest of the site of sample D28 (Fig. 1).

The forearc basement is inferred to be an extensive arc-volcanic complex, which has been exposed by tectonic erosion (Beccaluva and others, 1980; Hussong and Uyeda, 1981; Bloomer, 1983). The DSDP sections indicate that the volcanic rocks are likely late Eocene or early Oligocene in age (Ellis, 1981) and are among the oldest volcanic rocks associated with subduction in the Mariana region (Meijer and others, 1983). It should be noted that the boninitic samples are restricted to the older arc-volcanic units. Only tholeiitic and calcalkaline rocks have been sampled from the Palau-Kyushu Ridge, the West Mariana Ridge, the younger formations on Guam, and the active arc (Stern, 1979; Dixon and Batiza, 1979; Mattey and others, 1980; Wood and others, 1980; Stern, 1981; Meijer and Reagan, 1981; Reagan and Meijer, 1984). Thermal or chemical conditions unique to the early phases of subduction may be required to produce boninite melts. Where they do occur, however, the boninites often are associated with arc tholeiites. It is this association that may yield some clues to the origin of the tholeiitic suite.

ANALYTICAL METHODS

Major and trace element compositions reported in Table 1 were analyzed, except as noted, by standard x-ray fluorescence techniques at the Scripps Institution of Oceanography. Na_2O and Cr concentrations were made by standard atomic absorption analysis, Fe_2O_3/FeO by titration, and H_2O^+ by a gravimetric technique (after Johnson and Shephard, 1978). Rare-earth element analyses are of bulk-rock powders by isotope dilution, measured at the U.S. Geological Survey laboratory in Denver. Rare-earth analyses are estimated to be accurate to ±5 percent. Details of the analytic technique are discussed in Divis (1975).

GEOCHEMICAL CHARACTERISTICS

Two principal groups can be defined on the basis of chemical data (Tables 1, 2), each displaying trends consistent with changes in liquid composition as controlled by crystal fractiona-

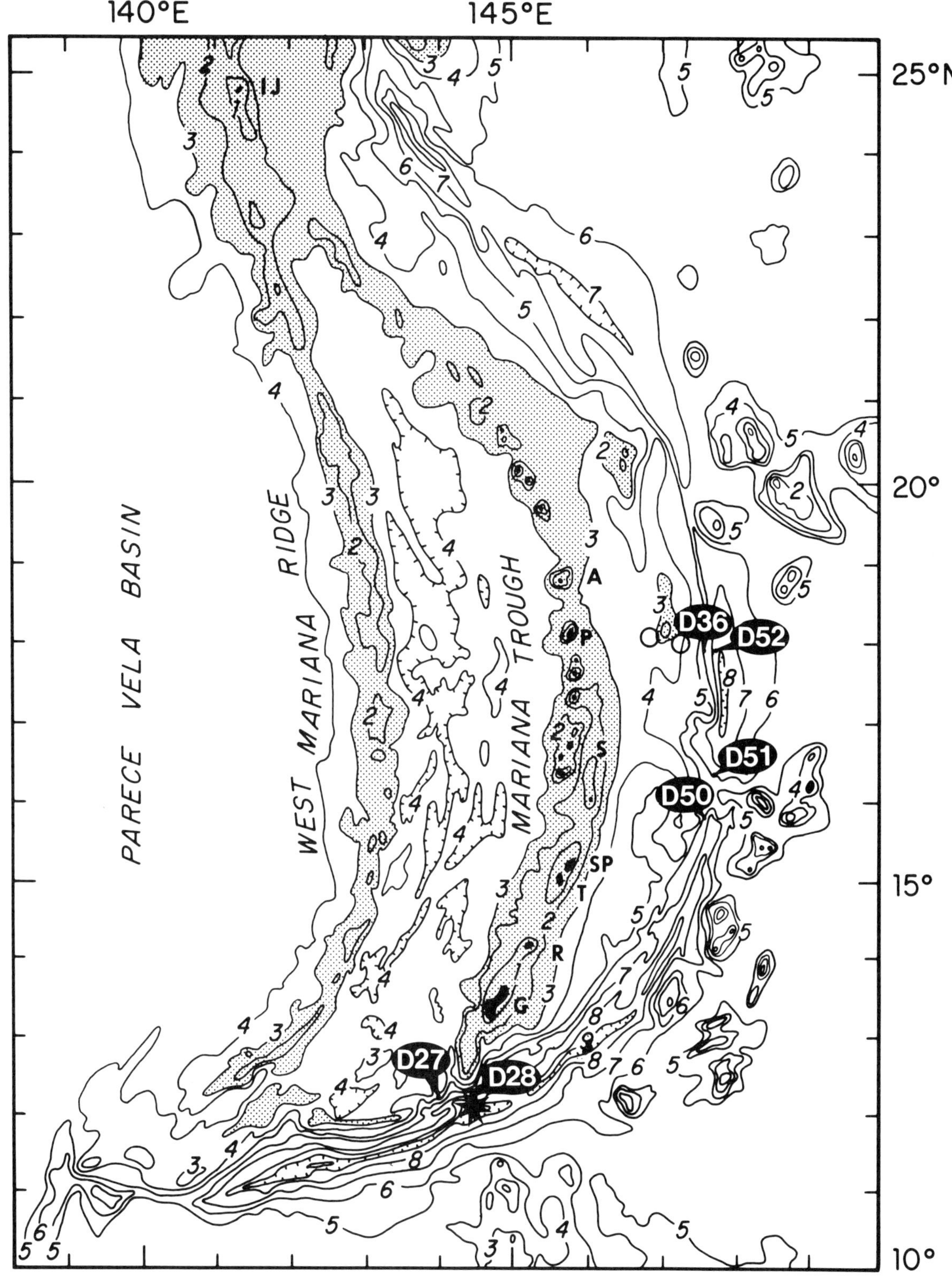

Figure 1. Location of dredge sites discussed in this paper. G, Guam; SP, Saipan; A, Agrigan; P, Pagan. Star south of D28 marks site of *Dmitry Mendeleev* dredge from which boninites were recovered (Beccaluva and others, 1980); open circles at 18°N, west of D36, indicate DSDP Sites 458 and 459. Palau-Kyushu Ridge runs north-south at 135°E.

TABLE 1. REPRESENTATIVE ANALYSES OF VOLCANIC ROCKS FROM THE MARIANA FOREARC*

	Boninite Series Samples						Tholeiite Series Samples						
	50-14	50-23a	28-18	28-1	28-9	28-43	50-13	51-5	50-15	52-21	52-16	36-46	36-65
SiO_2	53.60	54.44	55.09	59.46	61.90	67.45	50.67	51.70	53.99	52.93	58.44	61.79	67.54
TiO_2	0.18	0.27	0.20	0.16	0.17	0.19	0.48	0.57	0.37	0.52	0.88	0.81	0.56
Al_2O_3	10.77	13.65	10.26	11.16	11.66	13.20	14.98	15.94	14.02	15.53	14.16	14.13	13.48
Fe_2O_3	2.09	2.59	2.16	1.37	1.32	1.36	2.14	3.13	2.66	4.72	3.43	1.47	1.28
FeO	6.19	5.15	5.86	5.90	4.78	3.83	6.02	6.87	5.42	5.30	7.22	8.11	5.33
MnO	0.15	0.13	0.14	0.13	0.11	0.09	0.19	0.15	0.18	0.20	0.23	0.14	0.13
MgO	15.53	10.42	15.21	11.26	8.92	3.55	9.18	8.06	11.22	6.49	2.28	1.64	0.81
CaO	6.98	8.16	5.10	4.97	4.13	3.97	9.00	8.07	7.69	5.35	5.91	5.88	4.00
Na_2O	1.00	1.44	1.51	2.19	3.48	3.78	3.42	2.19	1.53	2.86	3.46	3.62	3.34
K_2O	0.66	1.02	0.77	0.64	0.70	1.37	1.11	1.11	1.36	1.98	1.12	0.47	1.20
H_2O^+	2.60	3.09	3.47	2.92	1.90	1.40	3.07	2.25	3.38	3.46	2.10	nd	2.24
P_2O_5	0.05	0.05	0.05	0.04	0.05	0.06	0.06	0.07	0.06	0.06	0.11	0.12	0.16
Total	99.80	100.41	99.83	100.20	99.12	100.25	100.32	100.11	100.63	99.40	99.34	98.20	100.07
Sr	71	91	99	154	220	283	116	135	61	112	111	108	118
Zr	32	48	38	50	63	76	36	38	32	45	82	85	99
Rb	10	14	10	10	10	7	<6	<6	7	13	10	8	10
Ni	384	182	344	234	150	24	89	87	164	46	<8	<8	<8
Cr	876	509	850	808	486	98	205	70	64	50	<20	<20	<20
Ba	17	20	23	36	44	42	35	12	16	30	25	27	31
V	161	164	124	126	100	85	203	258	198	227	131	131	21
Y	7	9	5	5	nd	7	11	15	13	16	32	35	34
Nb	<5	<5	<5	6	nd	<5	8	<5	nd	7	<5	<5	<5

	Transitional Samples			Arc Tholeiite Sample	Estimated
	51-8	51-10	51-12	27-3	1σ error
SiO_2	55.22	52.63	57.32	55.33	0.61
TiO_2	0.40	0.31	0.37	0.94	0.03
Al_2O_3	14.45	16.41	14.27	14.78	0.23
Fe_2O_3	1.84	3.04	2.10	2.83	0.14
FeO	4.45	3.68	4.50	7.48	0.14
MnO	0.11	0.10	0.14	0.16	0.01
MgO	7.76	6.93	7.30	4.26	0.33
CaO	8.44	8.38	6.52	8.56	0.10
Na_2O	2.60	1.39	5.06	2.25	0.20
K_2O	0.76	1.78	0.37	0.35	0.05
H_2O^+	3.26	6.26	2.08	2.49	0.20
P_2O_5	0.09	0.07	0.06	0.12	0.01
Total	99.38	100.98	100.09	99.55	
Sr	198	105	129	140	5
Zr	60	46	49	64	6
Rb	<6	6	<6	7	3
Ni	134	60	64	14	4
Cr	313	51	223	29	10
Ba	17	<8	14	45	4
V	160	159	161	321	5
Y	15	10	14	30	2.5
Nb	5	5	nd	<5	2.5

Notes:
Oxides are in wt.%.
Elements are in ppm.
nd = not determined.

TABLE 2. RARE-EARTH ELEMENT CONCENTRATIONS*

Element	Sample (ppm)							
	28-1	28-43	50-14	51-5	51-8	36-46	36-65	27-3
Ce	4.01	7.02	2.90	4.24	6.24	8.90	11.76	7.46
Nd	2.18	3.96	1.59	3.36	4.39	7.41	8.15	6.60
Sm	0.58	1.00	0.46	1.14	1.34	2.64	2.62	2.28
Eu	0.20	0.30	0.15	0.41	0.47	0.90	0.81	0.85
Gd	0.71	0.98	0.58	1.60	1.66	3.55	3.98	3.26
Dy	.68	.86	.68	1.79	1.75	4.74	4.52	4.00
Er	0.45	0.48	0.49	1.28	1.18	3.53	3.18	2.60
Yb	nd	nd	nd	nd	1.26	3.50	3.62	2.63

*nd = not determined.

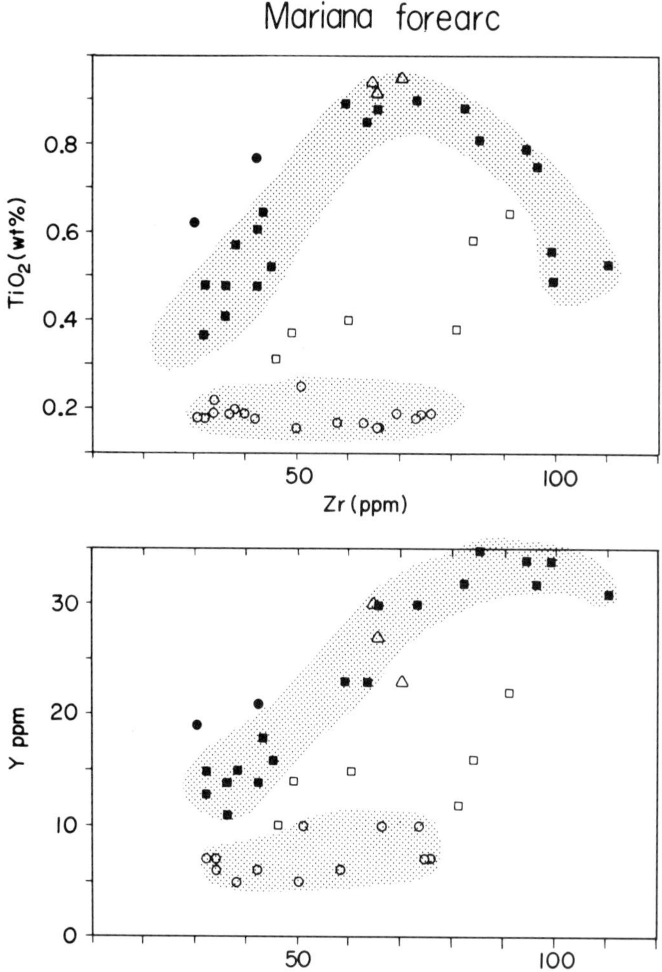

Figure 2. Chemical variation of boninite and tholeiite series rocks. Solid squares indicate tholeiitic basalts and derivatives (D36, D50, D51, D52); solid circles, D28 tholeiites; open circles, boninites and derivatives (D28, D50, D51); open squares, samples transitional between the two series (D51); triangles, more fractionated arc-tholeiites (D27) from southern Mariana Trench. Shaded areas enclose main trends for tholeiitic (solid squares) and boninitic series (open circles).

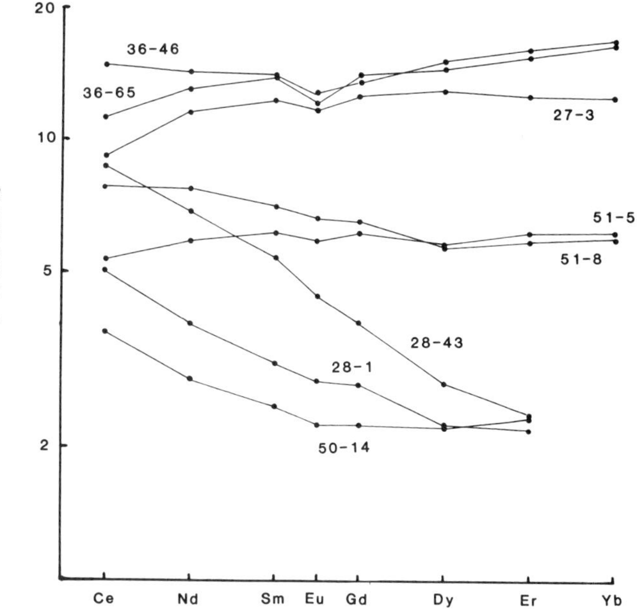

Figure 3. Rare-earth element patterns for dredged samples: 50-14, 28-1, and 28-43 are boninite series samples, 51-5 is tholeiitic basalt, 36-46 and 36-65 are dacites (part of tholeiitic series), 51-8 is transitional sample, 27-3 is a more fractionated arc-tholeiite from southern Mariana Trench.

tion (Fig. 2). The boninitic series includes samples containing 54–67 percent SiO_2. FeO, Ni, and Cr decrease, whereas SiO_2, Al_2O_3, Na_2O, and Sr increase, and TiO_2 stays relatively constant with increasing Zr. Zr is taken as an indication of the degree of fractionation, as it is both relatively immobile and highly incompatible in the mineral phases found in the boninitic samples. Typical of boninites, they have very low abundances of Ti, Y, and HREE (Table 2, Fig. 3). The chemical variations within this suite can be modeled by fractionation of olivine and orthopyroxene, joined in the later stages by clinopyroxene and plagioclase (Bloomer and Hawkins, in preparation). Two types can be identified among the least evolved boninites: one represented by D28-18 and D28-1 and one by D50-14 (Table 1). Both have high Ni, Cr, and MgO, and similar abundances of TiO_2, HREE, and Y, but have significantly different concentrations of Na_2O, Sr, CaO, and LREE (Tables 1, 2).

SiO_2 increases and Ni, Cr, and MgO decrease with increasing Zr in the tholeiitic samples (Table 1). There is an initial enrichment, characteristic of tholeiitic suites, in FeO, TiO_2, Y, and V with Zr (Fig. 2, Table 1), followed by a decrease that marks the onset of magnetite crystallization. Alteration in some of these samples precludes rigorous modeling of crystal fractionation. Qualitatively, olivine, plagioclase, clinopyroxene, and magnetite appear to have been the principal crystallizing phases. The tholeiitic samples are less depleted than the boninites in Ti, Y, and HREE, and include both slightly light rare-earth enriched and light rare-earth depleted samples (Fig. 3). This series also may include diverse parental magmas. Tholeiitic samples from D28 plot near, but consistently off, the main tholeiitic trend (Fig. 2). This suggests that they are part of a tholeiitic magma series, but one that evolved from a different parental composition.

A few samples have Ti, Y, and, to a lesser extent, HREE values, which fall between the boninite and tholeiite series samples (Figs. 2, 3, 4; Table 1). These may be part of a series transitional from the boninites to the tholeiites.

Some chemical characteristics of the least fractionated volcanic rocks from the forearc are compared to a primitive ocean-ridge basalt in Figure 4. A typical, more fractionated, arc tholeiite from the southern Mariana Trench (sample D27-3) is plotted for comparison. The origin of this sample is uncertain, as it was dredged from an intensely sheared and faulted portion of the

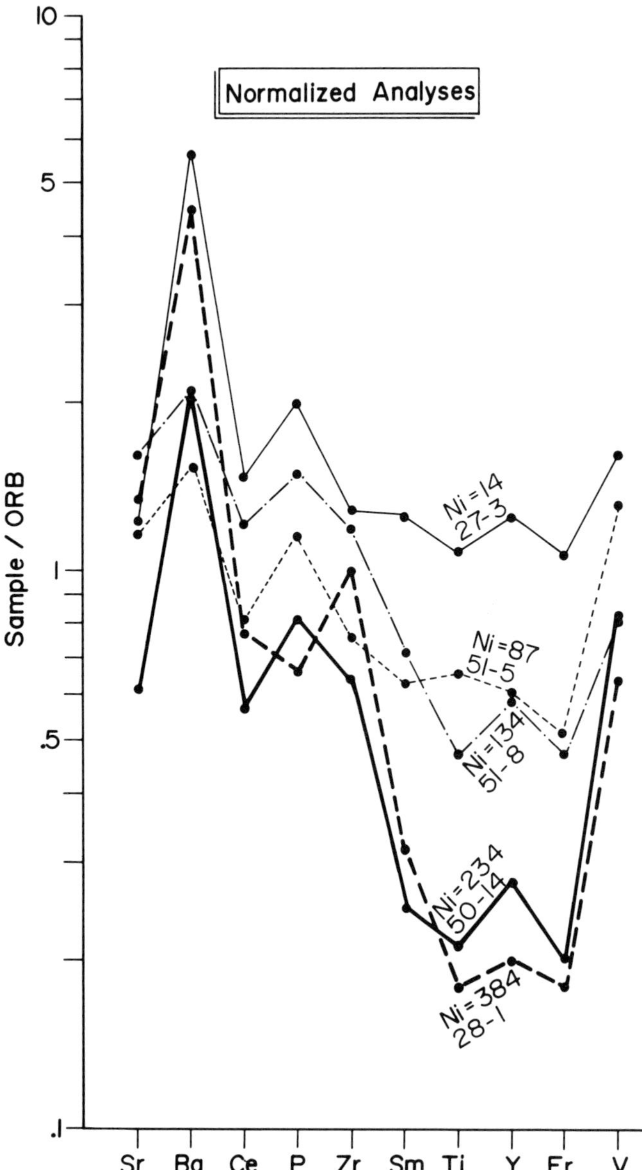

Figure 4. Element abundances in selected samples normalized to primitive ORB composition; samples as in Figure 3. Ni concentrations shown with sample numbers to indicate degree of fractionation. Normalizing values (in ppm) from Sun and others (1979): Sr = 114, Ba = 8, Ce = 5.12, P = 262, Zr = 50, Sm = 1.8, Ti = 5216, Y = 25, Er = 2.49, V = 196, Ni = 245.

trench slope, but it is identical to arc tholeiites from the Palau-Kyushu Ridge. It is used for comparison, as it was analyzed in the same laboratory as the forearc samples. The boninites are extremely depleted in Sm, Ti, Y, and Er relative to primitive ORB (Fig. 4). Least fractionated arc tholeiites are depleted in these elements as well, although not as greatly as are the boninites. These initial element depletions are obscured in more typical arc tholeiites, such as sample 27-3, which have been modified by fractionation (Fig. 4). Concentrations of Sr, Ba, Ce, and Zr in both boninites and tholeiites are enriched relative to Ti, Y and HREE, although the absolute concentrations may be higher or lower than those of ORB (Fig. 4). All of the samples show the enrichment in Ba characteristic of arc lavas.

Element abundances and ratios for less fractionated boninites and tholeiites are compared in Figures 5 and 6. Samples plotted include boninites with Ni and Cr concentrations above 100 ppm and SiO_2 concentrations of less than 60 percent, and tholeiites with Ni and Cr concentrations above 50 ppm and SiO_2 concentrations of less than 55 percent. Transitional samples meeting either criteria are included. All samples with Zr of less than 75 ppm (the onset of magnetite crystallization) are included in plots of rare-earth element data, because of the limited number of analyses. Ti/Zr increases from the boninites to the tholeiites (Fig. 5). This change in ratio is due principally to the increase of Ti in the tholeiites (Fig. 6, Table 1). Ba and Zr abundances show no consistent trends and have the same range of concentrations in both groups (Figs. 5, 6). Sr exhibits similar behavior. The few very high Ba and Sr values may result from low-temperature alteration. Ratios of highly incompatible elements such as Sr/Zr (Fig. 6), Ce/Zr (Fig. 5), and Ba/Zr show the same range in both tholeiitic and boninitic samples. Ratios of moderately incompatible elements, such as Y/Ti, are constant as well (Fig. 6). The degree of light rare-earth enrichment, as indicated by Ce/Sm, is inversely correlated with Ti/Zr, although Sm/Er is not (Fig. 5). The same relationship has been noted within other boninite suites (Jenner, 1981; Hickey and Frey, 1982; Cameron and others, 1983). In some boninite-series volcanics it has been noted that $^{143}Nd/^{144}Nd$ was positively correlated with the degree of light rare-earth enrichment (Hickey and Frey, 1982; Sharaskin and others, 1983; Cameron and others, 1983).

DISCUSSION

The least fractionated boninitic and tholeiitic samples from the Mariana forearc are depleted in Ti, Y, Sm, and HREE relative to primitive ORB; the tholeiites are depleted to a lesser extent than the boninites. Both have high ratios of alkali metals, alkaline earths, and LREE to the high field strength (HFS) and heavy rare-earth elements, although not necessarily high absolute abundances of those elements relative to primitive ORB. Concentrations of these incompatible elements vary independently of the degree of HFS cation and HREE depletion. These differences have been noted before in arc lavas (Perfit and others, 1980; Kay, 1980; McCulloch and Perfit, 1981) and indicate the arc lavas are not simply melts of mantle like that producing ORB.

Similarities in Sr, Zr, and Ba abundances, ratios of highly incompatible elements, and ratios of moderately incompatible elements, as well as the consistent variations of Ti, Ti/Zr, and Ce/Sm between the least fractionated boninites and tholeiites suggest that the processes or sources producing them may be related in some way. The high Ni, Cr, and MgO concentrations of the least fractionated samples indicate that the magmas parental

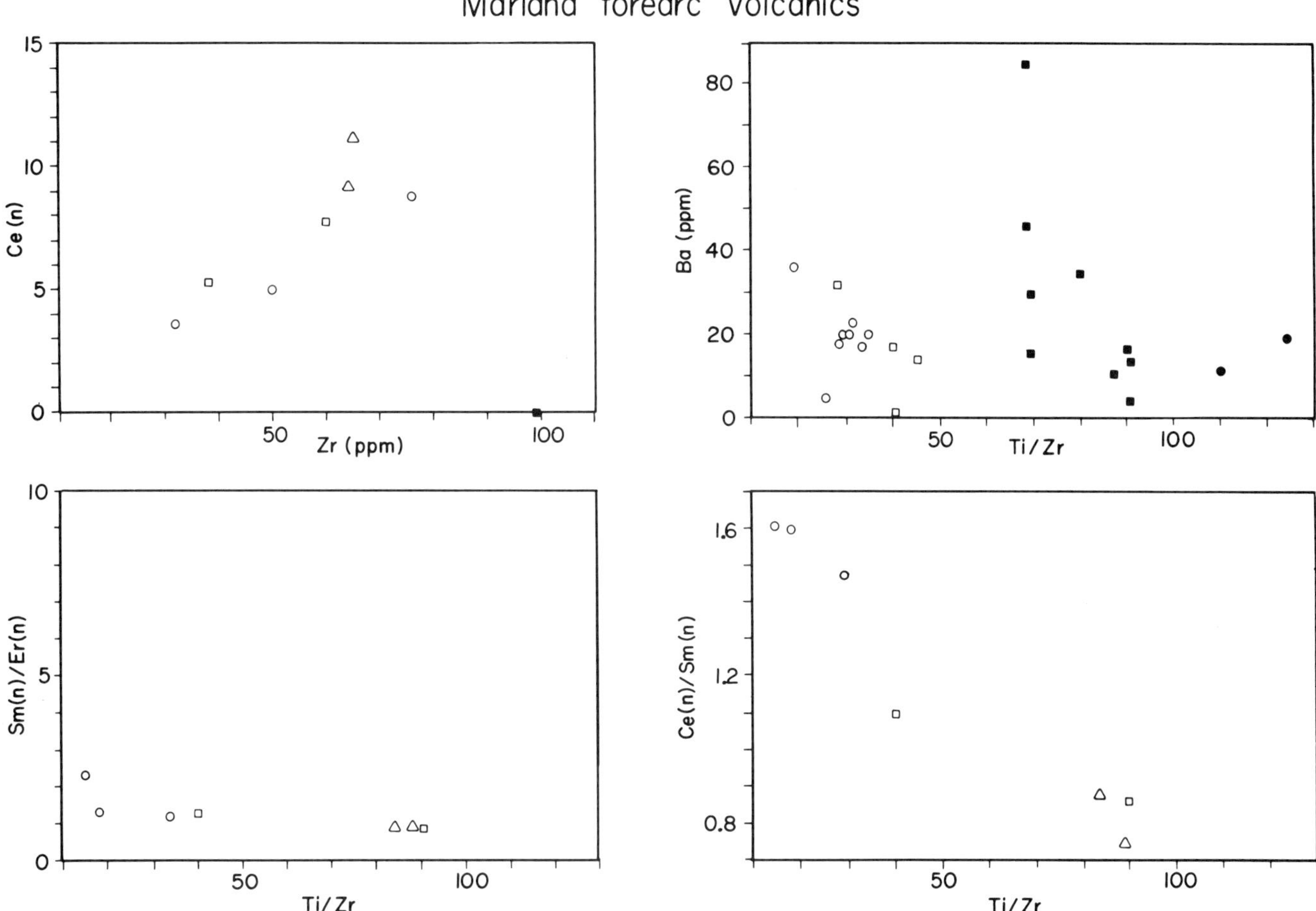

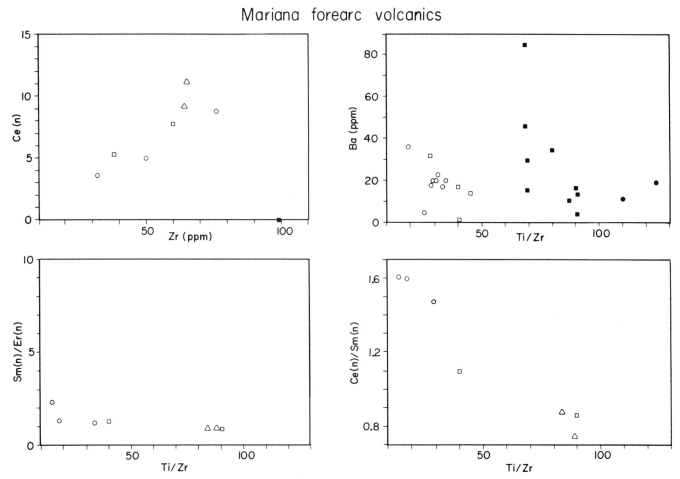

Figure 5. Comparison of selected chemical characteristics of least fractionated tholeiitic and boninitic samples. Symbols as in Figure 2.

to both series were melted from a peridotitic source. The decoupling of Sr, Ba, and LREE abundances from those of the HREE, Ti, and Y, and the similar incompatible element ratios in the two suites, show that their differences are not a result of crustal contamination. Most samples from the active and remnant arcs in the Marianas are interpreted to be derivatives of mantle melts (Dixon and Batiza, 1979; Stern, 1981; Meijer and Reagan, 1981). The boninites and tholeiites, however, cannot be derived by different degrees of melting of the same peridotitic source. Varying degrees of partial melting of a homogeneous source should produce higher ratios of moderately incompatible (Ti) to highly incompatible (Zr) elements with increasing partial melting. These should be correlated with decreasing abundances of Zr, Ba, Sr, Ti, and other incompatible elements. The boninites have the lowest Ti/Zr and Y/Zr values of the sampled rocks and therefore should be lower percentage partial melts than the tholeiites, if both are melted from the same source. The boninites' lower abundances of Ti, Y, Er, and Sm, their variable LREE contents, and their similar concentrations of Ba, Sr, and Zr when compared to the tholeiites, indicate that this is not the case.

Thus the fundamental differences between the magmas parental to the two rock series must be due to melting of distinct sources in the mantle. The occurrence of samples apparently transitional between the two series may reflect mixing of two distinct sources or sampling of a continuum of sources. It is simpler to constrain the nature of the boninite source that that of the tholeiites, because of the distinctive chemistry of the boninites. In the section below, the principal model for boninite petrogenesis is examined and an evaluation made of how that model may be related to the production of arc tholeiites.

Boninite Petrogenesis

The petrogenesis of boninite magmas has been discussed by a number of workers (Meijer, 1980; Jenner, 1981; Hickey and Frey, 1982; Crawford and others, 1981; Sharaskin and others, 1983; Cameron and others, 1983; Bloomer and Hawkins, in preparation), all of whom have reached substantially the same conclusions. This uniformity of opinion is due to the distinctive chemical characteristics of these rocks; there are a limited number

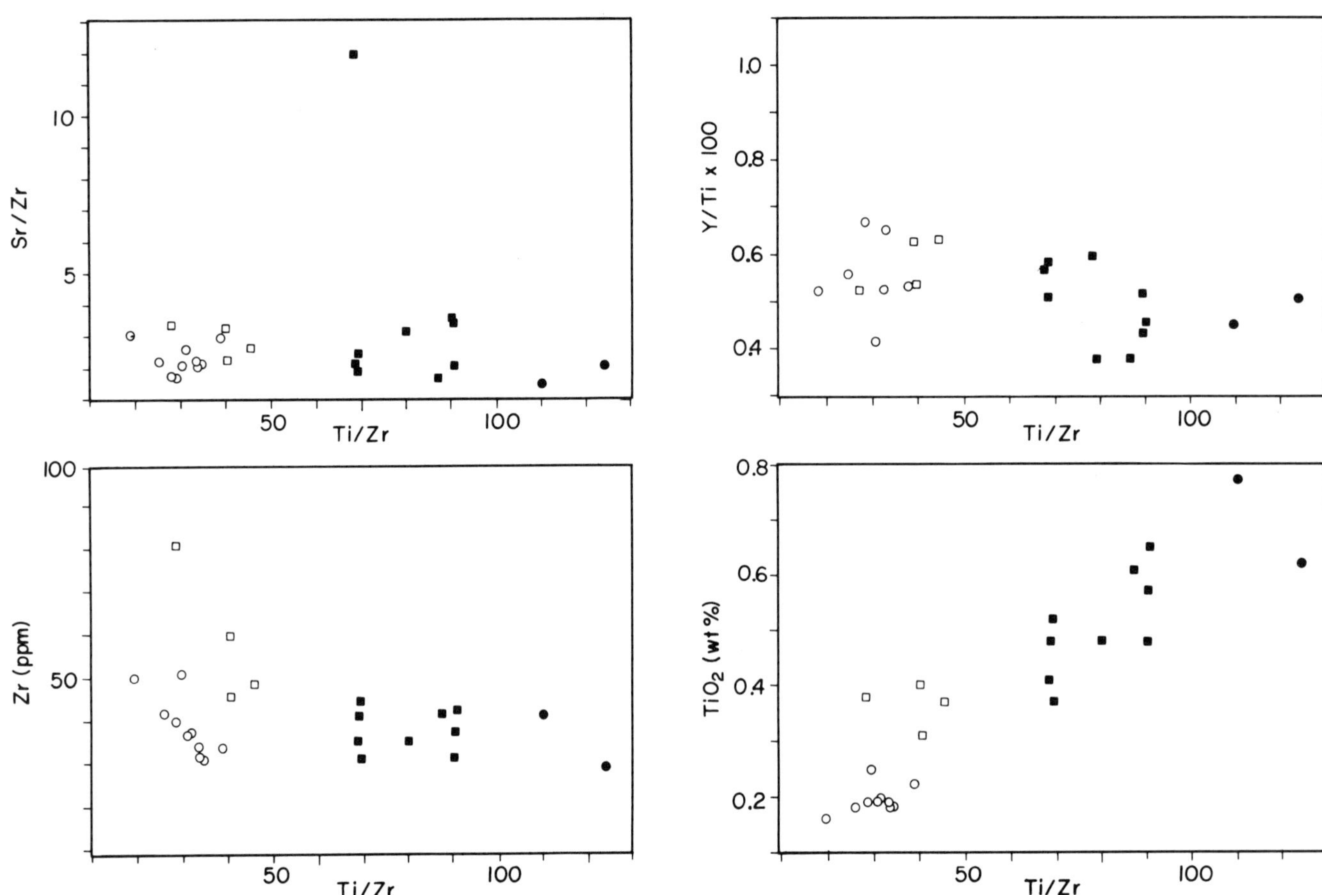

Figure 6. Comparison of selected chemical characteristics of least fractionated boninitic and tholeiitic samples. Symbols as in Figure 2.

of mechanisms for producing such a melt. Most models of boninite petrogenesis have three essential features:

1. Boninites are hydrous melts of mantle peridotites. Boninites are Si-saturated, yet have Ni, Cr, and MgO abundances indicating that they are at or near equilibrium with peridotitic materials. They are in bulk chemistry much like basalt-andesite compositions produced experimentally by hydrous melting of peridotite (Green, 1973).

2. Boninites are melts from a mantle that has been previously melted, that is, an ORB-mantle source from which ORB or arc-tholeiite magmas have been extracted. The extreme depletions in Ti, Y, and HREE, the high V/Ti, and the experimental evidence that these are moderate- to high-percentage partial melts favor this interpretation.

3. Prior to or concomitant with the melting event that produced the boninites, this depleted mantle source was metasomatized by a melt or fluid enriched in LREE, Ba, Sr, K, Rb, and Zr. Isotopic evidence indicates that this fluid (which differed in composition in different regions) had low $^{143}Nd/^{144}Nd$ values (Hickey and Frey, 1982; Sharaskin and others, 1983; Cameron and others, 1983). The mixing of different proportions of this fluid into the depleted mantle would be capable of producing boninitic magmas with various LREE enrichments, Ti/Zr, and $^{143}Nd/^{144}Nd$. The source of this metasomatic fluid is enigmatic. The low Nd isotopic signature suggests involvement of continental material, authigenic sediments, or an enriched mantle source (Hickey and Frey, 1982; Sharaskin and others, 1983; Cameron and others, 1983). There is no evidence of old continental material in the Mariana region, and Pb isotopic measurements on boninites (samples from D50 and D28) indicate that oceanic sediments have not contributed Pb to the magmas (Meijer and Hanan, 1981). A less depleted mantle, like that producing many intraplate basalts, might be the most likely source, at least in the Mariana region, for a fluid with such a low $^{143}Nd/^{144}Nd$ signature. Hickey and Frey (1982) have noted that Ba and Sr in some boninitic samples are not correlated with Ce/Sm, Ti/Zr, and Nd

TABLE 3. PARTITION COEFFICIENTS AND COMPOSITIONS (ppm) USED IN MODELING CALCULATIONS

Element	Partition Coefficient				Fertile Mantle	Subducted Slab	Boninite Fluid
	Olivine	Orthopyroxene	Clinopyroxene	Garnet			
Ti	0.02	0.11	0.30	0.30	1250	4763	0
V	0.04	2.00	1.50	8.00	97	143	0
Zr	0.01	0.03	0.12	0.50	12	52	289
Ni	7.00	4.00	4.50	5.10	2000	300	0
Cr	1.00	4.00	5.00	17.50	3000	600	0
Ba	0.003	0.01	0.01	0.02	1	10	173
Sr	0.003	0.02	0.1	0.02	22	105	704
Ce	0.002	0.008	0.15	0.07	0.798	3.81	27.5
Nd	0.002	0.015	0.25	0.3	0.964	4.48	11.8
Sm	0.002	0.02	0.30	0.9	0.428	1.96	2.3
Dy	0.002	0.05	0.35	1.0	0.763	3.33	1.5
Yb	0.002	0.15	0.35	5.0	0.523	0.215	0

References:
Langmuir and others, 1977; Sun and others, 1979; Tarney and others, 1978; Pearce and Norry, 1979; Irving, 1978; Irving and Frey, 1978; McCallum and Charette, 1978; Gill, 1978; Ray and others, 1983

isotopic ratio, and have suggested that a third component, perhaps from the subducted slab, may be involved in boninite petrogenesis as well.

Models involving mixing of depleted mantle sources and an incompatible element–enriched fluid, whatever its source, have great appeal in explaining the chemical variations in primitive arc magmas. Perfit and others (1980), Kay (1980), and McCulloch and Perfit (1981) have suggested such a process for generating Aleutian arc magmas, and Gill (1984) has presented evidence for tapping of both ocean-island and ocean-ridge basalt sources in arc lavas from Fiji. Models for boninite petrogenesis suggest that the depleted source involved in arc petrogenesis is depleted not only in the sense that an ORB-type mantle shows evidence of long-term depletion, but also that the source has been depleted of part of its basaltic component in a pre-arc melting event. The distinct chemical character of boninites allows a detailed estimate to be made of the metasomatic component that must be mixed into the depleted mantle prior to melting.

Boninites are an extreme type of arc lava, in that they are very depleted in high field strength and heavy rare-earth elements, yet are greatly enriched in incompatible elements relative to the HFS cations and HREE. Arc tholeiites associated with the boninites are similarly enriched in incompatible elements, and similarly, though not as strongly, depleted in HFS cations and HREE. It can be postulated that arc tholeiites are generated by a mechanism similar to that outlined above for boninites. The tholeiites represent melts of a depleted (once melted) ORB-type mantle, although one not as depleted as that producing boninites, that has been mixed with a metasomatic fluid like that affecting the boninite source. Hickey and Frey (1982) have suggested such a model, and discussed the relationship between arc tholeiites, boninites, a variously depleted mantle, and an enriched metasomatic fluid. The results of some simple mixing and melting models, examined in the next section, illustrate the effects of such a process and demonstrate how various degrees of depletion and proportions of metasomatic fluid change the resultant melt characteristics.

Partial-Melting Models

Two components are required for these models—an initial mantle composition and a metasomatic fluid. The degree of depletion of that initial mantle, the proportion of metasomatic fluid added, and the percentage of melting of the mixed source can then be varied. Models presented here start with a 59% olivine (OL), 16% clinopyroxene (CPX), 21% orthopyroxene (OPX), and 5% garnet (GA) mantle assemblage. Its chemical composition is that of a fertile ORB-source mantle (Table 3). This is melted in the proportions 15:35:30:20; garnet is exhausted at 25% melting. Residual mantle compositions are calculated for 25% batch melting with no melt trapped in the residue, 25% batch melting with 3% melt in the residue, 10% batch melting with no trapped melt, 10% batch melting with 3% trapped melt, and 15% batch melting with 3% trapped melt.

These residual compositions are mixed with different amounts of an incompatible element–enriched fluid. The composition of the metasomatic fluid can be estimated in two ways: a melt can be calculated using available partition coefficients and a postulated source, or the composition of the fluid required to match a specific boninite melt can be calculated and then used in other modeling.

The degree of melting of the mixed source, assumed to be an OL-OPX-CPX assemblage, is arbitrarily set at 20%, in accordance with evidence that boninites are high to moderate percentage partial melts (Green, 1973; Hickey and Frey, 1982). Small changes in the percentage of melt or mineral proportions

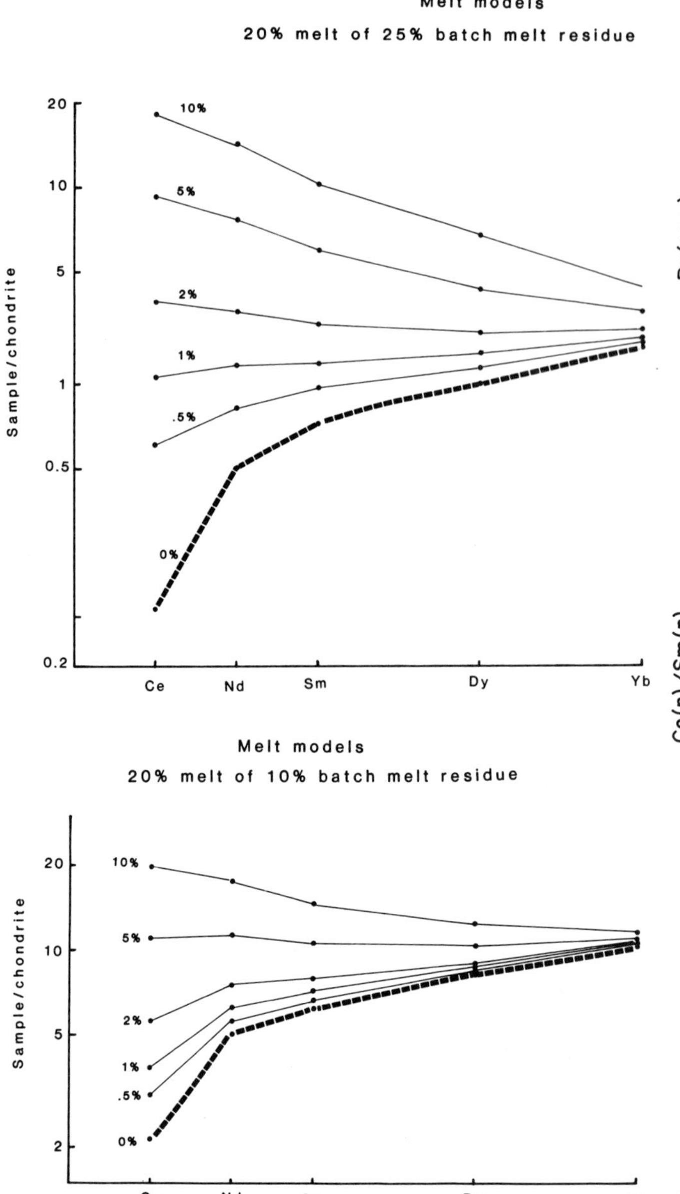

Figure 7. Calculated rare-earth element abundances for 20 percent partial melt of 25 percent batch-melted ORB-type mantle (top) and 10 percent melted ORB-type mantle with 3 percent trapped melt, each mixed with different proportions of melt derived from model subducted slab. Dotted line is pattern for melts derived from depleted source with no added fluid. Percentages indicate amount of slab melt mixed into depleted source prior to 20 percent melting.

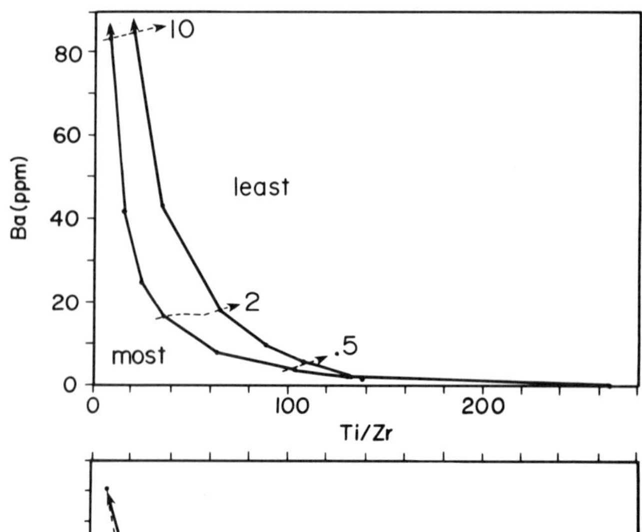

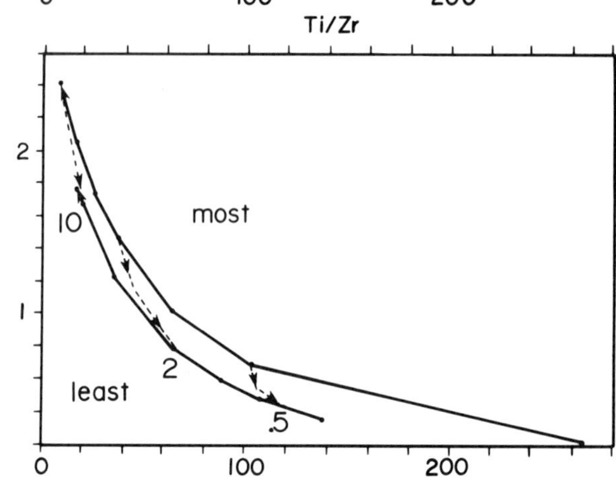

Figure 8. Melting models using same two depleted mantle compositions as in Figure 7, mixed with various proportions of fluid calculated to reproduce chemistry of boninite 50-14 (by 20 percent batch melt of mix of 25 percent depleted ORB-source mantle and 2 percent of calculated fluid). Melt compositions are calculated at 20 percent batch melting of mixed source. Solid lines indicate mixing of different proportions of fluid into same depleted mantle; arrows on solid lines indicate direction of increasing percentage of fluid. Side labeled "most" indicates line for 25 percent depleted ORB-source mantle; side labeled "least" indicates 10 percent depleted ORB-source mantle with 3 percent trapped melt. Dotted lines connect points calculated for mixes of same amount of fluid into different types of depleted mantle; these drawn for 0.5, 2, and 10 percent added fluid. Arrows indicate less depleted residual mantle compositions prior to addition of fluid.

have only small effects on the calculated melt compositions. Of most interest here are changes due to different degrees of depletion and proportions of added fluid.

The first set of calculations uses a metasomatic fluid derived by 3% partial melting of a CPX-GA source chemically like a primitive ORB (Table 3), to simulate a slab-derived melt. Some results for rare-earth elements are shown in Figure 7, for melts from mixtures of this fluid and the most and least depleted OL-OPX-CPX reservoirs. HREE abundances are controlled principally by the degree of depletion of the mantle and the Ce/Sm ratio by the percentage of fluid added (Fig. 7). The influence of the metasomatic fluid is much greater in the 25% melted residue than in the 10% residue. Two percent added fluid produces a boninite-like U-shaped pattern in the first case, and a LREE-depleted, arc-tholeiite–type pattern in the second (see samples 50-14, 51-5 in Fig. 3). V is buffered in most of the melts because

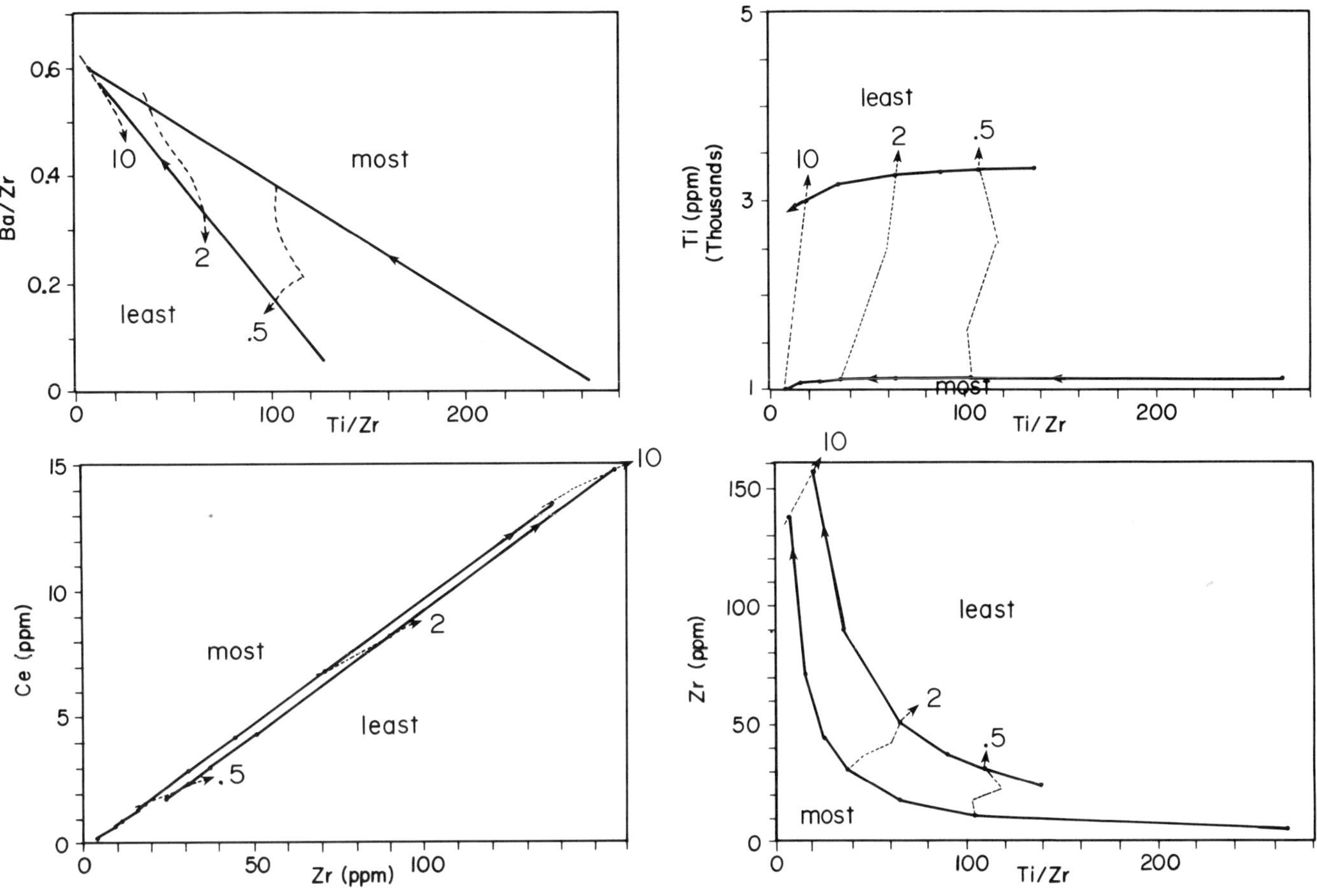

Figure 9. Results of mixing-melting calculations; symbols and labels as in Figure 8.

of its high partition coefficients in GA and CPX. Ba and Sr abundances in both models are controlled nearly completely by the amount of added fluid. Ba/Ce is somewhat higher in the models (5 to 13) than is typical of boninites and tholeiites (2.7 to 9). Sr/Ce is similar in the models (31.7 to 52.5) and in the samples (24 to 38). In all models using a slab-derived melt as the metasomatic component, too much Ti and too little Zr is added to adequately fit the boninite chemistry. Partition coefficients for both Ti and Zr are poorly constrained in garnet, and the Zr concentrations in the residual mentle can be varied significantly by changing the Zr partition coefficients. However, any reasonable partition coefficient for Ti still produces Ti concentrations far in excess of those observed in the boninites.

A second approach is to calculate the fluid composition required to reproduce boninite compositions. The initial modeling suggests that an ORB-type mantle depleted of 25% of its basaltic component mixed with 2% of a light rare-earth–enriched fluid, and then batch melted 20%, yields melts with boninite-like rare-earth element patterns. If these numbers are taken as a starting point, the composition of the requisite metasomatic fluid can be calculated (Table 3). This fluid was calculated to fit D50-14 exactly, assuming 20% melting of a mix of 2% metasomatic fluid and 98% of an ORB-type mantle, depleted of 25% of its basaltic component. It should be emphasized that this modeling is not intended to quantify the degrees of melting or mixing, but only to illustrate the element patterns that can be produced by such a process. The composition of the fluid depends, of course, on the partition coefficients and starting compositions selected. The calculated fluid composition, for the assumed conditions, is devoid of Ti and Y, and greatly enriched in LREE and Zr. It is in some ways similar to the metasomatic agent postulated to produce LREE enrichments observed in otherwise-depleted peridotites (Frey and Prinz, 1978).

The effects of varying the degree of depletion of the residual mantle and the proportion of the fluid in the mixture are illustrated in Figures 8 and 9. The model yields a strong inverse correlation between Ce_N/Sm_N and Ti/Zr, which vary with both the initial degree of depletion and the amount of added fluid (Fig.

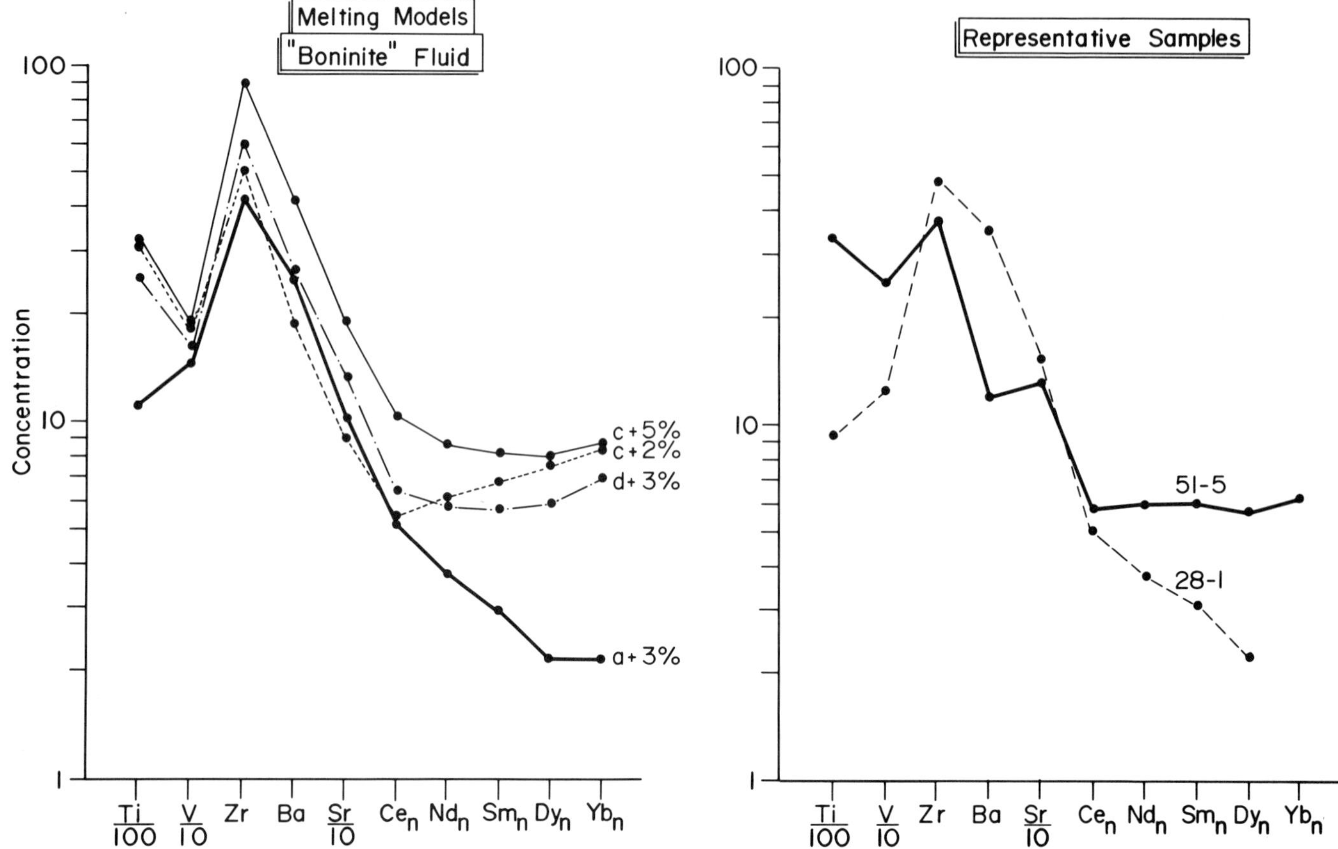

Figure 10. Results of melting-mixing models. Percentages indicate amount of calculated "boninite" fluid added to depleted source prior to 20 percent batch melting. Letters denote type of depleted source to which fluid added: a, residue of 25 percent batch melt of ORB source; c, residue of 10 percent batch melt of ORB source with 3 percent trapped melt; d, residue of 15 percent batch melt of ORB source with 3 percent trapped melt.

8). Ce/Zr is quite constant (Fig. 9), as are ratios of moderately incompatible elements not contained in the metasomatic fluid (Ti/Y). The abundance of Ti is principally controlled by the initial degree of depletion (Fig. 9); those of Ba and Zr by the amount of added fluid (Figs. 8, 9). Differently depleted mantle residues, mixed with the same proportion of fluid, yield melts with very similar concentrations of Ba and Zr (Fig. 8, 9).

Varying the parameters in this model can produce a wide range of element patterns (Fig. 10), all characterized by depletions of Ti, Y, and HREE relative to primitive ORB and by high ratios of Ba, Zr, and Sr to HFS and HREE cations. These are in general the patterns found in the boninites and tholeiites from the forearc. A 20% partial melt of a mix of 3% of this fluid and a 25% depleted mantle is very close to sample D28-1 (a distinctly different boninite composition than sample 50-14) in composition. A 20% partial melt of a 15% depleted ORB-source mantle, with 3% trapped fluid, mixed with 3% of the metasomatic fluid, produces a melt much like that of sample 51-5, a primitive tholeiitic basalt. This model does not reproduce all the characteristics of all the rock types exactly. One of the common failings in the calculation is that, for boninites other than sample 50-14 (by design fit exactly) and many of the tholeiites, Ba and Sr abundances are often too low in the calculated models, even when the LREE enrichments are well fit. Hickey and Frey (1982) noted that Ba and Sr were not clearly correlated with LREE enrichment, and suggested that there may be two metasomatic components involved in arc petrogenesis. The percentages and mineral assemblages used in these models are in no way intended to rigorously constrain the processes producing boninites and tholeiites. The modeling is intended rather to show that this approach—melting mixes of a variously depleted mantle and an incompatible element–enriched fluid—can produce a wide range of magma compositions, from boninite-like to arc tholeiite–like, and that it can produce many of the element trends and ratios observed in these two series. More realistic modeling requires an evaluation of major element and mineralogic depletions in a melted ORB source, and of their relations to boninitic and tholeiitic magmas.

This is a complicated and poorly constrained model, but it is not necessarily geologically unreasonable. Some portion of the upper 100 to 150 km of mantle (that most likely to be later involved in subduction zone volcanism) must have been partially depleted of its basaltic component during the production of oceanic crust. The extent of that melting and percent of trapped melt in that residue might vary significantly, particularly with depth in the mantle. The volume of depleted mantle, as well as its degree of depletion, are critically dependent on the percentage of melting required to produce ocean-ridge basalts. The restricted occurrence of boninites in the earliest arc units in the Marianas suggests that either extremely depleted upper mantle is rare, or that it cannot normally be tapped in arc volcanism. Unusual thermal or chemical conditions in the earliest phases of subduction may be required. LREE and alkali metal enrichments have been observed in otherwise depleted peridotites and have been attributed to the influence of a metasomatic component (e.g., Frey and Prinz, 1978). The contribution of such a fluid at subduction zones may be masked or diluted in part by fluids derived from the subducted slab. This depletion-enrichment type of model, suggested by analogy to bonitite-series rocks, may prove to be a useful way in which to examine the origin of arc-tholeiite magmas.

CONCLUSIONS

1. Boninite and arc-tholeiite series volcanic rocks are spatially related in the forearc of the Mariana Trench. Both series include relatively little fractionated samples; primitive arc-tholeiitic basalts have Ni = 60–160 ppm, Cr = 50–200 ppm, TiO_2 = 0.4–0.5 percent, Ba = 12–35 ppm, Sr = 60–130 ppm, Zr = 32–49 ppm, Y = 11–16 ppm, and Yb_N = 6.

2. Such primitive magmas are depleted in Ti, Y, Sm, and Er relative to ORB, but are not depleted in V. Boninitic samples are more extensively depleted than are the arc tholeiites. Concentrations of Ba, Sr, and Zr in the arc tholeiites are similar to or higher than those in primitive ORB, but ratios of these incompatible elements to Ti, Y, and HREE are much higher in the arc volcanic rocks. Boninites and arc tholeiites cannot be derived by different degrees of partial melting of a homogeneous source.

3. Least fractionated boninite and tholeiite samples have similar concentrations of Ba, Sr, and Zr and similar values of Ti/Y, Ti/Yb (moderately incompatible elements) and Ba/Zr, Ce/Zr (highly incompatible elements). Ti and Y concentrations and Ti/Zr increase from boninites to tholeiites. Ce/Sm varies antithetically with Ti/Zr.

4. Preliminary modeling suggests that many of these chemical characteristics can be produced by melting mixtures of variously depleted ORB-type and an incompatible element–enriched fluid. The calculated fluid composition has Ce/Sm of 12.0 and is enriched principally in alkali metals, alkaline earths, light rare-earth elements and Zr.

ACKNOWLEDGMENTS

The collection and analysis of these samples was supported by National Science Foundation grants OCE-78-16758, OCE-78-17823, OCE-80-19016 to J. W. Hawkins, Scripps Institution of Oceanography. Cheryl Burgess assisted with the drafting. Careful reviews by Fred Frey and Dave Vanko greatly improved the manuscript.

REFERENCES CITED

Beccaluva, L. G., Macciota, G., Savelli, C., Serri, G., and Zeda, O., 1980, Geochemistry and K/Ar ages of volcanics dredged in the Philippine Sea (Mariana, Yap, and Palau trenches and Parece Vela Basin), *in* Hayes, D. E., eds., The tectonic and geologic evolution of southeast Asian Sea and islands: American Geophysical Union Monograph 23, p. 247–270.

Bloomer, S. H., 1983, Distribution and origin of igneous rocks from the landward slope of the Mariana Trench; Implications for its structure and evolution: Journal of Geophysical Research, v. 88, p. 7411–7428.

Cameron, W. D., McCulloch, M. T., and Walker, D. A., 1983, Boninite petrogenesis; Chemical and Nd-Sr isotopic constraints: Earth and Planetary Science Letters, v. 56, p. 75–89.

Crawford, A. J., Beccaluva, L., and Serri, G., 1981, Tectonomagmatic evolution of the West-Philippine Mariana region and the origin of boninites: Earth and Planetary Science Letters, v. 54, p. 346–356.

DePaolo, D. J., and Wasserburg, G. J., 1977, The sources of island arcs as indicated by Nd and Sr isotopic studies: Geophysical Research Letters, v. 4, p. 465–468.

Divis, A. F., 1975, The geology and geochemistry of the Sierra Madre Mountains, Wyoming [Ph.D. thesis]: San Diego, University of California, 200 p.

Dixon, T. H., and Batiza, R., 1979, Petrology and chemistry of Recent lavas in the northern Marianas; Implications for the origin of island arc basalts: Contributions to Mineralogy and Petrology, v. 70, p. 167–181.

Ellis, C. H., 1981, Calcareous nannofossil biostratigraphy, Deep Sea Drilling Project Leg 60, *in* Hussong, D. M., Uyeda, S., and others, Initial reports of the Deep Sea Drilling Project, Vol. 60: Washington, D.C., U.S. Government Printing Office, p. 855–876.

Frey, F. A., and Prinz, M., 1978, Ultramafic inclusions from San Carlos, Arizona; Petrologic and geochemical data bearing on their petrogenesis: Earth and Planetary Science Letters, v. 38, p. 129–176.

Gill, J. B., 1978, Role of trace element partition coefficients in models of andesite genesis: Geochimica et Cosmochimica Acta, v. 42, p. 709–724.

——, 1984, Sr-Pb-Nd isotopic evidence that both MORB and OIB sources contribute to oceanic island arc magmas in Fiji: Earth and Planetary Science Letters, v. 68, p. 443–458.

Green, D. H., 1973, Experimental melting studies on a model upper mantle composition at high pressure under water-saturated and water-undersaturated conditions: Canadian Mineralogist, v. 14, p. 255–268.

Green, T. H., 1981, Experimental evidence for the role of accessory phases in magma genesis: Journal of Volcanology and Geothermal Research, v. 10, p. 405–422.

Hickey, R. L., and Frey, F. A., 1982, Geochemical characteristics of boninite series volcanics; Implications for their source: Geochimica et Cosmochimica Acta, v. 46, p. 209–2115.

Hussong, D. M., and Uyeda, S., 1981, Tectonic processes and the history of the Mariana Arc; A synthesis of the results of DSDP Leg 60, *in* Hussong, D. M., Uyeda, S., and others, Initial reports of the Deep Sea Drilling Project, Vol. 60: Washington, D.C., U.S. Government Printing Office, p. 909–929.

Irving, A. J., 1978, A review of experimental studies of crystal/liquid partitioning:

Geochimica et Cosmochimica Acta, v. 42, p. 743–770.

Irving, A. J., and Frey, F. A., 1978, Distribution of trace elements between garnet megacrysts and host volcanic liquids of kimberlitic to rhyolitic composition: Geochimica et Cosmochimica Acta, v. 42, p. 771–787.

Jakes, P., and Gill, J. B., 1970, Rare earth elements and the island arc tholeiite series: Earth and Planetary Science Letters, v. 9, p. 17–28.

Jenner, G. A., 1981, Geochemistry of high-Mg andesites from Cape Vogel, Papua, New Guinea: Chemical Geology, v. 33, p. 307–332.

Johnson, W. M., and Shephard, L. E., 1978, A gravimetric method for the determination of H_2O^+ in silicate rocks: Chemical Geology, v. 22, p. 341–346.

Kay, R. W., 1980, Volcanic arc magmas; Implications of a melting-mixing model for element recycling in the crust–upper mantle system: Journal of Geology, v. 8, p. 497–522.

—— , 1984, Elemental abundances relevant to identification of magma sources: Philosophical Transactions of the Royal Society of London, v. A310, p. 535–547.

Langmuir, C. H., Bender, J. F., Benie, A. E., Hanson, G. N., and Taylor, S. R., 1977, Petrogenesis of basalts from the FAMOUS area, Mid-Atlantic Ridge: Earth and Planetary Science Letters, v. 36, p. 133–156.

Mattey, D. P., Marsh, N. G., and Tarney, J., 1980, The geochemistry, mineralogy and petrology of basalts from the West Philippine and Parece Vela Basins and fromthe Palau-Kyushu and West Mariana Ridges, DSDP Leg 59, *in* Kroenke, L., Scott, R., and others, Initial reports of the Deep Sea Drilling Project, Vol. 59: Washington, D.C., U.S. Government Printing Office, p. 753–800.

McCallum, I. S., and Charette, M. P., 1978, Zv and Nb partition coefficient; Imphrations for the genesis of mare basalts, KREEP and sea floor basalts: Geochimica et Cosmochimica Acta, v. 41, p. 859–870.

McCulloch, M. T., and Perfit, M. R., 1981, $^{143}Nd/^{144}Nd$, $^{87}Sr/^{86}Sr$ and trace element constraints on the petrogenesis of Aleutian island arc magmas: Earth and Planetary Science Letters, v. 56, p. 167–179.

Meijer, A., 1980, Primitive arc volcanism and a boninite series; Examples from western Pacific island arcs, *in* Hayes, D. E., ed., The tectonic and geologic evolution of southeast Asian Seas and islands: American Geophysical Union Monograph 23, p. 269–282.

Meijer, A. and Hanan, B., 1981, Pb isotopic composition of boninite and related rocks from the Mariana and Bonin fore-arc regions [abs.]: EOS Transactions of the American Geophysical Union, v. 62, p. 408.

Meijer, A., and Reagan, M., 1981, Petrology and geochemistry of the island of Sarigan in the Mariana Arc; Calc-alkaline volcanism in an oceanic setting: Contributions to Mineralogy and Petrology, v. 77, p. 337–354.

Meijer, A., Anthony, E., and Reagan, M., 1981, Petrology of volcanic rocks from the fore-arc sites, *in* Hussong, D. M., Uyeda, S., and others, Initial reports of the Deep Sea Drilling Project, Vol. 60: Washington, D.C., U.S. Government Printing Office, p. 709–730.

Meijer, A., Reagan, M., Shafiquallah, M., Sutter, J., Damon, P., and Kling, S., 1983, Chronology of volcanic events in the eastern Philippine Sea, *in* Hayes, D. E., ed., The tectonic and geologic evolution of southeast Asian Seas and islands, Pt. 2: American Geophysical Union Monograph 27, p. 349–359.

Morris, J. D., and Hart, S. R., 1983, Isotopic and incompatible element constraints on the genesis of island arc volcanics from Cold Bay and Amak Island, Aleutians, and implications for mantle structure: Geochimica et Cosmochimica Acta, v. 47, p. 2015–2030.

Natland, J. H., and Tarney, J., 1981, Petrologic evolution of the Mariana Arc and back-arc system; A synthesis of drilling results in the southern Philippine Sea, *in* Hussong, D. M., Uyeda, S., and others, Initial reports of the Deep Sea Drilling Project, Vol. 60: Washington, D.C., U.S. Government Printing Office, p. 877–908.

Peace, J. A., and Norry, M. J., 1979, Petrogenetic implications of Ti, Zr, Y, and Nb variations in volcanic rocks: Contributions to Mineralogy and Petrology,v. 69, p. 33–47.

Perfit, M. R., Gust, D. A., Bence, A. E., Arculus, R. J., and Taylor, S. R., 1980, Chemical characteristics of island arc basalts; Implications for mantle sources: Chemical Geology, v. 30, p. 227–256.

Ray, G. L., Shimizu, N., and Hart, S. R., 1983, An ion microprobe study of the partitioning of trace elements between clinopyroxene and liquid in the system diopside-albite-anorthite: Geochimica et Cosmochimica Acta, v. 47, p. 2131–2140.

Reagan, M. K., and Meijer, A., 1984, Geology and geochemistry of early arc-volcanic rocks from Guam: Bulletin of the Geological Society of America, v. 95, p. 701–713.

Sharaskin, A. Y., Karpenko, S. F., Ljalikov, A. V., Zlobin, S. K., and Balashov, Y. A., 1983, Correlated $^{143}Nd/^{144}Nd$ and $^{87}Sr/^{86}Sr$ data on boninites from Mariana and Tonga arcs: Ofioliti, v. 8, p. 431–438.

Stern, R. J., 1979, On the origin of andesite in the northern Mariana Island arc; Implications from Agrigan: Contributions to Mineralogy and Petrology, v. 68, p. 207–219.

—— , 1981, A common mantle source for western Pacific island arcs and "hot-spot" magmas; Implications for layering in the upper mantle: Carnegie Institution of Washington Yearbook, v. 81, p. 455–462.

Sun, S., Nesbitt, R. W., and Sharaskin, A. Y., 1979, Geochemical characteristics of mid-ocean ridge basalts: Earth and Planetary Science Letters, v. 44, p. 119–138.

Tarney, J., Saunders, A. D., Weaver, S. D., Donellan, N.C.B., and Hendry, G. L., 1978, Minor element geochemistry of basalts from Leg 49, North Atlantic Ocean, *in* Luyendyk, B. P., Cann, J. R., and others, Initial reports of the Deep Sea Drilling Project, Vol. 49: Washington, D.C., U.S. Government Printing Office, p. 657–691.

Wood, D. A., Mattey, D. P., Joron, J. L., Marsh, N. G., Tarney, J., and Treuil, M., 1980, A geochemical study of 17 selected samples from basement cores recovered at sites 447, 448, 449, 450 and 451, Deep Sea Drilling Project Leg 59, *in* Kroenke, L., Scott, R., and others, Initial reports of the Deep Sea Drilling Project, Vol. 59, Washington, D.C., U.S. Government Printing Office, p. 743–752.

Wood, D. A., Marsh, N. G., Tarney, J., Joron, J.-L., Fryer, P., and Treuil, M., 1981, Geochemistry of igneous rocks recovered from a transect across the Mariana trough, arc, fore-arc, and trench, sites 453 through 461, Deep Sea Drilling Project Leg 60, *in* Hussong, D. M., Uyeda, S., and others, Initial reports of the Deep Sea Drilling Project, Vol. 60: Washington, D.C., U.S. Government Printing Office, p. 611–646.

Manuscript Accepted by the Society October 15, 1986

Printed in U.S.A.

Geological Society of America
Special Paper 215
1987

Alkaline and transitional subalkaline metabasalts in the Franciscan Complex mélange, California

John W. Shervais
Department of Geology, University of South Carolina, Columbia, South Carolina 29208
David L. Kimbrough
Department of Geological Sciences, University of California at Santa Barbara, Santa Barbara, California 93106

ABSTRACT

Metavolcanic rocks form an important component of the Franciscan complex in California and preserve evidence for the origin of oceanic crust during the late Mesozoic time. These rocks occur as tectonic inclusions within Franciscan mélange and, more rarely, as thrust klippen that rest on mélange. Four occurrences of Franciscan metavolcanic rock are studied here: Aliso Canyon, Avila Beach, Stonyford (Snow Mountain and Stony Creek complexes), and Paskenta. These rocks include enriched mid-ocean ridge basalt, transitional subalkaline to mildly alkaline basalt, and highly fractionated Fe-Ti basalt. They range in composition from TiO_2 = 1.35 to 2.94 wt.%, La = 13 to 37 ppm, Y $>$ 20 ppm, Ti/V = 22 to 66, and chondrite normalized La/Zr = 1.5 to 2.8. The more alkaline basalts are also high in Nb (10 to 47 ppm) and contain pyroxene that ranges from titan augite to aegerine augite. The high La/Zr ratios observed in all rocks, regardless of alkalinity, imply derivation from a similar, light element-enriched source region by varying degrees of partial melting.

Field and geochemical data indicate that these Franciscan metavolcanic rocks formed in a variety of tectonic settings. Alkaline diabase sills in Aliso Canyon that intrude thick sequences of ribbon chert suggest off-axis, intraplate volcanism. Transitional subalkaline to alkaline basalts of the Snow Mountain Complex represent seamount volcanism, as shown by MacPherson (1983) and others. Volcanic rocks from Paskenta and possibly Avila Beach occur as knockers in serpentinite matrix mélange and may have formed in a fracture zone setting. The Stony Creek Complex contains Fe-Ti basalts and occurs within the same serpentinite-matrix mélange as the Paskenta volcanics. The Stony Creek Complex may represent a small seamount associated with a fracture zone, or, alternatively, a propagating ridge segment.

The preservation of true oceanic crust is not, in general, favored by the geometry of subduction. Seamounts and other intraplate volcanics that are structurally detached from the underlying ocean crust may be preserved preferentially during subduction. In a similar fashion, fracture zones, aseismic ridges, and other structural discontinuities in ocean crust (e.g., the pseudofaults associated with propagating rifts) create zones of weakness that may fail during subduction, allowing laterally extensive slabs of ocean crust to be preserved within the mélange. It is not possible, however, to establish with any certainty whether these Franciscan metavolcanic rocks formed in a true ocean basin or within a back-arc/marginal basin–type setting.

INTRODUCTION

The Franciscan complex of California is an assembalge of diverse rock types that forms the basement terrane of the California Coast Ranges and also crops out locally in the western Transverse Ranges to the south (Bailey and others, 1964). Hsü (1969) interpreted the Franciscan complex as a mélange formed during the late Mesozoic and early Cenozoic convergence of the North American plate and oceanic plates of the Pacific basin. The mélange has a scaly clay, or, more rarely, serpentinite matrix, and contains tectonic inclusions of graywacke, serpentinized ultramafic rocks, greenstones, limestone, conglomerate, blueschist, and eclogite (Bailey and others, 1964). Recent studies have shown that the Franciscan complex has had a long and complex history, ranging from the mid-Jurassic to the Eocene (Blake, 1984, and references therein).

This study focuses on four locations where subaqueous volcanic and hypabyssal intrusive rocks occur within or associated with Franciscan complex mélange in western California (Fig. 1). The two southern occurrences (Avila Beach and Aliso Canyon) represent tectonic inclusions in fault-bounded outliers of Franciscan mélange that occur in the western Transverse Ranges of central California (Shervais and Kimbrough, 1984). The two northern occurrences (Stonyford and Paskenta) occur at the tectonic contact between the Franciscan complex and the coeval Great Valley sequence in the northern Coast Ranges, along the western margin of the Sacramento Valley. Because of their position adjacent to the Great Valley Sequence, these rocks were formerly correlated with the Coast Range Ophiolite (Bailey and others, 1970; Hopson and others, 1981). However, the Stonyford and Paskenta volcanic rocks are chemically and mineralogically more similar to Franciscan volcanic rocks, and are also associated with younger (Tithonian) radiolarian cherts (Blake and others, 1984; Hopson and others, 1981; McPherson, 1983; Shervais and Kimbrough, 1985a). The Stonyford and Paskenta volcanic rocks are considered here to be part of the Franciscan complex. Their structural position adjacent to the Great Valley Sequence, however, makes this correlation tentative.

Two main questions are addressed in this study: (1) in what tectonic setting did these Franciscan volcanic rocks originally form, and (2) what were the processes involved in their petrogenesis? Possible tectonic settings that we consider are mid-ocean ridge basalt (MORB) volcanism at a divergent plate boundary, island arc volcanism, and intraplate volcanism. The trace element systematics of these basalts can be modeled by approximately 7 to 17 percent partial melting of a similar light element–enriched mantle source region. The extent of light element enrichment in the mantle source region, about twice that of chondrite, is the same for all basalt types, regardless of alkalinity. This implies that metasomatic enrichment of the mantle is necessary but not sufficient to form alkaline basalts; small degrees of partial melting are also required. Our data are not consistent with a mixing origin for the transitional basalts. The Franciscan volcanic rocks from the four occurrences studied here are geochemically and petrologically distinct from ophiolitic volcanic rocks that occur at the base of the Great Valley Sequence (Shervais and Kimbrough, 1985a), and were not introduced into the mélange by tectonic erosion of the Coast Range Ophiolite. We conclude that these volcanic rocks are transitional subalkaline to alkaline basalts that represent off-axis, intraplate volcanism and/or enriched MORB formed at an oceanic spreading center or fracture zone. Either type of volcanism may have formed in a true "ocean basin" or in an arc-related back-arc basin.

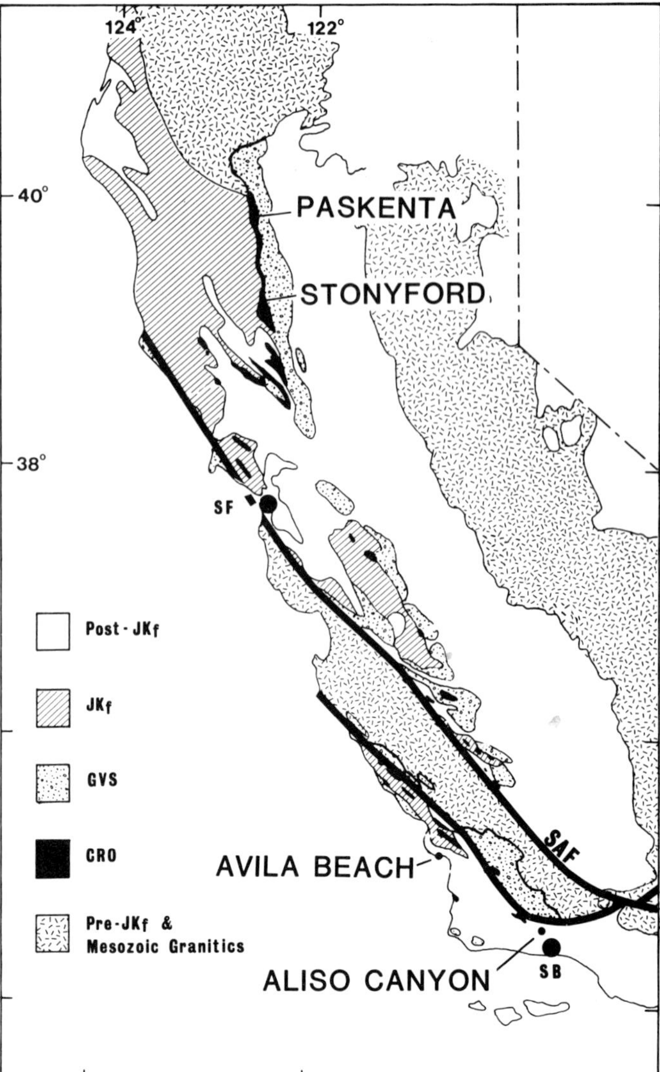

Figure 1. Generalized geologic map of northern and central California showing location of Franciscan metavolcanic rocks discussed in text. JKF indicates Franciscan Complex; GVS, Great Valley Sequence; CRO, Coast Range Ophiolite; SAF, San Andreas Fault; SB, Santa Barbara; SF, San Francisco.

FIELD RELATIONS

Four occurrences of Franciscan greenstone were studied (Fig. 1). Each of these occurrences features distinctive field relationships and is described separately, from south to north. Descriptions of the southern occurrences are based largely on our own field work; descriptions of the northern occurrences are summarized from the literature (Brown, 1964; Bailey and others, 1970; Hopson and others, 1981; MacPherson, 1983) and from our own reconnaissance studies.

Aliso Canyon (Santa Barbara County)

Aliso Canyon lies on the southern slopes of the San Rafael Range, north of Santa Barbara. The Franciscan complex is exposed here in a fault-bounded sliver south of the Little Pine fault, near its confluence with the Santa Ynez fault (Dibble, 1966). Franciscan rocks are overlain unconformably by Tertiary sedimentary rocks that range in age from Eocene to Miocene. The Franciscan complex beneath these sediments consists of mélange with a scaly clay matrix and tectonic inclusions of graywacke, siltstone, limestone, chert, diabase, pillow lava, and blueschist (Shervais, unpublished mapping; 1978, 1979; Schussler, 1981). The blueschist blocks are found only in the bed of the tiny creek that drains Aliso Canyon, but must have been derived locally because the creek heads against cliffs of massive sandstone in the overlying Tertiary section.

The largest tectonic inclusions in this area are a series of composite diabase-chert-graywacke knockers as much as 150 by 800 m in size that crop out discontinuously for about 4.5 km along strike (Fig. 2A). The chert is typical thinly bedded, red-green Franciscan chert with thin, shaly partings. The chert sections range up to about 50 m in thickness and appear to be conformably overlain by graywacke. Diabase intrudes the chert, as shown by locally discordant contacts, chert inclusions within diabase, and septa of massive, recrystallized white chert (as much as 2 m thick) sandwiched between adjacent diabase sills (Fig. 2B). The sill-like nature of the diabase is shown by the orientation of chert septa, which are flattened approximately parallel to bedding in the overlying ribbon chert, and by columnar jointing in the thicker sills, which is perpendicular to the bedding (Davis, 1969). The sills range in thickness from approximately 0.5 to 3 m, and are generally finer grained near the chert diabase contact and adjacent to chert septa deeper in the section. Thick, coarse-grained sills predominate at the base of the sections, which is a tectonic contact with the mélange matrix.

The single knocker of pillow lava in Aliso Canyon is small (≅20 by 25 m) but well exposed on three sides by a bend in the creek. It contains two distinctive flows: a fine-grained, aphyric to microphyric basalt, and a plagioclase megaphyric basalt containing blocky plagioclase megacrysts ≅1 cm across. The aphyric basalt appears to underlie the plagioclase megaphyric basalt, based on facing directions from pillows, but has been repeated by faulting along a small shear zone. The pillow lava has tops to the south, showing that it has been rotated relative to the composite diabase-chert-graywacke knockers described above, which have tops to the north.

Avila Beach (San Luis Obispo County)

Spectacular sea-cliff exposures of Franciscan greenstones occur just west of Avila Beach, near San Luis Obispo (Hall, 1973; Moore and Charlton, 1984). The pillow lavas are underlain by sheared serpentinite and probably represent a single large knocker in serpentinite-matrix mélange. The pillow lavas and serpentinite are both overlain by Cretaceous sandstones in a faulted angular unconformity (Hall, 1973). The pillow lavas strike approximately north-south and dip ≅50°E (Fig. 2C). Although exposed for about 1.5 km along the shoreline, less than 250 m of section is exposed because the shoreline here also trends north-south, approximately parallel to the strike of the pillows.

The most unusual aspect of this submarine volcanic sequence is the occurrence of thin sheet flows intercalated with massive and pillowed lava flows and associated hyaloclastite breccias. These sheet flows are similar to those observed on the Mid-Atlantic Ridge (Stakes and others, 1984), the East Pacific Rise (Ballard and others, 1979), and on seamounts in the eastern Pacific (Batiza and Vanko, 1984). The sheet flows in the Avila Beach section form individual layers 1 to 6 cm thick and occur in "stacks" up to 0.5 m thick (Fig. 2C,D). Sheet flows may merge laterally to form massive flows, or grade into hyaloclastite tuff-breccia (Fig. 2D). The locally observed stratigraphic succession (overlying pillow lava) of hyaloclastite tuff-breccia, sheet flows 1 to 6 cm, massive flows 20 to 50 cm thick, and pillow lava may represent an eruptive sequence or a single prograded lava flow (Fig. 2D). Moore and Charlton (1984) have suggested that the sheet flows form as successive stands of lava in a feeder pipe as it drains to feed more distal pillows. Ballard and others (1979) have suggested that sheet flows form when eruptions have rapid effusion rates.

Stonyford (Glenn, Colusa, and Lake Counties)

The Stonyford volcanics comprise a section 1.5 to 2 km thick of pillow lava, diabase, and pillow breccia, with minor intercalated thin flows and radiolarian chert (Brown, 1964; Hopson and others, 1981; MacPherson, 1983). The Stonyford volcanics occur in two structurally distinct massifs. Exposures centered around Stony Creek and Little Stony Creek are in fault contact with serpentinite that is adjacent to the base of the Great Valley Sequence near Stonyford. These exposures, which are in essence large mélange blocks within the serpentinite, are referred to here as the Stony Creek Volcanic complex. Thrust-faulted klippen centered around Snow Mountain and St. John's Mountain farther to the west comprise the Snow Mountain complex, which has been described in detail by MacPherson (1983). The Stony Creek and Snow Mountain complexes are similar stratigraphically; however, they exhibit minor but systematic chemical differ-

Figure 2. Field occurrence of Franciscan complex volcanic rocks in southern Coast Ranges: A, Composite knocker in scaly clay-matrix mélange, Aliso Canyon. Steep cliff is chert, underlain by diabase sills and overlain by graywacke. B, Septum of white, recrystallized chert ≅1.5 m thick between diabase sills, Aliso Canyon. Hammer indicates scale. C, Intercalated sheet flows and pillow lava, Avila Beach. Drill holes in pillow are 1.25 in. in diameter. D, Graded flow sequence, Avila Beach. Base of section is hyalocastite breccia, overlain successively by thin sheet flows, massive flows, and pillow lava. Note that sheet flows grade laterally into hyaloclastite (from right to left); hammer indicates scale.

ences which are discussed below. The term "Stonyford volcanics" is used informally to refer to the rocks of both complexes collectively (Shervais and Kimbrough, 1985b).

The Stonyford volcanics are the thickest accumulation of submarine volcanic rocks in the Coast Ranges (Bailey and others, 1964). Based on their thickness and composition, the Stonyford volcanics have been interpreted as a seamount by Bailey and others (1964), Bailey and Blake (1974), Hopson and others (1981), and MacPherson (1983). In particular, MacPherson (1983) has presented extensive geochemical, mineral, chemical, and stratigraphic data to support this interpretation for the Snow Mountain complex.

The stratigraphy of the Snow Mountain complex (MacPherson, 1983) is remarkably similar to that observed during submersible dives on young seamounts in the eastern Pacific (Lonsdale and Batiza, 1980; Batiza and others, 1984). Pillow lava dominates the lower volcanic unit, whereas hyaloclastites (pillow breccias) are more important at higher stratigraphic levels (MacPherson, 1983). These hyaloclastite deposits are inferred to form during deep-water phreatomagmatic eruptions, and by submarine fire fountaining (Lonsdale and Batiza, 1980; Batiza and others, 1984). Diabase dikes are common toward the base of the volcano, and peralkaline rhyolite is found in a structurally detached klippe that correlates with the upper volcanic unit of the Snow Mountain complex (MacPherson, 1983).

The lower and upper volcanic units of the Snow Mountain complex are separated by a chert horizon 0 to 3 thick (MacPherson, 1983). The Stony Creek complex contains chert intercalations 15 to 30 m thick that are conformably interbedded with the basalts (Brown, 1964; Hopson and others, 1981). Chert exposures along Stony Creek are notable for the occurrence locally of umbers rich in Fe-Mn oxide that have been prospected for manganese (Brown, 1964; Shervais, unpublished data). The upper part of the chert section in the Stony Creek complex contains radiolaria of Pessagno's radiolaria zone 2B, which are interpreted to be early Tithonian in age (Hopson and others, 1981).

The intercalated chert units within the Stonyford volcanics document a hiatus in volcanic activity and suggest that chert deposition may be favored on volcanic highs that are protected from clastic sedimentation by their elevation above the surrounding sea floor. The upper volcanics are conformably overlain by middle Tithonian graywacke and siltstone (Hopson and others, 1981; MacPherson, 1983). Presumably this marks the arrival of the seamount at a trench and the onset of its subduction (Hopson and others, 1981).

Paskenta (Tehama County)

Franciscan metavolcanic rocks form prominent tectonic inclusions in the serpentinite-matrix mélange that crops out near Paskenta (Hopson and others, 1981). This mélange, which is continuous with the serpentinite matrix mélange near Stonyford, separates South Fork Mountain Schist (a Franciscan blueschist facies terrane) on the west from Coast Range Ophiolite and Great Valley Sequence to the east (Hopson and others, 1981). Although previously correlated with the Coast Range Ophiolite (Bailey and others, 1970; Bailey and Blake, 1979; Hopson and others, 1981), the metavolcanic rocks in this mélange are petrographically and compositionally similar to those from Stonyford, Avila Beach, and Aliso Canyon, and are probably part of the Franciscan complex (Shervais and Kimbrough, 1985a, b).

Metavolcanic rock in the mélange consists mainly of pillow lava with subordinate massive flows, capped by red ribbon cherts of late Jurassic age (Hopson and others, 1981). Hopson and others (1981) showed that the mélange formed by the progressive break-up of large volcanic knockers by faulting and the injection of serpentine along fractures. Phipps (1984) described a sheared serpentinite unit south of Stonyford that is similar to the serpentinite-matrix mélange near Paskenta and occupies the same structural position as this mélange, between the Franciscan complex and Great Valley Sequence. The sheared serpentinite described by Phipps (1984) is cross-cut locally by mafic dike and sills and small diabase plugs (Phipps, 1984). These relationships imply that serpentinization and shearing of the ultramafic basement occurred *on the sea floor* prior to the end of volcanic activity. This evidence for early disruption of oceanic basement on the sea floor supports a fracture zone origin for the mélange.

AGE RELATIONS

Radiolarian cherts containing Late Jurassic (Tithonian) faunas are associated with Franciscan metavolcanic rocks at Stonyford, Paskenta, and Aliso Canyon. The cherts at Stonyford are intercalated with the volcanic flows and were more or less contemporaneous with volcanism (Hopson and others, 1981; MacPherson, 1983). The serpentinite-matrix mélange near Paskenta contains a few composite knockers in which Tithonian chert overlies pillow lava (Hopson and others, 1981; Blake and others, 1984). This provides a minimum age for the volcanic rocks. The diabase sills in Aliso Canyon intrude Tithonian chert, which provides a maximum age for the diabase (Schussler, 1981). Chert inclusions within the diabase have developed a distinct botryoidal surface texture (Davis, 1969; Schussler, 1981) that suggests they were included in the diabase magma prior to dewatering and lithification. If so, the diabase was intruded shortly after deposition.

Davis (1969) reported two K-Ar ages (on Franciscan diabase near Mt. Diablo) and one Rb-Sr age (from Aliso Canyon) of around 100 ± 10 Ma, significantly younger than the 150-Ma age of the Tithonian cherts. Davis (1969) interpreted these as crystallization ages. However, in view of the extensive metasomatic alteration and metamorphic recrystallization that has effected these rocks, it is more likely that radiometric ages of 100 ± 10 Ma are metamorphic, not igneous.

PETROGRAPHY AND MINERAL CHEMISTRY

Primary textural variations are well preserved in the metavolcanic rocks studied here, despite extensive alteration of the primary phases. Most of the lavas are aphyric or sparsely phyric clinopyroxene-plagioclase basalts that originally contained clinopyroxene, plagioclase, magnetite, ilmenite, glass, and in some cases, olivine as primary phases. Subhedral phenocrysts or megacrysts of plagioclase are rare in most samples, except for the upper pillow lava unit in Aliso Canyon, which contains abundant plagioclase megacrysts as much as 1 cm in diameter.

Textures vary as a direct function of cooling rate: sheet flows and pillow rims are vitrophyric to variolitic, pillow cores range from variolitic to intersertal, and diabase sills have subophitic to intergranular or intersertal textures. Some coarse-grained diabases from Aliso Canyon contain calcic pyroxene with exsolution lamellae of hypersthene (now chlorite). The trend of increasing grain size and decreasing glass content is typical of decreasing cooling rate in submarine basalts.

Pyroxene is commonly the only silicate to escape low-temperature alteration and metamorphism (Table 1). Pyroxenes from the Avila Beach and Paskenta lavas are neutral-colored, nonpleochroic augites that have low- to moderate-TiO_2 concentrations (TiO_2 = 0.3 to 2.2 wt. %). Pyroxenes from Aliso Canyon and Stonyford are pinkish, faintly pleochroic titanaugites (TiO_2 = 1.2 to 3.5 wt. %). There is a good correlation between cooling rate and quadrilateral components in these pyroxenes (Fig. 3a). Pyroxenes from the diabase sills in Aliso Canyon cooled slowly and have a salitic trend of Fe enrichment, almost parallel to the Di-Hd join. Pyroxenes from pillow lava cores at both Avila Beach and Aliso Canyon show little Fe enrichment and have Wo contents

TABLE 1. REPRESENTATIVE PYROXENE ANALYSES

	1	2	3	4	5	Weight % 6	7	8	9	10	11	12	13	14
SiO_2	51.33	51.28	44.61	49.95	46.75	47.47	49.89	50.89	52.42	50.57	51.55	49.88	50.18	49.10
Al_2O_3	0.92	1.00	7.09	2.62	5.49	5.74	4.67	2.41	2.11	4.32	2.01	2.90	4.02	5.61
Fe_2O_3*	30.14	28.54	3.27	1.60	3.16	1.89	1.45	1.95	1.12	1.28	1.78	1.89	1.53	1.47
FeO*	0.38	0.91	7.60	10.72	7.09	6.67	5.41	6.61	4.65	6.48	7.23	9.30	7.37	4.94
MnO	--	--	0.17	0.44	0.17	0.22	0.14	0.24	0.16	0.21	0.26	0.41	0.21	0.14
MgO	0.39	0.80	10.75	11.63	11.92	13.87	15.20	15.95	18.11	16.82	17.45	16.33	15.30	15.10
CaO	2.59	2.93	21.71	20.80	21.84	20.42	21.02	19.84	19.86	18.76	18.15	16.95	19.54	21.23
Na_2O	13.28	19.09	0.69	0.70	0.66	0.48	0.39	0.37	0.24	0.31	0.26	0.20	0.35	0.33
TiO_2	1.32	1.81	3.51	1.46	2.49	2.64	1.23	0.80	0.50	1.21	0.70	1.09	0.91	1.48
Cr_2O_3	--	--	0.0	0.0	0.00	0.30	0.38	0.06	0.51	0.00	0.32	0.21	0.04	0.43
Total	100.35	100.36	99.42	99.92	99.57	99.69	99.79	99.14	99.70	99.96	99.72	99.17	99.45	99.82
Cations per 6 Oxygens														
Si	1.963	1.959	1.702	1.894	1.770	1.776	1.845	1.900	1.921	1.860	1.908	1.873	1.869	1.814
Al-IV	0.036	0.040	0.297	0.105	0.229	0.223	0.154	0.099	0.078	0.139	0.087	0.126	0.130	0.185
Al-VI	0.005	0.004	0.021	0.012	0.015	0.029	0.049	0.006	0.012	0.047	0.000	0.001	0.046	0.059
Fe-3	0.867	0.820	0.094	0.045	0.090	0.053	0.040	0.054	0.030	0.035	0.049	0.053	0.042	0.040
Fe-2	0.012	0.028	0.242	0.339	0.224	0.208	0.167	0.206	0.142	0.199	0.223	0.292	0.229	0.152
Mn	--	--	0.005	0.014	0.005	0.006	0.004	0.007	0.004	0.006	0.008	0.012	0.006	0.004
Mg	0.022	0.045	0.611	0.657	0.673	0.773	0.838	0.888	0.989	0.922	0.963	0.914	0.849	0.831
Ca	0.106	0.119	0.887	0.845	0.886	0.818	0.833	0.793	0.779	0.739	0.719	0.682	0.780	0.840
Na	0.985	0.969	0.051	0.051	0.048	0.034	0.028	0.027	0.016	0.021	0.018	0.014	0.025	0.023
Ti	0.037	0.052	0.100	0.041	0.070	0.074	0.034	0.022	0.013	0.033	0.019	0.030	0.025	0.040
Cr	--	--	0.000	0.000	0.000	0.008	0.011	0.001	0.014	0.000	0.009	0.006	0.001	0.012
Sum	4.036	4.040	4.015	4.007	4.015	4.008	4.006	4.009	4.005	4.005	4.008	4.008	4.007	4.006

*Fe_2O_3 and FeO calculated by charge balance, after Papike and others (1976).

Note: Key to table column heads: 1. AC-2, aegerine-augite core; 2. AC-2, late magmatic aegerine-augite, core; 3. AC-2, large titanaugite with "opx" exsolution lamellae, core; 4. AC-2, rim of titanaugite in column 3; 5. AC-2 large titanaugite, core; 6. AC-3, augite microphenocrysts, core; 7. AC-4, augite microphenocryst, core; 8. AB-1, augite microphenocryst, core; 9. AB-2, augite microphenocryst, core; 10, AB-3, augite phenocryst core; 11. SF-12-4, large augite phenocryst, core; 12. SF-29-1, augite microphenocryst, core; 13. PK-57-1, augite phenocryst, core. 14. PK-9-1, augite phenocryst core.

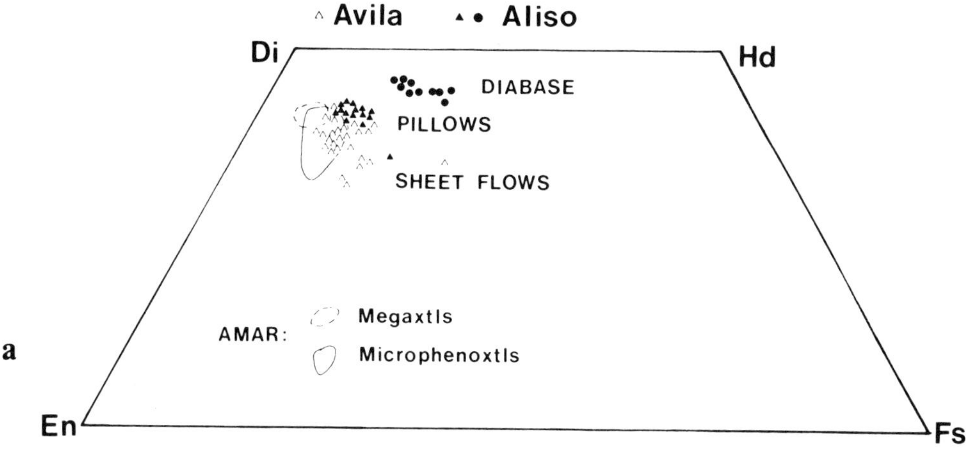

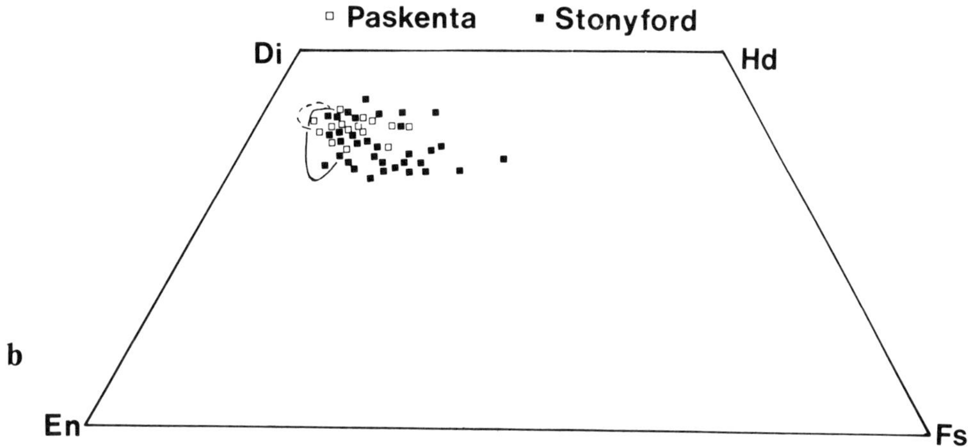

Figure 3. Pyroxene compositions projected onto pyroxene quadrilateral. Pillows and sheet flows from Avila Beach and Aliso Canyon show minor Fe enrichment, whereas pyroxenes from Aliso Canyon diabases follow salitic trend. Pillow lavas from Paskenta and Stonyford have pyroxenes that show moderate Fe enrichment. Data for Mid-Atlantic Ridge pyroxenes (Stakes and others, 1984) shown for comparison. a, Avila Beach, Aliso Canyon; b, Paskenta, Stonyford.

similar to pyroxene phenocrysts and megacrysts in glassy basalts from the Mid-Atlantic Ridge (Stakes and others, 1984). The lowest Wo contents are found in microphenocrysts from sheet flows and pillow rims (Fig. 3a) similar to microphenocrysts from the glassy basalts from the Mid-Atlantic Ridge (Stakes and others, 1984). Similar trends are exhibited by pyroxenes from the Stonyford and Paskenta volcanics (Fig. 3b). There is, however, no discernable correlation between minor element concentrations and cooling rate. The minor element concentrations of pyroxene are about the same in both cores and rims, and correlate roughly with bulk rock composition.

Aegerine-augite forms thin, discontinuous rims on titanaugite in many of the diabases and coarser grained basalts from Aliso Canyon and Stonyford. It also occurs as small euhedral prisms interstitial to plagioclase, or in "glass." This aegerine-augite is close to pure aegerine in composition, with <1.5 wt.% CaO and <1 wt.% Al_2O_3 (Table 1). These textural and compositional relationships suggest that the aegerine-augite formed by magmatic crystallization during the later stages of crystallization of an alkali-rich magma and is not metamorphic.

The other primary silicate phases are rarely preserved. Calcic plagioclase is most commonly altered to albite or oligoclase, which may be replaced subsequently by chlorite, pumpellyite, calcite, sausserite, smectites, or, more rarely, by prehnite (Aliso Canyon diabase only). Chlorite and calcite form pseudomorphs after olivine, whereas primary glass is altered to chlorite, pumpellyite, smectites, hematite, and/or calcite. Some primary ilmenite and Ti-magnetite are preserved in the coarser grained samples, but much is altered in part to sphene. Amygdules and vein fillings contain chlorite, pumpellyite, smectites, calcite, and rarely zeolites (Avila Beach only). In all cases, calcite is late in the paragenetic sequence of alteration minerals, filling the centers of veins and vesicles in which it occurs, or replacing earlier secondary phases in pseudomorphs. This is consistent with the common occurrence of pumpellyite (±prehnite), which is favored by low CO_2 activities in the hydrothermal solutions (Schiffman and Liou, 1982).

Representative analyses of chlorite and pumpellyite are shown in Figure 4a, b. The chlorites are more aluminous than chlorites from hydrothermally altered Mid-Atlantic Ridge basalts (Mottl, 1983), and have a wider range in Fe/Mg ratios (Fig. 4a). The pumpellyites are also very aluminous, similar to pumpellyites from high-pressure metamorphic terranes (Fig. 4b). These pumpellyites are distinct from those formed during zeolite facies hydrothermal alteration in the East Taiwan ophiolite (Lious, 1979).

There is no evidence in any of the metavolcanic samples studied here for the incipient formation of blueschist facies assemblages. The only sodic pyroxene found is aegerine-augite, which is magmatic in origin. The occurrence of chlorite, pumpellyite, prehnite, and smectites, and the absence (or rarity) of actinolite or zeolites implies subgreenschist facies metamorphism at moderate pressures, i.e., prehnite-pumpellyite facies.

GEOCHEMISTRY

Chemical analyses of aphyric or sparsely phyric volcanic rocks can be used to determine their magmatic affinities and to trace their fractionation histories. Whole-rock geochemical data for 16 samples of Franciscan metavolcanic rock are presented in Table 2. Ten major elements and 16 trace elements were determined by x-ray fluorescence using an automated Philips PW-1450 sequential spectrometer (Dietrich and others, 1976). Sample preparation techniques, instrumentation, and matrix correction methods are described by Dietrich and others (1976, 1981) and by Nisbet and others (1979). Ferrous iron and CO_2 were determined by wet chemistry (Aryanci, 1977), and H_2O by difference from the ignition loss. Additional data compiled from the literature are restricted mostly to major elements (Bailey and Blake, 1974), although MacPherson (1983) reports Cr and V as well. These data can be used to determine the primary magmatic history of the metavolcanic rocks, provided the effects of later alteration and metamorphism are taken into account.

Short descriptions of the analyzed samples are appended to Table 2. The samples from Aliso Canyon include two diabase sills, one aphyric pillow, and one plagioclase megaphyric pillow that is geochemically distinct from the other samples. The samples from Stonyford include three lavas from the Stony Creek complex and one from the Snow Mountain complex. The latter sample is compositionally similar to basalt analyses published by MacPherson (1983), but differs slightly from the analyses of Stony Creek complex lavas published here. These differences are further discussed below.

Alteration

All Franciscan metavolcanic rocks have undergone extensive low-temperature metamorphism, making interpretation of primary geochemical character difficult. In the samples studied here, structural water (100° to 1,050°C) ranges from 2.0 to 6.4 wt.%, CO_2 ranges from 0.1 to 3.6 wt.% (Table 2). The relative effects of this alteration are seen by comparing major and trace element concentrations between the crystalline core (AB-3) and "glassy" rim (AB-4) of a single lava pillow from Avila Beach (Table 2). The pillow rim is enriched in total Fe, Fe_2O_3/FeO, and K, and depleted in Mg, Ni, Cu, Sr, and S relative to the core.

Major Elements

Major element variations are compared in Figure 5 using MgO-variation plots. Before plotting, all analyses were recalculated to 100 percent after subtracting H_2O and $CaCO_3$. Also shown are fields for East Pacific Rise glass analyses from Melson and others (1975). The effects of low-temperature hydrothermal alteration are clear. Silica (not plotted), alumina (not plotted), MgO, and TiO_2 have concentrations in the same range as normal oceanic basalts, reflecting small net fluxes for these elements. Silica is apparently enriched in some samples after normalization, but this may be an artifact of the normalization procedure. CaO is strongly depleted relative to EPR glass analyses, whereas FeO*, Na_2O, and K_2O are all enriched (Fig. 5).

These data show that the metavolcanic rocks studied here are all more or less basaltic in composition, and are similar to MORB in major element geochemistry. Lavas from the Stony Creek complex are enriched in iron and are more similar to Ti-rich ferrobasalts ("Fe-Ti basalts") than to MORB. However, the usefulness of major elements in deciphering the petrogenesis of these rocks has been compromised by extensive hydrothermal alteration. Further efforts to characterize these rocks must rely on trace element data.

Trace Elements

Trace and minor elements that are immobile during hydrothermal alteration and low-temperature regional metamorphism have been used successfully as petrogenetic indicators in ophiolites and are used here to determine the magmatic affinities of Franciscan metavolcanic rocks. Of the elements listed in Table 2, the most useful are Ti, Zr, Nb, Y, Cr, V, Sc, and La. these elements are relatively immobile during low-temperature metamorphism, and may be used to calculate petrogenetic models.

All Franciscan metavolcanic rocks analyzed here have Ti/Zr ratios in the same range as MORB (Fig. 6). Concentrations of both elements are high, and several samples lie outside the range of normal MORB as defined by Pearce and Cann (1973), suggesting affinities to enriched MORB or ocean island tholeiites. Little resolving power is gained by adding either Y or Sr to form ternary Ti-Zr-Y or Ti-Zr-Sr plots (Pearce and Cann, 1973).

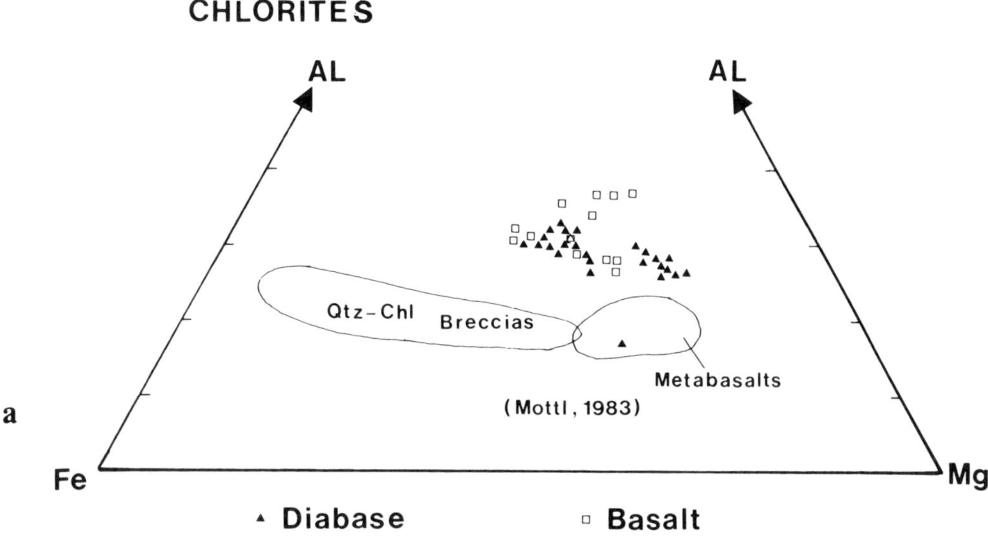

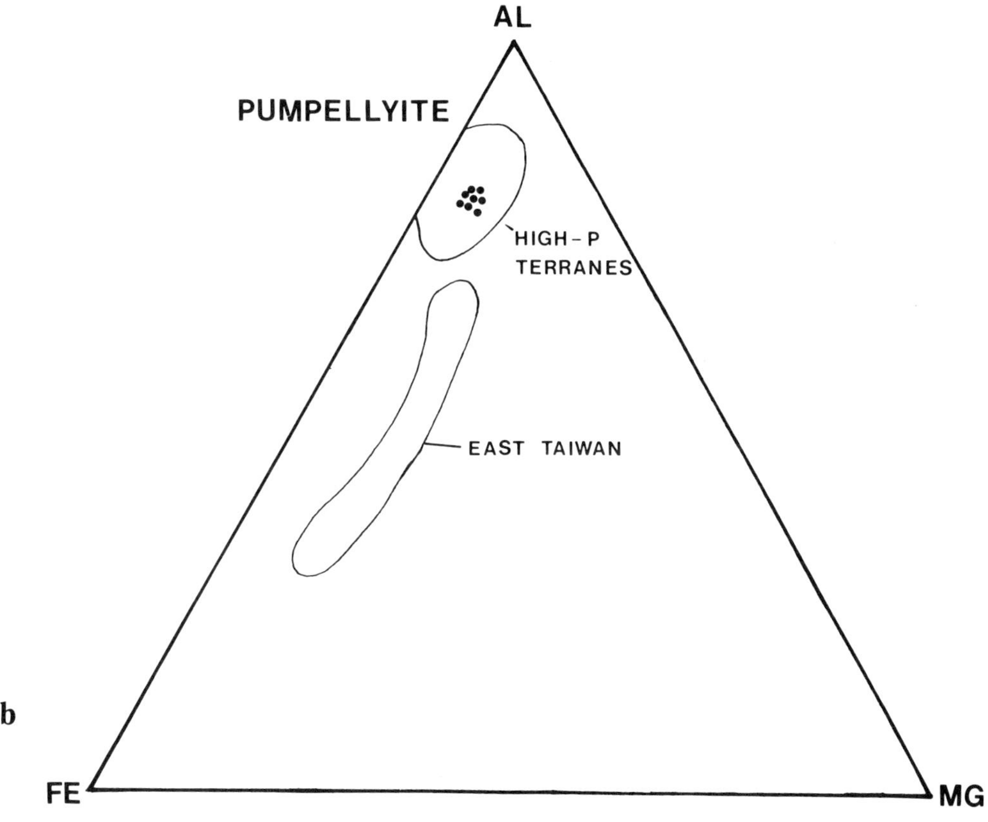

Figure 4. Compositions of chlorites (a) and pumpellyites (b) projected onto Al-Fe-Mg ternary. Chlorite compositions are more aluminous than in Mid-Atlantic Ridges metabasalts and quartz chlorite breccias (Mottl, 1983). Pumpellyites from Aliso Canyon are also aluminous, similar to pumpellyites from "high-P" terranes. Fields from Schiffman and Liou (1980) and Liou (1979).

TABLE 2. MAJOR AND TRACE ELEMENT ANALYSES OF INTRAPLATE METAVOLCANIC ROCKS FROM THE FRANCISCAN COMPLEX MÉLANGE*

	AC1	AC2	AC3	AC4	AB1	AB2	AB3	AB4	AB5	SF10	SF12	SF16	SF29	PK9	PK14	PK15
SiO_2	44.38	44.19	49.95	49.26	49.74	50.59	48.31	48.65	48.21	47.88	48.36	45.14	47.96	47.69	47.74	51.37
TiO_2	1.80	1.98	2.20	1.48	1.76	1.88	1.80	1.78	1.82	2.88	2.85	1.93	2.94	1.35	2.09	1.44
Al_2O_3	15.53	15.91	14.62	16.79	12.84	15.01	14.68	13.34	14.72	12.42	13.05	14.94	12.44	14.39	12.37	14.84
Fe_2O_3	3.69	4.23	5.20	4.18	6.27	4.01	4.21	9.18	3.31	7.51	6.60	3.98	6.83	3.62	5.60	3.14
FeO	5.75	6.95	3.85	3.35	4.80	4.00	4.65	4.40	4.55	8.50	7.60	5.00	8.05	6.50	7.40	7.50
MnO	0.31	0.17	0.12	0.12	0.24	0.16	0.18	0.21	0.20	0.22	0.25	0.17	0.24	0.17	0.18	0.16
MgO	8.37	7.28	3.87	4.09	5.60	5.55	7.29	5.54	6.37	5.27	5.20	7.49	5.60	8.00	4.87	6.36
CaO	5.61	9.28	8.45	10.05	9.17	8.87	9.53	9.06	10.80	7.38	10.51	10.26	7.79	11.35	11.95	5.77
Na_2O	2.37	2.48	4.98	4.81	2.79	2.59	3.72	3.49	3.37	3.48	1.95	1.62	3.46	2.47	2.87	3.98
K_2O	1.98	0.54	0.88	0.28	2.16	2.73	0.18	0.86	0.98	0.22	0.13	2.78	0.06	0.36	0.55	0.01
P_2O_5	0.35	0.23	0.50	0.25	0.23	0.25	0.17	0.17	0.17	0.30	0.29	0.26	0.29	0.13	0.21	0.14
H_2O+	5.33	6.37	2.92	1.99	2.76	2.43	3.19	2.77	2.84	2.13	1.63	4.33	2.95	3.14	2.59	3.67
CO_2	3.64	0.07	1.93	2.54	1.03	1.00	1.28	0.20	2.22	0.45	0.49	1.36	0.13	0.43	0.83	0.96
Total	99.11	99.68	99.47	99.19	99.39	99.07	99.19	99.65	99.56	98.64	98.91	99.26	98.73	99.60	99.25	99.34
Trace Elements (ppm)																
Nb	12	47	30	<3	<3	<3	<3	<3	<3	<3	<3	10	<3	<3	<3	<3
Zr	187	144	178	140	136	127	125	116	112	194	210	158	207	117	151	93
Y	25	38	27	7	27	29	28	29	24	51	55	21	57	23	39	21
Sr	379	307	223	375	195	184	338	201	365	247	132	111	197	80	83	69
Rb	19	20	13	<4	33	24	<4	16	<4	<4	<4	35	<4	<4	10	<4
Ni	42	79	25	53	19	24	39	25	39	23	45	83	49	66	32	9
Zn	58	74	56	55	80	73	74	80	79	125	123	58	171	83	99	74
Co	63	66	40	91	15	52	71	28	116	67	73	37	67	75	80	42
Cr	88	74	63	151	226	242	233	239	227	41	119	306	116	313	77	13
V	346	62	300	326	375	321	375	382	346	566	637	300	477	319	509	398
Sc	41	31	37	34	54	59	55	56	55	54	52	34	48	46	50	39
Ba	1660	161	208	157	75	52	46	46	38	72	50	68	55	23	89	24
La	27	13	30	15	21	22	16	19	17	32	37	23	33	12	25	11
S	1148	48	130	2067	237	176	781	201	1255	1820	1502	166	1674	110	74	856

*Abbreviations used: AC = Aliso Canyon; AB = Avila Beach; SF = Stonyford; PK = Paskenta.

Note: Key to sample numbers:

Aliso Canyon

AC-1 = Fine-grained diabase, subophitic texture.
AC-2 = Coarse-grained olivine diabase, subophitic texture with fresh titanaugite, aegerine-augite.
AC-3 = Aphyric pillow basalt, glassy texture with microphyric olivine, plagioclase, augite.
AC-4 = Plagioclase megaphyric pillow lava, large (≅ 1 cm) plagioclase megacrysts in fine-grained, intergranular to intersertal groundmass.

Avila Beach

AB-1 = Sheet flow, sparsely microphyric (clinopyroxene and plagioclase) basalt.
AB-2 = Sheet flow, aphyric basalt with rare clinopyroxene microphenocrysts.
AB-3 = Pillow lava, core; fine-grained granular to intersertal texture.
AB-4 = Pillow lava, rim of AB-3; glassy to fasciulitic texture, quench plagioclase clinopyroxene.
AB-5 = Pillow lava core; plagioclase phyric basalt, with augite microphenocrysts.

Stonyford

SF-10-1 = Sparsely phyric clinopyroxene basalt/diabase; coarse grained intergranular to hyalophitic texture. Pink titanaugite, with late aegerine-augite rims, prisms. Stony Creek, near diversion dam, Stony Creek complex.
SF-12-1 = Pyroxene-phyric basalt with sparse plagioclase phenocrysts. Intergranular texture. Stony Creek, near diversion dam, Stony Creek complex.
SF-16-1 = Coarse-grained intersertal basalt/diabase with titanaugite. Top of St. John's Mountain, Snow Mountain complex.
SF-29-1 = Aphyric to microphyric basalt, olivine, plagioclase, titanaugite microphenocrysts in variolitic groundmass. Little Stony Creek, Stony Creek complex.

Paskenta

PK-9-1 = Aphyric basalt with sparse augite microphenocrysts, variolitic texture.
PK-14-5 = Sparsely microphyric plagioclase + augite basalt, fine-grained groundmass with quench textures (variolitic, fasciulitic).
PK-57-1 = Sparsely microphyric clinopyroxene-plagioclase basalt, very fine grained, intersertal to hyalophitic groundmass.

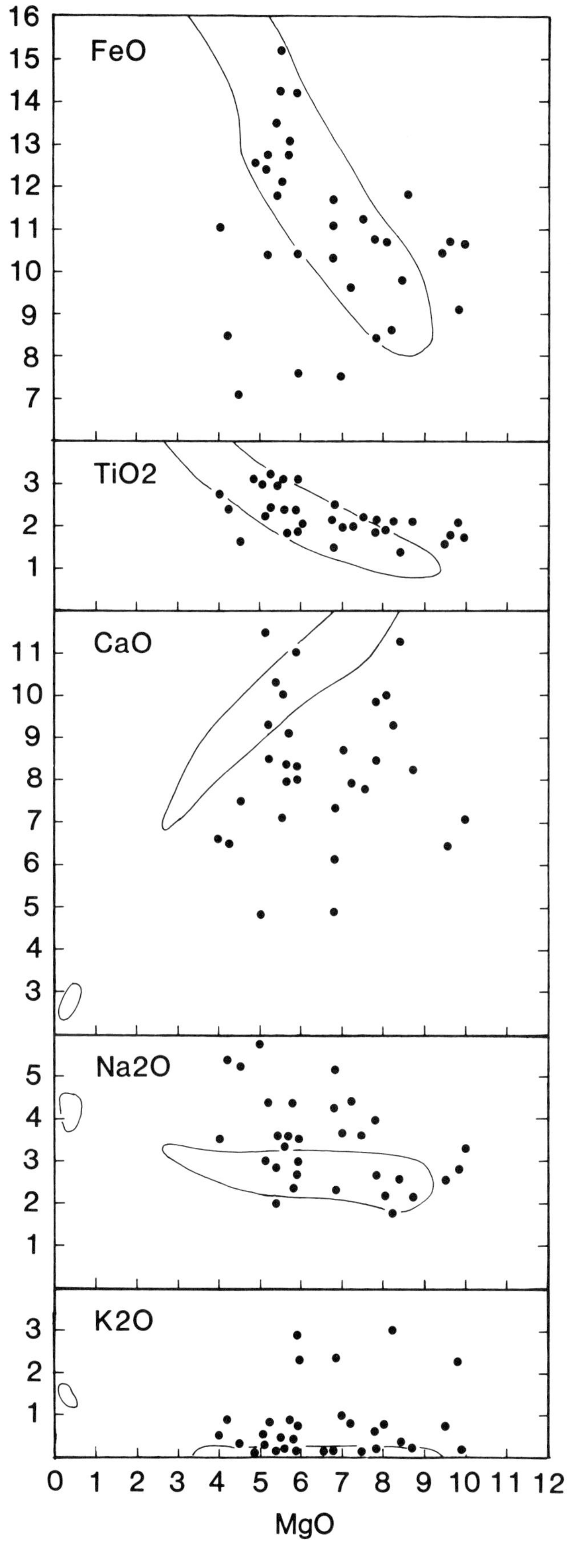

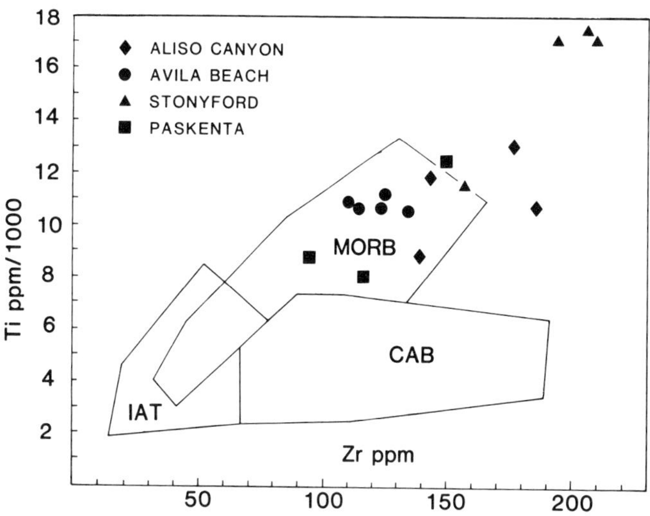

Figure 5. MgO variation plots for FeO, TiO_2, CaO, Na_2O, and K_2O in Franciscan metavolcanic rocks. Fields are for volcanic glasses from East Pacific Rise (Melson and others, 1976). Franciscan data from Bailey and others (1964), Bailey and Blake (1974), and this study.

Franciscan greenstones plot near the "ocean floor basalt"/"within plate basalt" boundary on the Ti-Zr-Y ternary (Fig. 7a), suggesting transitional magmatic affinities. Mostly they fall in the within plate basalt field in the Ti-Zr-Sr ternary (Fig. 7b), but form an array pointing toward the Sr apex at constant Ti/Zr, suggesting Sr mobility during alteration. In a plot Zr/Y versus Zr (Pearce, 1980), these samples again plot near the "ocean floor basalt"/"within plate basalt" boundary (Fig. 8), indicating an enriched tholeiite affinity.

Most of the samples analyzed here have Nb concentrations below the detection limit (3 ppm) and are probably enriched tholeiites transitional toward alkali basalts. Three of the four Aliso Canyon samples, however, and the only Snow Mountain volcanic for which we have data, have Nb = 10 to 47 ppm, suggesting more alkaline affinities. The only Aliso Canyon sample that has Nb < 3 ppm is the plagioclase megaphyric pillow basalt.

Elements that are equally incompatible during fractional crystallization and partial melting should behave coherently and preserve element/element ratios found in their source regions. The good linear correlation between La and Zr (Fig. 9) is unlikely to persist if either element was mobilized preferentially. Chondrite-normalized La/Zr ratios range from $\cong 1.5$ to 2.8, confirming an enriched source region for all of the basalts studied here, regardless of magmatic affinity.

Figure 6. Plot of Zr versus Ti (in parts per million) for Franciscan metavolcanic rocks. Fields from Pearce and Cann (1973): IAT indicates island arc tholeiite; MORB, mid-ocean ridge basalt; CAB, cal-alkaline basalt. Franciscan metavolcanic rocks have Ti/Zr ratios similar to MORB but with higher absolute concentrations.

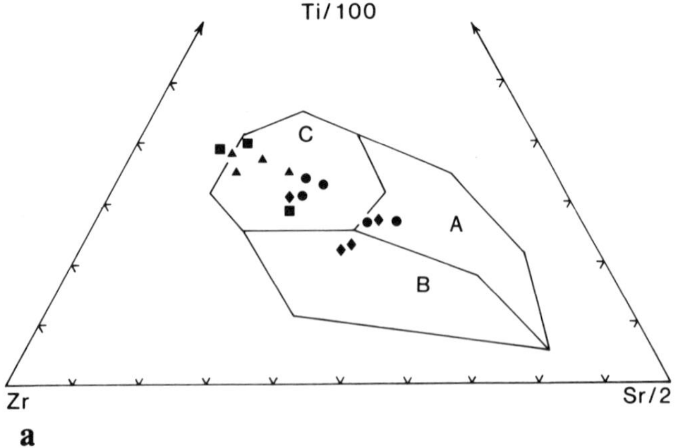

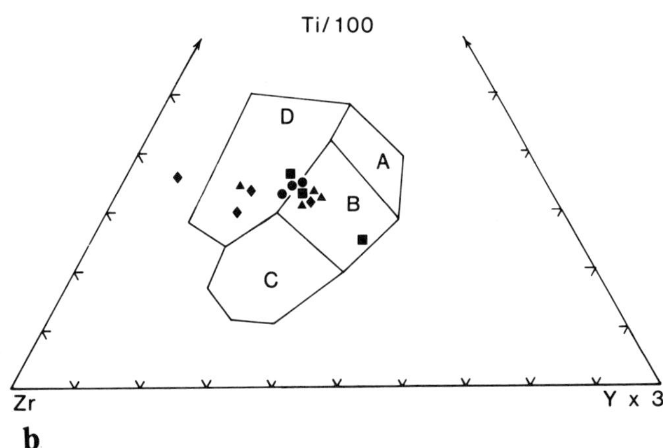

Figure 7. Plots of Ti-Zr-Y (a) and Ti-Zr-Sr (b) after Pearce and Cann (1973). Franciscan greenstones analyzed here have affinities to MORB or WPB (within plate basalts). Symbols same as in Figure 6.

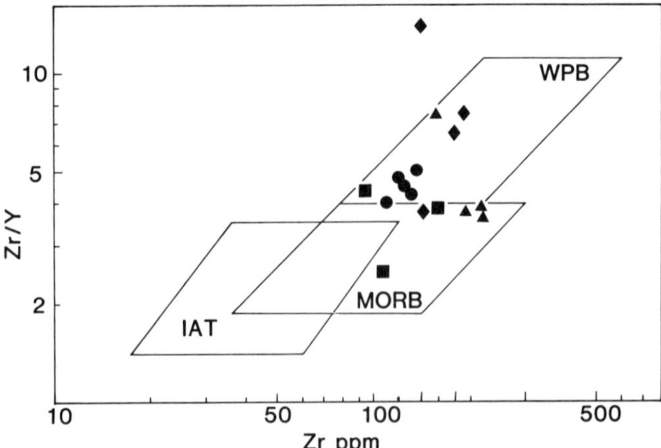

Figure 8. Plot of Zr/Y versus Zr for Franciscan metavolcanic rocks, with fields for IAT (island arc tholeiite), MORB, and WPB after Pearce (1980). Franciscan rocks plot mostly in WPB field, with some overlap into MORB field. Similar overlaps observed in data of Pearce (1980). Symbols as in Figure 6.

DISCUSSION

Magmatic Affinities of Franciscan Metavolcanics

The trace element data discussed above are consistent with the conclusion that the magmatic affinities of these Franciscan metavolcanic rocks are transitional between enriched subalkaline basalts and mildly alkaline basalts. This conclusion is supported by the high chondrite-normalized La/Zr ratios, moderate to high Ti/V ratios, and high concentrations (relative to MORB) of La, Ti, Zr, and, in some samples, Nb.

Shervais (1982) has shown that the fractionation of V from Ti during partial melting is a function of oxygen fugacity, and that Ti/V ratios of volcanic rocks increase in the order: island arc volcanics → MORB → alkali basalts. Mid-ocean ridge basalts, whether "normal" or "enriched," have Ti/V ≅ 20 to 50, whereas tholeiitic, transitional, and alkalic rocks from oceanic islands have somewhat higher ratios (Ti/V ≅ 40 to 100). Off-axis sills that intrude sediment in the Shikoku and Daito Basins (Nisterenko, 1980; Dick, 1982) are either tholeiitic (Ti/V ≅ 30 to 50) or transitional to alkaline (Ti/V ≅ 35 to 70).

Most of the Franciscan metavolcanic samples studied here have Ti/V ratios of between 22 and 66 (Fig. 10). The highest ratios are observed in lavas from the Snow Mountain complex (Ti/V = 30 to 66), which supports MacPherson's (1983) interpretation of these lavas as transitional to alkalic. Stonyford volcanics from the Stony Creek complex have Ti/V ratios similar to the Paskenta volcanic blocks, despite the more evolved compositions of the Stony Creek lavas (Fig. 10). The Stony Creek lavas are also richer in FeO* and Zr than any of the other samples studied here, suggesting similarities to Ti-rich ferrobasalts from the Galapagos rise (Sinton and others, 1983). The difference in Ti/V ratios between volcanic rocks from the Snow Mountain complex and Stony Creek complex implies that these massifs, although similar stratigraphically, are petrogenetically distinct. This is supported by the high Nb concentration of the Snow Mountain complex basalt analyzed here, which suggests more alkaline affinities for these rocks (Table 2, SF-16).

All of the samples from Avila Beach and Aliso Canyon have Ti/V ratios in the same range as MORB, except for AC-2, a coarse-grained diabase from Aliso Canyon (Fig. 10). The extremely low-V concentration of this sample may be due either to extensive hydrothermal alteration or to magnetite fractionation. Whereas the MORB-like Ti/V ratios of the Avila Beach lavas are consistent with their other geochemical characteristics, the low-Ti/V ratios of the Aliso Canyon samples is in conflict with their generally high-Nb concentrations. The reason for this conflict is not known; however, we believe that high-Nb concentra-

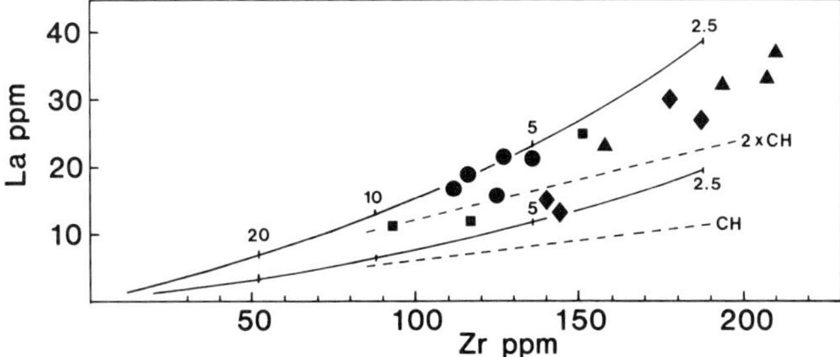

Figure 9. Plot of La versus Zr for Franciscan metavolcanic rocks, along with model melting curves (solid line) and constant La/Zr ratios in source (dashed line). Lower solid curve for melting source with La/Zr = chondritic, upper solid curve for melting source with La/Zr = two times chondritic. Fractional crystallization paths are tangential to melting curves and lie between melting curve and La/Zr ratio of its source composition. All Franciscan metavolcanic rocks plotted here lie between two melting curves and require source region enriched in La, with La/Zr ratios 1.5 to 2 times chondritic. Symbols same as in Figure 6; numbers on curves refer to percentage of melting.

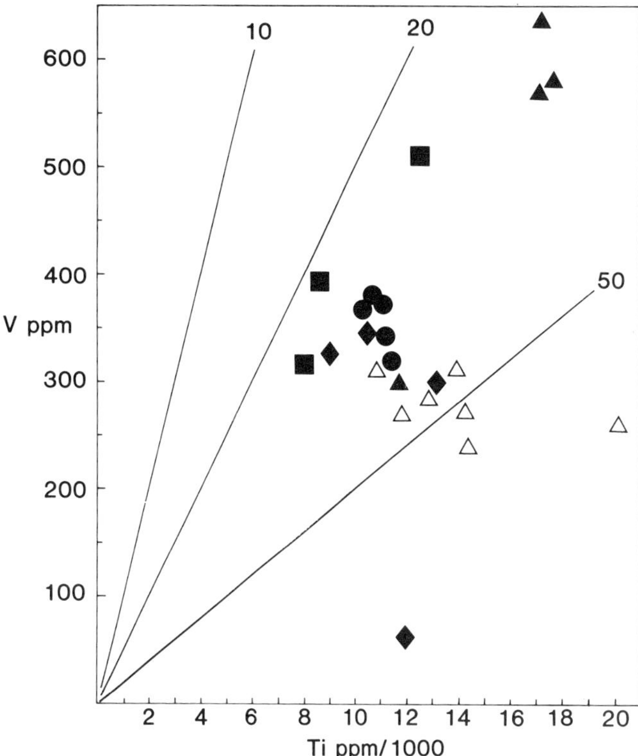

Figure 10. Plot of Ti versus V, after Shervais (1982). Range of Ti-V observed similar to MORB, transitional basalts, and mildly alkaline basalts. Island arc tholeiites (Ti/V <20) are absent. Solid symbols indicate this study (see Fig. 6); open triangles indicate data of MacPherson (1983).

tions are a more reliable indicator of alkaline affinities than the Ti/V ratios.

Confirmation of transitional alkaline-subalkaline affinities for these lavas comes from their pyroxene chemistry. Pyroxenes from all four locations plot in the "nonorogenic" field on the Ti + Cr versus Ca discrimination diagram (Fig. 11A,B) of Letterier and others, 1982). However, on a diagram of Ti versus Ca + Na diagram of (Fig. 11C,D), pyroxenes from the Aliso Canyon rocks have alkaline affinities, whereas those from Stonyford, Avila Beach, and Paskenta are subalkaline. The most enriched samples (high La, Zr, and/or Nb) are those from Aliso Canyon and Stonyford. These rocks are characterized by pinkish titanaugite, and, in the more slowly cooled samples, by late magmatic aegerine-augite. We conclude that the Aliso Canyon volcanics (excluding the plagioclase megaphyric pillow lava) are alkaline basalts, the Stonyford volcanics from the Snow Mountain complex are transitional subalkaline to mildly alkaline basalts, and the Stonyford volcanics from the Stony Creek complex are highly evolved, Ti-rich ferrobasalts ("Fe-Ti" basalts). Volcanic rocks from Paskenta, Avila Beach, and the plagioclase megaphyric pillow lava from Aliso Canyon are all enriched subalkaline tholeiites.

Petrogenesis

All of the Franciscan metavolcanics studied here have chondrite, normalized La/Zr ratios between 1.5 and 2.8 times that of chondrite, regardless of their magmatic affinities. This suggests that the tholeiitic, transitional, and alkaline basalts are all related to a common source. Partial melting models that assume a source with La/Zr ≅ chondritic cannot account for the observed trend in compositions (Fig. 9). Because mineral/liquid partition coefficients for La are smaller than those for Zr, partial melts of a chondritic source would lie on curve that is asymptotic to the La/Zr ratio of the source region. Fractional crystallization would increase both La and Zr without changing the parent magma ratio, creating trends that lie below and to the right of the melting curves in Figure 9.

Data for the Franciscan greenstones lie above the melting curve for a chondritic source, and below the curve for a source with La/Zr ≅ 2 times chondritic. An important point here is that both the "alkaline" and "transitional subalkaline" magma groups require more or less the same enrichment in their source regions. This argues against models that require variable enrichment ("metasomatism") of the mantle to create alkalic sources that are distinct from subalkaline source regions (e.g., Menzies and Murthy, 1980), or models based on mixing of tholeiitic MORB and more alkaline magmas (Batiza and Vanko, 1984).

The data considered here can be modeled more easily by variable degrees of partial melting of a similar source composition, as suggested by variations in Sc (a compatible transition metal) versus Zr (Fig. 12). It is immediately apparent in this diagram that the most alkaline samples (AC1, 2, 3; SF16) have the lowest Sc contents, whereas the least alkaline (Avila Beach)

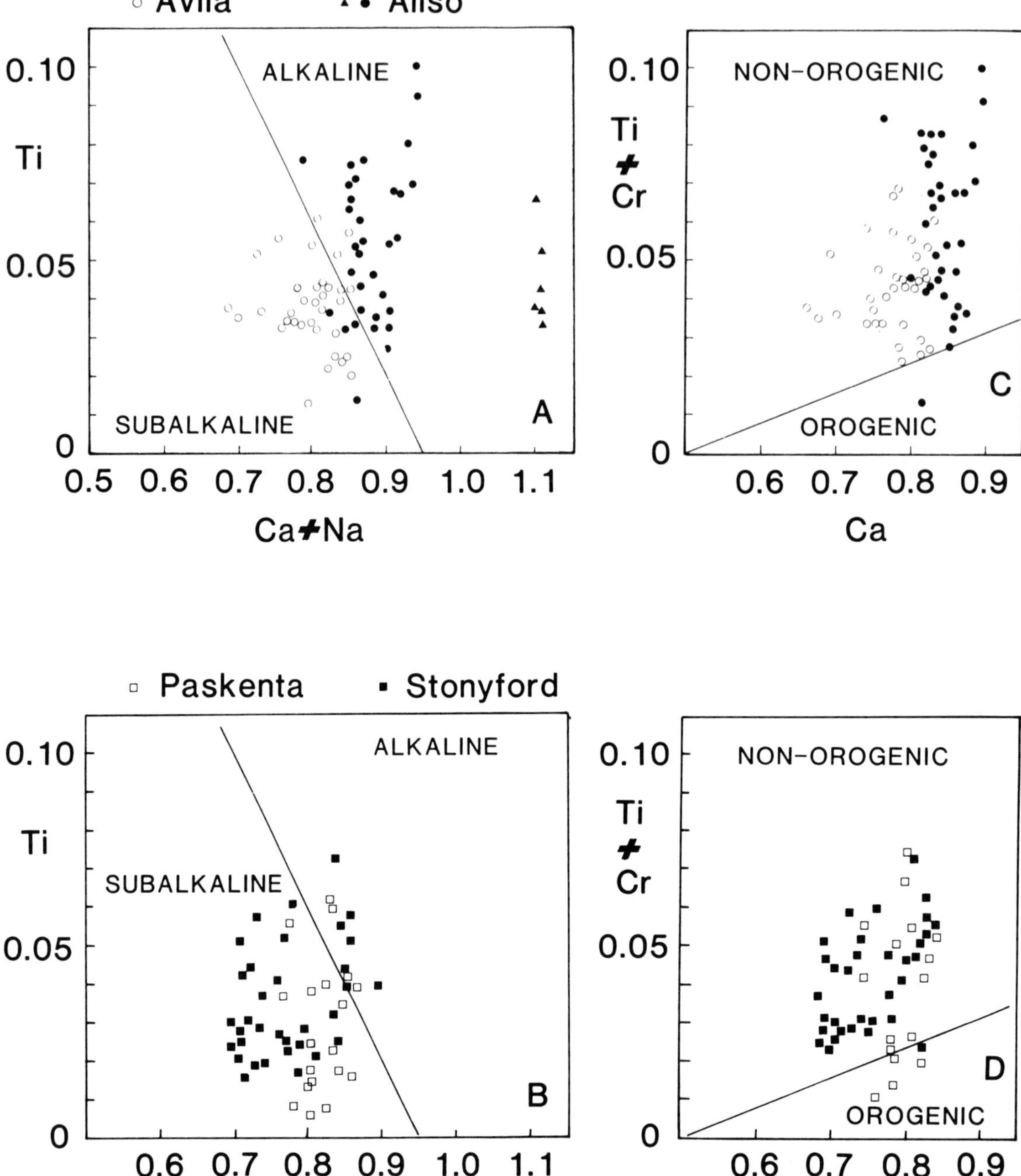

Figure 11. Pyroxene discrimination plots of Letterier and others (1982). Pyroxenes from all four locations are nonorogenic in character. Pyroxenes from Avila Beach, Paskenta, and Stonyford are subalkaline or transitional; those from Aliso Canyon are mildly alkaline. Alkaline character of most Aliso Canyon (and some Stonyford) basalts supported by occurrence of late magmatic aegerine-augite in these rocks (solid triangles).

have the highest. This correlation can be explained by the preferred partitioning of Sc into calcic pyroxene, and its exclusion from olivine and plagioclase. The melting model in Figure 12 shows that Sc is progressively enriched in the melt as clinopyroxene is eliminated from the refractory residue during nonmodal partial melting. Enriched tholeiites that form by 15 to 20 percent melting of an enriched mantle source contain *more* Sc than mildly alkaline basalts that form by ≅7 percent melting of the same source. Subsequent fractional crystallization of olivine ± plagioclase ($\bar{D}$ = 0.15) and olivine gabbro ($\bar{D}$ = 0.6–0.8) could create the trends observed in the data. This result would not be predicted by modal melting models. A detailed discussion of the partial melting models is presented in the Appendix.

The models discussed above suggest that compositional variations between these Franciscan greenstones can be explained by variable partial melting (7 to 20 percent) of a spinel lherzolite source that was enriched in light elements relative to chondritic values. This source is similar to that observed for oceanic islands and enriched MORB, but cannot be characterized further without isotopic data.

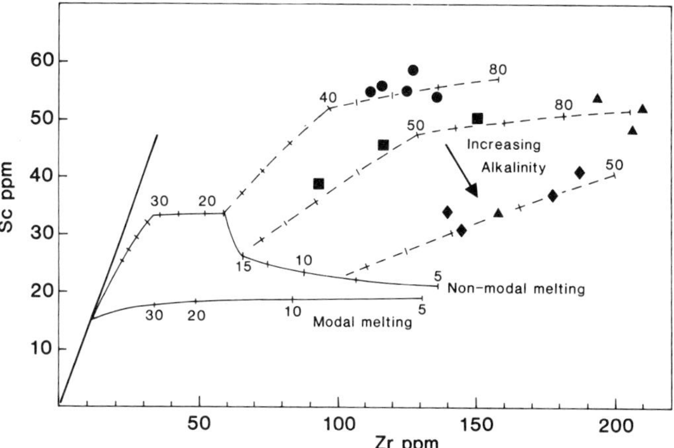

Figure 12. Plot of Sc versus Zr for Franciscan metavolcanic rocks. Also shown are model curves for partial melting (solid), fractional crystallization (dashed), and locus of chondritic Zr/Sc ratios (heavy solid line). Lower melting curve shows Sc-Zr variations during modal melting. Upper melting curve shows progressive enrichment of Sc in melt phase as percentage of melt increases during nonmodal melting, until clinopyroxene eliminated from residue. Fractional crystallization of melts formed by 7 to 20 percent nonmodal melting could create observed trends. Symbols as in Figure 6. Numbers inside diagram indicate percentage of melt (solid lines) or percentage of crystallization (dashed lines).

Tectonic Settings

Determination of the original tectonic setting of hydrothermally altered and structurally dismembered "oceanic" crust requires consideration of a wide range of geochemical, structural, and stratigraphic data. Geochemically, the Franciscan metavolcanic rocks studied here range in composition from enriched tholeiitic basalt to mildly alkaline basalt. Modern basalts with similar compositions are found in a variety of tectonic settings. In this section we address two main questions: did these lavas form at an oceanic spreading center or off-axis in an intraplate setting, and, in either case, do these lavas represent a major ocean basin, or a marginal, arc-related basin?

The Aliso Canyon sills are intrusive into chert and therefore formed during off-axis volcanic activity. The aphyric pillow lava found at this same location is geochemically similar and is probably related to this same off-axis volcanism. These rocks are also the most alkaline samples studied here, consistent with an intraplate origin.

The volcanic rocks at Avila Beach and Paskenta do not preserve unequivocal field evidence for their origin; however, the association of both sites with sheared serpentinite-matrix mélange is similar to fracture zone assemblages described by Saleeby (1982). Batiza and Vanko (1984) have noted the common occurrence of small seamounts near fracture zones in the Pacific Ocean Basin. This same association may be preserved at Paskenta and Avila Beach. The lavas from these sites are the least alkaline basalts studied here and may be similar to enriched MORB in composition.

The Stonyford volcanics preserve thick volcanic sections that are interpreted as seamounts, but whether these formed at a spreading center as axial seamounts or off-axis as intraplate seamounts is not known. The mildly alkaline transitional basalts of the Snow Mountain complex are equally common in either setting. The Ti-rich ferrobasalts of the Stony Creek complex are similar to ferrobasalts from the Galapagos Rise that are associated with propogating rift systems (Christie and Sinton, 1981; Sinton and others, 1983). The Stony Creek complex is enclosed within the same serpentinite-matrix mélange that contains the Paskenta volcanic knockers, and rocks from both localities may be related by fractional crystallization (e.g., Fig. 12). The geochemical relationship between the Stony Creek "Fe-Ti" basalts and the more normal Paskenta basalts is similar to that observed between Fe-Ti basalts erupted near the tip of a propagating rift and the normal MORB that erupts farther away (e.g., Christie and Sinton, 1981).

It is not possible, with the data presented here, to determine whether these Franciscan metavolcanic rocks formed in a true ocean basin or in a back-arc/marginal basins (e.g., the Shikoku Basin and Daito Basin; Nisterenko, 1980; Dick, 1982). It is clear, however, that a variety of tectonic settings are required to explain both the range in geochemistry of these lavas and the variety of structural and stratigraphic relationships that they display.

CONCLUSIONS

The geochemical data discussed above show that Franciscan greenstones from the four locations studied here include alkaline, transitional, and enriched tholeiite magma types. These magma types are probably related by increasing degrees of partial melting of a similar source region that was enriched in light elements (e.g., La/Zr ≅ two times chondrite). Mixing of tholeiitic basalts created by varying degrees of partial melting with alkalic magmas, as suggested by Batiza and Vanko (1984) for eastern Pacific seamounts, is not consistent with the uniform light element enrichment of all the basalts, regardless of magmatic affinity.

Geochemically similar basalts occur as tectonic inclusions in mélange or as thrust sheets associated with mélange in New Zealand (Grapes and Palmer, 1984), Newfoundland (Strong, 1974; Jamieson, 1977; Jacobi and Wasowski, 1985), Oregon (Snavely and others, 1968; Mullen, 1985), and the inner wall of the Marianas trench (Bloomer, 1983). Intraplate volcanic rocks and seamounts that are emplaced within or on top of abyssal sediments will be detached preferentially from the underlying crust during subduction, favoring their preservation within the tectonic mélange (e.g., Strong, 1974; Sporli, 1978; Bloomer, 1983; MacPherson, 1983). In a similar fashion, fracture zones, aseismic ridges, and other structural discontinuities in ocean crust (e.g., the pseudofaults associated with propagating rifts) create zones of weakness that may fail during subduction, allowing laterally extensive slabs of ocean crust to be preserved within the mélange. Preservation of true ocean crust (i.e., crust that formed at a spreading center) is not favored by the geometry of subduction, except in fracture zones and aseismic ridges.

Tectonic erosion of the upper plate, as postulated by Bloomer (1983) for the Mariana trench, has not been established for the Franciscan complex. Our data show that volcanic rocks of the overlying Coast Range Ophiolite are related to island arc volcanism, and are geochemically distinct from Franciscan volcanic rocks (Shervais and Kimbrough, 1985a).

The modern tectonic analogues discussed above show that a variety of tectonic settings are required to form the Franciscan metavolcanic rocks studied here. Although an oceanic origin may be likely for these rocks, their similarity to alkaline and transitional subalkaline sills in the Shikoku and Daito Basins (Nisterenko, 1980; Dick, 1982) suggests that a back-arc basin origin cannot be ruled out.

ACKNOWLEDGMENTS

The pyroxene and whole-rock analyses were made while one of us (J.W.S.) was a NATO postdoctoral fellow at the Institute for Crystallography and Petrography, Swiss Federal Institute of Technology (ETH Zurich), under the sponsorship of V. Trommsdorff. We especially thank V. Dietrich and B. Aryanci, who aided with the whole-rock analyses, and J. Sommerauer, who oversaw the electron microprobe analyses. C. A. Hopson (University of California at Santa Barbara) provided the samples from Stonyford and Paskenta, and valuable discussions on ophiolites in general and the Coast Ranges in particular. Careful reviews by S. H. Bloomer and R. D. Jacobi improved the manuscript significantly.

APPENDIX

The melting model used in Figures 9 and 12 is based on the non-modal melting equation of Shaw (1970). This equation is equivalent to the batch melting equation when D varies with changes in modal mineralogy during melting. Mineral/liquid partition coefficients for La, Zr, and Sc were taken from Weill and McKay (1975), Lindstrom (1976), McKay and Weill (1977), Grutzeck and others (1974), Nagasawa and others (1978), and McCallum and Charette (1978).

In Figures 9 (La versus Zr) two source regions are considered: one has La and Zr concentrations both at two times chondrite (i.e., La/Zr = chrondritic); the other has La = four times chondrite and Zr = two times chondrite (i.e., La/Zr = two times chondrite). Partial melting will increase La/Zr ratios somewhat, while fractional crystallization will have little or no effect (Fig. 9).

In Figure 12, both Zr and Sc are assumed to be two times chondrite in the source. A Sc concentration of two times chondrite (15 ppm) is consistent with concentrations observed by Frey and Green (1974) in the most pyroxene-rich spinel lherzolites that they analyzed. During melting, the Sc concentration of melt increases until clinopyroxene is eliminated from the residue ≅17 percent melt), then stays nearly constant until orthopyroxene is eliminated (≅33 percent melt). The Sc concentration of the melt decreases rapidly thereafter.

In both figures, the initial modal mineralogy is assumed to be olivine = 63 percent, orthopyroxene = 22 percent, clinopyroxene = 12 percent, spinel = 3 percent. This is consistent with most fertile pyroxene-rich lherzolites. The melting proportions were estimated to be olivine = 0.09, orthopyroxene = 0.20, clinopyroxene = 0.70, spinel = 0.01 before elimination of Cpx, and olivine = 0.30, orthopyroxene = 0.67, spinel = 0.03 after elimination of Cpx but prior to loss of Opx. These proportions are consistent with the results of high-pressure melting experiments on spinel peridotites, and result in modal mineralogies in the calculated refractory residues similar to modes observed in ultramafic xenoliths and alpine peridotites.

REFERENCES CITED

Aryanci, B., 1977, The major, minor, and trace element analysis of silicate rocks and minerals from a single sample solution: Schweizerische Mineralogische und Petrographische Mitteilugen, v. 57, p. 299–312.

Bailey, E. H., and Blake, M. C., 1974, Major chemical characteristics of Mesozoic Coast Range ophiolite in California: U.S. Geol. Survey Journal of Research, v. 2, p. 637–656.

Bailey, E. H., Irwin, W. P., and Jones, D. L., 1964, Franciscan and related rocks: California Division of Mines and Geology Bulletin 183, 177 p.

Bailey, E. H., Blake, M. C., and Jones, D. L., 1970, On-land Mesozoic oceanic crust in California Coast Ranges: U.S. Geological Survey Professional Paper 700-C, p. C70–81.

Ballard, R. D., Holcomb, R. T., and van Andel, T. H., 1979, The Galapagos Rift at 86°W, 3, Sheet flows, collapse pits, and lava lakes of the rift valley: Journal of Geophysical Research, v. 84, p. 5407–5422.

Batiza, R., and Vanko, D., 1984, Petrology of young Pacific seamounts: Journal of Geophysical Research, v. 89, p. 11235–11260.

Batiza, R., Fornari, D. J., Vanko, D. A., and Lonsdale, P., 1984, Craters, calderas, and hyaloclastites on young Pacific seamounts: Journal of Geophysical Research, v. 89, p. 8371–8390.

Blake, M. C., Jr., ed., 1984, Franciscan geology of northern California: Pacific Section, Society of Economic Paleontologists and Mineralogists, 254 p.

Blake, M. C., Jr., Jayko, A. S., and McLaughlin, R. L., 1984, Tectonostratigraphic terranes of the northern Coast Ranges, California, *in* Howell, D. G., ed., Tectonostratigraphic terranes of the Circum-Pacific: Circum-Pacific Council on Energy and Mineral Resources, Earth Science Series, v. 1.

Bloomer, S. H., 1983, Distribution and origin of igneous rocks from the landward slopes of the Mariana Trench: Journal of Geophysical Research, v. 88, p. 7411–7428.

Brown, R. D., 1964, Geologic map of the Stonyford quadrangle, Glen, Colusa, and Lake Counties, California: U.S. Geological Survey Mineralogical Investigations Map MF-279, scale 1:48,000.

Christie, D. M., and Sinton, J. M., 1981, Evolution of abyssal lavas along propagating segments of the Galapagos spreading center, Earth and Planetary Science Letters, v. 56, p. 321–335.

Davis, T., 1969, Strontium isotope and trace element geochemistry of igneous and metamorphic rocks in the Franciscan assemblage, California, [Ph.D. thesis]: University of California at Santa Barbara, 154 p.

Dibble, T. W., 1966, Geology of the central Santa Yvez Mountains, Santa Barbara County, California: California Division of Mines and Geology Bulletin 186, 99 pp.

Dick, H.J.B., 1982, The petrology of two back-arc basins of the northern Philippine Sea: American Journal of Science, v. 282, p. 644–700.

Dietrich, V., Oberhaenslt, R., and Walpen, P., 1976, Röntgenfluoreszenanalyse der Silikatgesteine: Institute für Kristallographie und Petrographie der ETH-Zürich.

Dietrich, V., Gansser, A., Sommerauer, J., and Cameron, E. E., 1981, Paleogene komatiites from Gorgona Island, East Pacific—A primary magma for ocean floor basalts? Geochemical Journal, v. 15, p. 141–161.

Frey, F. A., and Green, D. H., 1974, The mineralogy, geochemistry, and origin of lherzolite inclusions in Victorian basanites: Geochimica et Cosmochimica Acta, v. 38, p. 1023–1059.

Grapes, R., and Palmer K., 1984, Magma type and tectonic setting of metabasites, southern Alps, New Zealand, using immobile elements, New Zealand Journal of Geology and Geophysics, v. 27, p. 21–25.

Grutzeck, M. S., Kridelbaugh, S. J., and Weill, D. F., 1974, The distribution of Sr and the REE between diopside and silicate liquid: Geophysical Research Letters, v. 1, p. 273–275.

Hall, C. A., 1973, Geologic map of the Morro Bay South and Port San Luis Quadrangles, San Luis Obispo County, California: U.S. Geological Survey Miscellaneous Field Studies Map MF-511.

Hopson, C. A., Mattinson, J. M., and Pessagno, E. A., 1981, Coast Range Ophiolite, western California, *in* Ernst, W. G., ed., The geotectonic development of California, Ruby Volume I: Englewood Cliffs, New Jersey, Prentice-Hall, p. 418–510.

Hsu, K. J., 1968, Principles of mélange and their bearing on the Franciscan-Knoxville paradox: Geological society of America Bulletin, v. 79, p. 1063–1074.

Jacobi, R. D., and Wasowski, J. J., 1985, Geochemistry and plate tectonic significance of the volcanic rocks of the Summerford group, north-central Newfoundland: Geology, v. 13, p. 126–130.

Letterier, J., Maury, R. C., Thonon, P., Girard, D., and Marchal, M., 1982, Clinopyroxene compositions as a method of identification of the magmatic affinities of paleo-volcanic series: Earth and Planetary Science Letters, v. 59, p. 139–154.

Lindstrom, D. J., 1976, Experimental study of the partitioning of the transition metals between clinopyroxene and coexisting silicate liquids [Ph.D. thesis]: Eugene, University of Oregon, 180 p.

Liou, J. G., 1979, Zeolite facies metamorphism of basaltic rocks from the East Taiwan ophiolite: American Mineralogist, v. 64, p. 1–14.

Lonsdale, P., and Batiza, R., 1980, Hyaloclastite and lava flows on young seamounts examined by submersible: Geological Society of America Bulletin, v. 91, p. 545–554.

MacPherson, G. J., 1983, The Snow Mountain volcanic complex; An on-land seamount in the Franciscan Terrain, California: Journal of Geology, v. 91, p. 73–92.

McCallum, I. S., and Charette, M. P., 1978, Partition coefficients for Nb and Zr among armalcolite, ilmenite, clinopyroxene, and silicate melt: Geochimica et Cosmochimica Acta, v. 42.

McKay, G. A., and Weill, D. R., 1977, KREEP petrogenesis revisited: Proceedings, Lunar Science Conference, v. 8, p. 2339–2355.

Melson, W. G., Byerly, G. R., Nelen, J. A., O'Hearn, T., Wright, T. L., and Vallier, T., 1975, A catalog of the major element chemistry of abyssal volcanic glasses: Mineral Sciences Investigations, 1974-75, Smithsonian Contributions to the Earth Sciences, v. 19, p. 31–60.

Menzies, M., and Murthy, V. R., 1980, Mantle metasomatism as a precursor to the genesis of alkaline magmas: American Journal of Science, v. 280-A, pt. 2, p. 622–638.

Moore, J. G., and Charlton, D. W., 1984, Ultrathin lava layers exposed near San Luis Bay, California: Geology, v. 12, p. 542–545.

Mottl, M., 1983, Metabasalts, axial hotsprings, and the structure of hydrothermal systems at mid-ocean ridges: Geological Society of America Bulletin, v. 94, p. 161–180.

Mullen, E. D., 1985, Petrologic character of Permian and Triassic greenstones from the mélange terrane of eastern Oregon and their implications for terrane origin: Geology, v. 13, p. 131–134.

Nisbet, E. G., Dietrich, V. J., and Esenwein, A., 1979, Routine trace element determinations in silicate minerals and rocks by x-ray fluorescence: Fortschritte der Mineralogie, v. 57, p. 264–279.

Nisterenko, G. V., 1980, Petrochemistry and geochemistry of basalts in the Shikoku basin and Daito basin, Philippine Sea, *in* Initial reports of the Deep Sea Drilling Project, v. 58: Washington, D.C., U.S. Government Printing Office, p. 791–804.

Papike, J. J., Cameron, K. L., and Baldwin, K., 1974, Amphiboles and pyroxenes; Characterization of other than quadrilateral components and estimates of ferric iron from microprobe data: Geological Society of America Abstracts with Programs, v. 6, p. 1053–1054.

Pearce, J. A., 1980, Geochemical evidence for the genesis and eruptive setting of lavas from Tethyan ophiolites: Proceedings, International Ophiolite Symposium, Cyrus, 1979, p. 261–272.

Pearce, J. A., and Cann, J. R., 1973, Tectonic setting of basic volcanic rocks determined using trace element analysis: Earth and Planetary Science Letters, v. 19, p. 290–300.

Saleeby, J. B., 1982, Polygenetic ophiolite belt of the California Sierra Nevada; Geochronological and Tectonostratigraphic development: Journal of Geophysical Research, v. 87, p. 1803–1824.

Schiffman, P., and Liou, J. G., 1980, Synthesis and stability relations of Mg-Al pumpellyite, Ca4A15MgSi6021(OH): Journal of Petrology, v. 21, p. 441–474.

Schussler, S. A., 1981, Paleogene and Franciscan complex stratigraphy, southwestern San Rafael Mountains, Santa Barbara County, California [M.S. thesis]: University of California at Santa Barbara, 234 p.

Shaw, D. M., 1970, Trace element fractionation during anatexis: Geochimica et Cosmochimica Acta, v. 34, 237–243.

Shervais, J. W., 1982, Ti-V plots and the petrogenesis of modern and ophiolitic lavas: Earth and Planetary Science Letters, v. 59, p. 101–118.

Shervais, J. W., and Kimbrough, D. L., 1984 Alkaline and transitional subalkaline metabasalts; Seamount volcanism in the Franciscan Complex, southern California: EOS Transactions of the American Geophysical Union, v. 65, p. 1147.

Shervais, J. W., and Kimbrough, D. L., 1985a, Geochemical evidence for the tectonic setting of the Coast Range ophiolite; A composite island arc–oceanic crust terrane in western California: Geology, v. 13, p. 35–38.

Shervais, J. W., and Kimbrough, D. L., 1985b, Geochemical evidence for the tectonic setting of the Coast Range ophiolite: Reply, Geology, v. 13, p. 828–829.

Sinton, J. M., Wilson, D. S., Christie, D. M., Hey, R. N., and Delaney, J. R., 1983, Petrologic consequences of rift propagation on oceanic spreading ridges; Earth and Planetary Science Letters, v. 62, p. 193–207.

Snavely, P. D., MacLeod, N. S., and Wagner, H. C., 1968, Tholeitic and alkalic basalts of the Eocene Siletz River Volcanics, Oregon Coast Range: American Journal of Science, v. 266, p. 454–481.

Stakes, D. S., Shervais, J. W., and Hopson, C. A., 1984, The volcano-tectonic cycle of the Famous and AMAR valleys, Mid-Atlantic Ridge (36°47′N): Journal of Geophysical Research, v. 89, p. 6995–7028.

Strong, D. F., 1974, An off-axis alkali volcanic suite associated with the Bay of Islands ophiolite, Newfoundland: Earth and Planetary Science Letters, vol. 21, p. 301–309.

Weill, D. F., and McKay, G. A., 1975, The partitioning of Mg, Fe, Sr, Ce, SM, Eu, and Yb in lunar igneous systems: Proceedings, Lunar Science Conference, v. 6, p. 1143–1158.

MANUSCRIPT ACCEPTED BY THE SOCIETY OCTOBER 15, 1986

Printed in U.S.A.

Geological Society of America
Special Paper 215
1987

A reassessment of the diamondiferous Pamali Breccia, southeast Kalimantan, Indonesia: Intrusive kimberlite breccia or sedimentary conglomerate?

S. C. Bergman
ARCO Oil and Gas Company, Exploration and Production Research Center, PRC 3005, 2300 Plano Parkway, Plano, Texas 75075
*W. S. Turner**
Anaconda Indonesia Incorporated, Jakarta, Republic of Indonesia
*L. G. Krol**
Anaconda Minerals Company, DAT 1527, 555 Seventeenth Street, Denver, Colorado 80202

ABSTRACT

The "Pamali Breccia" is a small (<100-m diameter, 10- to 35-m thick), semicircular peridotite body that rests on the Bobaris ophiolite in southeast Kalimantan, Indonesia. Generally it has been regarded as a primary source rock for Borneo diamonds, which represents a tectonic breccia that is possibly related to a kimberlite. New geologic, geochemical (major and trace elements), and petrographic data indicate that the Pamali Breccia is, in fact, a sedimentary conglomerate with local Bobaris ophiolite provenance. The Pamali Breccia consists of well-rounded to subangular clasts (0.1 to 20 cm; mean, 0.7 cm), mostly serpentinites and pyroxenites, with clasts of gabbros and greenstones forming minor constituents.

Abundant sedimentary features characterize the Pamali Breccia and include well-sorted, matrix-supported, normally graded beds (coarse sand to cobble clast-size), and planar stratification with imbricated pebbles and cobbles. When corrected for near-surface, secondary calcite alteration and normalized to an anhydrous basis, whole-rock samples of the breccia and its component clasts and matrix display geochemical traits typical of ophiolites and other ultramafic rocks (SiO_2 = 42–45 wt. %, 9–12% FeO^T, 25–35% MgO, 3–12% CaO, 6–9% Al_2O_3, 0.1–1.3% Na_2O, <0.2% K_2O). Compatible trace element contents of the Pamali Breccia are more typical of ophiolites than kimberlites or lamproites and range from 600 to 2,000 ppm Ni, 1,400 to 4,000 ppm Cr, and 60 to 100 ppm Co. Large ion-lithophile (incompatible) element contents of the Pamali Breccia display the same features and are in the range of Rb = 10 to 30 ppm, Sr = <10 to 80 ppm, Y <10 ppm, Ce = 5 to 8 ppm, and Th <0.2 ppm. Various Pamali Breccia elemental ratios are also more characteristic of ophiolites than those of kimberlites or lamproites.

Thus the Pamali Breccia is not a kimberlite or lamproite. The primary source of the diamonds in the breccia may be the host ophiolite, or, alternatively, a distant kimberlite or lamproite, the unambiguous identification of which requires further study.

*Present addresses: Turner, Dominion Gold Mine, Ltd., S. Widjojo Centre, Second floor, Jl. Jend. Sudirman Kav. 57, Jakarta, Republic of Indonesia; Krol, 10040 East Grand Avenue, Englewood, Colorado 80111.

INTRODUCTION

Alluvial diamonds have been recovered in Kalimantan, which literally translated means "River of Gold and Diamonds," since the discovery of the first stones in the fifth century A.D. (Koolhoven, 1935; Wilson, 1982). Three major diamond provinces exist in Kalimantan: the Landak and Kapuas River basins in West Kalimantan (Witkamp, 1932), the Barito River basin in Central Kalimantan (Hovig, 1930), and the Martapura area in southeast Kalimantan (Halewijn, 1824; Horner, 1937; Schwaner, 1853; Koolhoven, 1935). Thus far, all known diamonds in Kalimantan (probably totaling less than 1 to 10 million carats) have been recovered from post-Cretaceous secondary sources; the primary source(s) of these alluvial "stones" has long been an enigma.

The Pamali Breccia is located on the southeast margin of the Bobaris peridotite massif in southeastern Kalimantan. It has been described as a diamond-bearing intrusive breccia pipe by several workers (e.g., Krol, 1919, 1920; Koolhoven, 1935; van Bemmelen, 1970). Although it has often received questionable connotations, more recently it has been cited as an example of a diamondiferous intrusive peridotite that occurs in a mobile tectonic setting (e.g., Hess, 1938, p. 328; Sobolev, 1951; Dawson, 1967, p. 248, 1980, p. 22; Carmichael and others, 1974, p. 523; Hutchison, 1975, p. 802; Bardet, 1977, v. 3, p. 141–145; Orlov, 1977, p. 158; Kaminskiy, 1980, p. 489; Legrand, 1980, p. 182; Trofimov, 1984, p. 80).

Nearly all diamondiferous kimberlites are restricted to stable continental regions that are underlain by Archean or Proterozoic cratons, which have not experienced much deformation in the last 1.5 b.y. (Dawson, 1980). The oldest exposed rocks in Borneo are Late Paleozoic/Early Mesozoic schists and granites of West and Central Kalimantan. Borneo is located in close proximity to the active Sunda arc. Therefore, the true nature of the Pamali Breccia is important with regard to current diamond-kimberlite issues. This paper attempts to accurately identify the Pamali Breccia by integrating major and trace element geochemical data with geologic field relations.

DIAMONDS IN BORNEO AND THEIR RELATIONSHIP TO OTHER ANOMALOUS DIAMOND OCCURRENCES

As reviewed by Koolhoven (1935), a variety of propositions has been advanced during the last 150 years for the true primary source rock* of diamonds in Kalimantan. Horner (1837) initially suggested that the "batu tatimahan" (literally, "friends of the diamond"; these include diaspore-corundum rocks, as well as associated minerals) was the primary source; Verbeek (1875) and Hooze (1893) indicated that the slates or veins and acid dikes within the crystalline slates were the primary source. Eruptive rocks or pegmatite were suggested as the source of diamonds by Gascuel (1901); Wing Easton (1904) suggested an olivine-bearing rock as the source. It was not until 1915 that van Es (cited in Krol, 1920) suggested that a kimberlite or peridotite breccia could be the primary source, based on what were then interpreted as analogous occurrences in South Africa. Krol (1919, 1920) indicated locations within the Bobaris peridotite where a faulted block of peridotite breccia could be the source of diamond, and Wing Easton (1925) concluded that, since there was only one possible primary source rock for diamonds, based on geologic and geochemical grounds, this breccia was kimberlite. When H.O.A. Meyer (personal communication cited in Kaminskiy, 1980) visited the Pamali Breccia locality in 1978, he did not recognize any "kimberlite indicator minerals," and further noted that the Pamali Breccia resembled an ordinary alpine ultramafite.

In an unpublished report to the Indonesian government, Stracke (1973) made special note of the lack of kimberlite indicator minerals (such as Cr-pyrope, Mg-ilmenite) in the diamondiferous sediments in southeast Kalimantan. Stracke reasoned that, if the diamonds were derived from a kimberlite source, the parent kimberlite must not be in the local vicinity. Nearly 10 years after Stracke's study, it was realized that lamproites are a second major primary source rock for diamonds (e.g., Atkinson and others, 1984; Jaques and others, 1984; Madigan, 1983). Because of the existence of a number of "anomalous" diamond occurrences (i.e., in rocks other than kimberlites or lamproites or in mobile belts: cf. Dawson, 1979; Nixon and others, 1986; P. H. Nixon and S. C. Bergman, in preparation), it is possible that several additional potential primary rocks for diamonds exist; witness the recent (albeit debatable) Russian reports of diamonds in harzburgites from ophiolites (Kaminskiy and Vaganov, 1976; Kaminskiy, 1980) and in an alkali basalt from the Onega Peninsula (Kaminskiy and others, 1974). Several other anomalous diamond occurrences exist in Russia: in the Ural picrites and peridotites (Kukharenko, 1955; Luk'yanova and others, 1978) and in the Sayan peridotites (Shestopalov, 1938). Gisolf (1923) reported the presence of diamond in olivine and anorthite-bearing bombs ejected from the Gunung Ruang volcano in the northern Celebes, and Van der Wegen (1963) discussed diamonds in serpentinites from Batang Pele (Indonesia). Twelvetrees (1914) reported finding several dozen diamonds in streams draining Bald Hill, a large serpentinite mass in Tasmania. Fan and Bai (1981) discovered diamonds 0.1 to 0.5 mm in diameter in harzburgite and dunite members of Tibetan alpine-type peridotites. Slodkevich (1982) reported graphite octahedra (retrograded diamonds?) in a garnet-clinopyroxene–bearing layered complex from the Beni Bouchera pluton in Morocco. Trofimov (1984) grouped the Pamali Breccia with the Beni Bouchera pluton in his "geosynclinal" group of diamond-bearing crustal structures.

*Prior to the mid-20th century realization that diamonds are derived from the Earth's upper mantle, the term "primary source of diamonds" had a totally different connotation than it has today. Strictly speaking, kimberlites or lamproites are not thought to be the most "primary sources of diamonds," but are merely the vehicles by which mantle-derived diamonds are transported to the surface (see, for example, Robinson, 1978; Meyer, 1985). The term "primary source" is used herein to refer to the igneous or metamorphic host rock of diamonds (i.e., kimberlite, lamproite, peridotite, eclogite, etc).

Although a detailed discussion of these occurrences is beyond the scope of this paper, it is noteworthy that many of the workers mentioned above have used the Pamali Breccia as an analogous occurrence to these supposed nonkimberlite or lamproite host rocks of diamonds. The Pamali Breccia may not be the "primary" source rock of the diamonds reported to have been obtained from it, yet it is becoming increasingly tenable, in view of the examples cited above, that the underlying Bobaris ophiolite may be the source of the Pamali diamonds (although a more distant unknown kimberlite/lamproite source cannot be yet ruled out).

PREVIOUS WORK

Geologic Setting

The geology of southeast Kalimantan is dominated by northeast-trending subduction complexes generally Late Cretaceous in age, which are overlain by Late Cretaceous sedimentary and volcanic rocks and in some areas by Tertiary sedimentary rocks. Tertiary intrusive rocks of intermediate to felsic composition occur throughout the area (Fig. 1; Hamilton, 1979; Geological Survey of Indonesia, 1970; Koolhoven, 1935; Krol, 1920). These subduction complexes represent younger additions to a mainly Paleozoic and Mesozoic craton that forms the basement terrain of western and central Borneo (Hamilton, 1979).

The subduction complexes form two prominent parallel mountain ranges in southeast Kalimantan, the Bobaris to the northwest, and the much higher Meratus Mountains to the southeast. The complexes comprise a wide variety of intercalated rock types that include peridotites, serpentinites, gabbros, greenschists, glaucophane schists, polymictic breccias, deep-ocean radiolarian cherts, and also clastic and carbonate sedimentary rocks (Hamilton, 1979; Koolhoven, 1935).

Bobaris Breccias

Several occurrences of mafic and ultramafic breccias have been reported to occur along the northwest and southeast margins of the Bobaris ophiolite massif (Fig. 1). Along the Riamterdjung River, a small tributary of the Haueran River (Fig. 1), a pipe-shaped "polymictic breccia" a few meters in diameter (Koolhoven, 1935) is described as containing clasts of serpentinite, peridotite, diabase, granite, epidote rock, schist, amphibolite and limestone in a matrix of serpentine and fine-grained material. Panning of the river gravels downstream from this breccia produced no diamonds (Koolhoven, 1935); however, diamonds have been found in the nearby Tampahalojo River (Krol cited in Koolhoven, 1935).

The Pamali Breccia occurs on the Ahim River (Fig. 1), a tributary of the Pamali. Initially diamonds were found in the stream sediments of the Pamali River, downstream from the Ahim, and were traced upstream to the Pamali Breccia body, which was subsequently proved to contain small diamonds (Koolhoven, 1935). No diamonds have ever been recovered in the Ahim River upstream from the Pamali Breccia, and seven test pits in this upstream area were found to be barren of diamonds (Koolhoven, 1935).

Diamonds were found in the 1930s in two of five test pits along the Pamali River above its confluence with the Ahim. Several breccias were found a short distance to the southwest of these diamondiferous pits but were thought to be tectonic breccias and distinctly different from those of the Pamali Breccia (Koolhoven, 1935). Koolhoven (1935) made mention of two other "breccia" localities: one in the Apui River and the other in the Takuti River (Fig. 1). Two small (<30 m) Apui River bodies are gabbroic breccias. The other breccias outcrop at three separate locations in the Takuti River region and are suggested by Koolhoven to be similar to those south of the Pamali River and thus tectonic in origin. According to Koolhoven (1935), no diamonds were found immediately downstream of the breccias in the Takuti River region.

In summary, the only "breccia" that shows some real association with diamonds is the so-called Pamali Breccia. A tectonic origin can easily explain the characteristics of the other breccia bodies.

Pamali Breccia

The Pamali Breccia crops out in a topographic low, bounded to the north and east by prominent hills of serpentinized peridotite and to the southeast by a ridge of hornblende diorite; to the south it is cut by the Ahim River (Fig. 2). Early Dutch drilling and test pitting delineated an oval-shaped body, approximately 300 m by 150 m in size, whose base was interpreted to be bounded by a curved plane striking N15°E and dipping 25° to the east. By drilling, the vertical extent of the body was determined to vary from 10 to 35 m. The presence of rolled-out serpentine and friction clay along a ±5-m-thick basal zone was interpreted by Koolhoven as evidence of a fault plane.

Koolhoven described the surface of the Pamali Breccia as weathered soil, thickness 0.5 to 2 m, that passes into a brecciated, serpentinized, fractured rock and that changes into a solid, serpentinized breccia with depth. A calcite-rich matrix was observed at the surface but was absent at depth in the drill holes and therefore considered to be supergene in origin. Koolhoven also noted the presence of orthopyroxene, often altered to uralitic hornblende, with chlorite, pyrite, chalcopyrite, and gold. The presence of hornblende with a slightly peculiar geochemistry(?) led him to speculate that the Pamali Breccia was not a true kimberlite, as asserted in early South African reports.

The Dutch found 13 small diamonds in a 10-ton bulk sample described as "weathered and deposited soil" from 17 shallow prospecting holes sunk in the top of the breccia (grade, 35 carats per 100 metric tons). In addition, two small diamonds were found in the rock excavated from test pits I and II. According to Koolhoven (1935), this established with certainty that the "breccia" was diamond-bearing, and, by implication, that this rock was a diamondiferous ultramafic intrusive breccia.

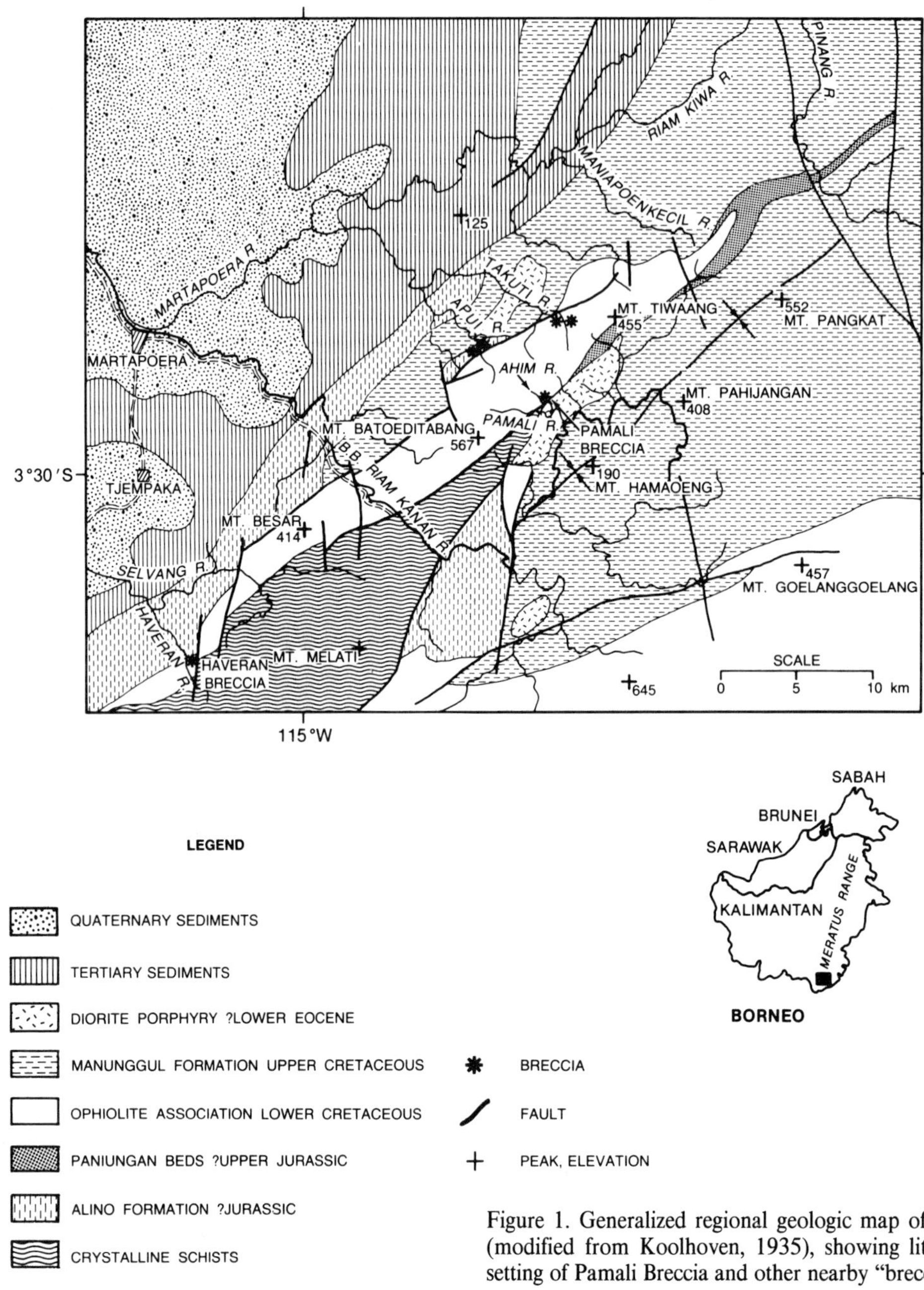

Figure 1. Generalized regional geologic map of southeast Kalimantan (modified from Koolhoven, 1935), showing lithologic and structural setting of Pamali Breccia and other nearby "breccia" occurrences.

RESULTS

During this reinvestigation of the Pamali Breccia, a number of the earlier test pits were reexcavated and inspected, and the streams that cross the body as well as the crests of the surrounding ridges were traversed. A new geologic sketch map was made by pace and compass traverse (Fig. 2), and a number of new pits were sunk (nos. 9 through 13 in Fig. 2). The deep pits on the southern side of Pamali River were also investigated, as these exposures were interpreted by Koolhoven (1935) to be peridotite breccias of tectonic origin, in contrast to the Pamali Breccia, which was at the same time considered to be of intrusive origin.

Field Relations

The outcrop area of the Pamali Breccia (see Fig. 2) is less extensive and more irregular in shape than that originally shown by Koolhoven. The west and northwest margins of the breccia are generally defined by a break in slope. The northwest margin is in fault contact with the massive peridotite. The eastern diorite

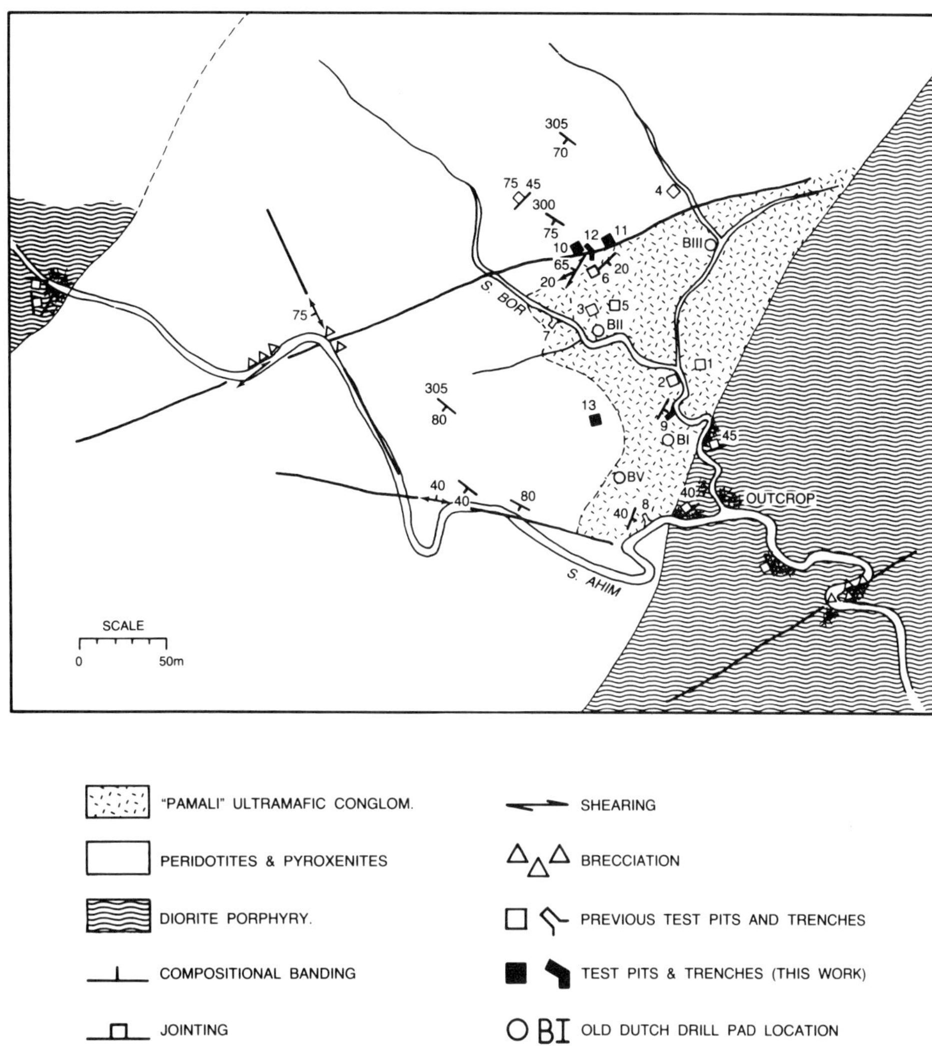

Figure 2. Geologic map of Pamali Breccia area showing underlying Bobaris ophiolite and intrusive diorite (this work).

contact proved difficult to excavate because of the steep slopes and thick soil cover, and consequently was not inspected.

The best exposures of the Pamali Breccia are found in the Ahim River, a few meters upstream from trench no. 8 (Fig. 2). At this location, graded bedding is clearly discernible, and the beds strike N25°E and dip 40° to the southeast. Stratification of a similar attitude, although less steeply dipping, was found at two other localities within the Pamali Breccia, at trench no. 9 and test pit no. 6.

The lithologic character of the Pamali Breccia varies a great deal. The exposures in the Ahim River consist of subrounded to subangular clasts (0.1–20 cm in size; mean, approximately 0.7 cm) in a calcite-rich polymodal matrix (Figs. 3, 4). Well-rounded, pebble-size clasts are common; sharp angular clasts are not, but when present are usually large. The clasts consist of 90 percent serpentinite, with minor greenstone chert, pyroxenite, and gabbro. Locally, the matrix material in the Ahim River outcrops is entirely replaced by calcite, but more typically it comprises a mixture of calcite and finer grained serpentinite clasts (less than 1–2 mm). This particular outcrop is the only place where calcite is observed in the Pamali Breccia. Where sorting is better developed, the fabric is most often clast-supported, and downdip pebble imbrication is discernible (see Fig. 4). Stratification is marked by a variety of individually well-sorted laminae, which typically are graded. Stratified conglomeratic layers (30 cm thick) are interstratified with and grade into pebbly sandstone layers (5–10 cm thick).

The rock exposed in test pit no. 6 is strikingly different from

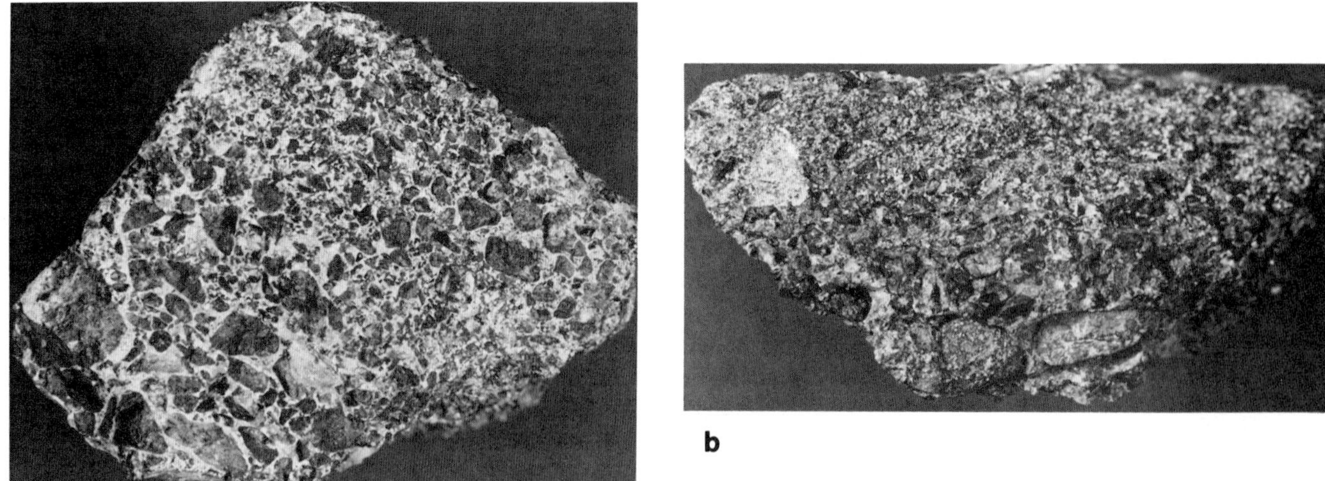

Figure 3. Hand-specimen photographs of Pamali Breccia samples from Ahim River exposure. a, Supergene calcite-rich matrix and sedimentary features including well-rounded clasts in clast-supported conglomerate framework (width of field, 35 cm). b, Normally graded sample (up is to the top of photograph) with well-developed stratification (width of field, 15 cm).

c

Figure 4. Outcrop photograph of Pamali Breccia at Ahim River exposure.

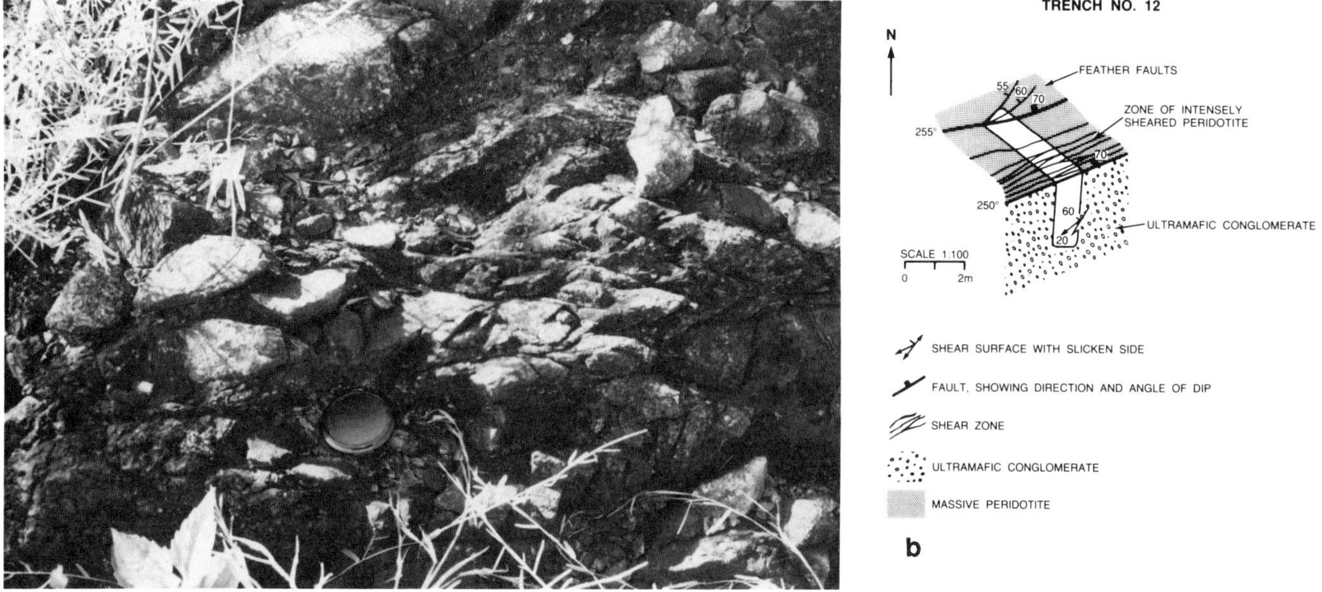

Figure 5. a, Outcrop photograph of nearby pyroxenite member of Bobaris ophiolite immediately underlying Pamali Breccia at its northwest margin, demonstrating well-developed, sheared fabric. b, Diagramatic sketch of trench 12, showing contact relations of Pamali Breccia and underlying pyroxenite.

that found at the two previously described locations. The clasts are much larger (ranging from large pebbles to cobbles) some reaching 35 cm long, and are typically subangular to angular. Pebble-size clasts are present in minor proportions (20–30%) and are usually subrounded to rounded.

The clast populations of the present-day Ahim and Bor River bedloads closely match that of the Pamali Breccia, especially when granule- to pebble-size clasts are compared. The streams' bedloads are dominated by serpentinized peridotite clasts, but also include clasts of banded gabbro, chert, gneissic granitoid, and mylonite. These clasts are typically subrounded to subangular, and similar in character to those of the Pamali Breccia.

The northwest margin of the Pamali Breccia was exposed by a 5-m trench (trench no. 12, Fig. 2) and found to be in fault contact with massive peridotite (see Fig. 5). This contact is very sharply defined and strikes N110°W and dips 70° to the north-northwest. Within about 1.5 m of the contact, the massive peridotite is sheared with increasing intensity as the faulted contact is approached. The shearing has completely obliterated the original igneous texture. Toward the northwest end of the trench, large blocks of serpentinized peridotite in which relict igneous textures can be seen are separated by discrete narrow faults 1 to 2 cm wide. Evidence of shallower angle feathering can be observed on one major fault (Fig. 5b).

In marked contrast to the peridotite, the Pamali Breccia in the southern end of the trench is characterized by very little faulting and disruption. The breccia is comprised of subangular to subrounded ultramafic and mafic clasts (size, 0.5–4 cm) in an extremely weathered, clayey matrix. A few of the clasts are up to 25 cm in size; these are usually less rounded than the smaller clasts. A small wedge of mainly well-sorted, clast-supported, sand- and granule-size material (about 1–6 mm), devoid of matrix, occurs near the northwest margin. Within the Pamali Breccia in this trench, there are a few minor faults that have slickensides plunging at about 20°. Small faults of similar orientation are found across the contact in the sheared peridotite.

We suggest that the major deformation and intense faulting now observed in the peridotite records an older event (pre–Pamali Breccia deposition), and that minor rejuvenation of this fault sometime after the deposition of the breccia caused its juxtaposition with the peridotite. In some areas at the surface, on the area mapped as Pamali Breccia, the tops of large subangular to angular boulders of massive peridotite can be seen poking through the soil profile. These are considered to be simply large boulders within Pamali Breccia body.

The field evidence alone gathered during this reinvestigation of the Pamali Breccia favors a reinterpretation of this rock as an ultramafic conglomerate of local Bobaris ophiolite provenance. The attitudes of the stratification observed in three separate locations, the lack of clast angularity, and the variation in clast size and lithology indicate that the breccia represents some sort of talus slope deposit that has been slightly modified by alluvial action.

TABLE 1. MAJOR ELEMENT COMPOSITIONS OF PAMALI BRECCIA AND ITS FRACTIONS WITH POTENTIALLY COMPARABLE ROCK TYPES*

(wt.%)	1	2	3	4	5	6	7	8	9	10	11	12
SiO_2	28.08	38.3	44.15	42.59	41.6	38.4	23.0	44.10	44.75	45.6	38.4 ± 7	52.5 ± 7
Al_2O_3	5.63	6.09	8.85	6.77	12.2	5.02	3.37	6.47	5.85	13.4	4.7 ± 2.7	9.0 ± 2.5
Fe_2O_3	2.75	6.01	4.17	6.68	1.81	3.68	3.18	6.10	4.29	1.98	---	---
FeO	5.27	4.47	8.00	4.97	6.0	5.3	3.60	6.90	6.18	6.58	---	---
(FeO*)	7.74	9.88	12.17	11.05	7.63	8.61	6.46	12.39	10.03	8.36	11.3 ± 3.5	6.8 ± 2.2
MnO	0.19	0.09	0.30	0.10	0.19	0.13	0.12	0.23	0.15	0.21	0.18 ± .09	0.10 ± .05
MgO	18.51	22.56	29.10	25.19	17.60	29.0	15.80	30.28	33.80	19.29	28.7 ± 7	12.3 ± 7
CaO	17.08	10.83	4.50	12.08	9.74	3.05	23.50	3.69	3.55	10.68	11.3 ± 7.6	6.1 ± 4
Na_2O	0.19	0.25	0.30	0.27	0.49	0.11	0.02	0.03	0.13	0.54	0.5 ± 0.9	1.4 ± 1.0
K_2O	0.11	<0.01	0.17	<0.01	0.82	0.04	0.05	0.10	0.05	0.90	1.4 ± 1.1	6.9 ± 2.8
TiO_2	0.20	0.17	0.41	0.19	0.36	0.14	0.11	0.20	0.16	0.39	2.6 ± 1.7	3.0 ± 1.7
P_2O_5	0.03	0.04	0.05	0.04	0.02	0.02	0.04	0.06	0.02	0.02	0.9 ± 0.7	1.3 ± 0.7
S	---	0.14	---	0.16	0.12	0.11	0.29	0.57	0.13	0.13	---	---
NiO	---	0.13	---	0.15	0.07	0.23	0.15	0.26	0.27	0.08	---	---
Cr_2O_3	---	0.56	---	0.62	0.20	0.58	0.50	0.97	0.68	0.22	---	---
H_2O^+	8.06	8.85	---	---	6.1	9.5	6.1	---	---	---	---	---
H_2O^-	2.82	1.42	---	---	0.6	3.3	2.5	---	---	---	---	---
CO_2	11.16	0.28	---	---	---	0.3	16.90	---	---	---	---	---
Totals	100.08	100.19	100.0	100.0	98.42	98.91	99.24	100.0	100.0	100.0	100.0	100.0

Key to column headings:

1. Pamali Breccia, carbonate-rich surface sample from the Ahim River exposure; analyst, I. den Haan (Koolhoven, 1935, analysis 1).
2. Pamali Breccia, 26-m-deep drill core from boring I in the east-central portion the body, approximately 130 m NNE of sample 1; analysts, H.W.V. Willems and Djokojuwonon (Koolhoven, 1935, analysis 2).
3. Analysis 1 normalized to calcite-free/volatile-free basis.
4. Analysis 2 normalized to calcite-free/volatile-free basis.
5. Large serpentinite clast (10 cm in diameter) in Ahim River exposure of Pamali Breccia (this study).
6. Small clasts (1.4-5.0 mm) in Ahim River exposure of Pamali Breccia (this study).
7. Pamali Breccia matrix (<0.355 mm) in Ahim river exposure (this study).
8. Analysis 7 normalized to calcite-free/volatile-free basis.
9. Analysis 6 normalized to calcite-free/volatile-free basis.
10. Analysis 5 normalized to calcite-free/volatile-free basis.
11. Average of 600 kimberlites normalized to calcite-free/volatile-free basis (Bergman, 1987).
12. Average of 300 lamproites, volatile-free basis (Bergman, 1987).

*See text for analytical techniques.

Geochemistry

Analytic Techniques. Major and trace element determinations were performed in Toronto in the laboratories of X-ray Assay Laboratories, Ltd., and Nuclear Activation Services, using standard analytic procedures. Wavelength-dispersive x-ray fluorescence spectroscopy was used for major elements as well as for Cl, S, Cr, Rb, Sr, Y, Zr, and Ba. H_2O, F, CO_2 and Fe^{2+} contents were determined using conventional wet methods. Instrumental neutron activation was used for Sc, Co, REE, Hf, Ta, U, and Th. Li contents were determined using atomic absorption spectroscopy; direct-current plasma spectroscopy was used for Ni and V.

Analytic precision and accuracy were verified using well-characterized standard rock powders. For major elements, we estimate relative errors of less than 1 to 3 percent. Minor and trace elements with concentrations close to detection limits for these rocks (e.g., Cl, Rb, Sr, Y, Zr, Ba and REE) have much higher relative errors (10 to 50 percent). Trace elements that occur in relatively high concentrations (e.g., Ni, Cr, Co, Sc) have smaller relative errors (5 to 10 percent).

Major and trace element systematics of various samples and of several different fractions of the breccia are presented in Tables 1 and 2. Two samples analyzed by Koolhoven (1935) include a calcite-rich, near-surface sample exposed on the Ahim River and a nearly calcite-free, 26-m-deep drill core sample. Samples analyzed in the present study include various fractions of the Ahim River exposure: a large serpentinite clast, a group of smaller clasts (1.4–5 mm), and the matrix of the Pamali Breccia that passed a 0.355-mm screen.

When normalized to 100 percent on a calcite-free, anhydrous basis, all samples and fractions display comparable major element compositions (analyses 3, 4, 8, 10, 11). These analyses are all typical of ultramafic rocks, particularly those from ophiolites (e.g., Coleman, 1977), in that they contain SiO_2 = 42–45 wt. %, FeO* = 9–12%, MgO = 35%, CaO = 3–12%, Al_2O_3 = 6–9%, and total alkalis (Na_2O + K_2O) = 0.1–1.3 wt. %. Compatible elements, such as Ni, Cr, Sc, and Co, are in the range of 600 to 2,000 ppm, 1,400 to 4,000 ppm, 12 to 40 ppm, and 60 to 100 ppm, respectively, which are also characteristic of ultramafic peridotites from ophiolites. Note that major element chemical data on the Pamali Breccia discussed by Burgath and Wilher Simandjuntak (1983) share chemical features with Koolhoven's (1935) and our present data. The relatively low alkali contents (analyses 1 and 2: $Na_2O + K_2O < 0.4$ wt. %; matrix <0.13 wt. %), depletion in P_2O_5 (typically <0.05 wt. %) and TiO_2 (always <0.4 wt. %),

TABLE 2. TRACE ELEMENT COMPOSITION OF PAMALI BRECCIA, ITS FRACTIONS, AND POTENTIALLY COMPARABLE ROCK TYPES

Element (ppm)	5	6	7	Avg. K* (n = 10-100) 12	Avg. L* (n = 30-180) 13
Li	66	38	30	25	31
Rb	30	20	10	65	272
Sr	80	<10	60	740	1530
Cl	<50	100	<50	---	---
F	140	170	180	---	---
V	210	200	190	120	123
Co	62	100	83	77	37
Sc	40	20	12	15	17
Ni	570	1800	1200	1050	420
Cr	1380	3950	3430	1100	580
Y	10	<10	<10	22	27
Zr	<10	<10	<10	250	922
Ba	600	<10	70	1000	5120
La	0.5	3.0	0.7	150	250
Ce	5	8	7	200	400
Nd	<3	3	<3	85	207
Sm	1.0	0.3	0.3	13	24
Eu	0.76	0.47	0.53	3.0	4.8
Tb	0.36	0.10	0.18	1.0	1.4
Dy	1.7	0.7	0.5	5.0	6.3
Yb	1.0	0.34	0.34	1.2	1.7
Lu	0.13	0.05	0.05	0.16	0.23
Hf	0.2	0.4	<0.2	7	39
Ta	<0.5	<0.5	<0.5	9	47
Th	<0.2	<0.2	<0.2	16	46
U	0.1	0.1	0.1	3.1	4.9

*K = Kimberlite; L = Lamproite (Bergman, 1987).

Note: For key to column headings, see Table 1. See text for analytic techniques.

and depleted large ion lithophile element (LILE) contents (Table 2) demonstrate a lack of kimberlite or lamproite affinity; rather, all are characteristic of ophiolitic peridotites. Light rare-earth element contents of the Pamali Breccia are several orders of magnitude below those of typical kimberlites or lamproites (Fig. 6). Although these patterns are not smooth, the detection limits for La, Ce, Sm, Eu, Yb, and Lu using this INA technique (see above) are 0.5 to 1 order of magnitude below the concentrations determined. Some type of surface fractionation related to alteration may alternatively explain the anomalous patterns.

Despite the relatively uniform major element character of the Pamali Breccia and its fractions, several minor and trace elements are extremely variable in their abundances. For instance, K, Sr, and Ba contents vary by one to two orders of magnitude, and while several elemental ratios are highly variable (K/Rb = 17–230, La/Lu = 4–60, Ti/Ba = 4–>84), others are remarkably uniform (Cr/Ni = 2.2–2.9). Distinct differences exist in compositional ratios between the breccia and kimberlites or lamproites; for instance, Pamali Breccia fractions possess Cr/Ni, Rb/Sr, and Th/U ratios of $\approx$2, 0.2 to >2, and <2, respectively, whereas these other rocks are characterized by respective ratios of $\approx$1, 0.1 $\pm$ 0.2, and 5 to 10. Therefore, Pamali Breccia geochemical data indicate an ophiolite affinity rather than a kimberlite or lamproite affinity.

Petrography

In thin section, the Pamali Breccia (Ahim River exposure) consists of well-rounded to subangular clasts of six lithologies, the most common being serpentinite and pyroxenite (Fig. 7). The matrix is largely formed by polycrystalline calcite; ultramafic matrix material is only rarely observed. The serpentinization textures of the Pamali Breccia clasts are similar to those typically observed in ophiolites. Pyroxenite, norite, chert, and gabbro clasts are sometimes extremely fresh. An anomalous "hornblende" was reported by Koolhoven (1935), but we were unable to locate this phase. The only abundant sheet silicates observed in the matrix or clasts of the breccia were chlorite and serpentine; no phlogopite or biotite was recognized. Whereas chromite occurs as small matrix grains and as larger clasts, its texture is typical of chromite seams in ophiolites.

DISCUSSION

Several workers have discussed the origin of ultramafic breccias apparently similar to the Pamali Breccia. Some of the more notable occurrences of ultramafic breccias include eastern Papua–New Guinea (Musa Valley: Green, 1961), the Philippines (Zambales Range: Gonzales and others, 1957), Cuba (Camaguey district: Flint and others, 1948); Cyprus (Troodos Range: Wilson, 1959), and Newfoundland (Bay of Islands: Smith, 1953). Despite some similarities, the Pamali Breccia is generally distinct from these other occurrences in texture, lithology, and local contact relations.

As Green (1961) noted, the origin of these ultramafic breccias may be related to tectonic, volcanic, plutonic, sedimentary, talus brecciation, or solution brecciation mechanisms. Most of these can be eliminated at the outset for the Pamali Breccia because of a lack of substantiating data; only the three most likely petrogenetic hypotheses are discussed here: (1) That the Pamali Breccia is of kimberlite or lamproite affinity and represents a faulted portion of a pipe whose near-surface portions have been reworked by surface sedimentary processes. The diamonds could be derived from the upper mantle or from more shallow diamondiferous sedimentary rocks. (2) That the Pamali Breccia is a sedimentary conglomerate that was deposited on a sheared ophiolite. The clasts are of local ophiolite parentage, and the diamonds are alluvial, derived from the adjacent ophiolite or some more conventional kimberlite or lamproite source. (3) That the Pamali Breccia is an ultramafic tectonic breccia of fault-related origin that contains diamonds as a result of contamination by near-surface rocks or mantle derivation.

Several features argue against the view that the Pamali Breccia is related to kimberlites or lamproites:

1. The major and trace element geochemistry of the Pamali Breccia, despite its ultramafic character, is atypical of kimberlites or lamproites because the breccia possesses LILE contents several orders of magnitude lower than kimberlites and lamproites. Whereas many of these elements are extremely mobile in a

typical rain forest, near-surface environment (such as K, Na, Li, Rb, Sr, Ca, U, and Th), some of the more immobile LIL and HFS (high field strength) elements that are typically enriched in kimberlites and related rocks (such as P, Ti, REE, Hf, Zr, etc.) are severely depleted in Pamali Breccia.

2. Whereas sedimentary-like textures occur in the near-surface crater facies or extrusive portions of some kimberlites (Dawson, 1980) and lamproites (Bergman, in press), these are totally different than those observed in the breccia.

3. Although generally ultramafic, the mineralogic character of the Pamali Breccia is different from typical kimberlites or lamproites because phlogopite, ilmenite, pyrope, or K-rich phases, so commonly found in kimberlites and related rocks, are absent in the breccia.

4. Instead of the Pamali Breccia being floored by a fault as Koolhoven (1935) suggested, we propose, on the basis of the contacts exposed at the northwest margin, that it was deposited on a previously sheared erosional surface of the Bobaris ophiolite. It is possible—although extremely improbable—that the Pamali Breccia represents a heretofore-unreported primary occurrence of diamonds and member of the kimberlite-lamproite clan.

Many features displayed by the Pamali Breccia support the hypothesis that the breccia represents a conglomerate with local Bobaris ophiolite parentage. These include:

- the shallow dipping uniform attitude and contact relations of the Pamali Breccia with the underlying Bobaris ophiolite;
- well-developed sedimentary structures including parallel laminations, graded bedding, and imbricated cobbles and pebbles;
- well-rounded to subangular clasts of local ophiolite provenance, none with an apparently exotic origin;
- the major and trace element geochemistry of the Pamali Breccia, typical of ophiolites; and
- the similarity of the Pamali Breccia clast population and the clast population of present-day bedloads from the Bor and Ahim Rivers.

In fact, no obvious features argue against this conglomerate hypothesis. A tectonic origin of the Pamali Breccia is supported only by the sheared nature of the underlying ophiolite and the fragmental and angular nature of the basal portion of the breccia. However, we interpret these relations to have resulted from tectonic activity that postdates the deposition of the breccia.

It is tempting to suggest that the Pamali Breccia represents a local variant of the Upper Cretaceous Manunggul Formation. The Manunggul contains both sedimentary and volcanic facies, the former of which predominates in the western exposures and near the base of the section, and the latter of which predominates in the east and in the upper part of the section. The sedimentary facies include stratified conglomerate, sandstone, marl, mudstone, and limestone lithologies, whereas the volcanic facies are predominantly breccias, tuffs, lavas, and other felsic to intermediate eruptive rocks. This hypothesis, however, requires much more detailed stratigraphic and geochronologic investigation. Stracke (1973) and Koolhoven (1935) noted that streams generally become diamond-bearing after crossing the Manunggul Formation. The basal conglomerate member of the Manunggul Formation was typically diamondiferous where it overlies the Bobaris ophiolite but barren where it overlies the Meratus Range (Stracke, 1937).

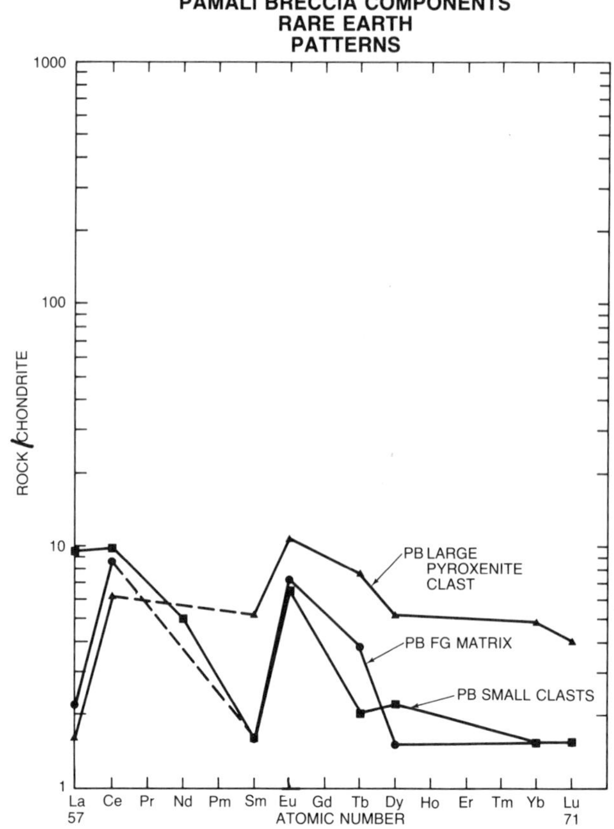

Figure 6. Chondrite-normalized rare-earth element patterns of three Pamali Breccia samples or fractions analyzed in this work.

As noted above, there has been an increasing number of reports of diamonds associated with alpine-type ultramafic rocks (such as those from China, the Urals, and Morocco: e.g., Nixon and others, 1986). Whereas it is always possible that these diamonds may be multicycle and ultimately related to a kimberlite or lamproite source, there is a growing body of evidence in favor of an alpine-type peridotite source rock for some diamonds. Until evidence to the contrary is advanced, we prefer a model in which the diamonds of the Pamali Breccia are derived from the underlying Bobaris ophiolite. This model requires that diamonds contained in ophiolites must behave in a metastable manner during their relatively slow (compared with kimberlites), but high-temperature, ascent from the upper mantle. In this regard, the experimental work by Cull and Meyer (1986) and that in other similar studies is of paramount importance.

a

b

Figure 7. Photomicrographs of Pamali Breccia (Ahim River exposure). a, Well-rounded pyroxenite clasts. b, Serpentinite clasts in calcite-rich matrix. Crossed polars; width of field, 1 cm.

CONCLUSION

On the basis of field relations, geochemistry, and petrology, the Pamali Breccia most likely represents a local fluvial/alluvial sedimentary conglomeratic deposit that was derived from the erosion of the adjacent Bobaris ophiolite and may represent a local facies of the Manunggul Formation. The diamonds in the breccia are therefore alluvial in origin, and the primary source of the southeast Kalimantan diamonds remains an enigma. Although these findings have been disappointing for some explorationists, we hope that they will remove some of the mysteries surrounding "anomalous" diamond occurrences in nature.

At this point, it seems invaluable to recall the contention of H. H. Hess (1955, p. 394) with regard to distinguishing between alpine serpentinites (ophiolites) and other members of the peridotite clan:

> Alnoites, kimberlites, mica peridotites, and related rock types can hardly be confused with alpine ultramafics because they differ conspicuously in mineralogy, chemistry, and tectonic setting. Considering the superfluity of names with which petrology is burdened it is most unfortunate that some name other than mica peridotite is not used for this dike rock. Gross errors have probably resulted from applying conclusions drawn from facts related to mica peridotites to alpine peridotites and vice versa. If mica peridotite had been called humptydumptyite, these probably would not have arisen.

ACKNOWLEDGMENTS

This study was supported by the Anaconda Minerals Company (a former division of the Atlantic Richfield Corporation) in conjunction with the exploration for and research on the primary source(s) of diamonds in Kalimantan. We appreciate the field assistance provided by the staff of P. T. Aneka Tambang (Banjarbaru office, southern Kalimantan, especially Augustinus Pasalli and Budi Susilo, and the Jakarta office, especially R. Tobing and J. Dahlan) and several local inhabitants of the Riam Kanan area. We thank Mike Covey (Exxon Research), Joe McGowen (ARCO Research), Maggie B. Mersmann, and Lucy N. Busby-Spera for their advice on the sedimentologic interpretations of hand-specimens of the Pamali Breccia. H.O.A. Meyer, I.S.E. McCallum, I. Munoz, and J. Pasteris provided constructive reviews for which we are most appreciative. The manuscript was typed by Sue Epperson.

REFERENCES CITED

Atkinson, W. J., Hughes, F. E., and Smith, C. B., 1984, A review of the kimberlitic rocks of western Australia, *in* Kornprobst, J., ed., Kimberlites and related rocks: Development in Petrology, Amsterdam, Elsevier, v. 9, p. 195–225.

Bardet, M. G., 1977, Geologie du Diamant, Bureau de Recherches Geologiques et Minères, Paris, Memoir 83, v. 3.

Bergman, S. C., 1987, Lamproites and other potassium-rich igneous rocks; A review of their occurrence, mineralogy, and geochemistry, *in* Fitton, J. G., and Upton, B. F., eds., Alkaline igneous rocks, Geological Society of London Special Paper (in press).

Burgath, K. P., and Wilher Simandjuntak, H. R., 1983, Investigation of ophiolite suites and their mineral possibilities in southeast Kalimantan (Meratus and Bobaris ophiolite zones): Bandung, Directorate of Mineral Resources, Field Activity Report 24.9.1983-30.11.1983.

Carmichael, I.S.E., Turner, F. J., and Verhoogen, J., 1974, Igneous petrology: New York, McGraw-Hill Book Co., 739 p.

Coleman, R. G., 1977, Ophiolites; Ancient oceanic lithosphere?: Berlin, Springer-Verlag, 229 p.

Cull, F. A., and Meyer, H.O.A., 1986, Oxidation of diamond at high temperature and 1 atm. total pressure with controlled oxygen fugacity. Fourth International Kimberlite Conference, Extended Abstracts. Geological Society of Australia, Abstracts, no. 16, p. 377–379.

Dawson, J. B., 1967, A review of the geology of kimberlite, *in* Wyllie, P. J., ed., Ultramafic and related rocks: New York, John Wiley & Sons, p. 241–251.

—— , 1979, New aspects of diamond geology, *in* Field, J. E., ed., Diamonds: New York, Academic Press, p. 539–554.

—— , 1980, Kimberlites and their xenoliths: Berlin, Springer-Verlag, 252 p.

Fan, C., and Bai, W., 1981, The discovery of Alpine-type diamond-bearing ultrabasic intrusion in Zizang (Tibet): Geological Review, v. 27, no. 35, p. 455–457 (in Chinese).

Flint, D. E., de Albear, J. F., Guild, P. W., 1948, Geology and chromite deposits of the Camaguey District, Camaguey Province, Cuba: U.S. Geological Survey Bulletin 954B, p. 39–62.

Gascuel, 1901, Les gisements diamantiferes de Borneo: Annales des Mines, v. 20, p. 5–23 (in French).

Geological Survey of Indonesia, 1970, Geologic map of southeast Kalimantan, Bandung.

Gisolf, W. F., 1923, On the occurrence of diamond as an accessory mineral in olivine- and anorthite-bearing bombs, occurring in basaltic lava, ejected by the volcano Gunung Ruang (Sangir-Archipelago, north of the Celebes): Proceedings, Koninklijke Nederlandse Akademie van Wetenschappen, 31, 78, p. 510–512 (in Dutch).

Gonzales, M. L., Peoples, J. W., Fernandez, N. S., and Victorio, V., 1957, The ultramafic and mafic rocks of Zambales Range, Luzon, Philippines: Geological Society of America Bulletin, v. 68, p. 1736.

Green, D. H., 1961, Ultramafic breccias from the Musa Valley, E. Papua: Geological Magazine, v. 98, p. 1–26.

Halewijn, M. J., 1824, Description of the diamond mines at Sungei Runti (Bandjermasin) in 1824: T.V.N.I. 1838, pt. II, p. 81–84 (in Dutch).

Hamilton, W., 1979, Tectonics of the Indonesian region: U.S. Geological Survey Professional Paper 1078.

Hess, H. H., 1938, A primary peridotite magma: American Journal of Science, 5th ser., v. 35, p. 321–344.

—— , 1955, Serpentinites, orogeny, and epeirogeny: Geological Society of America Special Paper 62, p. 391–408.

Hooze, J. A., 1893, Topographical, geological, mineralogical, and technical descriptions of part of the district of Martapura (south and east Borneo): Jaarboek Mijnwesen, p. 5 (in Dutch).

Horner, L., 1837, Report on a geological exploration in the southeastern part of Borneo: Verhandlingen Bat. Gen., v. 17, p. 87 (in Dutch).

Hovig, P., 1930, Diamonds in central Borneo: Jaarboek Mijnwezen (1930),

Hutchinson, C. S., 1975, Ophiolite in southeast Asia: Geological Society of America Bulletin, v. 86, p. 797–806.

Jaques, A. L., Lewis, J. D., Smith, C. B., Gregory, G. P., Ferguson, J., Chappell, B. W., and McCulloch, M. T., 1984, The diamond-bearing ultrapotassic (lamproitic) rocks of the west Kimberley region, western Australia, *in* Kornprobst, J., ed., Kimberlites and related rocks; Developments in petrology, v. 9: Amsterdam, Elsevier, p. 225–255.

Kaminskiy, F. V., 1980, The authenticity and regularity of occurrences of diamonds in alkaline-basaltoid and ultrabasic (other than kimberlite) rock: Vsesoyuznoye Mineralogicheskoye Obshchestvo, Zapiski, Leningrad, v. 109, no. 4, p. 488–493 (in Russian).

Kaminskiy, F. V., and Vagnov, V. I., 1976, Petrological conditions for diamond occurrence in Alpine-type ultrabasic rocks: International Geology Review, v. 19, no. 10, p. 1151–1162.

Kaminskiy, F. V., Klyuyev, Y. A., Konstantinovskiy, A. A., Piotrovskiy, S. V., Sochneva, E. G., and Yuzhakov, V. M., 1974, Evidence of diamonds in alkaline basaltoids in the northern part of the Russian platform: Doklady Akademia Nauk SSSR, Earth Sciences Section, v. 222, p. 939–941.

Koolhoven, W.C.B., 1935, The primary occurrence of South Borneo diamonds: Geologisch-Mijnbouwkundig Genootschap voor Nederland en Kolonien, verhandelingen, Geologische serie 9, p. 189–232 (in Dutch).

Krol, L. H., 1919, The diamonds of Borneo, their deposits and workings. Amsterdam, De Mijningenieur, p. 707–709.

——, 1920, Contribution to the knowledge of the origin and the dispersion of the diamond-bearing deposits of S.E. Borneo: Jaarboek van het Mijnwezen 1920, p. 250–304 (in Dutch).

Kukharenko, A. A., 1955, The diamonds of the Ural Mountains: Gosgeoltekhizdat (in Russian).

Legrand, J., 1980, Diamonds; myth, magic and reality: New York, Crown Publishers.

Luk'yanova, L. I., Smirnov, Yu. D., Zil'berman, A. M., Chernysheva, E. M., and Mikhailovskaya, L. N., 1978, Finds of diamonds in the picrites of the Urals: Vsesoyuznoye Mineralogicheskoye Obshchestvo, Zapiski, Leningrad, v. 107, p. 580–585 (in Russian).

Madigan, R., 1983, Diamond exploration in Australia: Indiaqua, v. 35, p. 27–38.

Meyer, H.O.A., 1985, Genesis of diamond; A mantle saga: American Mineralogist, v. 70, p. 344–356.

Nixon, P. H., Davies, G. R., Slodkevich, V. V., and Bergman, S. C., 1986, Graphite pseudomorphs after diamond in the eclogite-peridotite massif of Beni Bousera, Morocco, and a review of anomalous diamond occurrences. Fourth International Kimberlite Conference, Extended Abstracts. Geological Society of Australia, Abstracts, no. 16, p. 412–414.

Orlov, Y. L., 1977, The mineralogy of diamond: New York, John Wiley & Sons.

Robinson, D. N., 1978, The characteristics of natural diamond and their interpretation: Minerals Science and Engineering, v. 10, p. 55–72.

Schwaner, C.A.L.M., 1853, Borneo, pt. I (in Dutch).

Shestopalov, M. F., 1938, The ultrabasic mass of the Kitoisk Alps of Eastern Sayan and the related deposits, Tr: TSNIL Kamnei-samotsvetov Gostresta "Russkie Samotsvety."

Slodkevich, V. V., 1982, Graphite octahedra at Beni Boucera, N. Morocco, associated with garnet/clinopyroxene layered complex: Vsesoyuznoye Mineralogicheskoye Obshchestvo. Zapiski, Leningrad, v. 111, no. 1, p. 13–33 (in Russian; for translation, see International Geology Review, v. 25, no. 5, p. 497–514).

Sobolev, V. S., 1951, Geology of the diamond fields of Africa, Australia, Borneo, and North America: Moscow, Geogeoltekhizdat (in Russian).

Stracke, K. J., 1973, Report on the diamond potential of the P. N. Aneka Tambang diamond mining unit, Kalimantan Selatan, Indonesia: Unpublished report to P. N. Aneka Tambang (Bandung).

Trofimov, V. S., 1939, Diamond-bearing bed rock other than kimberlites: Soviet Geologiya, v. 9, no. 4–5, p. 40–59 (in Russian).

——, 1984, Diamond concentration in crustal structures: Doklady Academia Nauk SSSR, Earth Science Section, v. 273, no. 1–6, p. 78–81.

Twelvetrees, W. H., 1914, The Ball Hill Osmiridium field: Tasmania Geological Survey Bulletin, v. 17.

Van Bemmelen, R. W., 1970, The geology of Indonesia: The Hague, Martinus Nijhoff, 2nd ed., 2 vols.

Van der Wegen, G., 1963, The geology of the Island of Waigeo: Geologie en Mijnbouw, v. 42, p. 3–12 (in Dutch).

Verbeek, R.D.M., 1875, The geological description of the subdistricts Riam Kiwa and Riam Kanan in the south and east districts of Borneo: Jaarbuch Mijnwesen, v. I (in Dutch).

Wilson, A. N., 1982, Diamonds; From birth to eternity. Santa Monica, California, Gemological Institute of America, 450 p.

Wilson, R.A.M., 1959, Geology of the Xeros-Troodos area: Geological Survey of Cyprus, Memoir 1, p. 1–135.

Wing Easton, N., 1904, Geological synopsis of west Borneo; Differences and resemblances with central and southeastern Borneo: Report Geological Section Netherlands, Geologisch Mijnbouwkundig Genootschap. I 1912–1914, p. 179 (in Dutch).

——, 1925, The origin of the diamond: Amsterdam, De Mijningenieur, p. 44 (in Dutch).

Witkamp, H., 1932, Diamond deposits of Landak: Amsterdam, De Mijningenieur, v. 3, no. 13, p. 43–55 (in Dutch).

Manuscript Accepted by the Society October 15, 1986

Printed in U.S.A.

Geological Society of America
Special Paper 215
1987

A note on newly discovered kimberlites in Riley County, Kansas

William L. Mansker
INEX, 8704 Gutierrez N.E., Albuquerque, New Mexico 87111
Bill D. Richards
North Idaho College, 1000 West Garden Avenue, Coeur D'Alene, Idaho 83814
George P. Cole
Cominco American Incorporated, 831 East Glendale, Sparks, Nevada 89431

ABSTRACT

Five kimberlites have been discovered recently in Riley County, Kansas, within an area of six previously known kimberlites. Only one of the new kimberlites crops out; the others are covered by glacial and Recent colluvium.

The new kimberlites exhibit petrographic and magnetic similarities to the known kimberlites. Tuffaceous, crater-facies kimberlites (e.g., Fancy Creek, perhaps Baldwin Creek) are typically single phase, exhibit relatively low magnetic contrast (500–1,000 gammas), and tend to occur as topographic lows. Diatreme-facies kimberlites (e.g., Lonetree A, perhaps Lonetree B, perhaps Baldwin Creek) are brecciated and may be single or multiphase. They contain an abundance of crustal and mantle xenoliths, exhibit an intermediate magnetic contrast (2,000–3,000 gammas), and generally show slight positive relief beneath colluvial cover. Hypabyssal-facies kimberlites (e.g., Swede Creek) are single-phase, finer grained rocks containing smaller xenoliths and few kimberlite indicator minerals (e.g., pyrope garnet, Mg-ilmenite, chrome diopside); they generally exhibit slight to moderate positive relief, and are characterized by high magnetic contrast (3,500–6,000 gammas). Although the Baldwin Creek occurrence has not been confirmed by sampling, it is most likely of diatreme- or crater-facies, based on magnetic contrast comparisons.

Pervasive carbonatization and late-stage carbonate-magnetite alteration are characteristic of the diatreme- and hypabyssal-facies kimberlites. Fluid inclusion homogenization temperatures for late-stage carbonate veins yield relatively low temperatures (240–250°C).

Comparison of magnetic characteristics for most of the kimberlites in Riley County yields a correlation with kimberlite facies and may provide a technique for tentatively classifying other possible covered kimberlites.

INTRODUCTION

Kimberlites in Riley County, Kansas, were first discovered prior to 1920; new kimberlites have been recognized sporadically and studied since then. Prior to recent exploration efforts by mineral exploration companies, the most recent kimberlite discovery was in 1969, when Brookins (1970) recognized a circular feature in northern Riley County as a kimberlite crater (the Winkler crater). A chronology of kimberlite discoveries and investigations prior to 1969 was given by Brookins (1970). Although the general geologic setting and the accessibility are favorable for exploration, a relatively complex geomorphic terrain has overprinted the area. Kansan glaciation left residual effects in the form of outwash and colluvial cover in the middle to southern portions of the county and a northward-thickening veneer of till and colluvium. Consequently, kimberlite discoveries in Riley County have until now been limited to those occurrences exposed at the surface or those that manifest unusual topographic features (e.g., Winkler crater) in the glacial terrain.

Worldwide, kimberlites typically occur in clusters of numerous pipes, diatremes, and dikes. Therefore, it is not surprising that additional kimberlites have been found in the area. Through

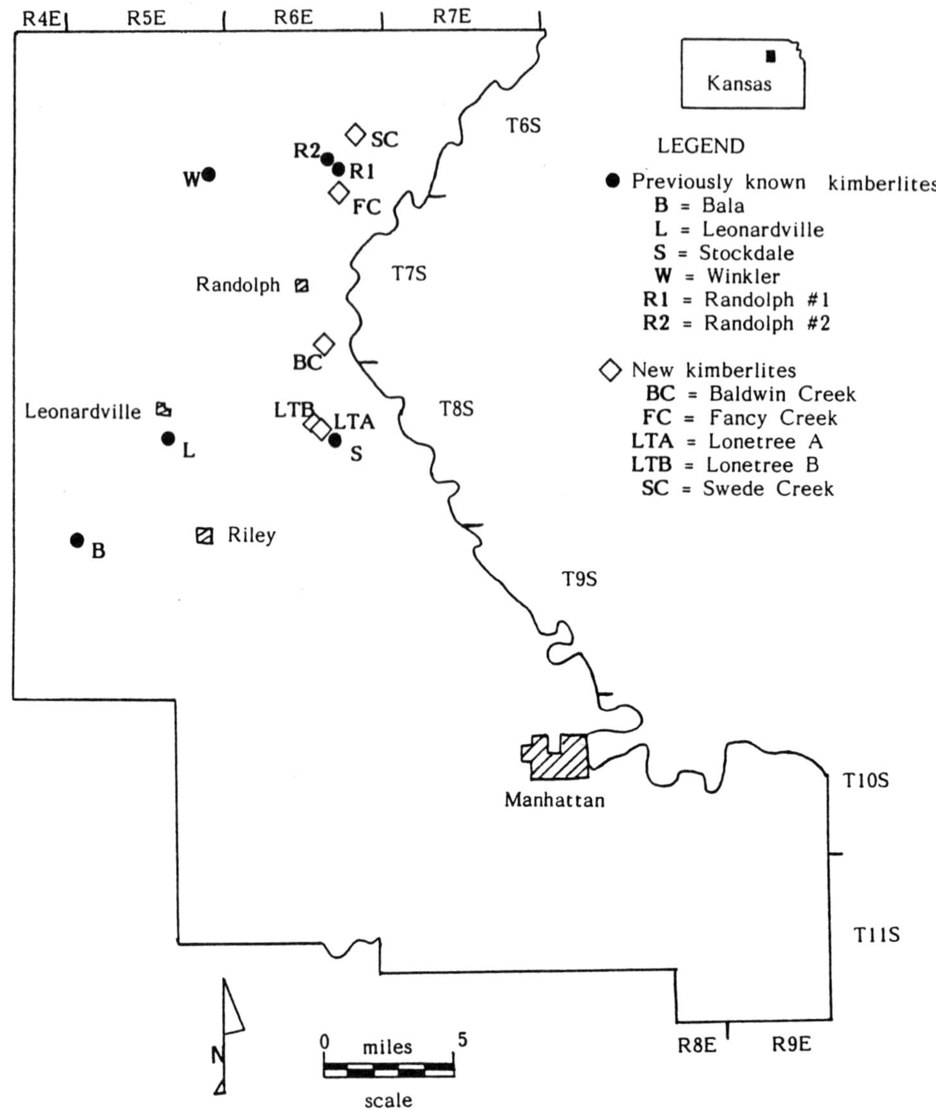

Figure 1. Map of Riley County, Kansas, showing locations of known and newly discovered kimberlites.

the efforts of mineral exploration groups, five new kimberlites have been discovered since 1979 (Fig. 1), and form the basis for this report.

GEOLOGIC SETTING AND AGE

The kimberlites in Riley County are situated along the southeast flank of the Abilene anticline–Irving syncline, although Winkler crater lies directly on the axis of the anticline. In a more regional setting, the kimberlite group is situated near the southeast flank of the Midcontinent Geophysical Anomaly (MGA). The MGA, underlain by Proterozoic mafic rocks (Steeples and Bickford, 1981), generally has been interpreted as an aborted, paleorift structure (see, for example, Goldich and others, 1961). Faults in the area are subtle features showing little vertical displacement. Mesoscopic (displacements about 1 m) post-late Paleozoic normal faults are apparent in several roadcuts throughout the area. Evidence of significant Cretaceous strike-slip displacement is apparent at certain localities within the county; where northwest-yielding, low-angle reverse faults show displacements of 10 m or greater (E. Wohler, personal communication). Isopach maps of various Paleozoic rock units indicate that some faults and fractures in the area persist at depth and may be related to basement structures. Exploration in the area has not determined whether kimberlite emplacement was directly fault-controlled.

The kimberlites are stratigraphically post–Lower Permian in age, intruding and enclosing xenoliths of limestones and shales of the Chase Group formations. Although the kimberlites have been dated radiometrically at 95 and 380 Ma (Brookins, 1970), the younger age value is in accordance with intrusive country rock relations. Cretaceous xenoliths have not been observed in any of the kimberlites, which may not be significant, as Cretaceous rocks were not deposited in the immediate area of the kimberlites (a topographic high during Cretaceous). Brookins (1970) has suggested an even younger age than Cretaceous, but no supporting evidence has been forthcoming. The kimberlites are definitely pre-Pleistocene in age, as demonstrated by the redistribution of kimberlite indicator minerals in glacial colluvium and areal drainages, and by glacial colluvial fill in Winkler and Fancy Creek craters.

DESCRIPTION

The following descriptions of the newly discovered kimberlites are based on local geologic setting, surficial expression, and preliminary hand-specimen and petrographic observations. Kimberlite classifications are tentative, pending more detailed petrographic study, and are descriptively based on discussions presented by Mitchell (1970), Skinner and Clement (1979), and Dawson (1980).

Of the five kimberlites, only Fancy Creek crops out. The other kimberlites are completely covered by glacial and Recent colluvium. Swede Creek, Lonetree A, and Lonetree B were exposed temporarily during exploration activities, then backfilled. The fifth occurrence, Baldwin Creek, has not been exposed, and no samples are presently available.

Swede Creek

Swede Creek is the northernmost kimberlite thus far discovered in Riley County. It is obscured by approximately 3 to 5 m of glacial till (minor) and colluvium, and no evidence of the kimberlite is observed on the surface. In subsurface exposure, the kimberlite exhibits positive relief and is calculated (from groundmagnetics and drilling) to have an apparent diameter of 67 m (220 ft), encompassing about 0.35 hectares (0.9 acres).

Hand-specimens of the rock are gray-green and medium to fine grained; contain relict, serpentinized, and oxidized olivine phenocrysts (or xenocrysts); and are moderately magnetic. No typical kimberlite indicator minerals and only a few small xenoliths are noted. A fine network of carbonate veins, some with limonitic stains, pervade the samples. Along the southernmost contact with shale and limestone country rock, a fine-grained, serpentine-rich skarn is present. On weathered surfaces, the rock exhibits a sandy appearance in response to differential weathering out of serpentinized and oxidized components.

Microscopically, the rock contains ubiquitous, rounded and euhedral, serpentinized and carbonatized pseudomorphs after olivine. The groundmass is well indurated with cloudy, mosaic carbonate, carbonate-magnetite veins, and magnetite-free carbonate veins. Magnetite is disseminated throughout the groundmass as anhedral specks and discrete anhedral and less common subhedral grains. Rare, rounded ilmenite macrocrysts consistently exhibit spongey magnetite reaction rims (Fig. 2). Magnetite is also commonly observed within grain boundaries of olivine replacement pseudomorphs, in association with carbonate veins, and rarely, as pure magnetite veins. Late-stage carbonate veins appear to be magnetite-free. Phlogopite is only a trace constituent, occurring as barely resolvable grains in proximity to some serpentine-carbonate pseudomorphs. Apatite is a minor constituent in the groundmass but is more abundant near the kimberlite-skarn contact. Rare carbonated laths occur in the groundmass, and are perhaps pseudomorphic after original melilite(?). Heavy mineral separates of the kimberlite contain fresh pyrope garnet, ilmenite, and rare chrome diopside in the –0.5-mm-size fraction, although none was observed in thin section. The Swede Creek kimberlite is similar to the Bala and Randolph 1 and 2 kimberlites described by Brookins (1970), is a nonmicaceous kimberlite, and is tentatively classified as an "opaque–mineral-rich" calcite-serpentine kimberlite (after Skinner and Clement, 1979). The kimberlite is considered to be of hypabyssal facies, as no diatreme or crater facies characteristics (e.g., brecciation, lapilli) are observed.

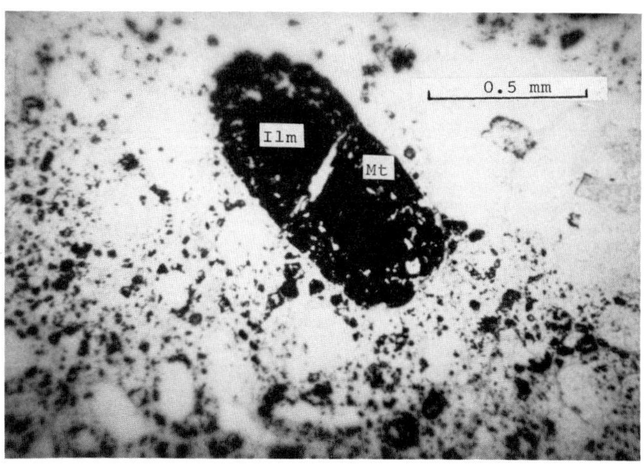

Figure 2. Magnetite reaction rim on ilmenite macrocryst, and view of the "opaque-rich" groundmass of Swede Creek kimberlite.

Lonetree A and B

The two Lonetree kimberlites are located approximately 1 km northwest of the Stockdale kimberlite (Fig. 1). Aligned along a northwest trend, the three intrusions likely share a common structural control at depth. Both Lonetree kimberlites are covered by postglacial colluvium and soil, and exhibit no anomalous topographic expression. A pond excavated in the drainage west of Lonetree A and south of Lonetree B contains decomposed kimberlite and abundant kimberlite indicator minerals. (It should be noted that Hanson (1969) attributed the abundance of indicators in this drainage to stream reversals associated with postglacial

Figure 3. Low-altitude, oblique view to southwest of Fancy Creek crater at approximately its widest point (230 m). Note crater rim of limestone and pond in center of crater.

drainage of the Stockdale kimberlite.) Lonetree A is covered by 1 to 3 m of colluvium, exhibits slightly positive relief in the subsurface, and intrudes thin beds of limestone and shale along the southeast contact. No pyrometasomatic effects are observed. Beneath the colluvium, the surface of the kimberlite is weathered to a depth of about 1 m. Coarse (up to 2 cm) carbonate veins appear to be late emplacement features. The intrusion is estimated to be 110 m in diameter, encompassing about 1.0 hectares (2.4 acres). Both megascopically and in thin section, Lonetree A is similar to the Stockdale kimberlite, containing abundant xenoliths and kimberlite minerals in a somewhat brecciated matrix. It differs from Stockdale in that carbonate veins in Lonetree A appear to be relatively magnetite-free, although magnetite is contained in the carbonated groundmass matrix. Lonetree B, located about 200 m northwest of Lonetree A, has an inferred diameter of about 10 m. Beneath a colluvial cover of 3 m, the kimberlite is decomposed to a yellow clay that contains residual kimberlite fragments, possible lapilli, and abundant kimberlite minerals. Rounded ilmenite and garnet megacrysts as large as 3 cm have been recovered from the kimberlite. The kimberlite fragments exhibit a brecciated appearance, laced with thin (1–2 mm) carbonate veins. No coarse carbonate veins have been observed in Lonetree B.

Lonetree A is diatreme facies, based on overall textural and petrographic observations similar to the diatreme facies Stockdale kimberlite. Although spatially associated with Lonetree A, Lonetree B exhibits textural differences. Lonetree B is tentatively classed as diatreme facies, pending a more detailed petrographic study. Confirmation of lapilli may indicate it to be of upper diatreme- or crater-facies. Both the Lonetree kimberlites are micaceous varieties, and in the scheme of Skinner and Clement (1979), they are tentatively classified as serpentine-calcite-(±phlogopite) kimberlites.

Abundant fluid inclusions populate the carbonate veins of the Lonetree and Stockdale kimberlites. Preliminary study of selected primary inclusions in the carbonate veins of Lonetree A and Stockdale kimberlites has yielded homogenization temperatures of 240 to 250°C.

Fancy Creek

Fancy Creek crater occurs about 1.5 km south of the Randolph no. 1 kimberlite, occupying a topographic low formed by a west-flowing drainage. The crater rim is elliptical in an east-west direction, 230 by 280 m, encompassing about 6.1 hectares (15 acres); it is breached on the west by the superposed drainage (Fig. 3). Well-exposed members of the Barnston Limestone Formation (i.e., Ft. Riley L.S., Oketo shale, Florence L.S.) form the rim and inward tapering walls of the crater. Pyrometasomatic effects are not apparent at exposed intrusive contacts. The actual kimberlite exposure is smaller than the crater outline, as determined by tracing the limits of kimberlite indicator minerals within the crater (Fig. 4), and occupies 3.5 to 4.0 hectares (9 to 10 acres). At least 75 percent of the kimberlite is variably covered (2 m or less) by silty, red-brown postglacial colluvium and soil. Because this colluvial remnant, as well as the kimberlite, occupies a topographic low, it may be inferred that the crater was partially excavated prior to, or during, Kansan glaciation. This observation is corroborated by numerous cut-and-fill channels observed beneath the colluvial cover, exposed during exploration. A pond excavated in the crater is totally within kimberlite and the pond dam is awash with residual xenoliths and kimberlite indicator minerals.

Fancy Creek kimberlite is similar in many respects to the Winkler kimberlite. Both occur as breached, partially eroded craters and are relatively large anomalous topographic features. Both occurrences appear to be highly brecciated and comminuted tuffaceous kimberlite of crater facies. Pyroclastic layering observed in the Winkler kimberlite is not apparent in Fancy Creek. Fancy Creek has a greater abundance of more deeply derived (crustal and upper mantle) xenoliths. These differences indicate that Fancy Creek may represent a slightly lower-level crater facies than Winkler.

The Fancy Creek xenolith suite is dominated by subangular to rounded Paleozoic rock fragments up to 0.5 m in size. Basement and crustal xenoliths are generally rounded, as large as 8 to 10 cm, and include granite and diorite (less common), slight to moderately metasomatized metagabbros, amphibolites, and basalts. One garnet peridotite xenolith, 12 cm across, has been recovered (Fig. 5). It is extensively altered but contains fresh pyrope garnets and chrome diopsides. Major mineral constituents of the kimberlite are ilmenite, garnet, chrome diopside, and phlogopite. Olivine, which was a major primary constituent, is completely serpentinized and carbonatized, and contains some minerals that have been altered to clay. Similar to Winkler kimberlite, Fancy Creek kimberlite is a micaceous variety, although—typical of crater facies kimberlites—the degree of

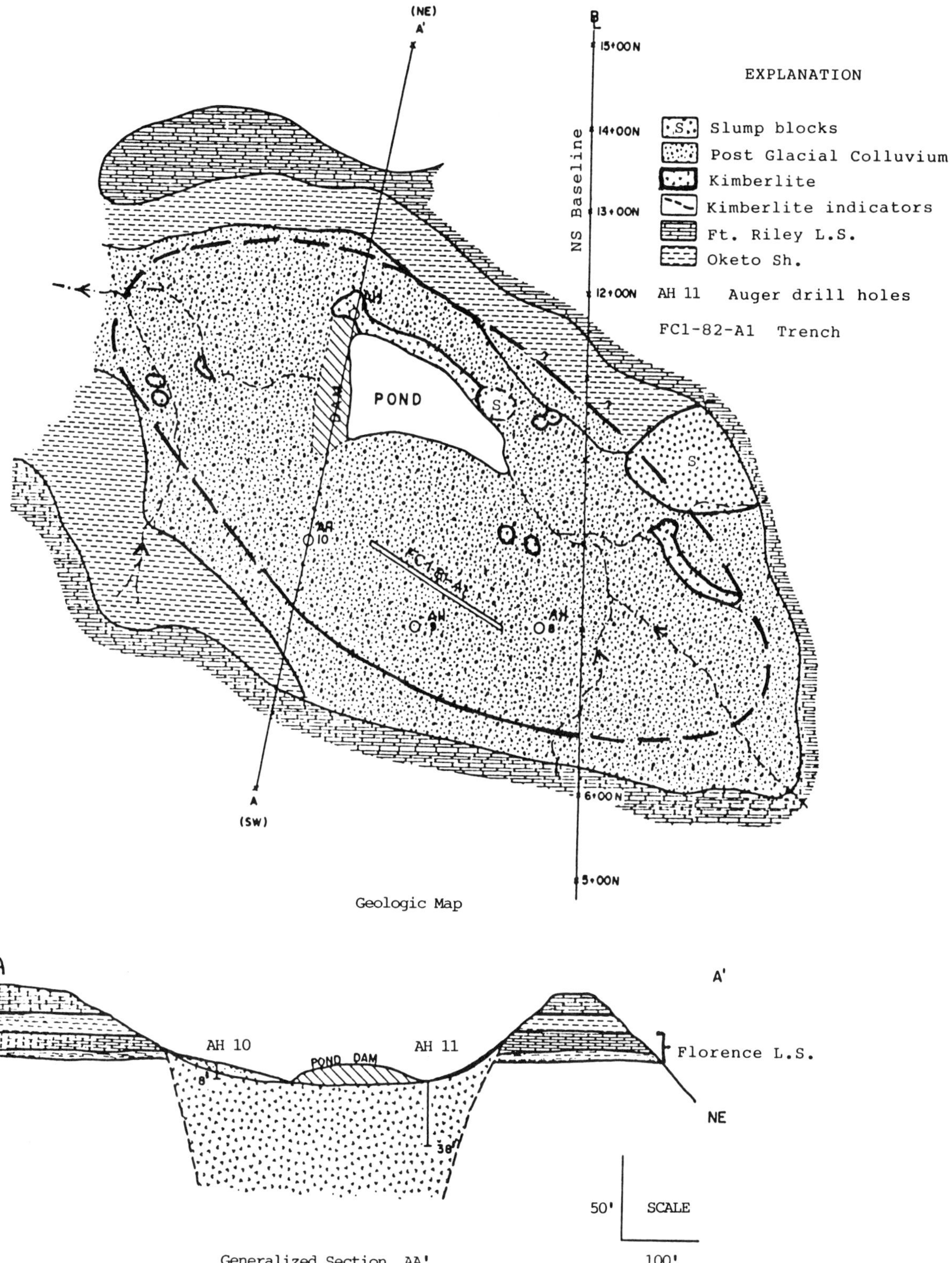

Figure 4. Generalized geologic map and cross section (A-A′) of Fancy Creek kimberlite. Note that limit of kimberlite indicators (dashed line) is approximately outline of kimberlite.

Figure 5. Close-up view of altered garnet peridotite xenolith (approximately 12 cm across) in place in Fancy Creek kimberlite. Note coin to indicate scale.

decomposition makes it difficult to place in the classification of Skinner and Clements (1979).

Baldwin Creek

The Baldwin Creek occurrence has not been sampled to confirm its kimberlitic nature. It is completely obscured by Baldwin Creek flood plain alluvium. The geophysical anomaly is a "predicted" kimberlite, based on comparison of its magnetic signature with that of known kimberlites in Riley County. Furthermore, magnetic intensity and contrast of the anomaly permit a tentative classification of diatreme- or possibly crater-facies. The size of the Baldwin Creek occurrence can be estimated from groundmagnetic data, indicating a northwest-elongated body about 100 m in its longest dimension, or two smaller bodies approximately 30 m apart.

MAGNETIC CHARACTERISTICS

All kimberlite occurrences in Riley County offer magnetic responses of varying magnitude and signature, and exhibit northwest-trending magnetic inclinations that dip to the southeast. Magnetic inclinations of the new kimberlites mimic those of previously known kimberlites (Brookins, 1970), substantiating kimberlite emplacement along northwest-trending zones of weakness throughout the area.

Integration of magnetic data with other attributes of the kimberlites (e.g., size, facies) permits some generalizations regarding evaluation of other kimberlites, yet undiscovered, which may not be exposed at the surface. Figure 6 illustrates relative aeromagnetic anomalies for the Lonetree-Stockdale cluster, Swede Creek, and Fancy Creek kimberlites. The illustration shows that

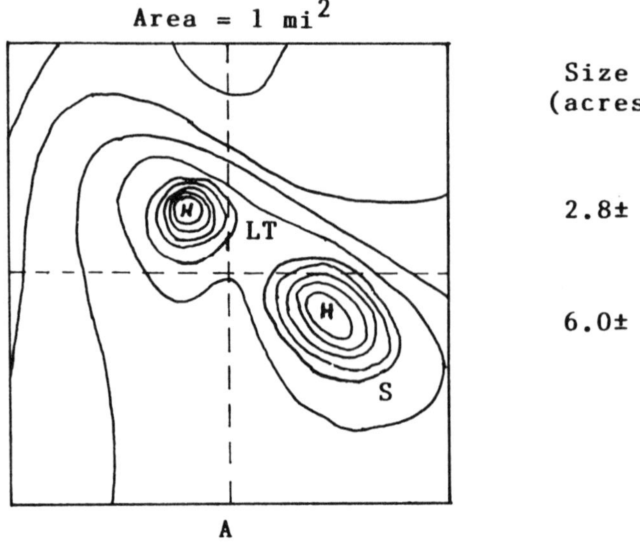

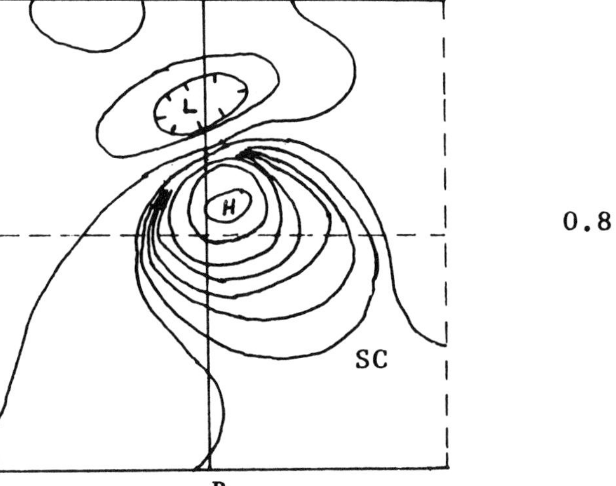

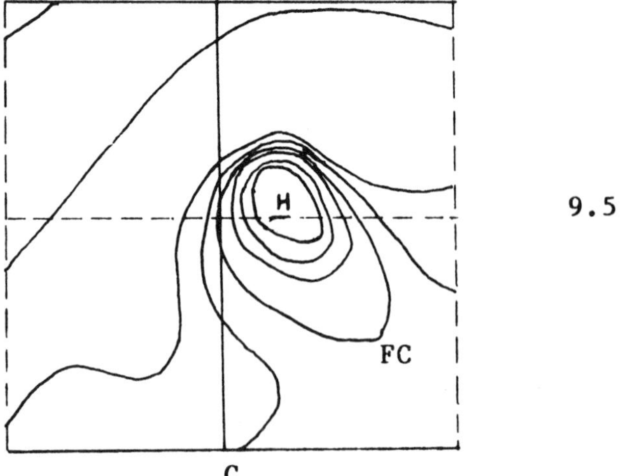

Figure 6. Comparison of aeromagnetic anomalies exhibited by several Riley County kimberlites, compared with actual kimberlite size. Kimberlite abbreviations same as in Figure 1.

TABLE 1. MAGNETIC CONTRAST OF KIMBERLITES IN RILEY COUNTY, KANSAS

Site	Groundmagnetic Data High-background (gammas)	Groundmagnetic Data East-West Distance (m)	Aeromagnetic Data High-background (gammas)	Aeromagnetic Data East-West Distance (m)	Size (acres)
Stockdale	3,000	76	60	402	6.0
Lonetree	2,168	152	60	302	2.8
Baldwin Creek	750	146	50	654	---
Fancy Creek	1,066	229	50	704	9.5
Randolph No. 1	6,321	98	30	905	0.9
Swede Creek	5,794	134	160	805	0.8
Winkler	500	274	60	352	18.5
Leonardville	2,000	137	60	704	>8.0
Bala*	>3,500	53	200	503	>9.0

*Groundmagnetic data for Bala from Dreyer (1947, Fig. 2).

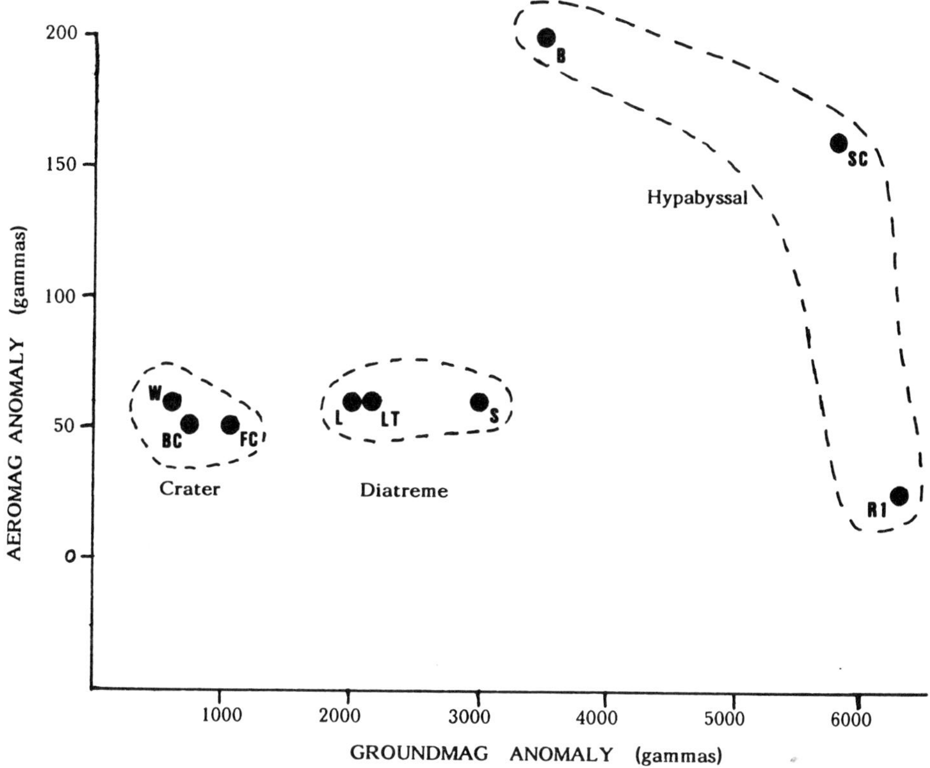

Figure 7. Graphic comparison of aeromagnetic and groundmagnetic anomalies, illustrating correlation with kimberlite facies. Kimberlite abbreviations same as in Figure 1.

the magnitude and extent of aeromagnetic anomalies does not necessarily indicate potential size of the kimberlite. Table 1 is a compilation of all aeromagnetic and groundmagnetic data derived for the kimberlites of Riley County. Comparison of the data indicates that groundmagnetic contrast of the kimberlites correlates well with kimberlite facies. Crater-facies kimberlites (e.g., Winkler, Fancy Creek, perhaps Baldwin Creek) yield the lowest contrast, with groundmagnetic intensities ranging from about 500 to 1000 gammas. Diatreme-facies kimberlites (e.g., Stockdale, Leonardville, Lonetree A and B) yield intermediate contrasts, with intensities ranging from about 2,000 to 3,000 gammas. Hypabyssal kimberlites (e.g., Bala, Randolph no. 1, Swede Creek) range from 3,500 to 6,300 gammas. The data also show that almost no correlation exists between the extent of the aeromagnetic anomalies (i.e., magnitude of east-west magnetic disturbance) and the actual size of the kimberlites. Extent of the groundmagnetic anomalies correlates somewhat better, but not precisely; this is primarily due to the asymmetry of some kimberlites (e.g., Leonardville and Bala). Figure 7 is a representation of the magnitude of aeromagnetic anomalies versus groundmagnetic anomalies for the kimberlites. Relatively close clustering of data is apparent for crater-facies and diatreme-facies kimberlites. Data for the hypabyssal kimberlites form an isolated, more widely scattered cluster. Numerous factors may be invoked to account for differences in magnetic contrast among the kimberlites. Mention of such factors is made here, but not discussed in detail. They include: (1) modal abundance of primary mafic minerals; (2) degree of serpentinization (olivine → serpentine + magnetite); (3) reduction-oxidation conditions prior to, during, and after (weathering) emplacement ($>fO_2 = >$magnetite); (4) size; (5) depth of cover; and (6) dilution via such factors as xenoliths, pyroclastic eruption, and epiclastic deposition.

SUMMARY

Five new kimberlites have been discovered in Riley County, Kansas. Four have been confirmed by sample investigation, and a fifth is "predicted" to be kimberlite, based on aeromagnetic and groundmagnetic characteristics. These geophysical characteristics and petrographic observations allow classification of the kimberlites as follows: (1) Swede Creek: hypabyssal-facies, "opaque-mineral–rich" calcite-serpentine kimberlite; (2) Lonetree A: diatreme-facies, serpentine-calcite (±phlogopite) kimberlite, micaceous variety; (3) Lonetree B: diatreme (possibly crater)-facies, serpentine-calcite kimberlite, micaceous variety; (4) Fancy Creek: crater-facies (kimberlite tuff), serpentine-calcite kimberlite (tentative), micaceous variety; (5) Baldwin Creek: diatreme- or crater-facies kimberlite (tentative).

Available data regarding magnetic characteristics of the kimberlites allow type-classification (e.g., whether hypabyssal-, diatreme-, or crater-facies), and may provide a technique for predicting kimberlite types should other kimberlites be discovered in the area.

ACKNOWLEDGMENTS

We thank D. G. Brookins (University of New Mexico); R. L. Cullers (Kansas State University); E. Wohler (Wamego, Kansas); P. Berendsen and D. Steeples (Kansas Geological Survey); L. L. Branch, D. E. Ladwig, M. B. Burrell, H. G. Coopersmith, and J. B. Lincoln (Cominco American, Inc.) for their expertise. Cominco American, Inc. is gratefully acknowledged for providing the opportunity to present the data and related information in this report. Appreciation is extended to the Department of Geology, University of New Mexico, for its analytic and petrographic facilities made available during this study. Special thanks are also due to Cecil, Miriam, and Cedric Pfaff (Randolph, Kansas) for their kind assistance during the exploration phase of this work.

REFERENCES CITED

Brookins, D. G., 1970, The kimberlites of Riley County, Kansas: Kansas Geological Survey Bulletin 200, 32 p.

Dawson, J. B., 1980, Kimberlites and their xenoliths: Berlin-Heidelberg, Springer-Verlag, 252 p.

Dreyer, R. M., 1947, Magnetic survey of the Bala intrusive, Riley County, Kansas: State Geological Survey of Kansas Bulletin 70, p. 21–28.

Goldich, S. S., Nier, A. D., Baadsgaard, H., Hoffman, J. H., and Krueger, H. W., 1961, The Precambrian geology and geochronology of Minnesota: Minnesota Geological Survey Bulletin 41, 193 p.

Hanson, C. G., 1969, Geochemical and mineralogical investigation of methods for detecting kimberlite in the area of the Stockdale kimberlite, Riley County, Kansas [M.S. thesis]: University Park, Pennsylvania State University 263 p.

Mitchell, R. H., 1970, Kimberlites and related rocks; A critical reappraisal: Journal of Geology, v. 78, p. 686–704.

Skinner, M.E.W., and Clement, R. C., 1979, Mineralogical classification of South African kimberlites, *in* Proceedings of the Second International Kimberlite Conference: American Geophysical Union, p. 129–139.

Steeples, D. W., and Bickford, M. E., 1981, Piggyback drilling in Kansas; An example for the Continental Scientific Drilling Program: EOS American Geophysical Union Transactions, v 62, p. 473–476.

MANUSCRIPT ACCEPTED BY THE SOCIETY OCTOBER 15, 1986

Geological Society of America
Special Paper 215
1987

Geology and petrography of the Twin Knobs #1 lamproite, Pike County, Arkansas

*Michael A. Waldman**
Long Lac Mineral Exploration (Texas), Inc., 2020 Airway Avenue, Fort Collins, Colorado 80524
*Tom E. McCandless**
Department of Geochemistry, University of Cape Town, Cape Town, Republic of South Africa 7700
*Hugo T. Dummett**
Nicor Mineral Ventures, 2341 South Friebus Avenue, Tucson, Arizona 85713

ABSTRACT

The Twin Knobs #1 diatreme (TK1), delineated in 1981, is located 3.8 km southeast of Murfreesboro, Pike County, Arkansas and is within the northeast-trending group of pipes, plugs, and dikes that comprise the Murfreesboro district. TK1 is an elongate, bilobate pipe, oriented north-south, and approximately 0.05 km^2 in surface area. The diatreme penetrated the Lower Cretaceous Trinity Formation, whereas the Upper Cretaceous Tokio Formation overlies portions of the pipe. These relationships suggest an emplacement age of early Late Cretaceous, which is contemporaneous with isotopic ages obtained at the Prairie Creek lamproite complex (97 to 106 Ma; Gogineni and others, 1978).

Rock types at TK1 include magmatic lamproites, sandy tuffs and breccias, and epiclastics. The magmatic breccias are porphyritic, with euhedral olivine, plus pyroxene and phlogopite phenocrysts set in a cryptocrystalline or glassy groundmass that also contains priderite, perovskite, and chromite. The sandy tuffs and epiclastics contain up to 30 percent quartz grains, in addition to lamproite and sedimentary rock fragments, set in a matrix of calcite, sericite, or fine-grained to cryptocrystalline glass.

The average diamond grade from bulk sampling of the phases is 0.17 carats per 100 metric tons. Diamond crystal habits include octahedra, tetrahexahedra, macles, and fragments. Megacrysts of picroilmenite, chrome spinel, clinopyroxene, orthopyroxene, garnet, and nodules of eclogite, peridotite, lherzolite, websterite, as well as crustal xenoliths, were recovered during sampling.

Whole-rock and trace element geochemistry strongly suggests that Twin Knobs #1 has more affinities to olivine lamproite than to kimberlite.

INTRODUCTION

The occurrence of an ultramafic body near Murfreesboro, Pike County, Arkansas, has been known since 1842, but it was not until 1889 that the Prairie Creek intrusion was classified as peridotite (Branner and Brackett, 1890). In 1906, a local farmer found two small diamonds on the surface of the Prairie Creek pipe; this discovery initiated diamond exploration in the district and the discovery of the smaller pipes and plugs to the northeast of Prairie Creek (Kimberlite Mine, American Mine, and Black Lick; Miser, 1913; Miser and Purdue, 1929). Diamond mining has been primarily confined to the Prairie Creek diatreme, which has been the only commercially operated diamond mine in the United States; in fact, until the discovery of the State Line kimberlites in Colorado and Wyoming, the diatreme was the only primary source rock of diamonds in the United States.

In 1979, Superior Minerals, as part of its diamond exploration program in the United States, began investigations of the Murfreesboro diatremes and exploration of the surrounding dis-

*Formerly with Superior Oil Company, Tucson, Arizona.

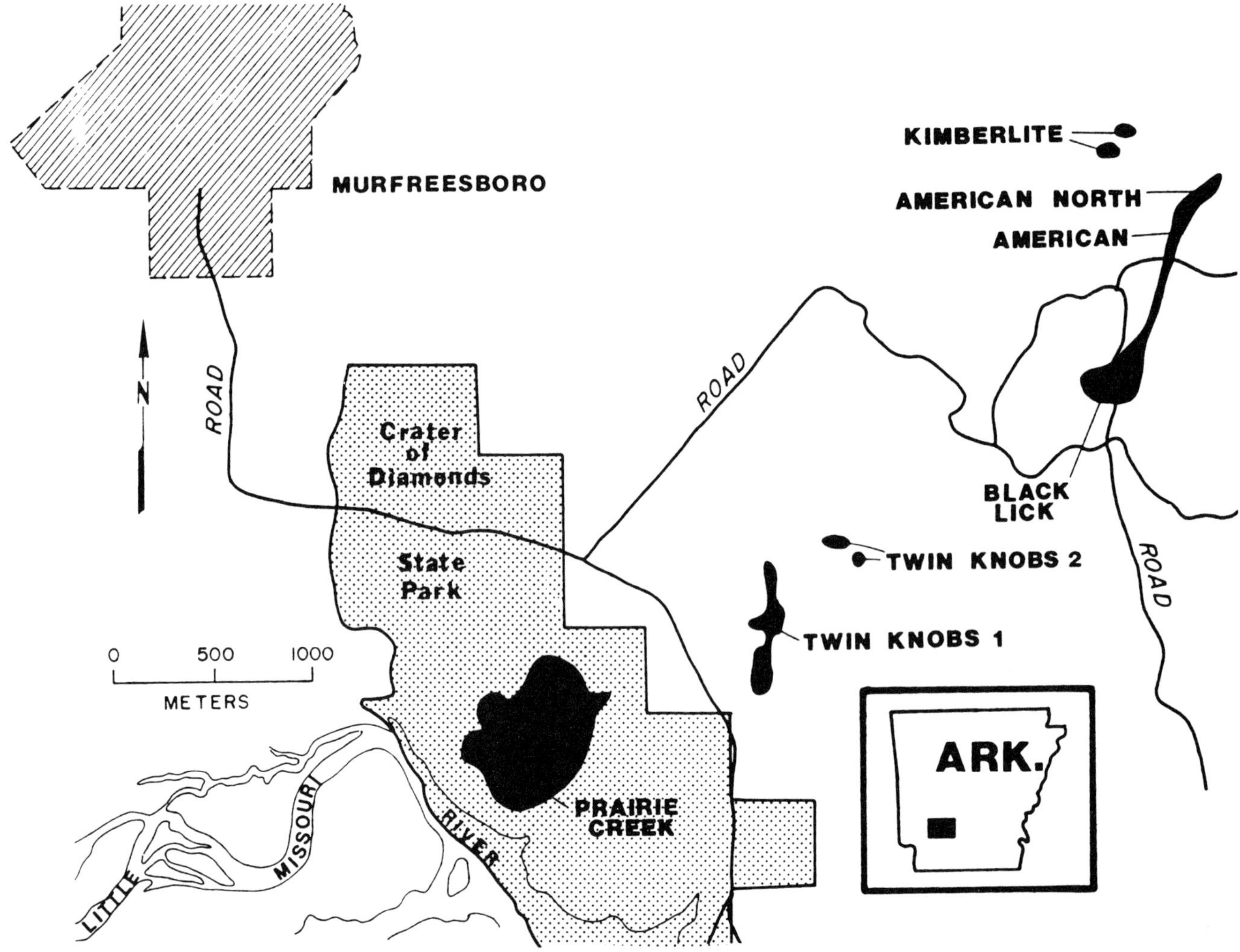

Figure 1. Location map of Twin Knobs #1 and adjacent lamproites. Extension of lamproite between Black Lick and American is inferred by ground magnetics.

trict for its diamond potential. Heavy mineral stream-sediment sampling, followed by detailed surveys with ground magnetometry and soil geochemistry, delineated the Twin Knobs #1 (TK1) pipe and two small satellite plugs to the northeast (Twin Knobs #2) in 1981 (Fig. 1). Miser, in 1912, had investigated waterlain "peridotite" tuffs found in a well on the old "Riley Place," which he suggested were derived from nearby vents (Ross and others, 1929). These peridotite tuffs may be tuffs on the southern portion of the TK1 pipe.

The ultramafic rocks of the Murfreesboro District were reclassified as kimberlite, because diamond occurred as an accessory mineral. However, many investigators recognized chemical and mineralogic differences between the classic kimberlites of South Africa and these diamond-bearing rocks of Arkansas (Bolivar, 1979; Gogineni and others, 1978; Lewis, 1977; Meyer, 1976; Mitchell and Lewis, 1983). The recent discovery of diamondiferous ultrapotassic rocks of the Kimberley region of Western Australia (Atkinson and others, 1984) and the petrologic work on the Murfreesboro District lamproites by Scott-Smith and Skinner (1984a, b) suggest that these rocks are, or at least have strong affinities to, olivine lamproites. The conclusion of our work is that TK1 also bears chemical and petrologic similarities to olivine lamproite.

REGIONAL SETTING

The Murfreesboro alkalic complex, which intrudes Gulf Coastal Plain sediments 6 to 8 km south of the boundary of the Ouachita Mountains Province, is composed of thick sequences of Paleozoic sandstones and shales that have been subjected to intense folding and thrust faulting. The Cretaceous-age Gulf Coastal Plain sediments consist of evaporites, limestones, claystones, sandstones, and coarse gravels, all gently dipping southward and unconformably overlying Paleozoic rocks. Detailed descriptions of the lithologies as well as structures in the region are found in Miser and Purdue (1929) and more recent works on the Ouachita Mountain and Gulf Coastal Plain regions.

Seismic data (Viele, 1979; Nelson and others, 1982) suggest that during the Carboniferous, there was a major continent-continent collision, analogous to the Himalayan and Appalachian types. The collision between the northward migrating Llanorian Plate and the southern margin of the North American Plate produced a thickened Paleozoic crust, as deep-water sediments in the Ouachita Basin were thrust and stacked over miogeosynclinal shelf sediments.

The predominant northeast trend of the Murfreesboro district diatremes parallels the trend of the long-active Mississippi Embayment rift system. This northeast trend is not evident in the Late Paleozoic rocks north of the Murfreesboro area, where the structural pattern is major east-west–trending thrust faults with associated minor, near-vertical, north-south–trending cross-faults. The TK1, Kimberlite, and the Black Lick–American dike complex are oriented north-south to north-northeast. The lamproite melt is thought to have risen along deep-seated structures associated with the Early Paleozoic rift system, but near-surface diatreme emplacement was controlled by the vertical, north-oriented cross-faults. It has also been suggested that Mesozoic plutonism in the region is due to extensional events within the rift system (Kane and others, 1981).

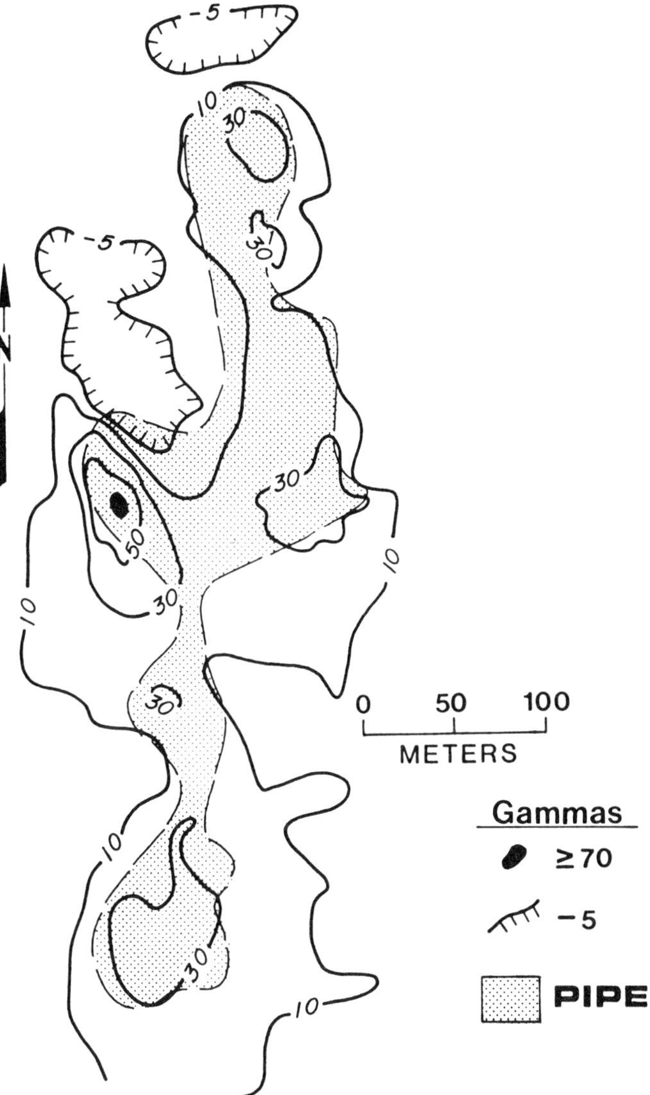

Figure 2. Ground magnetic survey map of Twin Knobs #1.

EXPLORATION METHODS AND RESULTS

Procedures used in exploring for diamondiferous diatremes in southwestern Arkansas consisted of closely spaced (200 m), low-level (100 m) airborne magnetometer and electromagnetic surveys (EM), in conjunction with stream sampling for kimberlitic indicator minerals (e.g., picroilmenite, chrome diopside, pyrope garnet, and chromite), followed by detailed ground geophysics and rock and soil geochemistry. TK1, as well as all known lamproites in the district, were detected by airborne magnetics and EM. TK2, too small to be detected by either, was located by ground magnetics. Orientation sampling for heavy minerals in streams draining the known lamproites, as well as heavy mineral studies of the lamproitic rocks, indicated a paucity of typical "kimberlitic" suite minerals. However, there were a sufficient number of indicator minerals in concentrates from stream silts to identify potential pipe locations.

Geophysical Surveys

Ground surveys with a Geometrics proton procession magnetometer and a Geonics EM-16 with an EM-16R resistometer were used successfully to help delineate the different phases of TK1 and its approximate boundaries. The use of gravity techniques did not delineate either, due to the minimal density contrast between the very abundant sandy tuffs and the country rock (Trinity Formation). Although the magnetic susceptibility contrast between all the Murfreesboro lamproites (except the hypabyssal phase at Prairie Creek) and the Trinity Formation is relatively low, detailed ground magnetics were useful in differentiating among the more magnetically susceptible breccias, the less magnetic sandy tuffs, and the nearly nonmagnetic Trinity Formation at TK1 (Fig. 2).

Soil Geochemistry

Soil sampling of the B horizon at TK1 was done on a 100- by 100-ft grid covering approximately 0.28 km^2 and totaling 304 samples. The –80 mesh-size fraction was analyzed for nickel, chromium, cobalt, and niobium by Skyline Labs (Tucson, Arizona), using standard acid digestion for AA (Ni, Cr, Co) and emission spectroscopy (Nb). These elements have proved to be sufficiently distinctive for detecting soil anomalies over ultramafic bodies, provided that background values have been determined for the surrounding country rocks and their soils. In southwestern Arkansas, soils developed over Trinity Formation and Tokio gravels average 14–17 ppm nickel, 10–25 ppm chromium, 12–15 ppm cobalt, and 10–20 ppm niobium. Threshold values for soils over TK1 are 95 ppm nickel, 225 ppm chromium, 17 ppm cobalt, and 23 ppm niobium.

The nickel dispersion pattern is well restricted and helped to define the pipe boundary (Fig. 3a). Chromium, on the other hand, tends to be more dispersed downslope to the west, particularly west of the primary breccias in the west-central part of the diatreme (Fig. 3b). Cobalt and niobium, although only very weakly anomalous, help to define the shape of the pipe (Fig. 3c,d).

Soil profiles are moderately well developed from depths of 0.5 m in the southern portion of the pipe to 2.3 m in the northern part. Soil profiles typically consist of organic material (0–10 cm) that overlies tan, sandy to silty clay (0.5–1.0 m) that overlies very weathered, black to green to red lamproite (1.0–2.5 m). Trace element values (Ni, Cr, Co, Nb) increase markedly below the contact of the tan, sandy-clay horizon (Fig. 4).

Bulk Testing

A total of 170.08 metric tons of olivine lamproite breccia and sandy, lamproite tuffs and breccias from five exploration pits were processed at the Superior Minerals diamond recovery plant at Sloan, 42 km north of Fort Collins, Colorado. Each pit was processed separately to determine grade variations of the different rock types. The plant circuit is designed to recover all stones larger than 0.5 mm via heavy media separation, grease table, jig bed, and wet sortex. Test diamonds from Prairie Creek lamproite and Sloan kimberlite were injected into the circuit periodically to ensure that the plant was operating at maximum effectiveness. All test stones, plus test density beads, were recovered and indicated a recovery rate of more than 95 percent.

Diamonds

Seventeen small diamonds (0.5–2.0 mm), weighing a total of 0.29 carats, were recovered, with 15 diamonds (0.26 ct, 88%) recovered from the olivine lamproite breccias. Two diamonds (0.03 ct, 12%) were recovered from sandy tuffs. Diamond crystal habits include octahedra (70%), tetrahexahedra (24%), and macles (6%). Distorted crystals are not common, although most showed broken surfaces. Light brown and white are the dominant colors. Octahedra exceed tetrahexahedra by a ratio of 3:1, which is in contrast to a predominance of tetrahexahedral forms at Prairie Creek (Robinson, 1980; T. E. McCandless, unpublished data). Hall and Smith (1984) reported abundant tetrahexahedra from the Western Australian lamproites, and a slight decrease in the ratio of tetrahexahedra to octahedra with decreasing size. The preponderence of resorbed tetrahexahedral forms is suggestive of a strongly corrosive environment.

GEOLOGY

The TK1 complex comprises a group of small plugs of magmatic olivine lamproite intruded into sandy lapilli tuffs and breccias, and is partially overlain by epiclastic crater facies rocks (Fig. 5). All the rocks observed are strongly altered and weathered, as exposures were limited to the exploration pits, which averaged 5 m in depth. In addition, the rock types tend to be gradational, with no sharp vertical or lateral contacts. The surface of the pipe is covered by a sandy clay soil that is light brown to dark brown and 0.1 to 2 m deep. In some cases the contact between the base of the soil horizon and weathered lamproite is marked by a thin, discontinuous gravel or cobble bed.

The magmatic olivine lamproite is light to dark greenish-black in color, and exhibits porphyritic textures due to the presence of macrocrystic olivine or its pseudomorphs. The magmatic lamproite grades quickly into breccia, with well-rounded xenoliths of Paleozoic shales, sandstones, and quartzites, as well as xenoliths of olivine lamproite.

The sandy lapilli tuffs and tuff-breccias are red to green, quartz-rich intercalated units exhibiting very weak bedding, in places dipping 10 to 20° to the west. They contain angular to subrounded grains of quartz, xenoliths of Lower Cretaceous Trinity Formation claystone, and lapilli of white clay and magmatic olivine lamproite. Occasionally, large blocks of claystone and Paleozoic sandstone as large as 1m across are observed in the lapilli tuff-breccias. Other lithic fragments include angular to rounded siltstone, quartzite, argillite, marl, and sandy limestone.

The epiclastic facies rocks occur as thin (2–4 cm) bedded units on the eastern side of the pipe. They are composed of quartz and clays, with a few small, bleached phlogopite flanks and fragments of very altered and weathered magmatic lamproite in a matrix composed largely of calcite and clays.

Xenoliths of mantle and lower crustal material, as well as megacrysts of ilmenite, chromite, chrome diopside, orthopyroxene, and garnet, have been recovered from TK1 during the bulk processing phase. The xenoliths include rutile websterite, garnet websterite, plagioclase websterite, eclogite, ilmenite-pyroxene intergrowths, and garnet lherzolite. Crustal xenoliths include gabbro, amphibolite, pegmatite, granite, granite gneiss, syenite, shale, sandstone, and quartzite.

The pipe is intruded into the Lower Cretaceous Trinity Formation and is partially overlain by the Upper Cretaceous Tokio Formation. This stratigraphic relationship suggests an age of very early Late Cretaceous, which corresponds to the K-Ar dates for the nearby Prairie Creek diatreme (97 ± 2 and 106 ± 3 Ma; Gogineni and others, 1978) and the ages suggested for the American, Kimberlite, and Black Lick bodies to the northeast (early Late Cretaceous; Miser and Purdue, 1929).

Field relationships suggest that the blocky tuff-breccias were

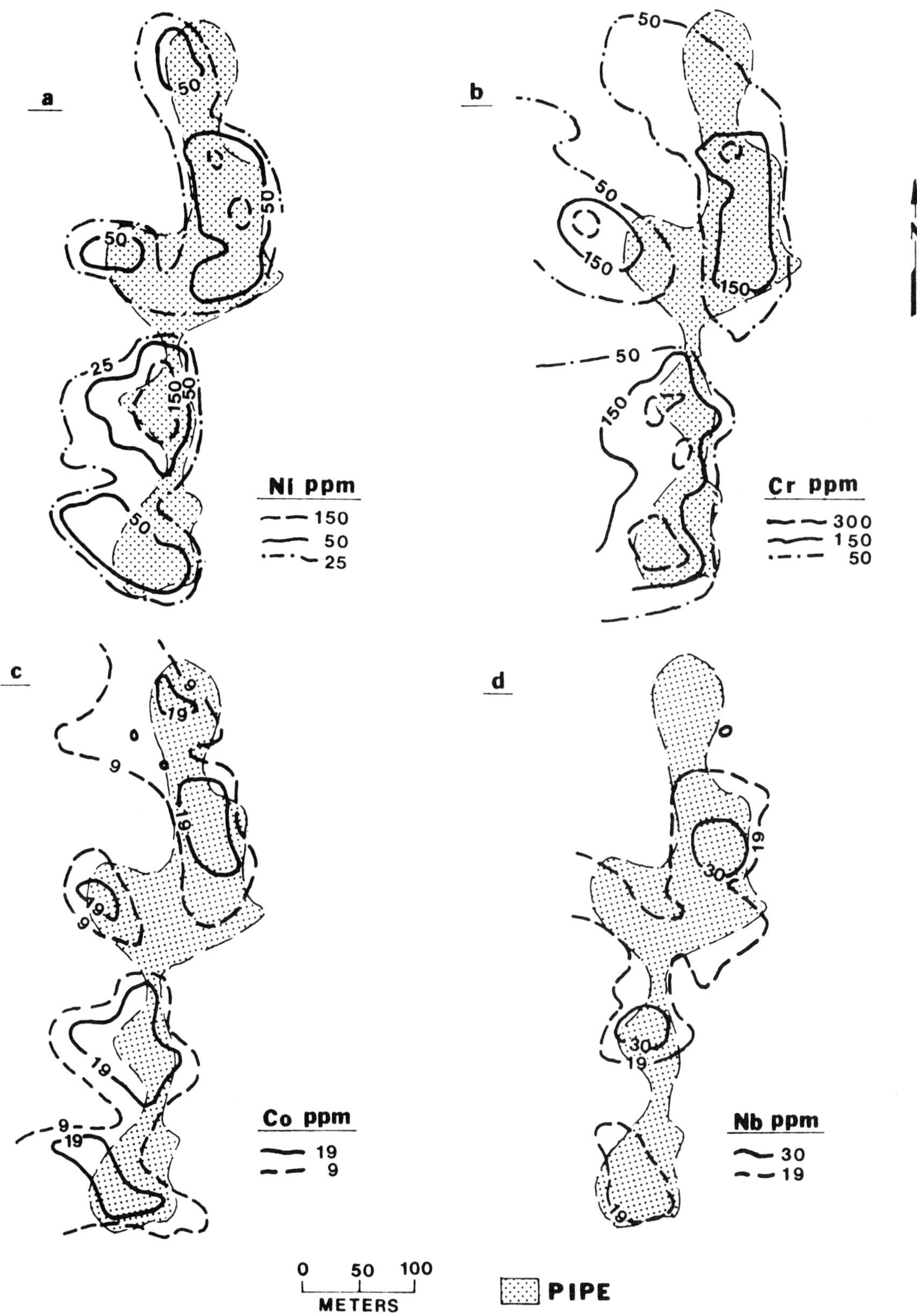

Figure 3 (a, b, c, d). Geochemical dispersion pattern of nickel, chromium, cobalt, and niobium in soils over Twin Knobs #1.

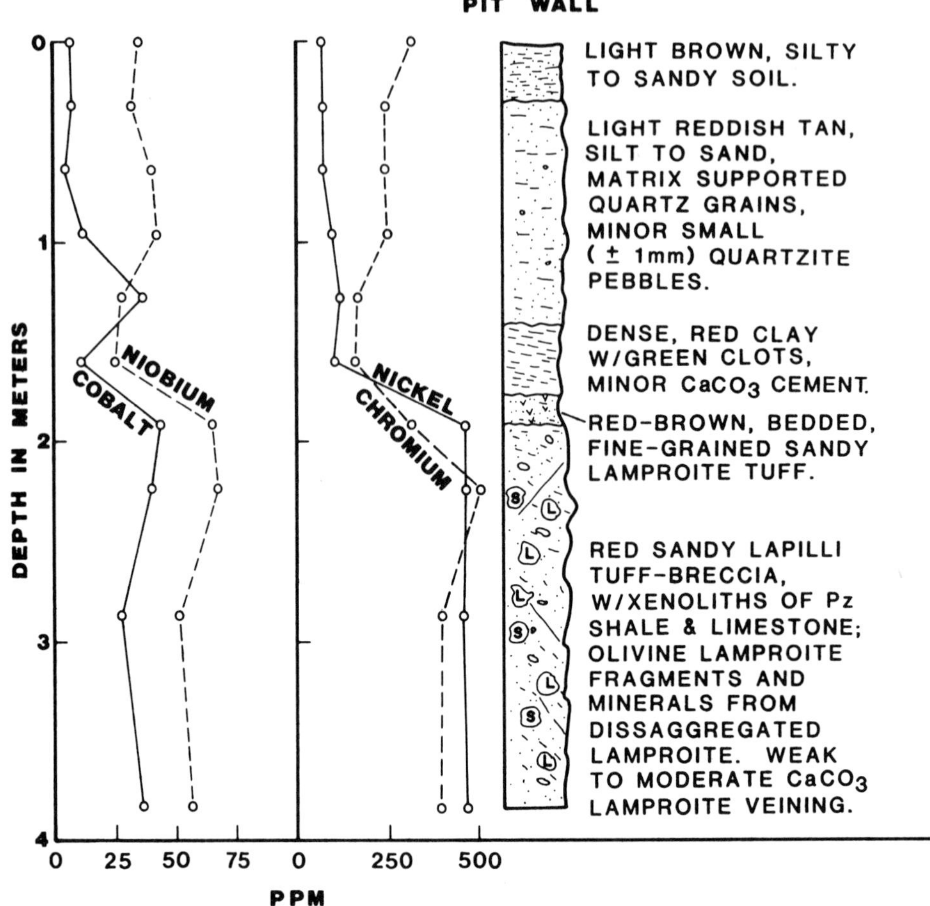

Figure 4. Profile of pit 1405, exhibiting changing trace element content with depth.

deposited as airfall ejecta in and around an elongated north-south–oriented fissure vent (crater facies), whereas the finer grained, more laminated or weakly bedded sandy tuffs were deposited in very shallow water, such as a small pond formed after subsidence and compaction of the rocks within the vent. Some mixing or reworking of the sandy units is evidenced by small (1–2 mm), sandy breccia fragments and detrital quartz. The quartz grains may have been derived in part, however, from the disaggregation of quartz-rich sedimentary rocks through which the lamproite passed. The tuffs and tuff-breccia units are cut by degassing vents and are intruded by magmatic lamproite. The epiclastic sediments were formed by the intermixing of weathered lamproite and fluvial detritus, and are postmagmatic lamproite emplacement. The preservation of the bedded epiclastic sediments suggests that only the uppermost portion of TK1 (75–150 m) has been eroded away.

In cross section, the shape of the TK1 diatreme is similar to the Western Australian lamproites, e.g., the champagne glass (Atkinson and others, 1984), rather than the typical kimberlite (Hawthorne, 1975). The paired high-low signatures of the ground magnetics and the lack of gravity contrast suggest that very small plugs of magmatic, olivine lamproite intruded the tuffs and breccias. Since the depth to very competent Paleozoic rocks beneath the present-day surface is no more than 100 m, the crater zone of tuffs and tuff-breccias is inferred to neck-down rapidly to magmatic lamproite root zones.

PETROGRAPHY

Lithologic units are not distinctly separable, as mentioned previously, due to extensive weathering and a lack of fresh samples. However, three major rock types can be distinguished in hand-specimens and from thin sections.

Magmatic Olivine Lamproite

Magmatic olivine lamproite contains euhedral to subhedral olivine phenocrysts, commonly exhibiting multiple growth aggregates up to 5 mm in diameter. Larger subrounded olivine macrocrysts are also present, but not common. Shale and sandstone xenoliths, detrital quartz, and feldspar also occur in varying amounts. All are set in a groundmass of glass that is heavily disseminated with perovskite and lesser amounts of priderite and spinel. The olivines are all pseudomorphed by serpentine, phlogopite, or calcite, and the glass is devitrified and partially replaced by fine, scaly micaceous minerals, possibly illite. These features are similar to those observed in olivine lamproite from Arkansas (Scott-Smith and Skinner, 1984a,b) and in Western Australia (Jaques and others, 1984). Calcite with or without barite veinlets cuts through the rock. Potassium richterite, commonly found in lamproites, was not observed in thin section, but may be unrecognizable due to the intensive alteration present.

It is likely that other magmatic rock types occur at TK1, although they were not encountered in the exploration pits. A single lamproite fragment from pit 2005 consists of 20 percent subhedral to rounded olivine phenocrysts, and 15 percent phlogopite laths as phenocrysts and microphenocrysts. The larger olivine phenocrysts are rimmed by serpentine and perovskite, whereas the smaller olivine microphenocrysts have no rims. The narrow laths of phlogopite are randomly oriented and the larger phlogopite laths poikilitically enclose perovskite, spinel, olivine, pyroxene, and priderite. These minerals also occur as microphenocrysts and groundmass minerals. The remainder of the groundmass consists of calcite and glass, now partially devitrified and altering to illite. Included in this fragment are ilmenite, pyroxene, olivine macrocrysts, and an altered micaceous peridotite. No additional samples of this phase were found.

Breccias and Tuffs

Tuffs and tuff-breccias are common at TK1 and are extremely variable in composition. Some are autolithic, while the majority contain up to 10 percent detrital quartz, feldspar, or country rock xenoliths. Other material includes garnet, chromite, clinopyroxene, orthopyroxene, ilmenite megacrysts, and mantle and lower crustal xenoliths. Magmatic lamproite fragments typically constitute as much as 20 percent of the breccias. The matrix is composed of euhedral to subhedral olivine microphenocrysts with perovskite, priderite, and spinel set in a glassy groundmass. The rocks are extensively altered, with olivine replaced by talc, illite or calcite; perovskite altering to cloudy leucoxene; and the glass devitrified and altered to talc or illite.

The sandy tuffs contain 15 to 30 percent angular to subangular quartz grains, along with lapilli of magmatic lamproite, shale, siltstone, and mantle and crustal xenoliths, set in a fine-grained matrix of serpentine, illite, and devitrified glass. Groundmass perovskite and phlogopite microphenocrysts are rarely seen in the matrix. As with the breccias, the sandy tuffs grade into sandy tuff-breccias in thin section. The presence of sandy breccia fragments in the tuffs suggests that some reworking of the tuffs has occurred at TK1.

Sandy Epiclastics

Sandy epiclastics are composed of as much as 35 percent silt- to sand-size, angular to subrounded quartz grains, detrital opaques, and rare lamproite fragments. The matrix contains up to 45 percent calcite cement, with variable amounts of illite and sericite. An occasional phlogopite lath or small lamproite fragment is the only physical evidence of its epiclastic origin.

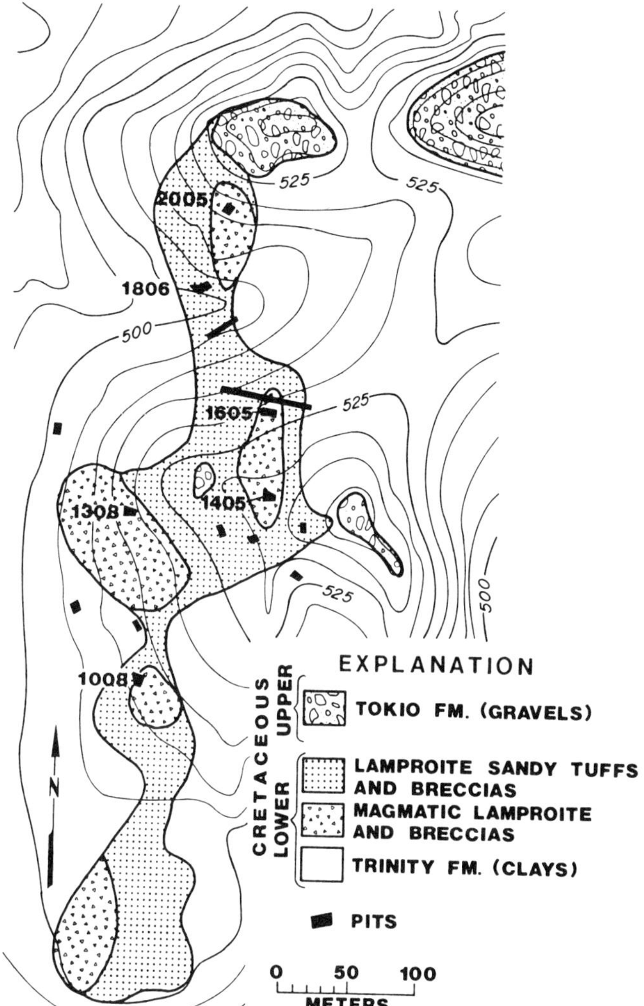

Figure 5. Geology of Twin Knobs #1 lamproite. Phases delineated from combined pit and trench exposures, soil geochemistry, and ground magnetics. Thin epiclastics on eastern portion of pipe omitted for clarity.

WHOLE-ROCK GEOCHEMISTRY

Extensive alteration, contamination from disaggregated country rock sediments, and mixing of lamproitic rock units at TK1 make detailed geochemical comparisons to other lamproites difficult. No unaltered samples of the magmatic phase were obtainable for comparison to the hypabyssal phase at Prairie Creek, which Scott-Smith and Skinner (1984a) have suggested have olivine lamproite affinities. However, some of the better preserved autolithic breccias at TK1 are chemically similar to brec-

TABLE 1. WHOLE-ROCK ANALYSES OF TWIN KNOBS #1 BRECCIAS FROM PITS 1308 AND 2005, COMPARED WITH REPRESENTATIVE ANALYSES OF OLIVINE LAMPROITES FROM ARKANSAS AND WESTERN AUSTRALIA

	Average Kimberlite (*)	Sample Number							
		TK 1 1308 (**)	TK 1 2005 (**)	PC 44 (**)	PC 59 (**)	AM 1 (**)	KI 1 (**)	WA (**)	ELL (***)
Number of Analyses		1	1	1	1	1	1	3	3
SiO_2 (wt.%)	35.2	41.5	41.5	42.0	40.0	44.0	43.0	41.2	41.5
TiO_2	2.3	2.2	3.8	2.2	3.4	2.8	3.1	3.0	2.7
Al_2O_3	4.4	5.7	8.2	4.0	5.0	5.8	5.2	3.6	3.5
FeO_t	9.8	8.0	15.6	8.6	9.9	9.0	9.7	9.2	8.6
MnO	0.1	0.1	0.4	0.1	0.1	0.1	0.1	0.1	0.1
MgO	27.9	14.9	13.1	21.8	19.0	17.8	19.1	23.7	26.0
CO	7.6	8.6	0.5	2.9	6.7	6.8	6.6	4.5	4.8
Na_2O	0.3	1.2	0.3	0.5	0.6	1.2	1.5	0.4	0.4
K_2O	1.0	0.5	1.2	1.9	2.1	3.5	2.7	3.0	3.9
P_2O_5	0.7	0.9	0.4	1.2	0.3	1.0	1.2	0.9	0.8
CO_2	3.3	3.2	0.2	0.6	0.3	0.5	0.2	---	0.1
H_2O^+	---	---	---	---	---	---	---	---	5.4
LOI	7.4	12.4	14.0	13.3	12.0	7.4	7.2	15.0	2.1
Total	100.0	101.2	99.1	99.1	99.4	99.9	99.6	98.8	100.1
K_2O/Al_2O_3	0.23	0.44	0.15	0.48	0.42	0.60	0.562	0.83	1.11
Al_2O_3/CaO	0.58	0.66	16.40	1.38	0.75	0.85	0.79	0.80	0.73
K_2O/Na_2O	3.33	2.08	4.00	3.80	3.50	2.92	1.80	7.50	9.75
MgO/K_2O	27.90	5.96	10.92	11.47	9.05	5.09	7.07	7.90	6.67
Ni (ppm)	710-1600	660	920	860	1120	740	860	1170	1190
Cr	550-1600	90	700	1000	1500	900	1000	1324	1450
Co	35-130	33	51	49	55	43	42	77	78
Nb	32-450	110	81	110	160	140	170	176	147
B	137-1970	8290	2850	3910	5130	1630	8320	12190	8680
Rb	0-350	40	11	110	175	59	90	406	439
Zr	84-700	505	430	495	780	665	740	650	617
Sr	40-1900	770	225	710	1040	880	1300	1062	1063
Rb/Zr	0.0-0.50	0.08	0.03	0.22	0.22	0.09	0.12	0.62	0.71
Rb/Sr	0.0-0.18	0.05	0.05	0.15	0.17	0.07	0.07	0.38	0.41
Ni/Co	15	20.00	18.04	17.55	20.36	17.21	20.48	15.19	15.26
Cr/Ni	1.2	1.36	0.76	1.16	1.34	1.22	1.16	1.13	1.22

*Average kimberlite (Dawson, 1967).
**Analyses by Bondar-Clegg, Vancouver, B.C.
***Average of three Ellendale olivine lamproites (Jaques and others, 1984).

Note: Analyses of oxides, Cr, Co, and Ni made by atomic absorption after total acid digestion; analyses of Rb, Sr, Zr, Ba, and Nb made by Bondar-Clegg.

Key to column headings:
PC = Prairie Creek
AM = American ol. lamproite
KI = Kimberlite ol. lamproite
WA = Average of three Ellendale ol. lamproites
ELL = Average of three Ellendale ol. lamproites

cias from Prairie Creek and to some of the olivine lamproites from Western Australia (Table 1).

The Twin Knobs #1 breccias are similar to olivine lamproites with respect to niobium and zirconium contents (Fig. 6), as well as to other trace element contents. TK1 breccias plot in an area shared by other Arkansas lamproite breccias and by some of the olivine lamproites from Western Australia. Rb:Sr ratios for TK1 rocks (0.05:0.56) are very similar to Rb:Sr ratios from Prairie Creek (0.04:0.44), but are considerably less than those from the Western Australian lamproites (0.26:0.56; Jaques and others, 1984). The other trace elements from Prairie Creek and TK1 are also very similar (Table 1).

XENOCRYST MINERALS

The minerals analyzed from TK1 are largely those recovered from concentrates during bulk sampling. The lower size cutoff of 0.5 mm did not allow for the recovery of primary groundmass minerals from the lamproite, particularly the spinels and other oxides that appeared unaltered in thin section. It is

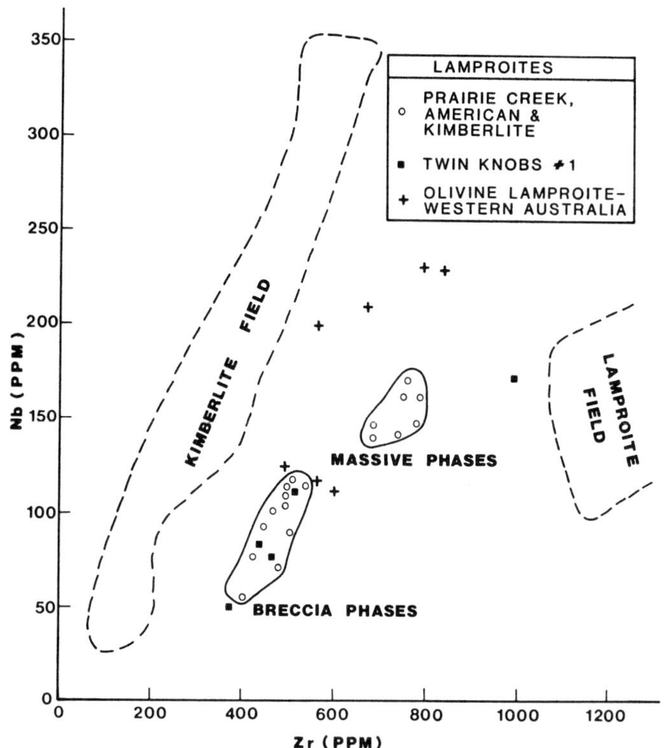

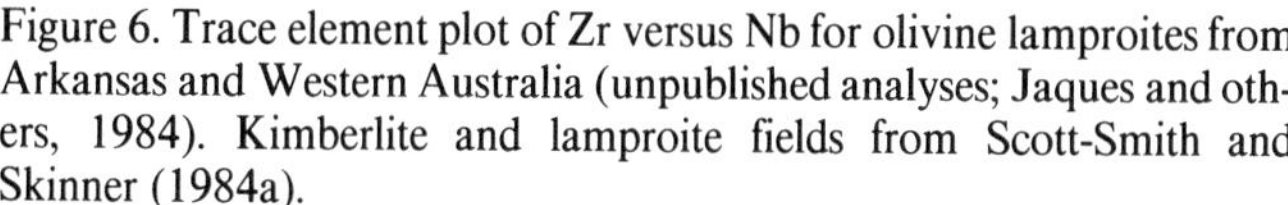
Figure 6. Trace element plot of Zr versus Nb for olivine lamproites from Arkansas and Western Australia (unpublished analyses; Jaques and others, 1984). Kimberlite and lamproite fields from Scott-Smith and Skinner (1984a).

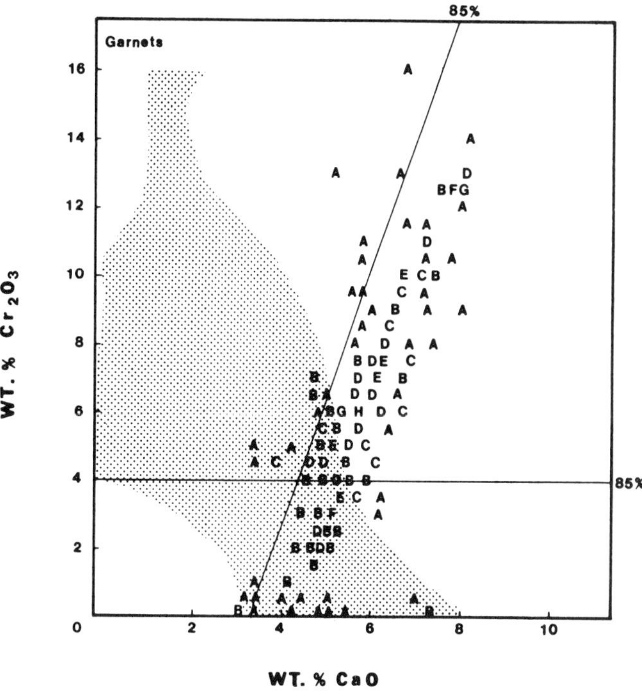

Figure 7. Plot of Cr_2O_3 versus CaO for pyrope garnets from Twin Knobs #1 concentrates. Shaded area represents diamond inclusion composition pyropes (Gurney, 1984). High Cr_2O_3-low CaO pyropes plotted in area above and to left of 85 percent lines are G10 pyropes (after Gurney, 1984). Letters represent numbers of observations, e.g., A = 1, B = 2.

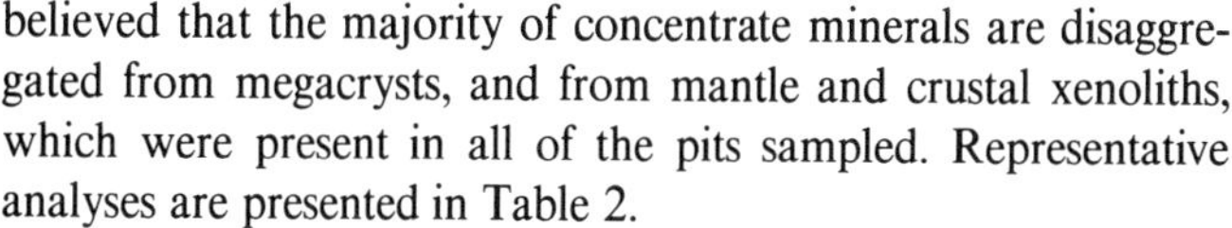
believed that the majority of concentrate minerals are disaggregated from megacrysts, and from mantle and crustal xenoliths, which were present in all of the pits sampled. Representative analyses are presented in Table 2.

Garnets

Garnets recovered from TK1 concentrates are largely lherzolitic chrome pyropes (groups 9 and 11; Dawson and Stephens, 1975, 1976). Gurney (1984) has shown that the presence in kimberlite of pyrope with CaO and Cr_2O_3 contents similar to pyrope included in diamond (G10 pyropes) indicates that a kimberlite will be diamondiferous. G10 pyropes are low-calcium, high-chromium pyropes and are best illustrated on a CaO-Cr_2O_3 plot, e.g., those pyropes that plot in the region above the +4.0% Cr_2O_3 line and left of the diagonal line marked 85% in Figure 7. This relation appears to hold true for the TK1 lamproite as well, as 10.6% of the +4.0% Cr_2O_3 pyropes fall within the G10 region.

Low-chrome pyrope and pyrope-almandine garnets are also present in the concentrates. They contain 0.00 to 2.96% Cr_2O_3, with 13.6 to 21.5% MgO (groups 3 and 6; Dawson and Stephens, 1975).

Pyroxenes

The clinopyroxenes are low in TiO_2 (<0.50%) with Ca/Ca + Mg ratios ranging from 0.36 to 0.50. Cr_2O_3 contents range from 0.35 to 2.66%. TK1 clinopyroxenes have similar Al_2O_3 contents (0.7–4.9% Al_2O_3) to Prairie Creek clinopyroxenes (0.4–5.6% Al_2O_3), but are lower in Al_2O_3 content than Prairie Creek groundmass clinopyroxenes (Scott-Smith and Skinner, 1984a) and both groundmass and xenolith clinopyroxenes from Western Australian lamproites (Jaques and others, 1984). The low TiO_2 and high Cr_2O_3 contents of the clinopyroxenes from TK1 precludes the likelihood that they are from the lamproitic groundmass. With respect to Ca:Mg:Fe ratios, TK1 clinopyroxenes are similar to chrome diopsides derived from lherzolites (Nixon and Boyd, 1973), and fall within the group 5 classification of Stephens and Dawson (1977) for clinopyroxenes from both sheared and granular garnet lherzolites. The diopside solvus of Mori and Green (1975), as shown in Figure 8, indicates a temperature range of 1,000 to 1,450°C at 30 kbar for the chrome diopsides. The orthopyroxenes are aluminous enstatites, with 0.74–2.69% Al_2O_3, 0.23–1.22% Cr_2O_3, and Mg/Mg + Fe ratios ranging from 0.914 to 0.935.

TABLE 2. REPRESENTATIVE ANALYSES OF MINERALS FROM TWIN KNOBS #1 CONCENTRATE

	Garnet			Pyroxene		Ilmenite	Spinel	
Sample No.	100819	130802	160524	2005151	2005160	1008135	160572	100881
SiO_2 (wt.%)	41.40	40.90	40.30	53.90	56.40	0.00	0.20	0.33
TiO_2	0.21	0.07	0.47	0.11	0.10	52.70	0.53	4.23
Al_2O_3	20.20	22.70	14.70	1.97	1.91	0.75	10.90	8.99
Cr_2O_3	4.42	0.19	11.10	1.52	0.50	2.18	58.00	46.50
FeO_t	7.39	14.20	6.64	2.66	5.14	31.10	16.90	25.90
MnO	0.41	0.42	0.41	0.33	0.00	0.27	0.35	0.25
MgO	20.70	17.80	19.10	18.50	34.20	12.70	12.30	13.30
CaO	4.78	3.04	7.18	19.50	0.96	0.00	0.10	0.00
Na_2O	0.00	0.00	0.11	1.52	0.74	0.00	0.00	0.00
Totals	99.50	99.40	100.01	100.01	100.00	99.80	99.30	99.50
Number of Cations for n Oxygens								
n =	12	12	12	6	6	3	4	4
Si	2.983	2.952	2.972	1.951	1.945	---	0.007	0.011
Ti	0.011	0.004	0.027	0.003	0.003	0.926	0.013	0.106
Al	1.714	1.932	1.280	0.084	0.077	0.021	0.421	0.351
Cr	0.251	0.012	0.647	0.043	0.014	0.040	1.503	1.228
Fe^{+3}	---	---	---	---	---	---	---	---
Fe^{+2}	0.446	0.859	0.408	0.080	0.149	0.608	0.463	0.722
Mn	0.026	0.026	0.027	0.011	---	0.006	0.010	0.008
Mg	2.225	1.917	2.101	0.998	1.758	0.442	0.600	0.662
Ca	0.360	0.234	0.567	0.757	0.035	---	0.004	---
Na	---	---	0.018	0.109	0.050	---	---	---
$(Mg/Mg + Fe^{+2})$	83.3	69.1	83.7	92.6	92.2	42.1	56.4	47.8
(Cr/Cr + Al)	12.8	0.6	33.6	---	---	---	8.1	77.6
(Ca/Ca + Mg)	13.9	10.9	21.3	43.1	2.0	---	---	---

Note: Data reduction by Bence-Albee method; <2σ variance of probe standards (Smithsonian Institution).

Ilmenites

The majority of ilmenites analyzed from TK1 concentrates are picroilmenites, with 7.1–15.4% MgO, and 0.1–12.0% Cr_2O_3. Picroilmenite is rare at Prairie Creek, and is reportedly absent in olivine lamproites at Ellendale and Argyle, Western Australia (Atkinson and others, 1984). Megacrysts of picroilmenite (1–2 cm) are common at TK1 and are suspected to be the main source of the ilmenites in the concentrate as well.

Ilmenites plotted on the Cr_2O_3 versus MgO diagram form a distinctive "J" pattern with low Cr_2O_3 to moderate MgO ilmenites on the hook and low to high Cr_2O_3–high MgO ilmenites forming the left limb (Fig. 9). The relationship is similar to ilmenites from Prairie Creek lamproite and the Sloan kimberlites (Eggler and others, 1979; M. A. Waldman, unpublished data).

Spinels

Less than 32% of the 255 spinels analyzed have TiO_2 over 1.0% (maximum, 6.09%). The high-titanium spinels have similar chromium contents (30.0–62.3% Cr_2O_3) to the low-titanium spinels, but are less aluminum-rich (4.3–20.3% versus 5.1–29.4% Al_2O_3). The TK1 spinels are largely magnesiochromites, with Mg/Mg + Fe ratios ranging from 0.33 to 0.67. The spinels are probably fragments from megacrysts or nodules, although some of the more titanium-rich varieties may be from the lamproite itself. One chromite grain analysis from TK1 is similar in composition to chromites found as diamond inclusions, e.g., Cr_2O_3 > 62%, Al_2O_3 < 7%, MgO >10% and Cr/Cr + Al >80% (Sobolev, 1974; Fig. 10).

DISCUSSION AND CONCLUSIONS

Our investigations have shown that the Twin Knobs #1 ultramafic rocks have many geologic, geochemical, mineralogic, and petrographic features similar to those described for the Prairie Creek complex, recently classified as olivine lamproite, and the olivine lamproites of Western Australia. Mineralogic and petrographic features typical of olivine lamproites include the presence of glass in the groundmass of the pyroclastic and magmatic rocks,

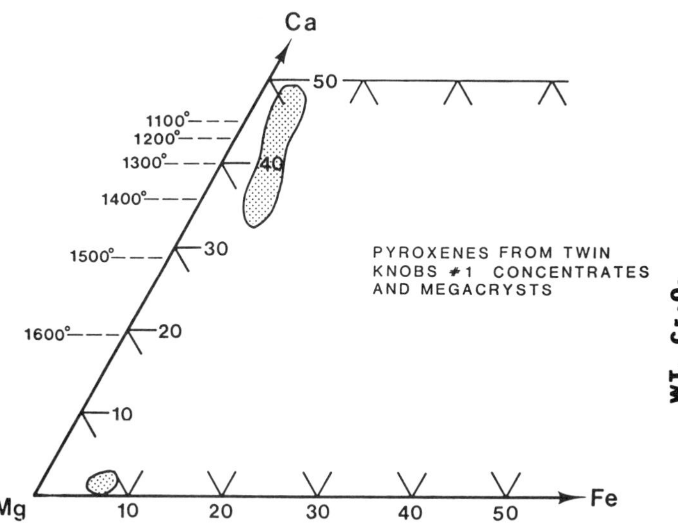

Figure 8. Plot of Ca:Mg:Fe ratios for pyroxenes from Twin Knobs #1 concentrates and megacrysts. Temperatures taken from diopside solvus of Mori and Green (1975).

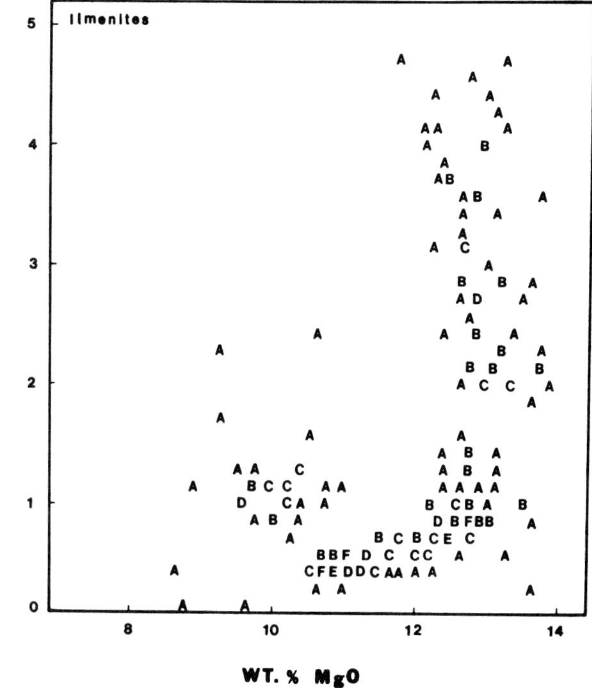

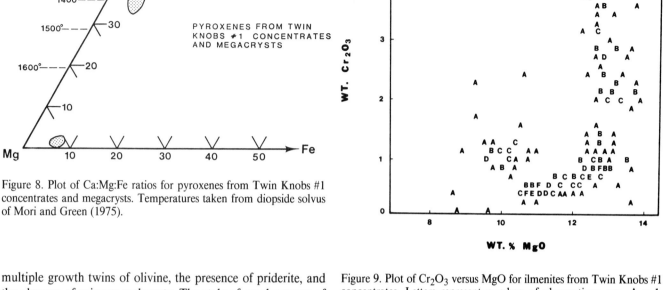

Figure 9. Plot of Cr_2O_3 versus MgO for ilmenites from Twin Knobs #1 concentrates. Letters represent numbers of observations, e.g., A = 1, B = 2.

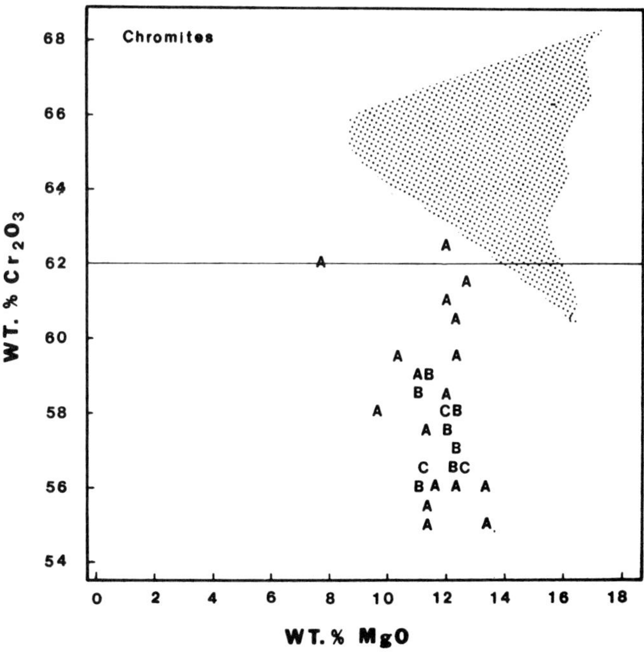

Figure 10. Plot of Cr_2O_3 versus MgO for high chromium spinels from Twin Knobs #1 concentrates. Letters represent numbers of observations, e.g., A = 1, B = 2. Shaded area represents 90 percent of spinel compositions from diamond inclusion chromites.

multiple growth twins of olivine, the presence of priderite, and the absence of primary carbonate. The style of emplacement of the TK1 rocks, with the magmatic phase intruding the older pyroclastics very near or at the surface, and the champagne glass shape of TK1 are typical of lamproites.

There are some features, however, that are typically kimberlitic and atypical of lamproites. TK1 contains a relative abundance of picroilmenite xenocrysts and megacrysts, potassium richterite has not been observed in thin section, and chrome diopside geochemistry is similar to those derived from kimberlites. Chemically, K_2O/Na_2O (2.08–4.00) and K_2O/Al_2O_3 (0.15–0.44) ratios at TK1 are within the kimberlitic range (average, 0.23 and 3.33) and are less than those of the more potassic-rich Western Australian lamproites. TK1 rocks have higher Al_2O_3 and lower MgO contents than either kimberlites or Western Australian olivine lamproites. The high Al_2O_3 suggests contamination from inclusion of the alumina-rich, country-rock sediments. The Murfreesboro District rocks are not closely associated with leucite-bearing lamproites, as are the olivine lamproites of Western Australia.

The diamond population from TK1 is not sufficiently large to make any broad comparisons, but the diamonds are typical in form and color to those found in both lamproites and kimberlites. The presence of typical "kimberlitic" indicator minerals in TK1 and the other pipes in the Murfreesboro district, albeit less abundant than in kimberlites, means that similar exploration techniques can be used in diamond exploration, whether the diamond host rocks are kimberlites or lamproites.

Although TK1 has some features similar to kimberlite, it is our conclusion that these rocks have a closer affinity to olivine lamproites. The chemical and mineralogic similarities to Prairie

Creek and the other intrusive rocks in the Murfreesboro district are considered to be comagmatic, as inferred by Scott-Smith and Skinner (1984b).

ACKNOWLEDGMENTS

We thank the Superior Oil Company for its permission to publish this work. We also thank all those individuals in Superior's diamond exploration program who worked so hard to gather the data; without their help this paper could not have been written. Special thanks to Chris B. Smith, C.R.A. Exploration, Ltd., for his most thoughtful comments. Critical reviews by Steve Bergman and Susan C. Erikson are also gratefully acknowledged.

REFERENCES CITED

Atkinson, W. J., Hughes, F. E., and Smith, C. B., 1984, A review of the kimberlitic rocks of Western Australia, *in* Kornprobst, J., ed., Kimberlites and related rocks: Proceedings of the Third International Kimberlite Conference, v. 1: Washington, D.C., American Geophysical Union, p. 195–224.

Bolivar, S. L., and Brookins, D. G., 1979, Geophysical and Rb-Sr study of the Prairie Creek, Ak Kimberlite, *in* Proceedings of the Second International Kimberlite Conference, October, 1977, v. 1: Washington, D.C., American Geophysical Union, p. 289–299.

Branner, J. C., and Brackett, R. N., 1890, The peridotite of Pike County, Arkansas: American Journal of Science, 3rd ser. v. 38, p. 50–59.

Dawson, J. B., 1967, Geochemistry and origin of kimberlite, *in* Wyllie, P. J., ed., Ultramafic and related rocks: New York, John Wiley and Sons, p. 269–278.

Dawson, J. B., and Stephens, W. E., 1975, Statistical classification of garnets from kimberlites and associated xenoliths: Journal of Geology, v. 83, p. 589–607.

—— , 1976, Statistical classification of garnets from kimberlites and associated xenoliths; Addendum: Journal of Geology, v. 84, p. 495–496.

Eggler, D. H., McCallum, M. E., and Smith, C. B., 1979, Megacryst assemblages in kimberlite from northern Colorado and southern Wyoming; Petrology, geothermometry-barometry, and aerial distribution, *in* Proceedings of the Second International Kimberlite Conference, October, 1977, v. 2: Washington, D.C., American Geophysical Union, p. 213–226.

Gogineni, S., Melton, C. E., and Giardini, A. A., 1978, Some petrological aspects of the Prairie Creek diamond-bearing kimberlite diatreme, Arkansas: Contributions to Mineralogy and Petrology, v. 66, p. 251–261.

Gurney, J. J., 1984, A correlation between garnets and diamonds in kimberlites, *in* Glover, J. E., and Harris, P. G., eds., Kimberlite occurrence and origin; A basis for conceptual models in exploration: Perth, University of Western Australia Publication 8, p. 143–166.

Hall, A. E., and Smith, C. B., 1984, Lamproite diamonds: Are they different?, *in* Glover, J. E., and Harris, P. B., eds., Kimberlite occurrence and origin: A basis of conceptual models in exploration: Perth, University of Western Australia Publication 8, p. 167–212.

Hawthorne, J. B., 1975, Model of a kimberlite pipe: Physics and Chemistry of the Earth, v. 9, p. 1–15.

Jaques, A. L., Lewis, J. D., Smith, C. B., Gregory, G. P., Ferguson, J., Chappell, B. W., and McCulloch, M. T., 1984, The diamond-bearing ultrapotassic (lamproitic) rocks of the West Kimberley region, Western Australia, *in* Kornprobst, J., ed., Kimberlites and related rocks: Proceedings of the Third International Kimberlite Conference, v. 1: Washington, D.C., American Geophysical Union, p. 225–254.

Kane, M. F., Hildenbrand, J. D., and Hendricks, J. D., 1981, Model for the tectonic evolution of the Mississippi embayment and its contemporary seismicity: Geology, v. 9, p. 563–568.

Lewis, R. D., 1977, Mineralogy, petrology, and geophysical aspects of Prairie Creek kimberlite near Murfreesboro, Arkansas [M.S. thesis]: W. Lafayette, Indiana, Purdue University, 161 p.

Meyer, H.A.O., 1976, The kimberlites of the continental United States; A review: Journal of Geology, v. 84, p. 377–403.

Miser, H. D., 1913, New areas of diamond-bearing peridotite in Arkansas: U.S. Geological Survey Bulletin 540-U, Contributions to Economic Geology, 1912, pt. 1, p. 534–546.

Miser, H. D., and Purdue, A. H., 1929, Geology of the DeQueen and Caddo Gap quadrangles, Arkansas: U.S. Geological Survey Bulletin 808, 191 p.

Mitchell, R. H., and Lewis, R. D., 1983, Priderite-bearing xenoliths from the Prairie Creek mica peridotite, Arkansas: Canadian Mineralogist, v. 21, p. 59–64.

Mori, T., and Green, D. H., 1975, Pyroxenes in the system $Mg_2Si_2O_6CaMgSi_2O_6$ at high pressure: Earth and Planetary Science Letters, v. 26, p. 277–288.

Nelson, D. K., Lillie, R. J., de Voogd, B., Brewer, J. A., Oliver, J. E., Kaufman, S., and Brown, L., 1982, COCORP seismic reflection profiling in the Ouachita Mountains of western Arkansas; Geometry and geologic interpretation: Tectonics, v. 1, no. 5, p. 413–430.

Nixon, P. H., and Boyd, F. R., 1973, Petrogenesis of the granular and sheared ultrabasic nodule suite in kimberlite, *in* Nixon, P. H., ed., Lesotho kimberlites: Maseru, Lesotho, Lesotho National Development Corporation, p. 48–56.

Robinson, D. N., 1980, Surface textures and other features of diamonds [Ph.D. thesis]: University of Cape Town.

Ross, C. S., Miser, H. D., and Stephenson, L. W., 1929, Waterlaid volcanics of early upper Cretaceous age in southwestern Arkansas, southern Oklahoma, and northeastern Texas: U.S. Geological Survey Professional Paper 154F, p. 175–202.

Scott-Smith, B. H., and Skinner, E.M.W., 1984a, A new look at Prairie Creek, Arkansas, 1984, *in* Kornprobst, J., ed., Kimberlites and related rocks: Proceedings of the Third International Kimberlite Conference, v. 1: Washington, D.C., American Geophysical Union, p. 255–284.

—— , 1984b, Kimberlite and American mines, near Prairie Creek, Arkansas, *in* Kornprobst, J., ed., Kimberlites III: Annales Scientifiques L'Universite de Clermont-Ferrand II, v. 74, pt. 1, p. 27–36.

Sobolev, N. V., 1974, Deep-seated inclusions in kimberlites and the problem of composition of the upper mantle: Washington, D.C., American Geophysical Union, 279 p.

Stephens, W. E., and Dawson, J. B., 1977, Statistical comparison between pyroxenes from kimberlite and their associated xenoliths: Journal of Geology, v. 85, p. 433–449.

Viele, G. W., 1979, Geologic map and cross section, eastern Ouachita Mountains, Arkansas: Map summary: Geological Society of America Bulletin, v. 90, pt. I, p. 1036–1099.

Manuscript Accepted by the Society October 15, 1986

Printed in U.S.A.

Geological Society of America
Special Paper 215
1987

The Cretaceous Arkansas alkalic province; A summary of petrology and geochemistry

*Ellen Mullen Morris**
Department of Geology, University of Arkansas, Fayetteville, Arkansas 72701

ABSTRACT

The Arkansas alkalic province consists of Cretaceous intrusions that range from lamproite to syenite in composition and are parallel to the trend of the Tertiary onlap through central Arkansas. Ages obtained by K-Ar and fission track methods become younger toward the northeast, and rocks generally become more leucocratic. However, mineral compositions show little evolution throughout the belt. Whole-rock major and trace element data indicate that fractionation played a minor role in the transition from mafic to syenitic intrusions. Separate mantle sources and potassic metasomatism are important factors in the development of the Arkansas alkalic province.

INTRODUCTION

The Cretaceous alkalic rocks of Arkansas (Fig. 1) are a diverse but petrologically related assemblage that varies from diamondiferous lamproite to quartz syenite. The total outcrop area is slightly less than 40 km^2, and occurs mostly in a 15-km-wide band along the northern edge of the Mississippi Embayment. Along this northeast-trending outcrop belt the igneous rocks become more leucocratic, as well as younger, from the Prairie Creek lamproite in the southwest to syenites near Little Rock, Arkansas, at the northeast end of the trend.

Seven major intrusions or intrusive complexes have been identified within Arkansas. Each seems to be a discrete body or magmatic system. The lamproite of the Prairie Creek complex near Murfreesboro, dated at 97 to 106 Ma (Zartman, 1977), is diamodiferous peridotite that intrudes co-magmatic breccia and water-laid tuff. Similar rocks occur in small bodies and dikes within Howard and Pike Counties, and were reported 60 km northward in Scott County by Miser and Ross (1923). The Magnet Cove intrusion, 85 km northeast of Prairie Creek, is a renowned ring-dike complex, 101 Ma in age and 12 km^2 in area, with a core of carbonatite and ijolite, a middle ring of phonolite and trachyte, and an outer ring of diverse syenites including garnet-pyroxenite syenite and pseudoleucite syenite.

Two related coeval intrusions—the V intrusive and Potash Sulphur Springs—are about 10 km west of Magnet Cove. The V intrusive, 0.2 km^2 in area, consists of cross-cutting dikes that vary from early sannaite to late nepheline syenite. The Potash Sulphur Springs intrusion, covering approximately 1.5 km^2, is crudely zoned. It has a rim of early syenites and a later core of lamprophyre dikes and carbonatite, and has been dated at 101 Ma (Zartman and Howard, 1985).

Lamprophyre dikes known as the Benton dike swarm are concentrated in a 100-km^2 area to the northeast of Magnet Cove. The dikes, 1 to 2 m wide and as much as 2 km in mapped length, trend generally east-west. They are mostly minettes and monchiquites, with syenite or other, more felsic, lithologies composing about 10 percent of the total number of dikes mapped.

Syenite is the most abundant rock type of the Arkansas alkalic province. Subsurface and gravity data suggest that syenite exposed near Little Rock extends for 300 km^2 beneath Tertiary cover and constitutes a pluton of major dimensions. Nepheline syenites with trachytic textures occur in Saline County. The Saline County syenite has a fission track age of 89 Ma (Morris and Eby, 1986). Laterization of these rocks during the early Cenozoic produced the well-known bauxite deposits of Arkansas. Granite Mountain, near Little Rock, consists of syenite, nepheline syenite, and minor quartz syenite. Late biotites from Granite Mountain have yielded a K-Ar age of 87 to 91 Ma (Zartman, 1977). Fission track dates of apatite and sphenes from the major portion of Granite Mountain rocks yield a similar age of 86–87 Ma, whereas rare olivine-bearing syenite from this pluton has a mean fission-track age of 95 Ma (Morris and Eby, 1986).

Although the principal exposures of igneous rocks in Arkan-

*Present address: Department of Geology, Sul Ross State University, Box C130, Alpine, Texas 79832.

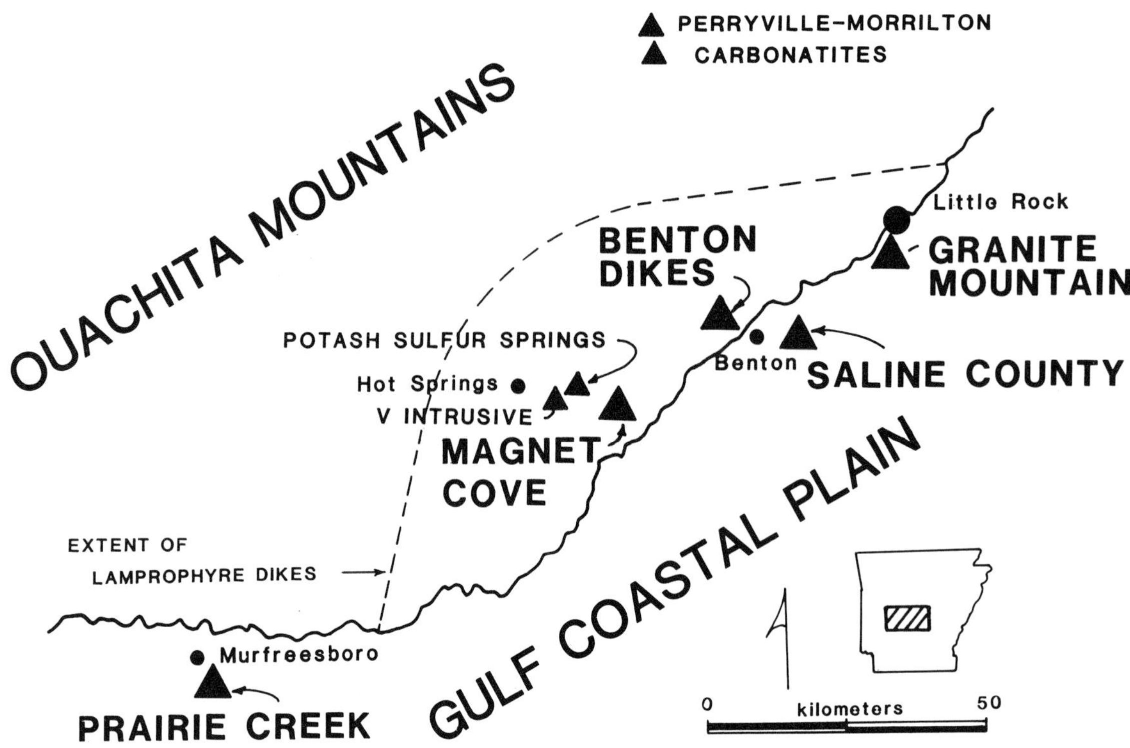

Figure 1. Location of Arkansas Cretaceous igneous intrusions.

sas are along the northeasterly trend summarized above, important alkaline rocks occur near Perryville and Morrilton, 70 to 80 km north of Magnet Cove. These rocks, dated at 96 Ma on K-Ar of phlogopite, are carbonatite breccias that contain olivine, phlogopite, apatite, ilmenite, and calcite as principal minerals. Altered granitoid, gneissic and/or sedimentary xenoliths are major components of some breccias. Very limited exposures of phlogopite-perovskite rocks (allikites) occur even farther to the north, 40 km north of Russellville. These rocks are geochemically distinct from the rocks of the principal Arkansas alkalic province.

This paper, broad in scope, summarizes previous work and presents new data with preliminary interpretations. It is meant to be a beginning toward the comprehensive understanding of the petrology of the Arkansas alkalic province.

PRAIRIE CREEK COMPLEX

The alkalic peridotite complex of Prairie Creek is significant not only because it contains diamonds, but also because it is a key to the genesis of the Arkansas alkalic province. Because the Prairie Creek rocks are the most mafic and undeniably mantle-derived, they may best indicate the nature of the mantle source(s) that produced the broad spectrum of alkalic rocks found along the western edge of the Mississippi Embayment in Arkansas.

Lithologies and Mineralogy of the Prairie Creek Complex

Three lithologic units are recognized in the Prairie Creek complex. The most abundant is hypabyssal peridotite, which comprises slightly more than half the exposure. The rock's groundmass consists of glass now substantially altered to serpentine and/or chlorite, and poikilitic phlogopite. Olivine occurs principally as embayed macrocrysts with serpentine reaction rims. Diopside macrocrysts are less corroded and less abundant. Small (<1 mm) crystals of diopside, apatite, Cr-spinel, and perovskite are dispersed throughout the groundmass. Amphibole (potassic richterite), priderite, and both high-Cr and low-Cr garnet are rare phases (Lewis and others, 1976). Ilmenite and orthopyroxene have not been reported from the Prairie Creek locality.

Mineral compositions of the hypabyssal peridotite (Table 1) are characteristic of alkali-rich ultramafic rocks, but somewhat different than kimberlites. Olivine is very magnesian (Fo_{86-92}) (Fig. 2) and overlaps the compositional field for olivine macrocrysts in kimberlites. Two generations of olivine have been recognized (Scott-Smith and Skinner, 1984a): a euhedral form and a much more common embayed morphology of larger macrocrysts (>1 mm). Both generations have the same compositional range. Diopside is very calcic (Ca_{50}, Fe_4, Mg_{46}) and plots above the field for groundmass diopside of kimberlite on the pyroxene quadrilateral (Fig. 3). It is comparable to clinopyroxene of some carbonatite complexes. The poikilitic phlogopite has a high content of SiO_2 and TiO_2, but low Al_2O_3 compared to most kimberlites (Fig. 4). Its pleochroic scheme is characteristic of low-pressure crystallization rather than the high-pressure scheme described for most kimberlites by Farmer and Boettcher (1981).

TABLE 1. MINERAL COMPOSITIONS: PRAIRIE CREEK PERIDOTITE

Oxide	Olivine						Diopside			Phlogopite	
SiO_2	42.3	42.2	41.7	41.3	42.2		53.0	53.0	53.2	42.9	43.0
TiO_2	0.0	0.0	0.0	0.0	0.0		1.2	1.6	0.9	6.2	6.4
Al_2O_3	0.0	0.0	0.0	0.0	0.0		0.3	0.3	0.3	8.0	8.7
FeO	8.3	8.1	8.5	8.3	7.9		3.3	3.0	2.6	6.8	6.5
MgO	49.6	48.8	49.1	49.1	50.3		16.7	16.6	16.6	0.0	0.0
CaO	0.1	0.1	0.2	0.1	0.1		25.1	25.3	25.1	22.3	22.3
Na_2O	0.0	0.0	0.0	0.0	0.0		0.3	0.3	0.4	0.6	0.5
K_2O	0.0	0.0	0.3	0.0	0.0		0.1	0.0	0.0	9.5	9.5
NiO	0.38	0.39	0.34	0.41	0.37		0.03	0.02	0.03	0.0	0.0
Cr_2O_3	0.04	0.05	0.05	0.05	0.04		0.46	0.35	0.20	0.1	0.1
Total	100.7	99.6	100.2	99.3	100.9		100.5	100.5	99.3	96.4	97.0
Fo	86	86	85	86	86	Ca	50	50	50		
						Mg	46	46	46		
						Fe	4	4	4		

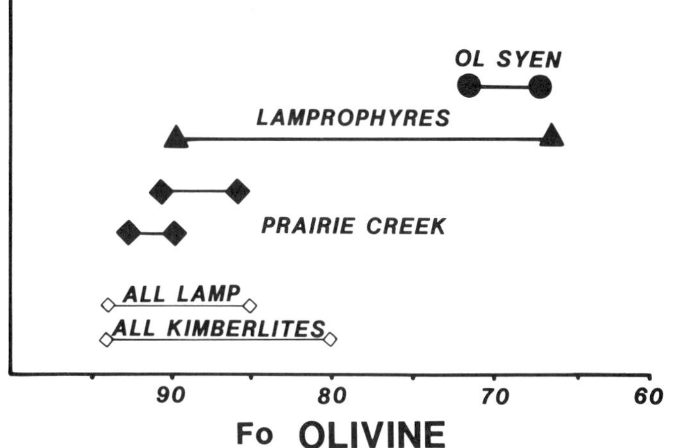

Figure 2. Olivine compositions for Cretaceous igneous rocks of Arkansas. Lower bar for Prairie Creek lamproite gives data from Scott-Smith and Skinner (1984a); upper bar for Prairie Creek lamproite is newly reported in this paper.

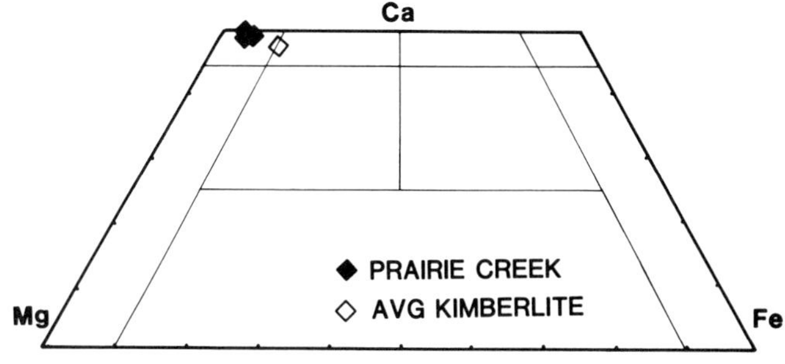

Figure 3. Compositions of clinopyroxene macrocrysts and microphenocrysts, Prairie Creek complex.

Phlogopite composition and occurrence as a groundmass phase indicate that it crystallized during emplacement.

The breccia unit constitutes about 40 percent of the Prairie Creek exposure. It is intruded by a dike of hypabyssal peridotite, so it is somewhat earlier than the peridotite. Because breccia is substantially altered and deeply weathered, its original character is difficult to determine. In general, it is an easily fragmented rock that contains a highly variable combination of juvenile lapilli, ultramafic autoliths, crustal xenoliths, and olivine macrocrysts in a serpentine and chlorite matrix. Bolivar (1984) divided the breccia into six units based on the degree of alteration, and speculated that some of those units represented different phases of intrusion.

The tuffs are the finest grained and have the most altered lithology. They are principally blue in color and contain rock fragments, lapilli, and a matrix rich in chlorite. Quartz grains, probably derived from the underlying Jackfork sandstone, are a principal constituent of some tuffs. Graded bedding in tuff units along the eastern limit of exposure suggests a waterlaid origin.

Other Diamondiferous Peridotites, South-Central Arkansas

Diamondiferous peridotites occur in several other localities near Prairie Creek. The best known of these are the Black Lick and American mines, which were prospected for only a short time. Both are highly altered hypabyssal peridotite. Mineralogy of the American Mine is similar to that of Prairie Creek. However, the rock at Black Lick contains 5 to 10 percent modal secondary nepheline as fillings in vugs, probably indicating a magma less rich in Mg and Si than that at Prairie Creek. A recently discov-

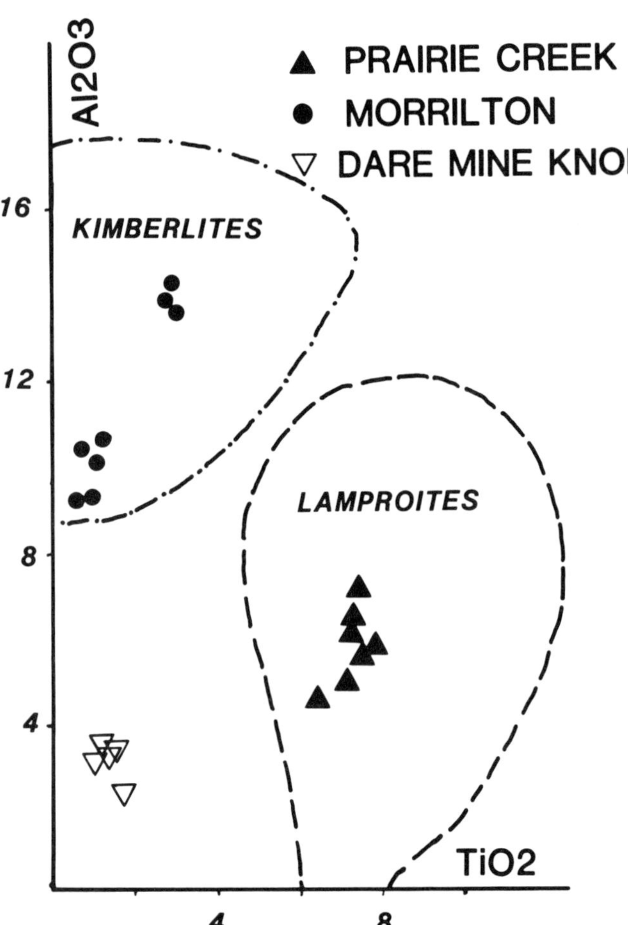

Figure 4. Plot of Al_2O_3 versus TiO_2 for phlogopites from Cretaceous alkalic rocks of Arkansas.

TABLE 2. MAJOR ELEMENT ANALYSES: ULTRAMAFIC ROCKS AND CARBONATITE CRETACEOUS ARKANSAS ALKALIC PROVINCE*

Oxide (wt.%)	1	2a	2b	3	4	5
SiO_2	12.66	49.90	29.96	39.46	42.10	38.40
TiO_2	0.96	2.48	3.15	3.61	4.13	2.36
Al_2O_3	2.53	3.45	6.73	3.53	5.49	3.47
FeO*	6.40	7.73	14.05	8.78	8.88	8.65
MgO	5.13	27.50	8.30	26.67	14.80	26.31
CaO	35.62	4.20	18.14	5.14	7.69	4.60
Na_2O	0.22	0.28	2.97	0.29	1.78	0.73
K_2O	0.23	4.00	0.47	2.56	5.78	3.07
P_2O_5	4.23	1.01	0.70	0.29	0.21	0.81
CO_2	---	5.00	---	0.21	---	0.18
H_2O^+	---	0.54	17.12	7.70	723	8.10
LOI	28.27	---	---	---	---	---
Total	97.25	106.09	99.10	97.24	100.09	96.68

*Key to column heads:
1. Carbonatite, Morrilton.
2a. Lamproite, Prairie Creek (Bolivar, 1984; $^{87/86}Sr$ = 0.7066).
2b. Carbonatite, Perryville.
3. Lamproite, Prairie Creek (Scott-Smith and Skinner, 1984).
4. Allikite, Dare Mine Knob
5. Lamproite, Prairie Creek (Scott-Smith and Skinner, 1984).

ered diatreme, Twin Knobs #1 (TK1), is located 2 mi southwest of Murfreesboro and covers approximately 13 acres. It consists of tuffs, epiclastics, and intrusive breccia (Waldman and others, 1985). Breccias at TK1 have yielded small, low-grade diamonds.

Geochemistry of the Prairie Creek Complex

In general, the major element geochemistry of the Prairie Creek hypabyssal peridotite is similar to kimberlites. Major element analyses (Table 2) indicate that the rocks are depleted in SiO_2, have 2 to 3 wt. % TiO_2, are Mg-rich, and contain about 10 wt. % H_2O with a significant amount of CO_2.

Trace element data are broadly similar to kimberlites, with some significant differences. A plot of Nb versus Zr (Fig. 5) indicates that the Prairie Creek samples are intermediate between kimberlites and lamproites (Scott-Smith and Skinner, 1984a). Rare-earth elements (REE) are light rare-earth (LREE) enriched, but less enriched than true kimberlites (Bolivar, 1984) (Fig. 6). The $^{87/86}Sr$ ratio of 0.707 reported by Scott-Smith and Skinner (1984a) is comparable with data from kimberlites and other mantle-derived, uncontaminated alkalic rocks.

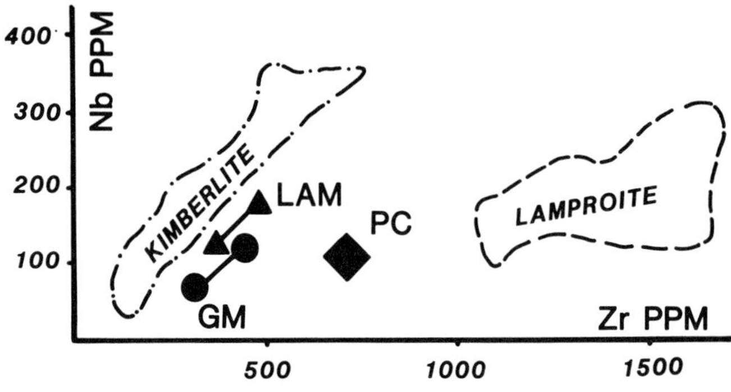

Figure 5. Plot of Nb versus Zr for Cretaceous alkalic rocks of Arkansas. GM indicates Granite Mountain; LAM, lamprophyres of Benton dikes; PC, Prairie Creek complex lamproite.

Classification of the Prairie Creek Complex

The classification best fitting the rocks of Prairie Creek seems to be lamproite, as recognized by Scott-Smith and Skinner (1984b). Lamproites, defined by Wade and Prider (1940) and Mitchell (1970), are potassium, titanium, and magnesium-rich lamprophyric rocks characterized by the presence of diopside,

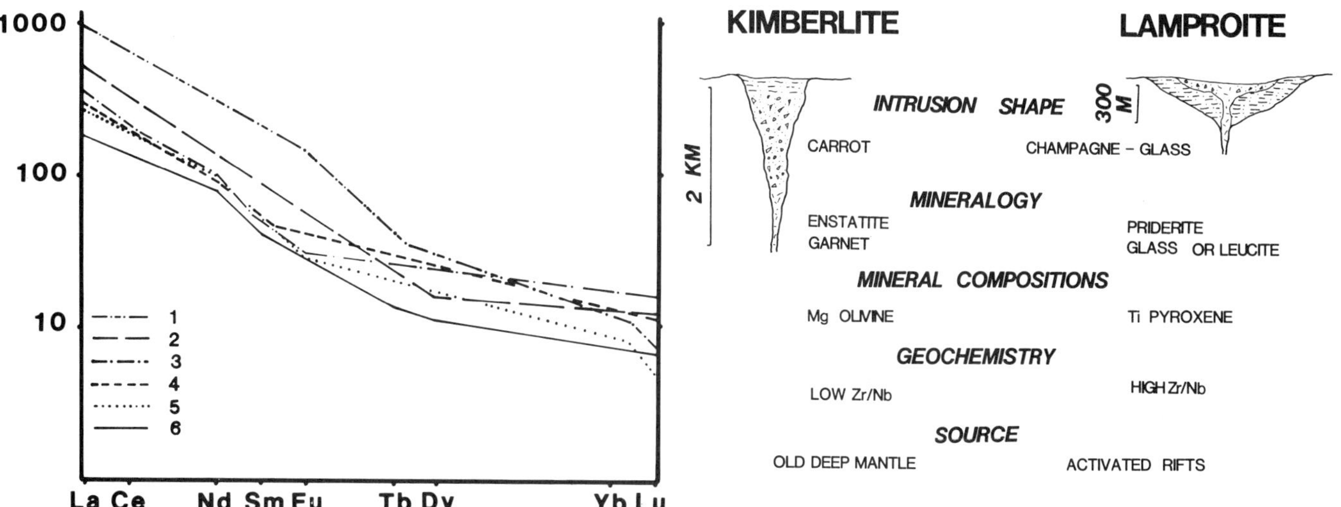

Figure 6. Chondrite-normalized rare-earth element plot for Cretaceous alkalic rocks of Arkansas. 1 indicates allikite, Dare Mine Knob; 2, lamproite, Prairie Creek complex; 3, light gray syenite, Granite Mountain; 4, olivine syenite, Granite Mountain; 5, minette, Benton dikes; 6, monchiquite, Benton dikes.

Figure 7. Generalized chart illustrating differences between lamproite and kimberlite.

amphibole, Ti-rich and Al-poor phlogopite, and olivine, with diamond, sanidine, leucite, priderite, perovskite, Cr-spinel, and apatite as possible accessories. Garnet, orthopyroxene, and ilmenite are absent or extremely rare. The only garnets recovered from Prairie Creek are near almandine in composition. Glass is a groundmass phase in lamproites, but is not reported from kimberlites. Furthermore, lamproite intrusions have a flared, "champagne glass" cross section, whereas kimberlites have a narrow, carrot-shaped, pipe-like morphology (Fig. 7). The mineralogy described for Prairie Creek is that of a lamproite. The cross section of Prairie Creek determined by limited drilling is more like the lamproite. The diamond-bearing intrusion at Prairie Creek is properly classified as lamproite rather than kimberlite.

PERRYVILLE-MORRILTON CARBONATITES

Thin carbonatite dikes, mostly less than 1 m in width, are present near Perryville, Morrilton, and Oppelo, 70 to 80 km north of Magnet Cove. These dikes consist of a carbonate matrix with phlogopite, perovskite, amphibole, and diopside phenocrysts, magnetite and diopside xenocrysts, and a variety of xenoliths. They are extremely undersaturated rocks with highly variable compositions (Table 2). McCormick and Heathcote (1979) reported two generations of carbonatite in the Morrilton dike: an early sovite, which is found as xenoliths in later alvikite. Intimate mixing of minerals such as phlogopite and apatite with brecciated shale and sandstone at the Perryville and Oppelo localities indicates explosive emplacement of at least some of these northern carbonatites.

Mineral compositions of these rocks suggest mantle origin. Phlogopites are Mg-rich, Ti-poor, and plot in fields near kimberlite phlogopites (Fig. 3). The Morrilton-Perryville rocks also contain Al-rich mica with very different compositions than Prairie Creek phlogopite. Most micas are strongly embayed, and are more likely xenocrysts than phenocrysts because of this extreme reaction. However, compositions within any one dike are fairly homogeneous. Apatite occurs as phenocrysts as large as 50 mm long, and is mostly carbonate apatite with high birefringence and a trace of SiO_2 (1 to 2 wt.%). Amphibole is rare, and is a kaersutite with 3 to 4 wt.% TiO_2. Clinopyroxenes reported by McCormick and Heathcote (1986) are Mg-enriched diopside with greater MgO in xenocrysts (Di83 Hd17) than phenocrysts (Di78 Hd22).

The Perryville-Morrilton carbonatites contain sedimentary (shale, sandstone) and igneous (granitoid, granite gneiss?) xenoliths, some of which are 0.5 to 1 m in diameter and are a major component of these dikes. Ultramafic xenoliths have not been reported. The original mineralogy of granitoid clasts is obscured by alteration of feldspar and mafic minerals to epidote and calcite. The faintly gneissic texture of many xenoliths has caused speculation that they represent lower crust (McCormick and Heathcote, 1986). The mineralogy of the carbonatites themselves indicate that the dikes had a deep mantle source.

MAGNET COVE COMPLEX

The Magnet Cove complex is located 20 km east of Hot Springs, Arkansas, and is exposed over an area of about 12 km^2. It is a classical alkalic ring-dike complex and intrudes the deformed Paleozoic sedimentary rocks of the Ouachita Mountains.

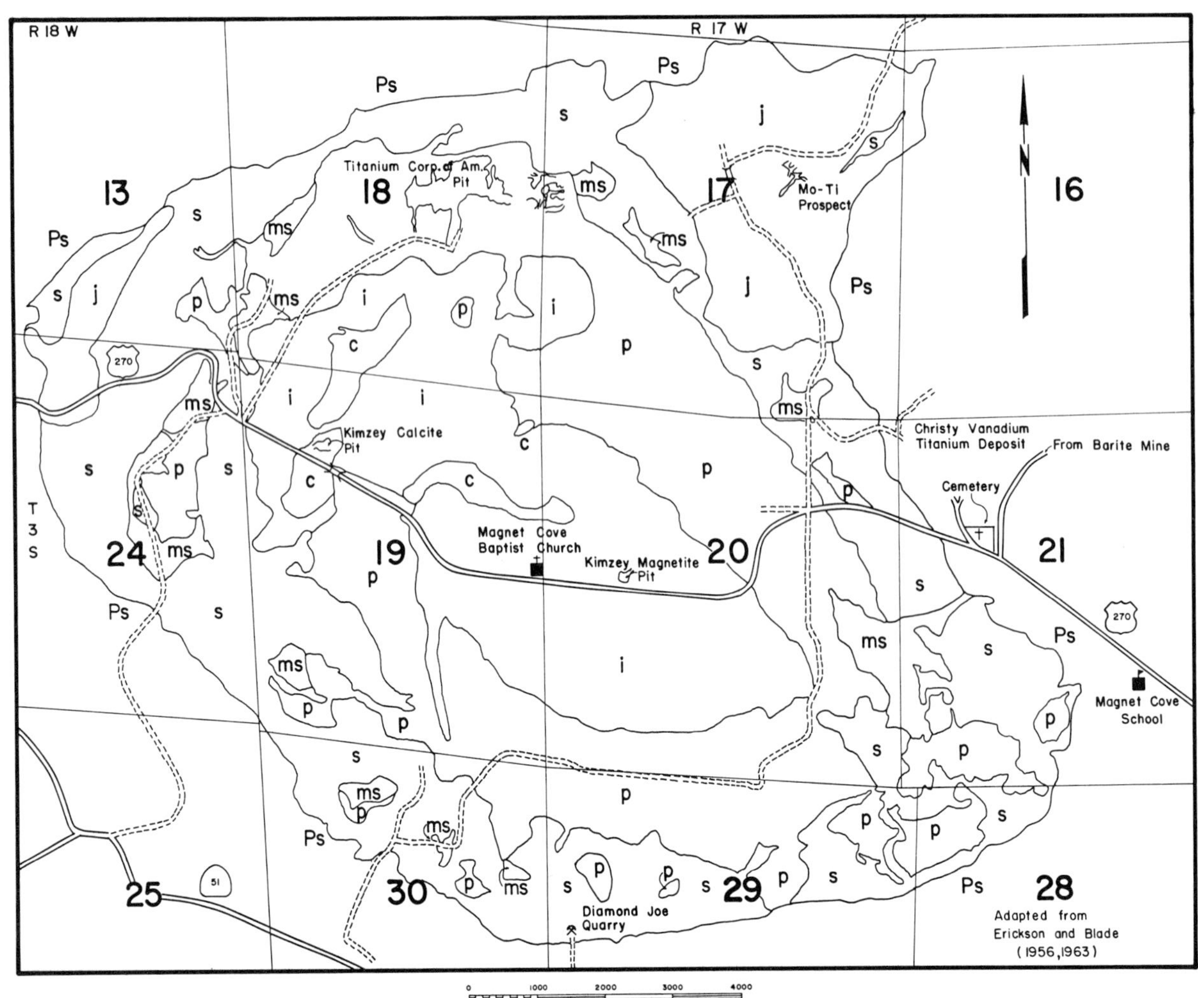

Figure 8. Geologic map of Magnet Cove complex: c indicates carbonatite; i, ijolite; j, jacuparangite; p, phonolite; s, syenite; ms, metasedimentary rocks (roof pendants and contact aureole); Ps, Paleozoic sedimentary rocks.

The only thorough investigations of the complex have been by Williams (1891) and Erickson and Blade (1963). Magnet Cove has been dated at 95 to 97 Ma, based on K-Ar ages of biotite separates (Zartman, 1977).

Lithologies and Mineralogy of the Magnet Cove Complex

Six principal rock types have been recognized at Magnet Cove (Fig. 8). The core of the complex consists of ijolite and low-alkali carbonatite. The carbonatite contains calcite, green or brown biotite, monticellite, and rare pyroxene or amphibole. High-Nb perovskite, magnetite, apatite, and Zr-garnet (kimzeyite) are the principal accessory minerals. The carbonatite is considered by Erickson and Blade to be the last major intrusion, due to the high-Nb content of its perovskite and the inclusions of ijolite xenoliths within carbonatite. The ijolite is a coarse- to medium-grained rock with porphyritic to xenomorphic granular textures. It contains varying amounts of diopside, biotite, nepheline, and garnet, with sphene and apatite as common accessory minerals. The lithology varies locally from ijolite to melteigite.

The middle, U-shaped ring, consisting of trachyte and phonolite, is considered by Erickson and Blade (1963) to be the first part of the complex emplaced. These rocks are very fine grained, commonly altered and weathered, and were mapped as metashale by Williams (1891). Their mineralogy consists of very fine grained orthoclase or sanidine, with varying amounts of nepheline, sodalite, hornblende, diopside, biotite, sphene, and apatite. Plagioclase (andesine) constitutes less than 10 percent of the trachyte. All middle-ring rocks are nepheline-normative.

The outer ring of the Magnet Cove complex is mostly nepheline syenite, with garnet pseudoleucite syenite, jacupirangite, and sphene-rich pyroxenite as subordinate units. Roof pendants and partly melted country-rock xenoliths are common in rocks of this zone. Syenites contain orthoclase, nepheline, zoned pyroxenes

TABLE 3. WHOLE-ROCK GEOCHEMISTRY OF MAGNET COVE ROCKS*

Oxide (wt. %)	1	2	3	4	5
SiO_2	1.09	36.89	50.48	35.42	47.31
TiO_2	0.10	3.32	2.21	4.05	0.88
Al_2O_3	0.33	13.28	20.98	9.21	20.10
Fe_2O_3	0.42	7.58	2.29	8.94	3.57
FeO	0.32	3.71	5.16	7.17	2.62
MnO	0.26	0.33	0.29	0.29	0.30
MgO	1.05	4.32	1.56	7.77	0.89
CaO	53.37	14.80	5.11	20.83	6.67
Na_2O	0.00	5.29	5.96	1.47	7.98
K_2O	0.16	4.62	3.39	0.62	6.50
BaO	0.00	0.18	0.18	0.00	0.36
P_2O_5	2.00	0.99	0.38	2.23	0.21
H_2O^-	0.04	0.14	0.11	0.11	0.10
H_2O^+	0.12	1.11	0.99	1.05	1.42
CO_2	39.41	1.54	0.31	0.11	1.07
SO_3	0.02	0.80	0.01	0.12	0.01
Cl	0.00	0.41	0.00	0.02	0.04
F	0.15	0.40	0.35	0.17	0.19
Total	99.74	100.56	100.20	99.96	100.28

*Key to column heads:
1. Carbonatite (L304).
2. Ijolite (MC-217.
3. Trachyte-phonolite (MC-227).
4. Jacupirangite (MC-173).
5. Garnet pseudo-leucite syenite (MC-121).

Source: Erickson and Blade (1963).

(diopside core, titanaugite intermediate zones, and aegerine-augite to acmite rims), hornblende, alkali amphiboles, apatite, magnetite, and variable amounts of sphene or melanite garnet. Sodalite is a late-magmatic to deuteric mineral.

The rocks of the outer ring are well exposed in the Diamond Jo quarry at the south end of the complex. At this locality, garnet-pyroxene syenite, ijolite, nepheline syenite, and pseudoleucite syenite occur in a syenite to phonolite matrix. All lithologies are strongly nepheline-normative (Erickson and Blade, 1963).

Jacupirangite constitutes a very small proportion of the outer ring and occurs in two localities. Its principal minerals are magnetite with perovskite rims, and pyroxene with a salite core rimmed by titanaugite. Garnet, sphene, apatite, and aegerine-augite occur in veins and as late phases in the rocks. The feldspathoidal pyroxenites are similar to jacupirangite except that they contain less magnetite.

Abundant late-stage dikes cross-cut the ring structure of the Magnet Cove complex. They have a wide range of lithologies, including tinguaite, analcime-olivine melagabbro, sodalite trachyte porphyry, and a variety of pegmatites. No systematic progression of compositions is apparent. Similar dikes are present in the surrounding country rock, some as much as 2 km in length.

Geochemistry of the Magnet Cove Complex

The Magnet Cove complex is an archetypal ring dike complex. Its major and trace element geochemistry (Table 3) shows extreme silica undersaturation. The rocks are enriched in volatiles, alkalis, and CaO, as well as V, Nb, Zr, and most large ion lithophile (LIL) elements. They are depleted in Cr, Ni, and other siderophiles, suggesting that fractionation played some role in the evolution of the complex, or that the source region underwent only very limited partial melting.

The geochemical relations between Magnet Cove units are not simple. Periodic intrusion of a single, fractionating, melanocratic phonolite magma was proposed as a model by Erickson and Blade (1963). Some evidence supports this idea: Nb/Zr, for example, plots as a straight line. However, points on this plot have considerable scatter, as do points on AFM and SiO_2/oxide diagrams.

The overall model proposed by Erickson and Blade (1963) for the sequence of intrusion is (1) middle-ring phonolite and trachyte, (2) jacupirangite, (3) outer-ring syenites, (4) ijolite, and (5) carbonatite. The model is based on progressive desilication, with phonolite and trachyte as "chilled margin" and the final carbonatite being the least siliceous and most undersaturated. Although some geochemical relations support this model, more detailed studies, including consideration of liquid immiscibility and separate sources, are needed to fully comprehend the petrology of the Magnet Cove complex.

V INTRUSIVE

The V intrusive, 10 km southwest of Magnet Cove and on the west shore of Lake Catherine, occupies an area of approximately 0.5 km^2. It consists of at least 40 cross-cutting dikes that have V-shaped outcrop patterns on existing maps. The dikes vary from 10 to 500 m in length and range from malignite to syenite in composition. Biotites have yielded fission track ages of 102 to 97 Ma (Owens, 1968; G. N. Eby, written communication, 1987).

Lithologies and Mineralogy of the V Intrusive

The relative ages and the sequence of intrusion of the dikes are difficult to define because of limited exposure. On the basis of field mapping, projected trends, and magnetic data, Owens (1968) suggested the sequence (from oldest to youngest) is malignite → melteigite → microijolite → tinguaite porphyry → sannaite → melanite nepheline syenite → pyroxene nepheline syenite → nepheline syenite → phonolite → syenite → syenite aplite → veins of fluorite, pyrite, and silica.

Mineral compositions of rocks within the V intrusive are apparently similar to those of the Magnet Cove complex. Olivine (Fo_{70}) occurs in V intrusive sannaite as large, embayed phenocrysts. Clinopyroxenes have complex zoning, with diopside cores, thin titanaugite outer bands, and aegerine-augite rims that replace and corrode both diopside and titanaugite. Amphiboles have not been reported in rocks of the V intrusive, although Owens (1968) suggests that aegerine replaces amphibole in syenites. Biotite is a very late, possibly subsolidus phase that has TiO_2-enriched rims in mafic rocks. Accessory minerals reported include sphene, perovskite, rutile, zircon, magnetite, ilmentite, fluorite, apatite, and calcite (Owens, 1968).

Geochemistry of the V intrusive

Geochemical investigations of the V intrusive are limited. Four analyses by Van Buren (1977) plot on a scattered horizontal trend on an AFM diagram, and V/TiO_2 also scatters somewhat. Information is too scarce for a conclusion regarding petrologic relations among these rocks. The model proposed by Owens (1968) is similar to the model proposed for the Magnet Cove complex: sequential intrusions from a fractionating magma. More data is needed, however, for a full understanding of the petrology of the V intrusive.

POTASH SULFUR SPRINGS

The Potash Sulfur Springs complex is located 10 km east of Magnet Cove and 5 km north of the V intrusive. It covers about 1 km^2 in southwest Garland County. Zircons from syenite have been dated at 101 Ma (Zartman and Howard, 1985), which is in excellent agreement with ages for the Magnet Cove complex and the V intrusive.

Lithologies and mineralogy of the Potash Sulfur Springs complex.

The Potash Sulfur Springs complex may be subdivided into rim, intermediate zone, core, and crosscutting dikes. Syenites compose the rim and extend into the core. The major constituents of these syenites are alkalic feldspar (orthoclase and sodic orthoclase) and nepheline. Minor minerals include biotite, aegerine, aegerine-augite, melanite, sphene, apatite, and minerals of the sodalite group. More mafic rocks, including ijolite, occur in the core and the intermediate zones. The pyroxene in these rocks is augite, mantled and replaced by aegerine. Olivine is similar to olivine of V intrusive sannaite, and is replaced by serpentine and zeolites. Melanite and biotite are late-stage minerals. Alkali feldspar and nepheline are also late and generally anhedral. Sodalite minerals, calcite, sphene, and ilmenite are the usual accessories.

Carbonatite is present in a single exposure near the core of Potash Sulphur Springs. Ijolite occurs as a small, arcuate body in the core area along with phonolite, tinguaite, and malignite.

Geochemistry of the Potash Sulfur Springs Complex

Geochemical trends for the Potash Sulfur Springs complex show a fairly broad scatter, but are generally consistent with its derivation from a single, fractionating magma. Complete major element analyses are not available. Plots of V/TiO_2 and Fe_2O_3/TiO_2 show an indistinct fractionation trend. However, similar patterns could be produced by contamination or alteration.

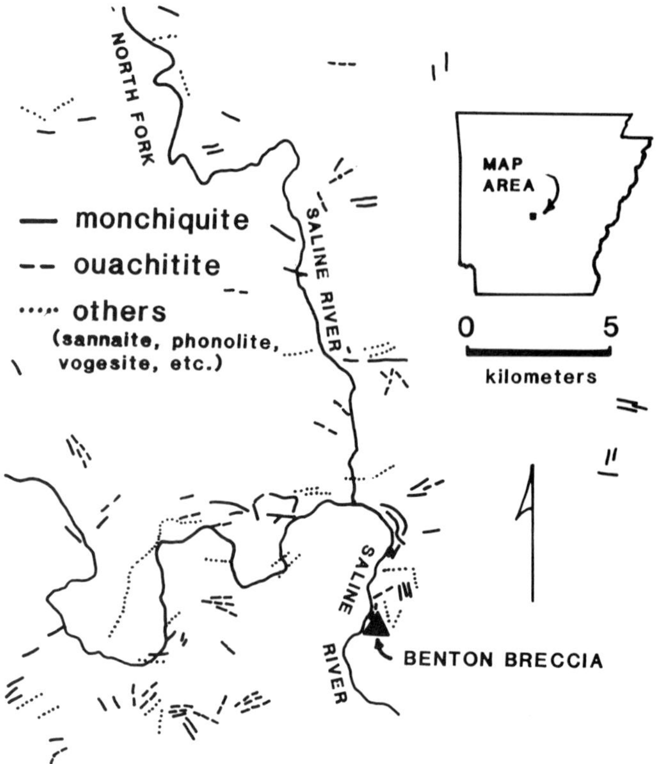

Figure 9. Locality map showing rock types and orientations of lamprophyre dikes, Benton dike swarm. Map after Valdovinos, 1968.

The model proposed by Howard (1974) for the origin of the complex suggests early emplacement of syenite, followed by intrusion of lamprophyres and culminating with intrusion of the carbonatite and ijolite core, a sequence similar to Magnet Cove. In this scenario, syenites are intruded as fractionated cupolas at the top of a large, alkali olivine basalt magma chamber. This same fractionating chamber yielded the later, diverse lamprophyres. The carbonatite intruded last and developed as the end-product of fractionation of carbonated nepheline syenite magma (Howard, 1974). Fenitization of the country rock was associated with this last stage of intrusion (Heathcote, 1974).

BENTON DIKE SWARM

Lamprophyre dikes occur throughout much of the central and eastern Ouachitas. They are abundant within a 350-km^2 area north of Benton (Fig. 9). The dikes are generally 1 to 2 m wide and vary from 100 m to 2 km in mapped length. Most trend east-west. Exposure of the dikes is poor; felsic varieties especially weather rapidly, and unless uncovered by excavation or stream erosion, the lamprophyre dikes are nearly impossible to find. There appears to be no systematic distribution of dike composi-

TABLE 4. REPRESENTATIVE MINERAL ANALYSES: LAMPROPHYRES, BENTON DIKE SWARM*

Oxides (wt.%)	1	2	3	4	5	6
SiO_2	39.36	48.48	44.07	39.86	35.58	33.72
TiO_2	0.03	1.52	3.85	5.45	5.26	5.87
Al_2O_3	0.03	5.86	7.87	14.06	16.56	16.83
FeO*	13.14	6.40	8.27	9.74	9.31	15.37
MnO	0.16	0.13	0.12	0.12	0.09	0.26
MgO	47.68	14.32	12.17	13.90	18.05	13.41
CaO	0.27	22.71	23.12	12.41	0.04	0.03
Na_2O	0.02	0.58	0.44	1.78	0.62	0.63
K_2O	0.01	0.00	0.00	2.00	9.23	8.74
Cr_2O_3	0.05	0.01	0.03	0.00	0.00	0.09
NiO	0.09	0.03	0.02	0.01	0.00	0.00
Total	100.8	100.3	99.9	99.3	94.6	94.8

*Key to column heads:
1. Olivine (SM-1).
2. Diopside core (A-102).
3. Titanaugite rim (A-102).
4. Kaersutite amphibole (SM-1).
5. Biotite core (A-66).
6. Biotite rim (A-66).

tions, size, orientation (Valdovinos, 1968), or geochemistry (Robinson, 1977).

Fission track ages for the Benton dike swarm are similar to ages for intrusions to the southwest. Sphene from sannaite yielded a date of 100 Ma; apatite from camptonite yielded a date of 99 Ma (G. N. Eby, written communication, 1987).

Figure 10. Photomicrograph of small, zoned clinopyroxene phenocryst, minette of Benton dike swarm. Bar = 1 mm; crossed nicols.

Lithologies and Mineralogy of the Benton Dike Swarm

The dikes of the Benton dike swarm vary greatly in lithology. Most are feldspathoidal monchiquites, ouachitites, or minettes. Camptonite, sannaite, and syenite have also been reported (Valdovinos, 1968). A breccia pipe approximately 100 m in diameter (the "Benton Complex") is associated with the dike swarm and may represent an ultravulcanian event (Peterson, 1972).

Although modal mineralogy varies from dike to dike, mineral compositions investigated to date are quite similar (Table 4). Olivine is magnesian, Fo_{87-82}. It is embayed, unzoned, and partly to completely altered. Pyroxenes are mostly euhedral and are zoned on two scales: an abrupt change from calcic diopside or salite to thin titanaugite rims, and a fine-scale, subtle, oscillatory variation in Mg/(Fe + Mg) (Fig. 10). Pyroxene compositions are salite to diopside (Fig. 11), and become relatively more calcic from core to rim. The abrupt transition to titanaugite may signal a rapid decrease in availability of SiO_2 and/or a change of magmatic conditions out of the diopside stability field. It is a nearly universal occurrence in lamprophyre dikes.

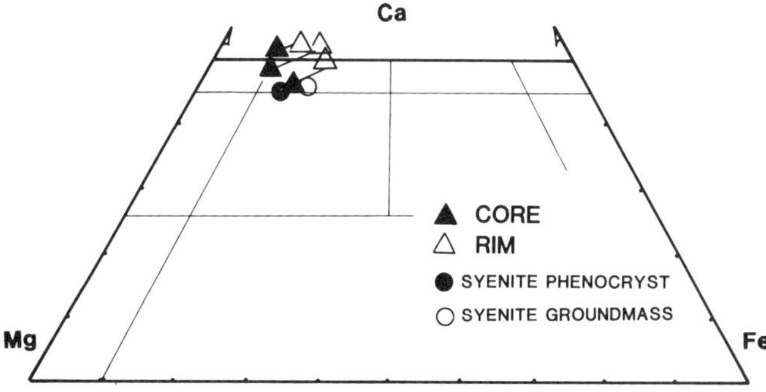

Figure 11. Compositions of clinopyroxenes in lamprophyres and syenite. Triangles indicate Benton dike swarm lamprophyres; circles, Granite Mountain dark gray syenite.

Amphiboles in the dikes are usually euhedral kaersutite, with thin, TiO_2-enriched rims. Mg-biotite or Fe-phlogopite is

TABLE 5. WHOLE-ROCK GEOCHEMICAL ANALYSES: BENTON DIKE SWARM

Oxides (wt. %)	1	2	3	4
SiO_2	41.3	35.8	39.9	36.1
TiO_2	3.4	3.9	2.6	3.3
Al_2O_3	12.4	11.9	14.7	9.8
FeO*	11.6	12.7	11.3	11.7
MnO	0.2	0.2	0.2	0.2
MgO	9.7	6.1	4.6	7.9
CaO	13.4	13.1	10.1	15.6
Na_2O	2.3	1.9	2.8	1.1
K_2O	1.5	3.2	3.6	3.2
P_2O_5	0.5	0.7	0.8	0.6
LOI	2.5	8.7	8.7	9.23
Total	99.0	98.5	99.8	99.1
Rb	50	90	110	86
Ba	990	940	na	1600
Sr	730	980	1250	1000
La	63	96	na	96
Yb	1.7	2	na	2

*Key to column heads:
1. Melamonchiquite (SM-1).
2. Ouachatite (A-202).
3. Monchiquite (A-66).
4. Minette (A-45).

euhedral to subhedral, zoned with Mg-rich cores, and, like the amphibole, has TiO_2-enriched rims.

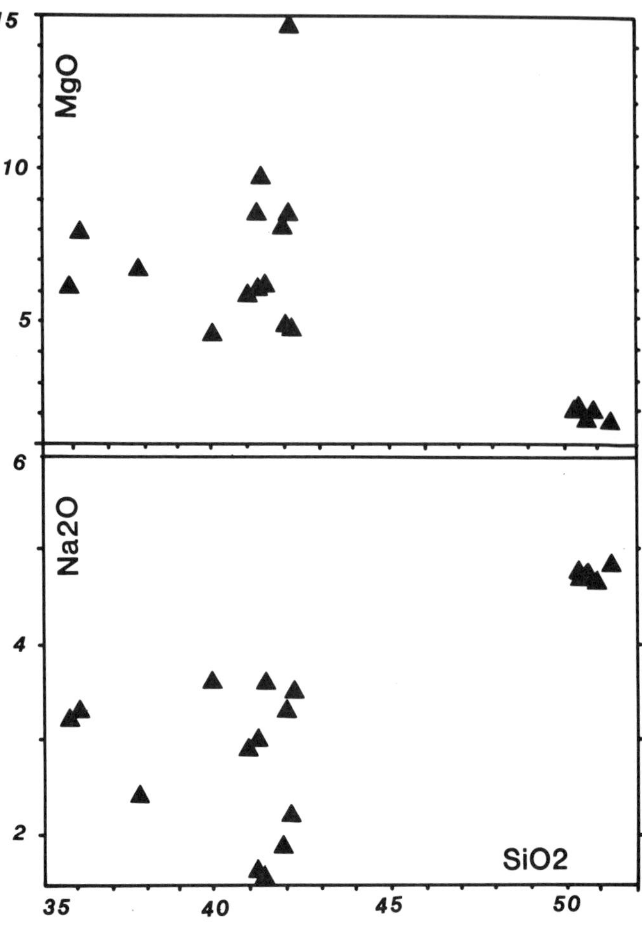

Figure 12. Oxide-silica plots for lamprophyres of Benton dike swarm.

Geochemistry of the Benton Dike Swarm

Major element geochemistry (Table 5) shows that rocks of the Benton dike swarm are strongly undersaturated, enriched in CaO and TiO_2, and relatively low in alkalis. They are olivine-nepheline–normative.

Trace element abundances suggest a relatively unfractionated magma: Cr = 100 to 230 ppm, LREE are 100 to 200 times chondritic abundances, and K/Rb ratio is 250 to 300. The MgO/(MgO + FeO) ratio for analyzed lamprophyres is 0.46 to 0.20. These indicators are similar to slightly fractionated oceanic alkali olivine basalts.

Strontium isotope data indicate that the lamprophyres are mantle-derived, with limited crustal influence. The $^{87/86}Sr$ ratio for biotite separates from a fresh monchiquite is 0.7043. This value is similar to other mantle-derived lamprophyres worldwide, and much lower than the whole-rock value of 0.707 reported for Prairie Creek by Scott-Şmith and Skinner (1984a).

The questions of the origin, magmatic history, and interrelations of these dikes are not yet resolved. Previous work (Van Buren, 1968; Robison, 1977; Steele and Robison, 1976) suggested fractional crystallization and sequential emplacement of a single-source, hydrous basaltic magma, with the addition of alkalis by fenitization. However, scatter on geochemical plots (AFM, Nb/Zr, Sr/Th, Rb/Th, etc.) (Figs. 12, 13) as well as preliminary modeling of new data indicate that other factors including multiple sources and magma mixing must be considered.

SYENITES

Syenites are exposed at two major localities: the Saline County syenite near the town of Bauxite, and Granite Mountain just south of Little Rock (Fig. 14). With the exception of olivine syenites from Granite Mountain that have a mean fission track age of 95 Ma (data from apatite counts), the age of the rocks based on both K-Ar dates of biotite separates and fission track ages of apatite and sphene are 89 to 87 Ma (Morris and Eby, 1986).

Drill-hole and gravity data indicate that syenite underlies much of the area between these major exposures (Gordon and others, 1958), and may represent either a single large pluton or multiple, smaller intrusions. Little information is available regarding the syenites of Saline County due to their poor exposure and high degree of alteration. The following discussion focuses on new data from Granite Mountain.

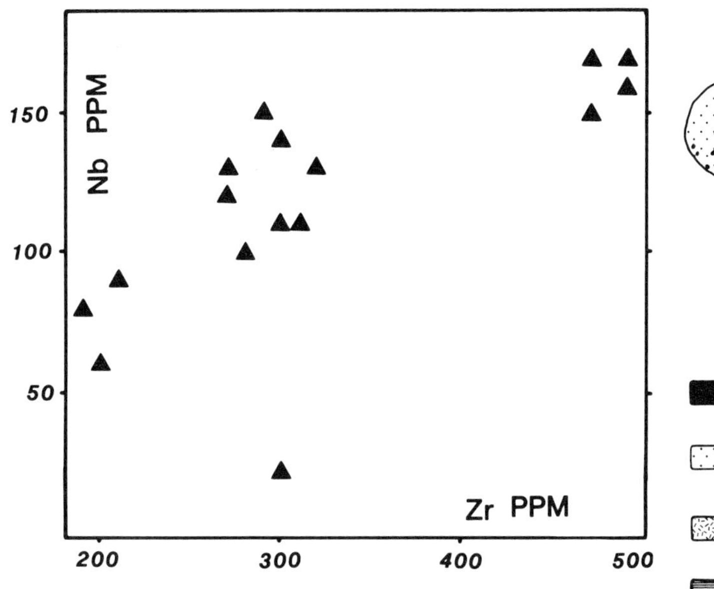

Figure 13. Plot of Nb versus Zr for lamprophyres of Benton dike swarm.

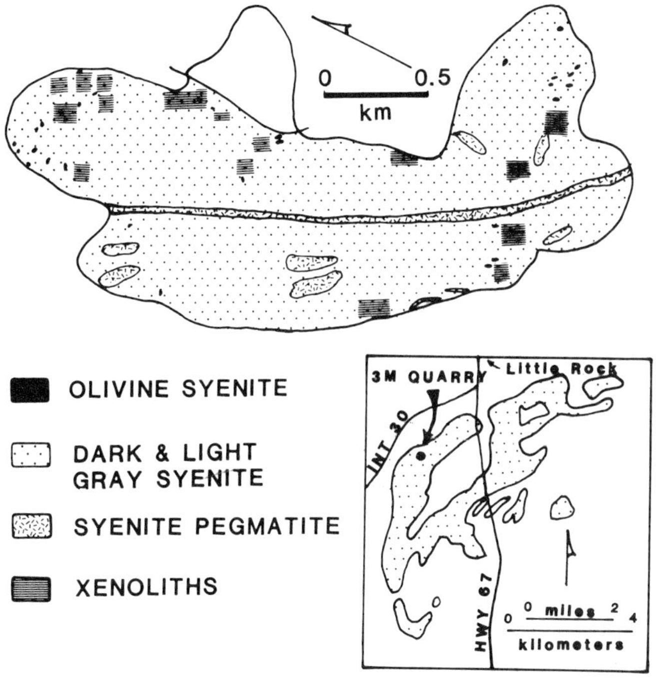

Figure 14. Map showing distribution of rock types in 3M quarry. Inset map shows location of quarry in Granite Mountain pluton.

Lithologies of Granite Mountain

The principal exposures of syenite in Arkansas are those of Granite Mountain, about 12 km^2 in area, just south of Little Rock. The pluton consists of medium- to coarse-grained syenite that may be subdivided into five principal lithologies: olivine syenite, dark gray syenite (commonly referred to as pulaskite), light gray syenite, syenite pegmatites, and quartz syenite. The major portion, perhaps 85 percent of Granite Mountain, consists of dark and light gray syenite intimately mixed and co-mingled. Contacts between syenite and the sedimentary country rock are sharp, and the syenite's 10-m-wide chilled margin retains a flow alignment of alkali feldspar phenocrysts. In addition to the syenite, at least three distinct types of xenoliths are present in this pluton: sedimentary country rock, lamprophyre (minette), and ultramafic cumulates, high in sulfide, which represent residual cumulate clots from the early magma.

The earliest and most "primitive" syenite of the Granite Mountain pluton is a very dark gray, medium-grained rock that contains 1- to 2-modal percent olivine (Fo_{70}), 10 to 15 percent euhedral diopside, 2 to 5 percent biotite, a trace of ilmenite, and at least 80 percent alkali feldspar, some of which contains patchy plagioclase (An_{30}) cores. This rock might best be classified as shonkinite or leucoshonkinite. It constitutes less than 1 percent of the portion of the Granite Mountain pluton mapped in detail (Fig. 14), and occurs as irregularly distributed "blobs" usually less than 1 m in diameter.

The dark gray syenite that is also known as pulaskite is nearly identical in appearance to the olivine syenite, but is a slightly lighter gray color. It is the principal rock of Granite Mountain. This rock contains no olivine, ilmenite, or relict plagioclase. The mineralogy is alkali feldspar (anorthoclase to orthoclase), salite, biotite, nepheline, sphene, and magnetite; transition from the olivine syenite to this dark gray pulaskite is gradational over several centimeters.

Lighter color gray syenite occurs as veins and blobs within the dark gray syenite, and is intimately mixed on a scale of centimeters in some locations (Fig. 15). It contains alkali feldspar (orthoclase to micro-perthite), biotite, salite, amphibole, sphene, magnetite, nepheline, and analcime. The lighter color of this rock is due to the greater hydration and alteration of its feldspar. Clinopyroxene is corroded and sometimes mantled by aegerine. Miariolitic cavities and vugs are apparent.

Coarse, faintly pink syenite pegmatite forms discrete, sharply bounded veins and dikes up to 20 m wide and 1,000 m long. The feldspars contain abundant fluid inclusions and have an overall composition of sodic anorthoclase. Acmite, zeolites, and a variety of rare minerals are present. Field relations and fission track data from sphene (Morris and Eby, 1986) indicate that the pegmatite was the last rock emplaced at Granite Mountain.

Quartz syenites are rare, and restricted to the westernmost outcrops and to a dike in the north-central part of the pluton. There are few data on these rocks, whose siliceous nature may be due to assimilation of local sandstones.

Mineralogy of Granite Mountain

Olivine (Fo_{70}), (Table 6) occurs only in the olivine syenite (shonkinite). It is embayed and unzoned, occurs both as isolated

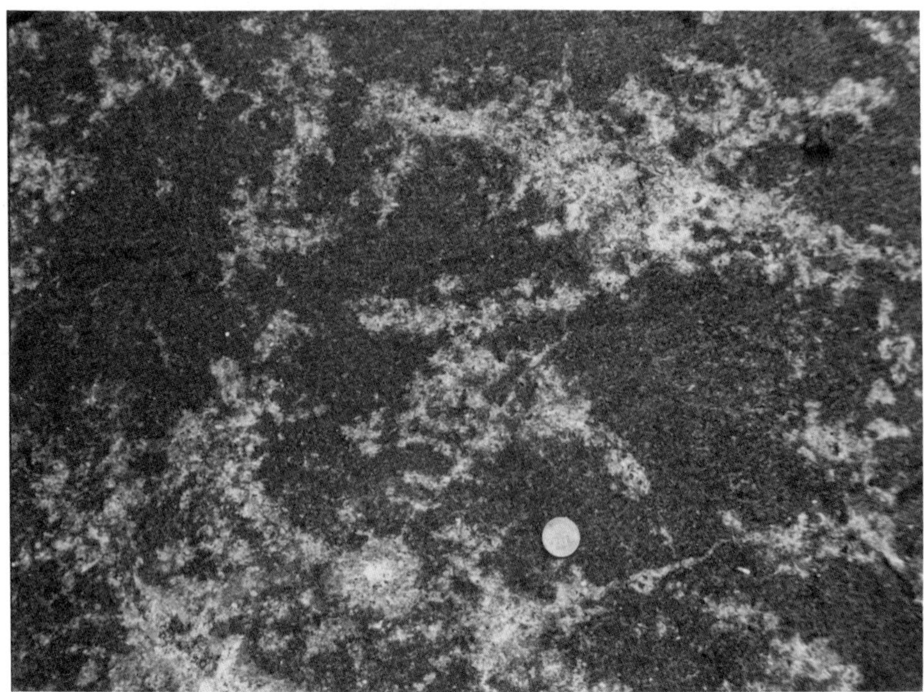

Figure 15. Dark gray syenite mottled by light gray syenite, 3M quarry, Granite Mountain.

TABLE 6. REPRESENTATIVE MAFIC MINERAL ANALYSES: GRANITE MOUNTINE OLIVINE SYENITE

Analysis Oxide (wt.%)	Olivine	Phenocryst Pyroxene	Groundmass Pyroxene	Late, Euhedral Biotite
SiO_2	37.13	50.11	49.70	37.65
TiO_2	0.00	0.83	1.70	6.52
Al_2O_3	0.00	2.64	3.18	13.62
FeO*	27.83	7.40	6.95	15.10
MnO	2.52	0.54	0.55	0.48
MgO	31.81	14.11	13.71	13.42
CaO	0.30	21.42	22.38	0.91
Na_2O	0.00	1.04	1.00	0.79
K_2O	0.00	0.00	0.05	8.21
Cr_2O_3	0.03	0.01	0.04	0.00
NiO	0.05	0.02	0.02	0.00
Total	99.6	99.8	99.2	96.7
Fo	70			
Ca		46	48	
Mg		42	41	
Fe		12	11	

crystals and in clusters, and constitutes as much as 2 modal percent of the rock (Fig. 16). The olivine has a markedly high content of CaO (as much as 0.30 wt.%), probably indicating near-surface crystallization (Simpkin and Smith, 1970) and also reflecting the alkalic character of the magma. Olivines contain virtually no NiO. The olivines observed in syenites are fresh and unaltered. They contain or partly enclose large, inclusion-rich apatite crystals as much as 2 mm in length. The olivines are commonly surrounded by calcic diopside and biotite, and are replaced by biotite as the olivine syenite grades into the gray syenite (pulaskite).

Clinopyroxene occurs as euhedral phenocrysts, 1 to 5 mm in diameter, in both olivine syenite and the dark gray syenite (Fig. 17). Small (0.10 mm) pyroxene laths are present in the

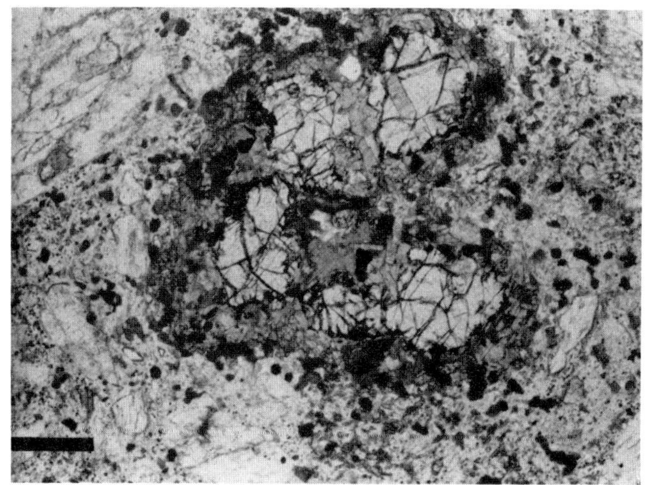

Figure 16. Photomicrograph of olivine in Granite Mountain olivine syenite. Biotite and salite at rim; apatite is euhedral crystal enclosed within olivine. Bar = 1 mm; PPL.

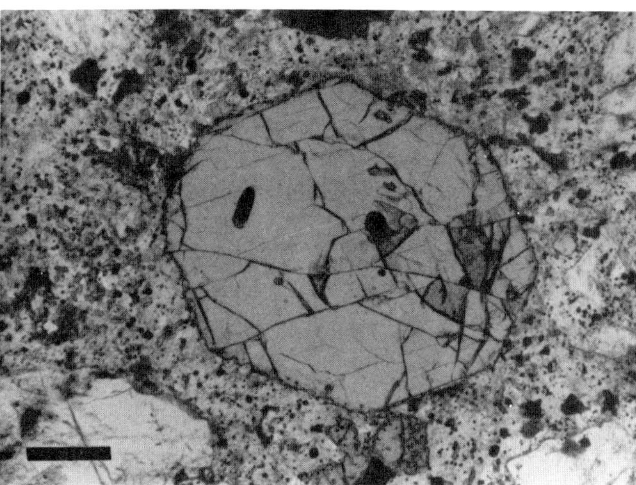

Figure 17. Euhedral augite in Granite Mountain olivine syenite. Bar = 1 mm; PPL.

"groundmass" of porphyritic syenite at the pluton's rim. Pyroxenes are Mg-salite. Phenocrysts are Ca_{46}, Mg_{43}, Fe_{11}. Zoning is absent. Groundmass pyroxenes have almost identical compositions, but are slightly more calcium- and iron-enriched (Ca_{48}, Mg_{40}, Fe_{12}) (Fig. 11). The clinopyroxenes of Granite Mountain follow a short trend of Ca and Fe enrichment similar to that evident in the lamprophyres. However, they lack the titanaugite rim common in lamprophyre and Magnet Cove complex clinopyroxenes. The clinopyroxenes in the light gray syenite and in syenite pegmatites are similar in composition to clinopyroxene in the early syenite. These rocks also contain aegerine in fibrous clusters and as mantles on some diopsides.

Amphibole is absent in the olivine syenite and uncommon in the dark gray syenite, but present in virtually all light gray syenite. Kaersutite constitutes 1 to 2 percent of dark gray syenite when present, and may form a thin rim on salite. In light gray syenite it is euhedral as well as mantling salite. In pegmatites, kaersutite is rimmed by green amphibole or acmite.

Biotite is uncommon in olivine syenite, and much more abundant in the dark gray and light gray syenite, where it constitutes as much as 15 percent of the rock. It is late, usually euhedral, and developed in at least two generations. Early biotite is Mg-rich and, where present as single crystals, is embayed. Later biotite rims the Mg-rich phase and forms euhedral, nonembayed crystals with Fe = Mg. Biotite contains 6 to 7 wt.% TiO_2 and about 13 wt.% Al_2O_3.

More than any other mineral, feldspar demonstrates the complex history of Granite Mountain syenites. Plagioclase (An_{20-30}) (Fig. 18; Table 7) is progressively replaced by anorthoclase and then orthoclase as the potassium content of the rock increases. Significant structural changes occur at the same time, from homogeneous plagioclase to orthoclase and to microperthite.

Figure 18. Plagioclase replaced by K-feldspar, Granite Mountain olivine syenite. Bar = 1 mm; crossed nicols.

Feldspar compositions plot along a continuous trend from andesine through anorthoclase to orthoclase/albite (Fig. 19). Feldspars in olivine syenite are plagioclase corroded and replaced by anorthoclase. Feldspar of the dark gray syenite is predominantly orthoclase with as much as 10 wt.% CaO. Feldspars in the light gray syenites are orthoclase to microperthite with about 0.1 wt.% CaO and slightly more sodic compositions. In the feldspars of the syenite pegmatites, CaO is virtually absent and feldspar has exsolved into end-member albite and orthoclase (microperthite) (Fig. 20). Albite constitutes about 70 percent of the total microperthite in the pegmatites. Overall, the feldspars suggest that potassic metasomatism was a major process in the petrology of Granite Mountain.

TABLE 7. REPRESENTATIVE FELDSPAR ANALYSES: GRANITE MOUNTAIN SYENITES

	Olivine Syenite			Dark Gray Syenite	Light Gray Syenite	
Oxide (wt.%)	BP-8 Core	BP-8 Rim	BP-10C	86-17	86-2	86-2
SiO_2	59.31	63.88	61.05	66.19	69.10	64.48
Al_2O_3	25.47	20.28	24.76	18.50	19.13	17.78
FeO*	0.25	0.38	0.42	0.22	0.11	0.00
CaO	6.11	0.87	5.05	0.29	0.08	0.00
BaO	0.60	0.12	0.31	0.34	0.04	0.20
Na_2O	6.63	4.66	7.43	6.31	11.39	0.28
K_2O	0.63	8.83	1.56	7.91	0.18	16.42
Total	99.00	99.12	100.58	99.76	100.03	99.16
An	33	05	24	0	0	0
Ab	63	42	69	55	99	1
Or	04	53	07	45	1	99

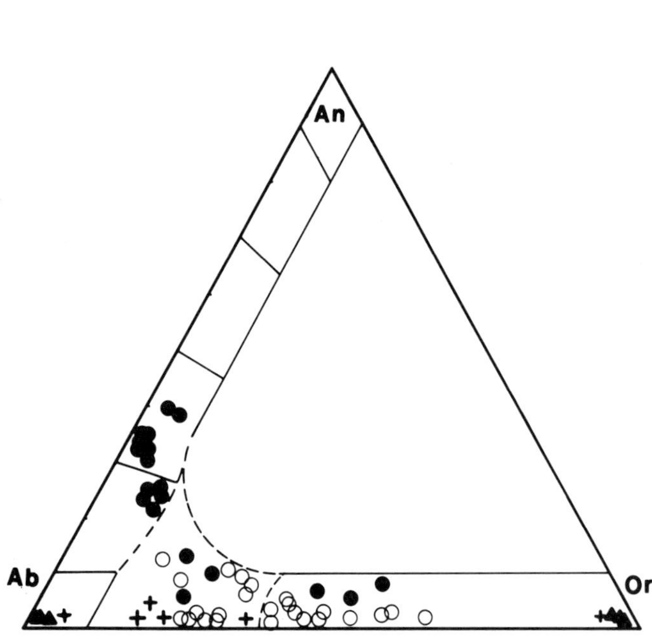

Figure 19. Feldspar compositions, Granite Mountain syenite. Filled circles indicate olivine syenite; open circles, dark gray syenite; crosses, light gray syenite; triangles, pegmatite.

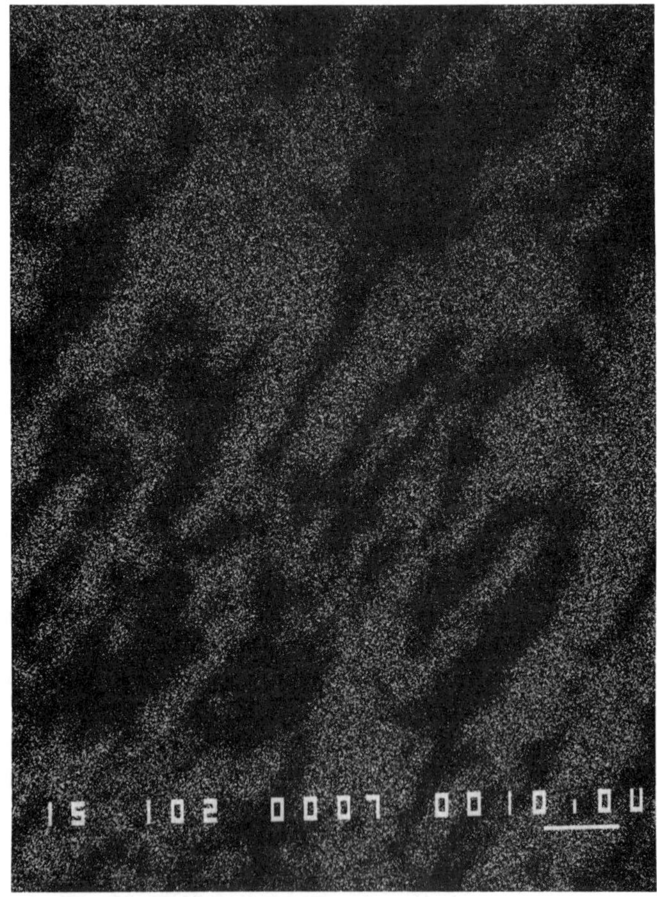

Figure 20. Back-scattered electron SEM dot map of feldspar in pegmatite, Granite Mountain. Dark areas indicate end-member orthoclase (Or_{99} Ab_1 An_0); light areas, albite (Or_1 Ab_{99} An_0).

TABLE 8. WHOLE-ROCK CEOCHEMICAL ANALYSES: GRANITE MOUNTAIN SYENITES

Oxide (wt. %)	Olivine-bearing syenite		Dark Gray Syenite		Light Gray Syenite
	BP-8	BP-10	BP-10A	BP-15	86-2
SiO_2	58.0	57.3	60.0	59.7	58.9
TiO_2	1.2	1.2	1.0	1.0	0.5
Al_2O_3	19.6	19.7	18.9	18.7	18.7
FeO*	4.0	4.0	3.1	3.3	3.7
MnO	0.14	0.14	0.18	0.17	0.28
MgO	1.4	1.4	1.0	0.9	0.5
CaO	3.8	3.8	1.9	1.9	0.8
Na_2O	6.1	6.2	6.2	6.1	7.6
K_2O	4.3	4.6	6.5	6.4	6.4
P_2O_5	0.41	0.40	0.20	0.20	0.04
LOI	0.6	0.9	1.00	1.00	2.54
Total	99.8	99.9	100.1	99.5	100.1
Rb	50	100	140	100	260
Ba	4300	4000	1400	1500	250
Sr	1600	1700	500	500	240
Nb	70	70	120	130	320
La	103	97	119	116	189
Sm	10.7	10.2	11.6	11.6	11.3
Yb	3	3	4	4	8.5
Hf	8	6	10	12	26
Th	8	7	12	12	36
U	2.4	2.2	3.8	3.8	14.6

Geochemistry of the Granite Mountain Syenite

Major element data from Granite Mountain syenites (Table 8) suggest that they are undersaturated, slightly fractionated rocks that have been altered or metasomatized. All samples are nepheline and olivine normative. On an AFM diagram (Fig. 21) syenites plot near the alkali apex. However, their MgO/(MgO + FeO*) ratio is 0.24 to 0.26, suggesting that fractionation in these rocks is not as extreme as their position on an AFM plot would indicate.

Trace element data also indicate that the syenites are not strongly fractionated rocks, and suggest that metasomatic addition of alkalis was important. The Rb/Sr ratio for olivine syenite is 0.04 to 0.06. Dark gray syenite that contains no plagioclase or olivine (and have K-feldspar and biotite clots in their place) have an Rb/Sr ratio of 0.19 to 0.22. These samples were all collected along a 100-m-long quarry wall. Samples with radically different Rb/Sr, collected within 2 m of each other, are nearly indistinguishable in hand sample. Similarly, abundances of U, Th, and Hf are nearly twice as high in gray syenites as they are in olivine syenite, although they are still very low for crustally derived magmas. Overall, Granite Mountain syenites are not greatly LREE-enriched, and the gray syenites are only slightly more LREE-enriched than their olivine-bearing cousins (Fig. 6), suggesting that extensive fractionation did not occur.

Isotopic compositions of the Granite Mountain syenites vary significantly. Olivine syenite has an $^{87/86}Sr$ ratio of 0.7039. The whole-rock ratio of dark gray syenite is 0.7051, and the ratio obtained from a feldspar separate of the light gray syenite is 0.7059. Thus it seems that the rocks have an increasing crustal component as they become more hydrous and more potassic, with the principal crustal influence noted in the dark gray syenite.

Systematic changes in trace element abundances support this interpretation. For example, Zr increases from 200 ppm in olivine syenites to 1,200 ppm in light gray syenite. Ba decreases from 4,000 ppm in olivine syenite, to 1,500 ppm in dark gray syenite and 250 ppm in light gray syenite. The uranium content of sphene and apatite increases from olivine syenite through the syenite pegmatites. Uranium is of low abundance (1 ppm) in apatite from olivine syenite, and increases to 8 ppm in apatite from pegmatite. Sphene in dark gray syenite contains 4 ppm uranium and increases to 14 ppm in light gray syenite. (Saline County syenites contain significantly more U in the same minerals.)

Plots of trace element ratios such as Nb/Zr, Rb/Zr, K_2O/Rb and others that should show strong linear correlations with fractionation instead have increasing scatter for the gray syenites and pegmatites. Whatever the initial magma composition, the syenites of Granite Mountain have undergone processes that have mobilized LIL and added a late, hydrous, crustal component to a rather mafic, mantle-derived parent magma.

SUMMARY

Petrologic and geochemical relations among rocks of the Arkansas alkalic province are not simple. On first examination, the province might be characterized as a series of intrusions from a single fractionating magma. Analyses from all intrusive bodies fall on a curved trend of low iron enrichment on an AFM diagram, and oxide/silica plots reveal broad linear correlations.

However, close scrutiny of geochemical and mineralogic data indicates that fractionation cannot be the sole mechanism for genesis of the Arkansas alkalic province. Trends on fractionation diagrams are not smooth. Each intrusion has its own field or subfield. Relations of elements that co-vary during fractionation are not linear, but have parallel or subparallel tends for each intrusive body. Trace element abundances, including REE, cannot be modeled from a single magma.

Mineralogic data also suggest that fractionation is a minor process overall. Olivine and clinopyroxene in most minettes have compositions similar to olivine and pyroxene in Granite Mountain syenites. Little change in mineral compositions occurs, even within intrusions. Where dramatic compositional differences do develop, such as in the feldspars of Granite Mountain, textural evidence indicates the action of metasomatic rather than magmatic agents.

On the basis of the data discussed above, a reasonable conclusion is that the Arkansas alkalic province originated as a number of discrete mafic intrusions from separate mantle sources. Potassium and other crustally derived metasomatism was greatest in the syenites of Granite Mountain and Saline County. Indeed, the late date for these rocks relative to the rest of the province suggests a separate event for creation of the later syenites.

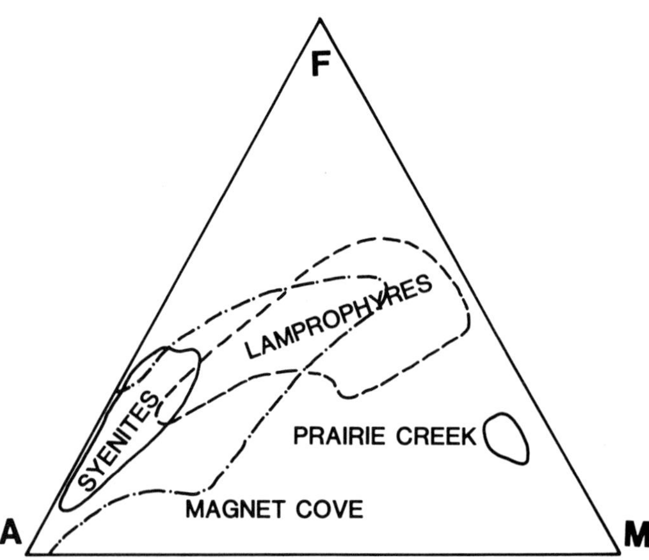

Figure 21. Generalized AFM plot for all rocks of Cretaceous Arkansas alkalic province.

ACKNOWLEDGMENTS

This work was partially supported by the U.S. Geological Survey, contract A-0194, and by ARCO Petroleum. The 3M Corporation allowed extensive access to its property. The comments and consideration of J. M. Howard, M. Ross, J. Lowery, and G. Dumont contributed positively to the research and resulting manuscript.

REFERENCES CITED

Beardsley, R., 1983, Modal analysis of the Granite Mountain Pulaskite, Pulaski County, Arkansas [M.S. thesis]: Fayetteville, University of Arkansas, 60 p.

Bolivar, S. L., 1984, An overview of the Prairie Creek intrusion, Arkansas: Society of Mining Engineers, preprint 84-346, 13 p.

Erickson, R. L., and Blade, L. V., 1963, Geochemistry and petrology of the alkalic igneous complex at Magnet Cove, Arkansas: U.S. Geological Survey Professional Paper 425, 94 p.

Gordon, M., Tracey, J., and Ellis, M., 1958, Geology of the Arkansas bauxite region: U.S. Geological Survey Professional Paper 299.

Heathcote, R. C., 1974, Fenitization of the Arkansas novaculite and adjacent intrusive, Garland County, Arkansas [M.S. thesis]: Fayetteville, University of Arkansas, 58 p.

Howard, J. M., 1974, Transition element geochemistry of the Potash Sulphur Spring intrusive complex, Garland County, Arkansas [M.S. thesis]: Fayetteville, University of Arkansas, 188 p.

Jackson, K. C., 1978, Arkansas syenites; Fenitized crustal material?: Geological Society of America Abstracts with Programs, v. 10, p. 7–8.

Lewis, R. D., Meyer, H.O.A., Bolivar, S. L., and Brookins, D. G., 1976, Mineralogy of the diamond-bearing "kimberlite," Murfreesboro, Arkansas: EOS American Geophysical Union Transactions, v. 57, p. 761.

McCormick, G., and Heathcote, R., 1979, Mineralogy of the Morrilton alvikite dike, Conway County, Arkansas: Geological Society of America Abstracts with Programs, v. 11, p. 163.

——, 1986, Mineral chemistry and petrogenesis of carbonatite intrusions, Perry and Conway Counties, Arkansas: American Mineralogist (in press).

Miser, H. O., and Ross, C. S., 1923, A peridotite dike in Scott County, Arkansas: U.S. Geological Survey Bulletin 735, p. 279–322.

Mitchell, R. H., 1970, Kimberlite and related rocks; A critical reappraisal: Journal of Geology, v. 78, p. 686–704.

Morris, E. M., and Eby, G. N., 1986, Petrologic and age relations in Granite Mountain syenites: Geological Society of America Abstracts with Programs, v. 18, p. 256.

Owens, D. R., 1968, Bedrock geology of the "V" intrusive, Garland County, Arkansas [M.S. thesis]: Fayetteville, University of Arkansas, 96 p.

Peterson, R. J., 1972, Bedrock geology of the Benton complex, Saline County, Arkansas [M.S. thesis]: Fayetteville, University of Arkansas, 81 p.

Pollack, D. W., 1965, The Potash Sulphur Springs alkali complex, Garland County, Arkansas (abs.): Mining Engineering, v. 17, p. 45–46.

Robison, E. C., 1977, Geochemistry of lamprophyric rocks of the eastern Ouachita Mountains, Arkansas [M.S. thesis]: Fayetteville, University of Arkansas, 147 p.

Scott-Smith, B. H., and Skinner, E.M.W., 1984a, A new look at Prairie Creek, Arkansas, *in* Kornprobst, J., ed., Kimberlites I; Kimberlites and related rocks: Amsterdam, Elsevier, p. 255–283.

——, 1984b, Diamondiferous lamproites: Journal of Geology, v. 92, p. 433–438.

Simkin, T., and Smith, J. V., 1970, Minor-element distribution in olivine: Journal of Geology, v. 78, p. 304–325.

Simmen, L. E., 1955, Petrography of the "blue" syenite of Granite Mountain, Pulaski County, Arkansas [M.S. thesis]: Fayetteville, University of Arkansas, 37 p.

Steele, K. F., and Robison, E. C., 1976, Chemical relationships of lamprophyre, central Arkansas: EOS American Geophysical Union Transactions, v. 57, p. 1018.

Tilton, G. R., and Kwon, S. T., 1985, Isotopic relations in Arkansas Cretaceous alkalic complexes: Geological Society of America Abstracts with Programs, v. 17, p. 194.

Valdovinos, D. L., 1968, Petrography of some lamprophyres of the eastern Ouachita Mountains of Arkansas [M.S. thesis]: Fayetteville, University of Arkansas, 146 p.

Van Buren, W. M., 1968, Geochemistry of syenite in Arkansas [M.S. thesis]: Fayetteville, University of Arkansas, 100 p.

Wade, A., and Prider, R. T., 1940, The leucite-bearing rocks of the West Kimberley area, western Australia: Journal of the Geological Society of London, v. 96, p. 39–98.

Waldman, M. A., McCandless, T. E., and Dummett, H. T., 1985, Geology and mineralogy of the Twin Knobs #1 lamproite, Pike County, Arkansas: Geological Society of America Abstracts with Programs, v. 17, p. 196.

Williams, J. F., 1891, The igneous rocks of Arkansas: Arkansas Geological Survey Annual Report 1890, v. 2, 457 p.

Zartman, R.E., 1977, Geochronology of some alkalic rock provinces in eastern and central United States: Annual Review of Earth and Planetary Sciences, v. 5, p. 257–286.

Zartman, R. E., and Howard, J. M., 1985, U-Th-Pb ages of large zircon crystals from the Potash Sulphur Springs igneous complex, Garland County, Arkansas: Geological Society of America Abstracts with Programs, v. 17, p. 198.

Manuscript Accepted by the Society October 15, 1986

Printed in U.S.A.

Geological Society of America
Special Paper 215
1987

Uranium-lead age of large zircon crystals from the Potash Sulfur Springs igneous complex, Garland County, Arkansas

Robert E. Zartman
U.S. Geological Survey, Denver Federal Center, Denver, Colorado 80225
J. Michael Howard
Arkansas Geological Commission, Little Rock, Arkansas 72204

ABSTRACT

Euhedral zircon crystals several millimeters in size were collected from feldspathoidal syenite constituting the central part of the Potash Sulfur Springs igneous complex, Garland County, Arkansas. Found loose in the overlying residual soil, the zircons displayed simple tetragonal dipyramid habits with small prism faces. The largest crystal weighed 78 mg and was internally adamantine except for some secondarily(?) corroded and iron-stained domains. Following crushing, a sample containing only completely clear fragments (A-1) and one containing some incipiently corroded surfaces (A-2) were picked from this crystal. Selected fragments from two other somewhat smaller, more translucent crystals constituted a third sample (BC-1).

Ages (Ma)

Sample	$^{206}Pb/^{238}U$	$^{207}Pb/^{235}U$	$^{207}Pb/^{206}Pb$	$^{208}Pb/^{232}Th$
A-1	98±1	98±1	102±4	99±1
A-2	92±1	93±1	99±6	92±1
BC-1	90±1	91±1	101±7	90±1

Sample A-1 yields concordant U-Th-Pb ages, demonstrating that it has behaved as a closed system subsequent to crystallization. Good agreement in $^{207}Pb/^{206}Pb$ ages of all samples further confirms this result, although some internal discordance within samples A-2 and BC-1 reveals them to be slightly open systems from which radiogenic lead recently has leaked. Current time scales place our preferred concordia intercept age of 100 ± 2 Ma for the Potash Sulfur Springs igneous complex near the end of the Albian Stage of the Early Cretaceous. Previously published ages for the nearby Magnet Cove Complex are in good agreement with the results of this study, which attest to the contemporaneity of the two plutons.

GEOLOGIC SETTING

During a field trip for the University of Arkansas at Fayetteville in 1972 by one of us (J.M.H.), three large, euhedral zircon crystals were collected from the residual soil in the central portion of the Potash Sulfur Springs igneous complex. It was recognized that the zircons held promise of being ideally suited for U-Pb age determination. The alkalic Potash Sulfur Springs intrusion is located in the W½,Sec.17, and E½,Sec.18,T.3S.,R.18W., Garland County, Arkansas. This igneous body intrudes the Devonian to Mississippian Arkansas Novaculite and the Mississippian Stanley Shale; it is poorly exposed on the nose of a complexly faulted, plunging anticline on the southern side of the Zigzag Mountains. The various lithologic units of the complex crop out as a roughly circular, discordant stock over an area of approximately 2 km^2 (Fig. 1). Previous workers who have contributed to knowledge of

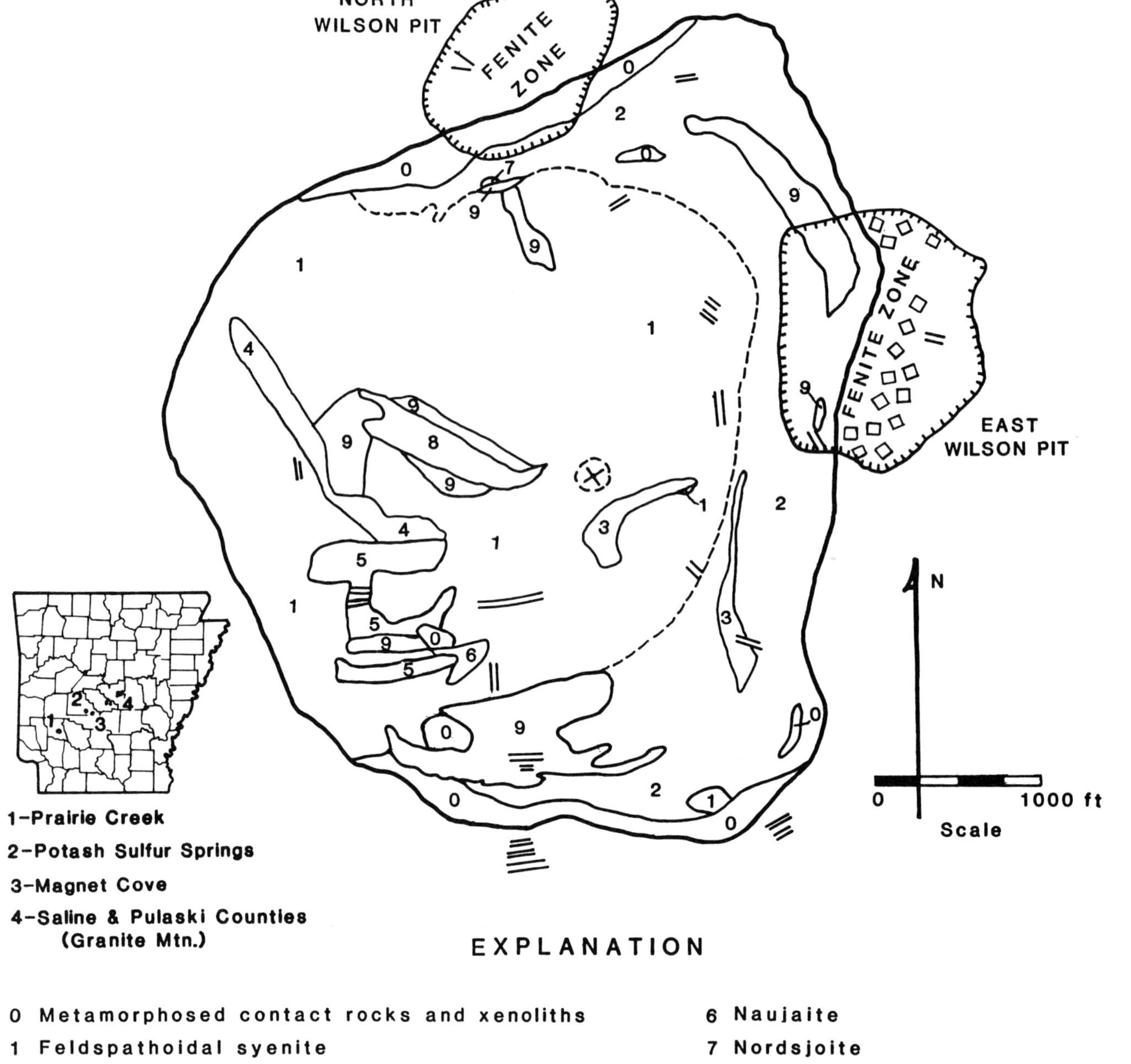

Figure 1. Geologic map of Potash Sulfur Springs igneous complex. Modified from Howard (1974).

TABLE 1. URANIUM-THORIUM-LEAD ISOTOPIC AGES OF ZIRCON FROM POTASH SULFUR SPRINGS IGNEOUS COMPLEX*

Sample designation	Concentration (ppm)			Isotopic composition of lead atom %				Age (Ma)			
	U	Th	Pb	^{204}Pb	^{206}Pb	^{207}Pb	^{208}Pb	$^{206}Pb/^{238}U$	$^{207}Pb/^{235}U$	$^{207}Pb/^{206}Pb$	$^{208}Pb/^{232}Pb$
A-1	834.9	1,438.4	17.97	0.0117	61.28	3.117	35.59	98±1	98±1	102±4	99±1
A-2	847.0	1,519.0	17.99	0.0591	59.72	3.737	36.49	92±1	93±1	99±6	92±1
BC-1	906.2	1,688.5	19.28	0.0725	58.51	3.877	37.54	90±1	91±1	101±7	90±1

Note: Errors, rounded to nearest million years, are estimates of 95 percent confidence level based on routine laboratory performance.

*Decay constants: ^{238}U = 1.55125 x $10^{-10}yr^{-1}$; ^{235}U = 9.8485 x $10^{-10}yr^{-1}$; ^{232}Th = 4.9375 x $10^{-11}yr^{-1}$; $^{238}U/^{235}U$ = 137.88. Isotopic composition of common lead assumed to be $^{204}Pb:^{206}Pb:^{207}Pb:^{208}Pb$ = 1:19.06:15.55:38.73, similar to that of galena from the Magnet Cove Complex (G. R. Tilton, written communication, 1985).

the Potash Sulfur Springs igneous complex include Williams (1891), Purdue and Miser (1923), Stroud (1951), Baxter (1952), Erickson and Blade (1963), Milton (1965, 1984), Owens (1967), Hollingsworth (1967), and Howard (1974).

Separated by 7 km from the renowned alkalic complex at Magnet Cove, this western satellite contains many of the same peculiar rock types, and it is almost certainly coeval with the Magnet Cove Complex. The major lithologies of the Potash Sulfur Springs intrusion consist of leucopulaskite, pulaskite, feldspathoidal syenite (sodalite nepheline syenite, nepheline sodalite syenite), ijolite, malignite, fasanite, naujaite, shonkinite, nordsjoite, and carbonatite (classification system of Johannsen, 1938). Additionally, nepheline sodalite syenite, nepheline syenite, ouachitite, and monchiquite dikes crosscut the intrusion and adjacent country rock.

Feldspathoidal syenite is the most abundant igneous rock type cropping out at the present erosional level. The unit consists of orthoclase (two generations in some instances), aegerine-augite, nepheline, a sodalite-group mineral (nosean, or its alteration products) ±euhedral zoned melanite, apatite, sphene, opaques, and zircon. Although Howard (1974) did not observe zircon in thin sections of feldspathoidal syenite, the large zircon crystals were collected from residual soils developed immediately above this rock type. Unfortunately, the collection site has subsequently been covered with mine waste dumps.

U-PB ZIRCON ANALYSES

The three zircons used in this study were simple tetragonal dipyramids with small prism faces; the *c* axis was shorter than the *a* axis in the approximate proportion of $c{:}a$ = 0.9:1. The largest crystal weighed 78 mg and measured 4 mm across in the direction of its *a* axis; it is designated herein as crystal A. Under the binocular microscope, crystal A was characterized internally by clear, adamantine areas separated by brown, iron-stained domains that seemed to be sites of secondary corrosion. The very clearest fragments obtained by hand-picking of the crushed crystal constitute sample A-1; hand-picked fragments containing some incipiently corroded internal surfaces constitute sample A-2. The other two crystals, designated here in as crystals B and C, were somewhat smaller in size, weighing 51 and 44 mg, respectively, and displayed more internal discoloration and translucency than crystal A. Sample BC-1 consists of hand-picked fragments of the least corroded portions of the two crushed smaller crystals.

All three samples were analyzed isotopically, and the resultant U-Th-Pb ages are reported in Table 1 and plotted on a concordia diagram in Figure 2. Sample A-1, containing the freshest fragments, yields concordant ages of 100 ± 2 Ma, demonstrating that this material has been a closed system subsequent to crystallization. The good agreement in $^{207}Pb/^{206}Pb$ ages of samples A-1, A-2, and BC-1 further confirms our interpretation, although internal discordancy, as much as 10 percent for sample BC-1, within the latter two samples reveals them to have been slightly open systems from which some of the radiogenic lead recently has leaked. This phenomenon of recent lead loss is common for zircon; it is probably best explained by a combination of accumulated radioactive damage or crystal defects and the accessibility of pore fluids to the degraded lattice sites as the rock experiences dilatancy under near-surface conditions (Goldich and Mudrey, 1972). Such recent removal of radiogenic lead (or, less likely, introduction of uranium) will result in a lowering of daughter-parent ratio but will not appreciably affect $^{207}Pb/^{206}Pb$ ages. Although the average $^{207}Pb/^{206}Pb$ age need not be the same as the concordia intercept age derived from a best-fit line to the individual analyses, no resolvable distinction between them

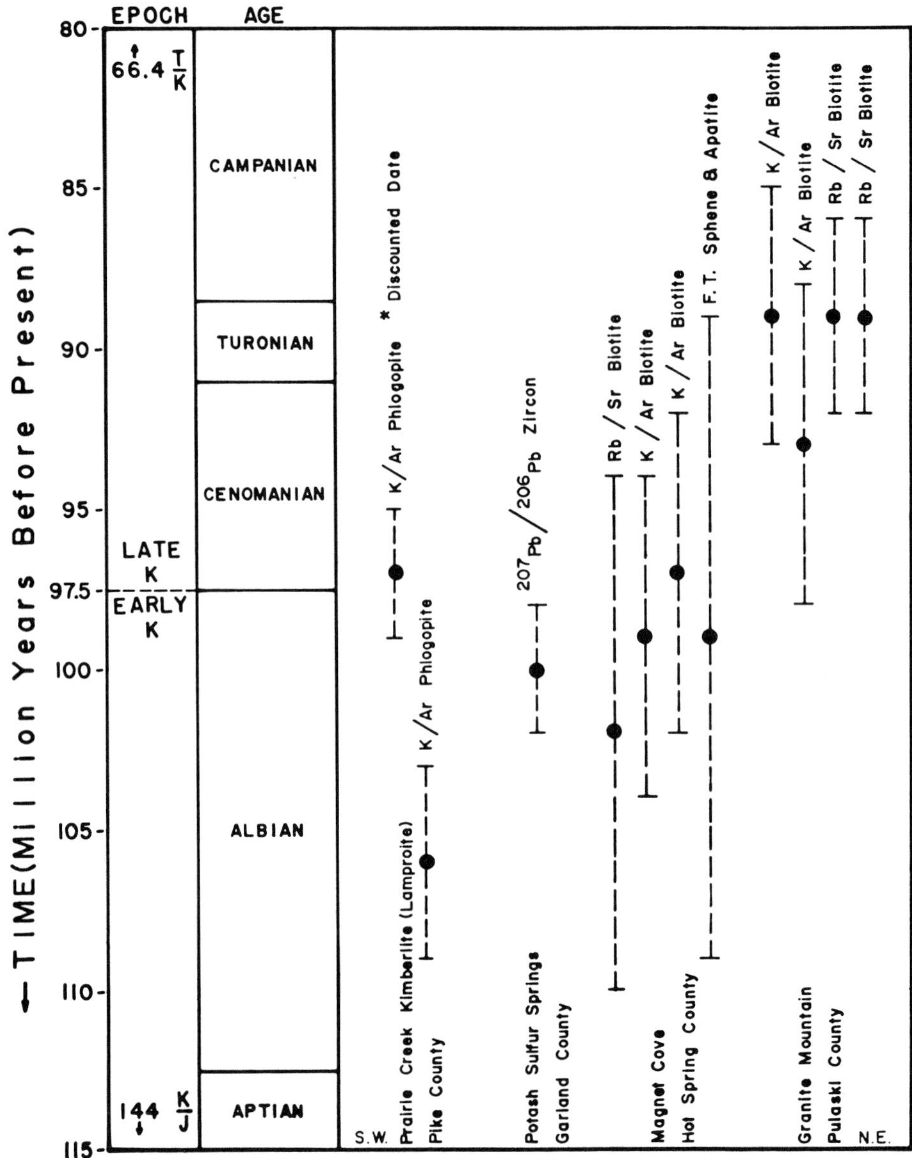

Figure 2. Uranium-lead concordia diagram of zircon from Potash Sulfur Springs igneous complex.

can be made in the present case. Most currently accepted time scales (see, for example, Palmer, 1983) place our preferred concordia intercept age of 100 ± 2 Ma for the Potash Sulfur Springs igneous complex near the end of the Albian Stage of the Early Cretaceous.

DISCUSSION

Previous efforts to isotopically date the alkalic igneous complex at Magnet Cove and the nepheline syenite from Granite Mountains near Little Rock have been reported by Zartman and others (1967). Two biotite K-Ar ages of 97 ± 5 and 99 ± 5 Ma and one biotite Rb-Sr age of 102 ± 8 Ma (recalculated to the decay constants recommended by the International Union of Geological Sciences Subcommission on Geochronology; Steiger and Jäger, 1977) for the Magnet Cove Complex are in excellent agreement with the results of the present study, confirming the contemporaneity of the two plutons. Two biotite K-Ar ages of 89 ± 4 and 93 ± 5 Ma and two biotite Rb-Sr ages of 89 ± 3 and 89 ± 3 Ma for the nepheline syenite from Granite Mountain appear to associate that body with the resolvably younger Turonian or Cenomanian Stage of the Late Cretaceous. Zartman (1977) also has obtained two phlogopite K-Ar ages of 97 ± 2 and 106 ± 3 Ma for the Prairie Creek kimberlite at Murfreesboro. A closer approach to stoichiometric potassium content and lack of vermiculate intergrowth for the phlogopite giving the older age argue in favor of that result, which would place the diatreme in the Albian, with its emplacement prior to the Potash Sulfur Springs and Magnet Cove bodies. In good agreement with the isotopic ages, but having an uncertainty too large to give the time resolution sought here, is the average of an apatite and a sphene fission track age of 99 ± 10 Ma for the Magnet Cove Complex reported by Naeser and Faul (1969). Figure 3 presents all published radiometric ages for Arkansas Cretaceous igneous rocks. When arranged geographically, it is apparent that the intrusion sequence of these presumed genetically related rocks began in the southwest and proceeded to the northeast over a time span of approximately 15 m.y., straddling the Early Cretaceous–Late Cretaceous boundary. It could be speculated from this trend that some mantle instability, which controlled the site of magmatic activity, moved across the area with a velocity of approximately 1 cm/yr. Farther to the west and east, other possibly related igneous bodies might lie buried under younger Coastal Plain sedimentary rocks. Yet without better evidence of such a "track," we hesitate to propose any specific mechanism, such as crustal migration over a mantle hotspot, for magma generation at this time.

Not included in Figure 3 are any ages for the ubiquitous lamprophyric rocks that would seem to bear close spatial and temporal relationship to the other igneous rocks. One unpublished age of 83 ± 8 Ma obtained by the K-Ar method on biotite from a kersantite(?) or ouachitite(?) from the Dare Mine, Polk County, however, has been reported to the Arkansas Geological Commission (R. E. Dennison, written communication, 1976). Some authors (Howard, 1974; Howard and Steele, 1975; Robinson, 1976) have contended that the lamprophyres are not only coeval with the major alkalic intrusions, but that they are derived from the same parental mantle source. When ages become available for other lamprophyres of the Ouachita Mountains, we predict they will be found to span essentially the same time range as do the major intrusions.

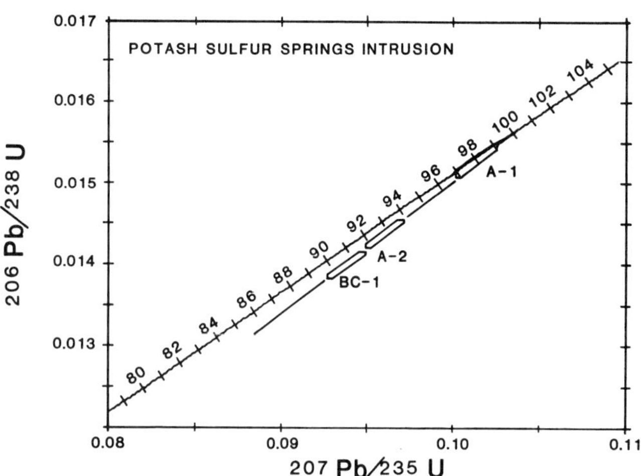

Figure 3. Radiometric ages for Arkansas Cretaceous igneous rocks. Individual ages discussed in text; geologic time scale from Palmer (1983).

ACKNOWLEDGMENTS

This study was undertaken after one of us (R.E.Z.) visited the headquarters of the Arkansas Geological Commission in Little Rock at the invitation of the State Geologist, Norman F. Williams. The zircon crystals mentioned in this paper were discovered by Don Owens of Umetco (formerly Union Carbide), John McFarland of the Arkansas Geological Commission, and one of the authors (J.M.H.).

REFERENCES CITED

Baxter, J. W., 1952, Potash Sulfur Springs igneous complex near Hot Springs, Garland County, Arkansas [M.S. thesis]: Little Rock, University of Arkansas, 36 p.

Erickson, R. L., and Blade, L. V., 1963, Geochemistry and petrology of the alkalic igneous complex at Magnet Cove, Arkansas: U.S. Geological Survey Professional Paper 425, 95 p.

Goldich, S. S., and Mudrey, M. G., Jr., 1972, Dilatancy model for discordant U-Pb zircon ages, *in* Contributions to recent geochemistry and analytical chemistry (A. P. Winograd Volume): Moscow, Nauka, p. 415–419.

Hollingsworth, J. S., 1967, Geology of the Wilson Springs vanadium deposits: Field Trip Guide Book, Central Arkansas Economic Geology and Petrology: Arkansas Geological Commission, p. 22–28.

Howard, J. M., 1974, Transition element geochemistry and petrography of the Potash Sulfur Springs intrusive complex, Garland County, Arkasas [M.S. thesis]: Little Rock, University of Arkansas, 118 p.

Howard, J. M., and Steele, K. F., 1975, Transition element geochemistry of the Potash Sulfur springs intrusion, Garland County, Arkansas: Geological Society of America Abstracts with Programs, v. 7, p. 174.

Johannsen, A., 1938, A descriptive petrography of the igneous rocks: University of Chicago Press, v. 4, p. 523.

Milton, C., 1965, Mineralogy of the "V" intrusive Garland County, Arkansas: Unpublished report to the Arkansas Geological Commission, 20 p.

——, 1984, Miserite; A review of world occurrences with a note on intergrown wollastonite: Contributions to the Geology of Arkansas, vol. II, Arkansas Geological Commission Miscellaneous Publication 18-B, p. 97–114.

Naeser, C. W., and Faul, H., 1969, Fission-track annealing in apatite and sphene: Journal of Geophysical Research, v. 74, no. 2, p. 705–710.

Owens, D. R., 1967, Bedrock geology of the "V" intrusive, Garland County, Arkansas [M.S. thesis]: Little Rock, University of Arkansas, 93 p.

Palmer, A. R., 1983, The decade of North American geology 1983 geologic time scale: Geology, v. 11, p. 503–504.

Purdue, A. H., and Miser, H. D., 1923, Description of the Hot Springs quadrangle, Arkansas: U.S. Geological Survey Atlas, Folio 215.

Robinson, E. C., 1976, Geochemistry of lamprophyric rocks of the eastern Ouachita Mountains, Arkansas [M.S. thesis]: Little Rock, University of Arkansas, 147 p.

Steiger, R. H., and Jäger, E., 1977, Subcommission on geochronology; Convention on the use of decay constants in geo- and cosmochronology: Earth and Planetary Science Letters, v. 36, p. 359–362.

Stroud, R. B., 1951, The areal distribution of radioactivity in the Potash Sulfur

Springs complex [M.S. thesis]: Little Rock, University of Arkansas, 42 p.
Williams, J. F., 1891, The igneous rocks of Arkansas: Arkansas Geological Survey Annual Report, 1890, v. 2, 457 p.
Zartman, R. E., 1977, Geochronology of some alkalic rock provinces in eastern and central United States: Annual Review, Earth and Planetary Science, 5, p. 257–286.
Zartman, R. E., Brock, M. R., Heyl, A. V., and Thomas, H. H., 1967, K-Ar and Rb-Sr ages of some alkalic intrusive rocks from central and eastern United States: American Journal of Science, v. 265, p. 848–870.

MANUSCRIPT ACCEPTED BY THE SOCIETY OCTOBER 15, 1986

Printed in U.S.A.

Geological Society of America
Special Paper 215
1987

Isotopic relationships in Arkansas Cretaceous alkalic complexes

G. R. Tilton, Sung Tack Kwon, and Donna Martin Frost
Department of Geological Sciences, University of California at Santa Barbara, Santa Barbara, California 93106

ABSTRACT

Isotopic data are reported from carbonatite, syenite, and galena. The average initial values for four carbonate samples from Magnet Cove, Hot Springs County, Arkansas, give ϵ (Nd) = +3.8 ± 0.3 and $^{87}Sr/^{86}Sr$ = 0.70363, based on an age of 100 m.y. for the complexes. Two carbonates from Potash Sulfur Springs yield lower ϵ (Nd) values of +2.4 and 3.0. Lead in the carbonate samples plots within the field defined by midocean-ridge basalts (MORB) but is not isotopically uniform. The least radiogenic sample yields $^{206}Pb/^{204}Pb$ = 18.98, $^{207}Pb/^{204}Pb$ = 15.53, and $^{208}Pb/^{204}Pb$ = 38.90. A galena sample at Magnet Cove has similar, but slightly more radiogenic, ratios. The Pb, Sr, and Nd data for the Magnet Cove samples indicate an origin in large ion-lithophile (LIL) element–depleted mantle, with little or no crustal interaction with the magmas.

Whole-rock samples of three syenites from the Little Rock district have average ϵ (Nd) = +2.6 at 100 m.y., but have indeterminate initial $^{87}Sr/^{86}Sr$ ratios due to postemplacement alteration. Lead ratios from seven of nine syenite feldspars are similar to the carbonatite leads, averaging $^{206}Pb/^{204}Pb$ = 19.20, $^{207}Pb/^{104}Pb$ = 15.59, and $^{208}Pb/^{204}Pb$ = 38.96, but two samples exhibit extensive contamination from crustal lead, as defined by galena samples from central Arkansas. Variations in Pb isotopic composition are documented for samples from a single syenite quarry.

The carbonatite data generally agree with parallel data from the Oka, Quebec, carbonatite complex, although Nd is somewhat less radiogenic and Sr is somewhat more radiogenic than the Oka values. The $^{206}Pb/^{204}Pb$ ratio at Magnet Cove is lower than the lowest value at Oka (19.65), although the $^{207}Pb/^{204}Pb$ ratios are identical within error limits. All the isotope data indicate that the source for the Magnet Cove carbonatite is slightly less depleted in LIL elements than is the Oka source, probably illustrating mantle compositional heterogeneity. The Arkansas data provide additional evidence that alkaline complexes can yield information on the geochemical character of subcontinental mantle.

INTRODUCTION

This project is part of a continuing study of the secular geochemical evolution of the Earth's mantle, which in particular seeks to define the evolution of the large ion-lithophile (LIL) element–depleted mantle. In the past, most information concerning depleted mantle has come from studies of oceanic rocks; however, with oceanic rocks and associated ophiolite complexes, the record can be traced back only about 500 m.y. (e.g., Jacobsen and Wasserburg, 1979). Therefore, alternate land-based methods must be sought if information is to be obtained for the Proterozoic and Archaean eras. By determining the isotopic composition of initial Sr in a suite of alkaline complexes, including carbonatite and syenite from the Canadian Shield, with ages ranging from 0.12 to 2.68 b.y. (AE), Bell and others (1982) have shown that carbonatite and alkaline syenite complexes appear to provide suitable "windows" for that purpose. When the $^{87}Sr/^{86}Sr$ ratios of the samples were plotted as a function of age, a few samples plotted on a development line for "bulk earth" (as defined, for example, by DePaolo and Wasserburg [1976]) but most yielded lower ratios, indicating that the Sr isotopic signature developed in LIL element–depleted mantle. Moreover, the depleted source appeared to have existed under large areas of the Superior and Grenville Provinces over approximately the last 3 AE (1 AE = 10^9 yr). More recently, Bell and Blenkinsop (1985) have verified the Sr observations with Nd measurements on some of the same samples. In a follow-up study on the Oka carbonatite, the youngest suite studied by Bell and others (1982), Grünenfelder and others (1985) found that Pb isotope data similarly support derivation of much of the Pb in the complex from LIL element–depleted mantle. However, some silicates in okaite from the complex show evidence of contamination with crustal lead. On the other hand, Sr ratios are uniform for the complex, indicating negligible contamination. The authors ascribed this difference in behavior to the lower ratio of Pb compared to Sr concentrations in the complex rocks relative to the country rocks, making Pb a more sensitive indicator of contamination. These data indicate that the magmas of alkaline complexes can traverse crustal rocks and preserve a record of the isotopic composition of the mantle source rocks, which in turn provides an indication of the geochemical characteristics of the sources.

The present investigations were undertaken to compare Pb, Sr, and Nd isotopic data from the Cretaceous Arkansas carbonatites and syenite magmas with data from the Oka carbonatite complex for which Eby (1984) has given a mean age of 118 m.y. An age of 99 ± 2 m.y. has been reported for zircon from the Potash Sulfur Springs carbonatite complex (Zartman and Howard, 1985). Zartman and others (1967) measured K-Ar and Rb-Sr ages of 95 to 99 m.y. on biotite from Magnet Cove and 86 to 91 m.y. on biotite from nepheline syenite at Little Rock. The only initial ratio information known to us is a single $^{143}Nd/^{144}Nd$ ratio of 0.512967 for "Magnet Cove Carbonatite" (Basu and Tatsumoto, 1980), a value in the range of many MORB samples. More data are needed, since an extensive data base from rocks of young complexes is essential if alkalic rocks are to be used to trace mantle evolution back into the Precambrian era. It would be futile to attempt measurements on older complexes if young alkalic rocks do not provide suitable mantle tracers.

ANALYTICAL PROCEDURES

Carbonates

Interior portions of single crystals were prepared by crushing. After careful hand-picking of samples of about 50 mg, the surfaces were etched in dilute HCl and the solutions discarded. The samples were then dissolved in cold 2M HCl, and any remaining residues discarded. Standard ion exchange procedures were used for chemical purification, which involved anion exchange for U, Th, and Pb, as described in Grünenfelder and others (1985), cation exchange in a 2.5-M HCl medium for Sr, and cation exchange with 2-methyl lactic acid for the final purification of Nd (Eugster and others, 1970). U, Th, and Pb were determined by a mixed ^{205}Pb-^{230}Th-^{235}U spike, Sr and Rb by a mixed ^{87}Rb-^{84}Sr spike, and Sm and Nd by a mixed ^{150}Nd-^{149}Sm spike.

Syenites

For Pb analyses, large feldspar crystals were collected by hand-picking, and the surfaces were washed in hot 2M HCl. Crushed whole-rock samples were used for the Sr and Nd analyses. The samples (approximately 50 mg) were then dissolved in HF-$HClO_4$ mixtures and processed by the same ion exchange procedures described above for the carbonates.

Chemical processing was carried out in laminar flow hoods at all times. Isotope ratios were measured on a Finnigan MAT 261 fixed multiple collector mass spectrometer, which allowed the simultaneous collection of masses 86, 87, and 88; or 143, 144, and 146; or 204, 206, 207, and 208. Sr and Pb data were collected in the static mode, but it was necessary to switch to three positions to obtain sufficient data for Nd, for which masses 143 through 148 were monitored. Sr and Nd were measured by double-filament techniques; Pb was run on single filaments by the phosphoric acid–silica gel method. Runs were made on samples that typically contained 0.1 to 0.5 μg of element.

Based on replicate analyses of known standards, the $^{204}Pb/^{206}Pb$ ratios are accurate within ±0.08%; $^{207}Pb/^{206}Pb$ and $^{208}Pb/^{206}Pb$ ratios to 0.04% standard errors for a single set of measurements. Two sigma mean errors for normalized $^{87}Sr/^{86}Sr$ ratios average 0.004% of the measured ratio; the normalized $^{143}Nd/^{144}Nd$ ratio also yields a 2-σ mean error of ±0.004% for a single sample. The Pb isotopic ratios have been normalized on the basis of replicate measurements on NBS standard 981, using the newer value of 2.1671 for $^{208}Pb/^{206}Pb$ reported by Todt and others (1984). We obtain a value of 0.71028 by the static mode for Sr standard NBS 987 when $^{86}Sr/^{88}Sr$ is normalized to 0.1194. We adjusted the ratio to 0.71025, the value obtained by double- and single-collector runs. For the La Jolla Nd standard we obtain 0.51188 for $^{143}Nd/^{144}Nd$ when $^{146}Nd/^{144}Nd$ is normalized to 0.72190. This ratio has been adjusted to 0.51185 to bring it into agreement with most other laboratories.

RESULTS

The analytical data are given in Tables 1 through 3. (For sample locations, see the Appendix.) The concentration of Pb in carbonatite carbonates at Magnet Cove and Potash Sulfur Springs is approximately 1 to 2 ppm, a level probably considerably lower than that in the country rocks. The U and Th concentrations are likewise low, however, causing corrections for the in situ production of radiogenic Pb from the decay of U and Th not to exceed 1.2% of the measured ratios for any of the samples. The Th/U ratios are variable and mostly lower than the average value of 3.8 for terrestrial igneous rocks.

In contrast to the Pb data, the Sr concentration is higher by more than an order of magnitude over that expected in country rocks. Sm and Nd have rather low concentrations in the calcite samples. The Nd concentrations in carbonates from Potash Sulfur Springs are about equal to average shale, 37 ppm (Haskin and Frey, 1966), but are lower by over a factor of two at Magnet Cove. Sm follows a similar pattern and is substantially lower than the shale value of 7.0 ppm (Haskin and Frey, 1966) at both localities. Whole-rock carbonatites typically have Sm and Nd abundances about 5 to 10 times those in average shale (Haskin

TABLE 1. Sr and Nd ISOTOPIC COMPOSITION OF ARKANSAS ALKALIC COMPLEXES*

Sample[1]	Sr (ppm)	Rb (ppm)	$^{87}Rb/^{86}Sr$	$^{87}Sr/^{86}Sr$ [2]	Nd (ppm)	Sm (ppm)	$^{147}Sm/^{144}Nd$	$^{143}Nd/^{144}Nd$ [3]	Nd[2]	Sr[4]
Whole Rock										
SY-2-1A	136.2	349.7	6.987	0.708991 (35)	150.1	21.1	0.0848	0.512673 (45)	2.1	
SY-2-1B	41.28	315.4	20.82	0.719800 (40)	147.3	22.5	0.0922	0.512704 (19)	2.6	
SY-2-1E	447.9	241.4	1.374	0.705131 (28)	77.91	11.8	0.0916	0.512710 (28)	2.8	
Calcite										
CA-3-1-I	12079	5.38	0.00129	0.703730 (22)	36.84	4.10	0.0672	0.512705 (28)	3.0	-12.0
CA-3-1-II	9899	4.66	0.00128	0.703781 (17)	33.63	3.72	0.0668	0.512674 (25)	2.4	-11.4
CA-4-1A	5161	6.84	0.00361	0.703674 (32)	12.73	2.06	0.0979	0.512780 (12)	4.0	-12.9
CA-4-1B	5405	9.74	0.00490	0.703554 (32)	12.14	1.75	0.0870	0.512780 (22)	4.0	-14.6
CA-4-1D	5186	5.73	0.00300	0.703588 (42)	15.21	2.26	0.0900	0.512753 (22)	3.6	-14.1
MC-6	5671	0.823	0.00040	0.703703 (42)	12.56	1.94	0.0934	0.512746 (49)	3.4	-12.4

*All concentration data measured by isotope dilution method; their errors are estimated to be about 1%.
$^{87}Sr/^{86}Sr$: normalized using $^{86}Sr/^{88}Sr$ = 0.1194, and adjusted to NBS 987 of 0.71025.
$^{143}Nd/^{144}Nd$: normalized using $^{146}Nd/^{144}Nd$ = 0.7219, and adjusted to 0.51185 for the La Jolla Nd standard.
[1] "Whole Rock" refers to powdered sample of total rock without mineral separation. "Calcite" refers to mineral ($CaCO_3$) separated from carbonatite total rock.
[2] Errors of 2 sigma mean shown in parentheses representing last significant digits.
[3] Calculated for assumed age of 100 Ma, using "bulk Earth" value of $^{143}Nd/^{144}Nd$ = 0.512638 and $^{147}Sm/^{144}Nd$ = 0.1967, after Jacobsen and Wasserburg (1984).
[4] Calculated for assumed age of 100 Ma, using "bulk Earth" value of $^{87}Sr/^{86}Sr$ = 0.7047 and $^{87}Rb/^{86}Sr$ = 0.08521.

TABLE 2. U-Th-Pb DATA FOR MINERAL SEPARATES

	Observed Ratios			Corrected for in situ decay*			Concentration (ppm)				
	$^{206}PB/^{204}PB$	$^{207}PB/^{204}PB$	$^{208}PB/^{204}PB$	$^{206}PB/^{204}PB$	$^{207}PB/^{204}PB$	$^{207}PB/^{204}PB$	Pb	U	Th	$^{238}U/^{204}Pb$	$^{232}Th/^{238}U$
Calcite											
CA-3-1	19.304	15.559	39.001	19.296	15.559	38.979	2.28	0.0185	0.153	0.526	8.57
CA-4-1A	19.047	15.577	38.706	19.041	15.577	38.705	1.64	0.0097	0.003	3.81	0.32
CA-4-1B	19.141	15.564	39.743	19.053	15.560	38.660	1.70	0.148	0.428	5.61	2.98
CA-4-1D	19.372	15.600	38.893	19.145	15.589	38.818	1.25	0.280	0.284	14.49	1.05
MC-6	19.125	15.539	38.908	18.975	15.532	38.901	1.45	0.215	0.0305	9.54	0.15
Potassium Feldspar											
SY-1-1	19.347	15.596	39.127	19.231	15.590	38.989	15.37	1.75	6.39	7.40	3.76
SY-1-2	19.557	15.615	39.401	19.201	15.598	38.936	13.75	4.79	19.09	22.8	4.12
SY-1-5	19.163	15.581	39.001	19.106	15.578	38.880	12.78	0.73	4.66	3.68	6.63
SY-2-1A	18.618	15.578	39.452	18.605	15.578	38.439	19.21	0.272	0.89	0.90	3.39
SY-2-1B	19.314	15.601	39.125	19.225	15.597	39.011	6.11	0.53	0.08	5.66	4.03
SY-2-1C	19.380	15.598	39.191	19.203	15.589	38.962	11.34	1.97	7.79	11.28	4.09
SY-2-1E	19.346	15.567	39.087	19.200	15.560	38.901	10.74	1.54	6.02	9.31	4.03
SY-3-3	20.216	15.660	39.453	18.691	15.586	38.840	5.42	8.01	9.82	97.5	1.27
MC-21	19.332	15.597	39.071	19.226	15.591	39.064	9.28	0.97	0.186	6.76	0.12

*Assumed age = 100 Ma (0.10 AE)
$\lambda(^{238}U) = 1.55125 \times 10^{-10}\ yr^{-1}$
$\lambda(^{235}U) = 9.8485 \times 10^{-10}\ yr^{-1}$
$\lambda(^{232}Th) = 0.49475 \times 10^{-10}\ yr^{-1}$
$^{232}U/^{235}U$ = 137.88

TABLE 3. ISOTOPIC COMPOSITION OF LEAD FROM ARKANSAS GALENAS

Sample	$^{206}Pb/^{204}Pb$	$^{207}Pb/^{204}Pb$	$^{208}Pb/^{204}Pb$
Magnet Cove			
1. Titanium Corporation of America Pit, Hot Springs County	19.058	15.549	38.734
Central Arkansas			
2. Kellogg Mine, North Little Rock	18.400	15.604	38.329
3. Preston Sweedon's Prospect, Pike County	18.641	15.637	38.519
4. Waterloo Mine, Montgomery County	18.163	15.542	38.115
5. Point Cedar, Hot Springs County	18.422	15.631	38.341
6. Bellah Mine, Sevier County	18.433	15.645	38.390
7. Pulaski County	18.322	15.566	38.217
8. Dare Mine, Pope County	18.414	15.638	38.399
9. Brewer-Robins Prospect, Pope County	18.327	15.571	38.174
10. Housley Mine, Pope County	18.434	15.634	38.350
North Arkansas			
11. North Pole Mine, Newton County	21.955	15.951	41.153
12. Brewer Mine, Newton County	21.941	15.942	41.010
13. Ponca City Mine, Newton County	22.114	15.944	40.993
14. Pigeon Roost Mine, Marion County	21.936	15.929	41.083

and others, 1966), values we have also found in calcite from several Canadian carbonatites we analyzed. The low-calcite Sm and Nd abundances at Magnet Cove may indicate that the rare-earth elements are concentrated in other minerals for which we have no concentration data.

From the syenites we have analyzed the potassium feldspars for Pb and whole rocks for Sr and Nd. As feldspar is very abundant in the rocks, in some cases exceeding 90%, the total rock concentrations are probably not very different from the feldspar values. Pb, Sm, Nd, and Rb are considerably more abundant in the syenites compared to the carbonates. Only Sr has a lower concentration, as expected.

DISCUSSION

Carbonatites

The mean $^{87}Sr/^{86}Sr$ ratio for the four Magnet Cove carbonate samples in Table 1 is 0.70363. The high-Sr concentrations in the calcites make in situ decay corrections unnecessary for these samples. The Sr ratio is higher than that found at Oka, 0.70331 (Grünenfelder and others, 1985), but still much lower than estimated "bulk earth" values of 0.7045 to 0.7050. Thus, these Sr data indicate an LIL element–depleted affinity for the carbonate sources. The Sr data are best interpreted in conjunction with the corresponding Nd data below.

The mean $^{143}Nd/^{144}Nd$ ratio for the carbonates is 0.51277, corresponding to a present-day ϵ (Nd) of +2.6. Here ϵ (Nd) is defined by the equation:

$$\epsilon\,(\mathrm{Nd}) = \left[\frac{R_x}{R_E} - 1 \right] \times 10^4,$$

where R_x is the $^{143}Nd/^{144}Nd$ ratio in the calcite and R_E is the ratio for average earth, obtained from the value for average chondrites (Jacobsen and Wasserburg, 1980, 1984). When corrected for in situ decay, the ratio becomes 0.51271, which yields ϵ (Nd) = +3.8 $\pm$ 0.3 (1σ) relative to the chondritic reservoir at 100 Ma, again indicating an LIL element–depleted source for the carbonates.

The Sr and Nd data are shown in a correlation diagram in Figure 1. Since the figure contains young oceanic rocks, as well as the 0.1 AE Arkansas rocks, we use the ϵ_T presentation for both Nd and Sr, where ϵ_T is a measure of the difference between the $^{143}Nd/^{144}Nd$ or $^{87}Sr/^{86}Sr$ ratio in a given sample and the ratio in average earth at time *T*, in this case the time at which the magmas were formed. This allows comparison of samples of differing age in a common diagram. The numerical values for the parameters used in the formulations are given in the footnotes to Table 1. This comparison with the oceanic data shows that the Magnet Cove carbonate ratios agree well with the field of Bouvet Island, hence, no crustal component is required to explain the data.

None of our $^{143}Nd/^{144}Nd$ ratios are as high as that reported by Basu and Tatsumoto (1980). Further measurements would be necessary to clarify the cause of this difference.

The Nd and Sr isotope ratios from the two Potash Sulfur Springs carbonates differ from those of at Magnet Cove, and probably from each other as well. Present data cannot distinguish whether this is due to crustal contamination or to mixing of isotopically heterogeneous mantle sources.

Pb data for five carbonate samples are shown in Figure 2. The samples plot generally within the MORB–ocean island fields in both parts of the figure, again indicating an LIL element–depleted affinity for the sources. Four Kimsey Quarry samples at

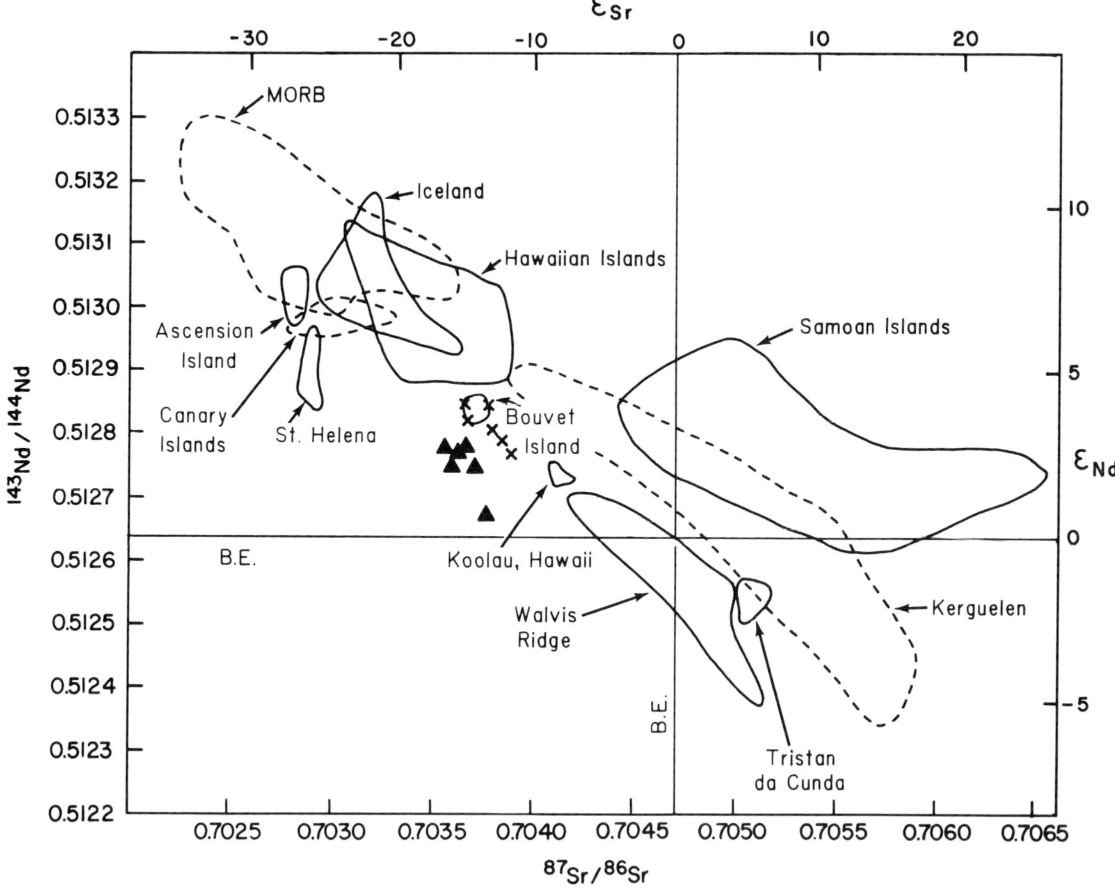

Figure 1. Nd-Sr isotope correlation diagram for Arkansas carbonatites. Triangles indicate present-day ratios for samples; crosses (x), initial ratios for age of 100 m.y., plotted on epsilon scales; B.E., estimated values for present-day "bulk earth." Oceanic data from the literature.

Magnet Cove, plus the Magnet Cove galena (Table 3), plot approximately along a line with a slope of 0.29 in Figure 2b. This suggests some kind of mixing, as the line corresponds to a formal Pb/Pb age of 3.26 AE, which seems too old to have geological significance. The implications of the possible mixing trend are discussed further under the galena data cited below. In contrast to the Nd-Sr data, the Pb ratios for the Magnet Cove carbonates differ from those at Bouvet Island (summary in Tatsumoto, 1978), particularly the $^{207}Pb/^{204}$ ratios, which are higher at Bouvet. The Potash Sulfur Springs carbonate plots off the Kimsey trend in the lower diagram, but as only one sample was measured at that locality, it is not clear how much the Pb isotopic compositions may differ between thc two localities.

The Magnet Cove carbonatite Sr, Nd, and Pb data agree generally with parallel data from the Oka carbonatite complex, although all three elements indicate a source slightly less depleted in LIL elements for the Arkansas rocks. We attribute the difference to regional mantle heterogeneity in the subcontinental mantle, analogous to the regional heterogeneities observed in suboceanic mantle.

Syenites

Nd data for whole-rock samples from three syenites are given in Table 1. The $^{143}Nd/^{144}Nd$ ratios average 0.51264 when corrected for in situ decay, a value similar to one of the initial ratios for Potash Sulfur Springs carbonate, but slightly lower than those at Magnet Cove. The corresponding ϵ(Nd) for the syenite initial ratio referred to the chondrite reservoir at 0.1 AE is +2.5. An attempt to determine the initial $^{87}Sr/^{86}Sr$ ratio for the feldspar samples by means of an isochron diagram failed, due either to nonuniform initial $^{87}Sr/^{86}Sr$ ratios or to open system behavior for the Rb-Sr system. Calculations of initial ratios for samples 2-1A and 2-1B from observed data yield 0.6900 and 0.6990, respectively, showing that postemplacement disturbance of the Rb-Sr system must be the major cause of the spread in the data. In view of the evidence for crustal contamination for Pb, discussed below, the initial Sr ratios may have been variable as well.

The initial Pb ratios for seven of the nine syenite feldspars are quite uniform and plot generally with the carbonate ratios in Figure 2. The mean ratios for all samples, excluding samples

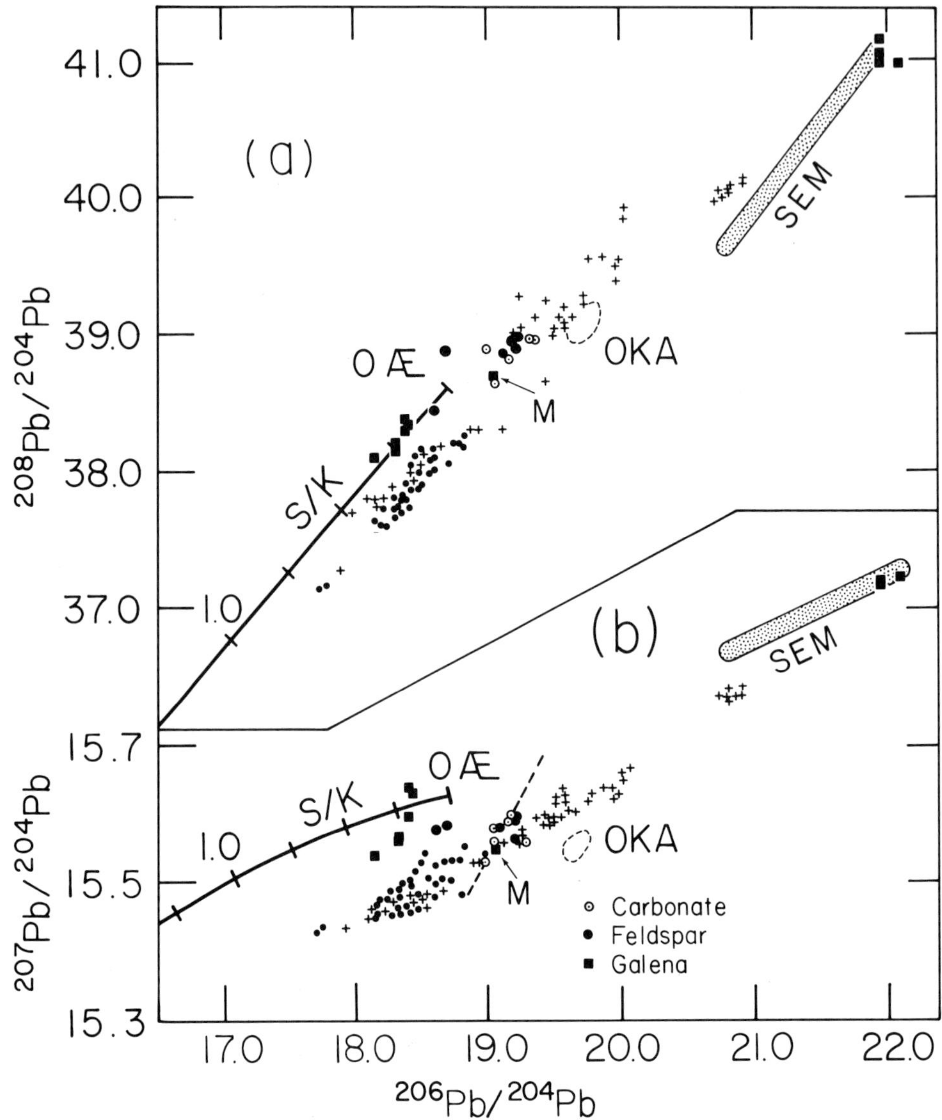

Figure 2. Lead isotope correlation diagram for Arkansas alkalic rocks and galenas. Small circles indicate MORB from Atlantic and Pacific Oceans; crosses, ocean island volcanic rocks; dotted circular fields, sövite and okaite samples from Oka. M, Magnet Cove galena; SEM, field for southeastern Missouri galenas, after Doe and Delevaux (1972); S/K, lead evolution model of Stacey and Kramers (1975). AE scale refers to model ages in 10^9 yr.

2-1A and 3-3, are $^{206}Pb/^{204}Pb = 19.199 \pm 0.016$ standard error; $^{207}Pb/^{204}Pb = 15.587 \pm 0.005$; $^{208}Pb/^{204}Pb = 38.963 \pm 0.024$. These ratios plot along the Magnet Cove calcite trend in Figure 2. Samples 2-1A and 3-3 have much lower initial $^{206}Pb/^{204}Pb$ and $^{208}Pb/^{204}Pb$ ratios than any of the other samples. Sample 2-1A also has a different isotopic composition than the other three samples from the same quarry. These observations suggest some form of contamination. To obtain further information on possible kinds of Pb contamination in the magmas, we analyzed several suites of galenas, which we assume to represent samples of large volumes of crustal material in most cases.

Galenas

The isotopic data are given in Table 3. The Magnet Cove galena ratios agree closely with the less radiogenic members of the carbonate suite, showing that the galena belongs to the carbonatite intrusive sequence. The northern Arkansas suite has ratios that are very similar to published data from southeast Missouri (Doe and Delevaux, 1972), which identifies the leads as members of the Mississippi Valley group. Doe and Delevaux (1972) showed that the source of the Missouri leads was probably in the Lamotte sandstone. The central Arkansas galenas, representing samples from six different counties, are rather uniform and show a distinctly different origin from the other two groups. The isotopic compositions of this group are appropriate for derivation from late Paleozoic or Mesozoic crustal sources; the lead ratios are, in fact, similar to those of the Appalachian Valley group that Heyl and others (1974) labeled "diagenetic(?)" in their analysis of central U.S. lead deposits.

The lead in samples 2-1A and 3-3, with the exception of the $^{208}Pb/^{204}Pb$ ratio of 3-3, can be accounted for by mixtures of lead from the central Arkansas galenas and from the average Pb cited above for the seven samples. Chen and others (1984) have similarly proposed crustal contamination of magmas from depleted mantle sources to explain Sr, Nd, and O isotopic relations within small alkaline intrusives in New England and southern Quebec. Eby (1985) also postulated crustal components in Monteregian Hills–White Mountains alkaline rocks on the basis of Sr and Pb isotopic data. It seems quite clear that syenitic rocks sometimes contain components from the crustal rocks with which they are associated.

There is no evidence for similar contamination in any of the remaining samples. The regression lines through the carbonates, Magnet Cove galena, and seven syenite feldspars in Figure 2 do not intersect the fields of the central Arkansas galenas, nor do they intersect the fields for the leads from northern Arkansas and southeast Missouri. We cannot specify accurately the components of the mixture, assuming that the trends in Figure 2 are real and not the result of too limited sampling. We believe, however, that this apparent mixing trend results from mantle processes rather than from crustal contamination. This view is based on the assumption that the central Arkansas galena data adequately characterize crustal Pb over large areas of the state in the Cretaceous era. If crustal contamination is a significant factor in the samples, it would be difficult to understand the approximately linear relation between $^{207}Pb/^{204}Pb$ and $^{206}Pb/^{204}Pb$ ratios in Figure 2b. A mantle mixing process can readily explain the ratios, with the higher $^{207}Pb/^{204}Pb$ end member having an isotopic composition similar to the Pb in the volcanic rocks at Bouvet Island.

Some carbon and oxygen isotopic data have been reported on seven samples of carbonate from Magnet Cove (Deines and Gold, 1973). In a plot of $\delta^{13}C$ versus $\delta^{18}O$, the Arkansas field overlaps the fields for complexes such as Oka and Alnö, which are generally considered to characterize uncontaminated igneous material. One sample, however, has a low $\delta^{13}C$ of –9 and a high $\delta^{18}O$ of +18, values characteristic of crustal rocks, especially sedimentary rocks. These data indicate that some parts of the complex may contain crustal material, in agreement with the Pb data discussed above, although other explanations of the C and O data are possible (Deines and Gold, 1973).

CONCLUSIONS

The Arkansas carbonatite complexes resemble the Oka, Quebec, complex in indicating derivation from an LIL element–depleted mantle source. The Sr and Nd isotopic compositions of the Magnet Cove carbonate samples are rather uniform and probably reflect the isotopic compositions of the mantle sources. The samples from Potash Sulfur Springs differ from those at Magnet Cove and indicate either derivation from a slightly more enriched mantle source or crustal contamination. The Arkansas carbonate leads have isotopic compositions that plot within the LIL element–depleted mantle field, as defined by oceanic data in Pb isotope correlation diagrams, and exhibit small variations that we tentatively attribute to mantle mixing processes. The small differences between the Oka and Arkansas Pb, Sr, and Nd ratios likely reflect a degree of mantle heterogeneity between the two localities.

Arkansas syenites likewise exhibit an affinity to an LIL-element–depleted mantle source. Nd is slightly less radiogenic than the corresponding ratios in the carbonatites from Magnet Cove, whereas Pb has, with two exceptions, similar isotopic compositions in both rock types. Pb in four syenite samples from a single quarry has variable isotopic compositions. Two of the samples contain a large component of lead characteristic of that observed in central Arkansas galenas that are thought represent crustal sources.

Overall, the results show that alkaline complexes can provide information about the geochemical characteristic of the underlying mantle, although careful attention must be given to deciphering the effects of contamination that may alter the mantle signature.

ACKNOWLEDGMENTS

We are indebted to the Arkansas Geological Commission, in particular to Michael Howard, for guiding the collection of samples for this study and for valuable discussions. We thank Mark Stein for assistance with the mass spectrometry, and David Crouch, who drafted the figures. We particularly thank B. A. Barreiro and K. A. Foland for constructive reviews, which had added substantially to the discussions. This project was supported by grant EAR 82-12931 (to G.R.T.) from the National Science Foundation.

APPENDIX. SAMPLE LOCATIONS AND DESCRIPTIONS

Syenites:

1-1: Blue pulaskite, roadcut on Highway 65, 2 mi south of intersection with Highway 70. Many small inclusions noted in outcrop. Collected sample was free of inclusions. No nepheline.

1-2: Pulaskite, abandoned 3M quarry by Holiday Inn near Little Rock airport. Few inclusions present, but fresher with fewer inclusions than sample 1-1.

1-5: Nepheline syenite, from cut on road into American Cyanamide Mine (Section 24), Benton quadrangle, near town of Bauxite. Nepheline filling small cavities between large orthoclase crystals.

2-1: Active 3M quarry, Little Rock quadrangle. A: Darker phase of pulaskite from center of east wall. B: Lighter phase at same location as A, possibly a more weathered phase of the rock; a few centimeter-sized darker inclusions present. C: From northeast corner of quarry; much sphene in sample. E: Dark pulaskite, collected about 30 m south of a 30-m-thick nepheline syenite dike along west wall of quarry; very fresh.

3-3: Syenite dike about 20-m wide in wall of Diamond Jo Quarry, near Potash Sulfur Springs, Nepheline with many inclusions. A sample nearly free of inclusions was obtained by hand-picking, but was not as pure as other syenite feldspars.

MC-21: Quarry along Arch Street pike, south Little Rock. Sample obtained from C. A. Hopson. No nepheline.

Carbonatites:

3-1: Potash Sulfur Springs, North Wilson Pit, abandoned Union Carbide quarry on north side of pluton. Samples collected from a meter-sized block from a carbonatite dike. Very pure.

4-1: Magnet Cove, Kimsey Quarry. Various samples from walls of quarry. All carbonates highly pure. Collected over a traverse of about 30 m.

MC-6: Magnet Cove, Kimsey Quarry. Sample obtained from C. A. Hopson.

Galenas:

Numbers correspond to sample numbers in Table 3.

Samples obtained from Michael Howard, Arkansas Geological Commission:

1. Magnet Cove. Float from separation tables of Titanium Corporation of America pit, Hot Springs County
2. Kellogg Mine, North Little Rock
3. Preston Sweeden's prospect, Pike County
4. Waterloo Mine, Silver City district, Montgomery County
5. Point Cedar, Hot Springs County
6. Bellah Mine, Sevier County
7. Ag-bearing galena, Little Rock, Pulaski County
8. Dare Mine, Pope County: replacement in fossils and conglomerate intergranular material
11. North Pole Mine, Newton County
12. Brewer Mine, Newton County
13. Ponca City, Newton County
14. Pigeon Roost Mine, Marion County

Samples from Ellen Mullen, University of Arkansas

9. Brewer-Robins prospect, Pope County: galena and sphalerite disseminated through very coarse sandstone.

10. Housely Pb-Zn mine, Cedar Point, Hot Springs County: veins in Devonian-Mississippian sediments.

REFERENCES CITED

Basu, A. R., and Tatsumoto, M., 1980, Nd-isotopes in selected mantle-derived rocks and minerals and their implications for mantle evolution: Contributions to Mineralogy and Petrology, v. 75, p. 43–54.

Bell, K., and Blenkinsop, J., 1985, Carbonatites; Clues to mantle evolution: Geological Society of America Abstracts with Programs, v. 17, p. 151.

Bell, K., Blenkinsop, J., Cole, T.J.S., and Menagh, D. P., 1982, Evidence from Sr isotopes for long-lived heterogeneities in the upper mantle: Nature, v. 298, p. 251–253.

Chen, Jiang-feng, Foland, K. A., Gilbert, L. A., and Randall, K. A., 1984, Depleted mantle sources for Mesozoic plutons of New England and Southern Quebec: Geological Society of America Abstracts with Programs, v. 16, p. 469.

Deines, P., and Gold, D. P., 1973, The isotopic composition of carbonatite and kimberlite carbonates and their bearing on the isotopic composition of deep seated carbon: Geochimica et Cosmochimica Acta, v. 37, p. 1709–1733.

DePaolo, D. J., and Wasserburg, G. J., 1976, Inferences about magma sources and mantle structure from variations of $^{143}Nd/^{144}Nd$: Geophysical Research Letters, v. 2, p. 743–746.

Doe, B. R., and Delevaux, M. H., 1972, Source of lead in southeast Missouri galena ores: Economic Geology, v. 67, p. 409–425.

Eby, G. N., 1984, Geochronology of the Monteregian Hills alkaline igneous province, Quebec: Geology, v. 12, p. 468–470.

——, 1985, Sr and Pb isotopes, U and Th chemistry of the alkaline Monteregian and White Mountain igneous provinces, eastern North America: Geochimica et Cosmochimica Acta, v. 49, p. 1143–1153.

Eugster, O., Tera, F., Burnett, D. S., and Wasserburg, G. J., 1970, The isotopic composition of Gd and neutron capture effects in some meteorites: Journal of Geophysical Research, v. 75, p. 2753–2768.

Grünenfelder, M. H., Tilton, G. R., Bell, K., and Blenkinsop, J., 1985, Lead and strontium isotope relationships in the Oka carbonatite complex, Quebec: Geochimica et Cosmochimica Acta, v. 50, p. 461–468.

Haskin, L. A., and Frey, F. A., 1966, Dispersed and not-so-rare earths: Science, v. 152, p. 299–314.

Haskin, L. A., Frey, F. A., Schmitt, R. A., and Smith, R. H., 1966, Meteoritic solar and terrestrial rare earth distributions: Physics and Chemistry of the Earth, v. 7, p. 167–321.

Heyl, A. V., Landis, G. P., and Zartman, R. E., 1974, Isotopic evidence for the origin of Mississippi Valley–type mineral deposits; A review: Economic Geology, v. 69, p. 992–1006.

Jacobsen, S. B., and Wasserburg, G. J., 1979, Nd and Sr isotope study of the Bay of Islands ophiolite complex and the evolution of the source of mid-ocean ridge basalts: Journal of Geophysical Research, v. 84, p. 7429–7445.

——, 1980, Sm-Nd isotopic evolution of chondrites: Earth and Planetary Science Letters, v. 50, p. 139–155.

——, 1984, Sm-Nd evolution of chondrites and achondrites, II: Earth and Planetary Science Letters, v. 67, p. 137–150.

Stacey, J. S., and Kramers, J. D., 1975, Approximation of terrestrial lead evolution by a two-stage model: Earth and Planetary Science Letters, v. 26, p. 207–221.

Tatsumoto, M., 1978, Isotopic composition of lead in oceanic basalt and its implication to mantle evolution: Earth and Planetary Science Letters, v. 38, p. 63–87.

Todt, W., Cliff, R. A., Hanser, A., and Hofmann, A. W., 1984, $^{202}Pb + ^{205}Pb$ spike for lead isotopic analysis [abs.]: Terra Congnita, v. 4, p. 209.

Zartman, R. E., and Howard, J. M., 1985, U-Th-Pb ages of large zircon crystals from the Potash Sulfur Springs igneous complex, Garland County, Arkansas: Geological Society of America Abstracts with Programs, v. 17, p. 198.

Zartman, R. E., Brock, M. R., Heyl, A. V., and Thomas, H. H., 1967, K-Ar and Rb-Sr ages of some alkalic intrusive rocks from central and eastern United States: American Journal of Science, v. 265, p. 848–870.

MANUSCRIPT ACCEPTED BY THE SOCIETY OCTOBER 15, 1986

Printed in U.S.A.

Geological Society of America
Special Paper 215
1987

Bibliochrony of igneous rocks of Arkansas with particular emphasis on diamonds

A.J.A. Janse
BP Minerals International Ltd., BP Minerals Australia, Box R1274, GPO Perth, Western Australia 6001
P. A. Sheahan
Konsult International, Inc., 44 Gemini Road, Willowdale, Ontario M2K 2G6, Canada

This is the second edition of the complete chronological reference catalogue (bibliochrony) of 428 records on alkaline and ultrabasic rocks in Arkansas.

Many of the records refer to the micaceous peridotites (also called kimberlites, olivine madupites, or olivine lamproites) near Murfreesboro as this occurrence (Prairie Creek or Crater of Diamonds) has been the subject of scientific and public interest for many years. We think that we have caught all the published articles in scientific journals, but there are probably omissions from obscure periodicals, society and outdoor sports magazines, newspapers and trade journals, and short references of a few lines in books of various kinds. The records include nine or more fields of information. The complete records are stored on BASIS as a private file on Infomat's VAX computer in Toronto. The labels (names of fields) have not been printed and the record may include from 5 to 10 lines in print in the following order:

DP Date
AU Author
TI Title
S1 Source journal
IN Book or proceedings
S3 Paginations
LOC Geographic location
LA Language

An example is given below:

1984 SCOTT SMITH, B.H.; SKINNER, E.M.W.
A NEW LOOK AT PRAIRIE CREEK, ARKANSAS.
THIRD KIMB. CONF.,
VOL. 1, PP. 255-284.
USA; GULF COAST; ARKANSAS; PIKE COUNTY
(ENG)
LAMPROITE; RELATED ROCKS; PETROGRAPHY; MINERAL CHEMISTRY; ANALYSES; WHOLE ROCK GEOCHEMISTRY; GEOCHRONOLOGY

The Arkansas reference catalogue forms part of a comprehensive database on diamonds and kimberlites worldwide, which has been developed by the authors within the British Petroleum Minerals International data system and is stored on a VAX computer. This database contains 17,000 bibliographic records at present and is continuously updated and augmented. Corrections and additions to the Arkansas database will be appreciated greatly by the authors.

BIBLIOCHRONY OF ARKANSAS

1835 FEATHERSTONHAUGH, G.W.
GEOLOGICAL REPORT OF AN EXAMINATION MADE IN 1834 OF THE ELEVATED COUNTRY BETWEEN THE MISSOURI AND RED RIVERS.
REPORT TO THE HOUSE OF REPRESENTATIVES, WASHINGTON,
97P.
USA; GULF COAST; ARKANSAS
(ENG)
MAGNET COVE; VEGETATION TYPES
COMPARES MAGNET COVE TO LOCALITIES IN VIRGINIA AND TENNESSEE.

1842 POWELL, W.B.
GEOLOGICAL REPORT UPON THE FOURCHE COVE AND ITS IMMEDIATE VICINITY.
LITTLE ROCK:
PRIVATELY PUBL. BY AUTHOR 22P.
USA; GULF COAST; ARKANSAS; PIKE COUNTY; PULASKI COUNTY
(ENG)
REGIONAL GEOLOGY; IGNEOUS INTRUSION
A FOOTNOTE ON P. 6 NOTES THE OCCURRENCE OF IGNEOUS ROCK.

1846 SHEPPARD, C.U.
ON THREE NEW MINERAL SPECIES FROM ARKANSAS AND THE DISCOVERY OF THE DIAMOND IN NORTH CAROLINA.
AMER. JOUR. SCIENCE,
SER. 2, VOL. 2, P. 253.
USA; GULF COAST; ARKANSAS; PIKE COUNTY; PULASKI COUNTY; SALINE COUNTY
(ENG)
MINERALOGY; PETROLOGY
A FOOTNOTE ON P. 253 NOTES THE OCCURRENCE OF A TRACHITIC PORPHYRY, WHICH PROBABLY REPRESENTS THE HARD MICACEOUS PERIDOTITE OR OLIVINE LAMPROITE.

1851 KING, H.
SOME REMARKS ON THE GEOLOGY OF THE STATE OF MISSOURI.
AMER. ASSOC. ADV. SCI.,
VOL. 5.
USA; GULF COAST; ARKANSAS
(ENG)
MAGNET COVE; LITHOLOGY; IRON ORE
DESCRIBES MAGNET COVE ROCKS AND NOTES LAMPROPHYRES(?) NEAR THE SALINE RIVER.

1852 BARNEY, J.
A SURVEY OF A ROUTE FOR A RAILROAD FROM THE VALLEY OF THE MISSISSIPPI RIVER TO THE PACIFIC OCEAN, COMMENCING AT ST. LOUIS.
EXECUTIVE DOCUMENTS,
NO. 49, P. 31.
USA; GULF COAST; ARKANSAS
(ENG)
FOURCHE MOUNTAIN, QUARRY STONE
MENTIONS USEABLE BUILDING STONE BEING PRESENT AT FOURCHE MOUNTAIN.

1860 OWEN, D.D.
SECOND REPORT OF A GEOLOGICAL RECONNAISSANCE OF THE MIDDLE AND SOUTHERN COUNTIES OF ARKANSAS MADE DURING THE YEARS 1859- 1860.
U.S.G.S. SECOND REPORT,
32P.
USA; GULF COAST; ARKANSAS
(ENG)
REGIONAL GEOLOGY

1878 KOENIG, G.A.
MINERALOGICAL NOTES.
ACAD. NAT. SCIENCES PHILADELPHIA PROC.,
PP. 292-293.
USA; GULF COAST; ARKANSAS; PIKE COUNTY
(ENG)

1886 HARVEY, F.L.
MINERALS AND ROCKS OF ARKANSAS, A CATALOG OF THE SPECIES.
GRANT AND FAIRES, PHILADELPHIA,
32P.
USA; GULF COAST; ARKANSAS
(ENG)
MINERALS; ROCK TYPES
HAND SPECIMEN DESCRIPTIONS OF IGNEOUS ROCKS.

1887 ROSENBUACH, H.
MIKROSKOPISCHE PHYSIOGRAPHIE DER MASSIGEN GESTEINE.
PRIVATELY PUBL. STUTTGART, GERMANY,
USA; GULF COAST; ARKANSAS
(GER)
MAGNET COVE; NEPHELINE SYENITE; PHONOLITE
FIRST PUBLISHED NOTE P. 631 OF PSEUDOMORPHS AFTER LEUCITE.

1888 COMSTOCK, T.G.
REPORT UPON THE PRELIMINARY EXAMINATION OF THE GEOLOGY OF THE WESTERN CENTRAL ARKANSAS.
ANNUAL REPORT OF THE GEOL. SURV. OF ARKANSAS FOR 1888,
VOL. 1.
USA; GULF COAST; ARKANSAS; HOT SPRING COUNTY; GARLAND COUNTY; OUACHITA MOUNTAINS
(ENG)
POTASH SULFUR SPRINGS; MINERAL COMPOSITION; MAGNET COVE
PP. 55-56 DISCUSSES POTASH SULFUR SPRINGS, PP. 81-83 MENTIONS MAGNET COVE.

1888 HILL, R.T.
THE NEOZOIC GEOLOGY OF SOUTHWESTERN ARKANSAS.
ARKANSAS GEOL. SURV. REPT. FOR 1888,
VOL. 2, PP. 56-61.
USA; GULF COAST; ARKANSAS
(ENG)
REGIONAL GEOLOGY

1889 BRANNER, J.C.
THE AGE OF CRYSTALLINE ROCKS OF ARKANSAS.
AMER. ASSOC. ADV. SCI.,
VOL. 37, P. 188. (ABSTR.).
USA; GULF COAST; ARKANSAS; PIKE COUNTY
(ENG)

1889 BRANNER, J.C.; BRACKETT, R.N.
THE PERIDOTITE OF PIKE COUNTY, ARKANSAS.
AMER. JOUR. SCI.,
SER. 3, VOL. 38, PP. 50-59. ALSO: REVIEW AMER. ASSOC. ADV. SCI. PROC., VOL. 37, PP. 188-189. ALSO: NEUES JAHRB. MIN., 1893 PP. 500-501.
USA; GULF COAST; ARKANSAS; PIKE COUNTY
(ENG)
MINERALOGY; PETROLOGY; PRAIRIE CREEK
THE CLASSICAL PAPER ON THE GEOLOGY OF THE CRATER OF DIAMONDS MINE.

1890 DAY, D.T.
MINERAL RESOURCES OF THE U.S. CALENDAR YEAR 1888.
DEPT. OF THE INTERIOR, U.S.G.S., WASHINGTON,
537P.
USA; GULF COAST; ARKANSAS
(ENG)
LITTLE ROCK; QUARRY STONE
DESCRIBES SYENITE AT LITTLE ROCK AND PRESENTS AN ANALYSIS BY THE U.S.G.S.

1891 GRISWOLD, L.S.
THE NOVACULITES OF ARKANSAS.
AMER. ASSOC. ADV. SCI.,
VOL. 39, PP. 248-250. EXTENDED ABSTRACT.
USA; GULF COAST; ARKANSAS
(ENG)

1891 **KEMP, J.F.**
THE BASIC DIKES OCCURRING OUTSIDE SYENITE AREAS OF ARKANSAS.
ARKANSAS GEOL. SURV. REPT. FOR 1890, IN: THE IGNEOUS ROCKS OF ARKANSAS, J.F. WILLIAMS EDITOR, CHAPTER 12, PP. 392-406.
USA; GULF COAST; ARKANSAS
(ENG)
MINERALOGY; PETROLOGY

1891 **KEMP, J.F.; WILLIAMS, J.F.**
TABULATION OF THE DIKES OF IGNEOUS ROCKS OF ARKANSAS.
ARKANSAS GEOL. SURV. REPT. FOR 1890, IN: THE IGNEOUS ROCKS OF ARKANSAS, J.F. WILLIAMS EDITOR, CHAPTER 13, PP. 407-432.
USA; GULF COAST; ARKANSAS;
(ENG)
MINERALOGY; PETROLOGY

1891 **WILLIAMS, J.F.**
DISTRIBUTION AND PETROGRAPHIC CHARACTER OF THE IGNEOUS ROCKS FROM PIKE COUNTY.
ARKANSAS GEOL. SURV. REPORT FOR 1890, IN: THE IGNEOUS ROCKS OF ARKANSAS, J.F. WILLIAMS EDITOR, CHAPTER 11, PP. 376-391.
USA; GULF COAST; ARKANSAS; PIKE COUNTY
(ENG)
MINERALOGY; PETROLOGY; PRAIRIE CREEK

1891 **WILLIAMS, J.F.**
THE IGNEOUS ROCKS OF ARKANSAS.
ARKANSAS GEOL. SURV. REPT. FOR 1890, 432P.
USA; GULF COAST; ARKANSAS
(ENG)
REGIONAL GEOLOGY; PETROLOGY; MINERALOGY

1892 **GRISWOLD, L.S.**
WHETSTONES AND THE NOVACULITES OF ARKANSAS.
ARKANSAS GEOL. SURV. ANN. REPT. FOR 1890, VOL. 3, 443P.
USA; GULF COAST; ARKANSAS
(ENG)

1894 **RUTLEY, F.**
ON THE ORIGIN OF CERTAIN NOVACULITES AND QUARTZITES.
GEOL. SOC. LONDON QUART. JOUR., VOL. 50, PP. 377-392. ALSO: ABSTR. IN GEOL. MAG., VOL. 4, NO. 1, PP. 232-233.
USA; GULF COAST; ARKANSAS
(ENG)
PETROLOGY

1898 **DERBY, A.G.**
ON THE ORIGIN OF CERTAIN SILICEOUS ROCKS, NOTES ON ARKANSAS NOVACULITE.
JOUR. GEOLOGY, VOL. 6, PP. 366-368.
USA; GULF COAST; ARKANSAS
(ENG)

1900 **WASHINGTON, H.S.**
IGNEOUS COMPLEX OF MAGNET COVE, ARKANSAS.
GEOL. SOC. AMER. BULL., VOL. 11, PP. 389-416.
USA; GULF COAST; ARKANSAS; HOT SPRING COUNTY
(ENG)
GEOLOGY

1901 **WASHINGTON, H.S.**
THE FOYAITE IJOLITE SERIES OF MAGNET COVE.
JOUR. GEOLOGY, VOL. 9, PP. 607-622; PP. 645-670.
USA; GULF COAST; ARKANSAS; HOT SPRING COUNTY
(ENG)
FOYAITE; IJOLITE

1906 **KANSAS CITY JEWELLER OPT.,**
A DIAMOND MINE IN ARKANSAS.
KANSAS CITY JEWELLER OPT., VOL. 6, OCTOBER P. 239.
USA; GULF COAST; ARKANSAS; PIKE COUNTY
(ENG)
NEWS ITEM

1906 **KANSAS CITY JEWELLER OPT.,**
THE FOURTH DIAMOND FOUND.
KANSAS CITY JEWELLER OPT., VOL. 6, NOVEMBER P. 268.
USA; GULF COAST; ARKANSAS; PIKE COUNTY
(ENG)
NEWS ITEM
J.W. HUDDLESTON HAS FOUND THE FOURTH DIAMOND ON HIS PROPERTY.

1906 **NASHVILLE NEWS,**
DIAMOND MINES OF ARKANSAS.
NASHVILLE NEWS SUPPLEMENT, AUGUST, 8P.
USA; GULF COAST; ARKANSAS
(ENG)
HISTORICAL REVIEW

1907 **AMERICAN DIAMOND COMPANY,**
PROSPECTUS.
TEXARKANA: AMERICAN DIAMOND COMPANY, 8P. 1 MAP.
USA; GULF COAST; ARKANSAS; PIKE COUNTY
(ENG)
QUOTES FROM VARIOUS AUTHORITIES. PHOTOGRAPHS OF THE FIRST 69 DIAMONDS FOUND IN ARKANSAS WITH DESCRIPTIONS.

1907 **JEWELLERS CIRC.,**
FURTHER DETAILS AS TO THE ARKANSAS DIAMONDS.
JEWELLERS CIRC., VOL. 55, NO. 2, AUGUST 7TH. P. 43; P. 45.
USA; GULF COAST; ARKANSAS; PIKE COUNTY
(ENG)
ILLUSTRATES AND DESCRIBES THE FIRST 69 DIAMONDS DISCOVERED IN PIKE COUNTY.

1907 JEWELLERS CIRC.,
DISCOVERY OF DIAMONDS IN ARKANSAS.
JEWELLERS CIRC.,
VOL. 55, NO. 2, AUGUST 14TH. PP. 63-64.
USA; GULF COAST; ARKANSAS; PIKE COUNTY
(ENG)
NEWS ITEM

1907 SCIENCE,
DIAMONDS IN ARKANSAS.
SCIENCE,
N.S. VOL. 26, AUGUST 16TH. P. 231.
USA; GULF COAST; ARKANSAS; PIKE COUNTY
(ENG)
NEWS ITEM

1907 MANUFACTURER JEWELLER,
AN AMERICAN DIAMOND MINE.
MANUFACTURER JEWELLER,
VOL. 41, SEPT. 19TH. P. 434.
USA; GULF COAST; ARKANSAS; PIKE COUNTY
(ENG)
NEWS ITEM

1907 ENG. MIN. JOUR.,
DIAMONDS IN ARKANSAS.
ENG. MIN. JOUR.,
VOL. 84, AUGUST 10TH. P. 270.
USA; GULF COAST; ARKANSAS; PIKE COUNTY
(ENG)
NEWS ITEM

1907 ENG. MIN. JOUR.,
DIAMONDS IN THE ORIGINAL MATRIX.
ENG. MIN. JOUR.,
VOL. 84, NOV. 2ND. P. 829.
AUSTRALIA; BINGARA; INVERELL; USA; GULF COAST; ARKANSAS
(ENG)
DIAMOND GENESIS

1907 AMER. GEOGRAPH. SOC. BULL.,
DIAMONDS IN PLACE IN THE UNITED STATES.
AMER. GEOGRAPH. SOC. BULL.,
VOL. 39, PP. 742-743.
USA; GULF COAST; ARKANSAS
(ENG)
DIAMOND OCCURRENCES

1907 SOUTH AFRICA MAGAZINE,
AN AMERICAN KIMBERLEY IN EMBRYO.
SOUTH AFRICA MAGAZINE,
VOL. 75, AUGUST 24TH. P. 487.
USA; GULF COAST; ARKANSAS; PIKE COUNTY
(ENG)
NEWS ITEM

1907 LITTLE ROCK DEMOCRAT,
DIAMONDS FOUND IN OZAN FIELD.
LITTLE ROCK DEMOCRAT,
MAY 22ND.
USA; GULF COAST; ARKANSAS; PIKE COUNTY
(ENG)
REPORTS ON THE FINDINGS OF SEVERAL DIAMONDS ON THE FARM OF D.W. GREN TWO MILES EAST OF OZAN, 15 MILES SOUTH OF THE KIMBERLEY DIAMOND MINES IN PIKE COUNTY.

1907 KUNZ, G.F.
DIAMONDS FOUND IN ARKANSAS.
MINERAL RESOURCES OF THE UNITED STATES FOR 1906,
PP. 1217-1220.
USA; GULF COAST; ARKANSAS; PIKE COUNTY
(ENG)
P. 1217 BY STERRETT, D.B.

1907 KUNZ, G.F.
THE OCCURRENCE OF DIAMOND IN NORTH AMERICA.
GEOL. SOC. AMER.,
VOL. 17, PP. 692-694.
USA; GULF COAST; ARKANSAS; WISCONSIN; GREAT LAKES
(ENG)
DIAMOND OCCURRENCE

1907 KUNZ, G.F.; WASHINGTON, H.S.
OCCURRENCE OF DIAMONDS IN ARKANSAS.
MINERAL RESOURCES OF THE UNITED STATES FOR 1906, PART 2, NONMETALS,
PP. 1247-1251. ALSO: SCI. AMER. SUPPL., OCT. 5TH. 1907 VOL. 64, PP. 211-212.
USA; GULF COAST; ARKANSAS
(ENG)
DIAMOND OCCURRENCE
REVIEW OF CURRENT ACTIVITIES FOR THE YEAR.

1907 KUNZ, G.F.; WASHINGTON, H.S.
NOTE ON FORMS OF ARKANSAS DIAMONDS.
AMER. JOUR. SCI.,
SER. 4, VOL. 24, PP. 275-276. ALSO: SOC. OURAL BULL., VOL. 28, P. 1, 1908 ABSTRACT BY O. CLERC. ALSO: CHEM. ABST. VOL. 1, P. 2546.; NEUES JAHRB. MIN. 1909 P. 164.
USA; GULF COAST; ARKANSAS
(ENG)
MORPHOLOGY
140 DIAMONDS FOUND FROM SIZES 1/64 TO 6 1/2 CARATS - WHITE MOST COMMON - SOME TINTED AND SOME INCLUSIONS.

1907 LANIER, R.S.
HAS ARKANSAS A DIAMOND FIELD?
REVIEW OF REVIEWS,
VOL. 36, P. 301.
USA; GULF COAST; ARKANSAS; PIKE COUNTY
(ENG)
HISTORY; GEOLOGY; NEWS ITEM

1907 SCHNEIDER, P.F.
A DESCRIPTIVE CATALOGUE OF THE DIAMOND BEARING ROCKS. TO ACCOMPANY THE EXHIBIT OF PERIDOTITE ROCK AT BUREAU OF MINES.
LITTLE ROCK: CENTRAL PRINTING CO.,
3P.
USA; GULF COAST; ARKANSAS; PIKE COUNTY; SOUTH AFRICA
(ENG)
PETROLOGY
THE PERIDOTITE EXHIBITED CAME FROM NEW YORK, KENTUCKY, ARKANSAS AND SOUTH AFRICA.

1907 **SCHNEIDER, P.F.**
A PRELIMINARY REPORT ON THE ARKANSAS DIAMOND FIELD.
LITTLE ROCK: CENTRAL PRINTING CO., 16P.
USA; GULF COAST; ARKANSAS; PIKE COUNTY
(ENG)
GEOLOGY; DIAMOND OCCURRENCE

1908 **LITTLE ROCK DEMOCRAT,**
BIG DIAMOND DEAL IS MADE.
LITTLE ROCK DEMOCRAT, NOV. 29TH.
USA; GULF COAST; ARKANSAS; PIKE COUNTY
(ENG)
NEWS ITEM
FORMATION OF THE OZARK DIAMOND MINING COMPANY WHICH ACQUIRED THE MAUNEY PROPERTIES.

1908 **STH. AFR. MIN. JOUR.,**
A NEW MENACE TO THE DIAMOND MARKET.
STH. AFR. MIN. JOUR., VOL. 6, PT. 1, MAY 23RD. P. 315.
USA; GULF COAST; ARKANSAS
(ENG)
NEWS ITEM
SOUTH AFRICA'S REACTION TO ARKANSAS DIAMOND FIND.

1908 **ZEIT. MINER.,**
DIAMANT FUNDE IN ARKANSAS.
[DIAMOND FINDS IN ARKANSAS.]
ZEIT. MINER., VOL. 2, JULY 31ST. PP. 188-191.
USA; GULF COAST; ARKANSAS; PIKE COUNTY
(GER)
FROM REPORT OF GERMAN CONSULATE AT ST. LOUIS, DATED APRIL 9TH. QUOTES KUNZ, SCHNEIDER AND HUDDLESTON.

1908 **MINING AND SCIENTIFIC PRESS,**
DIAMONDS IN ARKANSAS.
MINING AND SCIENTIFIC PRESS, VOL. 97, OCT. 31ST. P. 610.
USA; GULF COAST; ARKANSAS; PIKE COUNTY
(ENG)
NEWS ITEM
NOTES THE DISCOVERY OF ANOTHER OUTCROP OF PERIDOTITE, TWO AND HALF MILES FROM THE DIAMOND BEARING GROUP IN PIKE COUNTY.

1908 **STH. AFR. MIN. JOUR.,**
AMERICAN DIAMONDS.
STH. AFR. MIN. JOUR., VOL. 6, PT. 1, MAY 2ND. PP. 241-242.
USA; GULF COAST; ARKANSAS
(ENG)
NEWS ITEM

1908 **MANUFACTURER JEWELLER,**
ACTIVITY IN ARKANSAS DIAMOND FIELDS.
MANUFACTURER JEWELLER, VOL. 43, DEC. 17TH. P. 1076.
USA; GULF COAST; ARKANSAS; PIKE COUNTY
(ENG)
NEWS ITEM
NOTES THE FORMATION OF THE ARKANSAS DIAMOND MINING COMPANY AND THE OZARK DIAMOND MINING COMPANY.

1908 **ARKANSAS DIAMOND COMPANY**
A BRIEF ACCOUNT OF THE DISCOVERY AND INVESTIGATION AND THE OFFICIAL REPORTS OF GEOLOGIST AND MINING ENGINEER ON THE OCCURRENCE OF DIAMONDS IN PIKE COUNTY, ARKANSAS.
LITTLE ROCK: CENTRAL PUBLISHING CO., 37P.
USA; GULF COAST; ARKANSAS; PIKE COUNTY
(ENG)
GEOLOGY
GEOLOGICAL REPORT BY J.T. FULLER; H.S. WASHINGTON

1908 **BENEKE, A.**
NEUE DIAMANTENFUNDE IN DEN VEREINIGTEN STAETEN.
[NEW DIAMOND FINDS IN THE USA.]
ZEIT. MINER., VOL. 2, OCT. 30TH. PP. 265-268.
USA; GULF COAST; ARKANSAS; PIKE COUNTY
(GER)
REVIEW IN A GERMAN JOURNAL OF DIAMOND OCCURRENCES IN ARKANSAS.

1908 **FULLER, J.T.**
REPORT ON THE PROPERTY OF THE ARKANSAS DIAMOND COMPANY.
LITTLE ROCKS: CENTRAL PUBLISHING CO., IN: ARKANSAS DIAMOND COMPANY- A BRIEF ACCOUNT OF THE DISCOVERY AND INVESTIGATION AND THE OFFICIAL REPORTS OF THE GEOLOGIST AND MINING ENGINEER ON THE OCCURRENCE OF DIAMONDS IN PIKE COUNTY, ARKANSAS. PP. 9-29.
USA; GULF COAST; ARKANSAS; PIKE COUNTY
(ENG)
PROSPECTING; GEOLOGY

1908 **KUNZ, G.F.**
PRECIOUS STONES: DIAMOND.
THE MINERAL INDUSTRY DURING 1907, VOL. 16, PP. 792-804.
USA; GULF COAST; ARKANSAS; PIKE COUNTY
(ENG)
NOTES ON THE FULLER-BRANNER PROSPECTUS ON PRAIRIE CREEK. REVIEW OF CURRENT ACTIVITIES FOR THE YEAR.

1908 KUNZ, G.F.
DIAMONDS IN ARKANSAS.
MINES AND MINERALS,
VOL. 28, PP. 552-553. ALSO: MINING WORLD, VOL. 28, P. 443.
USA; GULF COAST; ARKANSAS; PIKE COUNTY
(ENG)
DIAMOND OCCURRENCE

1908 KUNZ, G.F.; WASHINGTON, H.S.
ON THE PERIDOTITE OF PIKE COUNTY, ARKANSAS AND THE OCCURRENCE OF DIAMOND THEREIN.
NEW YORK ACAD. SCI. ANNALS,
VOL. 18, P. 350.
USA; GULF COAST; ARKANSAS; PIKE COUNTY
(ENG)
GEOLOGY; PETROLOGY

1908 PURDUE, A.H.
A NEW DISCOVERY OF PERIDOTITE IN ARKANSAS.
ECON. GEOL.,
VOL. 3, PP. 525-528.
USA; GULF COAST; ARKANSAS; PIKE COUNTY
(ENG)
GEOLOGY

1908 SCHNEIDER, P.F.
GEOLOGY AND MINING IN ARKANSAS DIAMOND FIELD.
MINING WORLD,
VOL. 28, PP. 255-257.
USA; GULF COAST; ARKANSAS; PIKE COUNTY
(ENG)
HISTORY OF MINING METHODS

1908 SCHNEIDER, P.F.
A UNIQUE COLLECTION OF PERIDOTITE.
SCIENCE,
VOL. 28, PP. 92-93.
USA; GULF COAST; ARKANSAS; PIKE COUNTY
(ENG)
GEOLOGY; PETROLOGY

1908 STERRETT, D.B.
DIAMONDS IN THE UNITED STATES.
MINERAL RESOURCES OF THE UNITED STATES FOR 1907,
PP. 797-804.
USA; GULF COAST; ARKANSAS; PIKE COUNTY; KENTUCKY; APPALACHIA; CALIFORNIA; WEST COAST
(ENG)
DESCRIBES PROGRESS IN SEARCH FOR DIAMONDS. REVIEW OF CURRENT ACTIVITIES FOR THE YEAR.

1908 STIFFT, C.S.
THE ARKANSAS DIAMOND FIELD.
MANUFACTURER JEWELLER,
VOL. 43, OCT. 15TH. P. 602.
USA; GULF COAST; ARKANSAS; PIKE COUNTY
(ENG)
NEWS ITEM
THE VICE PRESIDENT OF ARKANSAS DIAMOND MINING COMPANY CORRECTS ERRORS IN NEWSPAPER ACCOUNTS.

1908 WASHINGTON, H.S.
REPORT ON THE ARKANSAS DIAMOND MINING COMPANY PROPERTY.
LITTLE ROCK: CENTRAL PUBLISHING CO., IN: ARKANSAS DIAMOND COMPANY- A BRIEF ACCOUNT OF THE DISCOVERY AND INVESTIGATION AND THE OFFICIAL REPORTS OF GEOLOGIST AND MINING ENGINEER ON THE OCCURRENCE OF DIAMONDS IN PIKE COUNTY, ARKANSAS.
PP. 30-37.
USA; GULF COAST; ARKANSAS; PIKE COUNTY
(ENG)
GEOLOGY; DIAMOND EVALUATION
LARGEST STONE 6 1/2 CARATS.

1908 WOODFORD, E.G.
REPORT ON THE PROPERTIES OF THE OZARK DIAMOND MINING COMPANY.
LITTLE ROCK: OZARK DIAMOND MINING COMPANY,
7P. 1 MAP.
USA; GULF COAST; ARKANSAS
(ENG)

1908 WOODFORD, E.G.
REPORT ON THE PROPERTIES OF THE OZARK DIAMOND MINING COMPANY.
LITTLE ROCK,
14P. UNPUBL.
USA; GULF COAST; ARKANSAS
(ENG)
GEOLOGY; PROSPECTUS

1909 ENG. MIN. JOUR.,
ARKANSAS - PIKE COUNTY.
ENG. MIN. JOUR.,
VOL. 87, APRIL 24, NO. 17, P. 870.
USA; GULF COAST; ARKANSAS; PIKE COUNTY
(ENG)
DIAMOND OCCURRENCE; NEWS ITEM

1909 STH. AFR. MIN. JOUR.,
MR. DRAPER AND THE ARKANSAS DIAMONDS.
STH. AFR. MIN. JOUR.,
VOL. 7, PT. 2, OCT. 9TH. P. 98.
USA; GULF COAST; ARKANSAS; PIKE COUNTY
(ENG)
NEWS ITEM
MR. DRAPER HAS CONDEMNED THE ARKANSAS DIAMOND FIELDS.

1909 COMMERCIAL APPEAL,
STATUS OF DIAMOND FIELDS OF PIKE COUNTY, ARKANSAS.
MEMPHIS, TENN.:COMMERCIAL APPEAL, MARCH 21ST.
USA; GULF COAST; ARKANSAS; PIKE COUNTY
(ENG)
NEWS ITEM
INTERESTING ACCOUNT OF THE DISCOVERY OF DIAMONDS IN ARKANSAS.

1909 BRANNER, J.C.
SOME FACTS AND CORRECTIONS REGARDING THE DIAMOND REGION OF ARKANSAS.
ENG. MIN. JOUR., VOL. 87, PP. 371-372.
USA; GULF COAST; ARKANSAS; PIKE COUNTY
(ENG)
GEOLOGY; HISTORY
BRANNER COMMENTS ON PAPER BY FULLER IN 1909.

1909 BRANNER, J.C.
BIBLIOGRAPHY OF THE GEOLOGY OF ARKANSAS.
ARKANSAS GEOL. SURV. BULL., PP. 97-164.
USA; GULF COAST; ARKANSAS
(ENG)
BIBLIOGRAPHY

1909 EBERLE, F.
THE ARKANSAS DIAMOND FIELDS.
MINING WORLD, VOL. 31, PP. 285-286.
USA; GULF COAST; ARKANSAS; PIKE COUNTY
(ENG)
GEOLOGY; DIAMOND OCCURRENCE

1909 FULLER, J.T.
DIAMOND MINE IN PIKE COUNTY, ARKANSAS.
ENG. MIN. JOUR., VOL. 87, PP. 152-155; PP. 616-617.
USA; GULF COAST; ARKANSAS; PIKE COUNTY
(ENG)
MINING; HISTORY; CATALOGUE OF DIAMOND OCCURRENCES
DISCUSSES CORRESPONDENCE REGARDING BRANNERS CRITICISM OF EARLIER ARTICLE.

1909 HARRIS, G.D.
MAGNETIC ROCKS. (A NOTE ON THE PERIDOTITE IN ARKANSAS.)
SCIENCE, VOL. 29, P. 384.
USA; GULF COAST; ARKANSAS
(ENG)
PETROLOGY

1909 KIMBERLEY TOWNSITE AND LAND COMPANY,
KIMBERLEY, THE DIAMOND CITY OF AMERICA.
NASHVILLE: THE NEWS PRINT, 16P.
USA; GULF COAST; ARKANSAS; PIKE COUNTY
(ENG)
INTERESTING INFORMATION ON THE EARLY HISTORY OF THE FIELDS.

1909 KUNZ, G.F.
DIAMONDS OF ARKANSAS.
THE MINERAL INDUSTRY DURING 1908, VOL. 17, PP. 734-735.
USA; GULF COAST; ARKANSAS
(ENG)
REVIEW OF CURRENT ACTIVITIES FOR THE YEAR.

1909 KUNZ, G.F.
DIAMONDS IN ARKANSAS.
ENG. MIN. JOUR., VOL. 87, P. 963.
USA; GULF COAST; ARKANSAS; PIKE COUNTY
(ENG)
DIAMOND OCCURRENCE; NEWS ITEM
COMMENTS ON FULLER-BRANNER PROSPECTUS.

1909 KUNZ, G.F.; WASHINGTON, H.S.
DIAMONDS IN ARKANSAS.
AMER. INST. MIN. ENG. TRANS., VOL. 39, PP. 169-176. ALSO A.I.M.E. BIMONTHLY BULL NO. 20, MARCH, PP. 187-194.; ENG. MAG. VOL. 35, PP. 113-115.
USA; GULF COAST; ARKANSAS; PIKE COUNTY
(ENG)
DIAMOND OCCURRENCE

1909 MCCOURT, W.E.
DIAMOND IN ARKANSAS.
SCIENCE, VOL. 30, JULY 30TH. P. 127. ALSO: ST. LOUIS ACAD. SCI. TRANS., VOL. 18, PP. LIX-LX.
USA; GULF COAST; ARKANSAS; PIKE COUNTY
(ENG)
DIAMOND OCCURRENCE
COLORS VARY WHITE, BLUE, YELLOW, AND SEVERAL BLACK.

1909 MILLAR, A.Q.
THE ARKANSAS DIAMOND FIELDS.
MINING AND SCIENTIFIC PRESS, VOL. 99, P. 534.
USA; GULF COAST; ARKANSAS
(ENG)
DIAMOND OCCURRENCE

1909 PEARSON, S.
AMERICAN DIAMONDS.
STH. AFR. MIN. JOUR., VOL. 7, PT. 1, JUNE 12TH. P. 412.
USA; GULF COAST; ARKANSAS; PIKE COUNTY; APPALACHIA; KENTUCKY
(ENG)
NEWS ITEM
BRIEF NOTES ON THE KENTUCKY AND ARKANSAS PERIDOTITE OCCURRENCES BY A FORMER SOUTH AFRICAN DIAMOND MINER WHO WORKED FOR TWELVE MONTHS ON THE ISON PIPES IN KENTUCKY. HE BELIEVES THERE ARE PLENTY OF DIAMONDS IN KENTUCKY.

1909 STERRETT, D.B.
DIAMOND IN THE UNITED STATES.
MINERAL RESOURCES OF THE UNITED STATES FOR 1908: PART 2, NONMETALS;
PP. 814-821.
USA; GULF COAST; ARKANSAS; PIKE COUNTY; SOUTH AFRICA; SOUTH WEST AFRICA; NAMIBIA; BRAZIL; GUYANA; INDIA; AUSTRALIA; WEST COAST; APPALACHIA; SOUTH AMERICA
(ENG)
REVIEW OF CURRENT ACTIVITIES FOR THE YEAR.
ORIGIN OF DIAMONDS ARKANSAS PP. 814-815.

1910 LONDON MINING JOURNAL,
ARKANSAS, U.S.A.
LONDON MINING JOURNAL,
VOL. 90, P. 553.
USA; GULF COAST; ARKANSAS; PIKE COUNTY
(ENG)
NEWS ITEM; DIAMOND OCCURRENCE

1910 MINING WORLD,
SOUTH AFRICA AND ARKANSAS DIAMONDS.
MINING WORLD,
VOL. 32, APRIL 9TH. P. 757.
USA; GULF COAST; ARKANSAS; SOUTH AFRICA
(ENG)
NEWS ITEM

1910 ARKANSAS DIAMOND MINING COMPANY,
INTERESTING FACTS, DIAMONDS AND DIAMOND MINING; COMPARISON OF THE SOUTH AFRICAN MINES WITH THE MINES OF PIKE COUNTY.
LITTLE ROCK: ARKANSAS DIAMOND MINING COMPANY,
12 P.
USA; GULF COAST; ARKANSAS; PIKE COUNTY
(ENG)
PROSPECTUS
CITES PROFITS MADE IN SOUTH AFRICA.

1910 COWAN, J.L.
DIAMOND MINES OF ARKANSAS.
MINING AND SCIENTIFIC PRESS,
VOL. 101, AUGUST 6TH. PP. 178-179. ALSO: STH. AFR. MIN. JOUR., VOL. 8, PT. 2, SEPT. 17TH. P. 54.
USA; GULF COAST; ARKANSAS; PIKE COUNTY
(ENG)
NEWS ITEM

1910 FOGG, F.P.
GENUINE DIAMOND MINES.
NATIONAL MAG.,
VOL. 32, SEPTEMBER PP. 604-615.
USA; GULF COAST; ARKANSAS; PIKE COUNTY
(ENG)
POPULAR ARTICLE

1910 FULLER, J.T.
THE ARKANSAS DIAMOND FIELD.
ENG. MIN. JOUR.,
VOL. 89, APRIL 9TH. PP. 767-768.
USA; GULF COAST; ARKANSAS; PIKE COUNTY
(ENG)
DIAMOND OCCURRENCE

1910 GARDINER, C.R.
NATIVE GEMS OF NORTH AMERICA.
JEWELLERS CIRC.,
VOL. 61, NO. 21, DEC. 21ST. PP. 75-77. ALSO: VOL. 61, NO. 22, DEC. 28TH. PP. 63 -65. MINING WORLD, VOL. 34, MARCH 18TH. PP. 593-595. 1911.
USA; GULF COAST; ARKANSAS; NORTH CAROLINA; APPALACHIA; CALIFORNIA; WEST COAST
(ENG)
NEWS ITEM
DIAMOND OCCURRENCES ARE REPORTED.

1910 GLENN, L.C.
REAL DIAMONDS IN ARKANSAS.
NASHVILLE AMERICAN,
MAY 7TH.
USA; GULF COAST; ARKANSAS
(ENG)
NEWS ITEM
INTERVIEW WITH PROF. GLENN OF VANDERBILT UNIV. WHO VISITED THE FIELDS.

1910 KOHR, H.F.
THOSE ARKANSAS DIAMONDS.
TECH. WORLD,
VOL. 13, MAY PP. 288-292.
USA; GULF COAST; ARKANSAS; PIKE COUNTY
(ENG)
GENERAL ARTICLE ON MINING ORGANIZATION.

1910 RANGE, P.
DIAMANTVORKOMMEN IN DEN VEREINIGTEN STAETTEN VON AMERIKA.
[DIAMOND OCCURRENCES IN THE UNITED STATES.]
DEUT. KOLONIALBLATT,
VOL. 21, PP. 942-943.
USA; GULF COAST; ARKANSAS
(GER)
DIAMOND OCCURRENCE

1910 STERRETT, D.B.
DIAMONDS IN THE UNITED STATES.
MINERAL RESOURCES OF THE UNITED STATES FOR 1909,
PT. 2, PP. 757-762.
USA; GULF COAST; ARKANSAS; CALIFORNIA; WEST COAST; BUTTE; INDIANA; APPALACHIA; NEW YORK;
(ENG)
REVIEW OF CURRENT ACTIVITIES FOR THE YEAR.
GRAYSON MCLEOD LUMBER CO. OWNS PERIDOTITE AT BLACK LICK ARKANSAS.

1911 LITTLE ROCK GAZETTE,
FOUND DIAMONDS 35 YEARS AGO.
WASHINGTON TELEGRAPH PUBLISHES EXTRACT FROM THE GAZETTE TELLING OF DISCOVERY OF FIRST GEM.
LITTLE ROCK GAZETTE,
SEPT. 10TH.
USA; GULF COAST; ARKANSAS; PIKE COUNTY
(ENG)
NEWS ITEM
ARTICLE FROM ARKANSAS GAZETTE WHICH WAS REPUBLISHED BY THE WASHINGTON, ARKANSAS TELEGRAPH OF APRIL 12, 1876. DESCRIBES A 1 1/2 CARAT DIAMOND AND LATER CUT IN NEW YORK.

1911 JEWELLERS CIRC.,
FURTHER DETAILS OF THE DISCOVERY OF THE LARGE DIAMOND AT PIKE COUNTY, ARKANSAS.
JEWELLERS CIRC.,
VOL. 62, NO. 22, JUNE 28TH, P. 57.
USA; GULF COAST; ARKANSAS; PIKE COUNTY
(ENG)
NEWS ITEM
DESCRIPTION OF THE BIGGEST DIAMOND YET REPORTED FROM PIKE COUNTY 10-13 CARATS.

1911 COWAN, J.L.
AMERICAN GEM MINES AND MINING.
MINES & MINERALS,
VOL. 32, SEPTEMBER PP. 103-105.
USA; GULF COAST; ARKANSAS
(ENG)
GIVES EARLY HISTORY OF ARKANSAS FINDS.

1911 FULLER, J.T.
THE ARKANSAS DIAMOND FIELD.
ENG. MIN. JOUR.,
VOL. 91, P. 6.
USA; GULF COAST; ARKANSAS; PIKE COUNTY
(ENG)
DIAMOND OCCURRENCE; NEWS ITEM

1911 KUNZ, G.F.
PRECIOUS STONES.
THE MINERAL INDUSTRY DURING 1910,
VOL. 19, PP. 563-589.
USA; GULF COAST; ARKANSAS: SOUTH AFRICA; SOUTH WEST AFRICA; NAMIBIA; ZAIRE; AUSTRALIA; CANADA; BRITISH COLUMBIA; ONTARIO; SOUTH AMERICA; BRITISH GUIANA; USA;
(ENG)
REVIEW OF CURRENT ACTIVITIES OF THE YEAR. ARKANSAS PP. 578-579.

1911 MILLAR, A.Q.
SUMMARY OF DIAMONDS AND DIAMOND MINES.
MINING WORLD,
VOL. 34, PP. 1125-1127; PP. 1188-1190.
USA; GULF COAST; ARKANSAS
(ENG)
DIAMOND OCCURRENCE; LOCATIONS; CATALOGUE

1911 RICHARD, L.M.
DIAMONDS.
DAHLONEGA NUGGET,
MAY 26TH.
USA; GULF COAST; ARKANSAS; APPALACHIA; GEORGIA
(ENG)
NEWS ITEM
CALLS ATTENTION TO THE SIMILARITY BETWEEN THE CHEMICAL COMPOSITION OF A ROCK MASS NEAR PORTER SPRINGS, GEORGIA AND THAT OF THE DIAMOND DEPOSITS OF ARKANSAS.

1911 STERRETT, D.B.
DIAMONDS IN THE UNITED STATES.
MINERAL RESOURCES OF THE UNITED STATES FOR 1910,
PT. 2, PP. 858-860.
USA; GULF COAST; ARKANSAS; WEST COAST; CALIFORNIA
(ENG)
REVIEW OF CURRENT ACTIVITIES OF THE YEAR.

1912 JEWELLERS CIRC.,
FINE FLAWLESS DIAMOND REPORTED TO HAVE BEEN FOUND AT MINE NEAR MURFREESBORO, ARKANSAS.
JEWELLERS CIRC.,
VOL. 64, NO. 18, JUNE 5TH. P. 67.
USA; GULF COAST; ARKANSAS; PIKE COUNTY
(ENG)
NEWS ITEM
DESCRIBES A FINE 3 1/2 CARAT STONE FOUND ON THE PROPERTY OF THE ARKANSAS DIAMOND COMPANY.

1912 LONDON MINING JOURNAL
ARKANSAS DIAMOND MINING.
LONDON MINING JOURNAL,
VOL. 7, DECEMBER P. 454.
USA; GULF COAST; ARKANSAS
(ENG)
CURRENT ACTIVITIES; NEWS ITEM

1912 FULLER, J.T.
THE ARKANSAS DIAMOND FIELD.
ENG. MIN. JOUR.,
VOL. 93, P. 6.
USA; GULF COAST; ARKANSAS; PIKE COUNTY
(ENG)
NEWS ITEM
SUMMARY OF YEAR'S PRODUCTION

1912 GLENN, L.C.
THE ARKANSAS DIAMOND BEARING PERIDOTITE AREA.
SCIENCE,
VOL. 35, P. 312. ALSO: NEUES JAHRB. 1914 BD. FEB. 23., P. 226.
USA; GULF COAST; ARKANSAS; PIKE COUNTY
(ENG)
GEOLOGY

1912 GLENN, L.C.
ARKANSAS DIAMOND BEARING PERIDOTITE AREA.
GEOL. SOC. AMER.,
VOL. 23, P. 726. (ABSTR.).
USA; GULF COAST; ARKANSAS; PIKE COUNTY
(ENG)
PRAIRIE CREEK; PETROLOGY

1912 KUNZ, G.F.
PRECIOUS STONES.
THE MINERAL INDUSTRY DURING 1911,
VOL. 20, PP. 624-644.
USA; GULF COAST; ARKANSAS; SOUTH AFRICA; SOUTH WEST AFRICA; NAMIBIA; AUSTRALIA; BRITISH GUIANA; BRAZIL; CANADA; BRITISH COLUMBIA; CHINA; ILLINOIS; APPALACHIA; CENTRAL STATES;
(ENG)
REVIEW OF CURRENT ACTIVITIES FOR THE YEAR. ARKANSAS; PP. 634-635.

1912 PURDUE, A.H.
ARKANSAS DIAMOND BEARING PERIDOTITE.
GEOL. SOC. AMER. BULL.,
VOL. 23, P. 726.
USA; GULF COAST; ARKANSAS; PIKE COUNTY
(ENG)
GEOLOGY
BRIEF DISCUSSION OF GLENN'S PAPER AND THE AGE OF THE PERIDOTITES.

1912 STERRETT, D.B.
DIAMONDS IN THE UNITED STATES.
MINERAL RESOURCES OF THE UNITED STATES FOR 1911,
PT. 2, P. 1047-1048.
USA; GULF COAST; ARKANSAS; TEXAS; CALIFORNIA; WEST COAST; ILLINOIS; GREAT LAKES
(ENG)
REVIEW OF CURRENT ACTIVITIES FOR THE YEAR.

1912 STOKES, R.S.G.
REPORT ON THE EXAMINATION OF PROPERTIES OF THE ARKANSAS DIAMOND MINING COMPANY AT MURFREESBORO.
COPY OF LETTER TO W.W. MEIN,
NOV. 2ND.
USA; GULF COAST; ARKANSAS; PIKE COUNTY
(ENG)
COMMENT ON PROSPECTS OF THE MINE AND RECORD OF TEST PITS.

1913 STH. AFR. MIN. JOUR.,
DIAMOND MINING IN THE U.S.A.
STH. AFR. MIN. JOUR.,
VOL. 22, PT. 1, FEB. 15TH. P. 761.
USA; GULF COAST; ARKANSAS; PIKE COUNTY
(ENG)
NEWS ITEM
W.W. MEIN, ENGINEER IN CHARGE OF ARKANSAS FIELDS PROPOSES TO TREAT 100,000 TONS AS SAMPLING. R. STOKES IS IN CHARGE OF THE WORK.

1913 ENG. MIN. JOUR.,
DIAMONDS IN ARKANSAS.
ENG. MIN. JOUR.,
VOL. 96, NO. 11, SEPT. 13TH. PP. 488-489.
USA; GULF COAST; ARKANSAS; PIKE COUNTY
(ENG)
DIAMOND OCCURRENCE NEWS ITEM
1375 STONES FOR A TOTAL OF 550 CARATS HAVE BEEN REPORTED TO DATE.

1913 ENG. MIN. JOUR.,
KIMBERLITE COMPANY'S CHARTER REVOKED.
ENG. MIN. JOUR.,
VOL. 96, OCT. 18TH. P. 751. NOV. 1, P. 855.
USA; GULF COAST; ARKANSAS; PIKE COUNTY
(ENG)
PROSPECTING; MINING ECONOMICS; TRADE; LAW; NEWS ITEM

1913 FULLER, J.T.
DIAMOND MINING IN ARKANSAS.
ENG. MIN. JOUR.,
VOL. 95, P. 75.
USA; GULF COAST; ARKANSAS; PIKE COUNTY
(ENG)
NEWS ITEM; MINING METHODS; CURRENT ACTIVITIES

1913 KUNZ, G.F.
PRECIOUS STONES.
THE MINERAL INDUSTRY DURING 1912,
VOL. 21, P. 712.
USA; GULF COAST; ARKANSAS; ROCKY MOUNTAINS; IDAHO
(ENG)
DIAMOND OCCURRENCE
REVIEW OF CURRENT ACTIVITIES FOR THE YEAR.

1913 STERRETT, D.B.
DIAMONDS IN THE UNITED STATES.
MINERAL RESOURCES OF THE UNITED STATES FOR 1912,
PT. 2, PP. 1037-1041.
USA; TEXAS; GULF COAST; ARKANSAS; INDIANA; GREAT LAKES; CALIFORNIA; WEST COAST
(ENG)
REVIEW OF CURRENT ACTIVITIES FOR THE YEAR.

1914 FULLER, J.T.
THE ARKANSAS DIAMOND FIELD IN 1913.
ENG. MIN JOUR.,
VOL. 97, JAN. 10TH. P. 52.
USA; GULF COAST; ARKANSAS; PIKE COUNTY
(ENG)
NEWS ITEM; MINING METHODS

1914 KUNZ, G.F.
DESCRIPTION OF ATTEMPT TO WORK THE AREA OF THE OZARK MINING COMPANY.
THE MINERAL INDUSTRY DURING 1913,
VOL. 22, PP. 640-641.
USA; GULF COAST; ARKANSAS; PIKE COUNTY
(ENG)
MINING METHODS
FIRST CLASS WASHING PLANT AT 100 LOADS PER HOUR OPERATED FOR THREE MONTHS FROM THE FIRST 'PIPE' DISCOVERED IN THE REGION. 5000 LOADS WERE WASHED AND RESULTS HAVE BEEN DISCOURAGING.

1914 MISER, H.D.
NEW AREAS OF DIAMOND BEARING PERIDOTITE IN ARKANSAS.
U.S.G.S. BULL. CONTRIB. OF ECONOMIC GEOLOGY, 1912, PT. 1,
NO. 540, PP. 534-546.
USA; GULF COAST; ARKANSAS; PIKE COUNTY
(ENG)
GEOLOGY; PROPERTY DESCRIPTIONS

1914 PRATT, J.H.
THE OCCURRENCE AND UTILIZATION OF CERTAIN MINERAL RESOURCES OF THE SOUTHERN STATES.
SCIENCE,
VOL. 39, P. 403. ALSO: ELISHA MITCHELL SCI. SOC. JOUR., VOL. 30, P. 1-25; PP. 90-115.
USA; GULF COAST; ARKANSAS; VIRGINIA; NORTH CAROLINA; GEORGIA; APPALACHIA;
(ENG)
DIAMOND OCCURRENCES

1914 SANFORD, S.; STONE, R.W.
USEFUL MINERALS OF THE U.S.
U.S.G.S BULL.,
NO. 585, 250P. P. 29; 58; 72; 98; 143; 194.
USA; GULF COAST; ARKANSAS; GEORGIA; INDIANA; MICHIGAN; NORTH CAROLINA; VIRGINIA; APPALACHIA; GREAT LAKES; CALIFORNIA; WEST COAST
(ENG)
DIAMOND OCCURRENCE

1914 SHIRAS, T.
DIAMONDS OF THE FIRST WATER FROM LITTLE KNOWN MINES OF ARKANSAS.
POPULAR MECH. REVIEW,
NOV. PP. 657-660.
USA; GULF COAST; ARKANSAS; PIKE COUNTY
(ENG)
POPULAR ACCOUNT OF DIAMONDS

1914 SHIRAS, T.
DIAMONDS MINED IN ARKANSAS CLAIMED TO GRADE WITH THE BEST THE WORLD PRODUCES.
THE KEYSTONE,
NOV. 15TH. 1P.
USA; GULF COAST; ARKANSAS; PIKE COUNTY
(ENG)
HISTORY; NEWS ITEM

1914 STERRETT, D.B.
DIAMONDS IN THE UNITED STATES.
MINERAL RESOURCES OF THE UNITED STATES FOR 1913,
PT. 2, PP. 663-666.
USA; GULF COAST; ARKANSAS; PIKE COUNTY; GREAT LAKES; INDIANA; APPALACHIA; NEW YORK
(ENG)
REVIEW OF CURRENT ACTIVITIES FOR THE YEAR.
OZARK MINING COMPANY FORMED IN ARKANSAS. FIRST CLASS WASHING PLANT AT 100 LOADS PER HOUR OPERATED FOR THREE MONTHS FROM THE FIRST 'PIPE' DISCOVERED IN THE REGION. 5000 LOADS WERE WASHED.

1915 STERRETT, D.B.
DIAMONDS IN THE UNITED STATES.
MINERAL RESOURCES OF THE UNITED STATES FOR 1914,
PT. 2, P. 320.
USA; GULF COAST; ARKANSAS; WEST COAST; CALIFORNIA
(ENG)
REVIEW OF CURRENT ACTIVITIES FOR THE YEAR.

1916 ENG. MIN. JOUR.,
PRECIOUS STONES IN 1915.
ENG. MIN. JOUR.,
VOL. 102, NO. 18, P. 800.
USA; GULF COAST; ARKANSAS; PIKE COUNTY; BRAZIL; SOUTH AFRICA
(ENG)
DIAMOND OCCURRENCE
REVIEW OF KUNZ MINERAL INDUSTRY REPT. VOL. 14. PIKE COUNTY IDLE.

1916 SCHALLER, W.T.
DIAMONDS IN THE UNITED STATES.
MINERAL RESOURCES OF THE UNITED STATES FOR 1915,
PT. 2, PP. 847-849.
USA; GULF COAST; ARKANSAS; WEST COAST; CALIFORNIA
(ENG)
REVIEW OF CURRENT ACTIVITIES FOR THE YEAR.

1917 KUNZ, G.F.
PRECIOUS STONES.
THE MINERAL INDUSTRY DURING 1916,
VOL. 26, PP. 593-594.
USA; GULF COAST; ARKANSAS; AUSTRALIA
(ENG)
REVIEW OF CURRENT ACTIVITIES FOR THE YEAR. STATUS OF DIAMOND PROPERTY IN ARKANSAS.

1917 SANFORD, S.; STONE, R.W.
USEFUL MINERALS OF THE U.S.
U.S.G.S. BULL.,
NO. 624, 412P.
USA; GULF COAST; ARKANSAS; APPALACHIA; GEORGIA; NORTH CAROLINA; VIRGINIA; GREAT LAKES; INDIANA; MICHIGAN; WEST COAST; CALIFORNIA
(ENG)
DIAMOND OCCURRENCES DISCUSSED P. 37; 54; 109; 137; 165; 231; 317.

1917 STERRETT, D.B.
DIAMONDS IN THE UNITED STATES.
MINERAL RESOURCES OF THE UNITED STATES FOR 1916,
PT. 2, PP. 892-893.
USA; GULF COAST; ARKANSAS; GREAT LAKES; INDIANA; WEST COAST; CALIFORNIA; BUTTE COUNTY
(ENG)
REVIEW OF CURRENT ACTIVITIES FOR THE YEAR.

1919 KUNZ, G.F.
PRECIOUS STONES.
THE MINERAL INDUSTRY DURING 1918,
VOL. 27, PP. 621-622.
USA; GULF COAST; ARKANSAS
(ENG)
REVIEW OF CURRENT ACTIVITIES FOR THE YEAR.

1920 FERGUSON, J.G.
OUTLINES OF GEOLOGY, SOILS AND MINERALS OF THE STATE OF ARKANSAS.
ARKANSAS STATE GEOL. SURVEY,
182P.
USA; GULF COAST; ARKANSAS
(ENG)
PP. 88-89 DIAMONDS - BRIEF GENERAL SUMMARY OF GEOLOGY AND DIAMOND OCCURRENCES.

1920 KUNZ, G.F.
ARKANSAS MINING INDUSTRY.
THE MINERAL INDUSTRY DURING 1919,
VOL. 28, PP. 603-604.
USA; GULF COAST; ARKANSAS
(ENG)
REVIEW OF CURRENT ACTIVITIES FOR THE YEAR.

1920 REYBURN, S.W.; ZIMMERMAN, S.H.
DIAMONDS IN ARKANSAS.
ENG. MIN. JOUR.,
VOL. 109, NO. 17, PP. 983-986.
USA; GULF COAST; ARKANSAS
(ENG)
DIAMOND OCCURRENCE; GEOLOGY

1920 STODDARD, B.H.
GEMS AND PRECIOUS STONES: DIAMOND.
MINERAL RESOURCES OF THE UNITED STATES FOR 1919: PART 2, NONMETALS;
PP. 166-171.
USA; GULF COAST; ARKANSAS; PIKE COUNTY; SOUTH AFRICA; GHANA; WEST AFRICA; CALIFORNIA; WEST COAST
(ENG)
REVIEW OF CURRENT ACTIVITIES FOR THE YEAR.
P. 170 DISCUSSES DIAMONDS IN ARKANSAS.

1921 KUNZ, G.F.
DIAMONDS OF ARKANSAS.
THE MINERAL INDUSTRY DURING 1920,
VOL. 29, PP. 216-217.
USA; GULF COAST; ARKANSAS
(ENG)
REVIEW OF CURRENT ACTIVITIES FOR THE YEAR.

1921 KUNZ, G.F.
DIAMONDS AT PIKE COUNTY, ARKANSAS.
GEOL. SOC. AMER. BULL.,
VOL. 32, P. 165.
USA; GULF COAST; ARKANSAS; PIKE COUNTY
(ENG)
LISTING OF A TITLE OF A PAPER PRESENTED AT AN ANNUAL MEETING.

1921 MUNDORFF, R.F.
DIAMOND MINING IN ARKANSAS.
THE KEYSTONE,
VOL. 48, NO. 6, FEBRUARY, P. 109; P. 111; P. 113.
USA; GULF COAST; ARKANSAS
(ENG)
HISTORY; NEWS ITEM

1921 STODDARD, B.H.
GEMS AND PRECIOUS STONES: DIAMOND.
MINERAL RESOURCES OF THE UNITED STATES FOR 1920: PART 2, NONMETALS;
PP. 216-217.
USA; GULF COAST; ARKANSAS
(ENG)
REVIEW OF CURRENT ACTIVITIES FOR THE YEAR. PRAIRIE CREEK MINE DISCUSSED.

1922 ENG. MIN. JOUR.,
DIAMONDS IN ARKANSAS.
ENG. MIN. JOUR.,
VOL. 113, JUNE 10TH. P. 990.
USA; GULF COAST; ARKANSAS
(ENG)
NEWS ITEM
BRIEF NOTES ON THE DIAMONDS AND POSSIBLE SOURCE.

1922 CURRENT OPINION,
DIAMOND MINING IS A FLOURISHING ARKANSAS INDUSTRY.
CURRENT OPINION,
VOL. 73, DECEMBER PP. 788-789.
USA; GULF COAST; ARKANSAS
(ENG)
NEWS ITEM

1922 **MISER, H.D.**
GEOLOGY AND GENERAL TOPOGRAPHIC FEATURES OF ARKANSAS.
LITTLE ROCK, ARKANSAS BUREAU OF MINES, PRIVATELY PUBL.
IN: MINERALS OF ARKANSAS, J.G. FERGUSON EDITOR,
PP. 11-14.
USA; GULF COAST; ARKANSAS
(ENG)
REGIONAL GEOLOGY

1922 **MISER, H.D.; ROSS, C.S.**
DIAMOND BEARING PERIDOTITE IN PIKE COUNTY, ARKANSAS.
ECON. GEOL.,
VOL. 17, PP. 662-674.
USA; GULF COAST; ARKANSAS; PIKE COUNTY
(ENG)
GEOLOGY
HISTORY OF INVESTIGATIONS, PETROGRAPHY, AND GEOLOGY COMPARED TO SOUTH AFRICAN DEPOSITS. DIAMONDS BRIEFLY DESCRIBED.

1923 **JEWELLERS CIRC.,**
DIAMONDS FROM ARKANSAS.
JEWELLERS CIRC.,
VOL. 86, NO. 18, JUNE 6TH. P. 65.
USA; GULF COAST; ARKANSAS; PIKE COUNTY
(ENG)
NEWS ITEM

1923 **INDIAN ENGINEERING,**
DIAMONDS IN THE UNITED STATES.
INDIAN ENGINEERING,
VOL. 74, AUGUST 11TH. P. 75.
USA; GULF COAST; ARKANSAS; PIKE COUNTY
(ENG)
NEWS ITEM

1923 **ENG. MIN. JOUR.,**
DIAMONDS FROM ARKANSAS.
ENG. MIN. JOUR.,
VOL. 115, JUNE 16TH, P. 1071.
USA; GULF COAST; ARKANSAS; PIKE COUNTY
(ENG)
6000 DIAMONDS REPORTED FROM AREA. MENTIONS RECENT U.S.G.S. BULL. 735-I BY MISER AND ROSS.

1923 **KUNZ, G.F.**
DIAMONDS IN ARKANSAS.
THE MINERAL INDUSTRY DURING 1922,
VOL. 31, PP. 603-605.
USA; GULF COAST; ARKANSAS; PIKE COUNTY
(ENG)
REVIEW OF CURRENT ACTIVITIES FOR THE YEAR.

1923 **MISER, H.D.; ROSS, C.S.**
PERIDOTITE DIKES IN SCOTT COUNTY, ARKANSAS.
U.S.G.S. BULL.,
NO. 735-H, PP. 271-278.
USA; GULF COAST; ARKANSAS; SCOTT COUNTY
(ENG)
GEOLOGY; PETROLOGY

1923 **MISER, H.D.; ROSS, C.S.**
DIAMOND BEARING PERIDOTITES IN PIKE COUNTY, ARKANSAS.
U.S.G.S. BULL.,
NO. 735-I, PP. 279-322.
USA; GULF COAST; ARKANSAS; PIKE COUNTY
(ENG)
GEOLOGY; PETROLOGY
PRAIRIE CREEK, AMERICAN AND BLACK LICK DISCUSSED. A COMPARISON OF THE GEOLOGY TO THE SOUTH AFRICAN DEPOSITS IS GIVEN AND BRIEF DIAMOND EVALUATION.

1923 **MISER, H.D.; ROSS, C.S.**
VOLCANIC ROCKS IN THE UPPER CRETACEOUS OF SOUTHWESTERN ARKANSAS AND SOUTHEAST OKLAHOMA.
AMER. JOUR. SCI.,
VOL. 9, FEB. PP. 113-126.
USA; GULF COAST; ARKANSAS
(ENG)
GEOLOGY

1923 **MITCHELL, G.J.**
DIAMOND DEPOSITS IN ARKANSAS.
ENG. MIN. JOUR.,
VOL. 116, NO. 17, AUG. 18TH. PP. 285-287.
USA; GULF COAST; ARKANSAS; PIKE COUNTY
(ENG)
COMMENTS ON A RECENT GEOLOGICAL STUDY OF A PROPERTY NEAR MURFREESBORO AND THE COMMERCIAL IMPORTANCE.

1923 **MUNDORFF, R.F.**
REAL DIAMONDS IN ARKANSAS.
ILLUSTRATED WORLD,
VOL. 38, PP. 708-710.
USA; GULF COAST; ARKANSAS; PIKE COUNTY
(ENG)
DIAMOND OCCURRENCE; NEWS ITEM

1924 **ARKANSAS BUREAU OF MINES, MANUFACTURERS AND AGRICULTURE,**
DIRECTORY OF ARKANSAS INDUSTRIES.
ARKANSAS BUREAU OF MINES, MANUFACTURERS AND AGRICULTURE,
174P.
USA; GULF COAST; ARKANSAS
(ENG)
PP. 32-33 GENERAL INFORMATION ON DIAMOND MINING.

1924 **SHIRAS, T.**
THE DESCRIPTION OF DIAMONDS IN ARKANSAS.
MANUFACTURERS RECORD,
JUNE 19TH. PP. 73-74.
USA; GULF COAST; ARKANSAS; PIKE COUNTY
(ENG)
DIAMOND MORPHOLOGY

1924 SHIRAS, T.
BRITISH AMERICAN COMPANY LEASES DIAMOND FIELDS IN ARKANSAS.
ENG. MIN. JOUR.,
VOL. 117, MAY 10TH. P. 779.
USA; GULF COAST; ARKANSAS; PIKE COUNTY
(ENG)
NEWS ITEM
PICTURE OF J. HUDDLESTONE ON MAUNEY TRACT.

1925 ANON.
DIAMOND FOUND IN ROSELAWN CEMETERY, LITTLE ROCK.
LETTER TO SAM SIGNED JOHN, IN G.F. KUNZ COLLECTION,
MAY 30TH.
USA; GULF COAST; ARKANSAS
(ENG)

1925 KUNZ, G.F.
DIAMOND IN ARKANSAS.
THE MINERAL INDUSTRY DURING 1924,
VOL. 33, P. 618.
USA; GULF COAST; ARKANSAS; PIKE COUNTY
(ENG)
A DIAMOND WAS FOUND EARLY IN 1925 BELOW THE SURFACE IN 18" IN SUBSOIL WHILE DIGGING A GRAVE IN LITTLE ROCK. DOES NOT HAVE THE SAME CHARACTERISTICS AS THOSE OF PIKE COUNTY. REVIEW OF CURRENT ACTIVITIES FOR THE YEAR.

1925 MISER, H.D.; ROSS, C.S.
VOLCANIC ROCKS IN THE UPPER CRETACEOUS OF SOUTHWESTERN ARKANSAS AND SOUTHEASTERN OKLAHOMA.
AMER. JOUR. SCI.,
5TH. SER. VOL. 9, PP. 113-126.
USA; GULF COAST; ARKANSAS; OKLAHOMA
(ENG)
REGIONAL GEOLOGY

1925 WILKES, W.N.
MINERALS OF ARKANSAS.
ARKANSAS BUR. MINES MANUF. AGRIC. BULL. FOR 1924,
127P.
USA; GULF COAST; ARKANSAS
(ENG)
DIAMOND
P. 41 DIAMOND MURFREESBORO HISTORY.

1926 ANON.
DIAMOND MINING AS DONE IN PIKE COUNTY.
1P.
USA; GULF COAST; ARKANSAS
(ENG)
MINING METHODS
CLIPPINGS OF A NEWS ITEM.

1926 ANON.
ARKANSAS DIAMOND MINES.
LETTER TO G.F. KUNZ,
DATED JULY 28TH.
USA; GULF COAST; ARKANSAS
(ENG)
ON HAND AT PRESENT TIME 14.64 CARATS IN CUT STONES. 3,819.78 CARATS IN ROUGH. "I RATHER DOUBT IF WE ARE ABLE TO GIVE THE TOTAL YIELD OF THE MINE. THE WEIGHT OF THE THREE LARGEST STONES MINED TO THE PRESENT TIME - ROUGH STONES 1 AT 40.25; 1 AT 8.96 AND 1 AT 7.03 CARATS."

1926 DACY, G.H.
AMERICA'S INFANT DIAMOND INDUSTRY.
COMPRESSED AIR MAG.,
VOL. 31, MARCH PP. 1553-1555.
USA; GULF COAST; ARKANSAS; PIKE COUNTY
(ENG)
NEWS ITEM

1927 MENTOR,
DIAMONDS IN ARKANSAS.
MENTOR,
SEPT., P. 65.
USA; GULF COAST; ARKANSAS; PIKE COUNTY
(ENG)
DIAMOND OCCURRENCE; NEWS ITEM

1927 BRANNER, G.C.
OUTLINES OF ARKANSAS MINERAL RESOURCES.
ARKANSAS STATE GEOL. SURV.,
352P.
USA; GULF COAST; ARKANSAS
(ENG)
PP. 124-133 EXCELLENT SUMMARY OF INFORMATION TO DATE. PP. 131-133 BIBLIOGRAPHY.

1927 MISER, H.D.; ROSS, C.S.
ARKANSAS DIAMOND INDUSTRY.
ABRASIVE INDUSTRY,
VOL. 8, SEPT. PP. 285-288.
USA; GULF COAST; ARKANSAS
(ENG)
EXCELLENT SUMMARY.

1929 BRANNER, G.C.
GEOLOGY OF AMERICAS DIAMOND FIELDS.
PAN. AMER. GEOL. (DES MOINES),
VOL. 51, NO. 5, PP. 339-353.
USA; GULF COAST; ARKANSAS; PIKE COUNTY; GREAT LAKES
(ENG)
DIAMOND OCCURRENCES; GEOLOGY; HISTORY

1929 CRONEIS, C.G.; BILLINGS, M.P.
NEW AREAS OF ALKALINE ROCKS IN CENTRAL ARKANSAS.
JOUR. GEOLOGY,
VOL. 37, NO. 6, PP. 542-561.
USA; GULF COAST; ARKANSAS
(ENG)
REGIONAL GEOLOGY

1929 HALTON, W.L.
MAGNET COVE ARKANSAS AND VICINITY.
AMER. MINERALOGIST,
VOL. 14, NO. 12, PP. 484-487.
USA; GULF COAST; ARKANSAS; HOT SPRING COUNTY
(ENG)
MINERALOGY

1929 HEILAND, C.A.
GEOPHYSICAL METHODS OF PROSPECTING: PRINCIPLES AND RECENT SUCCESSES.
COL. SCH. MINES QUART.,
VOL. 24, MARCH, PP. 1-163. (P. 45.).
USA; GULF COAST; ARKANSAS; PIKE COUNTY
(ENG)
GEOPHYSICS; PROSPECTING METHODS; KIMBERLITE
P. 45 TORSION BALANCE OVER PRESUMED KIMBERLITE AT MURFREESBORO.

1929 MISER, H.D.; PURDUE, A.H.
GEOLOGY OF THE DEQUEEN AND CADDO GAP QUADRANGLES, ARKANSAS.
U.S.G.S. BULL.,
NO. 808, 195P.
USA; GULF COAST; ARKANSAS
(ENG)
GEOLOGY
PP. 99-115 AREA OF INTEREST.

1929 ROSS, C.S.; MISER, H.D.; STEPHENSON, L.W.
WATER-LAID VOLCANICS OF EARLY UPPER CRETACEOUS AGE IN SOUTHWESTERN ARKANSAS, SOUTHERN OKLAHOMA AND NORTHEASTERN TEXAS.
U.S.G.S. PROF. PAPER,
NO. 154F, PP. 175-202.
USA; GULF COAST; ARKANSAS; TEXAS; OKLAHOMA
(ENG)
REGIONAL GEOLOGY; STRATIGRAPHY
DIAMONDS ON P.201 OCCUR IN WOODBINE AND TOKIO FORMATIONS. CHEMICAL COMPOSITION OF VOLCANIC ROCKS ON P. 187.

1930 CRONEIS, C.G.; BILLINGS, M.P.
IGNEOUS ROCKS IN CENTRAL ARKANSAS.
ARKANSAS GEOL. SURV. BULL.,
NO. 3, PP. 149-162.
USA; GULF COAST; ARKANSAS
(ENG)
REGIONAL GEOLOGY

1930 STEARN, N.H.
A GEOMAGNETIC SURVEY OF THE BAUXITE REGION IN CENTRAL ARKANSAS.
ARKANSAS GEOL. SURV. BULL.,
NO. 5, 16P.
USA; GULF COAST; ARKANSAS
(ENG)
GEOPHYSICS

1931 BLANK, E.W.
ARKANSAS HAS YIELDED SOME FINE DIAMONDS.
NATIONAL JEWELLER,
VOL. 27, JANUARY PP. 44-46.
USA; GULF COAST; ARKANSAS; PIKE COUNTY
(ENG)
GENERAL INFORMATION ON QUALITY OF THE DIAMONDS.

1931 CRONEIS, C.G.
TECTONICS OF ARKANSAS PALEOZOICS.
PAN. AMER. GEOL.,
VOL. 55, NO. 1, PP. 1-8.
USA; GULF COAST; ARKANSAS
(ENG)
GEOTECTONICS

1931 KUNZ, G.F.
DIAMONDS IN NORTH AMERICA.
GEOL. SOC. AMER. BULL.,
VOL. 42, NO. 1, PP. 221-222.
USA; GULF COAST; ARKANSAS; GREAT LAKES; CANADA
(ENG)
DIAMOND OCCURRENCE

1931 KUNZ, G.F.
DIAMONDS IN ARKANSAS.
THE MINERAL INDUSTRY DURING 1930,
VOL. 39, P. 522.
USA; GULF COAST; ARKANSAS; PIKE COUNTY
(ENG)
REVIEW OF CURRENT ACTIVITIES FOR THE YEAR. A PARCEL OF 3114 STONES WEIGHING 1485 CARATS AN AVERAGE OF 0.48 CARATS PER STONE- SHOWING UNUSUALLY HIGH PERCENTAGE OF WHITE STONES, AND ONE EXTRA WHITE OF 4.5 CARATS, OTHERS OF YELLOWISH BROWN BETWEEN 2 AND 3 CARATS.

1931 LANDES, K.K.
A PARAGENETIC CLASSIFICATION OF THE MAGNET COVE MINERALS.
AMER. MINERALOGIST,
VOL. 16, NO. 8, PP. 313-326.
USA; GULF COAST; ARKANSAS; HOT SPRING COUNTY
(ENG)
MINERALOGY

1932 ANON.
DIAMONDS... EXTRACT FROM ARKANSAS GEOLOGICAL SURVEY'S OUTLINES OF ARKANSAS MINERAL RESOURCES PUBLISHED IN 1927 AND PRODUCERS AND PRODUCTION FIGURES REVISED TO 1931.
UNKNOWN,
7P.
USA; GULF COAST; ARKANSAS; PIKE COUNTY
(ENG)
PRODUCTION; HISTORY; DIAMONDS; PROSPECTING

1932 STEARN, N.H.
PRACTICAL GEOMAGNETIC EXPLORATION WITH THE HOTCHKISS SUPERDIP.
AMER. INST. MIN. MET. TRANS.,
VOL. 97, PP. 195-199.
USA; GULF COAST; ARKANSAS; PIKE COUNTY
(ENG)
KIMBERLITE; GEOPHYSICS; GROUNDMAG; PRAIRIE CREEK; CRATER OF DIAMONDS

1934 HEILAND, C.A.
PRECIOUS STONES.
AMER. INST. MIN. ENG. TRANS.,
VOL. 110, P. 571.
USA; GULF COAST; ARKANSAS
(ENG)
DIAMOND OCCURRENCE

1935 FORTUNE,
DIAMONDS.
FORTUNE,
VOL. 11, NO. 5, MAY, PT. 1,PP. 66-74; NO. 6, PT. 2, PP. 96-107; P. 124; P. 126; PP. 128-139.
USA; GULF COAST; ARKANSAS; SOUTH AFRICA; EUROPE
(ENG)
POPULAR ACCOUNT OF HISTORY OF DIAMOND LOCATIONS.

1935 BLANK, E.W.
DIAMOND FINDS IN THE UNITED STATES.
ROCKS AND MINERALS,
VOL. 10, PP. 7-10; PP. 23-26; PP. 39-40.
USA; GULF COAST; ARKANSAS; GREAT LAKES
(ENG)
DIAMOND OCCURRENCE

1935 BROCK, C.L.
TITANIUM AT MAGNET COVE, ARKANSAS.
ROCKS AND MINERALS,
VOL. 10, NO. 11, NOVEMBER P. 169.
USA; GULF COAST; ARKANSAS; HOT SPRING COUNTY
(ENG)

1936 LITERARY DIGEST,
DIAMOND FINDS: NEW CAPITAL OBTAINED TO RE-OPEN MINE IN ARKANSAS GHOST TOWN.
LITERARY DIGEST,
VOL. 122, NOV. 14TH. PP. 9-10.
USA; GULF COAST; ARKANSAS; PIKE COUNTY
(ENG)
DIAMOND MINING; NEWS ITEM

1939 BRANNER, G.C.
WEALTH OF ARKANSAS.
ARKANSAS GEOL. SURV.,
135P.
USA; GULF COAST; ARKANSAS
(ENG)
DIAMOND INDUSTRY DISCUSSED BRIEFLY.

1939 BRANNER, G.C.
STATE MINERAL SURVEY OF ARKANSAS.
ECON. GEOL.,
VOL. 34, NO. 8, P. 941. (ABSTR.).
USA; GULF COAST; ARKANSAS
(ENG)
DIAMOND OCCURRENCE; GEOLOGY.
ABSTRACT ON THE BOOK WEALTH OF ARKANSAS BY BRANNER.

1940 HENSON, P.
ARKANSAS DIAMOND FIELD.
GEMS AND GEMOLOGY,
VOL. 3, NO. 7, PP. 109-112.
USA; GULF COAST; ARKANSAS; PIKE COUNTY
(ENG)
DIAMOND OCCURRENCE; GEOLOGY

1941 TIME,
DOMESTIC DIAMONDS.
TIME,
VOL. 38, JULY 21ST. P. 71.
USA; GULF COAST; ARKANSAS; PIKE COUNTY
(ENG)
NEWS ITEM

1941 GEMS AND GEMOLOGY,
ARKANSAS DIAMOND MINE CHANGES HANDS.
GEMS AND GEMOLOGY,
VOL. 4, FALL, P. 168.
USA; GULF COAST; ARKANSAS; PIKE COUNTY
(ENG)
NEWS ITEM

1941 MISER, H.D.; GLASS, J.J.
FLUORESCENT SODALITE AND HACKMANITE FROM MAGNET COVE, ARKANSAS.
AMER. MINERALOGIST,
VOL. 26, NO. 7, PP. 437-445.
USA; GULF COAST; ARKANSAS; HOT SPRING COUNTY
(ENG)
MINERALOGY

1942 BRANNER, G.C.
MINERAL RESOURCES OF ARKANSAS.
ARKANSAS GEOL. SURV. BULL.,
NO. 6, 101P.
USA; GULF COAST; ARKANSAS
(ENG)
DIAMOND OCCURRENCES
DIAMONDS PP. 55-57. BULLETIN REVISED AND REISSUED 1959, 84P.

1942 LUND, R.J.
INTRADEPARTMENTAL MEMORANDUM TO W.L. BOTT CONCERNING POSSIBLE SETTING UP OF A PROJECT ON ARKANSAS DIAMONDS.
U.S.G.S. MISC. MINERALS BRANCH,
3P. UNPUBL.
USA; GULF COAST; ARKANSAS; PIKE COUNTY
(ENG)
EVALUATION

1942 ROSENTHAL, E.
UNCLE SAM'S OWN DIAMONDS.
DIAMOND NEWS & STH. AFR. JEWELLER, VOL. 5, NO. 9, JUNE PP. 6-7.
USA; GULF COAST; ARKANSAS
(ENG)
DIAMONDS; NEWS ITEM

1942 VITT, G.N.
LETTER TO R.J. THOENEN CONCERNING PROPOSED PILOT PLANT AT ARKANSAS MINE.
NORTH AMERICAN DIAMOND CORP. (CHICAGO), 2P. UNPUBL.
USA; GULF COAST; ARKANSAS; PIKE COUNTY
(ENG)
EVALUATION; NEWS ITEM

1943 GEMS AND GEMOLOGY,
ARKANSAS DIAMOND MINE.
GEMS AND GEMOLOGY, VOL. 6, SPRING P. 72.
USA; GULF COAST; ARKANSAS; PIKE COUNTY
(ENG)
NEWS ITEM

1943 SCHENCK, E.
LETTER CONTAINING A STATEMENT OF VERIFICATION REGARDING DIAMONDS PRODUCED AT ARKANSAS DIAMOND MINE TO WAR PRODUCTION BOARD.
NEW YORK, 2P. UNPUBL.
USA; GULF COAST; ARKANSAS; PIKE COUNTY
(ENG)
PRODUCTION; POLITICS

1943 VITT, G.N.
MEMORANDUM ON DIAMOND CONTENT AND KINDS OF MURFREESBORO DIAMONDIFEROUS PERIDOTITE PIPE.
NORTH AMERICAN DIAMOND CORP. (CHICAGO), 8P. UNPUBL.
USA; GULF COAST; ARKANSAS; PIKE COUNTY
(ENG)
DIAMOND PRODUCTION; EVALUATION

1944 MINERAL RESOURCES OF THE UNITED STATES FOR 1943,
DIAMONDS IN ARKANSAS.
MINERAL RESOURCES OF THE UNITED STATES FOR 1943, P. 1567.
USA; GULF COAST; ARKANSAS
(ENG)
REVIEW OF CURRENT ACTIVITIES FOR THE YEAR.

1944 BALL, S.H.
DIAMONDS IN ARKANSAS.
THE DIAMOND INDUSTRY IN 1943, 19TH. ANNUAL REVIEW, PUBL. JEWELERS CIRC. KEYSTONE P.20
USA; GULF COAST; ARKANSAS;
(ENG)
DIAMOND PRODUCTION STATISTICS

1945 BUHLIS, R.
ARKANSAS DIAMONDS.
THE MINERALOGIST (PORTLAND, OREGON), VOL. 13, NO. 2, FEB. P. 46.
USA; GULF COAST; ARKANSAS; PIKE COUNTY
(ENG)
MINERALOGY

1945 SHOCKLEY, W.G.
SIX YEARS OF COLLECTING IN MAGNET COVE.
ARKANSAS MINERAL BULL., NO. 3, P. 1; PP. 3-8.
USA; GULF COAST; ARKANSAS; HOT SPRING COUNTY
(ENG)
MINERALOGY

1945 WHEELER, H.E.
DIAMONDS IN ARKANSAS.
ARKANSAS MINERAL BULL., NO. 4, PP. 1-3.
USA; GULF COAST; ARKANSAS
(ENG)
DIAMOND; GEOLOGY

1946 ARKANSAS MINERAL BULL.,
LIST OF GEM AND SEMI PRECIOUS STONES IN ARKANSAS.
ARKANSAS MINERAL BULL., NO. 5, PP. 3-8.
USA; GULF COAST; ARKANSAS
(ENG)
DIAMONDS

1946 ENG. MIN. JOUR.,
FORMER DIAMOND PRODUCING AREA IS UNDER MAGNETIC SURVEY.
ENG. MIN. JOUR., VOL. 147, P. 138.
USA; GULF COAST; ARKANSAS; PIKE COUNTY
(ENG)
GEOPHYSICS; KIMBERLITE; NEWS ITEM

1946 SCULLEY, F.J.
ARKANSAS STONE OR PURE NOVACULITE.
ROCKS AND MINERALS, VOL. 21, NO. 7, JULY, P. 424.
USA; GULF COAST; ARKANSAS
(ENG)

1946 SPENCER, R.V.
EXPLORATION OF THE MAGNET COVE RUTILE COMPANY PROPERTY, HOT SPRING COUNTY, ARKANSAS.
U.S.B.M. REPT. INV., NO. 3900, 23P.
USA; GULF COAST; ARKANSAS; HOT SPRING COUNTY
(ENG)
PROSPECTING

1946 WHEELER, H.E.
DIAMONDS IN ARKANSAS.
HOBBIES, VOL. 51, MAY, PP. 118-120.
USA; GULF COAST; ARKANSAS; PIKE COUNTY
(ENG)
NEWS ITEM ON PROSPECTING

1947 MINERAL RESOURCES OF THE UNITED STATES FOR 1946,
DIAMONDS IN THE UNITED STATES.
MINERAL RESOURCES OF THE UNITED STATES FOR 1946,
P. 547.
USA; GULF COAST; ARKANSAS; ROCKY MOUNTAINS; IDAHO
(ENG)
REVIEW OF CURRENT ACTIVITIES FOR THE YEAR.

1947 THE DIAMOND NEWS & STH. AFR. JEWELLER,
DIAMONDS IN THE UNITED STATES.
THE DIAMOND NEWS & STH. AFR. JEWELLER,
AUGUST, PP. 20-22.
USA; GULF COAST; ARKANSAS
(ENG)
DIAMOND OCCURRENCE; NEWS ITEM; HISTORY OF FINDS

1947 BALL, S.H.
NEW DIAMOND COMPANY STARTED IN ARKANSAS.
THE DIAMOND INDUSTRY IN 1946, S2 23RD. ANNUAL REVIEW, PUBL. JEWELERS CIRC. KEYSTONE,
USA; GULF COAST; ARKANSAS; PIKE COUNTY
(ENG)
INVESTMENT

1947 HOLBROOK, D.F.
A BROOKITE DEPOSIT IN HOT SPRING COUNTY, ARKANSAS.
ARKANSAS RESOURCES AND DEVEL. COMM. DIV. GEOLOGY BULL.,
NO. 11, 21P.
USA; GULF COAST; ARKANSAS; HOT SPRING COUNTY
(ENG)
BROOKITE; TITANIUM

1948 MINERAL RESOURCES OF THE UNITED STATES FOR 1947,
DIAMOND IN THE UNITED STATES.
MINERAL RESOURCES OF THE UNITED STATES FOR 1947,
P. 534.
USA; GULF COAST; ARKANSAS; WEST COAST; CALIFORNIA
(ENG)
REVIEW OF CURRENT ACTVITIES FOR THE YEAR.

1948 HOLBROOK, D.F.
MOLYBDENUM IN MAGNET COVE, ARKANSAS.
ARKANSAS RESOURCES AND DEVEL. COMM. GEOLOGY BULL.,
NO. 12, 16P.
USA; GULF COAST; ARKANSAS; HOT SPRING COUNTY
(ENG)
MOLYBDENUM

1948 SANSOM, W.J.
ARKANSAS DIAMOND MINE.
THE GEMMOLOGIST,
VOL. 17, PP. 58-65.
USA; GULF COAST; ARKANSAS; PIKE COUNTY
(ENG)
EVALUATION

1948 SHOCKLEY, W.G.
MINERAL COLLECTING IN MAGNET COVE, ARKANSAS.
ROCKS AND MINERALS,
VOL. 23, NO. 6, PP. 483-495.
USA; GULF COAST; ARKANSAS; HOT SPRING COUNTY
(ENG)
MINERALOGY

1949 GEMS AND GEMOLOGY,
ANOTHER ROUGH DIAMOND FOUND IN INDIANA. DIAMONDS IN ARKANSAS.
GEMS AND GEMOLOGY,
VOL. 12, WINTER, PP. 249-250.
USA; GULF COAST; ARKANSAS; GREAT LAKES; INDIANA
(ENG)
DIAMOND OCCURRENCES

1949 FRYKLUND, V.C.JR.
THE TITANIUM ORE DEPOSITS OF MAGNET COVE, HOT SPRING COUNTY, ARKANSAS.
PH.D. THESIS, UNIV. MINNESOTA,
USA; GULF COAST; ARKANSAS; HOT SPRING COUNTY
(ENG)
TITANIUM

1949 MOODY, C.L.
MESOZOIC IGNEOUS ROCKS OF NORTHERN GULF COASTAL PLAIN.
AMER. ASSOC. PETROL. GEOL. BULL.,
VOL. 33, NO. 8, PP. 1410-1428.
USA; GULF COAST; ARKANSAS
(ENG)
REGIONAL GEOLOGY

1949 REED, D.F.
INVESTIGATION OF CHRISTY TITANIUM DEPOSIT, HOT SPRING COUNTY, ARKANSAS.
U.S.B.M. REPT. INV.,
NO. 4592, 10P.
USA; GULF COAST; ARKANSAS; HOT SPRING COUNTY
(ENG)
TITANIUM

1949 REED, D.F.
INVESTIGATION OF MAGNET COVE RUTILE DEPOSIT, HOT SPRING COUNTY, ARKANSAS.
U.S.B.M. REPT. INV.,
NO. 4593, 9P.
USA; GULF COAST; ARKANSAS; HOT SPRING COUNTY
(ENG)
RUTILE; TITANIUM

1949 **THOENEN, J.R.; HILL, R.S.; HOWE, E.G.; RUNKE, S.M.**
INVESTIGATION OF THE PRAIRIE CREEK DIAMOND AREA, PIKE COUNTY, ARKANSAS.
U.S.B.M. REPT. INV.,
NO. 4549, 24P.
USA; GULF COAST; ARKANSAS; PIKE COUNTY
(ENG)
EVALUATION

1949 **WOOD, J.B.**
AMERICA'S 35 ACRES OF DIAMONDS.
NATIONAL'S BUSINESS,
VOL. 37, MARCH, P. 60.
USA; GULF COAST; ARKANSAS; PIKE COUNTY
(ENG)
PROSPECTING; NEWS ITEM

1950 **ALEXANDER, A.E.**
THE ARKANSAS DIAMOND CRYSTAL.
THE GEMMOLOGIST,
VOL. 19, FEBRUARY P. 29; DECEMBER P. 279.
USA; GULF COAST; ARKANSAS
(ENG)
DIAMOND OCCURRENCE

1950 **MINERAL RESOURCES OF THE UNITED STATES FOR 1949,**
DIAMOND IN THE UNITED STATES.
MINERAL RESOURCES OF THE UNITED STATES FOR 1949,
PP. 546-547.
USA; GULF COAST; ARKANSAS; GREAT LAKES; INDIANA
(ENG)
REVIEW OF CURRENT ACTIVITIES FOR THE YEAR.

1950 **BALL, S.H.**
MINING OPERATIONS START UP IN ARKANSAS.
THE DIAMOND INDUSTRY IN 1949, 26TH. ANNUAL REVIEW, PUBL. JEWELERS CIRC. KEYSTONE,
USA; GULF COAST; ARKANSAS; PIKE COUNTY
(ENG)
MINING METHODS

1950 **FRYKLUND, V.C.JR.; HOLBROOK, D.F.**
TITANIUM ORE DEPOSITS OF HOT SPRING COUNTY, ARKANSAS.
ARKANSAS RESOURCES AND DEVEL. COMM. DIV. GEOLOGY BULL.,
NO. 16, 173P.
USA; GULF COAST; ARKANSAS; HOT SPRING COUNTY
(ENG)
TITANIUM

1950 **SCHALLER, W.T.**
MISERITE FROM ARKANSAS- A RENAMING OF NATROXONOTLITE.
AMER. MINERALOGIST,
VOL. 35, NO. 9-10, PP. 911-921.
USA; GULF COAST; ARKANSAS; GARLAND COUNTY
(ENG)
MINERALOGY

1950 **THOENEN, J.R.; HILL, R.S.; HOWE, E.G.; RUNKE, S.M.**
INVESTIGATION OF THE PRAIRIE CREEK DIAMOND AREA, PIKE COUNTY.
EARTH SCI. DIGEST,
VOL. 4, NO. 6, PP. 3-8.
USA; GULF COAST; ARKANSAS; PIKE COUNTY
(ENG)
EVALUATION OF DEPOSIT AND MINING METHODS OUTLINED.

1951 **HERNDON, B.**
AMERICA'S ONLY DIAMOND MINE.
COLLIER'S,
AUG. 25TH., 5P.
USA; GULF COAST; ARKANSAS; PIKE COUNTY
(ENG)
POPULAR ACCOUNT OF THE HISTORY OF THE MINE.

1951 **HOWARD, D.L.**
DIAMOND MINES OF ARKANSAS.
LAPIDARY JOUR.,
VOL. 5, PP. 248-254; P. 256.
USA; GULF COAST; ARKANSAS; PIKE COUNTY
(ENG)
MINING METHODS

1953 **SHREVEPORT GEOL. SOC.**
UPPER AND LOWER CRETACEOUS OF SOUTHWESTERN ARKANSAS, CAMBRIAN-PENNSYLVANIAN OF THE OUACHITA MOUNTAINS AND MAGNET COVE.
GUIDEBOOK OF THE 19TH. ANN. FIELD TRIP, 36P.
USA; GULF COAST; ARKANSAS; HOT SPRING COUNTY; OUACHITA MOUNTAINS
(ENG)
GUIDEBOOK

1954 **THE GEMMOLOGIST,**
TRIAL EXPLORATION DRILLING IN ARKANSAS DIAMOND PIPE.
THE GEMMOLOGIST,
VOL. 24, AUGUST P. 147.
USA; GULF COAST; ARKANSAS; PIKE COUNTY
(ENG)
NEWS ITEM; PROSPECTTING

1954 **FRYKLUND, V.C.JR; HARNER, R.S.; KAISER, E.P.**
NIOBIUM (COLUMBIUM) AND TITANIUM AT MAGNET COVE AND POTASH SULFUR SPRINGS, ARKANSAS.
U.S.G.S. BULL.,
NO. 1015B, PP. 23-56.
USA; GULF COAST; ARKANSAS; HOT SPRING COUNTY; GARLAND COUNTY
(ENG)
NIOBIUM; COLUMBIUM; TITANIUM

1955 ROWE, R.B.
ASSOCIATION OF COLUMBIUM MINERALS AND ALKALINE ROCKS.
CAN. MIN. JOUR.,
VOL. 76, NO. 3, MARCH PP. 69-73.
USA; GULF COAST; ARKANSAS; HOT SPRING COUNTY; USSR; KOLA; CANADA; QUEBEC; SWEDEN
(ENG)
MAGNET COVE; BROOKITE; OKA

1956 SAN DIEGO UNION,
15 1/2 CARAT GEM IS NAMED STAR OF ARKANSAS.
SAN DIEGO UNION,
FRIDAY, MARCH 9TH.
USA; GULF COAST; ARKANSAS; PIKE COUNTY
(ENG)
NEWS ITEM; DIAMOND NOTABLE

1956 SAN DIEGO UNION,
HELP YOURSELF MINE: $15,000 DIAMOND HERS FOR FINDING.
SAN DIEGO UNION,
MAY 5TH. 1P.
USA; GULF COAST; ARKANSAS; PIKE COUNTY
(ENG)
NEWS ITEM; DIAMOND NOTABLE

1956 SAN DIEGO UNION,
$15,000. DIAMOND - HERS FOR THE FINDING.
SAN DIEGO UNION,
MARCH 5TH., 1/2P.
USA; GULF COAST; ARKANSAS; PIKE COUNTY
(ENG)
NEWS ITEM; DIAMOND NOTABLE

1956 JOHNSON, K.
ARKANSAS DIAMOND MINE.
COMMERCIAL APPEAL,
FEB. 26TH. SECT. 5, P. 1.
USA; GULF COAST; ARKANSAS; PIKE COUNTY
(ENG)
NEWS ITEM

1956 MCCORD, R.G.
FIFTY YEARS OF DREAMING AND DIGGING IN NORTH AMERICA'S ONLY DIAMOND MINE.
ARKANSAS DEMOCRAT SUNDAY MAGAZINE (LITTLE ROCK),
AUGUST 5TH. P. 1; P. 7; P. 9.
USA; GULF COAST; ARKANSAS; PIKE COUNTY
(ENG)
MINING; SAMPLING; DIAMOND; PRODUCTION; HISTORY

1956 ROBINSON, A.
HOW THEY CUT THE STAR OF ARKANSAS.
ARKANSAS DEMOCRAT (LITTLE ROCK), SUNDAY MAGAZINE,
SEPT. 30TH., PP. 1-7,8,9.
USA; GULF COAST; ARKANSAS; PIKE COUNTY
(ENG)
DIAMOND NOTABLE

1956 ST. CLAIR, J.Q
REPORT ON THE ARKANSAS DIAMOND PROPERTY.
COMPANY REPORT,
15P. UNPUBL.
USA; GULF COAST; ARKANSAS; PIKE COUNTY
(ENG)
EVALUATION; METHODS; MINING; PROSPECTUS

1957 MINERAL RESOURCES OF THE UNITED STATES FOR 1956,
DIAMONDS IN ARKANSAS.
MINERAL RESOURCES OF THE UNITED STATES FOR 1956,
P. 515.
USA; GULF COAST; ARKANSAS; PIKE COUNTY
(ENG)
REVIEW OF CURRENT ACTIVITIES FOR THE YEAR.
15.33 CARAT DIAMOND FOUND.

1957 THE GEMMOLOGIST,
THE IKE DIAMOND AND THE STAR OF ARKANSAS.
THE GEMMOLOGIST,
VOL. 27, JULY P. 132.
USA; GULF COAST; ARKANSAS; PIKE COUNTY
(ENG)
DIAMONDS NOTABLE

1957 BURGOON, J.R.
DIAMOND MINING IN ARKANSAS.
GEMS AND GEMOLOGY,
VOL. 20, PP. 355-362.
USA; GULF COAST; ARKANSAS; PIKE COUNTY
(ENG)
MINING; SAMPLING

1957 LEIPER, H.
ARKANSAS DIAMONDS.
JOUR. GEMOLOGY,
VOL. 6, NO. 2, PP. 63-71. ALSO: LAPIDARY JOUR., VOL. 11, PP. 4-6; P. 8; P. 10; P. 12.
USA; GULF COAST; ARKANSAS; PIKE COUNTY
(ENG)
MORPHOLOGY; DIAMOND OCCURRENCE

1957 MURPHY, A.G.
DIG YOUR OWN DIAMONDS.
THE AMERICAN WEEKLY,
JAN. 13TH. 2P.
USA; GULF COAST; ARKANSAS; PIKE COUNTY
(ENG)
NEWS ITEM

1958 MINERAL RESOURCES OF THE UNITED STATES FOR 1957,
DIAMONDS OF ARKANSAS.
MINERAL RESOURCES OF THE UNITED STATES FOR 1957,
P. 518.
USA; GULF COAST; ARKANSAS; PIKE COUNTY
(ENG)
REVIEW OF CURRENT ACTIVITIES FOR THE YEAR.

1958 GORDON, M.JR.; TRACEY, J.I.JR.; ELLIS, M.W.
GEOLOGY OF THE ARKANSAS BAUXITE REGION.
U.S.G.S. PROF. PAPER,
NO. 299, 268P. PP. 60-69 OF INTEREST.
USA; GULF COAST; ARKANSAS; PULASKI COUNTY; SALINE COUNTY
(ENG)
GEOLOGICAL DESCRIPTION OF IGNEOUS ROCKS PP. 60-69.

1958 MILTON, C.; BLADE, L.V.
PRELIMINARY NOTE ON KIMZEYITE, A NEW ZIRCONIAN GARNET.
SCIENCE,
VOL. 127, NO. 3310, P. 1343.
USA; GULF COAST; ARKANSAS;
(ENG)
MINERALOGY

1958 ROSE, H.J.JR.; BLADE, L.V.; ROSS, M.
EARTHY MONAZITE AT MAGNET COVE ARKANSAS.
AMER. MINERALOGIST,
VOL. 43, NO. 9-10, PP. 995-997.
USA; GULF COAST; ARKANSAS; HOT SPRING COUNTY
(ENG)
PETROGRAPHY

1959 THE EVENING STAR (WASHINGTON),
ARKANSAS CRATER STILL HOLDS DIAMOND TROVE.
THE EVENING STAR (WASHINGTON),
SEPT. 8TH., P. B5.
USA; GULF COAST; ARKANSAS; PIKE COUNTY
(ENG)
NEWS ITEM

1959 MINERAL RESOURCES OF THE UNITED STATES FOR 1958,
DIAMOND IN THE UNITED STATES.
MINERAL RESOURCES OF THE UNITED STATES FOR 1958,
P. 468.
USA; GULF COAST; ARKANSAS; GREAT LAKES; ILLINOIS; ROCKY MOUNTAINS; NEVADA
(ENG)
REVIEW OF CURRENT ACTIVITIES FOR THE YEAR.

1959 GEMS AND GEMOLOGY,
DIAMOND DIGGING IN ARKANSAS. A 3.65 CARAT DIAMOND.
GEMS AND GEMOLOGY,
VOL. 22, SUMMER P. 318; FALL P. 343.
USA; GULF COAST; ARKANSAS; PIKE COUNTY
(ENG)
DIAMOND NOTABLE

1959 GROSS, M.L.
THE INCREDIBLE AMERICAN DIAMOND MINE MYSTERY.
TRUE, THE MAN'S MAGAZINE,
SEPT. PP. 52-55; PP. 98-102.
USA; GULF COAST; ARKANSAS; PIKE COUNTY
(ENG)
DIAMOND; MINING; HISTORY

1959 HOWE, E.L.
DIAMONDS: FOR $1.50, IT'S FINDERS-KEEPERS AT YOU DIG 'EM GEM MINE. WHERE A STONE IS NEVER LEFT UNTURNED.
NEW YORK CITY,
SUNDAY NEWS, JULY 26TH., P. 70.
USA; GULF COAST; ARKANSAS; PIKE COUNTY
(ENG)
NEWS ITEM

1959 SINKANKAS, J.
DIAMONDS IN ARKANSAS.
IN: GEMSTONES OF NORTH AMERICA,
VOL. 1, PP. 34-38.
USA; GULF COAST; ARKANSAS; PIKE COUNTY
(ENG)
DIAMOND OCCURRENCE; HISTORY; LOCATION

1959 WILLIAMS, N.F.
DIAMONDS.
ARKANSAS GEOL. SURV. BULL.,
VOL. 6, PP. 60-61.
USA; GULF COAST; ARKANSAS; PIKE COUNTY
(ENG)
DIAMOND OCCURRENCE; GEOLOGY

1960 MINERAL RESOURCES OF THE UNITED STATES FOR 1959,
DIAMONDS OF ARKANSAS.
MINERAL RESOURCES OF THE UNITED STATES FOR 1959,
P. 472.
USA; GULF COAST; ARKANSAS; PIKE COUNTY
(ENG)
REVIEW OF CURRENT ACTIVITIES FOR THE YEAR.

1960 AUSTRALIAN GEMMOLOGIST,
DIAMOND MINING IN ARKANSAS.
AUSTRALIAN GEMMOLOGIST,
PP. 16-17.
USA; GULF COAST; ARKANSAS; PIKE COUNTY
(ENG)
NEWS ITEM

1960 BRANCH, H.
A 6.45 CARAT DIAMOND FROM ARKANSAS.
GEMS AND GEMOLOGY,
VOL. 23, SPRING, PP. 7-9.
USA; GULF COAST; ARKANSAS; PIKE COUNTY
(ENG)
DIAMOND NOTABLE

1960 POUGH, F.H.
DO IT YOURSELF DIAMOND MINING.
JEWELLERS CIRCULAR KEYSTONE,
VOL. 130, NO. 5, FEB., PP. 90-92; P. 116.
USA; GULF COAST; ARKANSAS; PIKE COUNTY
(ENG)
NEWS ITEM

1961 HARTWELL, J.W.; BRETT, B.A.
GEM STONES.
MINERALS YEARBOOK: METALS AND MINERALS, VOL. 1, PP. 585-596.
CANADA; USA; GULF COAST; ARKANSAS; PIKE COUNTY; USSR; BRAZIL; TANZANIA; CENTRAL AFRICA; KENYA; CAMEROUN; IVORY COAST; WEST AFRICA; SOUTH WEST AFRICA; NAMIBIA; SOUTH AFRICA
(ENG)
PRODUCTION; IMPORTS; REVIEW

1962 MINERAL RESOURCES OF THE UNITED STATES FOR 1961,
DIAMONDS OF ARKANSAS.
MINERAL RESOURCES OF THE UNITED STATES FOR 1961, P. 587.
USA; GULF COAST; ARKANSAS; PIKE COUNTY
(ENG)
REVIEW OF CURRENT ACTIVITIES FOR THE YEAR.

1962 HOLBROOK, D.F.
THE GEOLOGY OF MAGNET COVE.
MISSISSIPPI GEOL. SOC., IN: THE PALEOZOICS OF NORTHWEST ARKANSAS- GUIDEBOOK, 16TH. FIELD TRIP, PP. 10-11.
USA; GULF COAST; ARKANSAS; HOT SPRING COUNTY
(ENG)
GEOLOGY

1963 ERICKSON, R.L.; BLADE, L.V.
GEOCHEMISTRY AND PETROLOGY OF THE ALKALIC IGNEOUS COMPLEX OF MAGNET COVE, ARKANSAS.
U.S.G.S. PROF. PAPER, NO. 425, 95P.
USA; GULF COAST; ARKANSAS; HOT SPRING COUNTY
(ENG)
GEOCHEMISTRY; PETROLOGY

1963 KELLER, F.JR.; HENDERSON, J.R.; [ET AL.]
AEROMAGNETIC MAP OF THE MAGNET COVE AREA, HOT SPRING COUNTY, ARKANSAS.
U.S.G.S. MAP, NO. GP 409, 1: 24,000.
USA; GULF COAST; ARKANSAS; HOT SPRING COUNTY
(ENG)
GEOPHYSICS

1963 PETKOF, B.
GEM STONES.
MINERALS YEARBOOK: METALS AND MINERALS, VOL. 1, PP. 537-548.
USA; GULF COAST; ARKANSAS; PIKE COUNTY; BRAZIL; VENEZUELA; BELGIUM; INDIA; ANGOLA; EAST AFRICA; C.A.R.: GABON; GHANA; SOUTH AFRICA; SOUTH WEST AFRICA; NAMIBIA
(ENG)
INDUSTRIAL; REVIEW; IMPORTS; PRODUCTION

1963 STONE, C.G.; STERLING, P.J.
RELATIONSHIP OF IGNEOUS ACTIVITY TO MINERAL DEPOSITS IN ARKANSAS.
ARKANSAS ACAD. SCI. PROC., VOL. 17, P. 54. (ABSTR.).
USA; GULF COAST; ARKANSAS
(ENG)
ORIGIN

1964 STONE, C.G.; STERLING, P.J.
RELATIONSHIP OF IGNEOUS ACTIVITY TO MINERAL DEPOSITS IN ARKANSAS.
ARKANSAS GEOL. COM. MISC. PUBL., 55 P.
USA; GULF COAST; ARKANSAS; PIKE COUNTY
(ENG)
ORIGIN; MINERAL DEPOSITS

1965 COLLINS, H.F.
SUMMER SAFARI TO AMERICA'S DIAMOND FIELD.
THE DIAMOND NEWS & STH. AFR. JEWELLER, JANUARY, PP. 38-39.
USA; ARKANSAS; GULF COAST
(ENG)
HISTORY; GUIDEBOOK POPULAR ACCOUNT OF TRIP.

1965 POLLACK, D.W.
THE POTASH SULFUR SPRINGS ALKALI COMPLEX, GARLAND COUNTY, ARKANSAS.
MINING ENGINEERING, VOL. 17, PP. 45-46. (ABSTR.).
USA; GULF COAST; ARKANSAS; GARLAND COUNTY
(ENG)

1965 RUSSELL, J.
PLEASANT DREAM PAYS OFF.
THE TULSA TRIBUNE, AUGUST 25TH.
USA; GULF COAST; ARKANSAS; PIKE COUNTY
(ENG)
DIAMOND; PROPERTY; INVESTMENT; NEWS ITEM

1966 GITTINS, J.
SUMMARIES AND BIBLIOGRAPHIES OF CARBONATITE COMPLEXES.
INTERSCIENCE PUBL. IN: CARBONATITES, EDITORS, O.F. TUTTLE AND J. GITTINS, PP. 417-540.
USA; GULF COAST; ARKANSAS; ROCKY MOUNTAINS; MONTANA; COLORADO; CALIFORNIA; CANADA; QUEBEC; BRITISH COLUMBIA
(ENG)
BIBLIOGRAPHY

1966 LEIPER, H.
AMERICA'S ONLY DIAMOND BEARING PERIDOTITE PIPE.
LAPIDARY JOUR., VOL. 20, PP. 714-733.
USA; GULF COAST; ARKANSAS; PIKE COUNTY
(ENG)
POPULAR ACCOUNT OF HISTORY OF THE PRAIRIE CREEK AREA.

1966 **MILLAR, H.W.**
MILLAR'S CRATER OF DIAMONDS.
BROCHURE,
6P.
USA; GULF COAST; ARKANSAS; PIKE COUNTY
(ENG)
HISTORY; PRODUCTION

1966 **POWELL, J.L.; HURLEY, P.M.; FAIRBAIRN, H.W.**
THE STRONTIUM ISOTOPIC COMPOSITION AND ORIGIN OF CARBONATITES.
INTERSCIENCE PUBL.
IN: CARBONATITES, EDITORS O.F. TUTTLE AND J. GITTINS,
PP. 365-378.
USA; GULF COAST; ARKANSAS; HOT SPRING COUNTY; CANADA; QUEBEC; SOUTH AFRICA; TRANSVAAL
(ENG)
GEOCHRONOLOGY; SPITZKOP; OKA; MAGNET COVE

1966 **ROSS, M.**
DIAMONDS ARE DISCOVERED IN ARKANSAS- BUT THERE WERE A NUMBER OF EARLIER FINDS.
THE ARKANSAS GAZETTE,
OCTOBER 30TH.
USA; GULF COAST; ARKANSAS; PIKE COUNTY
(ENG)
NEWS ITEM; DIAMOND OCCURRENCE; HISTORY

1966 **STONE, C.G.**
GENERAL GEOLOGY OF THE EASTERN FRONTAL OUACHITA MOUNTAINS AND SOUTHEASTERN ARKANSAS VALLEY, ARKANSAS.
KANSAS GEOL. SOC. GUIDEBOOK, WICHITA KANSAS,
IN: FLYSCH FACIES AND STRUCTURE OF THE OUACHITA MOUNTAINS, KANSAS GEOL. SOC. 29TH. FIELD CONF.,
PP. 195-221.
USA; OUACHITA MOUNTAINS; ARKANSAS
(ENG)
STRUCTURE

1967 **HOLLINGSWORTH, J.S.**
GEOLOGY OF THE WILSON SPRINGS VANADIUM DEPOSITS.
ARKANSAS GEOL. COMM. CENTRAL ARKANSAS ECONOMIC GEOLOGY AND PETROLOGY- GEOL. SOC. AMERICA FIELD CONFERENCE GUIDEBOOK,
PP. 22-24.
USA; GULF COAST; ARKANSAS; GARLAND COUNTY
(ENG)
VANADIUM; POTASH SULFUR SPRINGS COMPLEX

1967 **OWENS, D.R.**
BEDROCK GEOLOGY OF THE 'V' INTRUSIVE, GARLAND COUNTY, ARKANSAS.
M.SC. THESIS, UNIV. ARKANSAS,
96P.
USA; GULF COAST; ARKANSAS; GARLAND COUNTY
(ENG)
PETROLOGY; STRUCTURE

1967 **UPTON, B.G.L.**
ALKALINE PYROXENITES.
JOHN WILEY & SONS. PUBL.
IN: ULTRAMAFIC AND RELATED ROCKS, EDITOR P.J. WYLLIE,
PP. 281-288.
USA; GULF COAST; ARKANSAS; HOT SPRING COUNTY
(ENG)
MAGNET COVE; PETROLOGY

1968 **VALDOVINOS, D.L.**
PETROGRAPHY OF SOME LAMPROPHYRES OF THE EASTERN OUACHITA MOUNTAINS OF ARKANSAS.
M.SC. THESIS, UNIV. ARKANSAS,
146P.
USA; OUACHITA MOUNTAINS; ARKANSAS
(ENG)
PETROLOGY

1969 **LAPIDARY JOUR.,**
OWNERS OF THE ARKANSAS DIAMOND PIPE.
LAPIDARY JOUR.,
VOL. 23, MAY PP. 366-372. OCT. PP. 970-973.
USA; GULF COAST; ARKANSAS; PIKE COUNTY
(ENG)
POPULAR ACCOUNT OF HISTORY OF OWNERSHIP.

1969 **CONWAY, C.M.; TAYLOR, H.P.JR.**
O18/O16 AND C13/C12 RATIOS OF COEXISTING MINERALS IN THE OKA AND MAGNET COVE CARBONATITE BODIES.
JOUR. GEOL.,
VOL. 77, NO. 5, PP. 618-626.
USA; GULF COAST; ARKANSAS; HOT SPRING COUNTY; CANADA; QUEBEC
(ENG)
GEOCHEMISTRY; GEOCHRONOLOGY

1969 **DEANE, E.**
MEMORIES OF ARKANSAS'S FAMOUS CRATER OF DIAMONDS.
ARKANSAS GAZETTE, SUNDAY EDITION,
JULY 6TH., 4P.
USA; GULF COAST; ARKANSAS; PIKE COUNTY
(ENG)
NEWS ITEM. HISTORY; DIAMOND

1969 GREGORY, G.P.
GEOCHEMICAL DISPERSION PATTERNS RELATED TO KIMBERLITE INTRUSIVES IN NORTH AMERICA,
PH.D. THESIS, UNIVERSITY OF LONDON, ROYAL SCHOOL OF MINES,
327P.
USA; GULF COAST; ARKANSAS
(ENG)
GEOCHEMISTRY

1969 GREGORY, G.P.; TOOMS, J.S.
GEOCHEMICAL PROSPECTING FOR KIMBERLITES.
COL. SCH. MINES QUART.,
VOL. 64, NO. 1, JANUARY PP. 265-304.
USA; GULF COAST; ARKANSAS
(ENG)
GEOCHEMISTRY; EVALUATION; PRAIRIE CREEK; MINERAL CHEMISTRY; SOIL MORPHOLOGY; MURFREESBORO; BUELL PARK; DRAINAGE SYSTEM; PROSPECTING; TECHNIQUES

1969 MANOYIAN, Z.
THE GEMSTERS.
OPA-LOCKA FLA: TODD-Z CORP.
144P.
USA; GULF COAST; ARKANSAS; PIKE COUNTY
(ENG)
ADVENTURE STORY OF DIAMOND DEPOSITS.

1969 SCHARON, L.; HSU I-CHI
PALEOMAGNETIC INVESTIGATIONS OF SOME ARKANSAS ALKALIC IGNEOUS ROCKS.
JOUR. GEOPHYS. RESEARCH,
VOL. 74, NO. 10, PP. 2774-2779.
USA; GULF COAST; ARKANSAS; HOT SPRING COUNTY
(ENG)
ABSOLUTE AGE

1970 MINERAL RESOURCES OF THE UNITED STATES FOR 1969,
ARKANSAS DIAMOND MINES UNDER ONE OWNER AND OPEN TO THE PUBLIC FREE.
MINERAL RESOURCES OF THE UNITED STATES FOR 1969,
P. 515.
USA; GULF COAST; ARKANSAS; PIKE COUNTY
(ENG)
REVIEW OF CURRENT ACTIVITIES FOR THE YEAR.

1972 TOWNER, J.
DIAMOND HUNTING IN MURFREESBORO, ARKANSAS.
LAPIDARY JOUR.,
VOL. 26, PP. 1268-1276.
USA; GULF COAST; ARKANSAS; PIKE COUNTY
(ENG)
HISTORY; PROSPECTING

1973 MINERAL RESOURCES OF THE UNITED STATES FOR 1972,
STATE OF ARKANSAS BUYS THE ARKANSAS DIAMOND PIPES FOR PARK.
MINERAL RESOURCES OF THE UNITED STATES FOR 1972,
P. 559.
USA; GULF COAST; ARKANSAS; PIKE COUNTY
(ENG)
REVIEW OF CURRENT ACTIVITIES FOR THE YEAR.

1973 HOWARD, J.M.; STEELE, K.F.; OWENS, D.R.
CHEMICALLY ROUNDED XENOLITHS IN AN ALKALIC DIKE, GARLAND COUNTY, ARKANSAS.
GEOL. SOC. AMER.,
VOL. 5, NO. 3, P. 263. (ABSTR.).
USA; GULF COAST; ARKANSAS; GARLAND COUNTY
(ENG)
PETROLOGY

1974 MINERAL RESOURCES OF THE UNITED STATES FOR 1973,
TWO GEM DIAMONDS FOUND IN ARKANSAS.
MINERAL RESOURCES OF THE UNITED STATES FOR 1973,
P. 548.
USA; GULF COAST; ARKANSAS; PIKE COUNTY
(ENG)
REVIEW OF CURRENT ACTIVITIES FOR THE YEAR. DIAMONDS NOTABLE.

1974 GIARDINI, A.A.; HURST, V.J.; MELTON, C.E.; STORMER, J.C.JR.
BIOTITE AS A PRIMARY INCLUSION IN DIAMOND: ITS NATURE AND SIGNIFICANCE.
AMERICAN MINERALOGIST,
VOL. 59, PP. 783-789.
USA; GULF COAST; ARKANSAS; PIKE COUNTY; SOUTH AFRICA
(ENG)
MINERAL CHEMISTRY

1974 HEATHCOTE, R.C.
FENITIZATION OF THE ARKANSAS NOVACULITE AND ADJACENT INTRUSIVE, GARLAND COUNTY, ARKANSAS.
M.S. THESIS, UNIV. ARKANSAS,
56P.
USA; GULF COAST; ARKANSAS; GARLAND COUNTY
(ENG)
ALTERATION

1974 HOTHEM, L.L.
ARKANSAS CRATER OF DIAMONDS.
GEMS AND MINERALS,
NO. 441, PP. 32-34.
USA; GULF COAST; ARKANSAS; PIKE COUNTY
(ENG)
NEWS ITEM. HISTORY; PROSPECTING

1974 HOWARD, J.M.
TRANSITION ELEMENT GEOCHEMISTRY AND PETROGRAPHY OF THE POTASH SULFUR SPRINGS INTRUSIVE COMPLEX, GARLAND COUNTY, ARKANSAS.
M.SC. THESIS, UNIV. ARKANSAS,
118P.
USA; GULF COAST; ARKANSAS; GARLAND COUNTY
(ENG)
GEOCHEMISTRY

1974 LANGFORD, R.E.
A STUDY OF THE ORIGIN OF ARKANSAS DIAMONDS BY MASS SPECTROMETRY.
PH.D. THESIS, UNIV. OF GEORGIA, ATHENS,
USA; GULF COAST; ARKANSAS; PIKE COUNTY
(ENG)
GENESIS; ISOTOPE CHEMISTRY

1975 SAN DIEGO UNION,
COUPLE'S HOBBY IS A REAL GEM.
SAN DIEGO UNION,
SUNDAY AUG. 3RD., P. A23.
USA; GULF COAST; ARKANSAS; PIKE COUNTY
(ENG)
NEWS ITEM

1975 GIARDINI, A.A.; MELTON, C.E.
THE NATURE OF CLOUD LIKE INCLUSIONS IN TWO ARKANSAS DIAMONDS.
AMERICAN MINERALOGIST,
VOL. 60, PP. 932-933.
USA; GULF COAST; ARKANSAS; PIKE COUNTY
(ENG)
MINERALOGY

1975 GIARDINI, A.A.; MELTON, C.E.
CHEMICAL DATA ON A COLORLESS ARKANSAS DIAMOND AND ITS BLACK AMORPHOUS C FE NI S INCLUSION.
AMERICAN MINERALOGIST,
VOL. 60, PP. 934-936.
USA; GULF COAST; ARKANSAS; PIKE COUNTY
(ENG)
MINERAL CHEMISTRY; AGE OF DIAMONDS

1975 HOWARD, J.M.; JACKSON, K.C.
PETROGRAPHY OF THE POTASH SULFUR SPRINGS INTRUSION, GARLAND COUNTY, ARKANSAS.
GEOL. SOC. AMER.,
VOL. 7, NO. 2, PP. 173-174. (ABSTR.).
USA; GULF COAST; ARKANSAS; GARLAND COUNTY
(ENG)
PETROGRAPHY

1975 HOWARD, J.M.; STEELE, K.F.
TRANSITION ELEMENT GEOCHEMISTRY OF THE POTASH SULFUR SPRINGS INTRUSION, GARLAND COUNTY, ARKANSAS.
GEOL. SOC. AMER.,
VOL. 7, NO. 2, P. 174. (ABSTR.).
USA; GULF COAST; ARKANSAS; GARLAND COUNTY
(ENG)
GEOCHEMISTRY

1975 HOWARD, J.M.; STEELE, K.F.
ORIGIN OF THE POTASH SULFUR SPRINGS INTRUSIVE COMPLEX, ARKANSAS.
GEOL. SOC. AMER.,
VOL. 7, NO. 4, P. 502. (ABSTR.).
USA; GULF COAST; ARKANSAS; GARLAND COUNTY
(ENG)
GEOLOGY

1975 JOHNSON, M.L.
CARBON AND OXYGEN ISOTOPE EVOLUTION IN THE MAGNET COVE COMPLEX, ARKANSAS.
M.SC. THESIS, RICE UNIVERSITY, HOUSTON, TEXAS,
63P.
USA; GULF COAST; ARKANSAS; HOT SPRING COUNTY
(ENG)

1975 MELTON, C.E.; GIARDINI, A.A.
EXPERIMENTAL RESULTS AND A THEORETICAL INTERPRETATION OF GASEOUS INCLUSIONS FOUND IN ARKANSAS NATURAL DIAMONDS.
AMERICAN MINERALOGIST,
VOL. 60, PP. 413-417.
USA; GULF COAST; ARKANSAS; PIKE COUNTY
(ENG)
MINERAL CHEMISTRY

1976 BOLIVAR, S.L.; BROOKINS, D.G.; LEWIS, R.D.; MEYER, H.O.A.
GEOPHYSICAL STUDIES OF THE PRAIRIE CREEK KIMBERLITE MURFREESBORO,ARKANSAS.
EOS,
VOL. 57, NO. 10, P. 762, (ABSTR.).
USA; GULF COAST; ARKANSAS; PIKE COUNTY
(ENG)
KIMBERLITE; GEOPHYSICS; GROUNDMAG; GRAVITY

1976 BROOKINS, D.G.; DELLA VALLE, R.S.; BOLIVAR, S.L.
URANIUM GEOCHEMISTRY OF SOME UNITED STATES KIMBERLITES.
EOS,
VOL. 57, NO. 10, P. V8, (ABSTR.).
USA; GULF COAST; ARKANSAS; CENTRAL STATES; ROCKY MOUNTAINS; NEBRASKA; COLORADO; WYOMING; KENTUCKY; TENNESSEE;
(ENG)

1976 HEATHCOTE, R.C.
FENITIZATION OF THE ARKANSAS NOVACULITE, GARLAND COUNTY, ARKANSAS.
GEOL. SOC. AMER.,
VOL. 8, NO. 6, P. 910. (ABSTR.).
USA; GULF COAST; ARKANSAS; GARLAND COUNTY
(ENG)
ALTERATION

1976 JACKSON, K.C.; STEELE, K.F.
NEW DATA ON SOME ARKANSAS IGNEOUS ROCKS.
GEOL. SOC. AMER.,
VOL. 8, NO. 1, PP. 25-26. (ABSTR.).
USA; GULF COAST; ARKANSAS; GARLAND COUNTY
(ENG)
GEOCHEMISTRY

1976 JOHNSON, M.; BAKER, D.R.
INTRUSIVE MODEL OF THE MAGNET COVE COMPLEX, ARKANSAS.
GEOL. SOC. AMER.,
VOL. 8, NO. 1, P. 26. (ABSTR.).
USA; GULF COAST; ARKANSAS; HOT SPRING COUNTY
(ENG)
GENESIS; STRUCTURE

1976 LEWIS, R.D.; MEYER, H.O.A.; BOLIVAR, S.L.; BROOKINS, D.G.
MINERALOGY OF THE DIAMOND BEARING 'KIMBERLITE' MURFREESBORO, ARKANSAS.
EOS,
VOL. 57, NO. 10, P. 761. (ABSTR.).
USA; GULF COAST; ARKANSAS; PIKE COUNTY
(ENG)
GEOCHRONOLOGY; ALTERATION; PETROGRAPHY; PEROVSKITE

1976 MILLAR, H.A.
IT WAS FINDERS-KEEPERS AT AMERICA'S ONLY DIAMOND MINE.
NEW YORK: CARLETON PRESS,
175P.
ARKANSAS; USA
(ENG)
HISTORY OF PRAIRIE CREEK MINE.

1976 NESBITT, B.E.
FLUID AND MAGMATIC INCLUSIONS IN THE CARBONATITE AT MAGNET COVE, ARKANSAS.
M.SC. THESIS, UNIV. MICHIGAN
USA; GULF COAST; ARKANSAS; HOT SPRING COUNTY
(ENG)
GEOCHEMISTRY; MINERAL CHEMISTRY

1976 ROBISON, E.C.
GEOCHEMISTRY OF LAMPROPHYRIC ROCKS OF THE EASTERN OUACHITA MOUNTAINS, ARKANSAS.
M.SC. THESIS, UNIV. ARKANSAS,
147P.
USA; OUACHITA MOUNTAINS; ARKANSAS
(ENG)
GEOCHEMISTRY

1976 SINKANKAS. J.
A BIBLIOGRAPHY OF NORTH AMERICAN GEMSTONE LITERATURE.
SEPERATELY PRINTED FROM VOLUME 2, GEMSTONES OF NORTH AMERICA, WITH APPENDIX 2, GEOGRAPHIC AND GEMSTONE INDEXES,
PP. 374-479.
USA; CENTRAL STATES; WEST COAST; ROCKY MOUNTAINS; APPALACHIA; GULF COAST; ARKANSAS
(ENG)
BIBLIOGRAPHY; DIAMOND OCCURRENCE
2661 ENTRIES BY LOCALITY AND GEMSTONE.

1976 SINKANKAS, J.
DIAMONDS IN ARKANSAS AND LOUISIANA.
IN: GEMSTONES OF NORTH AMERICA,
VOL. 2, PP. 2-5.
USA; GULF COAST; ARKANSAS; LOUISIANA
(ENG)
DIAMOND OCCURRENCE; CATALOGUE

1976 STEELE, K.F.; ROBISON, E.C.
CHEMICAL RELATIONSHIPS OF LAMPROPHYRE, CENTRAL ARKANSAS.
EOS,
VOL. 57, P. 1018. (ABSTR.).
USA; GULF COAST; ARKANSAS
(ENG)
GEOCHEMISTRY

1977 GEMS AND GEMOLOGY,
DIAMOND FOUND IN ARKANSAS.
GEMS AND GEMOLOGY,
VOL. 40, FALL P. 349.
USA; ARKANSAS; GULF COAST; PIKE COUNTY
(ENG)
NEWS ITEM. DIAMOND OCCURRENCE

1977 BOLIVAR, S.L.
GEOCHEMISTRY OF THE PRAIRIE CREEK, ARKANSAS AND ELLIOTT COUNTY, KENTUCKY INTRUSIONS.
PH.D. THESIS, UNIV. NEW MEXICO,
441P.
USA; GULF COAST; ARKANSAS; PIKE COUNTY; APPALACHIA; KENTUCKY; EASTERN KENT UCKY
(ENG)
GEOCHEMISTRY; LAMPROITE

1977 **DENISON, R.E.; BURKE, W.H.; OTTO, J.B.; HEATHERINGTON, E.A.**
AGE OF IGNEOUS AND METAMORPHIC ACTIVITY AFFECTING THE OUACHITA FOLDBELT.
ARKANSAS GEOL. COMM.,
IN: SYMPOSIUM ON THE GEOLOGY OF THE OUACHITA MOUNTAINS, VOL. 1, STRATIGRAPHY, SEDIMENTOLOGY, PETROGRAPHY, TECTONICS AND PALEONTOLOGY,
PP. 25-40.
USA; OUACHITA MOUNTAINS; ARKANSAS
(ENG)
STRUCTURE; GEOCHRONOLOGY

1977 **LEWIS, R.D.**
MINERALOGY, PETROLOGY AND GEOPHYSICAL ASPECTS OF THE PRAIRIE CREEK KIMBERLITE, NEAR MURFREESBORO, ARKANSAS.
M.SC. THESIS, PURDUE UNIV, WEST LAFAYETTE, INDIANA.
USA; GULF COAST; ARKANSAS
(ENG)
KIMBERLITE; GEOPHYSICS; LAMPROITE

1977 **LEWIS, R.D.; MEYER, H.O.A.**
DIAMOND BEARING KIMBERLITE OF PRAIRIE CREEK, MURFREESBORO, ARKANSAS.
INTERN. KIMB. CONF. SECOND EXTENDED ABSTRACT VOLUME,
USA; GULF COAST; ARKANSAS; PIKE COUNTY
(ENG)
PETROLOGY

1977 **MEYER, H.O.A.; LEWIS, R.D.; BOLIVAR, S.L.; BROOKINS, D.G.**
PRAIRIE CREEK KIMBERLITE, MUFREESBORO PIKE COUNTY, ARKANSAS.
INTERN. KIMB. CONF. SECOND, FIELD GUIDE, 14P.
USA; GULF COAST; ARKANSAS; PIKE COUNTY
(ENG)
PETROGRAPHY; MINERAL CHEMISTRY

1977 **NESBITT, B.E.; KELLY, W.C.**
MAGMATIC AND HYDROTHERMAL INCLUSIONS IN CARBONATITE OF THE MAGNET COVE COMPLEX, ARKANSAS.
CONTRIB. MIN. PETROL.,
VOL. 63, NO. 3, PP. 271-294.
USA; GULF COAST; ARKANSAS; HOT SPRING COUNTY
(ENG)
PETROLOGY; MINERAL CHEMISTRY

1977 **NEWTON, M.G.; MELTON, C.E.; GIARDINI, A.A.**
MINERAL INCLUSION IN AN ARKANSAS DIAMOND.
AMERICAN MINERALOGIST,
VOL. 62, NO. 5-6, PP. 583-586.
USA; GULF COAST; ARKANSAS; PIKE COUNTY
(ENG)
MINERALOGY; MURFREESBORO

1977 **ROBISON, E.C.; STEELE, K.F.; JACKSON, K.C.**
GEOCHEMISTRY OF LAMPROPHYRIC ROCKS, EASTERN OUACHITA MOUNTAINS, ARKANSAS.
GEOL. SOC. AMER.,
VOL. 9, NO. 1, PP. 69-70.
USA; OUACHITA MOUNTAINS; GULF COAST; ARKANSAS; GARLAND COUNTY; SALINE COUNTY
(ENG)
PETROLOGY; GEOCHEMISTRY

1977 **ROBISON, E.C.; STEELE, K.F.; JACKSON, K.C.**
GEOCHEMISTRY OF LAMPROPHYRIC ROCKS, EASTERN OUACHITA MOUNTAINS, ARKANSAS.
GEOL. SOC. AMER.,
VOL. 9, NO. 1, PP. 69-70. (ABSTR.).
USA; OUACHITA MOUNTAINS; ARKANSAS
(ENG)
GEOCHEMISTRY

1977 **STEELE, K.F.; ROBISON, E.C.**
CHEMICAL WEATHERING OF LAMPROPHYRIC ROCK, CENTRAL ARKANSAS.
ARKANSAS ACAD. SCI. PROC.,
VOL. 31, PP. 119-121.
USA; GULF COAST; ARKANSAS; SALINE COUNTY
(ENG)
PETROLOGY; GEOMORPHOLOGY

1977 **WAGNER, G.H.; STEELE, K.F.**
THE CHEMICAL COMPOSITION OF CARBONATITE IN CONWAY AND PERRY COUNTIES OF ARKANSAS.
ARKANSAS ACAD. SCI. PROC.,
VOL. 31, PP. 121-123.
USA; GULF COAST; ARKANSAS; CONWAY COUNTY; PERRY COUNTY
(ENG)
PETROLOGY

1977 **ZARTMAN, R.E.**
GEOCHRONOLOGY OF SOME ALKALIC ROCK PROVINCES IN EASTERN AND CENTRAL UNITED STATES.
ANN. REV. EARTH PLAN. SCI.,
VOL. 5, PP. 257-386.
USA; GULF COAST; ARKANSAS; APPALACHIA; KENTUCKY; EASTERN KENTUCKY; CENTRAL STATES
(ENG)
GEOCHRONOLOGY; RELATED ROCKS; KIMBERLITE

1978 **GOGINENI, S.V.; MELTON, C.E.; GIARDINI, A.A.**
SOME PETROLOGICAL ASPECTS OF THE PRAIRIE CREEK DIAMOND BEARING KIMBERLITE DIATREME, ARKANSAS.
CONTRIB. MIN. PETROL.,
VOL. 66, NO. 3, PP. 251-262.
USA; GULF COAST; ARKANSAS; PIKE COUNTY
(ENG)
PETROLOGY; LAMPROITE

1978 JACKSON, K.C.
ARKANSAS SYENITES, FENITIZED CRUSTAL MATERIAL?
GEOL. SOC. AMER.,
VOL. 10, NO. 1, PP. 7-8.
USA; GULF COAST; ARKANSAS; HOT SPRING COUNTY; GARLAND COUNTY PULASKI COUNTY
(ENG)
MAGNET COVE; POTASH SULFUR SPRINGS; PETROLOGY; GRANITE MOUNTAIN

1978 MORRISON, L.M.
CRATER OF DIAMONDS A SHORT HISTORY: 1906-1978.
LAPIDARY JOUR.,
VOL. 32, NO. 5, P. 1064; P. 1066; P. 1068; P. 1070; P. 1072.
USA; GULF COAST; ARKANSAS; PIKE COUNTY
(ENG)
HISTORY

1978 WAGNER, G.H.; HONIG, R.H.; JONES, M.D.
GEOCHEMISTRY OF A CARBONATITE IN MONTGOMERY COUNTY, ARKANSAS.
ARKANSAS ACAD. SCI. PROC.,
VOL. 32, PP. 93-94.
USA; GULF COAST; ARKANSAS; MONTGOMERY COUNTY
(ENG)
GEOCHEMISTRY

1979 BOLIVAR, S.L.; BROOKINS, D.G.
GEOPHYSICAL AND RB-SR STUDY OF THE PRAIRIE CREEK ARKANSAS KIMBERLITE.
INTERN. KIMB. CONF. SECOND PROC.,
VOL. 1, PP. 289-299.
USA; GULF COAST; ARKANSAS; PIKE COUNTY
(ENG)
KIMBERLITE; GEOPHYSICS; GROUNDMAG; GEOCHEMISTRY; LAMPROITE

1979 BROOKINS, D.G.; DELLA VALLE, R.S.; BOLIVAR, S.L.
SIGNIFICANCE OF URANIUM ABUNDANCE IN UNITED STATES KIMBERLITES.
INTERN. KIMB. CONF. SECOND PROC.,
VOL. 1, PP. 280-288.
USA; GULF COAST; ARKANSAS; APPALACHIA; KENTUCKY; CENTRAL STATES; KANSAS; STATE LINE; COLORADO; WYOMING; ROCKY MOUNTAINS
(ENG)
URANIUM

1979 HEATHCOTE, R.C.
TEMPORAL RELATIONSHIPS OF CARBONATITE AND FENITE AT POTASH SULFUR SPRINGS, ARKANSAS.
GEOL. SOC. AMER.,
VOL. 11, NO. 2, PP. 148-149. (ABSTR.).
USA; GULF COAST; ARKANSAS; GARLAND COUNTY
(ENG)
PETROLOGY

1979 MCCORMICK, G.; HEATHCOTE, R.
MINERALOGY OF THE MORRILTON ALVIKITE DIKE, CONWAY COUNTY, ARKANSAS.
GEOL. SOC. AMER.,
VOL. 11, P. 163. (ABSTR.).
USA; GULF COAST; ARKANSAS; CONWAY COUNTY
(ENG)

1979 PANTALEO, N.S.; NEWTON, G.S.; GOGINENI, S.V.; MELTON, C.E.
MINERAL INCLUSIONS IN FOUR ARKANSAS DIAMONDS: THEIR NATURE AND SIGNIFICANCE.
AMERICAN MINERALOGIST,
VOL. 64, NO. 9-10, PP. 1059-1062.
USA; GULF COAST; ARKANSAS; PIKE COUNTY
(ENG)
MINERALOGY; MINERAL CHEMISTRY

1979 STEELE, K.F.; JACKSON, K.C.; VAN BUREN, W.
GEOCHEMICAL COMPARISON OF ARKANSAS SYENITE.
GEOL. SOC. AMER.,
VOL. 11, NO. 2, P. 166. (ABSTR.).
USA; GULF COAST; ARKANSAS; GARLAND COUNTY; HOT SPRING COUNTY; PULASKI COUNTY; OUACHITA MOUNTAINS
(ENG)
MAGNET COVE; POTASH SULFUR SPRINGS; GEOCHEMISTRY; GRANITE MOUNTAIN

1979 STEELE, K.F.; WAGNER, G.H.
RELATIONSHIP OF THE MURFREESBORO KIMBERLITE AND OTHER IGNEOUS ROCKS OF ARKANSAS.
INTERN. KIMB. CONF. SECOND PROC.,
VOL. 1, PP. 393-399.
USA; GULF COAST; ARKANSAS; PIKE COUNTY; OUACHITA MOUNTAINS
(ENG)
PETROLOGY;

1979 VIELE, G.W.
GEOLOGICAL MAP AND CROSS SECTION, EASTERN OUACHITA MOUNTAINS, ARKANSAS.
GEOL. SOC. AMERICA,
NO. MC 28F, 7P.
USA; OUACHITA MOUNTAINS; ARKANSAS
(ENG)

1979 WATSON, K.D.
KIMBERLITES OF EASTERN NORTH AMERICA.
KREIGER PUBL.
IN: ULTRAMAFIC AND RELATED ROCKS, EDITOR P.J. WYLLIE,
PP. 312-323.
USA; GULF COAST; ARKANSAS; APPALACHIA; KENTUCKY; NEW YORK; PENNSYLVANIA; VIRGINIA; TENNESSEE; MISSOURI; KANSAS; CENTRAL STATES; CANADA; ONTARIO; QUEBEC
(ENG)
GEOCHEMISTRY; TECTONICS; GENESIS

1980 **THE NASHVILLE NEWS,**
MINING GIANT ENTERS DIAMOND HUNT, ANACONDA BUYING AREA MINERAL RIGHTS.
THE NASHVILLE NEWS,
AUGUST 9TH. NO. 64, P. 1; P. 5.
USA; GULF COAST; ARKANSAS; PIKE COUNTY
(ENG)
PROSPECTING; NEWS ITEM

1980 **BASU, A.R.; TATSUMOTO, M.**
ND-ISOTOPES IN SELECTED MANTLE-DERIVED ROCKS AND MINERALS AND THEIR IMPLICATIONS FOR MANTLE EVOLUTION.
CONTR. MIN. PETROL.,
VOL. 75, PP. 43-54.
SOUTH AFRICA; LESOTHO; USA; GULF COAST; ARKANSAS; HOT SPRING COUNTY; ROCKY MOUNTAINS; CALIFORNIA; HAWAII; GERMANY
(ENG)
KIMBERLITE; ALNOITE; CARBONATITE; PYROXENE; INCLUSIONS; XENOLITH; ECLOGITES; ROBERTS-VICTOR; MATSOKU KIMBERLITE
GEOCHEMISTRY OF MAGNET COVE DISCUSSED AS WELL AS DISH HILL.

1980 **HARNISH, A.**
ARKANSAS FINDS A 5.5 CARAT DIAMOND.
MURFREESBORO DIAMOND,
OCT. 30TH.
USA; GULF COAST; ARKANSAS; PIKE COUNTY
(ENG)
NEWS ITEM; DIAMOND NOTABLE

1980 **HARNISH, H.**
GEMS FROM THE DIAMOND MINE.
MURFREESBORO DIAMOND,
DEC. 11TH. 2P.
USA; GULF COAST; ARKANSAS; PIKE COUNTY
(ENG)
PROSPECTING; NEWS ITEM

1980 **MELTON, C.E.; GIARDINI, A.A.**
THE ISOTOPIC COMPOSITION OF ARGON INCLUDED IN AN ARKANSAS DIAMOND AND ITS SIGNIFICANCE.
GEOPHYS. RES. LETTERS,
VOL. 7, NO. 6, PP. 461-464.
USA; GULF COAST; ARKANSAS; PIKE COUNTY
(ENG)
ISOTOPE; INCLUSION; MINERAL CHEMISTRY

1980 **NISHIMORI, R.K.; POWELL, J.D.**
URANIUM IN CARBONATITES, U.S.A. FINAL REPORT.
N.U.R.E. REPT.,
NO. GJBX 147-80, 180P.
USA; GULF COAST; ARKANSAS; HOT SPRING COUNTY
(ENG)
MAGNET COVE

1980 **WOODS, C.L.**
UPDATE ON THE CRATER OF DIAMONDS STATE PARK.
GEMS AND MINERALS,
NO. 507, PP. 58-59; P. 61.
USA; GULF COAST; ARKANSAS; PIKE COUNTY
(ENG)
NEWS ITEM. HISTORY; PRODUCTION

1981 **MURFREESBORO DIAMOND,**
TOURISM COMMISSION REJECTS MINING OFFER AT CRATER.
MURFREESBORO DIAMOND,
APRIL 23RD. 2P.
USA; GULF COAST; ARKANSAS; PIKE COUNTY
(ENG)
PROSPECTING; NEWS ITEM

1981 **GEMS AND GEMOLOGY,**
SECOND LARGEST STONE DISCOVERED THIS YEAR, CRATER OF DIAMONDS.
GEMS AND GEMOLOGY,
FALL, P. 180.
USA; GULF COAST; ARKANSAS; PIKE COUNTY
(ENG)
NEWS ITEM; DIAMONDS NOTABLE

1981 **THE ARKANSAS GAZETTE,**
COMPANY ENCOURAGED BY TESTS FOR DIAMONDS.
THE ARKANSAS GAZETTE,
FEB. 12TH.
USA; GULF COAST; ARKANSAS; PIKE COUNTY
(ENG)
NEWS ITEM; EVALUATION; ANACONDA; PROSPECTING

1981 **ENG. MIN. JOUR.,**
ANACONDA'S PROPOSAL TURNED DOWN.
ENG. MIN. JOUR.,
VOL. 181, P. 178.
USA; GULF COAST; ARKANSAS; PIKE COUNTY
(ENG)
NEWS ITEM; DIAMOND; PROSPECTING

1981 **MURFREESBORO DIAMOND,**
COURTHOUSE RECORDS SHOW INTEREST IN EXPLORING FOR DIAMONDS.
MURFREESBORO DIAMOND,
JULY 22ND. 2P.
USA; GULF COAST; ARKANSAS; PIKE COUNTY
(ENG)
PROSPECTING; NEWS ITEM

1981 **ARKANSAS GAZETTE,**
DIAMONDS ARE STATE'S GOOD FRIEND.
ARKANSAS GAZETTE,
APRIL 19TH. 1P.
USA; GULF COAST; ARKANSAS; PIKE COUNTY
(ENG)
PROSPECTING; NEWS ITEM

1981 THE SENTINEL RECORD, HOT SPRINGS,
MINING OFFER IS REJECTED.
THE SENTINEL RECORD, HOT SPRINGS,
APRIL 17TH. 1P.
USA; GULF COAST; ARKANSAS; PIKE COUNTY
(ENG)
PROSPECTING; NEWS ITEM

1981 MURFREESBORO DIAMOND,
ANACONDA SEEKS MINING LEASE AT STATE PARK.
MURFREESBORO DIAMOND,
APRIL 9TH. 4P.
USA; GULF COAST; ARKANSAS; PIKE COUNTY
(ENG)
PROSPECTING; NEWS ITEM

1981 ARKANSAS GAZETTE,
HANDS OFF CRATER OF DIAMONDS.
ARKANSAS GAZETTE,
OCTOBER 7TH. 1P.
USA; GULF COAST; ARKANSAS; PIKE COUNTY
(ENG)
PROSPECTING; NEWS ITEM

1981 THE SENTINEL RECORD, HOT SPRINGS,
DIAMOND MINING SOUGHT BY FIRM.
THE SENTINEL RECORD, HOT SPRINGS,
APRIL 3RD. 1P.
USA; GULF COAST; ARKANSAS; PIKE COUNTY
(ENG)
PROSPECTING; NEWS ITEM

1981 BAUER, E.C.
LETTER TO THE EDITOR, AGREEING WITH REFUSAL OF MINING LEASE.
MURFREESBORO DIAMOND,
AUGUST 6TH. 1P.
USA; GULF COAST; ARKANSAS; PIKE COUNTY
(ENG)
PROSPECTING; NEWS ITEM

1981 BUSS, D.D.
INDUSTRY VIES WITH PROSPECTORS FOR RIGHTS TO STRIKE IT RICH AT CRATER OF DIAMONDS.
THE WALL STREET JOUR.,
AUGUST,
USA; GULF COAST; ARKANSAS; PIKE COUNTY
(ENG)
NEWS ITEM; PROSPECTING; INVESTMENT; ANACONDA

1981 DAVIES, A.J.
ARKANSAS DIAMOND LAKES COUNTRY- A ROCKHOUND'S END OF THE RAINBOW.
LAPIDARY JOUR.,
APRIL, PP. 428-432.
USA; GULF COAST; ARKANSAS; PIKE COUNTY
(ENG)
GUIDEBOOK; MINERALOGY

1981 EDICK, M.J.; BYERLY, G.R.;
POST PALEOZOIC IGNEOUS ACTIVITY IN THE SOUTHEASTERN UNITED STATES.
GEOL. SOC. AMER.,
VOL. 13, NO. 5, P. 236. (ABSTR.).
USA; GULF COAST; ARKANSAS
(ENG)
PETROLOGY

1981 GRIFFEE, C.
GEOLOGIST HITS REFUSAL TO LEASE DIAMOND AREA.
ARKANSAS GAZETTE,
MAY 1ST. 1P.
USA; GULF COAST; ARKANSAS; PIKE COUNTY
(ENG)
PROSPECTING; NEWS ITEM

1981 HARNISH, A.
GEMS FROM THE DIAMOND MINE.
MURFREESBORO DIAMOND,
JAN. 15TH.
USA; GULF COAST; ARKANSAS; PIKE COUNTY
(ENG)
NEWS ITEM; PRODUCTION STATISTICS

1981 HARNISH, H.
GEMS FROM THE DIAMOND MINE.
MURFREESBORO DIAMOND,
APRIL 16TH. 1P.
USA; GULF COAST; ARKANSAS; PIKE COUNTY
(ENG)
PROSPECTING; NEWS ITEM

1981 HAYS, S.
OFFER FOR DIAMOND MINE IS REJECTED.
THE ARKANSAS GAZETTE,
APRIL 17TH.
USA; GULF COAST; ARKANSAS; PIKE COUNTY
(ENG)
NEWS ITEM; INVESTMENT; ANACONDA

1981 HEATHCOTE, R.C.; OWENS, D.R.
FORMATION OF VANADIUM AT POTASH SULFUR SPRINGS, ARKANSAS.
GEOL. SOC. AMER.,
VOL. 13, NO. 7, P. 470. (ABSTR.).
USA; GULF COAST; ARKANSAS; GARLAND COUNTY
(ENG)
VANADIUM; ORIGIN

1981 HUDSON, S.
DIAMONDS FOR THE TAKING AT THE CRATER OF DIAMONDS STATE PARK.
LOST TREASURE,
VOL. 6, NO. 9, PP. 14-18.
USA; GULF COAST; ARKANSAS; PIKE COUNTY
(ENG)
NEWS ITEM; POPULAR ACCOUNT

1981 **MORSE, L.**
MURFREESBORO RESIDENTS FIGHT MINERAL RIGHTS DEAL.
ARKANSAS DEMOCRAT,
APRIL 13TH. 3P.
USA; GULF COAST; ARKANSAS; PIKE COUNTY
(ENG)
PROSPECTING; NEWS ITEM

1981 **OSWALD, M.**
ANACONDA SEEKING TO MINE DIAMONDS AT MURFREESBORO.
ARKANSAS GAZETTE,
APRIL 3RD. 2P.
USA; GULF COAST; ARKANSAS; PIKE COUNTY
(ENG)
PROSPECTING; NEWS ITEM

1981 **ROGERS, W.P.**
LETTER TO THE EDITOR IN FAVOR OF UNDERGROUND MINE.
MURFREESBORO DIAMOND,
JUNE 25TH, 2P.
USA; GULF COAST; ARKANSAS; PIKE COUNTY
(ENG)
PROSPECTING; NEWS ITEM

1981 **ROSS, D.**
JOTS IN JEWELS.
MURFREESBORO DIAMOND,
APRIL 9TH. 1P.
USA; GULF COAST; ARKANSAS; PIKE COUNTY
(ENG)
PROSPECTING; NEWS ITEM

1981 **SIMMONS, B.**
STATE GETS BID TO ALLOW DIAMOND MINE IN PARK.
ARKANSAS DEMOCRAT,
APRIL 3RD. 2P.
USA; GULF COAST; ARKANSAS; PIKE COUNTY
(ENG)
PROSPECTING; NEWS ITEM

1981 **WILCOX, J.D.; YOUNG, J.**
ARKANSAS INCREDIBLE DIAMOND MINE STORY.
SPECIALTY PRINTING COMPANY, NEWS SUPPLEMENT,
USA; GULF COAST; ARKANSAS; PIKE COUNTY
(ENG)
HISTORY

1982 **BEARDSLEY, R.H.**
MODAL ANALYSIS OF THE GRANITE MOUNTAIN PULASKITE, PULASKI COUNTY, ARKANSAS.
M.SC. THESIS, UNIV. ARKANSAS,
60P.
USA; GULF COAST; ARKANSAS; PULASKI COUNTY
(ENG)
SYENITE; PETROLOGY

1982 **BLOW, S.**
TOURISTS JOIN QUEST FOR GEMS.
THE DALLAS MORNING NEWS,
JULY 13TH. 2P.
USA; GULF COAST; ARKANSAS; PIKE COUNTY
(ENG)
PROSPECTING; NEWS ITEM

1982 **BOLIVAR, S.L.**
THE PRAIRIE CREEK KIMBERLITE, ARKANSAS.
ARKANSAS GEOL. COMM. MISC. PUBL.,
IN: CONTRIB. GEOL. OF ARKANSAS, EDITOR J.D. MCFARLAND III,
NO. 18, PP. 1-21.
USA; GULF COAST; ARKANSAS; PIKE COUNTY
(ENG)
GEOLOGY; PETROLOGY

1982 **BOLIVAR, S.L.**
KIMBERLITES; A PETROLOGIST'S BEST FRIEND.
EARTH SCIENCE,
VOL. 35, NO. 3, PP. 15-18.
USA; GULF COAST; ARKANSAS; PIKE COUNTY; APPALACHIA; KENTUCKY; ELLIOTT COUNTY; ROCKY MOUNTAINS; STATE LINE; COLORADO; WYOMING
(ENG)
POPULAR ACCOUNT.

1982 **GRIFFEE, C.**
PANEL UNANIMOUSLY REJECTS PLAN FOR BIDDING TO ALLOW GEM EXPLORATION IN PARK.
ARKANSAS GAZETTE,
OCTOBER 6TH. 1P.
USA; GULF COAST; ARKANSAS; PIKE COUNTY
(ENG)
PROSPECTING; NEWS ITEM

1982 **LEDGER, E.B.**
URANIUM CONTENT OF SELECTED ARKANSAS IGNEOUS ROCKS.
GEOL. SOC. AMER.,
VOL. 14, NO. 3, P. 115. (ABSTR.).
USA; GULF COAST; ARKANSAS; HOT SPRING COUNTY
(ENG)
GEOCHEMISTRY; PHONOLITE; LAMPROPHYRE

1982 **NELSON, D.K.; LILLIE, R.J.; DE VOOGD, B.; BREWER, J.A.; OLIVER, J.E.; KAUFMAN, S.; BROWN, L.**
COCORP SEISMIC REFLECTION PROFILING IN THE OUACHITA MOUNTAINS OF WESTERN ARKANSAS: GEOMETRY AND GEOLOGIC INTERPRETATION.
TECTONICS,
VOL. 1, NO. 5, PP. 413-430.
USA; GULF COAST; ARKANSAS
(ENG)

1982 **PARKHURST, D.**
GEMSTONES IN THE UNITED STATES.
CALIFORNIA MINING JOURNAL,
VOL. 52, NO. 4, DECEMBER, PP. 12-13.
USA; GULF COAST; ARKANSAS; PIKE COUNTY
(ENG)
DIAMOND OCCURRENCE
POPULAR ACCOUNT.

1982 **ROHN, K.H.**
SOUTH TO THE OZARKS.
JEWELRY MAKING GEMS AND MINERALS,
NO. 542, PP. 44-47; PP. 50-51.
USA; GULF COAST; ARKANSAS; PIKE COUNTY
(ENG)
DIAMOND OCCURRENCE

1982 **SCOTT SMITH, B.H.; SKINNER, E.M.W.**
A NEW LOOK AT PRAIRIE CREEK, ARKANSAS.
INTERN. KIMB. CONF. THIRD, TERRA COGNITA, ABSTRACT VOLUME,
VOL. 2, NO. 3, P. 210, (ABSTR.).
USA; GULF COAST; ARKANSAS; PIKE COUNTY
(ENG)
KIMBERLITE; BRECCIA; HYPABYSSAL; PERIDOTITE; LAMPROITE

1982 **STEELE, K.F.**
URANIUM AND OTHER ELEMENT ANALYSES OF IGNEOUS ROCKS OF ARKANSAS.
NTIS, DU PONT DE NEMOURS AND CO. AIKEN, S.C. SAVANNAH RIVER LAB. (MAY).
GJBX-129-82, DPST-8L-141-17, 14P. FICHE ONLY.
USA; GULF COAST; ARKANSAS; PIKE COUNTY; HOT SPRING COUNTY
(ENG)
GEOCHEMISTRY
76 SAMPLES BY NEUTRON ACTIVATION ANALYSIS FOR U, TH,NA,AL,SC, TI,V,MN,FE,LA,CE,SM,EU,DY,YB,LU,HF. GRANITE MOUNTAIN; BAUXITE; MAGNET COVE; MURFREESBORO

1982 **WESTMAN, B.J.**
".... IF DIAMONDS ARE NOT VIGOROUSLY SOUGHT, IT IS NOT LOGICAL TO ARGUE THAT DIAMONDS ARE RARE.."
CALIFORNIA MINING JOURNAL, ,
MARCH, PP. 4; PP. 6-8; P. 10.
USA; GULF COAST; ARKANSAS; WEST COAST; CALIFORNIA; ROCKY MOUNTAINS; STATE LINE; COLORADO; WYOMING
(ENG)
DIAMOND OCCURRENCES
HISTORY AND POPULAR ACCOUNT.

1983 **NATIONAL GEOGRAPHIC WORLD,**
DIGGING FOR DIAMONDS.
NATIONAL GEOGRAPHIC WORLD,
NO. 98, OCTOBER, PP. 32-33.
USA; ARKANSAS; GULF COAST; PIKE COUNTY
(ENG)
CRATER OF DIAMONDS;
ARTICLE PREPARED AS A POPULAR ACCOUNT FOR JUVENILES.

1983 **JEWELLERS CIRCULAR KEYSTONE ALMANAC,**
DIAMOND PARK PERKS.
JEWELLERS CIRCULAR KEYSTONE ALMANAC,
JUNE 21, P. 39.
USA; GULF COAST; ARKANSAS; PIKE COUNTY
(ENG)
CRATER OF DIAMONDS; PRODUCTION; SIZE; NEWS ITEM

1983 **BALES, J.R.; STEELE, K.F.**
A COMPARISON OF CARBONATITES AT MAGNET COVE AND POTASH SULFUR SPRINGS, ARKANSAS.
GEOL. SOC. AMER.,
VOL. 15, NO. 1, P. 7, (ABSTR.).
USA; GULF COAST; ARKANSAS; HOT SPRING COUNTY; GARLAND COUNTY
(ENG)
PETROLOGY; GEOCHEMISTRY; IJOLITE; MINERAL CHEMISTRY

1983 **BATH, T.P.**
IGNEOUS LAMINATION AND LAYERING IN THE NEPHELINE SYENITE QUARRY, SEC. 36, T1S, R 14W, SALINE COUNTY, ARKANSAS.
M.S. THESIS, UNIV. ARKANSAS,
111P.
USA; OUACHITA MOUNTAINS; ARKANSAS; SALINE COUNTY
(ENG)
SODALITE FOYAITE; PULASKITE; PETROGRAPHY

1983 **BRESHEARS, T.L.**
REGIONAL GRAVITY EFFECTS OF IGNEOUS INTRUSIONS, CENTRAL ARKANSAS.
M.SC. THESIS, UNIV. MISSOURI,
USA; GULF COAST; ARKANSAS; OUACHITA MOUNTAINS
(ENG)
GEOPHYSICS

1983 **BRESHEARS, T.L.**
REGIONAL GRAVITY EFFECTS OF IGNEOUS INTRUSIONS CENTRAL ARKANSAS.
MISSOURI ACAD. SCIENCE, TRANS.,
VOL. 17, P. 201. (ABSTR.).
USA; OUACHITA MOUNTAINS; ARKANSAS
(ENG)
GEOPHYSICS

1983 **BRISTOW, J.W.**
DIAMOND EXPLORATION-USA.
INDIAQUA,
NO. 34, 1983/1, PP. 27-32.
USA; GULF COAST; ARKANSAS; COLORADO; WYOMING; STATE LINE; ROCKY MOUNTAINS
(ENG)
TRAVELOGUE
POPULAR ACCOUNT.

1983 **MASTERSON, M.**
TOWN FINDS GEM IN DIAMOND MINE.
USA TODAY,
APRIL 6TH, WEDNESDAY, P. 3A.
USA; GULF COAST; ARKANSAS; PIKE COUNTY
(ENG)
NEWS ITEM; DIAMOND NOTABLE
CRATER OF DIAMONDS; 3.20 CARAT DIAMOND

1983 **MITCHELL, R.H.**
LAMPROITES: PETROGRAPHY AND MINERALOGY.
MANTLE METASOMATISM AND THE ORIGIN OF ULTRAPOTASSIC AND RELATED IGNEOUS ROCKS, SYMPOSIUM HELD AUGUST 31 TO SEPT. 3RD, UNIV. WESTERN ONTARIO, LONDON,
4P. (ABSTR.).
ITALY; UGANDA; WYOMING; USA; GULF COAST; ARKANSAS; MONTANA; AUSTRALIA; ROCKY MOUNTAINS
(ENG)

1983 **MITCHELL, R.H.; LEWIS, R.D.**
PRIDERITE BEARING XENOLITHS FROM THE PRAIRIE CREEK MICA, ARKANSAS.
CAN. MINERALOGIST,
VOL. 21, PP. 59-64.
USA; GULF COAST; ARKANSAS; PIKE COUNTY
(ENG)
PETROLOGY

1983 **PERRY, L.E.**
FIELD TRIP CRATER OF DIAMONDS. AMERICA'S ONE DIAMOND PIPE IS NOW AN ARKANSAS STATE PARK.
ROCK & GEM,
VOL. 13, NO. 2, FEBRUARY, PP. 22-25.
USA; GULF COAST; ARKANSAS; PIKE COUNTY
(ENG)
POPULAR ACCOUNT.

1983 **SCOTT SMITH, B.H.**
KIMBERLITES, LAMPROITES AND THEIR ORIGIN.
MANTLE METASOMATISM AND THE ORIGIN OF ULTRAPOTASSIC AND RELATED IGNEOUS ROCKS, SYMPOSIUM HELD AUGUST 31 TO SEPTEMBER 3RD, UNIV. WESTERN ONTARIO, LONDON,
1P. (ABSTR.).
USA; GULF COAST; ARKANSAS; AUSTRALIA; WESTERN AUSTRALIA
(ENG)
GENESIS; PETROGRAPHY

1983 **SCOTT SMITH, B.H.; SKINNER, E.M.W.**
KIMBERLITE AND AMERICAN MINES, NEAR PRAIRIE CREEK ARKANSAS.
ANN. SCI. DE L'UNIVERSITE DE CLERMONT FERRAND II,
NO. 74, PP. 27-36.
USA; GULF COAST; ARKANSAS; PIKE COUNTY
(ENG)
PETROGRAPHY; MINERAL CHEMISTRY; LAMPROITE

1984 **BERGMAN, S.C.**
LAMPROITES AND OTHER POTASSIUM RICH IGNEOUS ROCKS: A REVIEW OF THEIR OCCURRENCE, MINERALOGY AND GEOCHEMISTRY.
128P. 7 TABLES; 24 FIGS. 2 PLS.
USA; CANADA; GREENLAND; APPALACHIA; SOUTH CAROLINA; CENTRAL STATES; KANSAS; UTAH; WYOMING; ROCKY MOUNTAINS; GULF COAST; ARKANSAS; MONTANA; WEST COAST; CALIFORNIA; ARIZONA; COLORADO PLATEAU; COLORADO; ALASKA; MAINE; BRITISH COLUMBIA; YUKON; LABRADOR; AUSTRALIA; SPAIN; EUROPE; ITALY; ENGLAND; NORWAY; GERMANY; TURKEY; IRAN; BULGARIA; CZECHOSLOVAKIA; AFRICA; IVORY COAST; ZAMBIA; SOUTH AFRICA; UGANDA; ALGERIA; NIGERIA; ANTARCTICA; ASIA; INDONESIA; INDIA; KALIMANTAN; USSR; MONGOLIA; BRAZIL;
(ENG)
LAMPROITE; TERMINOLOGY; OCCURRENCES; GEOCHEMISTRY; MINERAL CHEMISTRY; PETROGENESIS

1984 **BOLIVAR, S.L.**
AN OVERVIEW OF THE PRAIRIE CREEK INTRUSION, ARKANSAS.
A.I.M.E. PREPRINT,
NO. 84-346, 12P.
USA; GULF COAST; ARKANSAS; PIKE COUNTY
(ENG)
LAMPROITE; GEOLOGY; GEOCHEMISTRY

1984 **COOPERSMITH, H.G.**
DIAMONDS IN NORTH AMERICA.
SME-AIME,
SYMPOSIUM OUTLINE FALL MEETING OCTOBER 24TH. P. 13. (ABSTR.)
USA; GULF COAST; ARKANSAS; CALIFORNIA; VIRGINIA; APPALACHIA; WEST COAST; S TATE LINE; COLORADO; WYOMING; ROCKY MOUNTAINS
(ENG)
ORIGIN; DISTRIBUTION
ABSTRACT ONLY. PAPER NOT PRESENTED.

1984 **GOLD, D.P.**
A DIAMOND EXPLORATION PHILOSOPHY FOR THE 1980'S. THE RECOGNITION OF NEW TARGET ROCKS AND TECTONIC SETTINGS FOR DIAMOND BEARING PIPES AND FISSURES CALLS FOR A REVISION IN EXPLORATION METHODS AND STRATEGY.
EARTH & MINERAL SCIENCES,
VOL. 53, NO. 4, SUMMER PP. 37-42.
USA; USSR; CANADA; TANZANIA; LESOTHO; SOUTH AFRICA; AUSTRALIA; WEST AFRICA; GULF COAST; ARKANSAS; WYOMING; STATE LINE; NORTHWEST TERRITORIES; SOLOMON ISLANDS; CHINA; BOTSWANA; BRAZIL; GREENLAND; MICHIGAN; GREAT LAKES; MONTANA; ROCKY MOUNTAINS; WISCONSIN; CALIFORNIA; NEW MEXICO; ONTARIO; BRITISH COLUMBIA; QUEBEC; LABRADOR; EAST AFRICA
(ENG)
BRIEF OVERVIEW OF EXPLORATION TECHNIQUES

1984 **MULLEN, E.D.**
ULTRAMAFIC PODS OF THE EASTERN OUACHITAS: OPHIOLITIC OR ALKALIC?
GEOL. SOC. AMER.,
VOL. 16, NO. 2, P. 110. (ABSTR.).
USA; OUACHITA MOUNTAINS; ARKANSAS
(ENG)
PETROLOGY

1984 **ROSS, M.**
ULTRAALKALIC ARFVEDSONITE AND ASSOCIATED RICHTERITE ACMITE, AND AEGERINE AUGITE IN QUARTZ SYENITE, MAGNET COVE ALKALIC IGNEOUS COMPLEX, ARKANSAS.
EOS,
P. 293. (ABSTR.).
USA; GULF COAST; ARKANSAS; HOT SPRING COUNTY
(ENG)
PETROGRAPHY

1984 **SCOTT SMITH, B.H.; SKINNER, E.M.W.**
A NEW LOOK AT PRAIRIE CREEK, ARKANSAS.
THIRD KIMB. CONF.,
VOL. 1, PP. 255-284.
USA; GULF COAST; ARKANSAS; PIKE COUNTY
(ENG)
LAMPROITE; RELATED ROCKS; PETROGRAPHY; MINERAL CHEMISTRY; ANALYSES; WHOLE ROCK GEOCHEMISTRY; GEOCHRONOLOGY

1984 **SCOTT SMITH, B.H.; SKINNER, E.M.W.**
DIAMONDIFEROUS LAMPROITES.
JOURNAL OF GEOLOGY,
VOL. 92, PP. 433-438.
USA; GULF COAST; ARKANSAS; PIKE COUNTY; AUSTRALIA; WESTERN AUSTRALIA
(ENG)
PETROLOGY; ARGYLE; PRAIRIE CREEK
DISCUSSION OF LAMPROITE TERMINOLOGY.

1984 **SHEDENHELM, W.R.C.**
DIAMONDS ARE FOREVER.
ROCK & GEM,
VOL. 14, NO. 2, PP. 36-39.
USA; GULF COAST; ARKANSAS; WEST COAST; CALIFORNIA
(ENG)
POPULAR ACCOUNT OF MARKETING.

1985 **FLOHR, M.J.K.; ROSS, M.**
NEPHELINE SYENITE, QUARTZ SYENITE AND IJOLITE FROM THE DIAMOND JO QUARRY, MAGNET COVE, ARKANSAS.
IN: ALKALIC ROCKS AND CARBONIFEROUS SANDSTONES OUACHITA MOUNTAINS NEW PERSPECTIVES.
A GUIDEBOOK FOR GSA REGIONAL MEETING, FAYETTEVILE, ARKANSAS APRIL 16-18, EDITORS R.C. MORRISS & E.D. MULLEN,
PP. 63-75.
USA; GULF COAST; ARKANSAS; HOT SPRING COUNTY
(ENG)
RELATED ROCKS

1985 **HOWARD, J.M.**
ALKALIC ROCKS OF ARKANSAS.
IN: ALKALIC ROCKS AND CARBONIFEROUS SANDSTONES OUACHITA MOUNTAINS NEW PERSPECTIVES.
A GUIDEBOOK FOR GSA REGIONAL MEETING, FAYETTEVILLE, ARKANSAS, APRIL 16-18, EDITORS R.C. MORRIS & E.D. MULLEN,
PP. 76-85.
USA; GULF COAST; ARKANSAS; PIKE COUNTY; HOT SPRING COUNTY
(ENG)
FIELD STOP GUIDEBOOK

1985 **KING, J.**
GEMS FROM THE CRATER.
MURFREESBORO DIAMOND,
APRIL 17, P. 11.
USA; GULF COAST; ARKANSAS; PIKE COUNTY
(ENG)
HISTORY; NEWS ITEM

1985 **MCFARLAND, J.D.; STOLARZ, T.**
THE GEOLOGIC HISTORY OF THE CRATER OF DIAMONDS STATE PARK.
ARKANSAS STATE PARKS,
11P.
USA; GULF COAST; ARKANSAS; PIKE COUNTY
(ENG)
HISTORY
POPULAR ACCOUNT.

1985 **MILTON, C.**
KASSITE CATI2O4 (OH)2 FROM MAGNET COVE, ARKANSAS.
GEOL. SOC. AMER.,
VOL. 17, NO. 3, P. 168. (ABSTR.).
USA; GULF COAST; ARKANSAS; HOT SPRING COUNTY; USSR
(ENG)
MINERALOGY; TERMINOLOGY

1985

F
A
F,
EL
PH
US
HC
(EI
OC
LAI
MA
BEN

1985 MUL
MIN
LAM
CEN
GEO
VOL.
USA;
COUI
(ENG
KERS

1985 MULL
PETRC
AND L
ALKAL
GEOL.
VOL. 1
USA; G
(ENG)
PETROL

1985 MURFRE
POLL FA
MURFRE
VOL. 10,
USA; GU
(ENG)
NEWS ITE

1985 THOMAS,
THE APPA
PALEOZO
MARGIN C
ANN. REV.
VOL. 13, P
USA; GULF
MOUNTAIN
GEOTECTO

1985 TILTON, G.R.; FROST, D.M.; KWON, SUNG TACK
ISOTOPIC RELATIONSHIPS IN ARKANSAS CRETACEOUS ALKALIC COMPLEXES.
GEOL. SOC. AMER.,
VOL. 17, NO. 3, P. 194. (ABSTR.).
USA; GULF COAST; ARKANSAS; HOT SPRING COUNTY; CANADA; QUEBEC
(ENG)
ISOTOPE
COMPARISON WITH MAGNET COVE AND OKA CARBONATITE, QUEBEC.

1985 TREIMAN, A.H.;
LOW ALKALI CARBONATITES IN ALKALINE COMPLEXES: SEPERATE MANTLE SOURCES FOR CARBONATE AND ALKALIS?
GEOL. SOC. AMER.,
VOL. 17, NO. 3, P. 194. (ABSTR.).
USA; GULF COAST; ARKANSAS; HOT SPRING COUNTY; CANADA; QUEBEC; KENYA; UGANDA
(ENG)
IJOLITE; CARBONATITE;
FORT PORTAL; OKA; KERIMISHI; MAGNET COVE

85 WALDMAN, M.A.; MCCANDLESS, T.E.; DUMMETT, H.T.
GEOLOGY AND MINERALOGY OF THE TWIN KNOBS # 1 LAMPROITE PIKE COUNTY, ARKANSAS.
GEOL. SOC. AMER.,
VOL. 17, NO. 3, P. 196. (ABSTR.).
USA; GULF COAST; ARKANSAS; PIKE COUNTY
(ENG)
GEOCHRONOLOGY; EVALUATION

WALDMAN, M.A.; MCCANDLESS, T.E.; DUMMETT, H.T.
GEOLOGY AND MINERALOGY OF THE TWIN KNOBS # 1 LAMPROITE PIKE COUNTY, ARKANSAS.
PREPRINT OF PAPER PRESENTED GEOL. SOC. AMER. SOUTHEASTERN SECTION, ARKANSAS MEETING,
17P. 12 FIGS. 1 TABLE
USA; GULF COAST; ARKANSAS; PIKE COUNTY
(ENG)
AMPROITE; PROSPECTING; GEOPHYSICS; EOCHEMISTRY; GEOLOGY; PETROGRAPHY; VHOLE ROCK GEOCHEMISTRY; XENOCRYST INERALOGY; GARNETS; PYROXENE; ILMENITE; PINEL

ARTMAN, R.E.; HOWARD, J.M.
TH PB AGES OF LARGE ZIRCON CRYSTALS OM THE POTASH SULFUR SPRINGS IGNEOUS MPLEX, GARLAND COUNTY, ARKANSAS.
OL. SOC. AMER.,
L. 17, P. 198. (ABSTR.).
A; GULF COAST; ARKANSAS; GARLAND UNTY
(ENG)
GEOCHRONOLOGY

AUTHOR INDEX

SUBJECT INDEX

DIAMOND FIELD
CLASSIC PAPER
BRANNER, J.C. 1889
SUMMARY TO DATE
BRANNER, G.C. 1927
GEOLOGY OF FIELDS
BRANNER, G.C. 1929
FULLER, J.T. 1908
GEOLOGY
FULLER, J.T. 1910
BRANNER COMMENTS
BRANNER, J.C. 1909
DIAMOND FIELDS
EBERLE, F. 1909
ARKANSAS FIELD
HENSON.P. 1940
DIAMOND FINDS
BLANK, E.W. 1935
LOCATIONS
SINKANKAS, J. 1959
WILLIAMS, N.F. 1959
MORPHOLOGY
KUNZ, G.F. 1907
SHIRAS, T. 1924
BLANK, E.W. 1931
KUNZ, G.F. 1931
PERIDOTITE AND DIAMOND
KUNZ, G.F. 1908
MISER, H.D. 1914
LEIPER, H. 1957
GENESIS
LANGFORD, R.E. 1974
DIAMOND INDUSTRY
MISER, H.D. 1927
CAPITAL RAISED
LITERARY DIGEST 1936
U.S.G.S. EVALUATION
LUND, R.J. 1942
NEW DIAMOND COMPANY
BALL, S.H. 1947
REVIEW
BALL, S.H. 1950
MARKETING
SHEDENHELM, W.R.C. 1984
DIAMONDS NOTABLE
SAN DIEGO UNION 1956
ROBINSON, A. 1956
GEMS & GEMOLOGY 1959
BRANCH, H. 1960
DIAMOND PRODUCTION
ANON. 1932
NORTH AMERICAN DIAMOND COMPANY
VITT, G.N. 1943
PRODUCTION FIGURES
WOODS, C.L. 1980
MILLAR, H.W. 1966
POLITICS
SCHENCK, E. 1943
STATISTICS
BALL, S.H. 1944
DIAMOND JO QUARRY
FLOHR, M.J.K. 1985
DIKES
GARLAND COUNTY
HOWARD, J.M. 1973
MAGNET COVE
MULLEN, E.D. 1985

EVALUATION
WASHINGTON, H.S. 1908
LETTER OF COMMENT
STOKES, R.S.G. 1912
COMMERCIAL ASPECT
MITCHELL, G.J. 1923
U.S.G.S. INTEREST
LUND, R.J. 1942
PILOT PLANT
VITT, G.N. 1942
DIAMOND CONTENT
VITT, G.N. 1943
LETTER/COMMENTARY
STOKES, R.S.G. 1912
MINE
SANSOM, W.J. 1948
THOENEN, J.R. 1949
1950
FENITIZATION
HEATHCOTE, R.C. 1974
1975
1976
1979
JACKSON, K.C. 1978
FICTION - ADVENTURE BOOK
PRAIRIE CREEK
MANOYIAN, Z. 1969
FOURCHE COVE
GEOLOGICAL REPORT -
POWELL, W.B. 1842
FOURCHE MOUNTAIN - QUARRY
BARNEY, J. 1852
FOYAITE
WASHINGTON, H.S. 1901
BATH, T.P. 1983
GARLAND COUNTY
**SEE ALSO* POTASH SULFUR SPRINGS
POLLACK, D.W. 1965
WILSON SPRING VANADIUM
HOLLINGSWORTH, J.S. 1967
V INTRUSIVE
OWENS, D.R. 1967
XENOLITHS IN ALKALIC DIKE
HOWARD, J.M. 1973
FENITIZATION NOVACULITE
HEATHCOTE, R.C. 1974
1976
GEOCHEMISTRY, PETROGRAPHY
POTASH SULFUR SPRINGS
HOWARD, J.M. 1974
1975
GEOCHEMISTRY IGNEOUS ROCKS.
JACKSON, K.C. 1976
VANADIUM
HEATHCOTE, R.C. 1981
PETROLOGY ALKALIC ROCKS
MULLEN, E.D. 1985
GENESIS
DIAMONDS
LANGFORD, R.E. 1974

MAGNET COVE
JOHNSON, M. 1976
KIMBERLITES
WATSON, K.D. 1979
GEOCHEMISTRY
**SEE ALSO* MINERAL CHEMISTRY
ANALYSIS SYENITE
DAY, D.T. 1890
ALKALIC MAGNET COVE
ERICKSON, R.L. 1963
POTASH SULFUR SPR.
HOWARD, J.M. 1974
1975
IGNEOUS ROCKS
JACKSON, K.C. 1976
CARBONATITE
NESBITT, B.E. 1976
BALES, J.R. 1983
LAMPROPHYRES
ROBISON, E.C. 1976
1977
STEELE, K.F. 1976
PRAIRIE CREEK RB SR
BOLIVAR, S.L. 1977
1979
KIMBERLITES
BROOKINS, D.G. 1976
WATSON, K.D. 1979
DISPERSION PATTERNS
GREGORY, G.P. 1969
PROSPECTING
GREGORY, G.P. 1969
MAGNET COVE
BASU, A.R. 1980
MAGNET COVE SYENITE
STEELE, K.F. 1979
LAMPROITES
BOLIVAR, S.L. 1984
SCOTT SMITH, B.H. 1984
BERGMAN, S.C. 1984
WALDMAN, M.A. 1985
GEOCHRONOLOGY
BRANNER, J.C. 1889
DENISON, R.E. 1977
BASU, A.R. 1980
MAGNET COVE
JOHNSON, M.L. 1975
ALKALIC ROCKS
ZARTMAN, R.E. 1977
POTASH SULFUR SPR.
ZARTMAN, R.E. 1985
GEOPHYSICS
MAGNETICS
HARRIS, G.D. 1909
HEILAND, C.A. 1929
ENG. MIN. JOUR. 1946
SURVEY BAUXITE
STEARN, N.H. 1930
HOTCHKISS
STEARN, N.H. 1932
PRAIRIE CREEK
BOLIVAR, S.L. 1976
1979
LEWIS, R.D. 1977
MAGNET COVE
KELLER, F. 1963
GRAVITY - IGNEOUS
BRESHEARS, T.L. 1983
TWIN KNOBS # 1
WALDMAN, M.A. 1985

MOLYBDENUM
HOLBROOK, D.F. 1948
CARBONATITE
NESBITT, B.E. 1976
GENESIS - INTRUSIVE
JOHNSON,M. 1976
MINERAL COLLECTING
SCHOCKLEY, W.G. 1948
CRETACEOUS GEOLOGY
SHREVEPORT 1953
EARTHY MONAZITE
ROSE, H.J. 1958
GEOLOGY
HOLBROOK, D.F. 1962
GEOCHEMISTRY & PETROLOGY
ERICKSON, R.L. 1963
ALKALINE PYROXENITE
UPTON, B.G.L. 1967
OXYGEN RATIOS
CONWAY, C.M. 1969
INCLUSIONS
NESBITT, B.E. 1977
GEOCHRONOLOGY
JOHNSON, M.L. 1975
GEOCHEMISTRY SYENITE
STEELE, K.F. 1979
URANIUM IN CARBONATITES
NISHIMORI, R.K. 1980
COMPARISON TO POTASH SULFUR SPRING
BALES, J.R. 1983
KASSITE
MILTON, C.E. 1985
PETROLOGY
MULLEN, E.D. 1985
COMPARISON TO OKA
TILTON, G.R. 1985
LOW ALKALI CARBONATITE
TREIMAN, A.H. 1985

MAGNETIC ROCKS (PERIDOTITE)
HARRIS, G.D. 1909

MICACEOUS PERIDOTITE
SHEPPARD, C.U. 1846

MINERAL CATALOGUE
HARVEY, F.L. 1886

MINERAL CHEMISTRY
**SEE ALSO* MINERALOGY
GIARDINI, A.A. 1975
MEYER, H.O.A. 1977
ARGON INCLUSIONS
MELTON, C.E. 1980
CARBONATITES
BALES, J.R. 1983
LAMPROITES
MITCHELL, R.H. 1983
SCOTT SMITH, B.H. 1983
1984
BERGMAN, S.C. 1984
TWIN KNOBS # 1
WALDMAN, M.A. 1985

MINERAL RESOURCES
SUMMARY OF DIAMONDS
BRANNER, G.C. 1927
ARKANSAS BULLETIN
BRANNER, G.C. 1942
REISSUED 1959

MINERALOGY
**SEE ALSO* MINERAL CHEMISTRY
SHEPPARD, C.U. 1846
KOENIG, G.A. 1878
MAGNET COVE
COMSTOCK, T.G. 1888
HALTON, W.L. 1929
LANDES, K.K. 1931
HACKMANITE SODALITE
MISER, H.D. 1941
MINERAL COLLECTING
SHOCKLEY, W.G. 1945
CRATER OF DIAMONDS
BRANNER, J.C. 1889
BASIC DIKES
KEMP, J.F. 1891
IGNEOUS ROCKS
WILLIAMS, J.F. 1891
DIAMOND MORPHOLOGY
KUNZ, G.F. 1907
DIAMOND MINERALOGY
BUHLIS, R. 1945
DIAMOND
ALEXANDER, A.E. 1950
DIAMOND BEARING
LEWIS, R.D. 1976
KIMZEYITE
MILTON, C. 1958
INCLUSIONS
GIARDENI, A.A. 1974
1975
MELTON, C.E. 1975
PANTALEO, S. 1979
PRAIRIE CREEK
LEWIS, R.D. 1977
MEYER, H.O.A. 1977
KASSITE
MILTON, C. 1985

MINING METHODS
SCHNEIDER, P.F. 1908
THOENEN, J.R. 1950
HOWARD, D.L. 1951
BURGOON, J.R. 1957

MISERITE
SCHALLER, W.T. 1950

MOLYBDENUM - MAGNET COVE
HOLBROOK, D.F. 1948

MONTGOMERY COUNTY
CARBONATITE
HEATHCOTE, R.C. 1979

MORPHOLOGY
DIAMONDS
KUNZ, G.F. 1907
COLOURS
MCCOURT, W.E. 1909

NEOZOIC GEOLOGY
SOUTH WEST ARKANSAS
HILL, R.T. 1888

NEPHELINE SYENITE
ROSENBUACH, H. 1887
DIAMOND JO QUARRY
FLOHR, M.J.K. 1985

NEWS ITEMS
DIAMOND MINE
KANSAS CITY JEWEL 1906
FOURTH DIAMOND
KANSAS CITY JEWEL 1906
DIAMOND MINE
NASHVILLE NEWS 1906
DETAILS
JEWELLERS CIRC 1907
DISCOVERY
JEWELLERS CIRC 1907
DIAMONDS
SCIENCE 1907
DIAMOND MINE
MANUFACT. JEWEL 1907
DIAMONDS
ENG. MIN. JOUR. 1907
DIAMONDS IN MATRIX
ENG. MIN. JOUR. 1907
DIAMOND OCCURRENCE
AMER. GEOGRAPH. 1907
AMERICAN KIMBERLEY
STH. AFRICA MAG. 1907
OZAN FIELD
LITTLE ROCK DEM. 1907
A DIAMOND FIELD?
LANIER, R.S. 1907
DIAMOND DEAL
LITTLE ROCK DEM. 1908
STH. AFR. REACTION
STH. AFR. MIN. J. 1908
GERMAN REPORT
ZEIT. MINER. 1908
DIAMONDS
MIN. SCI. PRESS 1908
AMERICAN DIAMONDS
STH. AFR. MIN. J. 1908
ARKANSAS DIAMONDS
MANUFACT. JEWEL 1908
NEW DIAMOND FINDS
BENEKE, A. 1908
DIAMOND FIELD
STIFFT, C.S. 1908
PIKE COUNTY
ENG. MIN. JOUR. 1909
MR. DRAPER
STH. AFR. MIN. J. 1909
STATUS OF PIKE CTY.
MEMPHIS TENN. COM. 1909
ARKANSAS FIELDS
EBERLE, F. 1909
DIAMONDS
KUNZ, G.F. 1909
DIAMOND FIELDS
MILLAR, A.Q. 1909
AMERICAN DIAMONDS
PEARSON, S. 1909
ARKANSAS USA
LOND. MIN. JOUR. 1910
STH. AFRICA AND ARK.
MINING WORLD 1910
DIAMOND MINES
COWAN, J.L. 1910
GENUINE MINES
FOGG, F.P.
NATIVE GEMS
GARDINER, C.R. 1910
REAL DIAMONDS
GLENN, L.C. 1910
THOSE ARKANSAS DIAM.
KOHR, H.F. 1910
DIAMONDS IN USA
RANGE, P. 1910
FOUND DIAMONDS 35 YRS
LITTLE ROCK GAZ. 1911

LARGE DIAMOND
JEWELLERS CIRC. 1911
AMERICAN GEM MINES
COWAN, J.L. 1911
ARKANSAS FIELD
FULLER, J.T. 1911
DIAMONDS
RICHARD, L.M. 1911
FLAWLESS DIAMOND
JEWELLERS CIRC. 1912
DIAMOND MINING
LOND. MIN. JOUR. 1912
YEAR'S PRODUCTION
FULLER, J.T. 1912
DIAMOND MINING U.S.A.
STH. AFR. MIN. J. 1913
DIAMONDS IN ARKANSAS
ENG. MIN. JOUR. 1913
COMPANY'S CHARTER
ENG. MIN. JOUR. 1913
DIAMOND MINING
FULLER, J.T. 1913
ARKANSAS DIELD IN 1913
FULLER, J.T. 1914
DIAMONDS OF FIRST WATER
SHIRAS, T. 1914
DIAMONDS IN ARKANSAS
REYBURN, S.W. 1920
DIAMOND MINING
MUNDORFF, R.F. 1921
DIAMONDS IN ARKANSAS
ENG. MIN. JOUR. 1922
FLOURISHING BUSINESS
CURRENT OPINION 1922
DIAMONDS FROM ARKANSAS
JEWELLER CIRC. 1923
DIAMONDS IN THE USA
INDIAN ENG. 1923
DIAMONDS FROM ARKANSAS
ENG. MIN. JOUR. 1923
DIAMOMD DEPOSITS ARK.
MITCHELL. G.J. 1923
REAL DIAMONDS IN ARK.
MUNDORFF, R.F. 1923
DIRECTORY OF ARKANSAS INDUSTRIES
ARKANSAS BUR. 1924
DESCRIPTION OF DIAMONDS
SHIRAS, T. 1924
DIAMOND ROSELAWN
ANON. 1925
DIAMOND MINING
ANON. 1926
ARKANSAS DIAMOND MINES
ANON. 1926
AMERICA'S INFANT DIAMOND INDUSTRY
DACY, G.H. 1926
DIAMONDS IN ARKANSAS
MENTOR 1927
YIELDED FINE DIAMONDS
BLANK, E.W. 1931
DIAMONDS, HISTORY
FORTUNE 1935
CAPITAL FOR MINE
LITERARY DIGEST 1936

ARKANSAS FIELD
HENSON, P. 1940
DOMESTIC DIAMONDS
TIME 1941
MINE CHANGES HANDS
GEMS & GEMOLOGY 1941
UNCLE SAM'S OWN
ROSENTHAL, E. 1942
ARKANSAS DIAMONDS
GEMS & GEMOLOGY 1943
MAGNETIC SURVEY
ENG. MIN. JOUR. 1946
DIAMONDS IN ARK.
WHEELER, H.E. 1946
ROUGH DIAMOND
GEMS & GEMOLOGY 1949
35 ACRES
WOOD, J.B. 1949
DIAMOND CRYSTAL
ALEXANDER, A.E. 1950
15 1/2 CARAT GEM
SAN DIEGO UNION 1956
ARKANSAS MINE
JOHNSON, E. 1956
FIFTY YEARS OF DREAM
MCCORD, R.G. 1956
STAR OF ARKANSAS
ROBINSON, A. 1956
IKE & STAR
GEMMOLOGIST 1957
DIG YOUR OWN
MURPHY, A.G. 1957
ARKANSAS CRATER
EVENING STAR 1959
DIAMOND DIGGING
GEMS & GEMOLOGY 1959
DIAMONDS FOR 1.50
HOWE, E.L. 1959
DIAMOND MINING
AUSTRAL. GEM. 1960
DO IT YOURSELF
POUGH, F.H. 1960
PLEASANT DREAM PAYS
RUSSELL, J. 1965
DIAMONDS DISCOVERED
ROSS, M. 1966
MEMORIES
DEANE, E. 1969
CRATER OF DIAMONDS
HOTHEM, L.L. 1974
COUPLES HOBBY
SAN DIEGO UNION 1975
MINING GIANT
NASHVILLE NEWS 1980
TOURISM COMMISSION
MURFREESBORO 1981
SECOND LARGEST STONE
GEMS & GEMOLOGY 1981
COMPANY ENCOURAGED
ARKANSAS GAZETTE 1981
ANACONDA'S PROPOSAL
ENG. MIN. JOUR. 1981
COURTHOUSE RECORDS
MURFREESBORO 1981
DIAMONDS ARE STATE'S
ARKANSAS GAZETTE 1981
MINING OFFER
SENTINEL RECORD 1981
ANACONDA SEEKS
MURFREESBORO 1981

HANDS OFF
ARKANSAS GAZETTE 1981
DIAMOND MINING
SENTINEL RECORD 1981
LETTER TO EDITOR
BAUER, E.C. 1981
INDUSTRY VIES
BUSS, D.D. 1981
GEOLOGIST HITS
GRIFFEE, C. 1981
GEMS FROM MINE
HARNISH, A. 1981
OFFER FOR DIAMOND
HAYS, S. 1981
MURFREESBORO
MORSE, L. 1981
ANACONDA SEEKING
OSWALD, M. 1981
LETTER TO EDITOR
ROGERS, W.P. 1981
JOTS IN JEWELS
ROSS, D. 1981
STATE GETS BID
SIMMONS, B 1981
ARKANSAS MINE STORY
WILCOX, J.D. 1981
TOURISTS JOIN
BLOW, S. 1982
PANEL REJECTS
GRIFFEE, C. 1982
DIGGING FOR DIAMONDS
NAT. GEOGRAPHIC 1983
DIAMOND PARK PERKS
JEWELLERS CIRC. 1983
TOWN FINDS GEM
MASTERSON, M. 1983
GEMS FROM THE CRATER
KING, J. 1985
POLL FAVORS
MURFREESBORO 1985

NIOBIUM
FRYKLUND, V.C. 1954

NORTH AMERICAN DIAMOND CORP.
VITT, G.N. 1943

NOVACULITE
GRISWOLD, L.S. 1891
1892
SCULLEY, F.J. 1946
RUTLEY, F. 1894
DERBY, A.G. 1898
FENITIZATION
HEATHCOTE, R.C. 1974
1976

OKLAHOMA VOLCANIC ROCKS
MISER, H.D. 1923
MISER, H.D. 1925
ROSS, C.S. 1929

OLIVINE LAMPROITE
SHEPPARD, C.U. 1846

ORIGIN OF IGNEOUS ACTIVITY
STONE, C.G. 1963
1964

ORIGIN OF DIAMOND
LANGFORD, R.E. 1974

OUACHITA MOUNTAINS GUIDEBOOK
SHREVEPORT 1953

LAMPROPHYRES
VALDOVINOS, D.L. 1968
REGIONAL GEOLOGY
STONE, C.G. 1966
NELSON, D.K. 1982
RELATIONSHIP TO KIMBERLITES
STEELE, K.F. 1979
NEPHELINE SYENITE
BATH, T.P. 1983
ULTRAMAFIC PODS
MULLEN, E.D. 1984
TECTONICS
THOMAS, W.A. 1985
ROBISON, E.C. 1976
1977
FOLDBELT
DENISON, R.E. 1977
GEOLOGIC MAP
VIELE, G.W. 1979
OZARK DIAMOND MINING CO.
WOODFORD, E.G. 1908
PERIDOTITE
BRANNER, J.C. 1889
KUNZ, G.F. 1908
PURDUE, A.H. 1908
SCHNEIDER, P.F. 1908
PETROLOGY
SHEPPARD, C.U. 1846
IGNEOUS ROCKS
HARVEY, F.L. 1886
WILLIAMS, J.F. 1891
CLASSIC PAPER
BRANNER, J.C. 1889
BASIC DIKES
KEMP, J.F. 1891
NOVACULITES
GRISWOLD, L.S. 1982
RUTLEY, F. 1894
DIAMOND BEARING ROCKS
SCHNEIDER, P.F. 1907
GLENN, L.C. 1912
MISER, H.D. 1914
MISER, H.D. 1922
1923
MAGNETIC ROCKS
HARRIS, G.D. 1909
PERRY COUNTY
WAGNER, G.H. 1977
PEROVSKITE
LEWIS, R.D. 1976
PETROLOGY
SHEPPARD, C.U. 1846
HARVEY, F.L. 1886
BRANNER, J.C. 1889
KEMP, J.F. 1891
WILLIAMS, J.F. 1891
RUTLEY, F. 1891
KUNZ, G.F. 1908
SCHNEIDER, P.F. 1908
HARRIS, G.D. 1909
GLENN, L.C. 1912
MISER, H.D. 1922
1923
PRAIRIE CREEK
LEWIS,R.D. 1977
CARBONATITE
HEATHCOTE, R.C. 1979

ALKALINE PYROX.
UPTON, B.G.L. 1967
LAMPROPHYRES
VALDOVINOS, D.L. 1968
POTASH SULFUR SPR.
HOWARD, J.M. 1974
1975
RELATIONSHIP OF IGNEOUS ROCKS
STEELE, K.F. 1979
IGNEOUS ACTIVITY
EDICK, M.J. 1981
PULASKITE
BEARDSLEY, R.H. 1982
BATH, T.P. 1983
PRAIRIE CREEK
BOLIVAR, S.L. 1982
LAMPROITES
SCOTT SMITH, B.H. 1982
1984
MITCHELL, R.H. 1983
BERGMAN, S.C. 1984
PRIDERITE
MITCHELL, R.H. 1983
PIKE COUNTY
**SEE ALSO*
PRAIRIE CREEK
CRATER OF DIAMONDS
NEWS ITEMS
SHEPPARD, C.U. 1846
AGE
BRANNER, J.C. 1889
PURDUE, A.H. 1912
PERIDOTITE
BRANNER, J.C. 1889
KUNZ, G.F. 1908
PURDUE, A.H. 1908
1912
DIAMOND BEARING
MISER, H.D. 1922
U.S.G.S. BULL.
MISER, H.D. 1923
DIAMONDS
MINERALOGY
BUHLIS, R. 1945
DIKES
MISER, H.D. 1923
GLENN, L.C. 1912
MISER, H.D. 1914
GEOPHYSICS
HARRIS, G.D. 1909
HEILAND, C.A. 1929
STEARN, N.H. 1932
IGNEOUS ROCKS
WILLIAMS, J.F. 1891
DIAMOND BEARING ROCKS
SCHNEIDER, P.F. 1907
INVESTIGATION AND REPORT
ARKANSAS DIAMOND CO 1908
REPORT ON DIAMONDS
FULLER, J.T. 1909
COMPARISON WITH SOUTH AFRICA
ARKANSAS DIAMOND CO 1910
NORTH AMERICAN DIAMOND CORP.
VITT, G.N. 1943

MINERALOGY
BUHLIS, R. 1945
RELATIONSHIP TO IGNEOUS ROCKS
STEELE, K.F. 1979
TWIN KNOBS # 1
WALDMAN, M.A. 1985
POTASH SULFUR SPRINGS
COMSTOCK,T.G. 1888
FRYKLUND, V.C. 1954
POLLACK, D.W. 1965
WAGNER, G.H. 1978
HEATHCOTE, R.C. 1981
GEOCHEMISTRY
STEELE, K.F. 1979
PETROGRAPHY
HOWARD, J.M. 1974
1975
VANADIUM
HOLLINGSWORTH, J.S. 1967
HEATHCOTE, R.C. 1981
COMPARISON TO MAGNET COVE
BALES, J.R. 1983
PETROLOGY
MULLEN, E.D. 1985
GEOCHEMISTRY
ZARTMAN, R.E. 1985
PRAIRIE CREEK
BRANNER, J.C. 1889
WILLIAMS, J.F. 1891
GLENN, L.C. 1912
MINERALOGY
LEWIS, R.D. 1977
MEYER, H.O.A. 1977
EVALUATION
THOENEN, J.R. 1949
GEOCHEMISTRY
BOLIVAR, S.L. 1977
GEOLOGY
BOLIVAR, S.L. 1982
1984
LAMPROITE
SCOTT SMITH, B.H. 1983
1984
PETROLOGY
MULLEN, E.D. 1984
PRIDERITE
MITCHELL, R.H. 1983
PROSPECTING
FULLER, J.T. 1908
MAGNET COVE RUTILE
SPENCER, R.V. 1946
GEMMOLOGIST 1954
SAMPLING
BURGOON, J.R. 1957
GEOCHEMISTRY
GREGORY, G.P. 1969
GENERAL
TOWNER, J. 1972
PROSPECTUS
ARKANSAS DIAMOND 1908
OZARK DIAMOND CO.
WOODFORD, E.G. 1908
ARKANSAS DIAMOND CO.
ARKANSAS DIAMOND 1908
1910
ST.CLAIR, J.Q. 1956

Printed in U.S.A.

Geological Society of America
Special Paper 215
1987

Late Cretaceous nephelinite to phonolite magmatism in the Balcones province, Texas

D. S. Barker
Department of Geological Sciences, University of Texas at Austin, Austin, Texas 78713-7909
R. H. Mitchell and D. McKay
Department of Geology, Lakehead University, Thunder Bay, Ontario P7B 5E1, Canada

ABSTRACT

Silica-undersaturated magmas intruded stable continental crust in southern Texas throughout Late Cretaceous time, forming at least 200 small volcanic centers and intrusive bodies in a shallow, epicontinental sea. Rock types, in order of decreasing abundance in outcrop, are melilite-olivine nephelinite, olivine nephelinite, alkali basalt, phonolite, and nepheline basanite. Melilite-olivine nephelinite, olivine nephelinite, and nepheline basanite contain xenoliths of spinel lherzolite, dunite, and harzburgite.

On the basis of Mg/(Mg + Fe) and Ni contents, some but not all analyzed samples of melilite-olivine nephelinite, olivine nephelinite, nepheline basanite, and alkali basalt could have crystallized from primitive mantle-derived liquids that underwent little fractionation. Preliminary trace-element data suggest that the four mafic types could all have formed from the same parent through extraction of varying amounts of liquid. Phonolites are strongly fractionated and peralkaline, and probably formed from olivine nephelinite magma, and possibly from the other mafic types, by removal of kaersutitic amphibole, olivine, and clinopyroxene at high pressure.

The bimodal Balcones suite resembles others in which mafic rocks and phonolites are associated, but it may be unique in the absence of coeval silica-oversaturated magmas and in the subordinance of alkali basalt.

INTRODUCTION

In the Balcones magmatic province of central and south Texas, small masses of nepheline-normative magmas formed at least 200 volcanic and shallow intrusive centers on and below the floor of a shallow epicontinental sea (Ewing and Caran, 1982). The province consists of two segments that extend from west of Uvalde to northeast of Austin, a distance exceeding 400 km (Fig. 1). The northwestern limit of magmatic activity roughly coincided with the inland margin of the buried Late Paleozoic Ouachita orogen, along a line that later became the Balcones zone of Miocene normal faults. Magmatism continued from approximately 100 to 66 Ma as sediments accumulated, without any unconformities of regional extent or contemporaneous faulting with significant displacement, on a stable platform. Most of the igneous centers or vents are not exposed, but have been discovered through gravity and magnetic surveys and confirmed by drilling. These buried bodies are of the same age range and rock types as those that are exposed, and are below the surface because of the southeastward regional dip of the enclosing Cretaceous sedimentary rocks. At least 60 buried volcanic centers have produced petroleum and natural gas, some from porous pyroclastic rocks and others from sedimentary strata draped over or abutting the volcanic piles.

Field relations of these small submarine volcanoes closely resemble those described at Surtsey by Kokelaar and Durant (1983). Pyroclastic cones rarely have relief exceeding a few hundred meters, and basal diameters generally are less than a kilometer. Near-vent accumulations are mostly lapilli, probably dominantly from phreatomagmatic explosions; some lapilli are cored by rock of contrasting vesicularity and degree of oxidation (Barker and Young, 1979). Bomb sags and cross-bedding indicative of surge deposits have also been noted.

Other deposits show rhythmic grading and are interpreted as air-fall. Still others, on the flanks of cones, are intimate mixtures of ash or lapilli with carbonate mud; they are formed by slumping.

Lava flows are generally only a few meters thick, and the

few examples that are exposed are of limited extent and show glassy tops but no pillow structures. At the Knippa traprock quarry, a columnar-jointed mass a few hundred meters in diameter within a pyroclastic rim is interpreted as a frozen lava lake (Ewing and Caran, 1982). Dikes are rare (probably because their wallrocks were poorly compacted and water saturated), only a few meters wide, and cannot be traced far because they are only exposed in stream beds and road-cuts. Sills and laccoliths (Lonsdale, 1927; Greenwood and Lynch, 1959) in the Uvalde segment achieve thicknesses of a few hundred meters and lateral extents exceeding a kilometer; these form resistant caps on hills and mesas, but contacts are rarely exposed. Other intrusive bodies, especially of the felsic rocks, appear to be plugs, some showing vertical flow lineation. In the vicinity of Uvalde many occurrences are merely low knobs poking through an extensive gravel blanket.

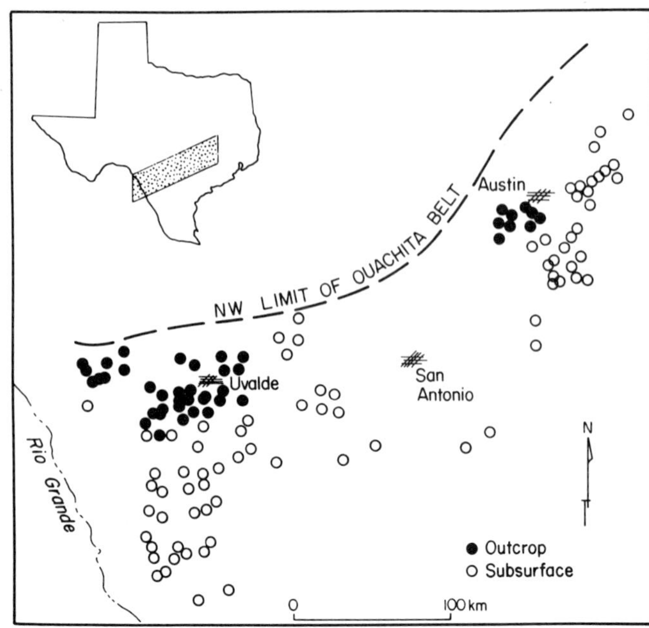

Figure 1. Extent of Balcones magmatic province, and distribution of exposed and buried volcanic and intrusive centers. Adapted from Ewing and Caran (1982).

PETROLOGY

Lonsdale (1927) provided petrographic descriptions and whole-rock major element analyses for many occurrences of igneous rocks in the Balcones province. Spencer (1969) recognized five igneous rock types, discriminated by modal compositions. In approximate order of decreasing abundance in outcrops, these are melilite-olivine nephelinite (hereafter abbreviated MON), olivine nephelinite (ON), alkali basalt (AB), phonolite (PH), and nepheline basanite (NB). All are porphyritic, containing 5 to 20 vol. % megacrysts.

Olivine and clinopyroxene are megacryst phases in all five types, joined by sphene, amphibole, nepheline, and alkali feldspar in PH. Melilite and locally perovskite are groundmass phases in MON. Plagioclase occurs in the groundmass of NB and more abundantly in AB, and alkali feldspar and analcime in PH. Clinopyroxene, apatite, and titanomagnetite are groundmass phases in all five types, and nepheline is absent from the groundmass only in a few samples of AB. Olivine megacrysts vary from euhedral to skeletal to anhedral and kink-banded (xenocrysts). Compositions cover the range Fo_{80-89} in all three habits in the four mafic rock types. Titanium and aluminum in clinopyroxene decrease, and sodium increases, with increasing differentiation index from MON to PH. Amphibole (kaersutite), so far found only in PH, forms xenocrysts with the atomic ratio 100 Mg/(Mg + Fe) varying from 35 to 71; both zoned grains and homogeneous grains of different compositions are present.

Figure 2 summarizes what is known of the distributions of the five rock types in the Uvalde segment. MON and PH have not been identified in the Austin segment, but the other three types appear to be intimately mingled at the same eruptive centers.

MON, ON, and NB carry xenoliths of spinel lherzolite, dunite, and harzburgite. These occur at several localities but have only been described from two (Cameron and Cameron, 1973; Barker and Young, 1979). All xenoliths studied so far belong to the chromian diopside suite of Wilshire and Shervais (1975).

The one felsic rock type, PH, is highest in Si, Al, Na, K, Rb, Y, Zr, and Nb, and lowest in Ti, Mg, Fe, Ca, P, and Ni. Some analyzed samples of PH are peralkaline, with the atomic ratio (Na + K)/Al reaching 1.14. The four mafic rock types fall into two groups: MON + ON are lowest in Si, Al, and Zr, but are highest in Ti, Mg, Ca, P, Ni, and Sr; Nb + AB are lowest in K and Nb, and are lower in Ni than MON + ON.

Values of 100 Mg/(Mg + Fe^{2+}) range from 73 to 78 in MON, 72 to 82 in ON, 68 to 72 in NB, 59 to 76 in AB, and 19 to 58 in PH. Major and trace element compositions and CIPW norms (Table 1) emphasize the distinctions between the three groups MON + ON, NB + AB, and PH while showing gradations within the two mafic groups. Analyzed samples of MON contain normative leucite and calcium orthosilicate (the latter indicating modal melilite). Although lacking melilite, ON in some samples contains small amounts of normative calcium orthosilicate and leucite, but other samples show albite and orthoclase in the norm. No sample of NB, AB, or PH contains normative leucite or calcium orthosilicate, but all analyzed Balcones rocks are nepheline-normative. Some PH samples are more mafic than the others, because of their higher mafic megacryst contents.

A plot of total alkalis versus silica (Fig. 3) shows that MON and ON cluster in one group, NB and AB in another, with PH separate from the mafic rocks. The coherent behavior of the four mafic rock types, and the distinctness of PH, are also shown on a plot of 100 Mg/(Mg + Fe^{2+}) versus Ni (Fig. 4), discussed below.

The MON in the Knippa quarry has been analyzed for

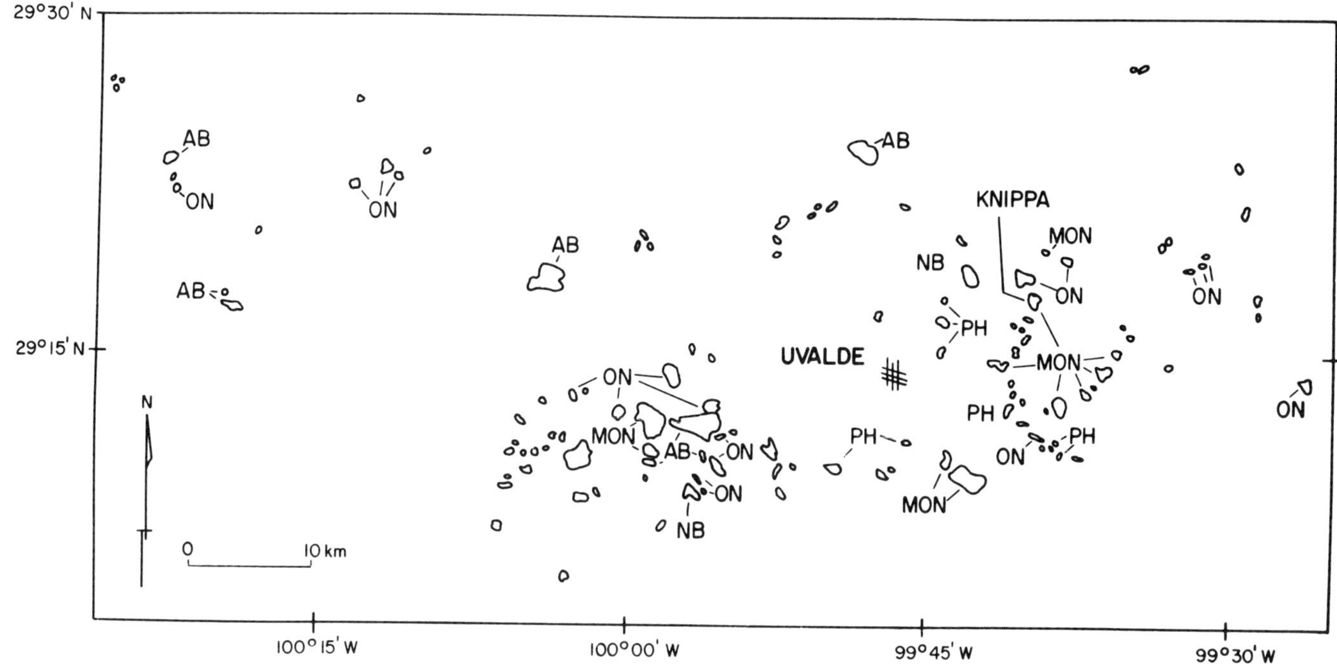

Figure 2. Distribution of five rock types in Uvalde area. MON, melilite-olivine nephelinite; ON, olivine nephelinite; NB, nepheline basanite; AB, alkali basalt; PH, phonolite. Outcrops for which no type is designated have not been sampled or sufficiently described to permit assignment of rock types. Outcrops of igneous rocks from Barnes (1974, 1977).

many trace elements; the results are compiled in Table 2, which also reports rare-earth element (REE) and other trace element analyses for samples from this and other Balcones localities by McKay (1984). Unfortunately, the Knippa data were gathered on different samples, and two distinct units of MON (one carrying ultramafic xenoliths and groundmass perovskite) are present in the quarry. Nevertheless, the trace element data for this locality deserve compilation here. Furthermore, the only published $^{87}Sr/^{86}Sr$ data for a sample from the Balcones province (Hedge, 1966) are from the Knippa MON, and recalculate to an initial ratio of 0.703.

GEOCHRONOLOGY

Burke and others (1969) reported a whole-rock K-Ar age of 67.5 ± 1.5 Ma for a sample (ON, NB, or AB?) from Pilot Knob in the Austin segment. For a sample of MON from the Uvalde segment, K-Ar ages are 70 ± 1.5 Ma for the whole rock, 73 ± 1.5 Ma for nepheline, and 69 ± 5 Ma for clinopyroxene (Burke and others, 1969). Baldwin and Adams (1971) gave whole-rock K-Ar ages for the following Balcones rocks: MON (five samples), 74 to 82 Ma; ON (seven samples), 73 to 86 Ma; NB + AB (eight samples), 71 to 81 Ma; and PH (two samples), 63 Ma. Associated pyroclastic rocks, intercalated with fossiliferous Upper Cretaceous limestones, indicate a wide time span of magmatism, from approximately 98 to 66 Ma, according to the DNAG time scale (Palmer, 1983). The low ages of the two felsic samples probably are due to argon loss from the analcime and alkali feldspar of the groundmass; K-Ar ages should be obtained on separated amphibole and alkali feldspar megacrysts in samples of Balcones PH.

COMPARISON WITH OTHER PROVINCES

The Balcones province contains a distinctly bimodal suite, lacking the intermediate types (hawaiite, mugearite, benmoreite) that could link PH to the mafic rocks. It resembles other provinces in which MON, ON, NB, AB, and PH are associated, including the Raton-Clayton volcanic field, New Mexico (Phelps and others, 1983), Oahu (Mansker, 1979; Clague and Frey, 1982; Wilkinson and Stolz, 1983), eastern Australia (Frey and others, 1978; Ferguson and Sheraton, 1979; Wass, 1980), Otago, New Zealand (Coombs and Wilkinson, 1969), Southeast Asia (Barr and Macdonald, 1981), East Africa (Williams, 1970; Aurisicchio and others, 1983), South Africa (McIver, 1981), East Greenland (Nielsen, 1980), the Canadian Arctic (Mitchell and Platt, 1984), the Laacher See district of Germany (Duda and Schmincke, 1978; Worner and others, 1983) and other parts of the Rhenish Massif (Fuchs and others, 1983), and several others mentioned by LeBas (1978; Fig. 1) and by Brey (1978). Among these, the Balcones province is unique in the absence of coeval silica-oversaturated magmas, the subordinance of alkali basalt, and the lack of faulting or crustal warping during magmatism.

TABLE 1. MAJOR AND TRACE ELEMENT DATA FOR IGNEOUS ROCKS OF THE BALCONES PROVINCE

Sample*	28	27	26	25	82021	24	82005	23	44845	22	21	20	19	18
Rock type	MON	MON	MON	MON	MON	MON	MON	MON	ON	MON	ON	ON	ON	ON
SiO_2	37.59	37.92	37.96	38.01	38.31	38.45	38.62	39.12	39.30	39.35	39.50	39.92	40.32	40.40
TiO_2	3.89	2.52	2.93	3.10	3.54	3.20	3.03	3.42	4.08	2.96	3.54	2.70	2.66	3.76
Al_2O_3	10.36	9.81	10.14	11.53	10.18	9.63	9.94	8.98	9.04	9.64	9.37	8.60	9.46	10.68
Fe_2O_3	4.97	4.40	3.69	3.57	4.28	3.22	4.56	3.28	4.70	3.36	5.55	4.40	4.75	3.95
FeO	7.77	7.49	7.59	8.25	7.72	8.16	7.30	8.87	9.35	8.16	6.70	8.00	7.48	7.89
MnO	0.21	0.18	0.22	0.19	0.19	0.18	0.17	0.18	0.18	0.17	0.18	0.24	0.25	0.07
MgO	14.30	14.38	14.69	12.40	15.20	15.18	14.90	16.66	14.86	14.94	15.21	20.17	18.12	13.94
CaO	12.51	15.85	16.28	15.49	13.16	15.21	14.98	12.54	13.38	14.93	12.62	10.68	10.55	12.89
Na_2O	3.15	2.51	2.18	3.11	2.63	2.78	2.38	2.36	1.35	3.06	2.72	1.91	2.62	3.35
K_2O	1.23	1.12	0.69	1.24	0.96	1.19	0.89	0.99	0.51	1.17	1.11	1.03	1.10	0.54
H_2O^+	---	1.95	1.82	1.16	2.20	1.48	1.47	2.20	1.87	1.30	2.00	1.45	1.25	1.35
H_2O^-	---	0.55	0.39	0.21	0.34	0.63	0.20	0.45	0.57	0.10	0.50	0.43	0.57	0.35
P_2O_5	---	1.13	1.13	1.08	0.75	0.85	0.84	0.83	0.82	0.92	0.74	0.51	0.68	0.77
CO_2	---	---	---	0.12	0.07	---	0.00	---	0.03	---	---	---	---	---
Total	95.98	99.81	99.71	99.46	99.53	100.16	99.28	99.88	100.04	100.06	99.74	100.04	99.81	100.04
Rb	---	---	---	---	53	---	17	---	37	---	25	---	25	---
Sr	---	---	---	---	982	---	917	---	1168	---	722	---	650	---
Y	---	---	---	---	21	---	19	---	23	---	18	---	11	---
Zr	---	---	---	---	291	---	247	---	332	---	230	---	180	---
Nb	---	---	---	---	72	---	63	---	69	---	71	---	51	---
Ni	---	---	---	---	379	---	287	---	367	---	376	---	603	---
or	---	---	---	---	---	---	---	---	3.09	---	---	---	2.70	2.28
ab	---	---	---	---	---	---	---	---	1.05	---	---	---	---	---
an	11.00	12.54	16.25	14.13	13.56	10.49	14.15	11.30	17.53	9.24	10.37	12.08	11.03	12.72
lc	5.97	5.33	3.28	5.86	4.59	5.62	4.22	4.72	---	5.49	5.29	4.86	3.09	0.76
ne	15.13	11.83	10.25	14.55	12.44	13.00	11.18	11.13	5.77	14.22	12.83	8.92	12.26	15.62
ac	---	---	---	---	---	---	---	---	---	---	---	---	---	---
ns	---	---	---	---	---	---	---	---	---	---	---	---	---	---
di-wo	13.92	12.29	12.88	11.56	14.92	12.12	15.01	15.96	18.80	14.22	19.42	15.04	15.82	19.70
di-en	11.28	9.21	9.81	8.45	11.69	9.12	11.76	11.98	14.23	10.63	15.69	11.67	12.16	15.25
di-fs	0.99	1.86	1.73	2.03	1.59	1.78	1.59	2.39	2.65	2.18	1.44	1.75	1.98	2.34
wo	---	---	---	---	---	---	---	---	---	---	---	---	---	---
ol-fo	18.25	19.35	19.42	16.17	19.18	20.63	18.40	21.51	16.61	18.98	16.32	27.69	23.77	14.06
ol-fa	1.76	4.31	3.78	4.29	2.87	4.43	2.75	4.72	3.40	4.29	1.65	4.58	4.26	2.38
cs	6.41	9.68	8.72	9.11	4.03	9.83	6.32	2.75	---	7.96	0.79	0.77	---	---
mt	7.55	5.99	5.49	5.28	6.40	4.76	6.73	4.89	6.98	4.94	7.52	6.21	6.16	5.82
il	7.74	4.92	5.71	6.01	6.94	6.20	5.90	6.68	7.94	5.70	6.92	5.23	5.16	7.26
ap	---	2.54	2.53	2.41	1.69	1.89	1.88	1.87	1.84	2.04	1.66	1.14	1.52	1.71
TTDI	21.10	17.16	13.53	20.42	17.03	18.62	15.40	15.84	9.91	19.71	18.12	13.78	18.05	18.65
wt.% an/(an+ab)	100	100	100	100	100	100	100	100	94.33	100	100	100	100	100
100 Mg/(Mg+Fe*), mol	68.55	69.12	70.59	65.85	70.07	70.99	69.96	71.53	66.11	70.42	69.86	75.04	73.32	68.47

*See explanation of column headings at end of table.

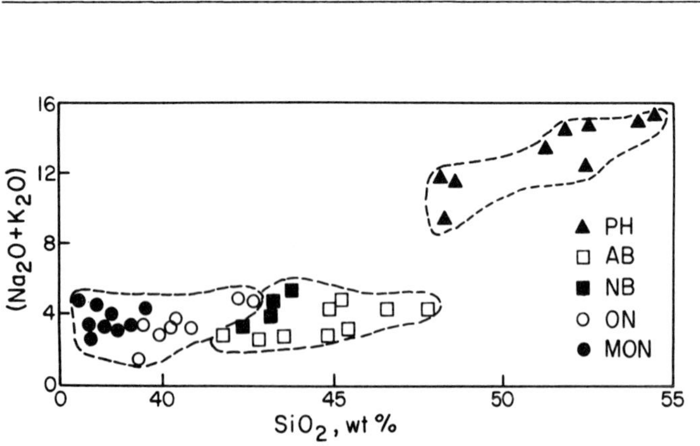

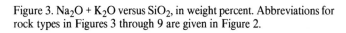

Figure 3. $Na_2O + K_2O$ versus SiO_2, in weight percent. Abbreviations for rock types in Figures 3 through 9 are given in Figure 2.

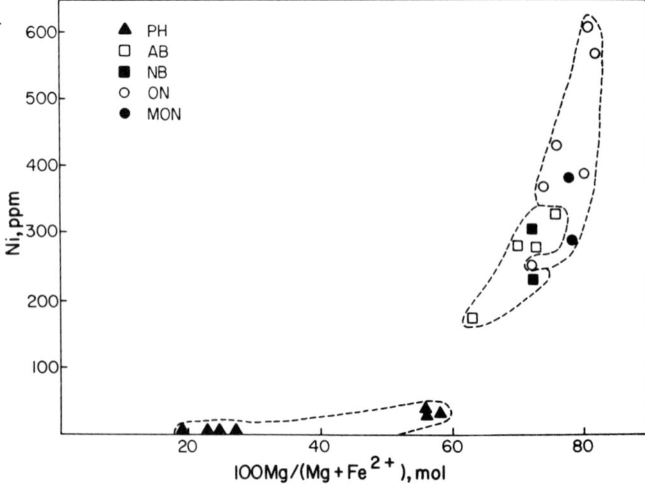

Figure 4. Atomic ratio 100 Mg/(Mg + Fe^{2+}) versus nickel (ppm).

TABLE 1 (CONTINUED)

Sample*	82012	44841	44837	17	16	44832	15	14	44847	13	12	11	10	9
Rock type	ON	AB	NB	ON	ON	AB	NB	NB	AB	NB	AB	AB	NB	AB
SiO_2	40.94	41.80	42.25	42.31	42.70	42.80	43.15	43.17	43.50	43.70	44.89	45.11	45.20	45.30
TiO_2	2.17	3.58	3.57	2.56	2.82	3.57	1.57	2.66	3.50	1.28	1.96	2.34	3.14	1.68
Al_2O_3	10.13	10.06	10.48	11.78	10.58	10.15	11.91	11.94	10.23	13.19	10.36	12.44	11.84	12.66
Fe_2O_3	3.77	5.48	3.79	2.47	3.72	2.71	3.30	2.45	3.70	3.19	1.80	2.67	2.34	2.93
FeO	7.14	7.65	8.85	8.00	7.77	9.90	9.18	8.40	8.82	7.61	10.15	9.36	8.95	10.54
MnO	0.17	0.18	0.17	0.18	0.18	0.17	0.17	0.18	0.17	0.17	0.16	0.22	0.16	---
MgO	17.78	13.29	12.60	11.36	13.90	13.10	12.82	12.10	13.25	11.05	15.96	11.56	10.77	8.39
CaO	12.34	12.77	11.59	10.53	10.93	12.27	10.67	10.81	11.97	10.49	10.46	10.61	10.70	11.86
Na_2O	2.68	2.10	2.70	3.98	3.71	2.16	3.11	3.82	2.05	3.98	2.04	3.05	3.12	2.66
K_2O	0.68	0.60	0.87	0.98	0.87	0.36	0.98	0.86	0.89	1.52	0.67	1.01	1.27	0.66
H_2O^+	1.24	2.07	1.89	4.85	1.35	2.26	2.15	2.45	1.39	2.75	0.74	0.78	1.30	1.79
H_2O^-	0.23	0.45	0.51	0.10	0.53	0.58	0.10	0.48	0.43	0.20	0.15	1.06	0.40	---
P_2O_5	0.60	0.68	0.66	0.64	0.64	0.62	0.60	0.62	0.61	0.68	0.30	0.51	0.61	0.89
CO_2	0.15	0.02	0.11	---	---	0.11	---	---	0.06	---	---	---	---	---
Total	100.02	100.73	100.04	99.74	99.70	100.76	99.71	99.94	100.57	99.81	99.64	100.72	99.80	99.35
Rb	21	37	24	36	34	20	---	---	34	51	---	---	---	---
Sr	703	712	670	890	863	643	---	---	568	908	---	---	---	---
Y	19	21	17	24	19	17	---	---	18	27	---	---	---	---
Zr	190	301	273	274	262	273	---	---	247	294	---	---	---	---
Nb	60	57	44	70	65	44	---	---	40	71	---	---	---	---
Ni	565	323	305	249	427	277	---	---	272	242	---	---	---	---
or	---	3.61	5.27	6.11	5.25	2.17	5.94	5.24	5.33	9.28	4.01	6.04	7.65	4.00
ab	---	7.92	8.60	8.25	6.69	10.87	7.65	8.71	10.64	5.93	11.51	13.79	14.67	19.52
an	13.83	16.56	14.26	12.01	9.86	17.32	16.06	13.29	16.30	14.09	17.35	17.47	14.83	21.17
lc	3.20	---	---	---	---	---	---	---	---	---	---	---	---	---
ne	12.49	5.52	8.03	14.78	13.76	4.23	10.49	13.33	3.76	15.63	3.23	6.67	6.63	1.92
ac	---	---	---	---	---	---	---	---	---	---	---	---	---	---
ns	---	---	---	---	---	---	---	---	---	---	---	---	---	---
di-wo	16.84	18.15	16.82	16.16	17.24	17.03	14.30	15.79	16.63	14.65	13.87	13.53	14.70	13.85
di-en	13.00	14.09	12.29	11.30	12.95	11.85	9.54	11.04	12.21	9.80	9.42	8.96	9.98	7.90
di-fs	2.05	2.10	2.95	3.50	2.57	3.76	3.71	3.42	2.84	3.76	3.37	3.59	3.58	5.35
wo	---	---	---	---	---	---	---	---	---	---	---	---	---	---
ol-fo	22.43	13.76	13.94	13.00	15.73	15.07	16.28	14.03	14.87	13.05	21.61	14.13	12.16	9.47
ol-fa	3.89	2.26	3.69	4.43	3.44	5.27	6.98	4.79	3.81	5.52	8.51	6.23	4.81	7.07
cs	1.26	---	---	---	---	---	---	---	---	---	---	---	---	---
mt	5.41	7.50	5.63	3.78	5.51	4.02	4.57	3.66	5.44	4.16	2.64	3.92	3.46	4.35
il	4.19	6.93	6.95	5.13	5.48	6.93	3.06	5.21	6.74	2.51	3.77	4.49	6.08	3.27
ap	1.33	1.51	1.48	1.48	1.43	1.39	1.35	1.40	1.35	1.53	0.66	1.13	1.36	1.99
TTDI	15.69	17.05	21.90	29.13	25.70	17.28	24.08	27.28	19.72	30.84	18.76	26.50	28.95	25.44
wt.% an/(an+ab)	100	67.65	62.38	59.29	59.60	61.43	67.73	60.42	60.51	70.36	60.11	55.87	50.28	52.02
100 Mg/(Mg+Fe*),mol	75.06	65.31	64.69	66.45	69.03	65.43	65.29	67.04	66.03	65.27	70.73	63.66	63.46	53.16

In spite of the large amount of effort expended on these magmatic provinces, there is no clearly prevailing opinion concerning the genetic relations. Brey (1978), on the basis of a literature survey and experiments, concluded that MON magmas are generated by a small degree of partial fusion at pressures exceeding 20 kbar, in the presence of high CO_2/H_2O, from lherzolites in which the residual phases are olivine, orthopyroxene, clinopyroxene, and garnet. Values of 100 Mg/(Mg + Fe total) between 65 and 69 indicate primary liquids; other values indicate loss or gain of olivine (Brey, 1978, p. 69). ON magmas, according to Brey and Green (1977, p. 161), arise from minimal partial fusion "in the presence of H_2O (no CO_2) alone." The addition of CO_2 will generate MON liquid instead. The results of Olafsson and Eggler (1983), however, imply that ON liquid is generated by partial fusion of amphibole-bearing peridotite with small amounts of CO_2. Yagi and Onuma (1978), on the basis of their experiments, concluded that MON magma is primary and fractionates to yield ON magma, which in turn can fractionate to NB magma. In the Raton-Clayton field, Phelps and others (1983) decided that MON, ON, and NB are all primary, from mantle that was metasomatized less than 1 b.y. ago. These mafic magmas were coeval with silica-oversaturated magmas and with phonolites, but Phelps and others did not treat the felsic feldspathoidal or the quartz-normative rocks.

MON magma may be primitive and may assimilate lherzolite to yield ON magma, according to Mansker (1979). Alternatively, MON, ON, NB, and AB may all be products of primitive liquids derived from metasomatized garnet lherzolite, according to Clague and Frey (1982), with the compositional differences reflecting varying degrees of fusion. This argument may involve

TABLE 1. (CONTINUED)

Sample	8	7	6	5	82009	82011	4	3	2	82008	82002	1
Rock type	AB	AB	PH	PH	PH	PH	PH	PH	PH	PH	PH	PH
SiO_2	46.61	47.51	48.13	48.23	48.66	51.01	51.25	51.97	52.51	52.64	54.07	54.42
TiO_2	2.92	0.51	1.74	2.00	1.84	1.22	0.80	0.80	0.80	0.51	0.38	0.40
Al_2O_3	12.66	15.52	18.44	17.43	17.85	19.00	18.07	18.79	18.21	19.60	19.92	20.76
Fe_2O_3	2.56	3.25	3.41	2.77	3.63	3.26	3.16	2.77	2.46	1.64	3.12	2.64
FeO	8.24	5.96	4.30	5.92	3.56	2.84	3.67	3.05	3.33	3.50	1.38	1.33
MnO	0.16	---	0.19	0.18	0.16	0.14	---	---	---	0.17	0.17	0.15
MgO	7.88	5.85	3.06	2.99	2.79	1.99	0.65	0.51	0.60	0.46	0.29	0.22
CaO	10.35	14.05	5.89	6.38	5.50	3.88	3.70	2.84	2.75	2.32	1.40	1.34
Na_2O	3.43	3.21	8.00	6.87	8.16	9.30	8.98	8.70	8.84	8.93	10.16	10.41
K_2O	0.88	1.32	3.80	2.78	3.68	4.26	4.70	5.50	5.78	3.46	4.96	4.89
H_2O^+	2.85	1.89	1.59	2.84	2.75	2.06	4.56	4.00	3.70	5.37	3.46	2.50
H_2O^-	1.17	---	0.18	0.54	0.18	0.17	0.45	0.65	0.40	0.71	0.20	0.22
P_2O_5	0.52	0.89	0.49	0.69	0.46	0.32	0.24	0.22	0.13	0.20	0.09	0.11
CO_2	---	---	---	---	0.13	0.00	---	---	---	0.24	0.02	---
Total	100.23	99.69	99.22	99.62	99.35	99.35	100.23	99.80	99.51	99.75	99.62	99.39
Rb	27	---	104	---	115	150	130	200	---	102	189	---
Sr	516	---	1323	---	1088	988	514	300	---	440	275	---
Y	21	---	29	---	22	29	33	36	---	30	33	---
Zr	180	---	430	---	427	537	795	812	---	722	730	---
Nb	38	---	96	---	91	102	109	143	---	101	120	---
Ni	173	---	37	---	31	28	2	6	---	3	4	---
or	5.40	7.96	23.04	4.89	22.59	25.92	29.19	34.17	35.80	21.86	30.59	29.91
ab	26.00	13.79	14.21	41.83	19.04	17.05	15.66	15.36	14.72	41.07	25.72	26.03
an	17.20	24.55	3.27	15.30	1.26	---	---	---	---	3.15	---	---
lc	---	---	---	---	---	---	---	---	---	---	---	---
ne	2.26	7.55	29.94	10.76	28.54	31.34	29.54	29.29	26.95	21.66	28.39	30.52
ac	---	---	---	---	---	4.74	6.99	7.00	6.98	---	5.68	5.69
ns	---	---	---	---	---	---	0.40	0.01	1.40	---	1.20	0.55
di-wo	13.63	16.99	9.79	5.64	9.74	7.68	7.37	5.55	5.60	3.35	2.77	2.56
di-en	8.86	9.48	6.52	3.10	7.22	4.49	1.51	1.19	1.35	0.96	0.44	0.40
di-fs	3.83	6.83	2.55	2.33	1.58	2.82	6.38	4.74	4.58	2.54	2.57	2.39
wo	---	---	---	---	0.27	---	---	---	---	---	---	---
ol-fo	8.59	3.78	0.91	3.37	---	0.58	0.13	0.10	0.15	0.21	0.22	0.12
ol-fa	3.86	3.00	0.39	2.80	---	0.40	0.62	0.44	0.56	0.62	1.46	0.78
cs	---	---	---	---	---	---	---	---	---	---	---	---
mt	3.86	2.98	4.82	4.26	5.03	1.74	---	---	---	3.10	---	---
il	5.76	0.99	3.39	4.03	3.63	2.44	1.60	1.60	1.59	0.98	0.75	0.79
ap	1.18	1.99	1.10	1.60	1.04	0.75	0.55	0.51	0.30	0.47	0.21	0.25
TTDI	33.66	29.31	67.19	57.48	70.17	74.31	74.40	78.83	77.47	84.59	84.69	86.46
wt.% an/(an+ab)	39.82	64.03	18.71	26.78	6.20	---	---	---	---	7.12	---	---
100 Mg/(Mg+Fe*),mol	57.12	53.99	42.54	38.78	42.15	39.21	15.10	14.09	16.17	15.32	10.99	9.57

some circularity, because nephelinite magmas are among the most likely candidates as the fluids responsible for mantle metasomatism (see, for example, Menzies and Murthy, 1980, and Wright, 1984, p. 3241–3244).

According to Mitchell and Platt (1984), the Bathurst Island suite in the Canadian Arctic is petrologically close to that of the Balcones province, except that the former also contains coeval tholeiite. In the bimodal suite of Bathurst Island, the most mafic lavas are interpreted as primitive, with Ni exceeding 250 ppm and with 100 Mg/(Mg + Fe total) ranging from 65 to 75. The sequence MON-ON-NB represents increasing degree of fusion, and the basalts may have formed by even more extensive melting, possibly from the same parent as the more silica-undersaturated magmas. The Bathurst Island model is therefore essentially the same as that proposed by Clague and Frey (1982) for the Oahu suite. Mitchell and Platt concluded that the phonolites formed by crystal-liquid fractionation from a nephelinite magma, not from basanite or basalt, because the phonolites lack the negative europium anomaly that would be imposed by fractionation from a parent containing modal plagioclase. Clague and Frey also found no negative Eu anomaly in rocks of the Oahu suite.

In contrast, Aurisicchio and others (1983) in eastern Kenya, and Wilkinson and Stolz (1983) on Oahu, concluded that phonolite must arise by low-pressure fractional crystallization of basanite magma. However, neither paper presents rare-earth data to document the presence or absence of a negative europium anomaly. Wilkinson and Stolz (1983, p. 363) find "no support for the proposal that olivine-melilite-nephelinite ultimately may fractionate to phonolite. Phonolitic differentiates can be generated by the low-pressure fractionation of alkali feldspar–bearing olivine

TABLE 1. (NOTE AND EXPLANATION OF COLUMN HEADS)

Note: Samples are from Uvalde County unless otherwise specified. Trace elements analyzed by x-ray fluorescence spectrometry by D. S. Barker and B. R. Pyle. Trace element data accompanying major element analyses from Spencer (1969) were obtained from Spencer's original samples; those accompanying analyses from Lonsdale (1927) and Clarke (1900) are from other samples from the same localities. CIPW norms, in wt.%, calculated volatile-free with Fe_2O_3 = TiO_2 + 1.5 (Irvine and Baragar, 1971). TTDI = differentiation index of Thornton and Tuttle (1960).

Explanation of column headings:

28	MON, Knippa Quarry (Carter, 1965).
27	MON, Houston Ranch (Spencer, 1969).
26	MON, "near Uvalde" (Clarke, 1900).
25	MON, Houston Ranch (Spencer, 1969).
82021	MON, Knippa Quarry. Major elements by G. K. Hoops.
24	MON, Houston Ranch (Spencer, 1969).
82005	MON, Taylor Hills, Major elements by G. K. Hoops.
23	MON, Langner Ranch (Spencer, 1969).
44845	ON, Pilot Knob, Travis County. Major elements by G. K. Hoops (Barker and Young, 1979).
22	MON, Taylor Hills (Spencer, 1969).
21	ON, Obi Hill (Spencer, 1969).
20	ON, Black Mountain, Kennedy Ranch (Clarke, 1900).
19	ON, Tom Nunn Hill, Allen Ranch (Clarke, 1900).
18	ON, Allen (Cantrell) Mountain, Allen Ranch (Spencer, 1969).
82012	ON, 3.3 mi west of Sabinal. Major elements by G. K. Hoops.
44841	AB, Pilot Knob, Travis County. Major elements by G. K. Hoops (Barker and Young, 1979).
44837	NB, Pilot Knob, Travis County. Major elements by G. K. Hoops (Barker and Young, 1979).
17	ON, Allen Ranch (Spencer, 1969).
16	ON, intrusion between Twin Peaks, Houston Ranch (Spencer, 1969).
44832	AB, Pilot Knob, Travis County. Major elements by G. K. Hoops (Barker and Young, 1979).
15	NB, intrusion 1/4 mi south of Sulphur Mountain, Allen Ranch (Spencer, 1969).
14	NB, Allen Ranch (Spencer, 1969).
44847	AB, Pilot Knob, Travis County. Major elements by G. K. Hoops (Barker and Young, 1979).
13	NB, Weymiller Butte, Allen Ranch (Spencer, 1969).
12	AB, Mustang Hill, Harris Ranch (Greenwood and Lynch, 1959).
11	AB, Pinto Mountain, Kinney County (Clarke, 1900).
10	NB, Gerdes Ranch (Spencer, 1969).
9	AB, Green Mountain (Lonsdale, 1927).
8	AB, Allen Ranch (Spencer, 1969).
7	AB, Green Mountain (Lonsdale, 1927).
6	PH, Inge Mountain (Lonsdale, 1927).
5	PH, near Big Mountain, Kennedy Ranch (Clarke, 1900).
82009	PH, Inge Mountain. Major elements by G. K. Hoops.
82011	PH, Inge Mountain. Major elements by G. K. Hoops.
4	PH, Connors Ranch (Lonsdale, 1927).
3	PH, Ange Siding (Lonsdale, 1927).
2	PH, 2 mi south of Black Water Hole, Houston Ranch (Lonsdale, 1927).
82008	PH, Connors Ranch. Major elements by G. K. Hoops.
82002	PH, Ange Siding. Major elements by G. K. Hoops.
1	PH, intrusion between Black and Big mountains, Kennedy Ranch (Clarke, 1900).

nephelinites but the relative volumes of the salic derivatives are minor. These volumetric and other constraints inherent in low-pressure fractionation models employing nephelinitic parents suggest that at least some phonolites may be products of deep crustal or upper mantle anatexis."

In the Laacher See district, Duda and Schmincke (1978, p. 22–27) found that basanites and nephelinites are not related by low-pressure fractionation. They concluded that nephelinite and basanite magmas were both primitive, and that basanite was parental to phonolite (see also Schmincke and others, 1983). Worner and Schmincke (1984) confirmed that nephelinites, as assumed parents of phonolites, do not yield satisfactory solutions in fractionation models, but separation of amphibole and clinopyroxene from basanite magma deep in the crust could produce phonolitic liquids.

At Otago, New Zealand, Coombs and Wilkinson (1969) identified three lines of descent by crystal fractionation: basanite-hawaiite-mugearite-benmoreite-phonolite; alkali basalt-hawaiite-mugearite-benmoreite-trachyte; and nephelinites.

Irving and Price (1981), in an overview of several occurrences of phonolite with more silica-undersaturated mafic associates, concluded that basanite fractionates to yield phonolite at high pressure; only one phonolite sample (from Oahu) showed a negative Eu anomaly, however.

In the Balcones province, Spencer (1969) concluded that ON magma was primary, and yielded MON, NB, and PH by crystal-liquid fractionation. AB formed from a separate liquid.

PETROGENESIS

Some, but not all, samples of Balcones MON, ON, and NB have $Mg/(Mg + Fe^{2+})$ ratios and Ni contents (Fig. 4) high enough to represent liquids equilibrated with upper mantle peridotite (Brey, 1978; Frey and others, 1978; Wass, 1980), and all three types contain ultramafic xenoliths. There is little doubt that at least one of these three most mafic magmas was derived directly by partial fusion of upper mantle rock, with some or all undergoing varying degrees of subsequent fractionation. AB has

TABLE 2. ADDITIONAL TRACE ELEMENT DATA, BALCONES ROCKS*

Analysis	Knippa Quarry/ MON (6?)	
Li	31.9	(c)
Sc	22.28, 24.75	(e)
Cr	394, 511	(e)
Co	76.8	(a)
	70, 69	(e)
Cu	15.0	(a)
Zn	125	(a)
Ga	20.2	(a)
Br (ppb)	1300	(a)
Rb	26	(a)
	27.9	(b)
	27.84	(c)
	20.5	(d)
Sr	1019	(b)
	1012	(c)
	710	(d)
In (ppb)	115	(a)
Cs (ppb)	745	(a)
	779	(c)
Ba	672	(b)
	645	(c)
	955, 1029	(e)
La	55	(b)
	59, 51	(e)
Ce	111	(b)
	122, 106	(e)
Nd	54	(b)
	43, 51	(e)
Sm	11.2	(b)
	12.1, 10.8	(e)
Eu	3.42	(b)
	3.77, 3.51	(e)
Gd	9.30	(b)
Tb	1.32, 1.11	(e)
Dy	6.18	(b)
Er	2.83	(b)
Yb	1.71	(b)
	1.77, 1.88	(e)
Lu	0.22	(b)
	0.24, 0.23	(e)
Hf	6.70, 6.93	(e)
Ta	6.00, 5.42	(e)
Au (ppb)	2.85	(a)
Tl (ppb)	20.3	(a)
Bi (ppb)	88.3	(a)
Th	7.85, 7.77	(e)

	Locality/Rock Type (no. of samples)						
Analysis	Taylor Hills/ MON (2)	Black Waterhole/ MON (1)	3.3 mi W of Sabinal/ ON (1)	Connor Ranch/ ON (1)	Connor Ranch/ PH (1)	Inge Mtn/ PH (1)	Ange Siding PH (1)
Sc	22.19	21.88	25.56	15.83	2.58	7.17	0.57
	26.29						
Cr	351	445	688	253	21	53	8
	532						
Co	61	64	92	49	22	28	28
	70						
Ba	653	682	629	690	1030	883	383
	432						
La	59	53	44	52	89	77	101
	47						
Ce	100	92	74	92	136	118	131
	81						
Nd	50	43	36	40	43	44	41
	39						
Sm	10.4	9.0	7.6	8.8	7.8	8.2	7.2
	8.5						
Eu	3.45	2.96	2.35	2.86	2.52	2.71	1.77
	2.63						
Tb	1.10	0.99	0.84	1.03	0.92	0.88	0.86
	0.93						
Yb	1.60	1.43	1.48	2.10	2.84	2.08	3.47
	1.57						
Lu	0.21	0.20	0.24	0.28	0.39	0.29	0.53
	0.22						
Hf	6.08	5.37	2.39	7.06	12.2	7.90	13.6
	5.21						
Ta	6.06	4.80	4.43	6.41	9.90	7.12	9.78
	5.83						
Th	6.65	5.94	4.49	5.62	13.6	10.6	24.1
	5.66						

*Letters in parentheses indicate references:
(a) Ganapathy and others (1970);
(b) Gast and others (1970);
(c) Tera and others (1970);
(d) Hedge (1966);
(e) McKay (1984).
All data in localities other than Knippe Quarry are from McKay (1984). Data expressed in parts per million unless otherwise specified.

not been found to contain ultramafic xenoliths, but it does have Mg/(Mg + Fe^{2+}) ratios and Ni contents overlapping those of the three most mafic types (Fig. 4). The lower average Ni contents of NB + AB, relative to MON + ON, are probably caused by olivine removal. The question of identifying a single primitive liquid to be the parent for all members of the Balcones suite seems moot, in the light of these data.

On a plot of niobium versus nickel (Fig. 5), NB and AB have lower Nb and Ni than do MON and ON. These two trace elements differ significantly in behavior; Ni is strongly partitioned into olivine relative to silicate liquids, but Nb is partitioned into liquid relative to any abundant crystalline phase. The coupled decreases in both elements must reflect two superimposed processes. At constant pressure, decreasing silica undersaturation from MON through ON and NB to AB should result if progressively larger liquid fractions were reached before each magma separated from its parent after partial fusion. Dilution of strongly incompatible elements (those with crystal/liquid partition coefficients very close to zero) would then be expected. More extensive partial fusion in the mafic sequence from MON and ON to NB and AB diluted niobium in the less silica-undersaturated liquids. Subsequently, fractional removal of olivine from ascending liquids removed nickel. PH samples show, relative to the mafic rocks, an increase in Nb and a decrease in Ni (Fig. 5), both attributable to crystal fractionation during generation of PH liquid.

All four mafic types show the same ranges for Zr and Rb (Fig. 6), but PH is much richer in Zr and Rb. Some analyzed Balcones phonolites are peralkaline, and peralkaline silicate liquids dissolve more Zr before becoming saturated with a zirconium-rich crystalline phase (Watson, 1979). Neither zircon nor eudialyte has yet been identified in PH.

A plot of Rb/Zr versus Rb (Fig. 7) is an example of a

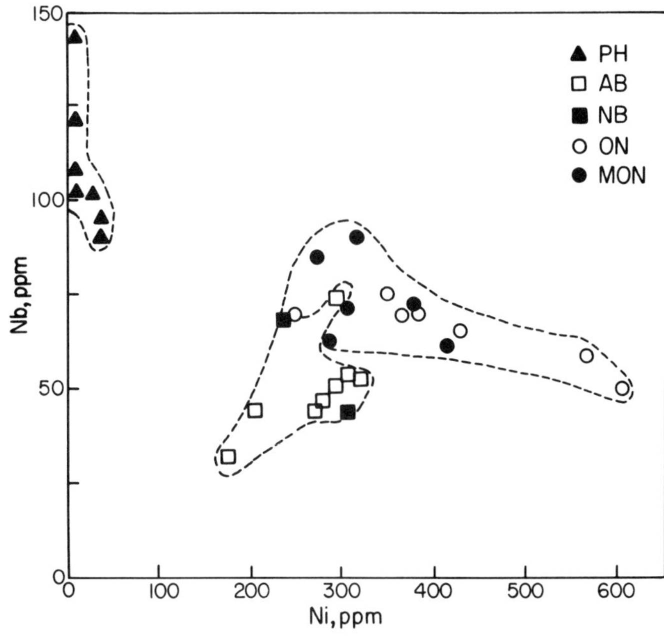

Figure 5. Niobium versus nickel (ppm).

diagram used by Minster and Allegre (1978, Fig. 2) employing the ratio of a strongly incompatible element to a less incompatible element versus the concentration of the strongly incompatible element. The four mafic types cluster in a steep trend typical of that produced by partial fusion, whereas PH samples show at least two trends with gentle slopes typical of fractional crystallization and diverging from the mafic trend at different values of Rb/Zr. These relationships suggest that phonolites may have formed by fractional crystallization from more than one mafic parent.

All four mafic types are possible parents, and interstitial glass in AB at Pilot Knob in the Austin segment does have a major element composition (Table 3) approximating that of PH (Barker and Young, 1979). However, PH is not known in the Austin segment, and in the Uvalde segment it forms isolated plugs, not spatially associated with any one of the other types (Fig. 2), and carries megacrysts of kaersutitic amphibole and xenoliths of Ouachita metasedimentary rocks, indicating derivation of PH at depth, not by filter-pressing of residual liquid from near-surface magma near its solidus.

Figure 8 shows chondrite-normalized rare-earth profiles for Balcones MON, ON, and PH. Profiles are linear for MON, slightly concave upward for ON, and more strongly concave upward for PH. For MON, ON, and PH, La concentrations are, respectively, 130 to 200, 130 to 150, and 210 to 260 times average chondrite; La/Yb ratios, chondrite-normalized, are, respectively, 16 to 22, 15 to 18, and 14 to 21. PH shows no europium anomaly, ruling out fractional removal of PH liquid from the plagioclase-bearing NB or AB magmas. Depletion in

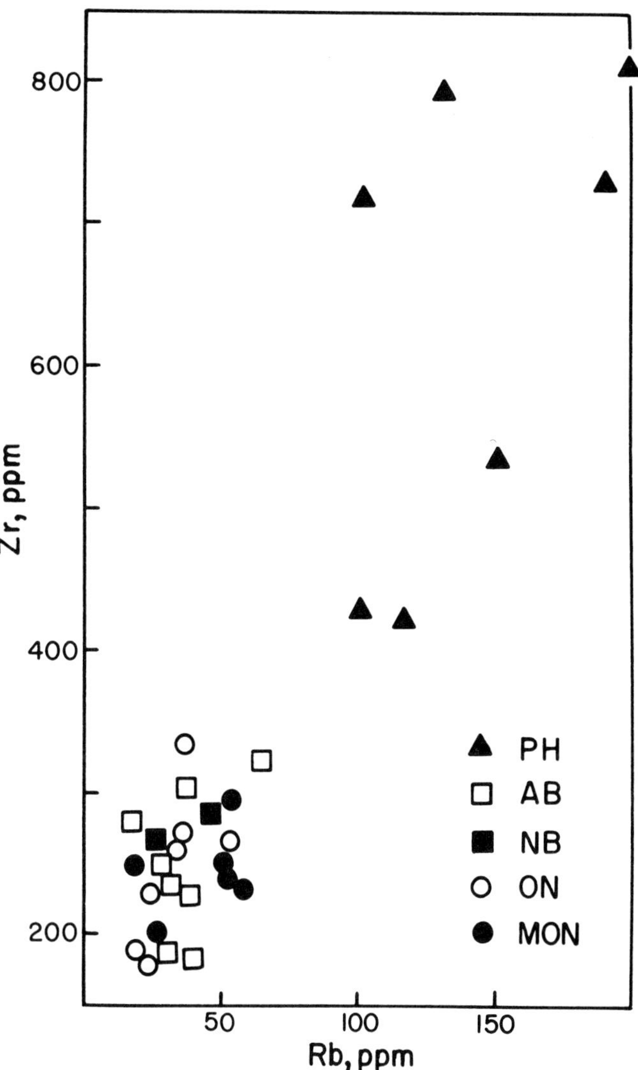

Figure 6. Zirconium versus rubidium (ppm).

middle rare-earth elements suggests that PH liquid may have formed by fractional removal of amphibole or sphene (both of which are megacryst phases in PH from either MON or ON.

A least-squares mixing model (LeMaitre, 1981) using either MON or ON as parents, PH as daughter, and amphibole, clinopyroxene, olivine, apatite, and sphene as fractionating phases, yields a best fit among all possible combinations with ON as parent, deriving 13 wt. % PH by removal of 31 wt. % amphibole, 34 wt. % clinopyroxene, and 22 wt. % olivine. The proportion of amphibole seems excessive.

Figure 9 shows concentrations of incompatible elements (more incompatible toward the left), normalized to chondrites with the exception of K, Rb, and P, using the format and normalizing values of Thompson and others (1983). The four mafic types form a coherent array, in contrast to PH, and are less enriched in Rb, K, Sr and Y than in Nb, P, and Ti. Average

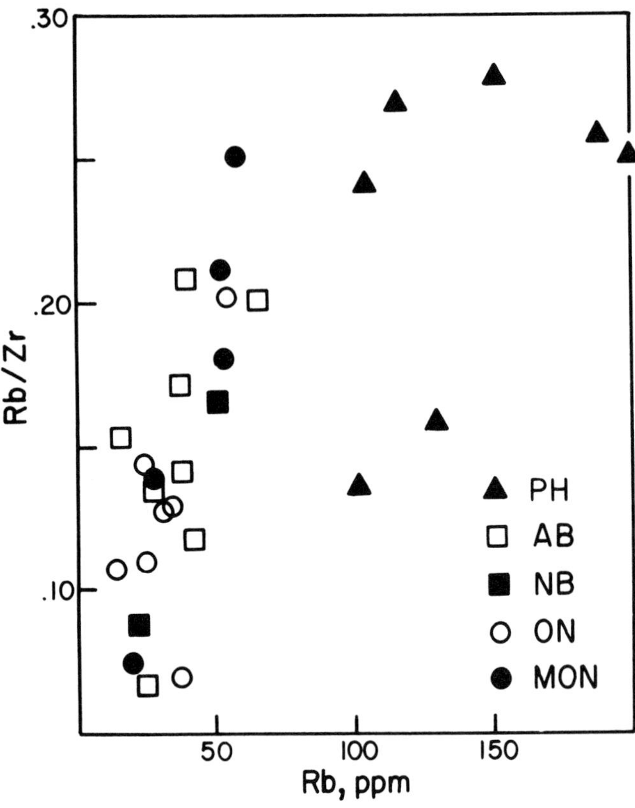

Figure 7. Rb/Zr versus rubidium (ppm).

TABLE 3. COMPARISON OF AVERAGE PHONOLITE COMPOSITION WITH INTERSTITIAL GLASS IN ALKALI BASALT*

	Phonolite (average of 10)	Glass in Alkali Basalt 44841	44832
SiO_2	53.9	56.8	58.7
TiO_2	1.10	1.29	1.24
Al_2O_3	19.8	22.8	22.3
FeO*	5.94	4.77	4.70
MnO	0.18	0.11	0.07
MgO	1.43	1.32	0.74
CaO	3.78	1.88	1.43
Na_2O	9.29	7.51	7.36
K_2O	4.60	3.22	3.44

*Phonolite average from data in Table 1. Glass compositions are microprobe analyses in two samples from Pilot Knob, Austin segment (Barker and Young, 1979, p. 15). All data normalized to 100 wt.%, volatile-free.

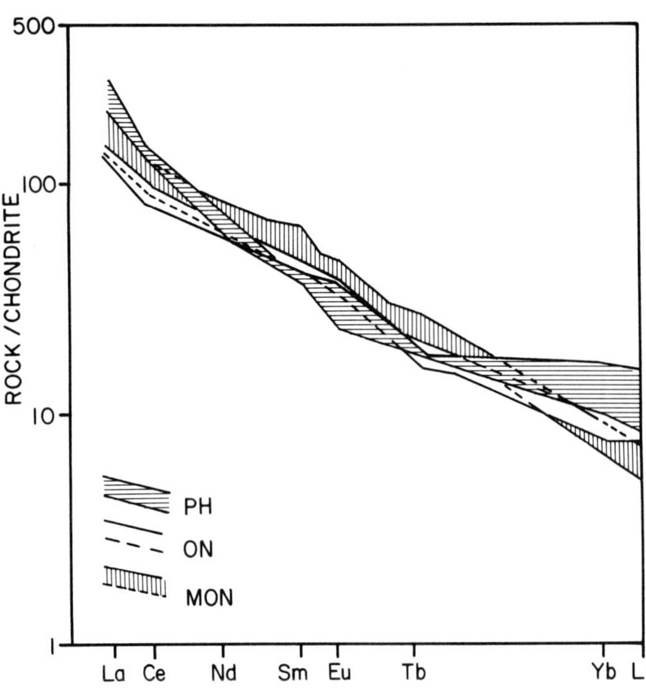

Figure 8. Chondrite-normalized rare-earth element profiles of melilite-olivine nephelinite (five samples), olivine nephelinite (two samples), and phonolite (three samples); data in Table 2.

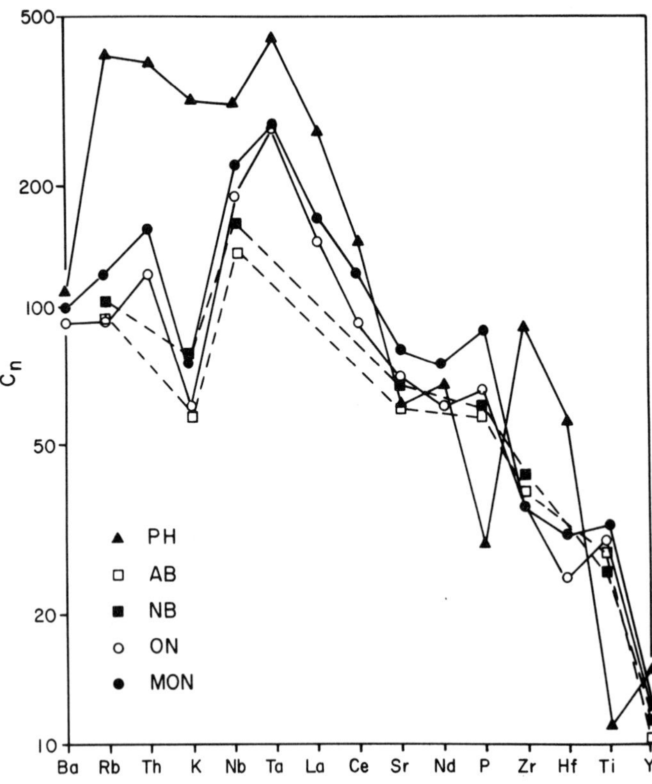

Figure 9. Incompatible elements normalized and arranged after Thompson and others (1983), for average of each Balcones rock type; data from Tables 1 and 2. For nepheline basanite and alkali basalt (dashed lines), data are available only for Rb, K, Nb, Sr, P, Zr, Ti, and Y. C_n indicates normalized value.

concentrations of Nb, Sr, and P decrease in the order MON, ON, NB, AB, but ratios among these elements (for example, P/Sr, P/Nb, and Sr/Nb) show only small variation from one rock type to another. Increasing degrees of partial fusion in the sequence from MON to AB can explain these relationships. Other incompatible elements, with the possible exception of Zr, also tend to decrease from MON to AB, but less regularly. Some of the dispersion is probably caused by averaging small numbers of analyses for each rock type. For all elements in Figure 9 except Zr, the ratio of concentration in MON to that in ON varies only from 1.04 to 1.37.

The strong coherence of all four mafic rock types in Figure 9 suggests that all were ultimately derived from one mantle source. We conclude that MON, ON, NB, and AB magmas were generated at increasing degrees of partial fusion; the decreases in incompatible elements, according to the data now available, contradict a model in which three of the mafic types are derived from the fourth by fractionation during ascent, although the decrease in Ni does suggest some removal of olivine.

CONCLUSIONS

A strongly bimodal suite of melilite-olivine nephelinite, olivine nephelinite, nepheline basanite, and alkali basalt as one group, with phonolite as the lone felsic type, characterizes the Balcones province. The igneous rocks show superimposed effects of partial fusion and fractional crystallization, and represent an extreme kind of magmatism that is perhaps symptomatic of stability in the overlying crust. The four mafic types could all have formed by varying (but small) degrees of fusion of an upper mantle source. Melilite-olivine nephelinite and olivine nephelinite form a chemically coherent subsuite, as do nepheline basanite and alkali basalt. Probably the latter pair represents liquids derived by a higher extent of fusion, which were modified by olivine removal. Phonolite probably is a product of extreme fractionation from one or more mafic parents.

Additional data needed to resolve the petrogenetic problems include Sr initial isotope ratios and more electron probe microanalyses of phenocryst and groundmass phases, including residual glasses (all in progress), trace element analyses of nepheline basanite and alkali basalt, and K-Ar ages, especially for phonolite.

ACKNOWLEDGMENTS

This research was supported by the Department of Geological Sciences, University of Texas at Austin, and by Lakehead University. Suggestions by G. Byerly, T. E. Ewing, W. L. Mansker, F. W. McDowell, H.-U. Schmincke, and D. Smith have been helpful, but we take full responsibility for any errors in fact or interpretation. The manuscript was improved through the suggestions of Calvin G. Barnes and an anonymous reviewer.

REFERENCES CITED

Aurisicchio, C., Brotzu, P., Morbidelli, L., Piccirillo, E. M., and Traversa, G., 1983, Basanite to peralkaline phonolite suite; Quantitative crystal fractionation model (Nyambeni range, East Kenya Plateau): Neues Jahrbuch für Mineralogie, Abhandlungen, v. 148, p. 113–140.

Baldwin, O. D., and Adams, J.A.S., 1971, K^{40}/Ar^{40} ages of the alkalic igneous rocks of the Balcones fault trend of Texas: Texas Journal of Science, v. 22, p. 223–231.

Barker, D. S., and Young, K. P., 1979, A marine Cretaceous nepheline basanite volcano at Austin, Texas: Texas Journal of Science, v. 31, p. 5–24.

Barnes, V. E., project director, 1974, San Antonio sheet, Geologic Atlas of Texas: Austin, University of Texas, Bureau of Economic Geology, scale 1:250,000, one sheet with explanatory text.

—— , 1977, Del Rio Sheet, Geologic Atlas of Texas: Austin, University of Texas, Bureau of Economic Geology: scale 1:250,000, one sheet with explanatory text.

Barr, S. M., and Macdonald, A. S., 1981, Geochemistry and geochronology of late Cenozoic basalts of Southeast Asia: Geological Society of America Bulletin, v. 92, pt. 2, p. 1069–1142.

Brey, G., 1978, Origin of olivine melilitites; Chemical and experimental constraints: Journal of Volcanology and Geothermal Research, v. 3, p. 61–88.

Brey, G., and Green, D. H., 1977, Systematic study of liquidus phase relations in olivine melilitite + H_2O + CO_2 at high pressures and petrogenesis of an olivine melilitite magma: Contributions to Mineralogy and Petrology, v. 61, p. 141–162.

Burke, W. H., Otto, J. B., and Denison, R. E., 1969, Potassium-argon dating of basaltic rocks: Journal of Geophysical Research, v. 74, p. 1082–1086.

Cameron, K. L., and Cameron, M., 1973, Mineralogy of ultramafic nodules from Knippa Quarry, near Uvalde, Texas: Geological Society of America Abstracts with Programs, v. 5, p. 566.

Carter, J. L., 1965, The origin of olivine bombs and related inclusions in basalt [Ph.D. thesis]: Houston, Rice University.

Clague, D. L., and Frey, F. A., 1982, Petrology and trace element geochemistry of the Honolulu Volcanics, Oahu; Implications for the oceanic mantle below Hawaii: Journal of Petrology, v. 23, p. 447–504.

Clarke, F. W., 1900, Analyses of rocks from the laboratory of the U.S. Geological Survey, 1880–1899: U.S. Geological Survey Bulletin 168.

Coombs, D. S., and Wilkinson, J.F.G., 1969, Lineages and fractionation trends in undersaturated volcanic rocks from the East Otago Volcanic Province (New Zealand) and related types: Journal of Petrology, v. 10, p. 440–501.

Duda, A., and Schmincke, H.-U., 1978, Quaternary basanites, melilite nephelinites and tephrites from the Laacher See area (Germany): Neues Jahrbuch für Mineralogie, Abhandlungen, v. 132, p. 1–33.

Ewing, T. E., and Caran, S. C., 1982, Late Cretaceous volcanism in south and central Texas; Stratigraphic, structural, and seismic models: Gulf Coast Association of Geological Societies Transactions, v. 32, p. 137–145.

Ferguson, J., and Sheraton, J. W., 1979, Petrogenesis of kimberlitic rocks and associated xenoliths of southeastern Australia, *in* Boyd, F. R., and Meyer, H.O.A., eds., Kimberlites, diatremes, and diamonds; Their geology, petrology, and geochemistry; Proceedings of the Second International Kimberlite Conference: Washington, D.C., American Geophysical Union, p. 140–160.

Frey, F. A., Green, D. H., and Roy, S. D., 1978, Integrated models of basalt petrogenesis; A study of quartz tholeiites to olivine melilitites from southeastern Australia utilizing geochemical and experimental petrological data: Journal of Petrology, v. 19, p. 463–513.

Fuchs, K., von Gehlen, K., Malzer, H., Murawski, H., and Semmel, A., eds., 1983, Plateau uplift: Berlin, Springer-Verlag, 411 p.

Ganapathy, R., Keays, R. R., Laul, J. C., and Anders, E., 1970, Trace elements in Apollo 11 lunar rocks; Implications for meteorite influx and origin of moon: Apollo 11 Lunar Science Conference Proceedings, v. 2, p. 1117–1142.

Gast, P. W., Hubbard, N. J., and Wiesmann, H., 1970, Chemical composition and petrogenesis of basalts from Tranquility Base: Apollo 11 Lunar Science Conference Proceedings, v. 2, p. 1143–1163.

Greenwood, R., and Lynch, V. M., 1959, Geology and gravimetry of the Mustang Hill laccolith, Uvalde County, Texas: Geological Society of America Bulletin, v. 70, p. 807–825.

Hedge, C. E., 1966, Variations in radiogenic strontium found in volcanic rocks: Journal of Geophysical Research, v. 71, p. 6119–6126.

Irvine, T. N., and Baragar, W.R.A., 1971, A guide to the chemical classification of the common volcanic rocks: Canadian Journal of Earth Sciences, v. 8, p. 523–548.

Irving, A. J., and Price, R. C., 1981, Geochemistry and evolution of lherzolite-bearing phonolitic lavas from Nigeria, Australia, East Germany, and New Zealand: Geochimica et Cosmochimica Acta, v. 45, p. 1309–1320.

Kokelaar, B. P., and Durant, G. P., 1983, The submarine eruption and erosion of Surtla (Surtey), Iceland: Journal of Volcanology and Geothermal Research, v. 19, p. 239–246.

LeBas, M. J., 1978, Nephelinite volcanism at plate interiors: Bulletin Volcanologique, v. 41, p. 459–462.

LeMaitre, R. W., 1981, GENMIX; A generalized petrological mixing model program: Computers and Geosciences, v. 7, p. 229–247.

Lonsdale, J. T., 1927, Igneous rocks of the Balcones fault region of Texas: University of Texas Bulletin 2744, 178 p.

Mansker, W. L., 1979, Petrogenesis of nephelinites and melilite nephelinites from Oahu, Hawaii [Ph.D. thesis]: Albuquerque, University of New Mexico.

McIver, J. R., 1981, Aspects of ultrabasic and basic alkaline intrusive rocks from Bitterfontein, South Africa: Contributions to Mineralogy and Petrology, v. 78, p. 1–11.

McKay, D., 1984, Application of ion exchange to neutron activation analysis of REE in geologic samples (a comparative study) [honours thesis]: Thunder Bay, Canada, Lakehead University.

Menzies, M., and Murthy, V. R., 1980, Mantle metasomatism as a precursor to the genesis of alkaline magmas; Isotopic evidence: American Journal of Science, v. 280-A, p. 622–638.

Minster, J. F., and Allegre, C. J., 1978, Systematic use of trace elements in igneous processes; Pt. III, Inverse problem of batch partial melting in volcanic suites: Contributions to Mineralogy and Petrology, v. 68, p. 37–52.

Mitchell, R. H., and Platt, R. G., 1984, The Freemans Cove volcanic suite; Field relations, petrochemistry, and tectonic setting of nephelinite-basanite volcanism associated with rifting in the Canadian Arctic Archipelago: Canadian Journal of Earth Sciences, v. 21, p. 428–436.

Nielsen, T.F.D., 1980, The petrology of a melilitolite, melteigite, carbonatite and syenite ring dyke system in the Gardiner complex, East Greenland: Lithos, v. 13, p. 181–197.

Olafsson, M., and Eggler, D. H., 1983, Phase relations of amphibole, amphibole-carbonate, and phlogopite-carbonate peridotite; Petrologic constraints on the asthenosphere: Earth and Planetary Science Letters, v. 64, p. 305–315.

Palmer, A. R., 1983, The Decade of North American Geology 1983 geologic time scale: Geology, v. 11, p. 503–504.

Phelps, D. W., Gust, D. A., and Wooden, J. L., 1983, Petrogenesis of the mafic feldspathoidal lavas of the Raton-Clayton volcanic field, New Mexico: Contributions to Mineralogy and Petrology, v. 84, p. 182–190.

Schmincke, H.-U., Lorenz, V., and Seck, H. A., 1983, The Quaternary Eifel volcanic fields, *in* Fuchs, K., von Gehlen, K., Malzer, H., Murawski, H., and Semmel, A., eds., Plateau uplift: Berlin, Springer-Verlag, p. 139–151.

Spencer, A. B., 1969, Alkalic igneous rocks of the Balcones Province, Texas: Journal of Petrology, v. 10, p. 272–306.

Tera, F., Eugster, O., Burnett, D. S., and Wasserburg, G. J., 1970, Comparative study of Li, Na, K, Rb, Cs, Ca, Sr, and Ba abundances in achondrites and in Apollo 11 lunar samples: Apollo 11 Lunar Science Conference Proceedings, v. 2, p. 1637–1657.

Thompson, R. N., Morrison, M. A., Dickin, A. P., and Hendry, G. L., 1983, Continental flood basalts; Arachnids rule OK?, *in* Hawkesworth, C. J., and Norry, M. J., eds., Continental basalts and mantle xenoliths: Nantwich, United Kingdom, Shiva Publishing, p. 158–185.

Thornton, C. P., and Tuttle, O. F., 1960, Chemistry of igneous rocks; 1. Differentiation index: American Journal of Science, v. 258, p. 664–684.

Wass, S. Y., 1980, Geochemistry and origin of xenolith-bearing and related alkali basaltic rocks from the Southern Highlands, New South Wales, Australia: American Journal of Science, v. 280-A, p. 639–666.

Watson, E. B., 1979, Zircon saturation in felsic liquids; Experimental results and applications to trace element geochemistry: Contributions to Mineralogy and Petrology, v. 70, p. 407–419.

Wilkinson, J.F.G., and Stolz, A. J., 1983, Low pressure fractionation of strongly undersaturated alkaline ultrabasic magma; The olivine-melilite-nephelinite at Moilili, Oahu, Hawaii: Contributions to Mineralogy and Petrology, v. 83, p. 363–374.

Williams, L.A.J., 1970, The volcanics of the Gregory Rift valley, East Africa: Bulletin Volcanologique, v. 34, p. 439–465.

Wilshire, H. G., and Shervais, J. W., 1975, Al-augite and Cr-diopside ultramafic xenoliths in basaltic rocks from western United States: Physics and Chemistry of the Earth, v. 9, p. 257–272.

Worner, G., and Schmincke, H.-U., 1984, Petrogenesis of the zoned Laacher See tephra: Journal of Petrology, v. 25, p. 836–851.

Worner, G., Beusen, J.-M., Duchateau, N., Gijbels, R., and Schmincke, H.-U., 1983, Trace element abundances and mineral/melt distribution coefficients in phonolites from the Laacher See Volcano (Germany): Contributions to Mineralogy and Petrology, v. 84, p. 152–173.

Wright, T. L., 1984, Origin of Hawaiian tholeiite; A metasomatic model: Journal of Geophysical Research, v. 89, p. 3233–3252.

Yagi, K., and Onuma, K., 1978, Genesis and differentiation of nephelinitic magma: Bulletin Volcanologique, v. 41, p. 466–472.

MANUSCRIPT ACCEPTED BY THE SOCIETY OCTOBER 15, 1986

Printed in U.S.A.

Geological Society of America
Special Paper 215
1987

Slickrock Mountain intrusive complex, Big Bend National Park, Texas

Elizabeth Hill and Calvin G. Barnes
Department of Geosciences, Texas Tech University, Lubbock, Texas 79409

ABSTRACT

Slickrock Mountain consists of two major sills intruded into Upper Cretaceous calcareous sandstone. The sills are an older Eocene quartz syenite approximately 200 m thick and a younger Oligocene alkali olivine gabbro approximately 100 m thick. Smaller mafic dikes and sills are abundant in the area. The quartz syenite contains fayalitic olivine and zoned alkali feldspar with plagioclase cores set in an intergranular groundmass of anorthoclase, sanidine, quartz, clinopyroxene, magnetite, and apatite. The smaller mafic intrusions commonly contain kaersutite megacrysts and an assemblage of crustal xenoliths that includes coarse-grained gabbro and partially fused feldspathic arenite. Major and trace element data do not support the derivation of the felsic compositions from the mafic ones by simple crystal fractionation at low pressure. The presence of kaersutite megacrysts and partially fused crustal xenoliths in nearby sills and dikes suggests that derivation of the Slickrock quartz syenite from some combination of fractionation and crustal assimilation at elevated pH_2O is possible. However, trace element concentrations indicate that the syenitic and gabbroic rocks are petrogenetically unrelated.

INTRODUCTION

Slickrock Mountain is one of numerous, small intrusive complexes in Big Bend National Park, Texas (Fig. 1). These intrusions lie within the northwest-southeast–trending Trans-Pecos magmatic province of Texas and New Mexico (Barker, 1977). The province is characterized by both silica-saturated and silica-undersaturated alkalic igneous rocks that range in composition from alkalic basalt to high-silica rhyolite and nepheline trachyte (Maxwell and others, 1967). Potassium-argon dating (McDowell, 1981) has determined that volcanic and intrusive activity in the Trans-Pecos began about 50 Ma and continued until 17 Ma. Activity peaked between 37 and 33 Ma.

The Trans-Pecos is considered by some to be the easternmost expression of volcanic activity associated with North American–Pacific convergence. This activity migrated eastward about 110 Ma and eventually reached Texas (Cameron and Cameron, 1985; Henry and Price, 1984). The province also has been interpreted as a southern extension of the Rio Grande Rift, as similar alkalic igneous activity is commonly associated with a rifting environment (Barker, 1979a). For example, in the Kenya Rift, as in the Trans-Pecos, silica-oversaturated and silica-undersaturated rock types coexist (Barker, 1977).

The presence of both silica-oversaturated and silica-undersaturated rocks in the Slickrock Mountain area and in the Christmas Mountains to the north (Fig. 1; Jungysuk, 1977) suggests a comagmatic relationship. This relationship could be due to the presence of two or more distinct parental magmas (Barker, 1979b), or it could be due to crystal-liquid fractionation, at least at a condition of high pressure.

The purpose of this study is to determine the magmatic history of the Slickrock Mountain intrusions and to evaluate the genetic association, if any, among them.

FIELD RELATIONS

Slickrock Mountain lies in the northern part of Big Bend National Park (Fig. 1). It is approximately 3.2 km southwest of the Christmas Mountains and due west of Croton Peak. The intrusive rocks of the area consist of a quartz syenite sill and numerous gabbroic sills and dikes. The quartz syenite is the largest intrusion; it underlies somewhat more than half the study area and forms Slickrock Mountain proper (Fig. 2). The syenite locally crops out beneath a sheet-like olivine gabbro body that is

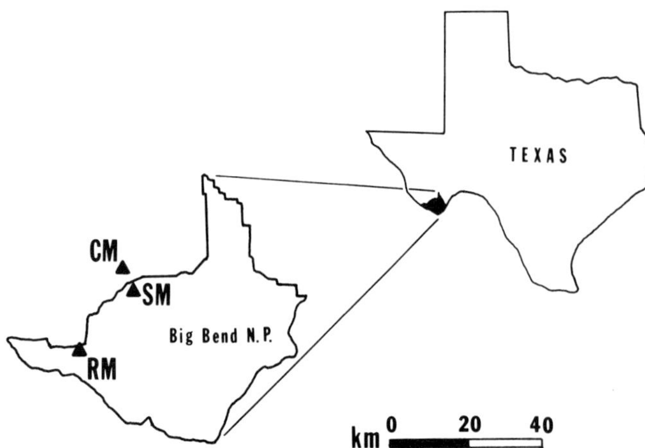

Figure 1. Location map showing outline of Big Bend National Park and location of following intrusive centers: SM, Slickrock Mountain; RM, Rattlesnake Mountain (Carman and others, 1975); CM, Christmas Mountains.

about one-third the size of the quartz syenite and trends east-west, perpendicular to the trend of the quartz syenite. Inferred crosscutting relations are described below. Another smaller gabbroic body lies adjacent to the Slickrock sill on the southwest corner of the area (Fig. 2). Other sills and dikes of gabbroic and syenitic composition crop out around the western and southern edge of Slickrock Mountain. The sills dip east to northeast about 12°.

The host sedimentary strata are part of the Upper Cretaceous (Gulfian) Aguja Formation. Locally, the Aguja is approximately 280 m thick and consists of three units (Maxwell and others, 1967): a basal sandstone that unconformably overlies the Cretaceous Pen Formation; a marine sandstone and claystone that grades upward into the third unit, a nonmarine sandstone and shale sequence with sparse lignite seams and fresh-water limestone (Maxwell and others, 1967).

The upper unit of the Aguja is host to the Slickrock Mountain intrusions. In the Slickrock Mountain area, the lowest exposed part of the upper Aguja is a calcareous claystone with lignite seams, yellow to gray in color, that crops out at the southwest base of Slickrock Mountain. The claystone grades into a sequence of alternating sandstone and sandy shale and then to quartz arenite that lies at the base of the quartz syenite. The lithology of the roof rocks of the quartz syenite was difficult to determine because erosion and colluvium have obscured this information. However, on the southeast side of Slickrock Mountain, a dark gray fossiliferous limestone is exposed near the top of the quartz syenite. The olivine gabbro intruded cross-bedded quartz arenite (Fig. 3). At the south side of this gabbro the roof rock consists of a resistant herringbone cross-bedded arenite. Very probably the olivine gabbro is a discordant body. It appears to dip more steeply and in a different direction than the sediments it intruded (south instead of east-northeast). Colluvial cover has concealed many contacts, but where they are exposed no visible contact metamorphism was observed. The thermal effects of the olivine gabbro on the host typically are seen as a red, indurated zone in the adjacent sandstone.

Crosscutting relations indicate that the quartz syenite is the oldest intrusion. In the southeastern part of the area, the quartz syenite is cut by olivine gabbro. The gabbro appears to extend beneath the syenite, as suggested by Maxwell and others (1967) (Fig. 3).

The intrusive relations between the large olivine gabbro sheet and the quartz syenite have been masked by colluvium. Evidence that gabbro was emplaced after the quartz syenite or intruded the quartz syenite can be inferred from the following observation. Minor faults cut the quartz syenite in the central part of the intrusion; however, the faults do not cut the adjacent olivine gabbro (Fig. 3). Petrographic evidence (discussed below) and evidence inferred from cross sections suggest that the quartz syenite was emplaced in at least two pulses of liquid. The order of intrusion among the olivine gabbro bodies is impossible to determine from field evidence. However, where cross-cutting relations were observed, the steeply dipping dikes in the area crosscut the syenite sill and olivine gabbro bodies.

Laughlin and others (1982) determined a K-Ar age of 43.5 ± 1.8 Ma for an analcite basalt dike with a northwest strike in the southeastern part of the area. This dike cuts quartz syenite, thus placing a minimum age on the syenite. C. Henry (written communication) has recently determined a K-Ar age of 28.1 ± 0.6 Ma for the olivine gabbro. Jungysuk (1977), however, described a quartz syenite–gabbro suite from the Christmas Mountains to the north (Fig. 1) that is very similar to the Slickrock Mountain suite. A gabbro from the Christmas Mountains yielded a K-Ar age of 41.5 ± 0.6 Ma (McDowell, 1979). These data suggest that both mafic and felsic magmas in the Christmas Mountains were contemporaneous with or slightly postdated Slickrock Mountain activity.

DESCRIPTION OF THE INTRUSIVE ROCKS

The intrusive rocks of the Slickrock Mountain area can be grouped into four petrographic types: quartz syenite, olivine gabbro, amphibole-bearing gabbro, and fine-grained syenites and diorites. Figure 4 shows that the syenitic rocks have a range of quartz content. We have used the term "quartz syenite" to refer to all of the syenites.

Quartz Syenite

The quartz syenite is porphyritic with fine- to medium-grained groundmass, is greenish-black on fresh surfaces, and weathers to brownish or grayish pink. The quartz syenite can be subdivided into three groups on the basis of groundmass texture. The bulk of the intrusion has medium-grained idiomorphic granular texture. In the northern lobe, grain size decreases inward to a fine-grained felsite. This unusual decrease suggests that the quartz

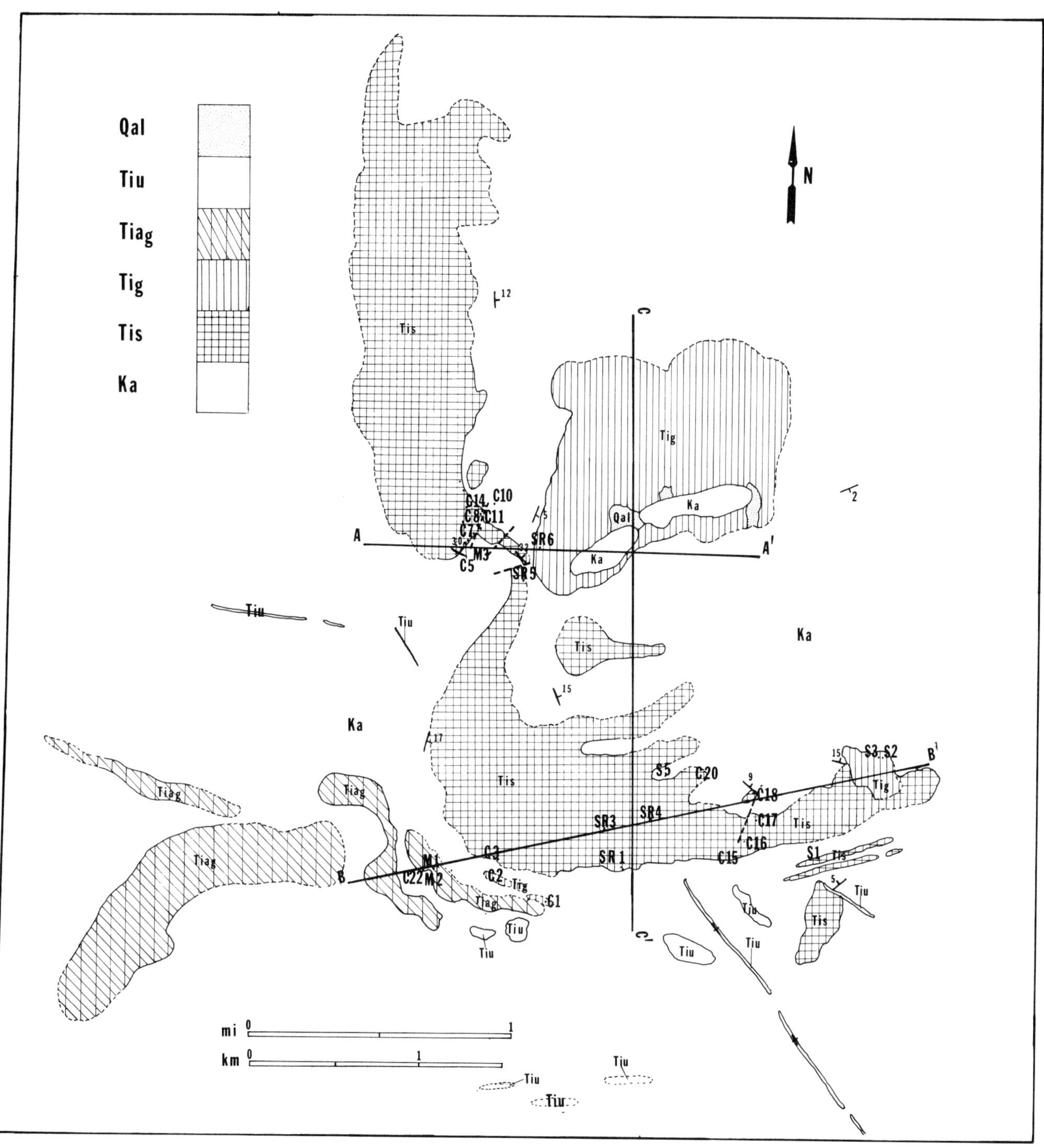

Figure 2. Geologic map; modified from Maxwell and others (1967). Ka, Cretaceous Aguja Formation; Tis, quartz syenite; Tig, olivine gabbro; Tiag, kaersutite gabbro; Tiu, intrusive rocks undivided; Qal, Quaternary alluvium. Note that Qal and Ka are shown only where they overlie an intrusion. Tiu units not included in this study. Numbers represent sample localities.

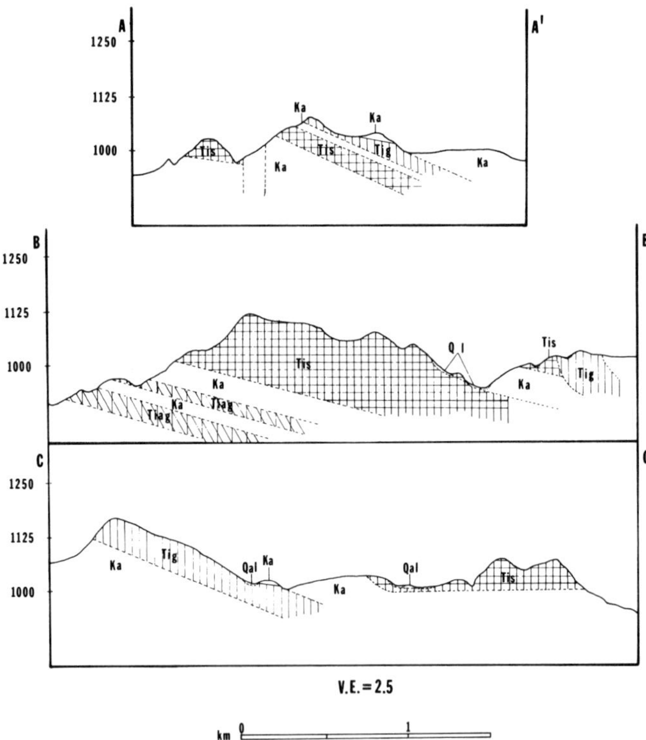

Figure 3. Cross sections. Note sill-like nature of intrusions and cross-cutting relation of olivine gabbro. Refer to Figure 2 for cross-section locations.

syenite was intruded in more than one pulse (see below). The upper few meters of the easternmost part of the southern lobe are characterized by sheaf-like alkali feldspar in the groundmass. This texture grades downward to idiomorphic granular, and may be the result of devitrification of an originally glassy groundmass. All three textural types contain feldspar and olivine phenocrysts. Clinopyroxene phenocrysts occur only in the fine-grained variety.

Feldspar phenocrysts are euhedral to subhedral and locally form glomerocrysts. The phenocrysts range in length from 1 mm to as much as 5 mm and show complex zoning. Patches of plagioclase (An_{26}; Table 1, Fig. 5) form the cores of the phenocrysts. The plagioclase is surrounded by grid-twinned anorthoclase that in turn is rimmed by sanidine, which gives the phenocrysts an equant shape. The cores commonly are resorbed and replaced by sanidine and have little or no remaining plagioclase.

Olivine phenocrysts (Fa_{86}; Table 2, Fig. 6) are euhedral to subhedral and range from 0.2 to 1.0 mm. Fresh olivine is uncommon in the quartz syenite; however, iddingsite pseudomorphs indicate that olivine was originally present in all quartz syenite samples.

Clinopyroxene (ferroaugite to hedenbergite; Table 3, Fig. 6) phenocrysts in the fine-grained quartz syenite are euhedral to subhedral and range from 0.5 to 1.2 mm in length. Equant titanomagnetite and ilmenite grains are microphenocryst phases only in the fine-grained textural type. Their average diameter is 0.2 mm.

The groundmass of the three textural types of quartz syenite consists of sanidine, magnetite, ilmenite, quartz, and acicular apatite. Quartz is more abundant in rocks from the southern and upper parts of the sill (Table 4). Magnetite and ilmenite com-

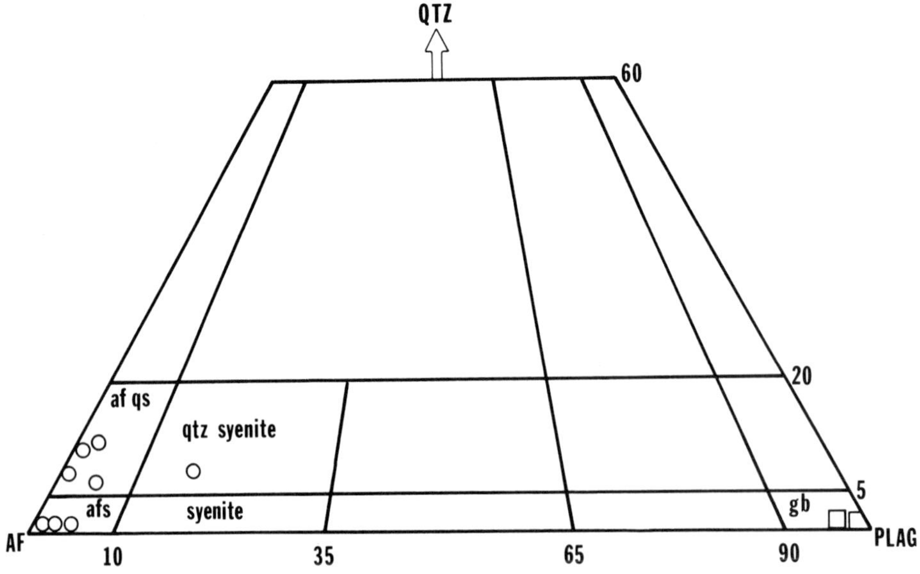

Figure 4. Quartz–alkali feldspar–plagioclase (Q-A-P) triangle showing modal composition of Slickrock Mountain suite according to IUGS classification (Streckeisen, 1976). Circles indicate quartz syenite; squares, olivine gabbro and kaersutite gabbro.

TABLE 1. REPRESENTATIVE FELDSPAR ANALYSES*

	Plagioclase						Anorthoclase		Sanidine		
	C5,3	C7	C8	SR2	M2,1	M2,1C	C7	C14	C14	C14,r	SR2
SiO_2	62.7	61.5	61.8	61.4	57.3	57.0	62.0	63.8	66.9	66.9	66.7
Al_2O_3	23.0	23.6	23.8	24.0	26.6	26.9	23.0	21.9	18.8	19.1	19.4
FeO	0.3	0.3	0.4	0.3	0.3	0.3	0.3	0.3	0.7	0.7	0.2
CaO	4.3	5.4	5.6	5.3	7.9	8.1	4.3	3.3	0.5	0.4	0.5
Na_2O	8.3	7.8	7.7	7.6	6.1	6.1	7.3	7.5	5.8	5.6	5.4
K_2O	1.3	0.9	0.9	1.0	0.6	0.6	2.4	2.9	7.0	7.2	7.6
BaO	0.4	0.2	0.2	0.1	0.1	0.1	0.2	0.3	0.1	0.1	0.1
SrO	nd	nd	nd	nd	0.6	0.6	nd	nd	nd	nd	nd
Total	100.3	99.7	100.4	99.7	99.5	99.7	99.5	100.0	99.85	100.0	99.9
Cations per 8 oxygens											
Si	2.787	2.747	2.746	2.742	2.587	2.575	2.781	2.845	3.000	2.997	2.987
Al	1.207	1.243	1.246	1.260	1.416	1.430	1.215	1.150	0.994	1.005	1.026
Fe	0.012	0.010	0.013	0.012	0.010	0.009	0.012	0.009	0.023	0.026	0.005
Ca	0.203	0.258	0.266	0.251	0.383	0.390	0.208	0.158	0.025	0.020	0.025
Na	0.711	0.676	0.659	0.655	0.535	0.529	0.638	0.646	0.505	0.483	0.468
K	0.073	0.051	0.052	0.058	0.036	0.032	0.134	0.162	0.401	0.409	0.437
Ba	0.005	0.002	0.002	0.001	0.000	0.001	0.004	0.004	0.025	0.000	0.000
Sr	nd	nd	nd	nd	0.016	0.014	nd	nd	nd	nd	nd
Total	4.997	4.986	4.983	4.978	4.982	4.979	4.991	4.974	4.950	4.940	4.947

*All samples are quartz syenite except M2 (kaersutite gabbro). Refer to text or Figure 2 for source, methods, and sample locations of data noted in this and following tables.

nd = not determined.

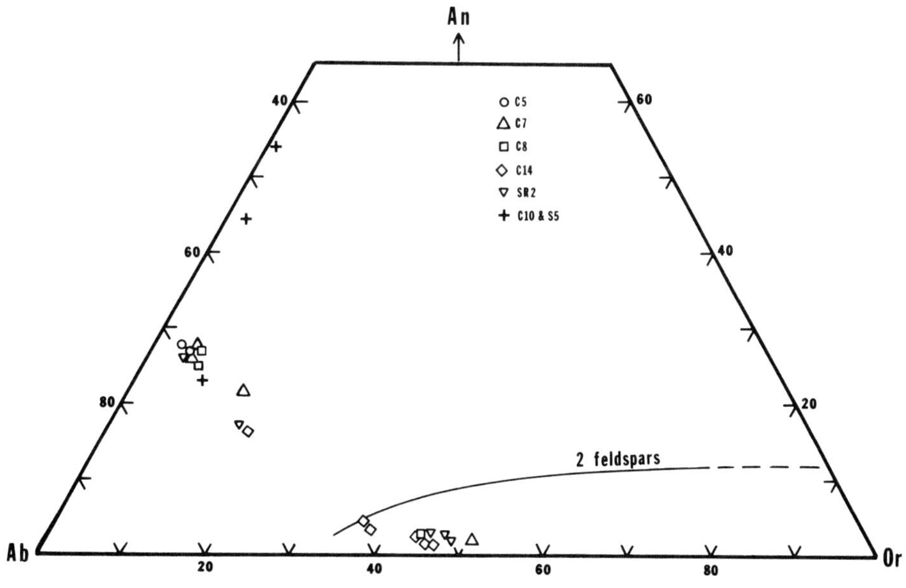

Figure 5. Feldspar compositions. C10 and S5 are olivine gabbro; all other samples are quartz syenite. Low-Ca plagioclase analyses from olivine gabbro from phenocryst rims; sanidine analyses from phenocryst rims and groundmass crystals.

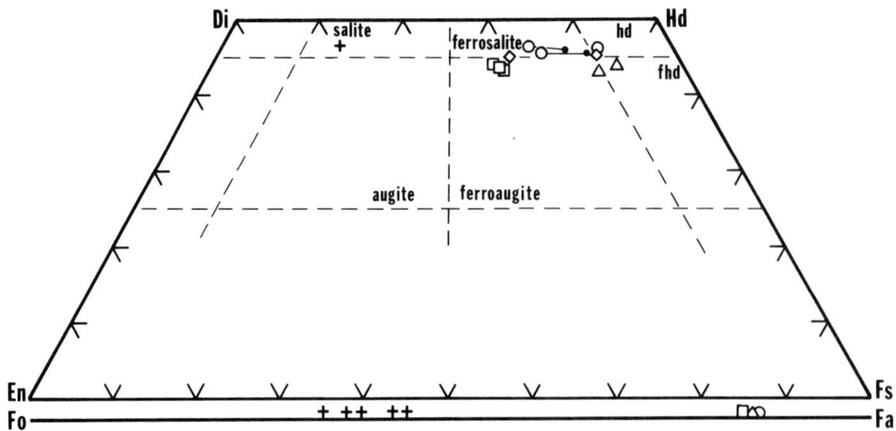

Figure 6. Olivine and clinopyroxene compositions. Refer to Figure 2 for sample localities. Symbols as in Figure 5.

TABLE 2. OLIVINE COMPOSITIONS*

	C8,1	C8,2	C5,2	C7,2
SiO_2	31.4	31.4	31.2	31.5
FeO	61.2	61.0	61.0	60.7
MnO	1.9	1.9	2.1	2.0
MgO	5.5	5.7	5.3	5.8
CaO	0.4	0.4	0.5	0.4
Total	100.4	100.4	100.1	100.4
Cations per 4 oxygens				
Si	1.008	1.009	1.008	1.009
Fe	1.648	1.640	1.646	1.631
Mn	0.052	0.052	0.058	0.054
Mg	0.263	0.271	0.256	0.276
Ca	0.012	0.012	0.017	0.013
Total	2.983	2.984	2.985	2.983

*All samples are quartz syenite.

monly occur as granules adjacent to olivine and clinopyroxene. Acicular apatite ranges in length from 0.05 to as much as 0.5 mm. Interstitial calcite typically occurs as replacements after feldspar and in miarolitic cavities. Locally, however, calcite is graphically intergrown with sanidine and quartz, suggesting a primary origin. Other secondary minerals are iddingsite after olivine, quartz, iron oxides, and clay minerals.

In the idiomorphic granular textural type, sanidine and anorthoclase rimmed by sanidine form equant crystals with average grain size from 0.2 to 0.4 mm. The grains exhibit sharp and consertal boundaries. Skeletal boxwork sanidine occurs in interstices where it is intergrown with quartz or deuteric(?) calcite. Groundmass clinopyroxene grains occur as subhedral prismatic and equant grains that range in length from 0.1 to 1 mm and as skeletal and swallow-tail crystals. Microprobe analyses show a compositional range from ferroaugite to hedenbergite (Fig. 6, Table 3). Interstitial micrographic intergrowths of quartz and alkali feldspar also occur in this textural type as do miarolitic cavities filled with calcite, quartz, and iron oxides.

The fine-grained textural type occurs in the center of the northern part of the quartz syenite. The groundmass ranges from seriate to felsitic, with grain size from 0.01 to 0.3 mm. In addition to equant and prismatic grains, clinopyroxene occurs as tiny rod-shaped crystals in the seriate texture and as microlites in felsitic samples. This fine-grained texture may represent a chilled margin, suggesting that the quartz syenite was emplaced in two pulses. One pulse emplaced the southern lobe and upper parts of the northern lobe of the quartz syenite (idiomorphic granular), and the second pulse emplaced the lower part of the northern lobe of the quartz syenite.

Veins that cross-cut the upper part of the sill show two textural types. One has a fine-grained seriate texture similar to the fine-grained texture described previously, but lacks phenocrysts. This fine-grained type contains groundmass biotite, sanidine, and rare plagioclase without sanidine or anorthoclase rims. Olivine is absent in these veins. The other type of vein (dike?) may or may not be related to the quartz syenite. It has a felted texture of plagioclase, acicular and dendritic magnetite, and intergranular calcite.

Olivine Gabbro

Dark gray olivine gabbro is slightly porphyritic with a very fine grained matrix in which patches of feldspar as much as 3 mm in diameter are distinguishable. Weathering surfaces of the olivine gabbro bodies range from dark brownish black to light gray.

The olivine gabbro is characterized by euhedral to anhedral olivine phenocrysts (Fo_{55-65}; Fig. 6, Table 2), ranging from 0.3 to 0.5 mm in length and set in a felted arrangement of seriate (0.1

TABLE 3. REPRESENTATIVE CLINOPYROXENE ANALYSES

	Olivine Gabbro				Quartz Syenite					Kaersutite Gabbro	
	S5,1	S5,2	C10,2	C10,3	C5,2	C5,2r	C7,1	C8,2	C14,2	M2,1	M2,2
SiO_2	48.8	50.6	50.6	50.5	48.9	49.1	47.6	49.9	49.9	51.8	51.5
TiO_2	2.4	1.5	1.6	1.7	0.6	0.4	1.0	0.6	0.5	0.4	0.4
Al_2O_3	4.4	2.9	2.7	2.9	0.9	0.7	1.3	1.5	1.2	3.0	3.2
MnO	0.2	0.2	0.2	0.2	0.9	1.1	1.0	0.7	0.8	0.5	0.4
FeO	8.6	9.0	8.8	8.5	22.3	25.2	26.4	20.0	20.1	10.6	11.5
MgO	13.4	13.4	13.7	13.7	5.2	3.1	3.5	7.3	6.7	12.4	12.5
CaO	21.9	21.9	21.8	21.6	20.5	20.0	19.2	19.6	20.8	21.0	19.9
Na_2O	0.6	0.5	0.5	0.5	0.3	0.4	0.4	0.4	0.3	0.8	0.7
Cr_2O_3	0.0	0.0	0.0	0.0	0.0	0.0	0.0	0.1	0.0	0.0	0.0
Total	100.3	100.0	99.9	99.6	99.6	100.0	100.4	100.0	100.3	100.5	100.1
Cations per 6 oxygens											
Si	1.826	1.894	1.895	1.892	1.958	1.983	1.926	1.954	1.959	1.935	1.930
Ti	0.065	0.041	0.046	0.048	0.019	0.013	0.029	0.018	0.015	0.010	0.012
Al	0.193	0.127	0.117	0.129	0.043	0.032	0.063	0.067	0.053	0.132	0.141
Mn	0.005	0.006	0.005	0.006	0.028	0.037	0.033	0.024	0.026	0.014	0.012
Fe	0.270	0.280	0.275	0.265	0.745	0.852	0.895	0.654	0.659	0.329	0.360
Mg	0.747	0.748	0.761	0.765	0.307	0.182	0.208	0.425	0.394	0.690	0.696
Ca	0.875	0.878	0.875	0.867	0.877	0.866	0.833	0.822	0.872	0.841	0.799
Na	0.042	0.038	0.036	0.038	0.023	0.028	0.028	0.032	0.025	0.055	0.047
Cr	0.001	0.000	0.000	0.000	0.001	0.000	0.000	0.001	0.000	0.000	0.001
Total	4.025	4.012	4.010	4.010	4.001	3.992	4.015	3.998	4.002	4.006	3.998

TABLE 4. REPRESENTATIVE OXIDE ANALYSES*

	Spinel Phase				Rhombohedral Phase				
	S5,1	S5,2	C10	C8,1	S5	C10.1	C10,4	C8,1	C8,2
SiO_2	0.0	0.0	0.1	0.1	0.0	0.0	0.0	0.0	0.0
TiO_2	20.6	20.5	22.0	18.8	49.9	48.8	48.7	47.9	48.5
Al_2O_3	1.4	2.2	3.6	0.9	0.1	0.1	0.1	0.1	0.0
Cr_2O_3	0.1	0.1	0.1	0.1	0.1	0.0	0.1	0.0	0.1
Fe_2O_3	41.4	39.6	35.4	44.3	10.1	10.6	10.3	11.6	10.2
FeO	35.6	35.8	36.4	34.5	36.7	34.9	35.4	37.0	37.9
MnO	0.8	0.7	0.7	1.0	1.0	0.9	0.8	1.2	1.2
MgO	1.2	1.2	2.5	0.1	1.6	2.1	1.9	0.1	0.1
Total	101.1	100.3	100.8	99.8	99.5	97.4	97.3	97.9	98.0
Cation									
Si	0.001	0.001	0.002	0.002	0.000	0.000	0.001	0.000	0.000
Ti	0.573	0.572	0.598	0.536	0.942	0.939	0.939	0.933	0.941
Al	0.060	0.097	0.151	0.040	0.003	0.002	0.004	0.001	0.001
Cr	0.001	0.002	0.002	0.001	0.001	0.000	0.001	0.000	0.001
Fe^{3+}	0.792	0.755	0.645	0.883	0.112	0.120	0.116	0.133	0.116
Fe^{2+}	1.483	1.489	1.447	1.502	0.861	0.840	0.851	0.903	0.911
Mn	0.021	0.021	0.019	0.030	0.021	0.018	0.016	0.026	0.026
Mg	0.056	0.063	0.135	0.006	0.060	0.081	0.071	0.004	0.004
Total	3.000	3.000	3.000	3.000	2.000	2.000	2.000	2.000	2.000

*FeO and Fe_2O_3 calculated according to Stormer (1983). C8 is quartz syenite; all others are olivine gabbro.

to 1.0 mm) plagioclase laths (An_{45-55}), magnetite, ilmenite, clinopyroxene, analcite, apatite, and biotite (Fig. 5, Table 1). Sparse plagioclase phenocrysts and glomerocrysts that consist of olivine, magnetite, ilmenite, clinopyroxene, and biotite are also present. Iddingsite is pseudomorphous after olivine phenocrysts. Other secondary minerals include hematite and chlorite. In the east-west–trending gabbro, clinopyroxene is ophitic to subophitic, ranges in diameter from 1.5 to 2.5 mm, and has a uniform salitic composition (Fig. 6, Table 3). Biotite occurs as coronas around magnetite and olivine. The gabbros in the southwestern and southeastern parts of the area contain anhedral clinopyroxene and prismatic biotite grains as much as 1 mm in diameter in a felted arrangement of plagioclase laths. Sparse miarolitic cavities are 3 to 4 mm in diameter and have a drusy rim of euhedral clinopyroxene and plagioclase with an inner zone of analcite and calcite.

Kaersutite Gabbro

A second type of gabbro occurs as sills and dikes within the southeastern part of the area. This type is characterized by olive gray color, kaersutitic amphibole megacrysts, quartz-rich xenoliths, and coarse-grained mafic crystal clots.

Kaersutitic amphibole (Table 5) occurs as resorbed phenocrysts as much as 5 cm long that are commonly rimmed by radially oriented titanaugite, plagioclase, magnetite, and biotite. Other phenocryst phases are plagioclase (Table 1), magnetite, and ilmenite. Plagioclase phenocrysts are subhedral to anhedral (2 to 4 mm long) with resorbed rims. Anhedral magnetite phenocrysts range from 1.5 to 5 mm in diameter and are embayed. The mafic crystal clots contain clinopyroxene (Table 3), olivine, and plagioclase in a cumulate(?) texture. Quartzofeldspathic xenoliths range in length from 2 mm to 15 cm. Individual grains within the xenoliths are anhedral and average 0.5 mm in diameter. Grain boundaries are diffuse, and grains are commonly separated by a zone of recrystallized intergranular glass(?), which suggests that the xenoliths were partially fused. Groundmass texture is a fine-grained subtrachytoidal arrangement of plagioclase laths, prismatic titanaugite, anhedral magnetite and ilmenite, biotite, and prismatic and acicular apatite. Secondary calcite locally has replaced resorbed plagioclase and filled sparse miarolitic cavities 2 to 5 mm in diameter.

ATTEMPTS AT GEOTHERMOMETRY AND OXYGEN BAROMETRY

If it is assumed that the olivine gabbro represents a liquid composition, then the FeO/MgO ratio of the gabbro and coexisting olivine can be used to estimate the temperature at which olivine was in equilibrium with the liquid (Roeder and Emslie, 1970). A silicate liquid with FeO/MgO values of gabbro samples S5 and SR6 would be in equilibrium with Fo_{83} olivine at 1,100°C. Olivine phenocrysts in S5 are Fo_{60}, suggesting that olivine has accumulated in the gabbro samples to such an extent that the gabbros cannot represent liquid compositions.

TABLE 5. KAERSUTITE MEGACRYST ANALYSES

	M2,1c	M2,2c	M2,1r
SiO_2	39.4	39.4	39.4
TiO_2	5.3	5.4	5.5
Al_2O_3	14.4	14.1	14.1
MnO	0.1	0.1	0.2
FeO	11.4	11.5	11.8
MgO	11.4	11.4	11.5
CaO	9.3	9.1	9.5
Na_2O	2.6	2.5	2.7
K_2O	1.0	1.0	1.0
Total	94.9	94.5	95.7
Cations per 23 oxygens			
Si	5.967	5.984	5.930
Ti	0.599	0.611	0.624
Al	2.562	2.530	2.514
Mn	0.013	0.009	0.018
Fe	1.434	1.455	1.487
Mg	2.575	2.579	2.589
Ca	1.500	1.477	1.526
Na	0.754	0.741	0.774
K	0.195	0.183	0.181
Total	15.601	15.568	15.643

Titanomagnetite and ilmenite are interstitial groundmass phases in the olivine gabbro and are microphenocryst phases in the fine-grained quartz syenite. The titanomagnetite typically shows very fine scale lamellar "exsolution" of ilmenite. We attempted to obtain the pre-exsolution composition of the titanomagnetite by averaging several 20μ-diameter spot analyses of individual grains (Table 4). Oxide analyses were recast according to the method of Stormer (1983), after which temperature and oxygen fugacity were estimated using the results of Spencer and Lindsley (1981). Olivine gabbro sample S5 yielded a temperature estimate of 850°C at log fO_2 = –14.3, and olivine gabbro C10 gave a temperature of 888°C at log fO_2 = –13.4. Inasmuch as no granule exsolution was observed, these temperatures appear to represent near-solidus crystallization of Fe-Ti oxides. Microphenocryst oxide pairs from quartz syenite sample C8 give temperature estimates in the range 820°C and log fO_2, from –13.9 to –14.8. All such results must be interpreted with caution, owing to the possibility that Ti mobility during formation of ilmenite lamellae modified the bulk composition of titanomagnetic grains. However, the fO_2 estimates for the quartz syenite sample are slightly lower than the FMQ oxygen buffer at 5×10^7 Pa (500 bars). These estimates are consistent with the presence of Fe-rich olivine and titanomagnetite as phenocryst phases and of quartz as a late-stage groundmass mineral.

CHEMICAL VARIATION

Whole-rock major- and trace-element analyses were obtained by x-ray fluorescence for 15 samples (9 samples from the

TABLE 6. REPRESENTATIVE MODES

Sample	M1	M2	SR6	S5	C2	S2
Plagioclase	48	17	63	60	66	68
Clinopyroxene	11	10	12	9	5	8
Olivine	1	0	7	7	14	3
Opaque minerals	9	12	8	7	7	4
Biotite	18	12	7	11	4	7
Apatite	2	7	1	3	3	2
Analcite	2	7	1	3	3	6
Feldspar phenocryst	2	4	0	0	0	0
Kaersutite phenocryst	4	23	0	0	0	0
Other*	3	7	2	3	3	6
Total	100	100	100	100	100	100

Sample	SR1	SR3	SR4A	SR5	C5	C7	C8**	C14
Feldspar phenocryst	4.9	5.9	1.0	1.0	0.0	4.0	0.0	10.3
Alkali feldspar	69.2	61.1	73.0	68.0	71.0	71.0	13.3	67.2
Clinopyroxene	0.0	0.0	0.0	0.0	8.0	4.0	0.5	5.8
Olivine phenocryst	0.0	0.0	0.0	0.0	8.0	4.0	0.6	2.5
Magnetite	9.6	11.7	10.0	8.0	5.0	9.0	2.7	3.8
Apatite	2.6	3.7	2.3	2.0	4.0	4.0	0.0	3.3
Quartz	5.0	8.9	9.0	6.0	1.0	0.0	0.0	0.0
Calcite	3.8	2.0	1.0	1.0	0.0	0.0	0.0	0.0
Amygdule	0.0	6.0	3.7	6.0	0.0	0.0	0.0	0.0
Other*	4.9	0.7	0.0	8.0	3.0	4.0	82.9	7.1
Total	100.0	100.0	100.0	100.0	100.0	100.0	100.0	100.0

*Includes hematite, iddingsite, and clay minerals.

**Groundmass minerals included in "Other."

quartz syenite, 5 from the olivine gabbro, and 1 from the kaersutite gabbro; see Table 7). Analyses of mineral phases from five quartz syenite samples and two olivine gabbro samples were carried out on a JEOL JXA-733 microprobe at Southern Methodist University (Tables 1 through 4). Typical analytical conditions were 15 KV, 20 NA, and a 10-μ diameter spot.

Figure 7 shows variation diagrams for the Slickrock Mountain suite plotted versus SiO_2. The suite is clearly bimodal, with a compositional gap between 53 and 63 wt. % SiO_2. In general, K_2O and Rb increase with increase in SiO_2, whereas CaO, MgO, Fe_2O_3, TiO_2, and Sr decrease. Al_2O_3 appears to reach a maximum and then decrease toward the quartz syenite; Na_2O shows little variation from mafic to felsic compositions. With the exception of sample C2, all of the analyzed gabbroic rocks are silica-undersaturated (nepheline normative). All of the syenitic samples are silica-oversaturated (normative quartz) and two contain normative corundum (Table 6). In these rocks, corundum in the norm may be the result of Ca or Na loss due to weathering.

DISCUSSION

In the Slickrock Mountain area, silica-undersaturated rocks crop out in close proximity to potassium-rich silica-oversaturated rocks. The presence of Eocene gabbroic rocks similar in composition to Slickrock gabbro in the Christmas Mountains (Fig. 1; Jungysuk, 1977) and elsewhere in the Trans-Pecos province (Barker, 1977) suggests—but certainly does not prove—a genetic relationship. One means of testing possible genetic relationships among the Slickrock Mountain suite is the use of least-squares mass-balance calculations (Bryan and others, 1969) that subtract observed phenocryst phases from a postulated parent to produce a daughter composition similar to the evolved rocks in the suite.

Numerous experimental studies (e.g., Kushiro, 1968; Jaques and Green, 1980) have shown that low-pressure crystal fractionation cannot produce a silica-oversaturated magma from a silica-undersaturated parent. Mass-balance calculations for the Slickrock Mountain rocks support this conclusion. No combination of olivine, pyroxene, feldspars, and oxide phases can be subtracted from an olivine gabbro composition to produce a quartz syenite.

The presence of kaersutite and intermediate plagioclase as megacrysts in late-stage dikes suggests the possibility of crystal fractionation at elevated water pressure. Mass-balance calculations in which kaersutite, plagioclase, and minor olivine, clinopyroxene, apatite, and oxide phases are subtracted from olivine gabbro to produce quartz syenite yield acceptable results (i.e., sum of the squares of residuals <1). However, the results of the mass-balance calculations are not compatible with the extreme range in Sr between gabbro and quartz syenite. The fraction of

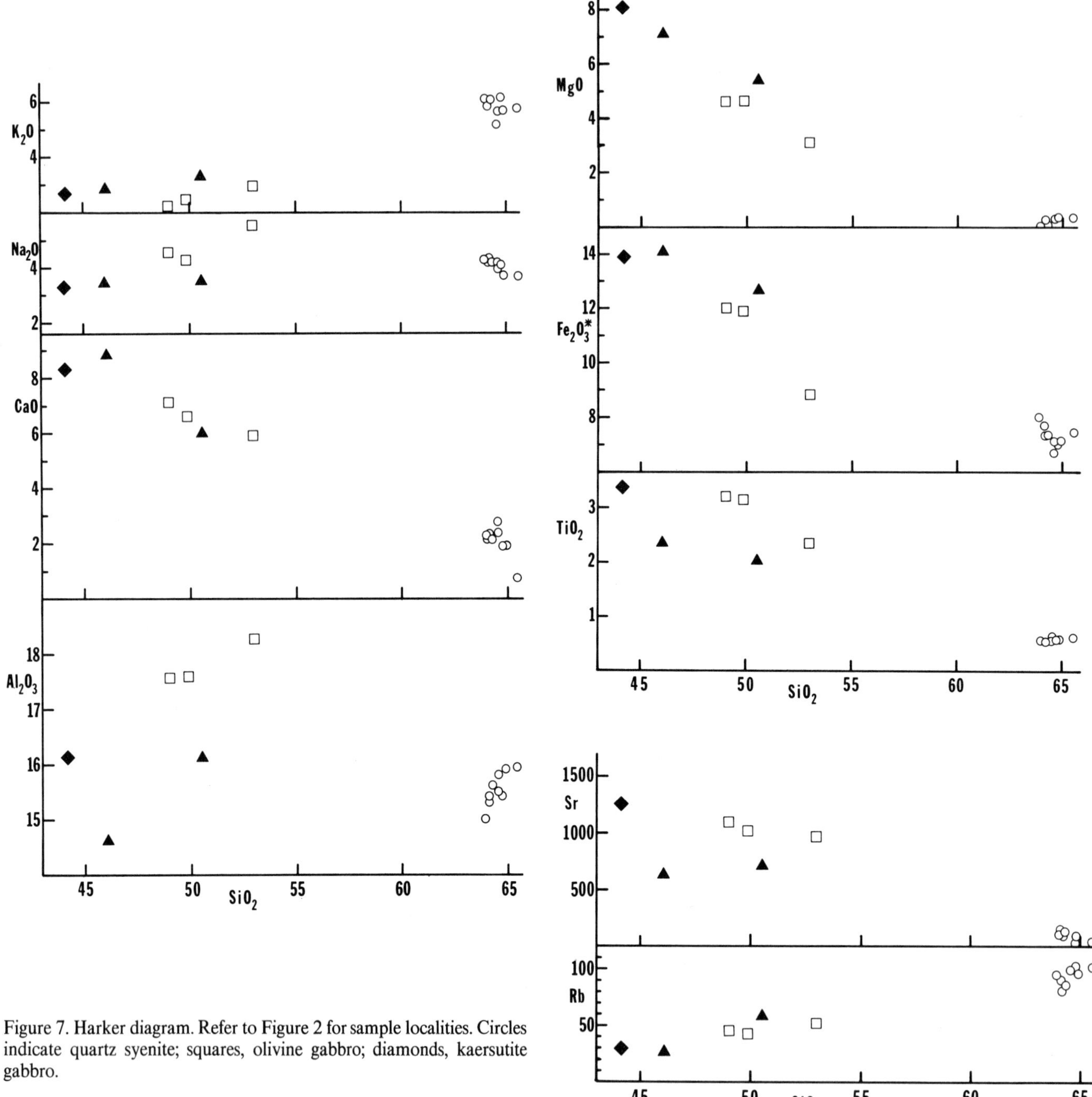

Figure 7. Harker diagram. Refer to Figure 2 for sample localities. Circles indicate quartz syenite; squares, olivine gabbro; diamonds, kaersutite gabbro.

TABLE 7. WHOLE-ROCK MAJOR AND TRACE ELEMENT ANALYSES (NORMALIZED VOLATILE-FREE) AND CIPW NORMS*

	M1	C20	SR6	S5	C2	S2	C5	C7	C8	C14	SR1	SR3	SR4A	SR5	S1
SiO_2	44.16	46.09	49.02	49.88	50.59	52.95	64.29	63.95	64.10	64.09	64.54	64.58	64.76	65.90	64.91
TiO_2	3.39	2.34	3.20	3.13	2.02	2.34	0.55	0.59	0.56	0.56	0.61	0.60	0.59	0.63	0.62
Al_2O_3	16.16	14.63	17.56	17.60	16.11	18.29	15.62	15.03	15.31	15.44	15.83	15.60	15.43	15.95	15.93
Fe_2O_3	1.67	1.70	1.44	1.42	1.52	1.05	0.88	0.96	0.92	0.88	0.80	0.85	0.84	0.89	0.85
FeO	11.09	11.34	9.60	9.49	10.11	7.04	5.87	6.39	6.13	5.86	5.32	5.67	5.60	5.89	5.69
MnO	0.16	0.27	0.16	0.16	0.20	0.14	0.19	0.22	0.20	0.19	0.17	0.19	0.18	0.20	0.21
MgO	8.10	7.09	4.61	4.59	5.64	3.06	0.08	0.05	0.08	0.26	0.29	0.23	0.33	0.28	0.35
CaO	8.39	8.82	7.07	6.60	6.00	5.88	2.16	2.28	2.31	2.21	2.78	2.35	1.86	0.71	1.88
Na_2O	3.37	3.46	4.52	4.23	3.53	5.51	4.11	4.29	4.30	4.23	3.90	4.16	4.09	3.66	3.71
K_2O	2.68	2.83	2.19	2.20	3.29	2.91	6.05	6.09	5.90	6.10	5.61	5.16	6.18	5.74	5.65
P_2O_5	0.82	0.66	0.64	0.69	0.99	0.83	0.18	0.17	0.19	0.19	0.51	0.16	0.14	0.15	0.20
Total	99.99	100.00	100.01	99.99	100.00	100.00	99.98	100.02	100.00	100.01	100.00	100.00	100.00	100.00	100.00
Trace element (ppm)															
Rb	29	27	45	42	58	52	119	95	80	105	97	96	106	101	95
Sr	1250	628	1097	1001	706	955	85	80	81	91	33	37	80	35	46
Normative minerals															
AP	1.75	1.42	1.40	1.51	2.18	1.82	0.42	0.40	0.44	0.44	0.35	0.37	0.33	0.36	0.47
IL	6.26	5.80	6.04	5.91	3.85	4.44	1.11	1.19	1.13	1.13	1.22	1.21	1.19	1.32	1.27
MT	2.36	2.42	2.08	2.05	2.21	1.52	1.35	1.47	1.41	1.35	1.22	1.30	1.29	1.42	1.32
OR	15.36	16.39	12.83	12.89	18.43	17.14	36.76	38.00	36.81	38.03	34.81	34.97	38.56	37.21	35.79
AB	8.88	11.53	29.29	33.89	29.86	37.94	6.66	38.34	38.42	37.78	34.66	37.14	36.55	33.98	33.66
AN	20.43	15.71	20.98	22.35	18.39	16.53	2.68	3.98	5.33	5.41	9.58	7.73	5.80	2.79	8.60
DI	12.61	19.48	8.15	4.77	4.13	6.01	3.33	5.32	4.28	3.83	3.07	2.74	2.36	0.00	0.00
HY	0.00	0.00	0.00	0.00	0.19	0.00	0.00	0.00	0.00	0.00	3.13	3.39	3.60	4.42	4.32
OL	22.16	17.94	14.58	15.77	19.04	9.98	0.00	0.00	0.00	0.00	0.00	0.00	0.00	0.00	0.00
NE	10.18	9.30	4.67	0.86	0.00	4.63	0.00	0.00	0.00	0.00	0.00	0.00	0.00	0.00	0.00
AC	0.00	0.00	0.00	0.00	0.00	0.00	0.00	0.00	0.00	0.00	0.00	0.00	0.00	0.00	0.00
C	0.00	0.00	0.00	0.00	0.00	0.00	0.00	0.00	0.00	0.00	0.00	0.00	0.00	3.08	0.85
Q	0.00	0.00	0.00	0.00	0.00	0.00	9.98	8.32	9.05	8.07	11.97	11.16	10.33	18.50	14.57
Total	99.99	100.00	100.00	99.99	100.00	100.00	99.98	100.02	100.00	100.01	100.00	100.00	100.00	100.00	100.00

*Iron oxides normalized so that $Fe_2O_3/Fe = 0.15$.

liquid remaining after removal of these phases is approximately 64 percent. Using bulk distribution coefficients of 2.0 for Sr (extreme) and 0.01 for Rb, the model predicts values of 500 to 600 ppm for Sr and 65 to 70 ppm for Rb in the quartz syenite. Inspection of Table 7 shows that these values do not fit those measured in the quartz syenite. Thus amphibole-dominated crystal fractionation can be ruled out by the trace element data. Similarly, steady-state fractionation plus assimilation of quartzofeldspathic crustal rocks (e.g., the partially melted xenoliths in the kaersutite gabbro) to produce quartz syenite should result in Sr concentrations far greater than those measured in the Slickrock Mountain syenite (e.g., DePaolo, 1981).

We are left with two possible explanations for the Slickrock Mountain suite. These are that the gabbros and quartz syenite are genetically unrelated (e.g., Barker, 1979b) or that the quartz syenite is the product of partial melting of lower crust with heat provided by alkali olivine basaltic magma. Our chemical data cannot constrain the second possibility, but refractory phases that are characteristic of S-type granites (White and Chappell, 1977) are absent in the quartz syenite.

Barker (1979b) argued for the presence of two distinct parental magma types in the Trans-Pecos, one silica-undersaturated and one silica-saturated. Our investigation of the Slickrock Mountain suite indicates that Barker's (1979b) proposal is the only one that is adequately supported by available chemical and petrographic data.

CONCLUSIONS

The Slickrock Mountain intrusive suite consists of a series of shallow dikes and sill-like bodies intruded into Cretaceous sedimentary rocks between about 43 and 28 Ma. The sequence of intrusion was from felsic to mafic: two pulses of quartz syenite, then kaersutite gabbro and fine-grained gabbro, then alkali olivine gabbro. No intermediate compositions occur in the area. Thus the Slickrock Mountain intrusive rocks form a distinctly bimodal suite.

Major element mass-balance calculations show that fractional crystallization of observed phenocryst phases cannot explain the compositional variations in the suite. Petrographic evidence suggests differentiation by crustal contamination or fractionation dominated by kaersutitic amphibole. Trace-element data rule out both possibilities. The most likely explanation for the Slickrock Mountain suite involves two parental magmas. The quartz syenite was derived from a silica-saturated parental composition, whereas the gabbros were derived from one or more silica-undersaturated magmas.

ACKNOWLEDGMENTS

We are indebted to the National Park Service for permission to sample at Big Bend National Park, and in particular to Mike Fleming for his assistance and advice. C. Henry kindly provided K-Ar data on the olivine gabbro. Reviews by D. Parker, J. Pasteris, E. Mullen, and an anonymous reviewer are also greatly appreciated. This project was supported by a new-faculty research grant from Texas Tech University.

REFERENCES CITED

Barker, D. S., 1977, Northern Trans-Pecos magmatic province; Introduction and comparison with the Kenya Rift: Geological Society of America Bulletin, v. 88, p. 1421–1427.

—— , 1979a, Cenozoic magmatism in the Trans-Pecos province; Relation to the Rio Grande Rift, *in* Riecker, R. E., ed., Rio Grande Rift; Tectonics and magmatism: Washington, D.C., American Geophysical Union, p. 382–392.

—— , 1979b, Magmatic evolution in the Trans-Pecos province, *in* Walton, A. W., and Henry, C. D., eds., Cenozoic geology of the Trans-Pecos volcanic field of Texas: Austin, University of Texas, Bureau of Economic Geology, Guidebook 19,, p. 4–9.

Bryan, W. B., Finger, L. W., and Chayes, F., 1969, Estimating proportions in petrographic mixing equations by least squares approximation: Science, v. 163, p. 926–927.

Cameron, K. L., and Cameron, M., 1985, Rare earth element, $^{87}Sr/^{86}Sr$, and $^{143}Nd/^{144}Nd$ compositions of Cenozoic orogenic dacites from Baja California, northwestern Mexico, and adjacent west Texas; Evidence for the predominance of a subcrustal component: Contributions to Mineralogy and Petrology, v. 91, p. 1–11.

Carman, M. F., Jr., Cameron, M., Gunn, B., Cameron, K., and Butler, J. C., 1975, Petrology of the Rattlesnake Mountain sill, Big Bend National Park, Texas: Geological Society of America Bulletin, v. 86, p. 177–193.

DePaolo, D. J., 1981, Trace element and isotopic effects of combined wallrock assimilation and fractional crystallization: Earth and Planetary Science Letters, v. 53, p. 189–202.

Henry, C. D., and Price, J. G., 1984, Variations in caldera development in the Tertiary volcanic field of Trans-Pecos Texas: Journal of Geophysical Research, v. 89, p. 8765–8786.

Jaques, A. L., and Green, D. H., 1980, Anhydrous melting of peridotite at 0-15 kb pressure and the genesis of the tholeiitic basalts: Contributions to Mineralogy and Petrology, v. 73, p. 287–310.

Jungysuk, N., 1977, Petrology of the Christmas Mountains igneous rocks, Trans-Pecos, Texas [M.S. thesis]: Austin, University of Texas, 178 p.

Kushiro, I., 1968, Compositions of magmas formed by partial zone melting of the earth's upper mantle: Journal of Geophysical Research, v. 73, p. 619–634.

Laughlin, A. W., Kress, V. C., and Aldrich, M. J., 1982, K-Ar ages of dike rocks, Big Bend National Park, Texas: Isochron/West, no. 35, p. 17–18.

Maxwell, R. A., Lonsdale, J. T., Hazzard, R. T., and Wilson, J. A., 1967, Geology of Big Bend National Park, Brewster County, Texas: Austin, University of Texas Publications 6711, 320 p.

McDowell, F. W., 1979, Potassium-argon dating in the Trans-Pecos Texas volcanic field, *in* Walton, A. W. and Henry, C. D., eds., Cenozoic geology of the Trans-Pecos volcanic field of Texas: Austin, University of Texas, Bureau of Economic Geology, Guidebook 19, p. 10–19.

—— , 1981, Chronology of Tertiary volcanism in Trans-Pecos Texas: Geological Society of America Abstracts with Programs, v. 13, p. 242.

Roeder, P. L., and Emslie, R. F., 1970, Olivine-liquid equilibrium: Contributions to Mineralogy and Petrology, v. 29, p. 275–289.

Spencer, K. J., and Lindsley, D. H., 1981, A solution model for coexisting iron-titanium oxides: American Mineralogist, v. 66, p. 1189–1201.

Stormer, J. C., 1983, The effects of recalculation on estimates of temperature and oxygen fugacity from analyses of multicomponent iron-titanium oxides: American Mineralogist, v. 68, p. 586–594.

Streckeisen, A., 1976, To each plutonic rock its proper name: Earth Science Reviews, v. 12, p. 1–33.

White, A.J.R., and Chappell, B. W., 1977, Ultrametamorphism and granitoid genesis: Tectonophysics, v. 43, p. 7–22.

MANUSCRIPT ACCEPTED BY THE SOCIETY OCTOBER 15, 1986

Printed in U.S.A.

Geological Society of America
Special Paper 215
1987

Geochemical comparison of alkaline volcanism in oceanic and continental settings; Clarion Island versus the eastern Trans-Pecos magmatic province

Dennis O. Nelson and Kerri L. Nelson
Department of Geology, Sul Ross State University, Alpine, Texas 79832

ABSTRACT

Clarion Island (CI) represents young (1–1.7 Ma) fracture zone-related volcanism in the east-central Pacific Ocean. The eastern Trans-Pecos magmatic province (TPMP) of west Texas represents continental volcanism that developed during waning Laramide compression, primarily 30 to 38 Ma. Both rock series are alkaline, with the CI rocks conforming to a trend of increased silica undersaturation, and the bulk of the TPMP rocks to a trend of increased oversaturation with respect to silica. A sodic series of alkali basalt–hawaiite–mugearite–benmoreite–trachyte occurs in both regions. On Clarion Island, the evolved rocks tend to be phonolitic. The TPMP contains an additional potassic series, alkali basalt–trachybasalt–tristanite–trachyte, as well as basanites. Modeling and petrography suggest that these sequences were produced by low-pressure crystal fractionation involving olivine, clinopyroxene, plagioclase, Ti-magnetite, amphibole, and apatite. Compositional trends for CI and TPMP rocks are very similar for many elements (e.g., Rb, Th, Ta, Eu, Sc, Ti, Ba, and REE), suggesting similar mechanisms of evolution, possibly similar sources, and an argument for minimal crustal assimilation for the TPMP alkali basalt–trachyte sequence. In addition to these silica-enrichment trends, both regions possess a trend toward silica depletion, possibly the result of fractionation of a high-pressure pyroxene-dominated assemblage.

Absent from CI, but important in the TPMP rocks, is the sequence of quartz trachyte to rhyolite/comendite, produced by fractionation of alkali feldspar, clinopyroxene, Ti-magnetite, amphibole, quartz, and apatite, modified by magma replenishment and assimilation of crustal rocks or their partial melts.

We regard the nature and thickness of the crust as important controls in the evolution of magmas in these two regions. Mafic magmas are similar in both regions. Low-pressure fractionation beneath Clarion Island's thin basaltic crust produced the alkali basalt–phonolite trend, whereas fractionation beneath the TPMP's thicker continental crust led to the alkali basalt–rhyolite trend.

The rocks of CI and the eastern TPMP exhibit nearly identical ranges of $(Sm/Nd)_n$. This variation, outside analytical uncertainty, is suggested to be the result of mixing of depleted and enriched mantle, produced during earlier melting or metasomatic events. The rocks of the eastern TPMP have Th/Ta and Rb/Ta ratios that extend from those characteristic of CI to much higher values. This variation is interpreted to have been produced by limited interaction of the ascending melt with the continental lithosphere.

In both CI and the eastern TPMP, incompatible elements such as Zr and Nb have nearly constant concentrations over a range of Mg numbers from 70 to 50. This indicates that such Mg numbers may represent primary melts of heterogeneous mantle comprising different proportions of depleted and enriched (veined) mantle.

INTRODUCTION

Basalts of alkaline affinity have erupted in a variety of geologic environments. Analytical data for these alkaline basalts clearly indicate that their source regions are distinct from that of the ocean-ridge basalts, the former being derived from mantle enriched in the incompatible elements, and the latter derived from a source region depleted in these constituents (for example, Kay and Gast, 1973; Sun and Hanson, 1975). Isotopic studies of Nd and Sr are consistent with the derivation of the ocean-ridge basalts from depleted mantle, but Nd data, coupled with the rare-earth element (REE) geochemistry of many of the alkaline basalts, appear conflicting. In particular, a common characteristic of the alkaline basalts is light rare-earth element (LREE) enrichment ($(La/Sm)_n > 1.0$), whereas the Nd isotopes indicate that the source regions of the alkaline magmas have had a time-averaged LREE–depleted character (e.g., DePaolo and Wasserburg, 1976; O'Nions and others, 1977). The conflict here arises from the difficulty in producing an LREE–enriched melt from an LREE–depleted source with reasonable (>1%) degrees of melting. Several workers (Frey and others, 1978; Menzies and Murthy, 1980; Wass and Rogers, 1980; Menzies and Wass, 1983) have suggested that mantle metasomatism by a H_2O–CO_2 fluid immediately (geologically) prior to magma generation could supply the required incompatible element-enriched character without influencing the isotopic compositions. Still others, partly to accommodate the restrictions of Pb isotopes, have called upon recycling of oceanic crust to produce the alkaline basalts of the sea floor (Hofmann and White, 1982; Chase, 1981).

Schwarzer and Rogers (1974) demonstrated that most alkali basalts have been derived from similar parental mantle. Since that time, the nature of the source region(s) of alkali basalts and the relative roles of the suboceanic and subcontinental lithosphere and continental and oceanic crusts have been much debated (Allegre and others, 1981; Futa and LeMasurier, 1983; Downes, 1984; Tarney and others, 1981; Fitton and Dunlop, 1985; McDonough and others, 1986).

A significant development in the debate has been the recognition, through combined isotopic techniques, of a heterogeneous source for the alkaline (and other) magmas. In particular, successful models have been constructed for the development of magmas from a source comprising a mixture of depleted, undepleted, and enriched mantle components (Tarney and others, 1981; Chen and Frey, 1983; Stille and others, 1983; Feigenson, 1984; Vollmer and others, 1984). Identification of these components has provided new insight into the processes of alkaline magma formation.

In this study we compare an alkaline suite of an oceanic island with one that erupted in a continental setting to try to constrain and compare their respective source regions, the relative impacts of the oceanic and continental lithosphere, the role of continental crust, and the major processes of magmatic evolution in the two regions. We report previously unpublished major and trace element data from Clarion Island (CI) in the east-central Pacific and compare it with data (Nelson and others, 1987a) from the eastern Trans-Pecos magmatic province (TPMP) of west Texas (Fig. 1).

ANALYTICAL TECHNIQUES

All analyses were performed using analytical facilities at Sul Ross State University. Major elements and Sr, Y, and Nb were analyzed by energy dispersive x-ray fluorescence (XRF) using pressed-powder pellets. Each sample was analyzed three times for 300 seconds; each analysis was fit to a standard curve using a linear regression program. Standards for the standard curve were analyzed concurrently with the samples. Precision of XRF analyses was determined by multiple analyses of U.S. Geological Survey standards and is better than 5 percent of the value for all major and trace elements except MgO, Sr, and Nb, which are within 8 percent of their respective values.

All other trace elements were determined by instrumental neutron activation analysis (INAA). Two splits of each sample were irradiated at the Nuclear Science Center, Texas A&M University, with U.S. Geological Survey and in-house standards for 6 hr at a neutron flux of approximately 8×10^{12} neutrons/cm^2/sec. Precision, based on 38 replicate analyses of our in-house standard, is as follows: better than 3 percent of the value for Lu, La, Ce, Hf, and Rb; better than 5 percent for Sm, Eu, Th, Yb, Cs, Sc, Zn, and Ta; better than 10 percent for Ba, Nd, Tb, Cr, Zr, and Co; and better than 12 percent for U. Corrections for peak interference were made when necessary.

REGIONAL GEOLOGY

Clarion Island

Clarion Island is the westernmost of the Revilla Gigedo Island group. It is located on the Clarion fracture zone, approximately 650 km west of the East Pacific Rise (Fig. 1). The island, 8 by 3 km in size, is situated between anomalies 5 and 6 in the Pacific sea floor (Sclater and others, 1971) and therefore is resting on oceanic crust that is approximately 10 to 15 m.y. old. Whole-rock K-Ar age determinations of exposed units on Clarion vary from 1.02 to 1.69 ± 0.04 Ma (J. Clark, J. Dymond, and L. Hogan, personal communication).

Areal geology, as well as lithologic and petrographic descriptions have been provided by Bryan (1967), who defined five major units on the island (Table 1). Bryan (1967) proposed two cycles of volcanic activity in the exposed rocks of Clarion Island. The first commenced with eruptions of basaltic lavas and terminated with eruptions of trachyte to quartz trachyte. Basaltic eruptions returned during the initial phases of the second cycle that ended with the emplacement of trachybasalt to trachyandesite. Evaluation of radiometric ages of the CI rocks (Table 1) indicates that the episodes of volcanic activity on the island were more complex than Bryan envisioned. There is considerable overlap between formations, and progressive compositional variation with time is not observed.

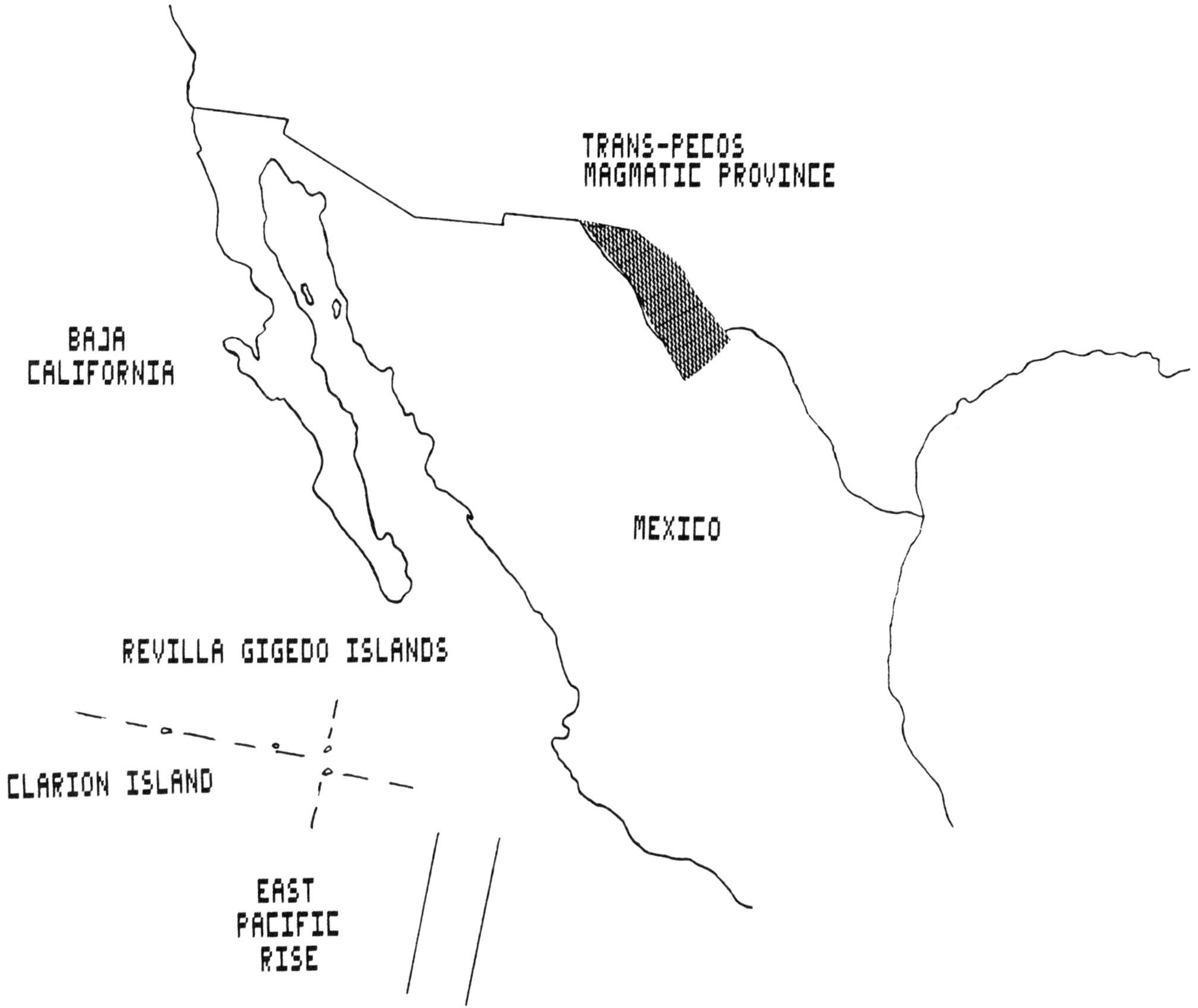

Figure 1. Index map showing locations of Clarion Island and Trans-Pecos magmatic province. Solid lines indicate East Pacific Rise; dashed lines, fracture zones.

Bryan (1967) reported seven major element analyses and norms of selected rocks of the CI group. The mafic rocks vary from nepheline- to hypersthene-normative. A trachyandesite and trachyte both yield normative quartz. However, this may be the result of the oxidized state of the rocks (Fe_2O_3/FeO = 1.77 and 1.34, respectively). Our analyses, assuming Fe_2O_3/FeO = 0.2, yield nepheline-normative compositions for most rocks, although one sample (CHB-23) contains 0.73 percent normative quartz. Bryan (1967) suggested that the various compositions may be related through a crystal fractionation model, and cited the common occurrence of "cumulate gabbro xenoliths" as supporting evidence for this model.

Trans-Pecos Magmatic Province

The TPMP is a significant belt of intracontinental alkaline magmatism that extends from New Mexico, through west Texas, into Coahuila, Mexico (Fig. 1). Barker (1977, 1979a, b) demonstrated that, in terms of major element chemistry, the province is similar to the Kenyan rift field. McDowell and Clabaugh (1979) and Barker (1979b) presented evidence suggesting that the TPMP was related to subduction processes that were active off western North America during much of the Tertiary (see Keith, 1978).

Magmatism in the TPMP extended from 48 to 17 Ma (McDowell and Clabaugh, 1979; McDowell, 1979; Parker and McDowell, 1979; Laughlin and others, 1982), with a major pulse of activity during 38 to 30 Ma. At least eight major eruptive centers have been described in the west Texas portion of the belt (Henry and Price, 1984).

McDowell and Clabaugh (1979) have delineated belts of distinct composition within and subparalleling the trend of the

TPMP, based on the atomic (Na + K) /Al ratio versus SiO_2. Barker (1977) referred to the western belt of the TPMP as metaluminous and the eastern belt as "more alkalic." These compositions appear to grade into rocks of more calcalkaline affinity to the west through Chihuahua toward the Sierra Madre Occidental (McDowell and Clabaugh, 1979; Cameron and others, 1980; Keller and others, 1982; Nelson and Nelson, 1986).

Barker and others (1977) have determined that the initial $^{87}Sr/^{86}Sr$ ratios of evolved intrusive phases from the northern TPMP vary from 0.7031 to 0.7120. Similarly, Kesler and others (1983) have analyzed intrusives along the southern extension of the belt in Coahuila, Mexico. The $^{87}Sr/^{86}Sr$ values of these rocks vary from 0.7074 to 0.7118. Both groups of authors suggest that assimilation of crust was an important part of the evolution of the evolved TPMP magmas.

Elsewhere, we have presented a model for the evolution of the rocks of the eastern TPMP (Nelson and others, 1984a, 1984b, 1987a). The salient points of this model are summarized below.

The model has been constructed from major and trace element data from two major volcanic terrains within the eastern TPMP. For simplicity, we refer to these terrains as the Big Bend and the Davis Mountain Centers, even though they extended over large areas and certainly comprised multiple vents. The rocks within these centers follow either a sodic trend of alkali basalt, including hypersthene-normative varieties, hawaiite, mugearite, trachyte, quartz trachyte, and rhyolite, or a potassic trend of similar evolved rocks, but having alkali basalts, trachybasalts, and trachyandesites in the mafic to intermediate range. The evolved rocks in both centers vary from metaluminous to peralkaline, and high-silica rhyolites (comendites) are abundant. Rocks as undersaturated as basanite and nephelinite have been found. Figure 2 illustrates that, within the eastern TPMP, a progressive variation from nepheline-normative to quartz-normative composition occurs. The utility of diagrams such as Figure 2 in interpreting the evolution of volcanic rocks has been discussed by Miyashiro (1978).

Chemical variation diagrams yield linear data arrays, with the two centers being indistinguishable on most plots. Scatter outside analytical uncertainty is evident on most diagrams, however, and this, when considered with the duration of magmatism in the belt, argues convincingly against a single magma chamber or even a single episode of magmatic activity. Marked changes in slopes in the trends of the data were used to delineate various stages of magmatic activity; specific element variations were then utilized to constrain magmatic processes using phenocryst phases and assuming progressive variations in the values of the partition coefficients. Four stages of magmatic evolution were identified; the major processes that were operating include crystal fractionation, combined fractionation-replenishment, and combined fractionation-assimilation. Thorium concentration was used as a monitor of magmatic evolution, and stages I through IV correspond to Th concentration ranges of 2 to 10, 10 to 22, 22 to 32, and 32 to 66 ppm, respectively. Although we have identified specific assemblages involved in the fractionation, we suggest that these assemblages and/or their compositions and relative proportions were changing continuously, even within a given stage.

TABLE 1. CLARION ISLAND VOLCANIC ROCK DATA*

Formation	Rock Type**	Age (m.y.)
Monte Gallegos	Trachyandesite	1.66±0.09
	Trachybasalt	1.02±0.04
	Basalt	1.21±0.02-1.29±0.02
Pico Tienda de la Campaña	Trachyte	1.69±0.02
Bahìa Sulphur	Basalt	1.10±0.015-1.52±0.02

*Clarion Island data from Bryan (1967).
**Bryan's (1967) terminology.
Whole-rock K-Ar age determinations from J.Clark, J. Dymond, and L. Hogan.

Stage I consists of the variation from alkali basalt to trachyte. The dominant process was crystal fractionation of a time-averaged assemblage comprising olivine (33%), clinopyroxene (18%), plagioclase (39.5%), amphibole (7%), apatite (2%), and magnetite (0.5%). The trachyte magmas represented 10 to 30 percent of the original mass of basaltic magma. Magma replenishment may have been operative here; however, the simultaneous variations of incompatible (e.g., Th and Rb) and compatible (e.g., Sr and Sc) elements limit replenishment to less than 10 percent. Crustal assimilation is not considered to have been important during this stage. Supporting this contention are the initial Sr ratios for Big Bend rocks of stage I lithologies reported by Lewis (1978). Rocks varying from hawaiite to benmoreite analyzed by Lewis (1978) varied in initial $^{87}Sr/^{86}Sr$ from 0.7019 to 0.7044. Within this range, a hawaiite (0.7043), a mugearite (0.7042), and a benmoreite (0.7044) are indistinguishable.

Stage II produced the variation from trachyte to quartz trachyte. Crystal fractionation—olivine (0.5%), clinopyroxene (4%), plagioclase (10%), anorthoclase (72%), amphibole (6.5%), apatite (3%), and magnetite (4%)—magma replenishment, and crustal assimilation acted in varying proportions during this evolution. Replenishment is indicated by unfractured, subhedral glomerocrysts from a magma of alkali basalt to hawaiite composition in more evolved (quartz trachyte) rocks. Crustal assimilation may be indicated by simultaneously low Sr and Th abundances, and by high (>+9‰) oxygen isotopic compositions of some of the rocks (Nelson and others, 1984b).

In stage III, quartz trachyte to rhyolite (comendite), magma replenishment became less important, and combined crystal fractionation-assimilation dominated. The time-averaged assemblage during stage III consisted of clinopyroxene (2%), plagioclase (2%) anorthoclase (37%), sanidine (45%), amphibole (7%), quartz (4%), apatite (1%), and magnetite (2%). Open-system fractionation is clearly indicated by the steady-state behavior of Ba

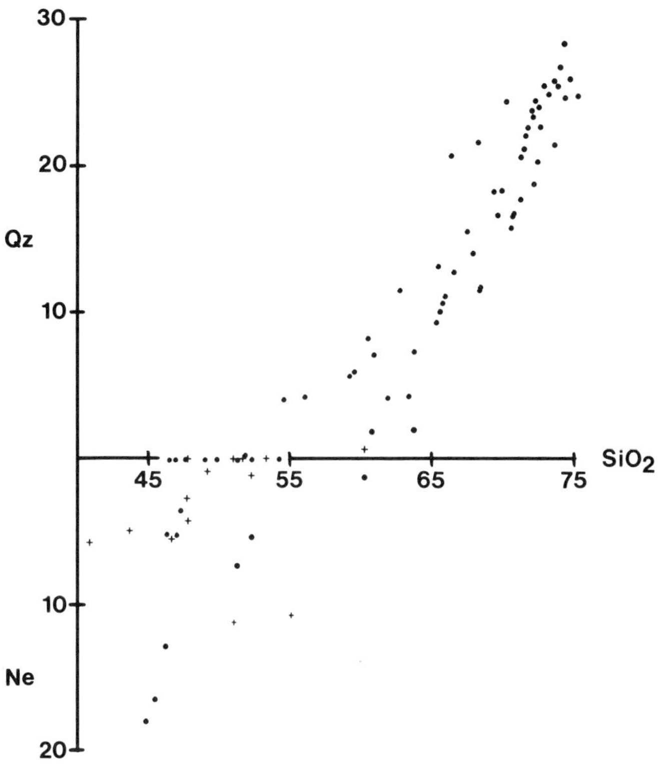

Figure 2. Weight percent of silica versus percentage of normative nepheline or normative quartz for eastern Trans-Pecos magmatic province (dots) and Clarion Island (plus signs).

and Sr. The development of deep Eu anomalies clearly requires high proportions of alkali feldspar in the fractionating assemblage. The concentration of Ba, however, instead of steadily declining as predicted by simple fractionation, remains constant (Fig. 4).

The prevalence of combined fractionation-assimilation continued through stage IV (rhyolite [comendite] to high-silica rhyolite [comendite]), as evidenced by the continued steady-state behavior for Ba and Sr and elevated $\delta^{18}O$ values of some rocks. The fractionation assemblage at this stage included clinopyroxene (0.3%), anorthoclase (33%), sanidine (45%), quartz (15%), amphibole (4.9%), apatite (1%), and magnetite (0.8%). Variable oxygen isotopic values (+7.3 to +10.7) may indicate a heterogeneous assimilant. Calculations indicate that the mass of these high-silica rhyolites is less than 5 percent of the original mafic magmas.

The age span of the eastern TPMP and the random nature of the exposed stratigraphy require that the evolution from stage I to IV occurred episodically. At any given time, separate chambers must have existed in various stages of the evolution. These chambers randomly(?) fed magmas to the surface.

We emphasize that this model represents a possible solution to the geochemical evolution of the eastern TPMP. We stress that, although the two eruptive centers—the Big Bend and Davis Mountains—evolved similarly, they did not do so identically.

GEOCHEMISTRY

Major and trace element compositions of selected Clarion Island rocks are given in Table 2. Normative mineralogies have been used to classify the rocks using the scheme of Irvine and Baragar (1971); norms were calculated assuming Fe_2O_3/FeO = 0.2. We make this assumption because we do not believe that the measured Fe_2O_3/FeO accurately reflects magmatic values. Oxidation of iron during and after emplacement can obviously have a significant and variable impact on this ratio. A narrow range of Fe_2O_3/FeO probably prevailed for the CI magmas and, although the 0.2 value may not be correct, it is certainly reasonable; the assummption of the constant value may provide a more accurate reflection of the relative norms. Most of the rocks plot on the sodic side of the An-Ab-Or diagram, and therefore the terms hawaiite-mugearite are used instead of trachybasalt-trachyandesite, as applied by Bryan (1967).

In contrast to rocks of the eastern TPMP (Fig. 2), the CI rocks do not display a trend of marked increase in degree of silica oversaturation. The rocks do define critically saturated and undersaturated trends, retaining nepheline in the norm through most of the evolved compositions (Fig. 2). Two of the intermediate compositions have sufficient normative nepheline to be classified as phonolites.

We will report the compositions and trends of the rocks of the eastern TPMP elsewhere (Nelson and others, 1987a). Here, to facilitate comparison of the CI and eastern TPMP rocks, we present chemical variation diagrams in which the CI data have been superimposed on the general trends of the Trans-Pecos rocks (stages I through IV) described above. Figure 3 (Th versus CaO, TiO_2, and SiO_2) and Figure 4 (Th versus Eu, Ba, and Sc) clearly indicate a marked similarity between CI and a portion of the eastern TPMP rocks. Not illustrated, but also displaying similar trends in the two regions, are plots of Th versus La, Rb, Zn, Sr, K, and others. The compositional variations observed in the CI rocks are virtually identical to stage I of the eastern TPMP evolution. The similarity extends to the rare-earth elements (REE). In Figure 5, the REE patterns of rocks of stage I in the eastern TPMP are compared with those of Clarion Island. The $(La/Sm)_n$ ratio of CI rocks (2.63 ± 0.6) and the eastern TPMP (2.71 ± 0.6) are identical, and REE concentration ranges, as indicated by Yb_n values, are also similar (Fig. 5).

DISCUSSION

Clarion Island and Stage I of TPMP

The marked similarities, both in concentrations and trends of most major and trace constituents between the stage I rocks of the eastern TPMP and of Clarion Island certainly suggest that the rocks evolved through similar processes. Similar arguments concerning continental versus oceanic alkaline volcanism have been made for the Cameroon line by Fitton and Dunlop (1985). Stage I processes in the eastern TPMP involved extensive crystal frac-

TABLE 2. MAJOR AND TRACE ELEMENT COMPOSITION OF SAMPLES FROM CLARION ISLAND*

	Bahia Sulphur			Monte Gallegos								Pico Tienda de la Campana		
	CHA-57	CHA-59	CHA-61	CHB-15	CHB-18	CHB-19	CHB-26	CHB-28	CHA-34	CHA-37	CHA-56	CHB-21	CHB-22	CHB-23
SiO_2	49.24	40.93	51.43	47.73	47.73	47.72	51.17	53.22	51.00	52.41	43.58	55.38	46.98	60.21
Al_2O_3	16.25	19.94	16.05	15.66	16.64	15.35	16.44	17.53	16.28	15.33	21.26	17.99	14.83	16.22
Fe_2O_3	2.42	2.37	2.33	2.15	2.62	1.98	2.04	2.38	2.34	2.67	2.32	1.50	2.71	1.49
FeO	9.98	9.79	9.62	8.86	10.81	8.18	8.40	9.85	9.66	10.92	9.57	7.52	11.21	7.43
MgO	5.36	4.71	3.25	6.74	4.30	4.96	2.56	1.77	2.68	3.75	3.81	0.94	6.44	1.14
CaO	8.89	8.66	6.63	8.91	8.30	9.02	5.10	6.12	6.13	6.61	8.81	2.68	9.16	3.67
Na_2O	4.11	3.44	4.61	4.00	4.23	3.93	6.44	4.81	4.32	5.20	3.75	7.25	4.13	5.95
K_2O	1.05	1.36	2.17	1.61	0.87	1.56	3.00	1.95	1.98	1.91	1.08	3.85	1.05	3.81
TiO_2	2.79	2.45	2.60	3.50	3.94	2.68	1.79	2.62	2.65	2.64	3.21	0.83	3.45	0.90
P_2O_5	0.68	1.55	0.83	0.72	1.07	0.64	0.88	1.04	0.93	0.91	0.81	0.22	0.46	0.81
MnO	0.19	0.16	0.17	0.16	0.19	0.15	0.23	0.19	0.15	0.19	0.18	0.26	0.17	0.25
Total	100.96	95.36	99.69	100.04	100.70	96.17	98.05	101.48	98.12	101.23	98.38	97.67	100.59	101.13
Rb	22.8	40.04	36.9	35.0	37.3	32.9	61.0	87.6	51.1	26.01	30.5	84.7	23.5	57.2
Sr	368	361	355	505	287	465	361	375	356	341	371	268	314	377
Zn	116	144	133	99	125	88	124	151	101	118	119	147	129	127
Y	26.38	29.92	35.83	29.08	49.96	29.31	46.57	59.63	35.60	34.57	30.70	55.71	24.05	60.04
Nb	21.60	19.48	25.98	28.57	14.67	24.80	38.90	60.53	26.77	28.08	17.76	62.63	14.38	59.23
Sm	6.99	6.93	9.79	8.12	8.26	7.19	10.34	12.42	9.59	9.51	7.47	13.21	5.11	13.70
Eu	2.50	2.17	3.31	2.58	2.74	2.38	3.03	3.30	2.75	2.75	2.50	4.65	1.94	3.90
Lu	0.35	0.32	0.50	0.31	0.74	0.28	0.56	0.56	0.44	0.46	0.36	0.82	0.26	0.80
Ba	412	405	590	593	457	547	1255	832	595	552	470	1431	368	1247
V	-	0.69	1.16	1.27	-	0.83	1.92	0.99	1.00	0.98	2.12	1.79	1.55	9.75
Th	2.51	2.48	5.04	4.86	2.34	3.99	7.25	7.03	4.81	4.65	2.74	10.07	1.91	9.75
Yb	2.60	2.47	3.66	2.25	5.21	2.24	4.53	4.42	3.45	3.52	2.64	6.01	1.99	6.1
Nd	28.67	29.37	39.35	49.38	25.48	33.08	52.36	45.74	34.79	34.29	27.98	61.17	25.74	53.29
La	23.97	24.22	40.33	44.0	23.07	36.42	58.10	60.87	38.94	38.11	26.26	69.55	15.69	70.35
Ce	53.46	78.55	84.23	83.90	78.69	70.81	112	138	106	50.9	53.85	147	33.24	116
Tb	1.12	1.55	1.49	1.14	1.57	1.09	1.76	2.17	1.57	1.13	1.23	2.05	0.85	1.98
Cr	120	20.85	18.89	208	19.55	200	21.09	2.63	23.55	117	75.67	-	216	28.99
Hf	5.14	8.22	8.54	6.27	7.87	5.69	11.31	14.64	7.58	4.82	5.0	16.04	4.22	10.24
Zr	339	321	317	272	341	279	528	570	271	294	226	672	-	556
Cs	-	0.29	-	0.64	0.27	-	0.46	0.82	1.13	0.21	-	4.25	-	0.26
Sc	24.81	16.82	17.20	25.04	16.13	26.27	11.49	10.21	16.48	23.27	24.57	10.08	3.08	12.50
Ta	2.21	3.24	3.48	3.35	3.17	2.62	4.13	5.55	1.91	2.12	2.13	6.47	1.75	4.08
Co	48.41	36.70	3.74	43.13	32.02	42.92	20.96	1.87	27.20	45.16	40.86	1.76	55.26	22.84
Classification:**	Hawaiite	Basanite	Hawaiite	Hawaiite	Hawaiite	Hawaiite	Phonolite	Hawaiite	Hawaiite	Hawaiite	Alkali basalt	Phonolite	Basanite	Mugearite

*Formation names from Bryan (1967).
**Procedure used techniques from Irvine and Baragar (1971).

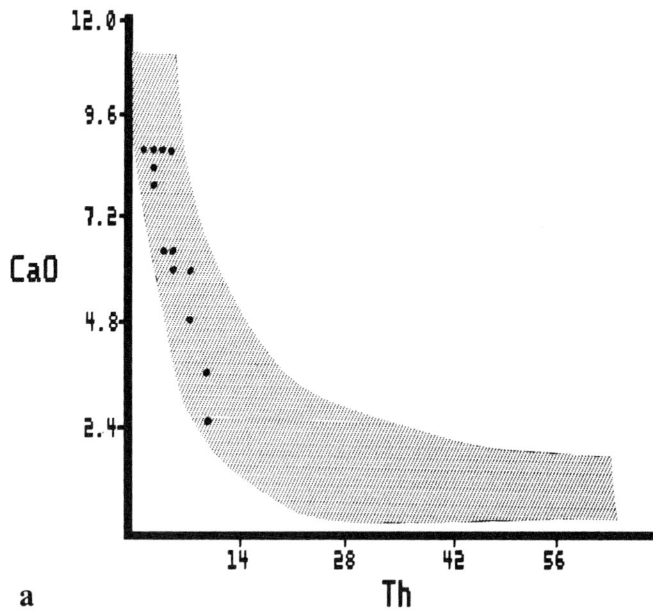

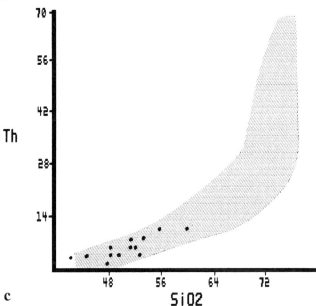

Figure 3. Plots of Th versus CaO (a), TiO_2 (b), and SiO_2 (c), comparing Clarion rocks (dots) with those of eastern TPMP, shown as shaded area.

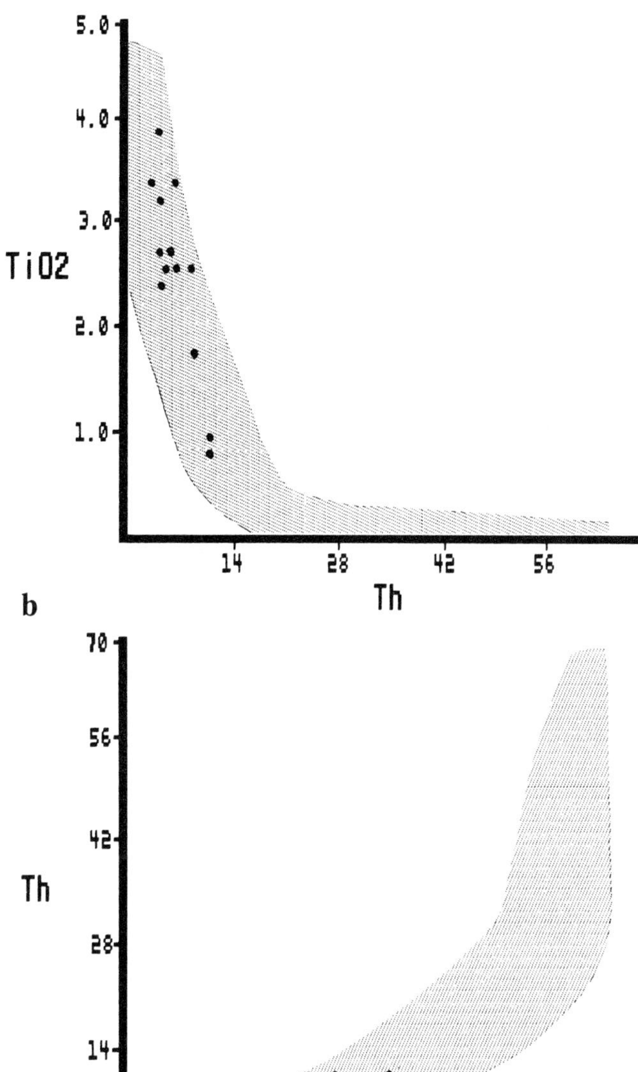

tionation of an assemblage that consisted of olivine, plagioclase, clinopyroxene, amphibole, magnetite, and apatite (Nelson and others, 1984a; Cameron and others, 1984). Figure 6 shows Harker diagrams for the CI rocks. Examination of Figures 3, 4, and 6 indicates that, for the bulk of the rocks analyzed, Al_2O_3, K, Nb, Rb, Th, Zr, and Ta increase with increasing SiO_2, whereas CaO, MgO, TiO_2, Sc, and Mg number decrease. These variations are consistent with removal of the above mineral phases.

In this regard, abundant phenocrysts of plagioclase and lesser pyroxene are found within the Bahia Sulphur basalt; Bryan (1967) reported the occurrences within this unit of "cognate" gabbroic xenoliths with mineralogies of plagioclase, olivine, augite, opaque minerals, and apatite; amphibole was not reported. These inclusions may well represent fragments of material crystallized against the walls of early chambers, only to be disrupted and brought to the surface by later magma pulses. The inclusions do preserve an open texture, perhaps supportive of a cumulate origin (Bryan, 1967), as does the crude layering observed in some inclusions. Based on our observations, the ratio of plagioclase to pyroxene of these inclusions varies markedly, indicating that the relative crystallization rates changed with time in the CI chambers. In our model for the TPMP, we suggest limited (<10%) replenishment of fresh magma during the fractionation event. In this regard, Bryan (1967, p. 1466) describes "segregations of trachytic compositions" observed in thin sections of a basalt within the Roco Monumento tuff unit. We have observed such segregations in hand-specimens from the Monte Gallegos unit. As we do not have compositional data on these segregations, we can only speculate that they may represent the incomplete homogenization of a limited mix of evolved and less-evolved magmas.

As indicated above, amphibole can be found in the rocks of the TPMP, particularly in rocks more evolved than hawaiite, although it has not been observed in the rocks of Clarion Island (Bryan, 1967). The lack of amphibole in Clarion rocks may be of significance in terms of developing the undersaturated trend, as is discussed below.

Stage O: Clarion Island

Further evaluation of Figure 6 indicates a possible compositional discontinuity in many of the Harker diagrams at a silica

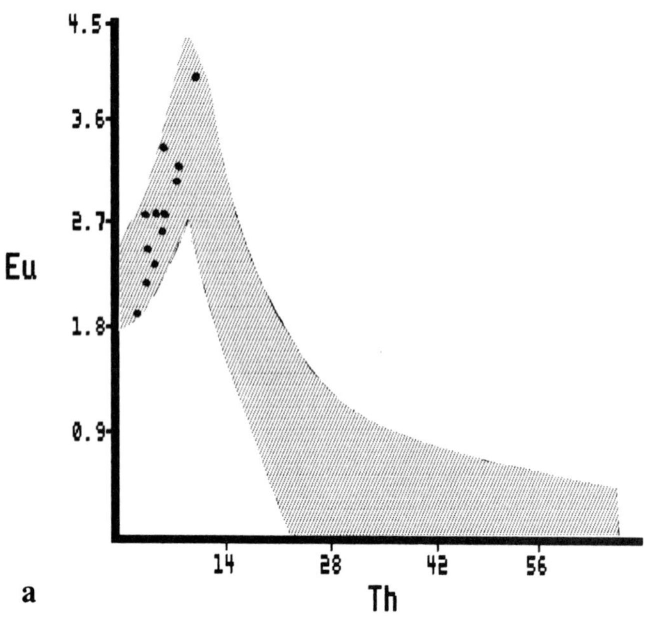

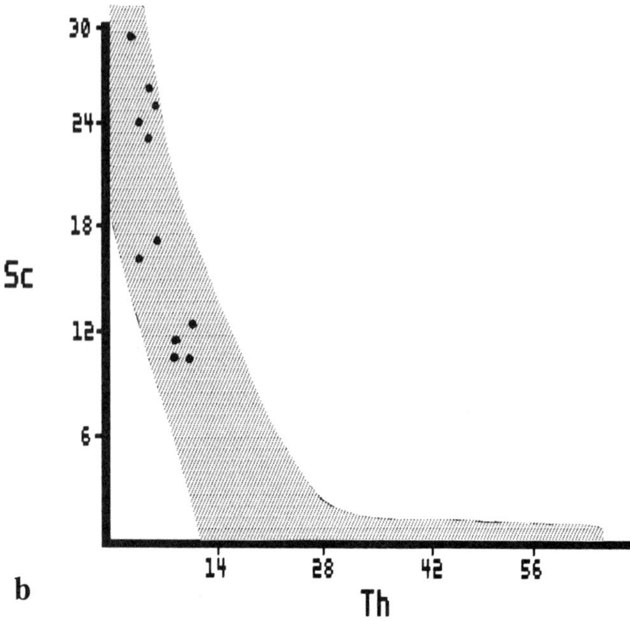

Figure 4. Plots of Th versus Eu (a), Sc (b), and Ba (c) for Clarion Island (dots) and eastern TPMP (shaded area).

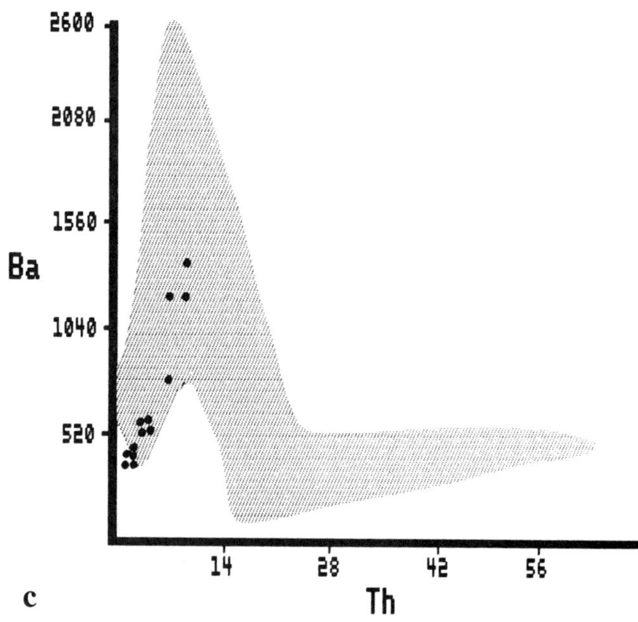

content of ~46 percent. Although this discontinuity is "defined" by only two samples (CHA-56 and CHA-59; Table 2), the departure of these two samples from the general trends of the CI rocks are well outside analytical uncertainty and, as discussed below, display considerable internal consistency. We therefore feel justified in comparing these samples to other CI rocks and in evaluating their compositions in terms of possible petrogenetic models. It is apparent, however, that the rocks are not on a single trend, and probably evolved independently.

In terms of most components, the CI rocks appear to become more evolved as the compositions move away from ~46 percent SiO_2 toward both higher and *lower* silica values. This seemingly paradoxical situation is particularly manifest in increasing incompatible element (e.g., K, Th, Rb, Nb, Zr, and Ta) and decreasing compatible element (e.g., Cr and Sc) abundances toward lower silica contents. Further, the Mg numbers of the lower silica rocks (<44%) are less (41.5–46.1) than those for many of the rocks in the silica range of 46 to 49 percent (e.g., Mg numbers vary up to 57.6 at SiO_2 = 47.7 percent).

In terms of the major components, we can summarize the variation of samples 56 and 59, relative to the general trend, as increasing Al_2O_3, K_2O, and Na_2O, as SiO_2, MgO, and CaO decrease. If these variations were produced by an episode of crystal fractionation not initially recognized in the TPMP (i.e., stage 0), then the inverse relation between CaO and Al_2O_3 argues against plagioclase as a significant component. Further, the decrease in SiO_2 suggests a high-silica assemblage, eliminating olivine and oxides(?) as principal phases in the assemblage. These variations, coupled with decreasing concentrations of MgO, CaO, Cr, and Sc, strongly suggest a pyroxene-dominated assemblage.

We envision, therefore, that some of the CI rocks have undergone a stage 0 fractionation event, dominated by orthopyroxene and clinopyroxene, but probably involving minor olivine (Presnall and others, 1978). We have not attempted to quantify this model because we do not have direct knowledge of the concentration of elements within the primary magma(s) of Clarion Island and because stage 0 trends are not clearly defined. For reasons discussed below, we do not feel justified in "adding in" appropriate amounts of the stage I assemblage to those lavas on the high-silica side to bring the Mg number to 68 to 70. By extrapolating incompatible element trends for stages 0 and I to a common intersection point, we arrive at the following concentration in the "primary" magma for selected elements: Th = 1.9, Rb = 3.0, K = 6,500, Ta = 1.4, and Zr = 80. From these concentrations, assuming all bulk distribution coefficients to be zero, we calculate that stage 0 rocks were produced by 30 to 60 percent fractionation.

Sc is more strongly partitioned into clinopyroxene than orthopyroxene; the opposite is true for Cr (see the compilation by Irving, 1978). Based on this and the observation that the relative stage 0 variation of Cr (210 to 21) is substantially greater than that observed for Sc (28 to 17), we suggest that the proportion of orthopyroxene in the stage 0 assemblage is greater than clinopyroxene.

Although there is some evidence that stage 0 may have operated in the eastern TPMP, the chemical trends discussed above for Clarion Island are not as clearly defined in the TPMP. Indirect evidence of such a high-pressure fractionation stage does exist in the form of megacrysts of orthopyroxene and aluminous augite in some basanite and alkali basalt intrusives in the eastern TPMP (Scheiffer and Nelson, 1981).

Oversaturated versus Undersaturated Trends

In spite of marked similarities in chemical trends in the rocks of Clarion Island and the eastern TPMP—leading to the suggestion that stage I fractionation is applicable to both regions—Figure 2 reveals a significant difference in trends of silica saturation. In particular, the eastern TPMP rocks follow a trend from nepheline-normative to quartz-normative within stage I and continuing through stage IV; the CI rocks remain critically saturated or become markedly undersaturated (ne >10 percent), during stage I. Here we wish to consider the origins of the oversaturated (leading to quartz trachytes to rhyolites) versus undersaturated (leading to phonolites) trends.

Three possible mechanisms may have been responsible for these different trends: (1) crustal assimilation, (2) the presence or absence of amphibole in the fractionating assemblage, and (3) migration of the thermal divide.

Crustal Assimilation. The potential for crustal assimilation is certainly greater for the TPMP than for Clarion Island, owing to the former's continental setting. The range of potential contaminants for the TPMP magmas is great, varying from "granitic" rocks, through carbonates. This virtually precludes quantitative evaluation of this process, although limited oxygen isotopic data of stage I rocks (Nelson and others, 1984b, 1987a) argue against significant assimilation of metasedimentary rocks. Further, highly oversaturated rocks can be found on many oceanic islands (e.g., Easter, Ascension, and Reunion Islands: Miyashiro, 1978). Thus it is not necessary to have crustal assimilation to produce a trend of increasing normative quartz. In this regard, Downes (1984) has demonstrated, through the use of Nd and Sr isotopes, that in the production of silica-saturated and silica-undersaturated magmas of the Massif Central of France, equal amounts of contamination were involved.

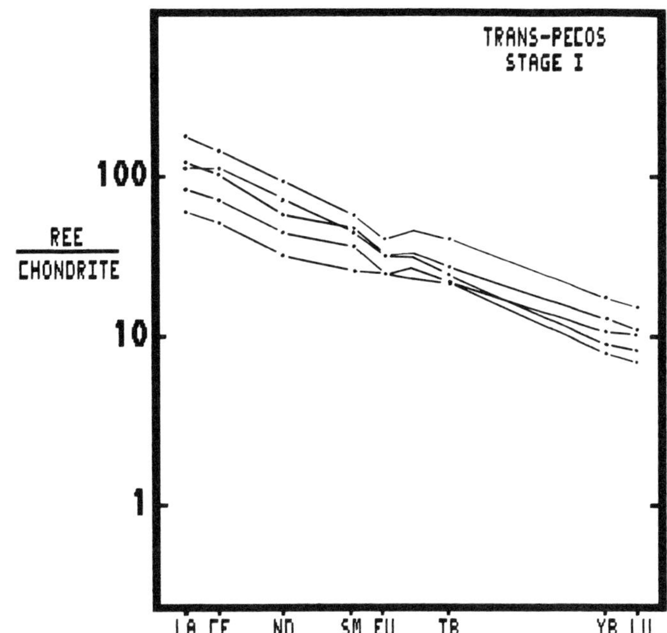

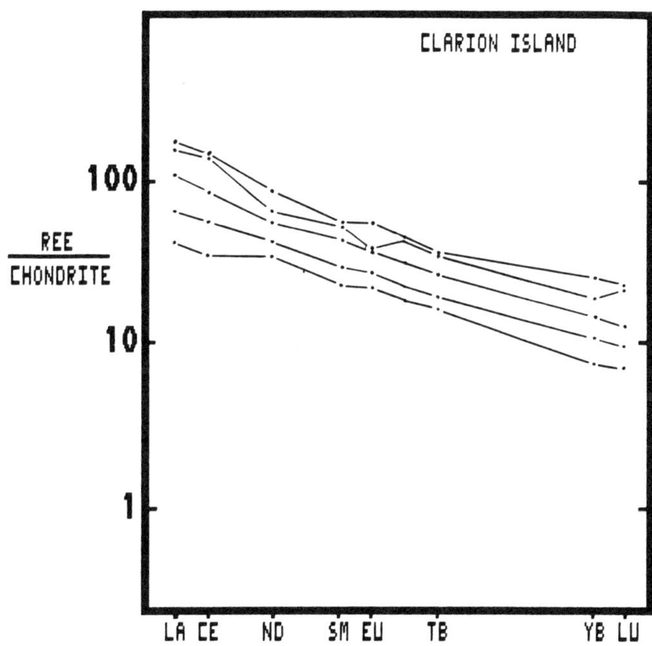

Figure 5. Rare-earth element diagrams for selected stage I rocks of eastern TPMP and rocks of Clarion Island. Analyses have been normalized to values for C-1 carbonaceous chondrites (Taylor and McLennan, 1981).

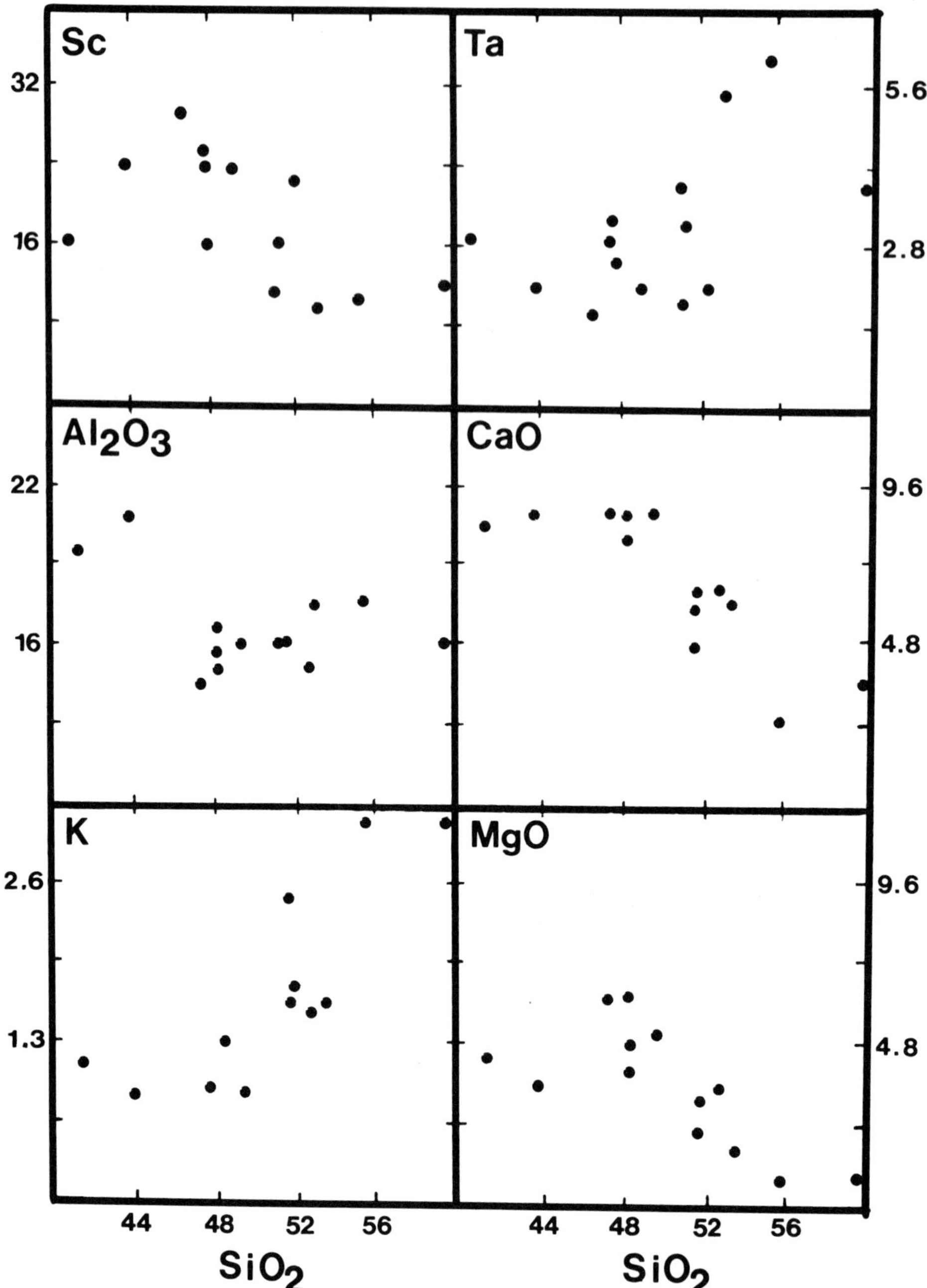

Figure 6. Harker diagrams for selected elements of Clarion Island rocks.

Presence or Absence of Amphibole. The production of silica-oversaturated melts by fractionation of amphibole from an originally nepheline-normative parent magma has been suggested by several authors (see the references in Downes, 1984). The low-silica content of the amphibole drives the residual liquid into the silica-saturated field. Amphibole occurs within the eastern TPMP rocks where rocks evolve from nepheline-normative to quartz-normative composition, but only in small amounts, and fractionation models allow amphibole to comprise no more than 7 percent of the total separating assemblage. This is probably insufficient to produce the trend toward quartz-normative compositions (e.g., Irving and Price, 1981). Further, in terms of the REE, amphibole exhibits maximum distribution coefficients in the middle REE range (Hanson, 1980; Irving and Price, 1981). As a consequence, extensive fractionation of amphibole, or reaction of the liquid with crustal or lithospheric amphibole, leads to a concave-upward REE pattern. Evaluation of the REE pattern of the evolved rocks of the Trans-Pecos indicates only a slight

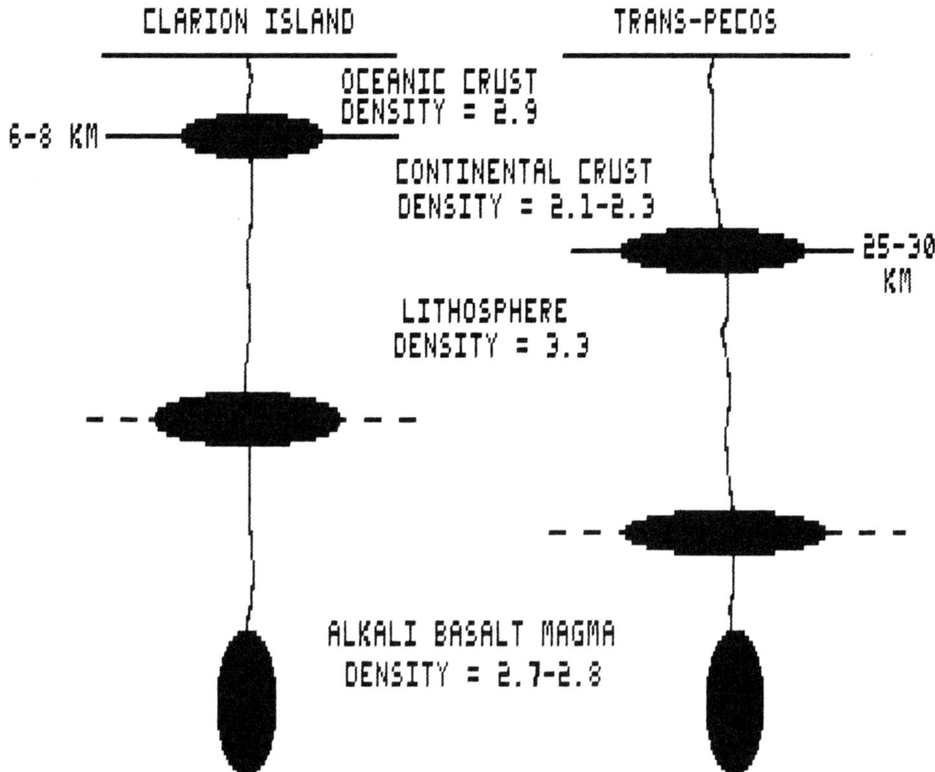

Figure 7. Schematic diagram depicting evolution of alkaline magmas below Clarion Island and Trans-Pecos magmatic province. Alkali basalts, produced in similar(?) manner under both regions, eventually pool at base of respective crusts. Alkali basalts fractionate at low pressures beneath Clarion Island, producing trend of silica undersaturation, whereas similar basalts fractionating under higher pressures beneath Trans-Pecos magmatic province produce trend of silica oversaturation (see text).

tendency for this development, which is consistent with only minor amphibole fractionation (Nelson and others, 1987a).

Migration of the Thermal Divide. Yoder and Tilley (1962) demonstrated that the thermal divide, broadly equivalent to the critical plane of undersaturation at low pressures, disappears at high pressures. Thompson and others (1972) and Miyashiro (1978) suggested that the evolution of quartz-normative from nepheline-normative magmas may be the result of the migration of this divide into the nepheline field. As a consequence, magmas originally nepheline-normative will evolve "down" the thermal gradient toward and eventually across the critical plane of undersaturation. Presnall and others (1978) evaluated the crystallization of "basalts" within the CaO-Al_2O_3-MgO-SiO_2 system at pressures up to 20 kbar. They determined that the position of the divide was pressure sensitive, primarily because of the changing composition of diopside, which became poorer in SiO_2 and richer in Al_2O_3 at higher pressure. At pressures greater than 4 kbar, the thermal divide resides within the nepheline-normative field.

We suggest that the migration of the thermal divide plays a critical role in the evolution of composition in the two regions. The situation is depicted in Figure 7. We regard the production of parental alkaline magmas in the two regions as broadly similar (see below). During ascent of the magmas, high-pressure fractionation (stage 0) may have affected both. Beneath the Trans-Pecos region, the thick continental crust acted as a density barrier, causing the pooling of the magmas at pressures exceeding 4 kbar. Under these conditions, the thermal divide lies within the nepheline field, and the originally nepheline-normative basaltic magma evolves toward and eventually across the critical plane of undersaturation. The alkali basalt to rhyolite trend is produced. An upper pressure of 9 kbar, however, is indicated by the requirement of plagioclase in the fractionating assemblage. In contrast, the most probable pooling place for the Clarion magmas would be at the base of, or within, the oceanic crust (Fig. 7). As a consequence of a pressure <4 kbar, the nepheline-normative magma(s) will lie to the nepheline side of the thermal divide and will therefore migrate *away* from the critical plane of undersaturation. The undersaturated trend of alkali basalt to phonolite results.

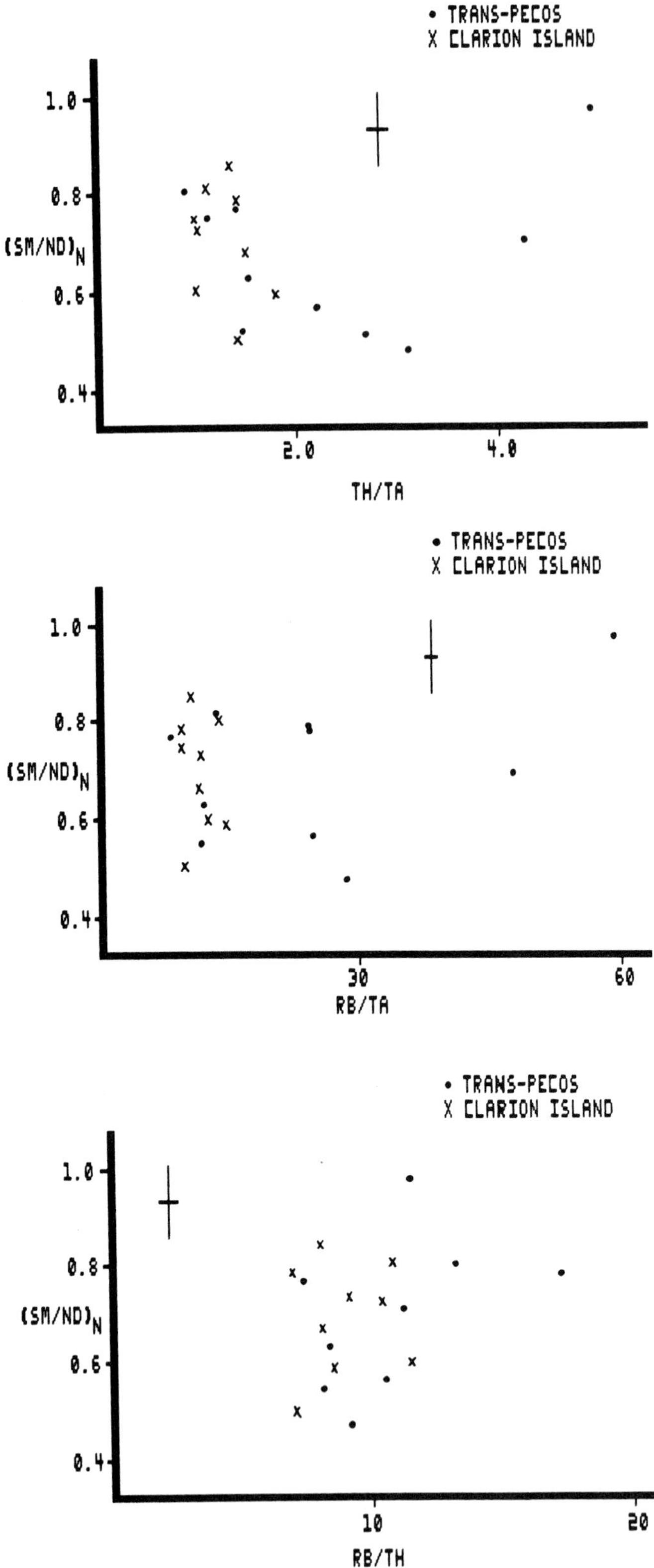

Figure 8. Chondrite-normalized Sm/Nd ratio plotted against Th/Ta, Rb/Ta, and Rb/Th ratios for rocks of Clarion Island and eastern Trans-Pecos magmatic province.

Note that these processes are not mutually exclusive. A variety of situations may arise through which magmas may fractionate at multiple levels (i.e., pressures) in both oceanic and continental regions; this gives rise to associations comprising both quartz- and nepheline-normative compositions (e.g., the Massif Central: Downes, 1984; the Cameroon line of West Africa: Fitton and Dunlop, 1985; and the eastern TPMP itself: Barker, 1977, 1979a, Nelson and others 1987b).

Implications for Mantle Source Regions. Abundant evidence, in the form of major and trace element compositions and of Sr and Nd isotopic compositions, suggests broad similarities for the source regions of alkali basalt, whether in continental or oceanic settings (e.g., Schwarzer and Rogers, 1974; Hanson, 1977; Allegre and others, 1981, Fitton and Dunlop, 1985; Sun and Hanson, 1975; Hanson, 1977). The implication of these models is that the principal influence on the composition of the magmas lies below the respective crust and lithosphere of the two environments (Sun and Hanson, 1975). The similarity of compositional trends and the lack of apparent crustal contamination observed in stage I of Clarion Island and the eastern TPMP are consistent with the above conclusions.

In an attempt to evaluate the suggestion of similar sources, we have evaluated the trace element compositions of flow units in both areas. Many authors have used the compositions of primary magmas to characterize the nature of the source regions. Although rocks with compositions appropriate for primary magmas do exist in the eastern TPMP (Nelson and others, 1987a), as yet there are none analyzed from Clarion Island (see below, however). Accordingly, we use ratios of elements whose bulk distribution coefficients are low and approximately equal. These ratios will therefore be relatively insensitive to fractionation and, if properly selected, to differing degrees of melting. We have chosen Sm/Nd, Rb/Th, Rb/Ta, and Th/Ta. Partial melting models, assuming a garnet lherzolite source, indicate that these ratios vary less than the analytical uncertainty for melts exceeding 1 percent. Variations of these ratios outside analytical uncertainty should thus reflect variations within the source region.

Figure 8 portrays $(Sm/Nd)_n$ plotted against the three remaining ratios. Examination of these diagrams indicates that real variations occur with all four ratios. Of particular importance is the observation that the source regions of Clarion Island and the eastern TPMP have nearly identical $(Sm/Nd)_n$ ratios, whereas in all three cases the rocks of the eastern TPMP have much broader ranges in Rb/Th, Rb/Ta, and Th/Ta, which extend from values characteristic of CI rocks to much higher values. Although not shown, the overlap of the compositions of rocks from Clarion Island and from stage I of the eastern TPMP is also apparent in terms of Zr/Nb and Zr/Y.

In all three diagrams, the CI rocks define relatively narrow trends that extend from $(Sm/Nd)_n$ of 0.5 to 0.85, implying that the sources for the CI magmas were LREE–enriched at the time of magma genesis. Similar arguments can be made for the eastern TPMP.

We suggest that the observed variations of $(Sm/Nd)_n$ reflect

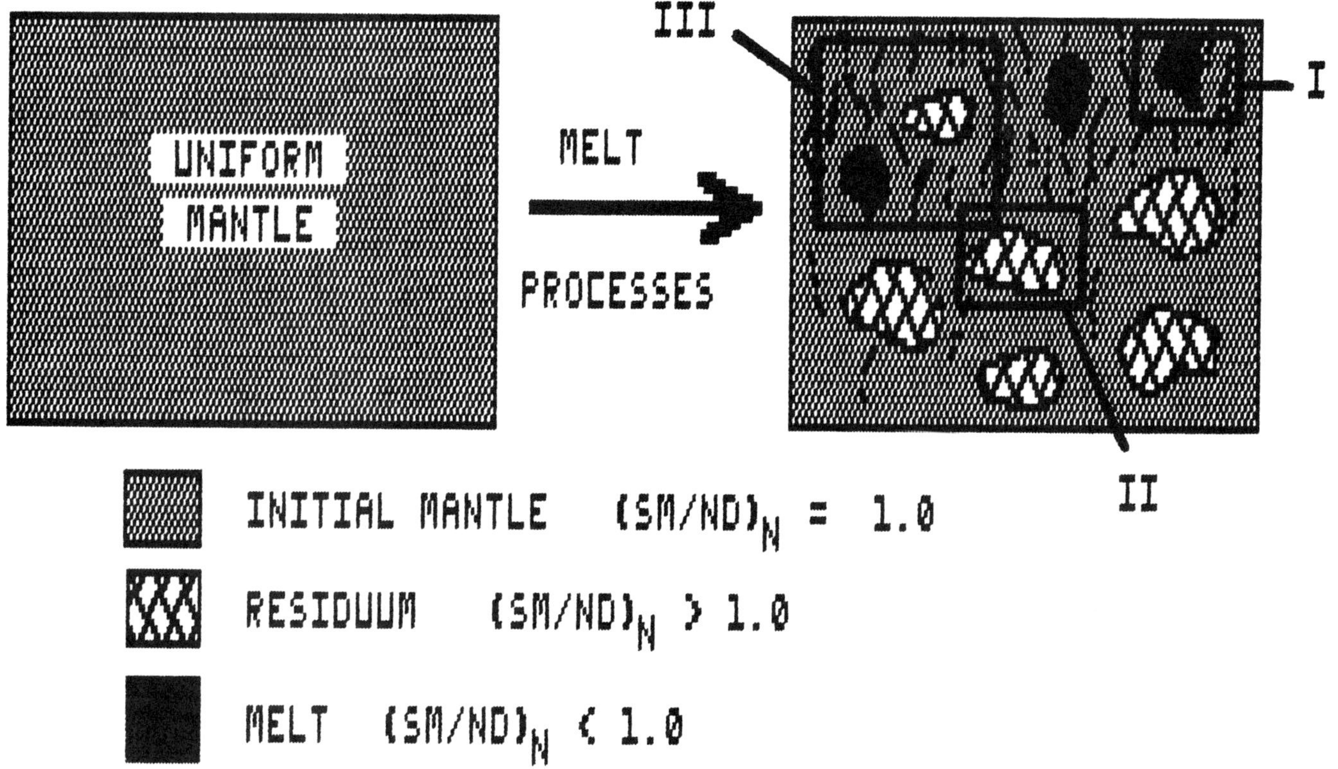

Figure 9. Schematic diagram depicting evolution of an originally homogeneous mantle into heterogeneous mantle through melting or metasomatic processes. Through these processes, original mantle evolves toward either enriched (i.e., veined with cumulates or trapped melt with Mg number of 68) or depleted (residuum after melt extraction with Mg number >72-75) mantle domains . Variable proportions of these domains contributing to melt, either as result of differing volumes undergoing melting or differing degrees of melting, will produce range of "primary" compositions.

magma production from a heterogeneous mantle. Preliminary isotopic data support this contention (Nelson and Gerlach, 1986). The range $(Sm/Nd)_n$ is probably the result of differing contributions from the compositional end members that define the heterogeneity. A variety of processes may act to produce these compositional end members and their co-mingling in a given volume of mantle.

Perhaps the simplest mechanism to produce heterogeneities in the mantle, and therefore variations in the Sm/Nd ratio, is variable partial melting and liquid extraction. During this process, as the result of solid-liquid partitioning, the liquid acquires a low Sm/Nd ratio, whereas the residuum acquires a higher value. The emplacement or incomplete extraction of the liquid may lead to a veined mantle. The veins could be either cumulates or trapped melt, and might possibly bear a genetic relationship to their host peridotite. Conversely, a genetic relation may not exist, and the host mantle might be undepleted, previously depleted, or previously and variably enriched. Further, the "fluids" involved might be magma, volatiles, or both (Schilling and others, 1980). Given these possibilities, the overlap of Clarion and the eastern TPMP in terms of Sm/Nd, La/Sm, Zr/Nb, Zr/Y, Rb/Ta, Th/Ta, etc., becomes even more striking. Figure 9 depicts a simplified model, similar to the SYS model of Zindler and others (1979), in which an originally homogeneous mantle becomes heterogeneous on a scale ranging from centimeters to tens of kilometers. Individual domains exist of residuum and veins or plutons of crystallized melt (Hanson, 1977; Tarney and others, 1981; Le Roex and others, 1983; Humphris and Thompson, 1983; Feigenson, 1984; Fodor and Vetter, 1984; Fitton and Dunlop, 1985). Two related phenomena, then, contribute to the variation in Sm/Nd. The first is the relative proportions of veins, residuum and original mantle (e.g., volumes I and II in Fig. 9). The second factor influencing the Sm/Nd value is the degree of melting and the relative scales of melting and heterogeneity (Hanson, 1977; Zindler and others, 1979; Tarney and others, 1980; Fodor and Vetter, 1984). If the volume of mantle undergoing melting and the degree of melting are large enough, the original heterogeneities will be "averaged out" (Cohen and O'Nions, 1982). Similar conclusions have been reached by Staudigel and others (1984) regarding the Loihi Seamount, the most recent manifestation of the hawaiian hot spot.

It is significant that the rocks of the eastern TPMP have

nearly identical $(Sm/Nd)_n$ to the CI rocks. Similar source regions are therefore implied. The question remains, however, as to why the eastern TPMP rocks have larger variations in Th/Ta, Rb/Ta, and Rb/Th ratios. We suggest that these variations result from interactions of the ascending melt with continental lithosphere. Although some workers have argued against interaction with the continental lithosphere (e.g., Allegre and others, 1981; Fitton and Dunlop, 1985), others have found support for such an interaction (McDonough and others, 1986).

Saunders and others (1980) and Tarney and others (1980) suggested that, during the evolution of the crust and subjacent lithosphere, fractionation of the large ion-lithophile elements (LILE) (e.g., Rb and Th) from the high field strength (HFS) elements (e.g., Ta) would occur as a result of either fluid transport away from the subduction zone, preferentially carrying the LILE, the presence of accessory phases such as ilmenite, rutile, zircon, and apatite in the oceanic crust concentrating the HFS ions, or fractionation of these mineral phases. As a consequence, the continental lithosphere should have higher Th/Ta and Rb/Ta ratios than its oceanic counterpart. During the passage of magma through the continental lithosphere, production of small amounts of crustal melt may occur. The extraction and subsequent mixing of these melts with the ascending magmas (McDonough and others, 1986) may produce the higher Th/Ta and Rb/Ta ratios observed in the eastern TPMP magmas. The higher Rb/Th ratios may reflect the presence of phlogopite in the subcontinental lithosphere (Mahoney and others, 1985) and its preferential incorporation in the extracted melt.

Mg Numbers of Primary Mantle Melts. As mentioned, none of the CI rocks we analyzed were primary melts. We reached that conclusion based on accepted compositional criteria for recognition of primary melts. One of the most important criterion used to identify primary magmas has been a Mg number of the bulk rock of ~68, the value calculated (Roeder and Emslie, 1970) to be in equilibrium with mantle olivine of composition Fo_{86-92} (e.g., Irving and Green, 1976). Wilkinson and Binns (1977) and Wilkinson (1982), however, suggested that substantial portions of the mantle might be more Fe-rich, and therefore primary melts with Mg numbers of less than 68 might be possible. The model for CI and TPMP sources presented above, constructed from major and trace elements, is consistent with the conclusion of Wilkinson and Binns (1977). The heterogeneities produced certainly would have variable Mg/Fe ratios (Fig. 9), ranging from high in the residuum to lower in the vein material, depending on whether it represents trapped melt or a "cumulate" precipitated in the vein walls (Irving, 1980). It follows that the melts produced during melting of this mantle would possess variable Mg numbers as a function of the relative proportions of depleted versus enriched mantle melted.

Figure 10 shows plots of Nb and Zr against Mg numbers for both the CI and eastern TPMP rocks having less than 51 percent SiO_2. Clearly, these incompatible elements remain virtually constant in their concentration throughout a range of Mg numbers of 70 to 50. These data are not easily reconcilable with a fractionation model involving a single parent magma. Rather, we believe the data in Figure 10 indicate that primary magmas with a range in Mg numbers (i.e., 70 to 50) were produced (see Fig. 9). The variations in Nb and Zr observed within a given small range of Mg numbers probably reflect either variable proportions of depleted and enriched mantle material, or minor fractionation of individual magmas.

CONCLUSIONS

Clarion Island and the eastern Trans-Pecos magmatic belt consist of a series of alkaline rocks that have erupted in oceanic and continental settings, respectively. Exposed rocks of Clarion Island are sodic in nature, and follow a trend of increasing silica undersaturation. Rocks vary from alkali basalt to phonolite. Rocks of the eastern TPMP follow both sodic and potassic trends, and evolve from nepheline-normative to quartz-normative compositions. Rock types vary from basanite to high-silica comendites. Four stages of evolution have been recognized in the eastern TPMP: stage I, alkali basalt to trachyte; stage II, trachyte to quartz trachyte; stage III, quartz trachyte to rhyolite/comendite; and stage IV, rhyolite/comendite to high-silica rhyolite/comendite. The rocks have evolved under open-system conditions, with magma replenishment recognized in stages I and II, and crustal assimilation in stages II through IV.

Trace element concentrations and trends of the CI rocks are very similar to stage I of the eastern TPMP, suggesting similar sources, melt levels, and mechanisms of evolution. In addition, a high-pressure, pyroxene-dominated (Opx >Cpx) fractionation scheme has been recognized on Clarion Island and possibly the TPMP. Incompatible- and compatible-element variation suggests an evolution of magmas to *lower* silica content as a result of this fractionation process.

The mechanism suggested to explain the normative quartz versus normative nepheline trends in the two regions is different depths of fractionation, as a result of variable crustal thicknesses. At the higher pressures at the base of the continental crust beneath the TPMP, a migration of the thermal divide into the field of normative nepheline is produced, causing the originally nepheline-normative liquids to migrate toward and eventually across the critical plane of silica undersaturation, leading to quartz-normative compositions. At lower pressures, at the base of the oceanic crust beneath Clarion Island, the thermal divide is broadly coincident with the critical plane of silica undersaturation, and the original liquids will migrate away from this plane toward increasing normative nepheline. Crustal contamination and amphibole fractionation may contribute to the quartz-normative trend, but apparently do not control it.

In terms of incompatible-element ratios—$(Sm/Nd)_n$, Rb/Ta, and Th/Ta—CI and eastern TPMP rocks overlap extensively, although the eastern TPMP rocks extend to higher values of Rb/Ta and Th/Ta. The variation in $(Sm/Nd)_n$ is interpreted to be a source effect, produced by mixing of depleted, undepleted, and enriched (veined) mantle. Differing contributions of these

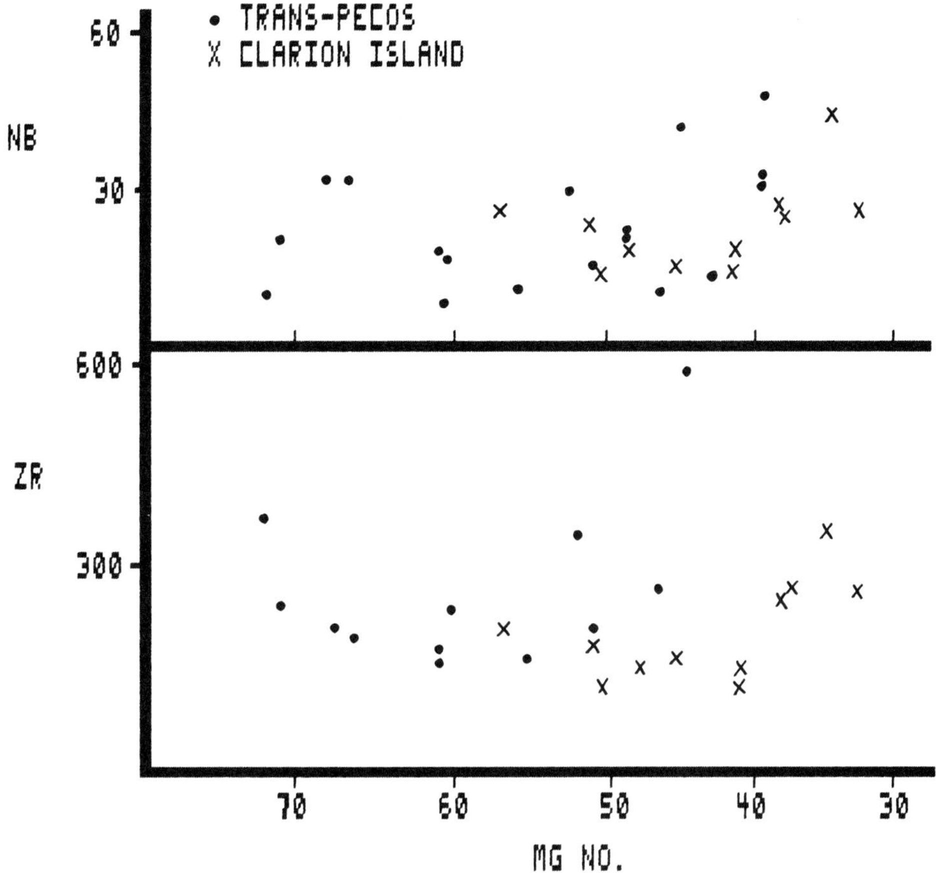

Figure 10. Nb and Zr plotted against Mg numbers for rocks of Clarion Island and stage I of eastern Trans-Pecos magmatic province. Note independence of Zr and Nb concentrations for Mg numbers in range of 70 to 50 for latter.

end members, through either differing melt domains or melt degree, produce the Sm/Nd variation. The higher Rb/Ta and Th/Ta ratios are believed to be the result of interaction of the TPMP magmas with the continental lithosphere during ascent.

For both CI and eastern TPMP rocks, variations of Zr and Nb are independent of Mg number in the range of 50 to 70 for the latter. This may be the result of the same mixing of sources that produced the Sm/Nd variation, and suggests that primary magmas with Mg numbers of less than 68 may be common.

Based on the compositional similarities of CI and eastern TPMP rocks described in this report, the compositions of alkaline magmas are in fact independent of the environment in which they erupt. This does not mean, however, that the respective sources are identical, only that the respective mantles consist of mixtures of similar enriched and depleted components. It is probably worth noting that the Sm/Nd characteristics are preserved independent of the volume of melt produced. Although we have not attempted to estimate the respective volumes of volcanic rocks in the two areas, it is certain that the TPMP exceeds Clarion Island by orders of magnitude. Because the controlling parameter is the relative proportion of enriched to depleted mantle, it is tempting to suggest that the first melts produced, primarily from the low melting temperature–enriched (veined) mantle, somehow control the amount of melt produced within the depleted mantle.

ACKNOWLEDGMENTS

We appreciate the constructive reviews of Dan Barker and Jill Pasteris. Funding for this project was in part derived from the Chihuahuan Desert research funds provided by Senate Bill 179 of the 68th Texas Legislature, General Appropriations Act of 1983, a Reactor-Sharing Grant from the Department of Energy through Texas A&M University, and the National Science Foundation (grant RII-8504808). We thank the National Park Service for its support and permission to collect samples from Big Bend National Park, and the Government of Mexico for its permission to visit and collect specimens on Clarion Island. The help and patience extended to one of us (D.O.N.) by the captain and crew of Oregon State University's oceanographic research vessel, the R.V. *Yaquina,* during a visit to Clarion Island in 1974, is much appreciated, as is the assistance of the other members of the Clarion landing party, particularly Jim Clark and John Toth who collected the CHB sample series. Jack Dymond resurrected the Clarion samples from storage at Oregon State and generously supplied unpublished K-Ar age determinations. Finally, we thank Socorro Brito for typing the manuscript.

REFERENCES CITED

Allegre, C. J., Duprè, B., Lambret, B., and Richard, P., 1981, The subcontinental versus suboceanic debate, I. Lead-neodymium-strontium isotopes in primary alkali bassalts from a shield area; The Ahaggar volcanic suite: Earth and Planetary Science Letters, v. 52, p. 85–92.

Barker, D. S., 1977, Northern Trans-Pecos magmatic province; Introduction and comparison with the Kenya Rift: Geological Society of America Bulletin, v. 88, p. 1421–1427.

——, 1979a, Magmatic evolution in the Trans-Pecos province, *in* Walton A. W., and Henry C. D., eds., Cenozoic geology of the Trans-Pecos volcanic field of Texas: Austin, University of Texas, Bureau of Economic Geology, Guidebook, v. 19, p. 4–9.

——, 1979b, Cenozoic magmatism in the Trans-Pecos province; Relation to the Rio Grande Rift, *in* Riecker R. E., ed., Rio Grande Rift; Tectonics and magmatism: Washington, D.C., American Geophysical Union, p. 382–392.

Barker, D. S., Long, L. E., Hoops, G. K., and Hodges, F. N., 1977, Petrology and Rb-Sr isotope geochemistry of intrusions in the Diablo Plateau, northern Trans-Pecos magmatic province, Texas and New Mexico: Geological Society of America Bulletin, v. 88, p. 1437–1446.

Bryan, W. B., 1967, Geology and petrology of Clarion Island, Mexico: Geological society of America Bulletin: v. 78, p. 1461–1476.

Cameron, K. L., Cameron, M., Bagby, W. C., Moll, E. J., and Drake, R. E., 1980, Petrologic characteristics of mid-Tertiary volcanic suites, Chihuahua, Mexico: Geology, v. 8, p. 87–91.

Chase, C. G., 1981, Oceanic island Pb; Two-stage histories and mantle evolution: Earth and Planetary Science Letters, v. 52, p. 277–284.

Chen, C.-Y., and Frey, F. A., 1983, Origin of Hawaiian tholeiite and alkali basalt: Nature, v. 302, p. 785–789.

Cohen, R. S., and O'Nions, R. K., 1982, The lead, neodymium, and strontium isotopic structure of ocean ridge basalts: Journal of Petrology, v. 23, p. 299–324.

DePaolo, D. J., and Wasserburg, G. J., 1976, Nd isotopic variations and petrogenetic models: Geophysical Research Letters, v. 3, p. 249–252.

Downes, H., 1984, Sr and Nd isotope geochemistry of coexisting alkaline magma series, Cantal, Massif Central, France: Earth and Planetary Science Letters, v. 69, p. 321–334.

Feigenson, M. D., 1984, Geochemistry of Kauai volcanics and a mixing model for the origin of Hawaiian alkali basalts: Contributions to Mineralogy and Petrology, v. 87, p. 109–119.

Fitton, J. G., and Dunlop, H. M., 1985, The Cameroon line, West Africa, and its bearing on the origin of oceanic and continental alkali basalt: Earth and Planetary Science Letters, v. 72, p. 23–38.

Fodor, R. V., and Vetter, S. K., 1984, Rift-zone magmatism; Petrology of basaltic rocks transitional from CFB to MORB, southeastern Brazil margin: Contributions to Mineralogy and Petrology, v. 88, p. 307–321.

Frey, F. A., Green, D. H., and Roy, S. D., 1978, Integrated models of basalt petrogenesis; A study of quartz tholeiites to olivine melilitites from southeastern Australia utilizing geochemical and experimental petrological data: Journal of Petrology, v. 19, p. 463–513.

Futa, K., and LeMasurier, W. E., 1983, Nd and Sr isotopic studies on Cenozoic mafic lavas from West Antarctica; Another source for continental alkali basalts: Contributions to Mineralogy and Petrology, v. 83, p. 38–44.

Hanson, G. N., 1977, Geochemical evolution of the suboceanic mantle: Journal of the Geological Society of London, v. 134, p. 235–253.

——, 1980, Rare earth elements in petrogenetic studies of igneous systems: Annual Review of Earth and Planetary Sciences, v. 8, p. 371–406.

Henry, C. D., and Price, J. G., 1984, Variations in caldera development in the Tertiary volcanic field of Trans-Pecos Texas: Journal of Geophysical Research, v. 89, p. 8765–8786.

Hofmann, A. W., and White, W. M., 1982, Mantle plumes from ancient oceanic crust: Earth and Planetary Science Letters, v. 57, p. 421–436.

Humphris, S. E., and Thompson, G., 1983, Geochemistry of rare earth elements in basalts from the Walvis ridge; Implications for its origin and evolution: Earth and Planetary Science Letters, v. 66, p. 223–242.

Irvine, T. N., and Baragar, W.R.A., 1971, A guide to the chemical classification of the common volcanic rocks: Canadian Journal of Earth Science, v. 8, p. 523–548.

Irving, A. J., 1978, A review of experimental studies of crystal/liquid trace element partitioning: Geochimica et Cosmochimica Acta, v. 42, p. 743–770.

——, 1980, Petrology and geochemistry of composite ultramafic xenoliths in alkalic basalts and implications for magmatic processes within the mantle: American Journal of Science, v. 280-A, p. 389–426.

Irving, A. J., and Green, D. H., 1976, Geochemistry and petrogenesis of the Newar Basalts of Victoria and South Australia: Journal of the Geological Society of Australia, v. 23, p. 45–66.

Irving, A. J., and Price, R. C., 1981, Geochemistry and evolution of lherzolite-bearing phonolite lavas from Nigeria, Australia, East Germany and New Zealand: Geochimica et Cosmochimica Acta, v. 45, p. 1309–1320.

Kay, R. W., and Gast, P. W., 1973, The rare earth content and origin of alkali-rich basalts: Journal of Geology, v. 81, p. 653–682.

Keith, S. B., 1978, Paleosubduction geometries inferred from Cretaceous and Tertiary magmatic patterns in southwestern North America: Geology, v. 6, p. 516–521.

Keller, P. C., Bockoven, N. T., and McDowell, F. W., 1982, Tertiary volcanic history of the Sierra del Gallego area, Chihuahua, Mexico: Geological Society of America Bulletin, v. 93, p. 303–314.

Kesler, S. E., Ruiz, J., and Jones, L. M., 1983, Strontium-isotopic geochemistry of fluorite mineralization (Coahuila, Mexico): Isotope Geoscience, v. 41, p. 65–76.

Laughlin, A. W., Kress, V. C., and Aldrich, M. J., 1982, K-Ar ages of dike rocks, Big Bend National Park, Texas: Isochron/West, v. 35, p. 17–18.

Le Roex, A. P., Dick, H.J.B., Erland, A. J., Reid, A. M., Frey, F. A., Hart, S. R., 1983, Geochemistry, mineralogy, and petrogenesis of lavas erupted along the southwest Indian Ridge between the Bouvet triple junction and 11 degrees east: Journal of Petrology, v. 24, p. 267–318.

Lewis, P. S., 1978, Igneous petrology and strontium isotope geochemistry of the Christmas Mountains, Brewster County, Texas [M.A. thesis]: Austin, University of Texas, 120 p.

Mahoney, J. J., Macdougall, J. D., Lugmair, G. W., Goplan, K., and Krishnamurthy, P., 1985, Origin of contemporaneous tholeiitic and K-rich alkalic lavas; A case study from the northern Deccan Plateau, India: Earth and Planetary Science Letters, v. 72, p. 39–53.

McDonough, W. F., McCulloch, M. T., and Sun, S. S., 1985, Isotopic and geochemical systematics of Tertiary-Recent basalts from southeastern Australia and implications for the evolution of the subcontinental mantle: Geochimica et Cosmochima Acta, v. 49, p. 2051–2067.

McDowell, F. W., 1979, Potassium-argon dating in the Trans-Pecos Texas volcanic field, *in* Walton, A. W., Henry, C. D., eds., Cenozoic geology of the Trans-Pecos volcanic field of Texas: Austin, University of Texas, Bureau of Economic Geology, Guidebook, v. 19, p. 10–19.

McDowell, F. W., and Clabaugh, S. E., 1979, Ignimbrites of the Sierra Madre Occidental and their relation to the tectonic history of western Mexico: Geological Society of America Special Paper 180, p. 113–124.

Menzies, M. A., and Murthy, V. R., 1980, Nd and Sr isotope geochemistry of hydrous mantle nodules and their host alkali basalts; Implications for local heterogeneities in metasomatically veined mantle: Earth and Planetary Science Letters, v. 46, p. 323–334.

Menzies, M. A., and Wass, S. Y., 1983, CO_2- and LREE-rich mantle below eastern Australia; A REE and isotopic study of alkaline magmas and apatite-rich mantle xenoliths from the Southern Highlands Privince, Australia: Earth and Planetary Science Letters, v. 65, p. 287–302.

Miyashiro, A., 1978, Nature of alkalic volcanic rock series: Contributions to Mineralogy and Petrology, v. 66, p. 91–104.

Nelson, D. O., and Gerlach, D. C., 1986, Sr, Nd, and Pb isotopic compositions of rocks of Clarion Island, east-central Pacific Ocean: Geological Society of America Abstracts with Programs, v. 18, p. 704.

Nelson, D. O., Nelson, K. L., and Mattison, G. D., 1984a, Reconnaissance geochemistry of the Trans-Pecos magmatic belt: Geological Society of America Abstracts with Programs, v. 16, p. 110.

——, 1984b, Implications of trace element and oxygen isotopic compositions for the origins of silicic rocks of the east-central Trans-Pecos magmatic province: Geological Society of America Abstracts with Programs, v. 16, p. 249.

Nelson, D. O., Nelson, K. L., Reeves, K. D., and Mattison, G. D., 1987a, Geochemistry of Tertiary alkaline rocks of the eastern Trans-Pecos magmatic province, Texas: Contributions to Mineralogy and Petrology (in press).

Nelson, D. O., Nelson, K. L., and Miner, R. S., 1987b, Geochemistry of plutons in the eastern Trans-Pecos magmatic province: Geological Society of America Abstracts with Programs, v. 19, p. 175.

Nelson, K. L., and Nelson, D. O., 1986. Geochemical evaluation of the magmatic evolution of the Van Horn Mountains Caldera and comparison with alkaline magmatism in the eastern Trans-Pecos magmatic belt, west Texas, *in* Price, J. G., Henry, C. D., Parker, D. F., and Barker, D. S., eds., Igneous geology of Trans-Pecos Texas; Field trip guide and research articles: Austin, University of Texas, Bureau of Economic Geology, Guidebook 23, p. 164–177.

O'Nions, R. K., Hamilton, P. S., and Evensen, N. M., 1977, Variations in $^{143}Nd/^{144}Nd$ and $^{87}Sr/^{86}Sr$ ratios in oceanic basalts: Earth and Planetary Science Letters, v. 34, p. 13–22.

Parker, D. F., and McDowell, F. W., 1979, K-Ar geochronology of Oligocene volcanic rocks, Davis and Barrilla Mountains, Texas: Geological Society of America Bulletin, v. 90, p. 1100–1110.

Presnall, D. C., Dixon, S. A., Dixon, J. R., O'Donnell, T. H., Brenner, N. L., Schrock, R. L., and Dycus, D. W., 1978, Liquidus phase relationships on the join diopside-forsterite-anorthite from 1 atm to 20 kbar; Their bearing on the generation and crystallization of basaltic magma: Contributions to Mineralogy and Petrology, v. 66, p. 203–220.

Roeder, P. L., and Emslie, R. F., 1970, Olivine-liquid equilibria: Contributions to Mineralogy and Petrology, v. 29, p. 275–289.

Saunders, A. D., Tarney, J., and Weaver, S. D., 1980, Transverse geochemical variations across the Antarctic Peninsula; Implications for the genesis of calc-alkaline magmas: Earth and Planetary Science Letters, v. 46, p. 344–360.

Schieffer, J. H., and Nelson, D. O., 1981, Petrology and geochemistry of megacrysts, xenoliths, and their host basalts from Terlinqua Mercury District of west Texas: Geological Society of America Abstracts with Programs, v. 13, p. 547.

Schilling, J. G., Bergeron, M. B., and Evans, R., 1980, Halogens in the mantle beneath the North Atlantic: Royal Society of London Philosophical Transactions, ser. A, v. 297, p. 147–178.

Schwarzer, R. R., and Rogers, J.J.W., 1974, A worldwide comparison of alkali olivine basalts and their differentiation trends: Earth and Planetary Science Letters, v. 23, p. 286–296.

Sclater, J. G., Anderson, R. N., and Bell, M. L., 1971, Elevation of ridges and evolution of the central eastern Pacific: Journal of Geophysical Research, v. 76, p. 7888–7915.

Staudigel, H., Zindler, A., Hart, S. R., Leslie, T., Chen, C.-Y., and Clague, D., 1984, The isotope systematics of a juvenile intraplate volcano; Pb, Nd, and Sr isotope ratios of basalts from Loihi Seamount, Hawaii: Earth and Planetary Science Letters, v. 69, p. 13–29.

Stille, P., Unruh, D. M., and Tatsumoto, M., 1983, Pb, Sr, Nd, and Hf isotopic evidence of multiple sources for Oahu, Hawaii basalts: Nature, v. 304, p. 25–29.

Sun, S. S., and Hanson, G. N., 1975, Evolution of the mantle; Geochemical evidence from alkali basalt: Geology, v. 3, p. 297–302.

Tarney, J., Wood, D. A., Saunders, A. D., Cann, J. R., and Varet, J., 1980, Nature of mantle heterogeneity in the North Atlantic; Evidence from deep sea drilling: Royal Society of London Philosophical Transactions, ser. A, v. 297, p. 179–202.

Taylor, S. R., and McLennan, S. M., 1981, The composition and evolution of the continental crust; Rare earth element evidence from sedimentary rocks: Royal Society of London Philosophical Transactions, ser. A, v. 301, p. 381–399.

Thompson, R. N., Esson, J., and Dunham, A. C., 1972, Major element chemical variation in the Eocene lavas of the Isle of Skye, Scotland: Journal of Petrology, v. 13, p. 219–253.

Vollmer, R., Ogden, P., Schilling, J.-G., Kingsley, R. H., and Waggoner, D. G., 1984, Nd and Sr isotopes in ultrapotassic volcanic rocks from the Leucite Hills, Wyoming: Contributions to Mineralogy and Petrology, v. 87, p. 359–368.

Wass, S. Y., and Rogers, N. W., 1980, Mantle metasomatism; Precursor to continental alkaline volcanism: Geochimica et Cosmochimica Acta, v. 44, p. 1811–1823.

Wilkinson, J.F.G., 1982, The genesis of mid-ocean ridge basalt: Earth-Science Reviews, v. 19, p. 1–57.

Wilkinson, J.F.G., and Binns, R. A., 1977, Relatively iron-rich lherzolite xenoliths of the Cr-diopside suite; A guide to the primary nature of anorogenic tholeiitic andesite magmas: Contributions to Mineralogy and Petrology, v. 65, p. 199–212.

Yoder, H. S., Jr., and Tilley, C. E., 1962, Origin of basalt magmas; An experimental study of natural and synthetic rock systems: Journal of Petrology, v. 3, p. 342–532.

Zindler, A., Hart, S. R., Frey, F. A., and Jakobsson, S. P., 1979, Nd and Sr isotope ratios and rare earth element abundances in Reykjanes Peninsula basalts: Evidence for mantle heterogeneity beneath Iceland: Earth and Planetary Science Letters, v. 45, p. 249–262

MANUSCRIPT ACCEPTED BY THE SOCIETY OCTOBER 15, 1986

Printed in U.S.A.

Geological Society of America
Special Paper 215
1987

Alkalic rocks of contrasting tectonic settings in Trans-Pecos Texas

Jonathan G. Price and Christopher D. Henry
Bureau of Economic Geology, University of Texas at Austin, Austin, Texas 78713
Daniel S. Barker
Department of Geological Sciences, University of Texas at Austin, Austin, Texas 78713
Don F. Parker
Department of Geology, Baylor University, Waco, Texas 76798

ABSTRACT

Alkalic rocks in Trans-Pecos Texas were emplaced in two distinctly different tectonic environments: one compressional (contractional) and one extensional. Rocks (Eocene and early Oligocene) of the older compressional environment can be divided into a western alkali-calcic belt and an eastern alkalic belt. The boundary between the two belts is parallel to the paleotrench that used to lie off the west coast of Mexico. The alkalic rocks were the most inland expression of subduction-generated volcanism. The predominance of east-striking dikes and veins and the orientation of en echelon dikes indicate igneous activity during residual compression remaining from Laramide deformation. As the dip of the subducting slab became gentler with time, calc-alkaline magmatism of Laramide age in Mexico graded eastward into the alkaline magmatism in Texas.

Widespread extension and normal faulting began about 24 Ma in the Texas portion of the Basin and Range province. Between 24 and 17 Ma, alkalic basalts were extruded and intruded at several localities, dominantly as north-northwest–striking dikes. Both nepheline- and hypersthene-normative basalts occur in the extensional environment.

Rocks of the compressional environment follow two major lines of differentiation: hypersthene-normative basalt to rhyolite and nepheline-normative basalt to phonolite. In contrast, rocks of the extensional environment are apparently limited to basalts. During contraction, magmas rising from the mantle probably formed chambers in which differentiation could occur. During extension, less differentiation occurred, either because tectonically dilated fractures permitted more direct rise of magma to the surface or because the volume of magma was too small.

Basalts of the two contrasting tectonic settings are broadly similar in alkalinity and silica saturation. The basalts of the extensional environment are, however, generally richer in magnesium than are the basalts of the compressional environment. This difference is not simply a matter of degree of differentiation but is probably related to the different pressure-temperature regimes of the mantle from which the basalts originated.

INTRODUCTION

Continental alkalic rocks occur primarily in regions of extension or continental rifting (Sorensen, 1974; Bailey, 1974). Although many authors have used the presence of peralkaline and alkalic rocks as evidence of extension (for example, Macdonald, 1975), occurrences in areas of compression or in orogenic belts are known (Sorensen, 1974; Barker, 1974). These two contrasting tectonic settings were superimposed in time in the Trans-Pecos region of Texas, as well as New Mexico and adjacent Chihuahua and Coahuila in Mexico. Alkalic rocks were emplaced during periods of both compression and extension.

The voluminous Eocene and early Oligocene alkaline magmatism of the region accompanied compressional deformation

TABLE 1. SUMMARY OF TECTONIC AND MAGMATIC EVENTS, TRANS-PECOS TEXAS

Age (Ma)	Tectonic Environment	Magmatic Events	Examples	Time Sequence (Ma)
~75-50	Laramide folding and thrusting	None in Trans-Pecos Texas; calc-alkaline magmatism to the west in Mexico		
48-32	Compression residual from Laramide deformation	Widespread, voluminous, compositionally diverse, alkali-calcic and alkalic magmatism	Nine Point Mesa hawaiite	47
			Christmas Mountains gabbro and rhyolites	42
			Van Horn Mountains caldera	38
			Marble Canyon stock	37
			Paisano Pass caldera	36
			Chinati Mountains caldera	33-32
31-25	Transitional period of early extension	Localized, compositionally diverse activity	Sierra Rica caldera complex	31-28
			Rattlesnake Mountain sill	28
			Bofecillos volcano	28-27
24-17	Basin and Range extension	Widespread, volumetrically minor alkalic basalt	Rim Rock dikes	24-18
			Black Gap flows and dikes	23-20
			Cox Mountain flow	17
16-0	Continued Basin and Range extension	None in Texas		

related to earlier Laramide folding and thrusting. Volumetrically minor Miocene alkalic basalts were emplaced during a later extensional period of Basin and Range deformation. In this paper we document and discuss petrologic differences between igneous rocks of the two contrasting tectonic environments, and we relate chemical variations between igneous suites to tectonic settings.

TERTIARY TECTONIC ENVIRONMENTS

Three tectonic environments characterized the Tertiary Period in the Trans-Pecos region (Table 1). First, Laramide crustal shortening began in Cretaceous time, peaked in the late Paleocene, and essentially ended in the early Eocene (Wilson, 1971). Second, widespread magmatism occurred in middle Eocene to early Oligocene time (McDowell, 1979) during a period of weak compression that was probably residual from the Laramide state of stress (Price and Henry, 1984). Third, Basin and Range extension, which was presaged by a shift in stress orientations approximately 31 Ma, began with region-wide normal faulting approximately 24 Ma and continues today. This section briefly discusses structural evidence for each of the three Tertiary tectonic environments. For more detailed descriptions, see Muehlberger (1980), Price and Henry (1984), and Henry and Price (1985; 1986b).

Laramide Deformation

Laramide deformation involved thrust faulting, folding, and monoclinal warping of Cretaceous and older rocks. Maximum principal compressive stress, σ_1, was directed east to northeast. Faults, fractures, folds, stylolites, and calcite twin lamellae found throughout the Trans-Pecos region provide evidence for east to east-northeast σ_1 (Muehlberger, 1980; Moustafa, 1983). Northeast compression has been documented locally (DeCamp, 1981; Berge, 1981; Price and others, 1985) and appears to have predated east to east-northeast compression.

Faults in the Precambrian basement, particularly those of the northwest- to west-northwest–trending Texas lineament, have been reactivated throughout the tectonic history of the region (Muehlberger, 1980). The directions of principal stress axes during Laramide deformation caused left-lateral strike slip on the basement faults of the Texas lineament (Muehlberger, 1980). These faults also controlled the terminations of some monoclines (Moustafa, 1983).

Convergence of the Farallon and North American plates is the generally accepted cause of Laramide deformation throughout North America (Dickinson, 1981). During the main pulse of Laramide deformation, from Late Cretaceous to early Eocene time, subduction-related igneous activity occurred mostly to the west of Texas, in Arizona, New Mexico, and northern Mexico (Coney and Reynolds, 1977). Magmatism in Trans-Pecos Texas did not begin until near the end of Laramide folding and thrusting, which occurred approximately 50 Ma.

Eocene and Early Oligocene Magmatism

Tertiary intrusive and extrusive rocks are widespread throughout Trans-Pecos Texas (Fig. 1). The earliest documented

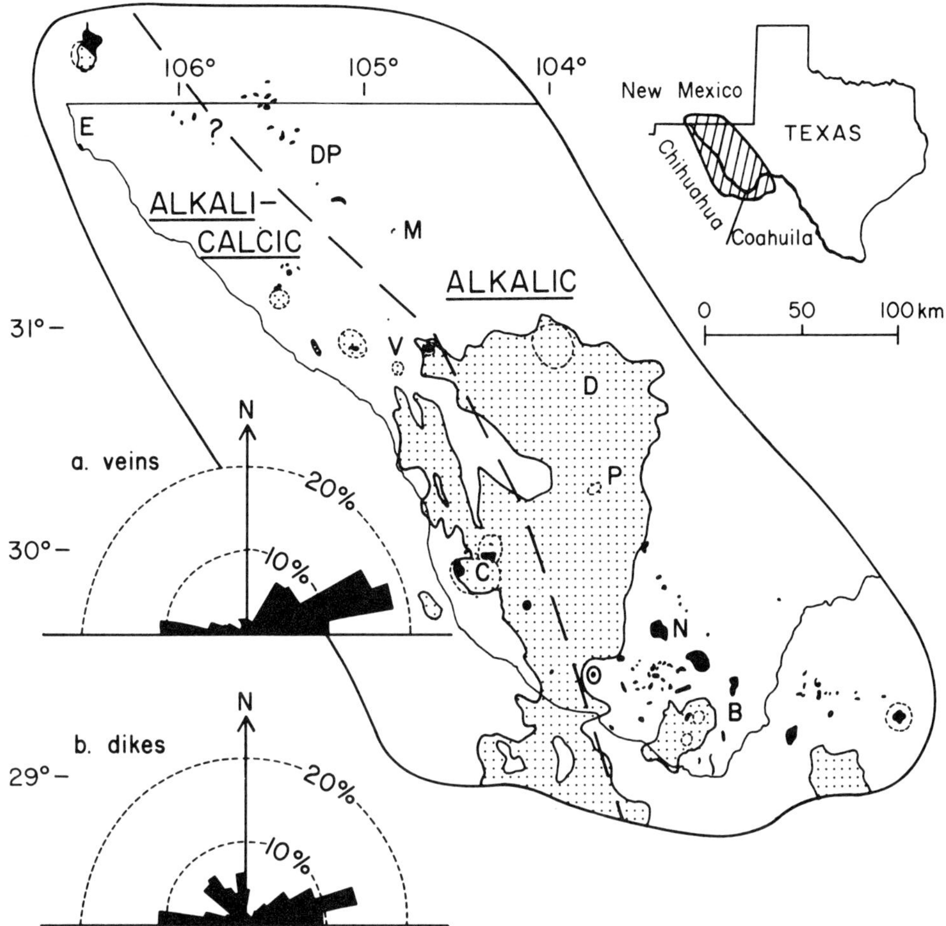

Figure 1. Eocene and early Oligocene igneous rocks in Trans-Pecos region. Heavy dashed line indicates caldera; dots, volcanic rock; solid, intrusion; B, Big Bend National Park; C, Chinati Mountains caldera; D, Davis Mountains; DP, Diablo Plateau; E, El Paso; M, Marble Canyon; N, Nine Point Mesa; P, Paisano Pass caldera; V, Van Horn Mountains caldera. Also shown are strike rosettes for veins (a) and dikes (b) that were emplaced during this period of igneous activity (from Price and others, 1985). Lengths of strike bars indicate percentages of total strike length: 74 km, representing 866 veins, and 175 km, representing 394 dikes.

igneous activity in the region was about 48 Ma (Hoffer, 1970; McDowell, 1979; Laughlin and others, 1982). The largest volumes of igneous rocks in the region were erupted from calderas between 38 and 32 Ma (Henry and Price, 1984). Laramide crustal shortening apparently ended before emplacement of the middle Tertiary igneous rocks. In general, the igneous rocks are not folded or thrust-faulted. Tilted volcanic rocks occur locally in and near calderas and near younger Basin and Range normal faults (Henry and Price, 1986b).

Orientations of dikes and veins that formed during the main phase of igneous activity (Fig. 1) indicate that least principal stress, σ_3, was oriented north to north-northwest. Price and Henry (1984) interpreted this orientation as evidence that σ_1 at this time was east to east-northeast, essentially the same as during Laramide deformation. At Glenn Draw in Big Bend National Park, a north-northwest–trending, left-lateral set of east-striking en echelon dikes indicate σ_1 in the same direction as Laramide deformation. Although these dikes have not been dated, similar basaltic dikes and other igneous rocks in Big Bend National Park fall within the time span, 47 to 33 Ma (McDowell, 1979; Laughlin and others, 1982).

Transitional Period of Early Extension

Orientations of dikes indicate that about 31 Ma, σ_3 shifted from the direction characteristic of Laramide compression to the direction characteristic of early Basin and Range extension. Basin and Range normal faulting did not begin in Texas, however, until about 24 Ma (Henry and Price, 1986b). Two volcanic centers were active during this transitional period of early extension. The

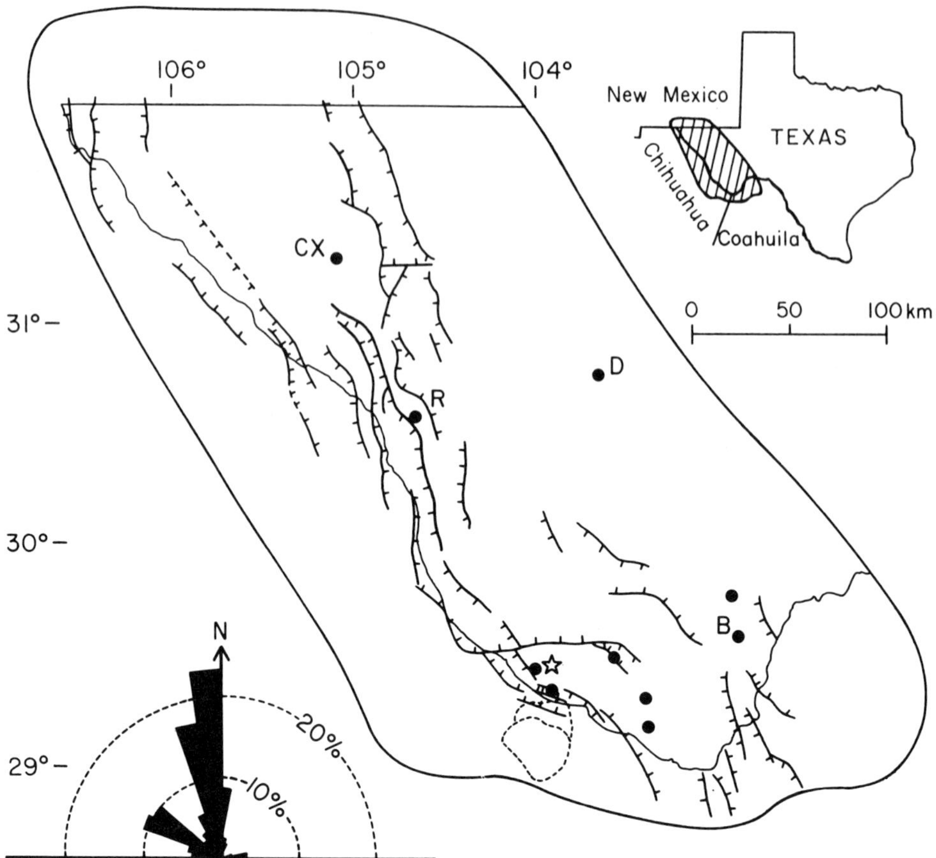

Figure 2. Basin and Range faults and late Oligocene to early Miocene mafic rocks in Trans-Pecos region (modified from Henry and others, 1985). Tick marks indicate major faults, generally with displacement greater than 300 m (ticks on downthrown side); heavy dashed line, Sierra Rica caldera complex; star, Bofecillos volcano; solid circles, 24- to 17-m.y.-old basalts; B, Black Gap; CX, Cox Mountain; D, Davis Mountains (localities noted by Parker, 1972); R, Rim Rock area. Also shown is strike rosette for Rim Rock dikes (from Price and others, 1985). Total strike length is 70 km, representing 480 dike segments.

Sierra Rica caldera complex in Chihuahua (Fig. 2) erupted 30 and 28 Ma (Chuchla, 1981; Gregory, 1981; Henry and Price, 1984). A nearby 31-Ma north-striking basaltic dike (Chuchla, 1981; Gregory, 1981) and a north-elongated granitic intrusion were emplaced with σ_3 oriented east to east-northeast. The predominantly north-northwest–striking basaltic dikes radiating from the 28- to 27-Ma Bofecillos volcano (Fig. 2) (McKnight, 1970; McDowell, 1979; Henry and Price, 1986b) indicate an east-northeast σ_3. The shift in σ_3 approximately 31 Ma presumably was accompanied by the beginning of regional extension.

Basin and Range Extension

Basaltic magmas intruded along Basin and Range normal faults and intercalated with early basin fill by 23 Ma (Dasch and others, 1969; Henry and Price, 1986b). Volumetrically minor basaltic magmatism was widespread in the Trans-Pecos region from about 24 to 17 Ma (Fig. 2) (Henry and Price, 1986b). Normal faulting probably began about 24 Ma and continued to the present. Extension in the Basin and Range province as a whole is attributed to a change from a convergent to a transform boundary along the western edge of North America (Dickinson, 1981; Zoback and others, 1981).

Stress orientations during early extension in Texas probably were similar to those throughout the province (Zoback and others, 1981); σ_3 was east-northeast. For example, the dominant strike of dikes in the Rim Rock area is north-northwest (Fig. 2). West-northwest–striking dikes, the subordinate trend in Figure 2, tend to be older than north-northwest–striking dikes. Dasch and others (1969) interpreted this change in strike with time as the result of a change in regional σ_3. Conversely, Henry and Price (1986b), using additional data from the Rim Rock area and other parts of the region, suggested that σ_3 was east-northeast throughout early Basin and Range extension. In the Rim Rock area, older northwest- and west-northwest–striking dikes were probably injected along preexisting faults of the Texas lineament. Once ex-

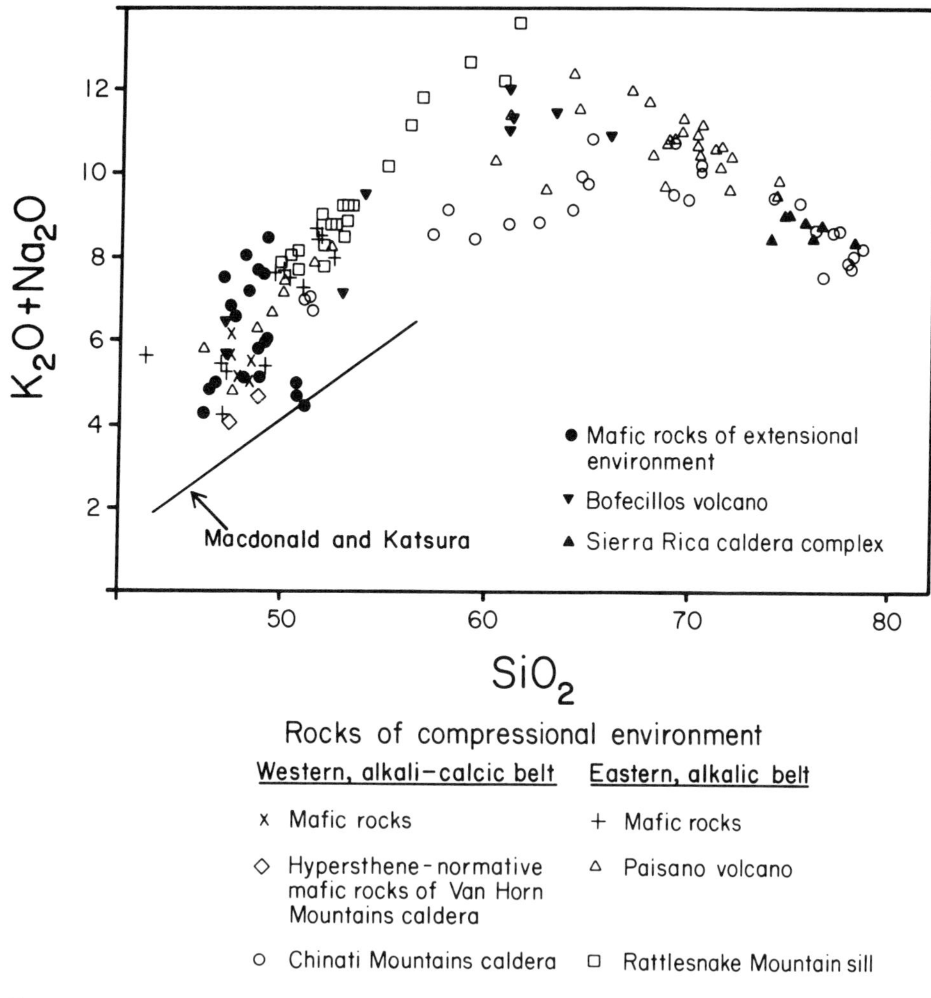

Figure 3. Alkali-silica plot (wt. %) comparing igneous rocks of contrasting tectonic environments in Trans-Pecos region. Data for Rattlesnake Mountain sill from Carman and others (1975). Included in mafic rocks of extensional environment are three samples of Dahl (1984) from north of Black Gap. Macdonald and Katsura (1964) line separating Hawaiian alkalic basalts (above) and tholeiitic basalts (below) shown for comparison.

tension was well underway, by approximately 21 Ma, new fractures that formed perpendicular to σ_3 were preferentially filled with dikes.

PETROLOGICAL DIFFERENCES BETWEEN IGNEOUS ROCKS EMPLACED DURING COMPRESSION AND EXTENSION

Alkaline igneous rocks were emplaced during two widely contrasting tectonic episodes in the Trans-Pecos region: Eocene and early Oligocene compression residual from Laramide deformation and Miocene Basin and Range extension. The Eocene and early Oligocene rocks can be separated into a western alkali-calcic or metaluminous (Barker, 1977) belt and an eastern alkalic belt (Fig. 1) on the basis of the Peacock (1931) alkali-lime index for suites of rocks. Volcanic rocks ranging from basalt to trachyte to rhyolite were erupted from numerous calderas in both belts (Henry and Price, 1984). Silica-undersaturated differentiates, phonolites and nepheline syenites, are restricted to the eastern alkalic belt. Although peralkaline varieties occur in both belts, the most strongly peralkaline rhyolites are in the eastern alkalic belt. Only basaltic rocks are recognized in the Miocene extension-related magmatism. The igneous rocks emplaced during the transitional period of early extension share characteristics of both the older compressional suites and the younger extensional basalts.

Comparison of Mafic Rocks: Similarities and Differences

Mafic rocks of the two contrasting tectonic environments are similar. The variation of total alkalies and silica in mafic rocks of the province (Fig. 3) shows the general alkaline character.

TABLE 2. REPRESENTATIVE ANALYSES OF MAFIC ROCKS FROM DIFFERENT TECTONIC ENVIRONMENTS IN TRANS-PECOS TEXAS
(Analyses in wt. %)

	Basin and Range Extension		Eocene to Early Oligocene Compression			
	Rim Rock Dike	Black Gap Lava	Western: Alkali-Calcic Belt Van Horn Mountains caldera		Eastern: Alkalic Belt Paisano Pass caldera	
	H82-83	77042	81-192	H84-10	SHEEP-3	PP-391
OXIDE						
SiO_2	46.64	49.59	47.06	46.80	46.06	48.72
TiO_2	2.37	1.69	2.27	3.59	2.88	2.75
Al_2O_3	16.51	15.35	13.89	16.32	17.29	16.39
Fe_2O_3	3.46	6.81	4.97	4.47	2.85	4.75
FeO	7.29	4.03	7.80	7.50	9.20	5.89
MnO	0.16	0.16	0.20	0.17	0.17	0.23
MgO	6.63	6.84	9.19	5.52	5.64	3.54
CaO	7.55	8.19	9.44	7.86	7.89	6.02
Na_2O	4.41	3.40	2.88	3.65	4.34	4.68
K_2O	2.34	0.87	1.18	1.94	1.00	2.62
P_2O_5	1.12	0.46	0.56	0.63	0.96	2.05
CO_2	0.20	0.20	0.12	0.13	0.00	0.12
H_2O^+	1.76	1.53	0.58	0.95	1.57	1.89
Total[1]	100.44	99.12	100.14	99.53	99.85	99.65
NORM						
or	13.83	5.32	6.97	11.46	5.91	15.42
ab	24.62	29.62	24.37	29.92	30.48	39.35
an	18.35	24.78	21.49	22.42	24.71	15.63
ne	6.88	----	----	0.52	3.34	----
di	8.48	11.31	16.73	9.32	6.60	----
hy	----	17.62	4.67	----	----	5.03
ol	13.96	2.15	12.24	9.88	15.25	4.67
mt	5.02	4.78	7.21	6.48	4.12	6.84
il	4.50	3.30	4.31	6.82	5.45	5.20
ap	2.45	1.04	1.22	1.38	2.10	4.46
cc	0.45	----[2]	0.27	0.30	0.00	0.27

[1]Analyses reported on H_2O^--free basis.
[2]Recalculated using correction for oxidized iron of Irvine and Baragar (1971).

References:
H82-83, Henry and Price (1986b)
77042, D. S. Barker (unpublished data)
81-192, Henry and Price (1986a)
H84-10, Henry and Price (1986a)
SHEEP-3, Goldich and Elms (1949)
PP-391, Parker (1983)

They range from nepheline- to hypersthene-normative. According to the chemical classifications of Barker (1979) and of Carmichael and others (1974), many of the rocks are hawaiites or trachybasalts. The relationships of total alkalies to silica for all tectonic environments and for both alkalic and alkalic-calcic belts overlap. Representative analyses of mafic rocks from the different environments are presented in Table 2.

Mafic rocks in the different tectonic environments are also similar petrographically. They contain a single titanium-rich clinopyroxene and olivine as phenocrysts and in the groundmass. Plagioclase is relatively sodic; rims or discrete grains of anorthoclase are common.

Although broadly similar, mafic rocks from the extensional environment and those from the two different belts of the compressional environment are distinguishable on some compositional plots (Figs. 4 through 6). Rocks emplaced during extension appear as a distinct group on an (Na_2O + K_2O)-total iron as FeO-MgO (AFM) diagram (Fig. 4) and generally contain more MgO (Fig. 5) and less TiO_2 (Fig. 6) than do rocks emplaced during compression. Although they overlap somewhat, rocks of the alkalic and alkali-calcic belts are also apparently distinct in MgO content (Fig. 5). Analyses of some MgO-rich rocks reported by Maxwell and others (1967) would plot with the extension-related basalts on AFM or MgO/SiO_2 diagrams. However, the rocks of Maxwell and others (1967) are probably olivine and pyroxene cumulates that do not reflect actual magmatic compositions.

Several components show clear correlations with SiO_2. The

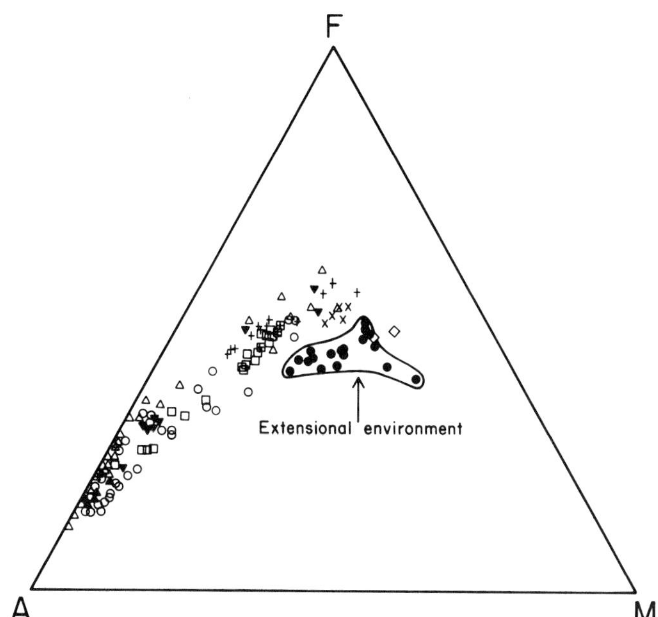

Figure 4. AFM diagram comparing rocks of contrasting tectonic environments; values in wt. %. A indicates $Na_2O + K_2O$; F, $FeO + Fe_2O_3$ recalculated as FeO; M, MgO. Symbols as in Figure 3.

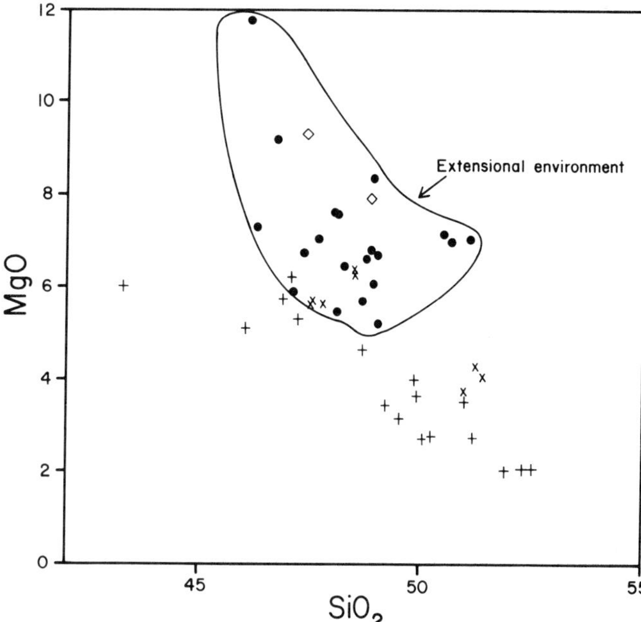

Figure 5. Magnesia-silica plot (wt. %) of mafic rocks. Symbols as in Figure 3.

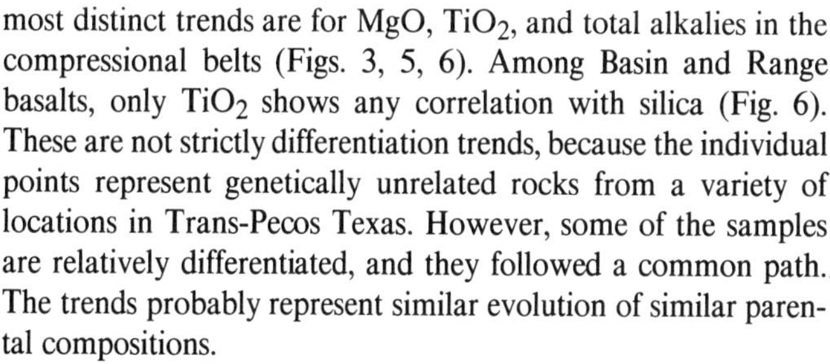

most distinct trends are for MgO, TiO_2, and total alkalies in the compressional belts (Figs. 3, 5, 6). Among Basin and Range basalts, only TiO_2 shows any correlation with silica (Fig. 6). These are not strictly differentiation trends, because the individual points represent genetically unrelated rocks from a variety of locations in Trans-Pecos Texas. However, some of the samples are relatively differentiated, and they followed a common path. The trends probably represent similar evolution of similar parental compositions.

Variations occur within belts and even within single eruptive centers within the belts. Two of six hawaiites from the Van Horn Mountains caldera (Henry and Price, 1986a) in the alkali-calcic belt are hypersthene-normative and have distinctly lower total alkalies and TiO_2 and higher MgO than do other basalts in the western belt (Figs. 3, 5, 6). The other four basalts from the Van Horn Mountains caldera are nepheline-normative and are distinctive in various major and minor element concentrations from the hypersthene-normative rocks (Henry and Price, 1986a).

Extension-related basalts vary widely in composition and in silica saturation (Fig. 7). Geographically related rocks are chemically similar. All the rocks are hawaiites but range from strongly nepheline-normative (7 to 11 percent ne in the Rim Rock dikes) to strongly hypersthene-normative (11 to 17 percent hy in the Black Gap lavas; Fig. 7).

Degree of Differentiation

Extension-related rocks show little evidence of differentiation. Limited differentiation in the extensional environment is

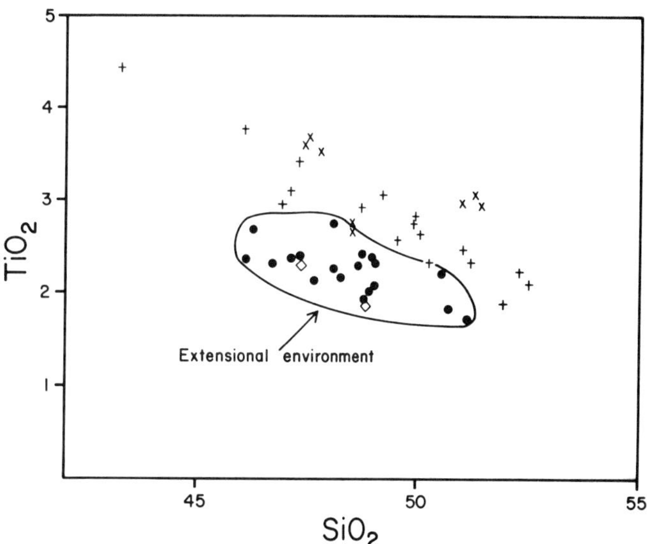

Figure 6. Titanium dioxide–silica plot (wt. %) of mafic rocks. Symbols as in Figure 3.

suggested by the compositional trends (Fig. 3) and by the presence of biotite and anorthoclase megacrysts in some basalts (Irving and Frey, 1984). Rocks from each area show a narrow range of compositions. Dasch and others (1969) indicated from proportions of xenocrysts that the Rim Rock dikes include phonolites and lamprophyres. Our petrographic and chemical data on the basaltic groundmasses do not support this conclusion (Henry and

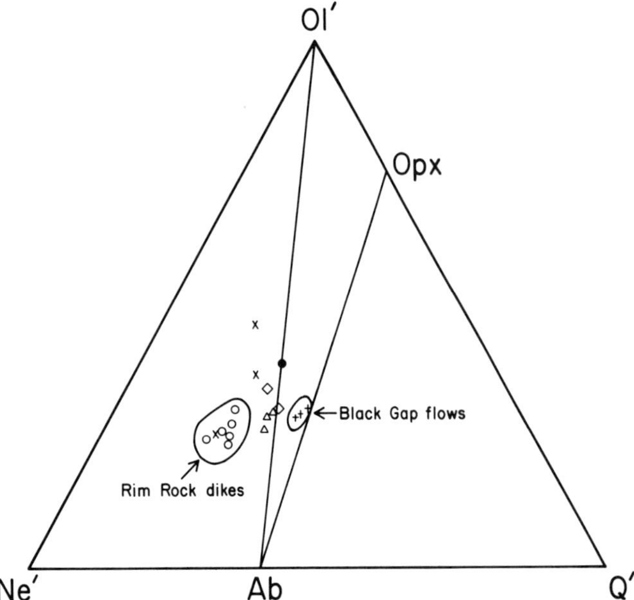

Figure 7. Normative nepheline-olivine-quartz plot (wt. %) of extension-related basalts. Values of Ne′ and Ol′ calculated according to Irvine and Baragar (1971) using CO_2-free analyses and Fe_2O_3 corrected for oxidation. Open circles indicate Rim Rock dikes; solid circle, Cox Mountain flow; diamonds, northern Bofecillos basalts; triangles, southern Bofecillos basalts; +, Black Gap flow; x, dikes and small plugs north of Black Gap.

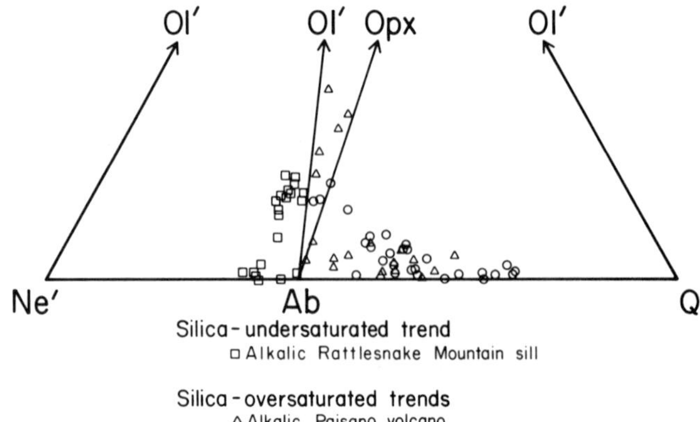

Figure 8. Normative nepheline-olivine-quartz plot of compression-related suites of rocks.

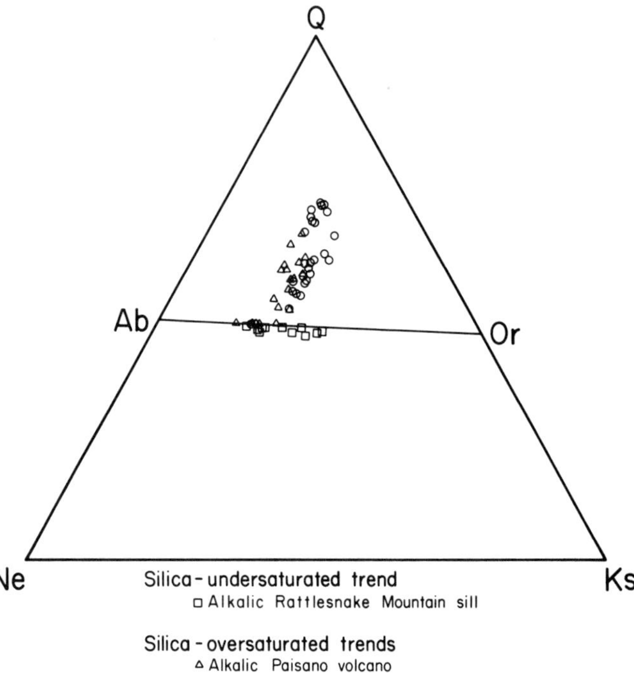

Figure 9. Normative nepheline-quartz-kalsilite plot (wt. %) of compression-related suites of rocks.

Price, 1986b). Because both nepheline- and hypersthene-normative rocks exist, differentiation presumably could have duplicated the phonolitic and rhyolitic trends seen in the older rocks.

In contrast, the older compression-related rocks are commonly strongly differentiated. Compositional trends in the Chinati Mountains caldera, Paisano Pass caldera, and Rattlesnake Mountain sill illustrate differentiation trends in the alkali-calcic and alkalic belts (Figs. 3, 4, 8, 9).

The Rattlesnake Mountain sill (Carman and others, 1975) is a mafic intrusion in Big Bend National Park in the eastern, alkalic belt. Although biotite from the sill yielded a K-Ar date of 28 Ma (M. F. Carman, Jr., personal communication, 1984; Henry and others, 1986), thereby suggesting that the sill formed during the transitional period from compression to extension, it is chemically similar to other intrusions of the compressional tectonic environment. Carman and others (1975) showed that a parental magma having approximately 50 percent SiO_2 differentiated at low pressure to produce small amounts of syenite having approximately 60 percent SiO_2. All analyzed rocks are nepheline normative (Figs. 8, 9). Rocks that show similar differentiation trends elsewhere in Texas commonly contain modal nepheline (Barker, 1977). Figures 3 and 4 illustrate the compositional range.

The Paisano Pass caldera is a small volcanic center 5-km in diameter, built on a trachytic shield volcano (Parker, 1976, 1983). Rocks of the area include both quartz- and nepheline-normative varieties, but only the oversaturated trend is well documented. Nepheline-normative trachyte and phonolite intrusions cut the volcanic rocks and were probably not related to the caldera reservoir. Nepheline-normative mafic rocks are older than the caldera-associated volcanic rocks. The most mafic rocks are commonly oxidized, so determination of silica saturation is difficult. Rhyolitic rocks are commonly peralkaline. Figures 3, 4, 8, and 9 illustrate the oversaturated trend, which Parker (1983) modeled by fractional crystallization. McDonough and Nelson

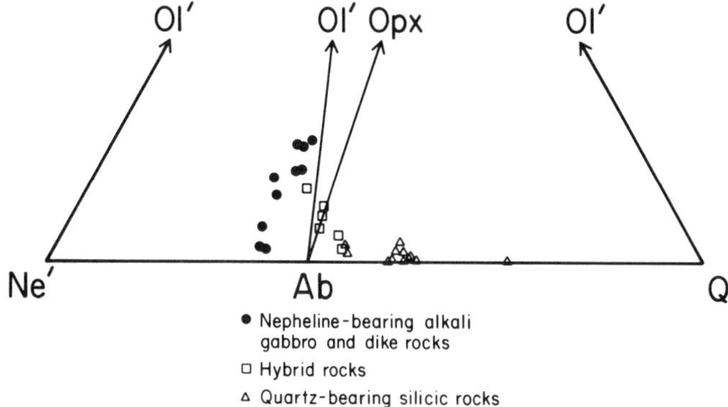

Figure 10. Normative nepheline-olivine-quartz plot of 29 intrusive rocks from Marble Canyon stock and nearby Cave Peak intrusions.

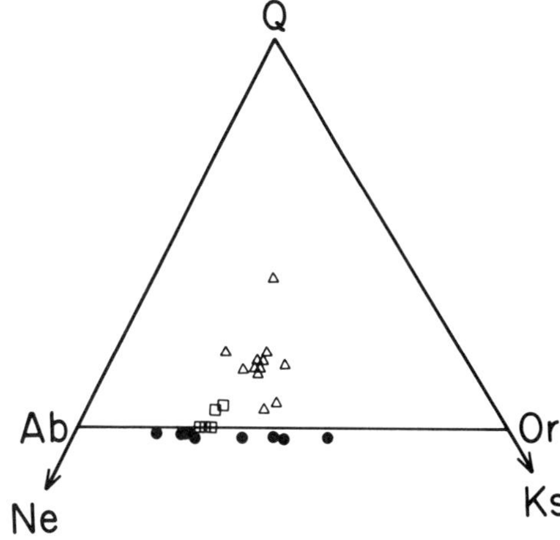

Figure 11. Normative nepheline-quartz-kalsilite plot of intrusive rocks from Marble Canyon area. Symbols as in Figure 10.

(1984) suggested that magma mixing was a minor process in the formation of the rocks of the Paisano volcano.

The Chinati Mountains caldera is the largest in Trans-Pecos Texas (Cepeda and Henry, 1983; Henry and Price, 1984). Erupted rocks range from abundant rhyolite to mafic trachyte and minor hypersthene-normative hawaiite (Figs. 3, 4, 8, 9). No nepheline-normative rocks have been identified. Peralkaline rhyolites represent minor late differentiates. Cameron and others (1982, 1983) attributed the sequence to fractional crystallization of a mantle-derived mafic magma.

Although broadly similar, the oversaturated differentiation trends in the alkalic and alkali-calcic belts are not identical. The Paisano Pass trend is enriched in alkalies (Fig. 3), particularly sodium (Fig. 9), and depleted in MgO (Fig. 4) relative to the Chinati Mountains trend. Both suites differentiate toward a minimum melt composition (Fig. 9), but rhyolite is more abundant in the Chinati Mountains suite. Peralkaline rocks are more abundant in the alkalic belt, however.

Interaction between the oversaturated and critically undersaturated trends is important in at least one locality in the alkalic belt. The Marble Canyon stock exhibits both a trend to nepheline syenite, similar to that at Rattlesnake Mountain, and a trend to quartz syenite (Figs. 10, 11). The least differentiated rocks are all nepheline normative. Preliminary analytical results indicate mixing of the nepheline-normative hawaiite magma with small amounts of a magma having a composition near minimum-melting granite (Price and others, 1986). The mixing probably resulted from assimilation of small amounts of crustal material during rise of the nepheline-normative magma. Fractional crystallization further modified the resulting mix.

Magmatism before about 38 or 39 Ma, when calderas first started to form, has not been well studied in terms of either age or compositional trends. The igneous rocks consist almost entirely of abundant but volumetrically minor intrusive rocks. Mafic to intermediate rocks seem proportionally more abundant than during the 38- to 32-Ma period of caldera formation. Nevertheless, the existence of all the compositional examples is well established. Nepheline-normative mafic rocks include an intrusion at Nine Point Mesa (47 Ma; Henry and others, 1986) and the Christmas Mountains gabbro (42 Ma; Henry and others, 1986), both in the eastern belt. Hypersthene-normative rocks include the main quartz syenite sill at Nine Point Mesa (≤47 Ma) and the Campus "Andesite" (48 Ma; Hoffer, 1970), actually a trachyte, in El Paso in the western belt. Rhyolites of this age occur in the Christmas Mountains (42 Ma; Henry and others, 1986) in the eastern belt and in the Chinati Mountains (43 Ma; Henry and others, 1986) in the western belt.

Magmatism during the early extensional or transitional period of 31 to 25 Ma is also compositionally diverse, but different from the preceding activity. The Bofecillos volcano (McKnight, 1970) and the Sierra Rica caldera complex (Chuchla, 1981; Gregory, 1981) in Chihuahua immediately south of the Bofecillos volcano were active during this time. Geographically, both lie in the western alkali-calcic belt. The Bofecillos volcano erupted lavas that were alkalic and mafic to intermediate from a central vent 28 to 27 Ma (Henry and others, 1986); no silicic rocks occur with it. Igneous rocks of the Sierra Rica caldera complex include two rhyolitic ash-flow tuffs and a granitic intrusion, all formed between 31 and 28 Ma (Gregory, 1981; Chuchla, 1981). The younger ash-flow tuff is interbedded with lava flows of the Bofecillos volcano but was clearly derived from the Sierra Rica. Intermediate to mafic lava flows, which characteristically fill the older calderas in Texas, are absent in the Sierra Rica complex (Henry and Price, 1984).

Compositional data (Fig. 3) show the distinctly different compositions of the two centers. The mafic to intermediate rocks of the Bofecillos volcano are compositionally more like those in the eastern belt (Fig. 4). The silicic rocks of the Sierra Rica complex are similar to silicic rocks in the western belt (Figs. 3, 4).

DISCUSSION

Basalts of Contrasting Tectonic Environments

Basaltic rocks of the two contrasting tectonic environments (compression and extension) are chemically distinct. The essential difference in major elements is in magnesium content. The extensional basalts are generally richer in magnesium than are the compressional basalts (Fig. 5). Differences in apparent differentiation trends (Fig. 4) illustrate that the extensional and compressional basalts are fundamentally different. The extensional rocks follow an apparent differentiation trend at relatively higher magnesium contents; that is, the extensional basalts are not simply less differentiated than are the compressional basalts.

Two types of basaltic rocks occur in each of the two contrasting tectonic environments: a nepheline-normative type and a hypersthene-normative type. In the contractional environment, differentiation of the nepheline-normative type yielded phonolites, and differentiation of the hypersthene-normative type yielded rhyolites. Locally, as at Marble Canyon, both types of magma occupied the same conduit, mixing probably occurred, and hybrid rocks developed. Both types occur near one another, but not obviously in the same conduit, in several areas: Davis Mountains, Diablo Plateau, Nine Point Mesa, and Big Bend National Park in the eastern alkalic belt and Van Horn Mountains caldera in the western alkali-calcic belt. Limited Sr-isotopic data (Dasch, 1969; Barker and others, 1977; Cameron and others, 1983) suggest mantle sources for the basaltic rocks, although small amounts of crustal contamination of the differentiates are possible and locally likely (Barker and others, 1977).

Comparison of the hypersthene- and nepheline-normative basalts from the Van Horn Mountains caldera (Table 2) suggests that assimilation of granitic material cannot explain all chemical differences. The hypersthene-normative basalts contain more Mg and Ca than do the nepheline-normative basalts. Incorporation of granitic material would increase Si content in the melt but would lower Mg and Ca. Crustal contamination may work for the Paisano Pass basalts (Table 2), for which Mg and Ca are higher in the nepheline-normative varieties than in the hypersthene-normative varieties.

The causes of the subtle differences in chemical compositions of basalts from the compressional and extensional environments are poorly understood. More Sr, Nd, Pb, and O isotopic studies, trace element investigations, and studies of xenoliths and xenocrysts are needed to assess the roles of depth of melting in the mantle, degrees of partial melting, polybaric fractional crystallization, and crustal contamination.

Tectonic Significance of Observed Petrologic Differences

Alkalinity versus Distance from Paleotrench. The north-northwestern trend of the boundary between western alkali-calcic and eastern alkalic rocks emplaced during the compressional tectonic environment in Trans-Pecos Texas (Fig. 1) parallels the paleotrench that was active off the western coast of North America (Dickinson, 1981). Using the potassium-depth relationships of Dickinson and Hatherton (1967), Keith (1978) postulated that subduction-related igneous activity in western North America ranged from calcic near the trench, through calc-alkaline and alkali-calcic, to alkalic farthest from the trench. Coney and Reynolds (1977) and Keith (1978) interpreted age relationships of these igneous rocks in terms of variable dips of the subducting slab. Throughout Laramide time, the dip of the subducting slab decreased, perhaps in response to an increased rate of plate convergence. The alkalic rocks that formed 48 to 32 m.y. ago during a waning stage of Laramide compression in Trans-Pecos Texas were probably generated when the subducting slab was at or near its shallowest angle. As the dip of the subducted slab decreased with time, magmatism swept eastward from Mexico into Texas. The Texas alkalic rocks were emplaced approximately 1,000 km from the trench.

Alkaline rocks may be common as the most inland expression of subduction-induced magmatism. The Trans-Pecos region is part of the broader North American Cordillera, along the eastern or inland part of which alkaline rocks are concentrated (Barker, 1974). Other regions of the world exhibit similar relationships between subduction and alkalic rocks. Stewart (1971) postulated that Neogene peralkaline rocks in eastern Peru are the inland result of Andean subduction. Butakova (1974) noted that many alkaline rocks of Siberia coincide temporally with calc-alkaline magmatism in nearby orogenic zones. For example, the alkaline mafic rocks of the Koryak-Kamchatka folded region may be related to Cenozoic subduction east of Siberia. Structural data, such as dike orientations and presence or absence of contemporaneous folding, are needed to determine whether the subduction-related alkaline rocks were emplaced in compressional tectonic environments, as in Trans-Pecos Texas, or in back-arc extensional environments. Also, these examples include a variety of alkaline rock types. If valid, the hypothesis of subduction-related origin must account for this diversity.

Degree of Differentiation versus Tectonic Setting. Degree of differentiation is clearly greater for the alkaline rocks emplaced during compression in the Trans-Pecos region than for those emplaced during extension (Fig. 3). The most silicic rock analyzed from the Basin and Range extensional environment contains less than 52 percent SiO_2. In contrast, the alkaline rocks from the compressional environment exhibit differentiation trends to either phonolites, with as much as 62 percent SiO_2, or rhyolites, with as much as 78 percent SiO_2.

Extension itself is a possible explanation for the lesser degree of differentiation in the extensional environment than in the compressional environment. During extension, rising magma would tend to reside a shorter time in the crust. During compression, fewer more poorly connected and less dilated tectonic fractures would be available for more-or-less direct rise of magma to the surface. Magma would rise more slowly and have more opportunity to form chambers in which differentiation could occur.

Some suites of alkalic rocks in continental rifts, notably in

East Africa, contain substantial volumes of rhyolites and phonolites (Bailey, 1974). Whether these felsic magmas were partial melts or products of crystal fractionation is controversial (Bailey, 1974; Barker, 1977). A possible explanation for the difference between these continental rift settings and the Trans-Pecos extensional environment is that the volume of mafic magma emplaced in the Trans-Pecos region was insufficient either to form magma chambers capable of yielding felsic differentiates or to melt substantial volumes of rock.

Volume considerations may also explain the observed differences in degree of differentiation between the Trans-Pecos compressional and extensional environments. In comparison to the compositionally diverse rocks of the compressional environment, the less differentiated rocks of the extensional environment are volumetrically minor. The area of exposed compression-related igneous rocks is between two and three orders of magnitude greater than the area of exposed extension-related rocks. Using the ratios of these areas to approximate ratios of total volumes and using the overall time spans for compression-related (48 to 32 Ma) and extension-related (24 to 17 Ma) rocks, we estimate that the average rate of magma production in the compressional environment was at least two orders of magnitude greater than that in the extensional environment.

CONCLUSIONS

Alkalic rocks occur in two contrasting tectonic environments, one compressional and one extensional, in the Tertiary magmatic province of the Trans-Pecos region. Rocks of the older (48 to 32 Ma) compressional environment can be divided into a western alkali-calcic belt and an eastern alkalic belt. The north-northwest–trending boundary separating the two belts parallels the trench that lay off the coast of northern Mexico at that time. The alkalic rocks were emplaced approximately 1,000 km from the trench and were the most inland expression of subduction-related volcanism.

Igneous rocks with characteristics of both the compressional and extensional environments were emplaced during a transitional period of early extension between 31 and 25 Ma. Widespread extension and normal faulting began about 24 Ma in the Texas portion of the Basin and Range province. Between 24 and 17 Ma, volumetrically minor alkalic basalts were intruded and extruded at localities scattered throughout the region.

Rocks of the compressional environment exhibit two major lines of differentiation depending on silica saturation: hypersthene-normative basalt to rhyolite and nepheline-normative basalt to phonolite. In contrast, rocks of the extensional environment are apparently limited to basalts. A direct link between degree of differentiation and state of stress is obvious. During compression, magmas had ample opportunity to form chambers in which differentiation could occur. During extension, less differentiation occurred, perhaps because dilated tectonic fractures permitted rapid rise of magma to the surface, or because the volume of magma was too small to yield felsic differentiates.

Basalts of the two contrasting tectonic environments are broadly similar in degrees of alkalinity and silica saturation. They are, however, chemically distinct. The extensional basalts are generally richer in magnesium than the compressional basalts. The difference is not a matter of degree of differentiation. The tectonic change from subduction-related compression to extension offers abundant possibilities for different conditions of magma generation. The processes, depths, pressure-temperature conditions, and source materials are uncertain for either tectonic environment. More petrologic and geochemical studies, especially of Sr, Nd, Pb, and O isotopes and selected trace elements, need to be undertaken to document differences between the basalts and to explain their origins and tectonic significance.

ACKNOWLEDGMENTS

Research support was provided by U.S. Bureau of Mines grants G1134148 and G1144148 to the Texas Mining and Mineral Resources Research Institute of the University of Texas at Austin. The manuscript was improved through review comments by Jules R. DuBar, John C. Stormer, and John A. Wolff. Publication was authorized by the Director of the Bureau of Economic Geology.

REFERENCES CITED

Bailey, D. K., 1974, Continental rifting and alkaline magmatism, *in* Sorensen, H., ed., The alkaline rocks: New York, John Wiley, p. 148–159.

Barker, D. S., 1974, Alkaline rocks of North America, *in* Sorensen, H., ed., The alkaline rocks: New York, John Wiley, p. 160–171.

—— , 1977, Northern Trans-Pecos magmatic province; Introduction and comparison with the Kenya rift: Geological Society of America Bulletin, v. 88, p. 1421–1427.

—— , 1979, Magmatic evolution in the Trans-Pecos province, *in* Walton, A. W., and Henry, C. D., eds., Cenozoic geology of the Trans-Pecos volcanic field of Texas: University of Texas at Austin, Bureau of Economic Geology, Guidebook 19, p. 4–9.

Barker, D. S., Long, L. E., Hoops, G. K., and Hodges, F. N., 1977, Petrology and Rb-Sr isotope geochemistry of intrusions in the Diablo Plateau, northern Trans-Pecos magmatic province, Texas and New Mexico: Geological Society of America Bulletin, v. 88, p. 1437–1446.

Berge, T.B.S., 1981, Structural evolution of the Malone Mountains, Hudspeth County, Texas [M.S. thesis]: University of Texas at Austin, 95 p.

Butakova, E. L., 1974, Regional distribution and tectonic relations of the alkaline rocks of Siberia, *in* Sorensen, H., ed., The alkaline rocks: New York, John Wiley, p. 172–189.

Cameron, M., Cameron, K. L., and Cepeda, J. C., 1982, Geochemistry of Oligocene igneous rocks from the Chinati Mountains, West Texas: Geological Society of America Abstracts with Programs, v. 14, no. 3, p. 107.

Cameron, M., Cameron, K. L., Sawlan, M., and Gunderson, R., 1983, Calc-alkalic volcanism in Chihuahua, Mexico; Its regional geochemical context, *in* Clark, K. F., and Goodell, P. C., eds., Geology and mineral resources of north-central Chihuahua: El Paso Geological Society, Guidebook for the 1983 Field Conference, p. 94–101.

Carman, M. F., Jr., Cameron, M., Gunn, B., Cameron, K. L., and Butler, J. C., 1975, Petrology of Rattlesnake Mountain sill, Big Bend National Park, Texas: Geological Society of America Bulletin, v. 86, p. 177–193.

Carmichael, I.S.E., Turner, F. J., and Verhoogen, J., 1974, Igneous petrology: New York, McGraw-Hill, 739 p.

Cepeda, J. C., and Henry, C. D., 1983, Oligocene volcanism and multiple caldera formation in the Chinati Mountains caldera, Presidio County, Texas: University of Texas at Austin, Bureau of Economic Geology, Report of Investigations no. 135, 32 p.

Chuchla, R. J., 1981, Reconnaissance geology of the Sierra Rica area, Chihuahua, Mexico [M.S. thesis]: University of Texas at Austin, 199 p.

Coney, P. J., and Reynolds, S. J., 1977, Cordilleran Benioff zones: Nature, v. 270, p. 403–406.

Dahl, D. A., 1984, Petrology and geochemistry of nepheline trachytes and phonolites in the Black Hills area, Brewster County, Trans-Pecos Texas [M.S. thesis]: Fort Worth, Texas Christian University, 103 p.

Dasch, E. J., 1969, Strontium isotope disequilibrium in a porphyritic, alkali basalt, and its bearing on magmatic processes: Journal of Geophysical Research, v. 74, p. 560–565.

Dasch, E. J., Armstrong, R. L., and Clabaugh, S. E., 1969, Age of Rim Rock dike swarm, Trans-Pecos, Texas: Geological Society of America Bulletin, v. 80, p. 1819–1824.

DeCamp, D. W., 1981, Structural geology of Mesa de Anguila, Big Bend National Park, Trans-Pecos Texas [M.S. thesis]: University of Texas at Austin, 185 p.

Dickinson, W. R., 1981, Plate tectonic evolution of the southern Cordillera: Arizona Geological Society Digest, v. 14, p. 113–135.

Dickinson, W. R., and Hatherton, T., 1967, Andesitic volcanism and seismicity around the Pacific: Science, v. 157, p. 801–803.

Goldich, S. S., and Elms, M. A., 1949, Stratigraphy and petrology of the Buck Hill Quadrangle, Texas: Geological Society of America Bulletin, v. 60, p. 1133–1182.

Gregory, J. L., 1981, Volcanic stratigraphy and K-Ar ages of the Manuel Benavides area, northeastern Chihuahua, Mexico, and correlations with the Trans-Pecos Texas volcanic province [M.S. thesis]: University of Texas at Austin, 78 p.

Henry, C. D., and Price, J. G., 1984, Variations in caldera development in the Tertiary volcanic field of Trans-Pecos Texas: Journal of Geophysical Research, v. 89, no. B10, p. 8765–8786.

Henry, C. D., and Price, J. G., 1985, Summary of the tectonic development of Trans-Pecos Texas: University of Texas at Austin, Bureau of Economic Geology, Notes to accompany Miscellaneous Map 36, 8 p.

Henry, C. D., and Price, J. G., 1986a, The Van Horn Mountains caldera; Geology and development of a small (10 km^2) ash-flow caldera: University of Texas at Austin, Bureau of Economic Geology, Report of Investigations no. 151, 46 p.

——, 1986b, Early Basin and Range development in Trans-Pecos Texas and adjacent Chihuahua; Magmatism and orientation, timing, and style of extension: Journal of Geophysical Research, v. 91, no. B6, p. 6213–6224.

Henry, C. D., Gluck, J. K., and Bockoven, N. T., 1985, Tectonic map of the Basin and Range province of Texas and adjacent Mexico: University of Texas at Austin, Bureau of Economic Geology, Miscellaneous Map 36, scale 1:500,000.

Henry, C. D., McDowell, F. W., Price, J. G., and Smyth, R. C., 1986, Compilation of potassium-argon ages of Tertiary igneous rocks, Trans-Pecos Texas: University of Texas at Austin, Bureau of Economic Geology, Geological Circular 86-2, 34 p.

Hoffer, J. M., 1970, Petrology and mineralogy of the Campus Andesite pluton, El Paso, Texas: Geological Society of America Bulletin, v. 81, p. 2129–2135.

Irvine, T. N., and Baragar, W.R.A., 1971, A guide to the chemical classification of the common volcanic rocks: Canadian Journal of Earth Sciences, v. 8, p. 523–548.

Irving, A. J., and Frey, F. A., 1984, Trace element abundances in megacrysts and their host basalts; Constraints on partition coefficients and megacryst genesis: Geochimica et Cosmochimica Acta, v. 48, p. 1201–1221.

Keith, S. B., 1978, Paleosubduction geometries inferred from Cretaceous and Tertiary magmatic patterns in southwestern North America: Geology, v. 6, p. 516–521.

Laughlin, A. W., Kress, V. C., and Aldrich, M. J., 1982, K-Ar ages of dike rocks, Big Bend National Park, Texas: Isochron/West, no. 35, p. 17–18.

Macdonald, G. A., and Katsura, T., 1964, Chemical composition of Hawaiian lavas: Journal of Petrology, v. 5, p. 82–133.

Macdonald, R., 1975, Tectonic settings and magma associations: Bulletin Volcanologique, v. 38, p. 575–593.

Maxwell, R. A., Lonsdale, J. T., Hazzard, R. T., and Wilson, J. A., 1967, Geology of Big Bend National Park, Brewster County, Texas: University of Texas at Austin, Bureau of Economic Geology, Publication 6711, 320 p.

McDonough, W. F., and Nelson, D. O., 1984, Geochemical constraints on magma processes in a peralkaline system; The Paisano volcano, west Texas: Geochimica et Cosmochimica Acta, p. 2243–2455.

McDowell, F. W., 1979, Potassium-argon dating in the Trans-Pecos Texas volcanic field, *in* Walton, A. W., and Henry, C. D., eds., Cenozoic geology of the Trans-Pecos volcanic field of Texas: University of Texas at Austin, Bureau of Economic Geology, Guidebook 19, p. 10–18.

McKnight, J. F., 1970, Geology of Bofecillos Mountains area, Trans-Pecos Texas: University of Texas at Austin, Bureau of Economic Geology, Geologic Quadrangle Map no. 37, scale 1:48,000, text 36 p.

Moustafa, A. R., 1983, Analysis of Laramide and younger deformation of a segment of the Big Bend region, Texas [Ph.D. thesis]: University of Texas at Austin, 157 p.

Muehlberger, W. R., 1980, Texas lineament revisited, *in* Dickerson, P. W., and Hoffer, J. M., eds., Trans-Pecos region, southwestern New Mexico and West Texas: Socorro, New Mexico Geological Society, 31st Field Conference Guidebook, p. 113–121.

Parker, D. F., Jr., 1972, Stratigraphy, petrography, and K-Ar geochronology of volcanic rocks, northeastern Davis Mountains, Trans-Pecos Texas [M.S. thesis]: University of Texas at Austin, 136 p.

——, 1976, Petrology and eruptive history of an Oligocene trachytic shield volcano, near Alpine, Texas [Ph.D. thesis]: University of Texas at Austin, 183 p.

——, 1983, Origin of the trachyte-quartz trachyte-peralkalic rhyolite suite of the Oligocene Paisano volcano, Trans-Pecos Texas: Geological Society of America Bulletin, v. 94, p. 614–629.

Peacock, M. A., 1931, Classification of igneous rock series: Journal of Geology, v. 39, p. 54–67.

Price, J. G., and Henry, C. D., 1984, Stress orientations during Oligocene volcanism in Trans-Pecos Texas; Timing the transition from Laramide compression to Basin and Range extension: Geology, v. 12, p. 238–241.

Price, J. G., Henry, C. D., Standen, A. R., and Posey, J. S., 1985, Origin of silver-copper-lead deposits in red-bed sequences of Trans-Pecos Texas; Tertiary; mineralization in Precambrian, Permian, and Cretaceous sandstones: University of Texas at Austin, Bureau of Economic Geology, Report of Investigations no. 145, 65 p.

Price, J. G., Henry, C. D., Barker, D. S., and Rubin, J. N., 1986, Petrology of the Marble Canyon stock, Culberson County, Texas, *in* Price, J. G., Henry, C. D., Parker, D. F., and Barker, D. S., eds., Igneous geology of Trans-Pecos Texas; Field trip guide and research articles: University of Texas at Austin, Bureau of Economic Geology, Guidebook 23, p. 353–360.

Sorensen, H., 1974, Regional distribution and tectonic relations; Introduction, *in* Sorensen, H., ed., The alkaline rocks: New York, John Wiley, p. 145–147.

Stewart, J. W., 1971, Neogene peralkaline igneous activity in eastern Peru: Geological Society of America Bulletin, v. 82, p. 2307–2312.

Wilson, J. A., 1971, Vertebrate biostratigraphy of Trans-Pecos Texas and northern Mexico, *in* Seewald, K., and Sundeen, D., eds., The geologic framework of the Chihuahua Tectonic belt: Midland, West Texas Geological Society, p. 157–166.

Zoback, M. L., Anderson, R. E., and Thompson, G. A., 1981, Cainozoic evolution of the state of stress and style of tectonism of the Basin and Range province of the western United States: Philosophical Transactions of the Royal Society of London, A, v. 300, p. 407–434.

Manuscript Accepted by the Society October 15, 1986

Printed in U.S.A.

Geological Society of America
Special Paper 215
1987

Petrology and geochemistry of xenolith-bearing alkalic basalts from the Geronimo Volcanic Field, southeast Arizona; Evidence for polybaric fractionation and implications for mantle heterogeneity

Pamela D. Kempton* and Michael A. Dungan
Southern Methodist University, Dallas, Texas 75275
Douglas P. Blanchard
NASA-Johnson Space Center, Houston, Texas 77058

ABSTRACT

Alkalic basalts and included ultramafic xenoliths from the Geronimo Volcanic Field (GVF), southeast Arizona, have been analyzed for their major and trace element compositions. Chemical, petrographic, and physiographic distinctions define two groups of lavas and a temporal evolution for the suite. Basalts on the uplifted flanks of the San Bernardino Valley (3.5 to 9 Ma) range in composition from alkali olivine basalt to hawaiite and mugearite. Most are slightly to moderately evolved (Mg value, 0.30 to 0.66). Major and trace element concentrations show well-defined linear correlations with Mg value, and incompatible trace element ratios are distinct (Zr/Nb > 4, La/Sm <5). They contain relatively few xenoliths, although megacrysts are common. Younger cinder cones and flows (0.25 to 3.0 Ma), which cover the valley floor, are predominantly basanites with lower Zr/Nb (<4) and greater light rare-earth element enrichment (La/Sm >6) than flank basalts. Valley basanites exhibit generally higher Mg values (0.52 to 0.68), many of which are within the range for liquids calculated to be in equilibrium with mantle olivine compositions. Major and trace element concentrations correlate poorly with Mg values in most cases. Mafic and ultramafic xenoliths, as much as 45 cm in diameter, are abundant.

Flank basalts record the effects of significant polybaric fractional crystallization that occurred as these magmas established conduits to the surface. Fractionation at moderate pressures involved the phases aluminous clinopyroxene + aluminous spinel ± olivine. The compositions and modal proportions of fractionated phases are similar to those observed in Type II (Al-augite series) xenoliths and megacrysts. Younger lavas (<3.5 Ma) were probably generated at greater depths in the mantle, but escaped to the surface through a crust "primed" by previous magma passage, undergoing less extensive fractional crystallization. The distinctive Zr/Nb ratios demand that the source areas of flank and valley basalts be distinct; these sources differ largely in the modal proportion of clinopyroxene relative to garnet and in the proportion of garnet + clinopyroxene relative to olivine and orthopyroxene. The compositional spectrum observed in the younger lavas reflects differences in the degree of partial melting.

*Present address: NASA-Johnson Space Center, SN2, Houston, Texas 77058

INTRODUCTION

The petrogenesis of alkalic basalts is perplexing because of the apparent necessity for deriving their alkali-enriched compositions by very small degrees of partial melting (<1%) from an apparently alkali-poor source material, i.e., garnet or spinel lherzolite (Kay and Gast, 1973; Sun and Hanson, 1975; Frey and others, 1978; Fitton and Dunlop, 1985). Because it has been assumed that such small volumes of melt could not escape from the mantle, numerous alternate mechanisms have been proposed to account for these alkali-rich compositions. A popular model at present is that of trace element enrichment occurring just prior to magma genesis via mantle metasomatism. The attractive aspect of this model is that it can account for the decoupling of trace element abundances and isotopic values characteristic of alkalic basalts (Frey and others, 1978; Menzies and Murthy, 1980). A second model (Hofmann and White, 1982; Ringwood, 1982) calls upon mixing melt components from several different sources to produce oceanic and continental alkali basalts (i.e., depleted mantle plus "streaks" of subducted oceanic or continental crust and/or lithospheric mantle, along with possible contribution from undepleted lower mantle below the 650-km seismic discontinuity). However, in modeling the behavior of partially melted rock, McKenzie (1984) suggested that melt migration will be sufficiently rapid to restrict the amount of melt present in the mantle to <3% such that even small volumes of partial melt can ascend through the mantle.

Although it can be argued that the presence of xenoliths precludes the involvement of substantial fractional crystallization in the petrogenesis of alkalic basalts, their presence does not rule out fractionation prior to xenolith entrainment; nor does the argument apply to those alkalic basalts that bear few or very small xenoliths. Therefore, the possibility that fractionation, magma mixing, and crustal or mantle contamination occur subsequent to magma genesis must also be considered if we are to evaluate the source area of these magmas.

Petrographic and geochemical observations of GVF alkalic basalts indicate a substantial diversity in lava compositions erupted closely in space and time. Such diversity negates the existence of a single parental magma and demands a spectrum of parental magmas to give rise to the range of lava compositions at GVF. Some of the inferred mineralogic and compositional characteristics of the mantle source(s) that contributed to this diversity are discussed on the basis of trace element data. The effects of moderate- and low-pressure fractional crystallization are superimposed on the diversity of primitive compositions at GVF; it is the discussion of this process that forms the principal subject of this paper.

PREVIOUS WORK

Previous work on Geronimo Volcanic Field basalts includes reconnaissance mapping by Cooper (1959), and petrologic studies by Lynch (1972, 1978) and Evans and Nash (1979). Evans and Nash employed thermodynamic calculations in an attempt to estimate the pressure (*P*) and temperature (*T*) of equilibration between basanite melt and spinel lherzolite xenoliths, megacrysts, and phenocrysts. Their results are compatible with a model in which the lava compositions are derived by 10 percent partial melting of pyrolite at a depth of 80 to 120 km, followed by fractionation of high-pressure megacrysts (40–50 km) and low-pressure phenocrysts. Their calculations assume that the incorporated ultramafic xenoliths and megacrysts were in equilibrium with the host magma at some *P* and *T* even though such inclusions in alkalic basalts are known to be largely accidental in origin, having been plucked from conduit walls by the ascending magma (Frey and Prinz, 1978; Wass, 1979a, Irving, 1980). Furthermore, several of the basalts chosen for modeling had evolved, rather than primitive, compositions. Thus, it is difficult to evaluate the significance of such calculations.

GEOLOGIC SETTING

The Geronimo Volcanic Field is located in southeastern Arizona on the eastern boundary of the Basin and Range province in the southwestern United States (Fig. 1). Structures analogous to the major tectonic features found throughout the Basin and Range province are observed, including (1) a northwest-trending system of reactivated Precambrian faults, (2) northeast-directed thrust faults, and (3) north-trending Basin

QUATERNARY

GVF Cinder Cones

GVF Lava Flows

GVF Tuffs

Alluvium

TERTIARY

Rhyolite and associated volcanics

Alluvium

MESOZOIC

Bisbee Formation

Andesite, associated pyroclastics and sediments

PALEOZOIC

Limestone and Dolomite

Thrust Faults

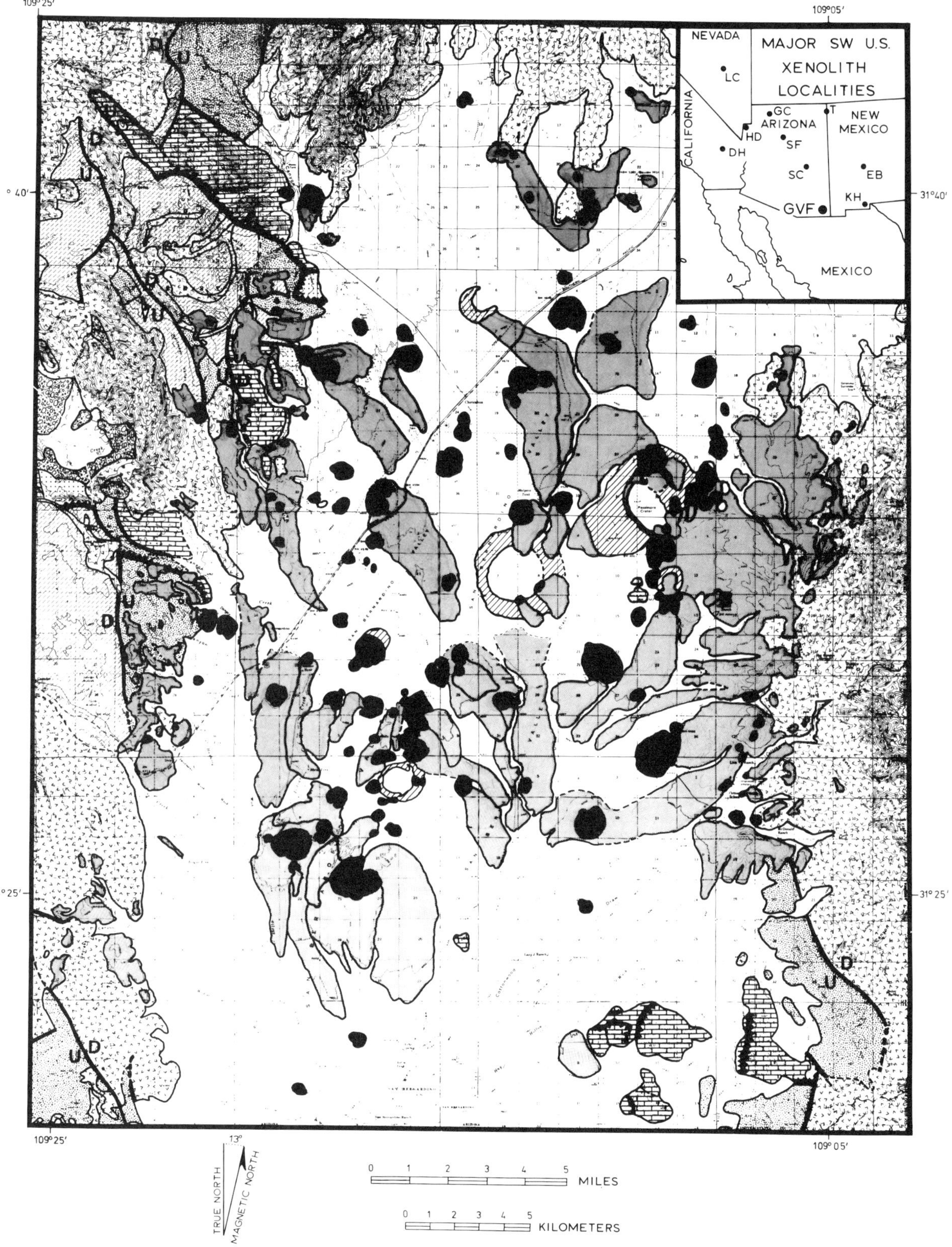

Figure 1. Geologic map of Geronimo Volcanic Field. Stratigraphy and distribution of rocks older than GVF volcanics generalized from Cooper (1959). Inset shows locations of major xenoliths localities in southwestern U.S. GVF indicates Geronimo Volcanic Field; LC, Lunar Crater; DH, Dish Hill; HD, Hoover Dam; GC, Grand Canyon; SF, San Francisco; SC, San Carlos; T, The Thumb; KH, Kilbourne Hole; EB, Elephant Butte. Explanation on facing page.

and Range normal faults. Northeast-directed Laramide thrust sheets were mapped by Cooper (1959) and later by Drewes (1981) in the flank mountains (Fig. 1).

The San Bernardino Valley in which the Geronimo Volcanics erupted is a graben bounded by north-northeast–oriented Basin and Range–type faults. On the east the valley is bounded by the Peloncilla Mountains and on the west by the Chiricahua, Pedregosa, and Perilla Mountains (north to south). The San Bernardino Valley has remained tectonically active into recent times (Drewes, 1981; Herd and McMasters, 1982). The most recent activity, in 1887, resulted in the formation of a high-angle fault scarp 76 km long along the southern extension of the valley (Herd and McMasters, 1982).

Although the San Bernardino Valley is several hundred kilometers in length, volcanic activity is for the most part restricted to that section of the graben offset by the intersection with older northwest-southeast–trending structures. This intersection produced a more highly fractured crust in this location, facilitating greater access of magma to the surface (Fig. 1; Lynch, 1978; Evans and Nash, 1979; Drewes, 1981).

As a result of normal faulting concurrent with volcanism, GVF basalts exposed on the flanks of the valley are older than those on the valley floor. K-Ar age determinations yield dates ranging from 3 Ma to as old as 9 Ma for the flank basalts (Table 1). Some of the oldest lavas contain anorthoclase megacrysts, and the potential for inheriting radiogenic argon from these inclusions makes the accuracy of the oldest dates uncertain. The flank lavas occur as plateau capping flows whose vents have been largely eroded, thus obscuring their eruptive style. The valley lavas range in age from 3.5 to 0.26 Ma. These basalts erupted to form cinder cones, maars, and associated flows. Many of the cinder cones overlie aprons of pyroclastic surge deposits, and some vents erupted entirely as phreatomagmatic tuffs, forming extensive tuff rings and large maar craters (Fig. 1).

TABLE 1. K/Ar CHRONOLOGY OF GERONIMO VOLCANICS

Sample No.	Group	Age* (Ma)
D5	F**	9.20 ± 0.13
D8	F	5.63 ± 0.08
D9	F	4.95 ± 0.07
D11	F	4.72 ± 0.07
D6	V	3.62 ± 0.14
D3	F	3.23 ± 0.04
D12	F	3.08 ± 0.06
D4	F	3.01 ± 0.04
D2	V	1.48 ± 0.02
D7	V	0.67 ± 0.01
D10	V	0.26 ± 0.02

*Dates determined by University of Washington Geochronology Laboratory, Seattle.

**F and V designate Flank and Valley basalts, respectively.

ANALYTIC TECHNIQUES

Fifty-nine samples of basalt were analyzed for major element composition, 33 for trace elements and 30 for rare-earth elements (REE). Prior to analysis, the samples were crushed to fragments of 1 cm or less and sorted to assure minimal contamination from xenolithic material. Samples were then ground to a powder in a tungsten-carbide shatterbox.

Major elements were determined using an automated Siemens x-ray fluorescence (XRF) spectrometer at the Johnson Space Center in Houston. Major elements, except sodium, were analyzed on lanthanum-bearing, lithium borate glass disks (Norrish and Hutton, 1969). Na was determined separately by instrumental neutron activation analysis (INAA). H_2O^+ content was measured coulometrically using a DuPont moisture analyzer in which the sample is mixed with a Pb oxide flux. In this procedure, dry gas is first passed over the sample to remove adsorbed surface water; after which the sample is run at 1,000°C for 15 min to analyze for structurally bound H_2O. FeO was determined titrimetrically using the modified cold acid digestion method of Wilson (Maxwell, 1968); Fe_2O_3 was obtained by difference from the XRF total iron value. Because the degree of secondary alteration and oxidation generally increases with age, CIPW norms and Mg values were calculated using molecular Mg/(Mg + Fe^{2+}), where molecular Fe has been proportioned as $Fe^{3+}/(Fe^{2+} + Fe^{3+}) = 0.1$ (Basaltic Volcanism Study Project, 1981).

Trace elements were determined on unfired, pressed-powder pellets by XRF analysis (Y, Sr, and Rb, with an Mo tube; Zr, Nb, Zn, Ni, Cr, and V with an Au tube) using a modification of the method of Norrish and Chappel (1967) on an automated Siemens SRS-2 X-ray fluorescence machine at the University of Massachusetts. The rare-earth elements (La, Ce, Nd, Sm, Eu, Tb, Yb, Lu) along with Hf, Pa(Th), Sc, Cr, Ni, Sr, Ba, and Rb were determined by INAA at the Johnson Space Center. Samples were analyzed using both thermal and epithermal neutron irradiation. U.S. Geological Survey standards (BCR, GSP-1, BHVO, DTS-1) were used for calibration. Further details of sample preparation and data reduction are described by Jacobs and others (1977).

Microprobe analyses were performed on a fully automated, three-spectrometer model JXA-733 JEOL electron microprobe at Southern Methodist University, using mineral standards at 15 kV accelerating potential and a 20-nA sample current. Raw data were corrected for dead time, background, absorption, fluorescence, and atomic number using the alpha correction procedure developed by Bence and Albee (1968).

PETROGRAPHY AND MINERAL CHEMISTRY

The two groups of lavas defined on the basis of contrasting physiographic distribution (graben flank versus valley floor) and age also exhibit distinct petrographic and compositional trends. Figure 2 shows that the older GVF lavas exposed on the flanks of the San Bernardino Valley include alkali olivine basalts and

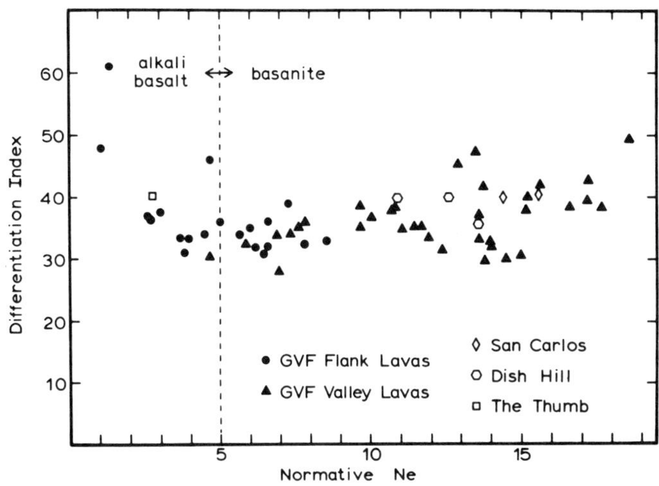

Figure 2. Plot of differentiation index versus normative nepheline content for GVF basalts. Dashed line indicates division between alkali olivine basalt and basanite based on normative nepheline. Data from other xenolith localities include Dish Hill (Wilshire and Trask, 1971); San Carlos (Frey and Prinz, 1978); and The Thumb (Ehrenberg, 1982).

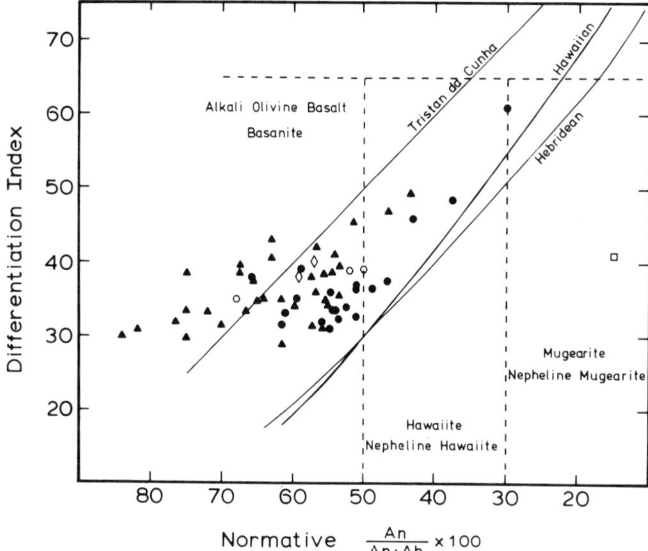

Figure 3. Normative plagioclase composition versus differentiation index. Symbols same as in Figure 2. Trends for lavas from Tristan da Cunha, Hawaii, and the Hebrides from Coombs and Wilkinson (1969).

minor amounts of basanite, ranging from 2 to 8 percent normative nepheline. In contrast, the younger lavas exposed on the valley floor are almost exclusively basanites (4 to 18 percent Ne). Although highly evolved compositions such as phonolite or trachyte have not been found at GVF, moderately fractionated compositions occur, i.e., hawaiite, nepheline hawaiite, and mugearite (Fig. 3). The discussion below focuses on comparing and contrasting the characteristics of these groups and the different processes responsible for their genesis.

Flank Alkali Olivine Basalts and Basanites

Flank alkali olivine basalts and basanites are fine-grained, porphyritic or microporphyritic rocks with groundmass textures ranging from intergranular to intersertal. Plagioclase phenocrysts (bytownite to labradorite) exhibit normal and slight oscillatory zoning (Fig. 4; Table 2). Clinopyroxene is generally confined to the groundmass, but may form small, sector-zoned microphenocrysts less than 0.5 mm in size. Larger clinopyroxene crystals typically have partially resorbed cores that are compositionally distinct from the remainder of the crystal (Fig. 5; Table 3). These cores are lower in Ti and Ca, but higher in Na, Cr, and Fe. Olivine microphenocrysts (Fo_{86} to Fo_{71}) are normally zoned and the rims are usually oxidized (Fig. 5). Aluminous Cr-spinel occurs rarely as inclusions in olivine phenocrysts in the more primitive basanites (Table 4). Groundmass phases include plagioclase, Ti-augite, olivine, magnetite + ilmenite, K-feldspar, and nepheline.

Variably resorbed xenocrysts of plagioclase and anorthoclase are common, but xenoliths are rare. Where present, feldspathic mafic granulites dominate over Type I spinel lherzolites and Type II aluminous clinopyroxenites (nomenclature of Frey and Prinz, 1978).

Flank Hawaiites and Mugearite

Three K-rich hawaiites (pk-G-63, D9, Qtz Hill) and one mugearite (pk-G-58) are known from GVF. These are fine-grained, aphyric, plagioclase-rich rocks. Olivine microphenocrysts occur in a trachytic groundmass of plagioclase laths. Small crystals of FeTi oxides, clinopyroxene, and olivine occupy the interstitial areas. Partially resorbed plagioclase megacrysts occur, but both megacrysts and xenocrysts are rare.

Groundmass plagioclase ranges in composition from oligoclase to labradorite in the hawaiites, and from oligoclase to andesine in the mugearite (Fig. 4; Table 2). Anorthoclase and sanidine have been identified as interstitial phases in the groundmass by electron microprobe analysis (Table 2). The groundmass anorthoclase compositions are similar to analyzed anorthoclase megacrysts from GVF (Fig. 4) but differ in their concentrations of Sr and Ba. Anorthoclase megacrysts exhibit higher concentrations of SrO (0.2–0.80 wt%) than groundmass anorthoclase (less than 0.05 wt.%), but BaO is detectable only in the groundmass anorthoclase (0.15–0.34 wt.%) (Table 2).

Olivine compositions range from Fo_{80} in the hawaiites to Fo_{50} in the mugearite (Fig. 5). Oxidation of both microphenocrysts and groundmass olivines is extensive, and the complete range of olivine compositions has not been established.

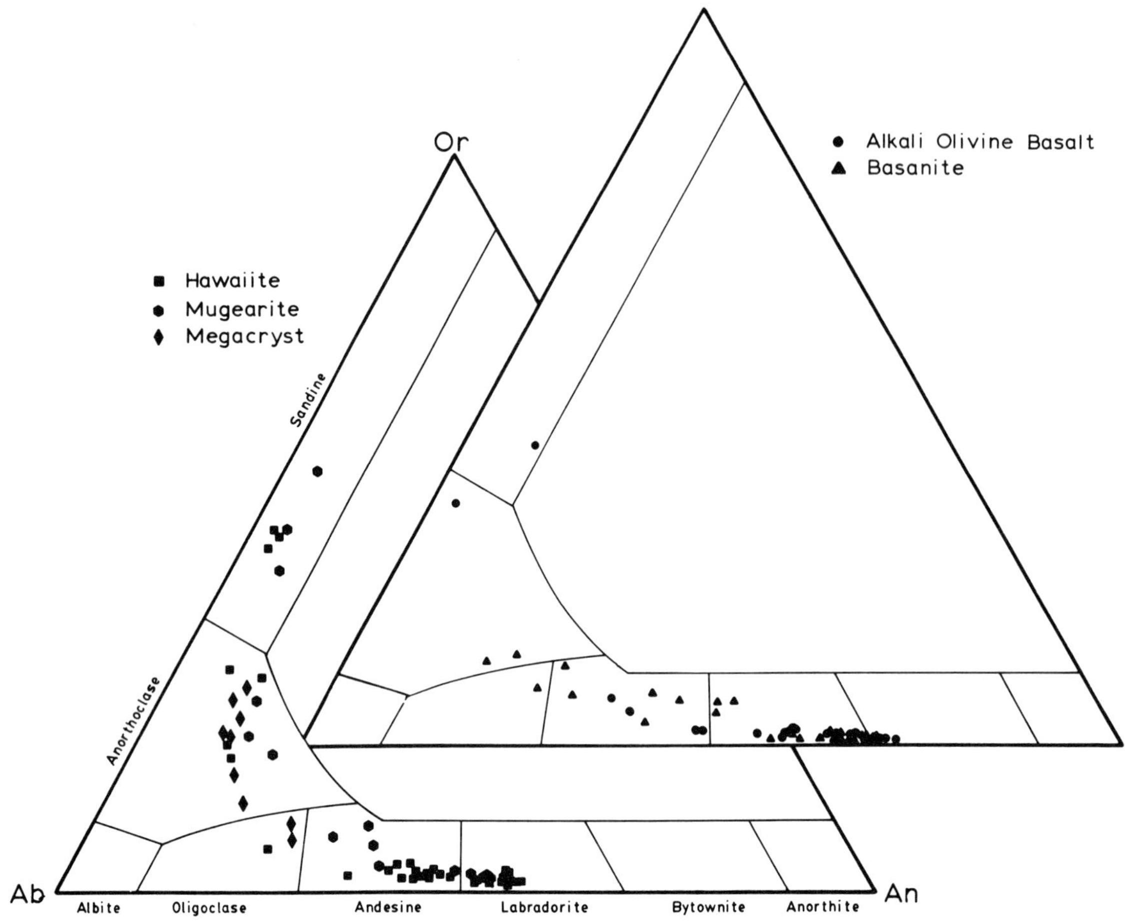

Figure 4. Microprobe analyses of phenocryst and groundmass feldspars and anorthoclase megacrysts from GVF basalts. Each data point in Figures 4 through 6 represents single analysis taken from traverses oriented to encompass maximum compositional zonation in each crystal.

Clinopyroxene is restricted to the groundmass, where it appears as minute (15–30 μm), pale-green rods. These pyroxenes are higher in Na, lower in Ti and Al, and comparable in Ca, Mg, and Fe contents to the titanaugites in less evolved GVF lavas (Table 3). They are similar to the Na-rich cores of some clinopyroxene phenocrysts in GVF valley lavas (Fig. 5) and other alkalic basalts (Wass, 1979b).

Valley Alkali Olivine Basalts and Basanites

The younger valley lavas are fine grained, with textures ranging from glassy to intersertal, less commonly intergranular. Groundmass phases include olivine, clinopyroxene, plagioclase, FeTi oxides ± phlogopite, amphibole, and nepheline. Phenocrysts or microphenocrysts of plagioclase (bytownite to labradorite), olivine (Fo_{72} to Fo_{90}), and titanaugite occur, but plagioclase is most abundant. Aluminous Cr-spinels are common as inclusions in the olivine phenocrysts of basanites with Mg values greater than about 0.63. These spinels are compositionally distinct from groundmass oxides, and from spinels in Types I and II xenoliths (Fig. 6).

Clinopyroxene is far more abundant modally, both as a groundmass and as a phenocryst phase, in valley lavas relative to flank lavas. Clinopyroxene phenocrysts are sector-zoned titanaugites. Compositional variations are analogous to sector-zoned titanaugites described by Wass (1973) for hawaiites from the Southern Highlands, New South Wales, Australia. Many clinopyroxene phenocrysts have xenocrystic cores that are compositionally distinct from the phenocryst rim and groundmass titanaugites. These cores may be derived from one of several sources: (1) Cr-diopside xenocrysts disrupted from Type I lherzolite xenoliths; (2) Al-augite xenocrysts disrupted from Type II clinopyroxenite xenoliths; or (3) green, Fe- and Na-rich pyroxenes of uncertain origin. Pyroxenes in the latter class, rich in jadeite component, are similar in occurrence and composition to pyroxenes described by Wass (1973). They are also essentially

TABLE 2. REPRESENTATIVE MICROPROBE ANALYSES OF FELDSPAR SAMPLES*

Analysis	Sample 22 Groundmass (1)	Sample 50 Groundmass (2)	Sample 63 Core (3)	Sample 63 Groundmass (4)	Sample 63 Groundmass (5)	Sample 21-M Megacryst (6)
Oxide (wt.%)						
SiO_2	50.94	50.43	58.46	63.38	65.64	63.91
TiO_2	0.15	0.10	0.17	0.08	0.16	0.09
Al_2O_3	30.44	21.23	25.63	21.75	19.96	21.99
FeO	0.59	0.49	0.49	0.27	0.30	0.12
CaO	13.50	14.41	7.62	2.80	0.84	2.71
SrO	0.27	0.07	0.10	0.04	n.d.	0.80
BaO	0.06	0.01	0.17	0.15	n.d.	0.00
Na_2O	3.70	3.35	6.80	7.57	5.08	7.46
K_2O	0.27	0.20	1.15	3.69	8.14	2.58
Total	99.91	100.29	100.58	99.73	100.11	99.66
Cations (calculated on the basis of 8 oxygens)						
Element						
Si	2.331	2.299	2.620	2.843	2.947	2.853
Ti	0.004	0.002	0.005	0.002	0.005	0.002
Al	1.641	1.677	1.353	1.149	1.055	1.156
Fe	0.021	0.019	0.017	0.009	0.010	0.004
Ca	0.662	0.703	0.365	0.134	0.039	0.129
Sr	0.006	0.001	0.001	0.000	---	0.020
Ba	0.000	0.000	0.002	0.002	---	0.000
Na	0.328	0.295	0.590	0.658	0.441	0.644
K	0.015	0.010	0.065	0.210	0.465	0.146
Total	5.007	5.006	5.019	5.008	4.962	4.955
An	0.659	0.697	0.358	0.134	0.041	0.140
Ab	0.327	0.293	0.578	0.657	0.467	0.701
Or	0.015	0.010	0.064	0.210	0.492	0.159

*Note: All analyses performed by P. Kempton. Microprobe analysis work done at Southern Methodist University; x-ray fluorescence majors and instrumental neutron activation analysis at NASA; x-ray fluorescence traces at University of Massachusetts.

Key to column heads:
1. Average composition of groundmass plagioclase in valley basanite pk-G-22.
2. Average composition of groundmass plagioclase in flank basanite pk-G-50.
3. Core composition of microphenocryst in flank bawaiite pk-G-63.
4. Groundmass anorthoclase in flank hawaiite pk-G-63.
5. Groundmass sanidine in flank hawaiite pk-G-63.
6. Anorthoclase megacryst in valley basanite pk-G-21.

equivalent to groundmass pyroxenes described previously for the GVF hawaiites and mugearite (Fig. 5). One GVF clinopyroxenite xenolith has the same petrographic and compositional characteristics as these pyroxenes (Fig. 5).

In contrast with the older flank lavas, many valley lavas contain abundant ultramafic xenoliths, up to 45 cm in diameter. Xenolith types include Type I spinel lherzolite, harzburgite, dunite, websterite; Type II clinopyroxenite, wehrlite, kaersutite peridotite; and mafic granulites. Megacrysts of olivine, clinopyroxene, orthopyroxene, apatite, anorthoclase, plagioclase, and kaersutite are common as well. Xenoliths and megacrysts show far less evidence of interaction with the host magma in valley basalts than do similar inclusions in flank lavas. Aspects of the petrology and geochemistry of the various GVF xenolith types have been described elsewhere (Kempton and others, 1984a; Kempton, 1984, 1985; Menzies and others, 1985).

Valley Nepheline Hawaiites

Nepheline hawaiite has been identified from two localities, Red Hill cinder cone and lava flow, pk-G-66. Both are aphyric with very fine-grained, intersertal textures, and contain microlites of plagioclase, olivine, minor titanaugite, and FeTi oxides set in glass. Although phenocrysts are absent; megacrysts and small xenoliths are common. No mineral chemistry has been determined.

BASALT CHEMISTRY

Representative GVF lavas were analyzed for major and trace elements; a typical group of these is presented in Tables 5 and 6. The lavas are entirely basaltic, with MgO ranging from 10.60 to 4.70 wt.%. Despite the broad compositional spectrum in

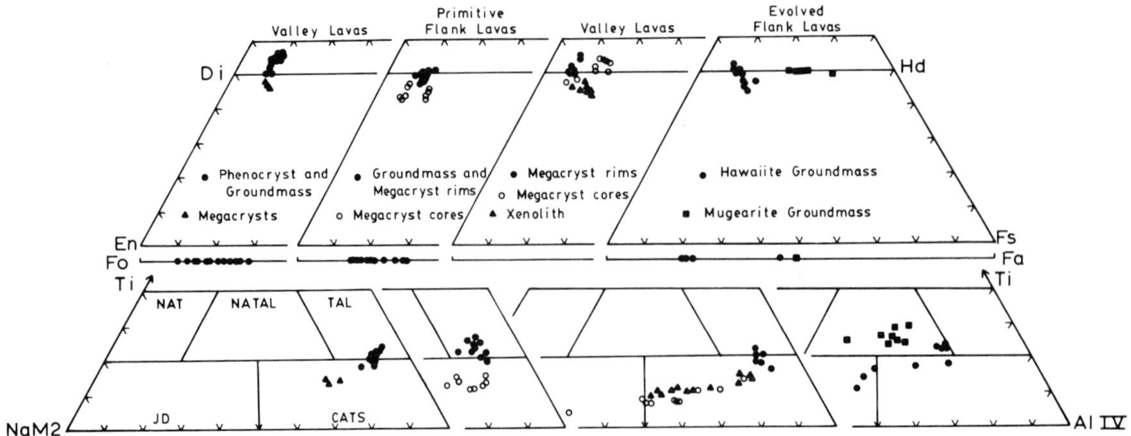

Figure 5. Clinopyroxene ("Quad" and "Others" components) and olivine microprobe analyses. Data summarized include (from left to right): primitive valley and flank lavas and Al-augite megacrysts; Na- and Fe-rich pyroxene megacrysts and xenoliths; and evolved flank lavas.

TABLE 3. REPRESENTATIVE MICROPROBE ANALYSES OF CLINOPYROXENE SAMPLES*

Analysis	D7-3 Core (1)	D7-3 Rim (2)	22-1 Core (3)	22-1 Rim (4)	22MD (5)	23-M3 (6)	56 Core (7)	56 Rim (8)	63 Groundmass (9)	58 Groundmass (10)
Oxide (wt.%)										
SiO_2	49.17	44.97	50.94	48.64	50.57	47.02	46.96	47.55	49.68	47.51
TiO_2	0.67	3.80	0.35	1.94	0.74	2.02	2.00	2.75	1.56	2.90
Al_2O_3	5.02	8.26	4.16	5.09	4.12	9.46	8.23	5.26	3.59	2.88
Cr_2O_3	0.04	0.05	0.04	0.08	0.03	0.00	0.08	0.08	0.04	0.03
FeO	11.10	8.11	12.07	7.09	10.93	7.46	9.74	7.66	9.25	16.59
MnO	0.25	0.20	0.64	0.14	0.17	0.20	0.21	0.11	0.30	0.48
MgO	10.99	11.80	9.66	13.98	12.82	12.88	12.84	13.10	12.65	6.84
CaO	21.96	23.08	20.03	22.76	19.92	19.75	18.95	22.70	21.30	20.89
Na_2O	1.31	0.34	2.31	0.44	1.09	1.34	1.09	0.50	1.02	1.44
Total	100.50	100.93	100.20	100.15	100.39	100.12	100.09	99.71	99.38	99.57
CATIONS (calculated on the basis of 6 oxygens)										
Si	1.856	1.683	1.927	1.812	1.892	1.742	1.756	1.789	1.876	1.865
Ti	0.019	0.106	0.010	0.053	0.020	0.055	0.056	0.077	0.043	0.085
Al^{IV}	0.144	0.317	0.073	0.188	0.108	0.258	0.244	0.211	0.124	0.132
Al^{VI}	0.078	0.046	0.112	0.035	0.073	0.155	0.118	0.022	0.035	0.000
Cr	0.000	0.001	0.001	0.002	0.000	0.000	0.002	0.002	0.001	0.000
Fe^{+2}	0.227	0.149	0.272	0.144	0.267	0.141	0.213	0.171	0.216	0.467
Fe^{+3}	0.123	0.104	0.109	0.076	0.074	0.089	0.091	0.069	0.076	0.074
Mn	0.007	0.005	0.020	0.003	0.004	0.005	0.006	0.003	0.009	0.015
Mg	0.618	0.658	0.544	0.776	0.714	0.711	0.715	0.734	0.711	0.399
Ca	0.887	0.925	0.812	0.908	0.798	0.784	0.759	0.914	0.861	0.878
Na	0.095	0.046	0.169	0.031	0.079	0.096	0.079	0.036	0.074	0.109
Total	4.053	4.041	4.049	4.027	4.028	4.037	4.039	4.028	4.026	4.029
QUAD										
WO	0.512	0.534	0.499	0.497	0.449	0.479	0.450	0.502	0.481	0.504
EN	0.357	0.380	0.334	0.424	0.401	0.435	0.424	0.404	0.398	0.229
FS	0.131	0.086	0.167	0.079	0.150	0.086	0.126	0.094	0.121	0.268
Others										
NAM_2	0.368	0.098	0.671	0.114	0.381	0.235	0.208	0.111	0.307	0.334
Al^{IV}	0.558	0.676	0.290	0.691	0.522	0.631	0.644	0.651	0.514	0.405
Ti	0.074	0.226	0.040	0.195	0.097	0.134	0.148	0.238	0.178	0.261

*Key to column heads:
1. Na-rich xenocrystic core in titanaugite microphenocryst.
2. Titanaugite microphenocryst in valley basanite D7.
3. Na-rich xenocrystic core in titanaugite phenocryst.
4. Titanaugite microphenocryst in valley basanite pk-G-22.
5. Green, Na-rich pyroxene in clinopyroxenite xenolith 22MD.
6. Al-augite megacryst in valley basanite pk-G-23B.
7-8. Phenocryst in flank alkali olivine basalt.
9. Na-rich groundmass clinopyroxene in flank hawaiite pk-G-63.
10. Na-rich groundmass clinopyroxene in flank mugearite pk-G-58.

TABLE 4. REPRESENTATIVE MICROPROBE ANALYSES OF SPINEL SAMPLES*

Analysis	D7 (1)	D7 (2)	50 (3)	50 (4)	50 (5)	22 Core (6)	22 Rim (7)
Oxide (wt.%)							
SiO_2	0.14	0.07	0.16	0.04	0.15	0.07	0.15
TiO_2	1.09	27.28	0.87	26.50	51.54	16.19	21.13
Al_2O_3	37.64	1.43	46.16	1.52	0.11	6.38	4.54
Cr_2O_3	20.73	0.25	12.98	0.19	0.12	9.89	1.25
Fe_2O_3**	1.025	15.61	8.66	17.42	3.70	22.35	23.52
FeO	15.88	53.52	14.06	52.78	41.72	40.28	44.17
MnO	0.24	1.05	0.21	0.78	0.92	0.75	0.82
MgO	14.58	1.37	16.36	1.73	1.81	4.13	4.10
CaO	0.06	n.d.	0.02	n.d.	n.d.	n.d.	n.d.
Total	100.62	100.57	99.48	100.96	100.07	100.03	99.69
CATIONS (calculated on the basis of 4 oxygens for magnetite and spinel; 3 oxygens for ilmenite)							
Si	0.003	0.002	0.004	0.001	0.003	0.006	0.017
Ti	0.023	0.752	0.018	0.726	0.966	0.431	0.570
Al	1.271	0.062	1.506	0.065	0.003	0.266	0.192
Cr	0.670	0.007	0.284	0.005	0.002	0.277	0.035
Fe^{+3}	0.221	0.430	0.181	0.477	0.069	0.595	0.634
Fe^{+2}	0.381	1.639	0.326	1.607	0.869	1.191	1.324
Mn	0.006	0.033	0.005	0.024	0.019	0.022	0.025
Mg	0.623	0.075	0.676	0.094	0.067	0.218	0.219
Ca	0.001	---	---	---	---	---	---
Total	2.999	3.000	3.000	2.999	1.998	3.006	3.016
$Mg/(Mg + Fe^{+2})$	0.621	0.044	0.675	0.055	0.072	0.155	0.142
Cr/(Cr +Al)	0.270	0.101	0.159	0.071	0.400	0.510	0.154

*Key to column heads:
1. Aluminous Cr-spinel inclusion in olivine phenocryst in valley basanite D7.
2. Groundmass titanomagnetite in valley basanite D7.
3. Aluminous Cr-spinel inclusion in olivine phenocryst in flank basanite pk-G-50.
4. Groundmass titanomagnetite in flank basanite pk-G-50.
5. Groundmass ilmenite in flank basanite pk-G-50.

6-7. Cr-rich, groundmass titanomagnetite in valley basanite pk-G-22.

**Fe_2O_3 calculated assuming stoichiometry and charge balance.

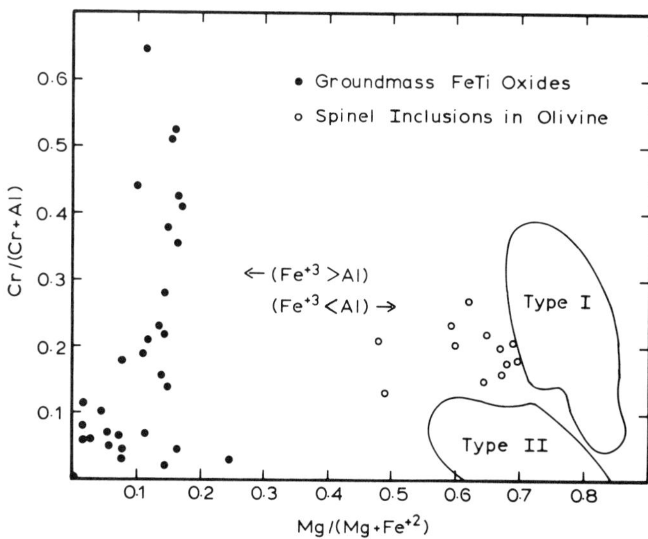

Figure 6. Microprobe analyses of GVF spinels summarized in terms of Cr/(Cr + Al) versus $Mg/(Mg + Fe^{+2})$. Groundmass oxides are typically magnetite with low Mg and Cr values, (cation proportion of $Fe^{+3} > Al$); Cr values higher than 0.1 are common only in primitive valley basanites. Spinel inclusions in olivine phenocrysts have higher Mg values ($Fe^{+3} < Al$) and higher concentrations of Cr and Al. Fields Types I and II enclose spinel compositions from GVF spinel lherzolite and Al-augite xenoliths, respectively.

TABLE 5. WHOLE-ROCK MAJOR ELEMENT CHEMICAL DATA AND CALCULATED CIPW NORMATIVE COMPOSITIONS OF GVF ALKALIC BASALTS*

Analysis	Alkali Olivine Basalt Samples				Basanite Samples						
	D5 F	D4 F	pk-G-51 F	SH-14 V	pk-G-22 V	82-1 V	D7 V	D2 V	QH8 V	RF26 V	pk-G-55 V
Oxide (wt.%)											
SiO_2	46.64	47.17	47.76	46.43	45.12	44.76	47.27	45.85	44.05	45.94	45.91
TiO_2	2.28	2.79	2.09	2.45	2.05	2.58	2.04	2.22	2.52	2.02	2.21
Al_2O_3	16.18	16.26	15.80	15.80	16.32	16.39	16.18	16.38	15.30	16.11	15.69
Fe_2O_3	4.93	1.69	2.92	6.90	2.68	2.84	2.00	1.56	2.32	4.03	2.90
FeO	5.73	8.82	7.82	4.29	7.13	8.52	7.49	8.69	7.90	6.36	6.86
MnO	0.19	0.18	0.16	0.21	0.22	0.22	0.19	0.19	0.18	0.24	0.17
MgO	7.98	7.03	7.96	8.07	8.88	6.88	7.97	7.95	9.81	6.98	10.03
CaO	9.04	9.03	9.64	10.00	8.56	8.72	8.69	9.23	10.39	8.32	10.27
Na_2O	3.33	3.98	3.46	3.07	4.46	4.65	4.78	4.42	3.57	4.60	3.75
K_2O	1.39	1.08	1.42	1.36	2.55	2.36	2.45	2.23	1.94	2.24	1.71
P_2O_5	0.39	0.48	0.49	0.60	0.42	0.53	0.50	0.62	0.47	0.67	0.44
H_2O^+	1.68	1.04	1.01		0.74	1.03	0.72	0.85	0.88	n.d.	0.60
Total	99.76	99.55	100.53	99.18	99.13	99.47	100.28	100.19	99.33	97.51	100.54
Or	8.38	6.48	8.43	8.10	15.32	14.17	14.54	13.24	11.65	13.58	10.11
Ab	21.32	24.94	21.06	19.13	5.74	8.20	11.81	9.59	3.88	14.45	8.92
An	25.59	23.67	23.50	25.53	17.26	17.15	15.53	18.41	20.31	17.12	20.94
Ne	4.01	5.01	4.53	3.83	17.67	17.21	15.61	15.21	14.52	13.80	12.37
Di	14.51	15.48	17.68	17.07	18.98	19.40	20.18	19.60	23.65	17.34	22.17
Ol	18.85	16.24	17.86	18.05	18.42	15.75	15.74	16.68	18.33	16.34	18.61
Mt	1.67	1.69	1.69	1.71	1.56	1.81	1.50	1.64	1.63	1.65	1.53
Ilm	4.42	5.38	3.99	4.69	3.96	4.98	3.89	4.24	4.86	3.93	4.20
Ap	0.87	1.06	1.08	1.32	0.93	1.18	1.10	1.36	1.04	1.50	0.96
Total	99.61	99.95	99.82	99.42	99.84	99.84	99.91	99.96	99.98	99.70	99.82
Mg value**	0.61	0.57	0.60	0.60	0.65	0.55	0.63	0.61	0.66	0.58	0.68
DI	33.7	36.4	34.02	31.06	38.7	39.6	42.0	38.0	30.0	41.8	31.4

*Physiographic occurrence designated as V (valley) or F (flank).

**Mg value calculated as $Mg/(Mg + 0.9Fe^{tot})$.

terms of silica saturation, the GVF rocks display a number of common chemical characteristics. They are all mildly to strongly alkaline and show uniformly high Al_2O_3 contents, ranging from 15.3 to 18.3 wt.%; CaO/Al_2O_3 is also low relative to chondrites (0.8), averaging 0.55.

Because Mg value is sensitive to the removal or addition of ferromagnesian phases, representative major and trace elements have been plotted against Mg value to best display the gross chemical variations of the GVF lava suite (Fig. 7). The salient features of this diagram include the following: (1) The flank lavas encompass a larger range and lower average Mg value than the valley lavas. (2) SiO_2 and Al_2O_3 show strong positive correlations with Mg value in the flank lavas, whereas CaO exhibits a strong negative correlation. The valley lavas show similar overall trends, but far more scatter in the data is apparent. (3) TiO_2 shows little correlation with Mg value in either group. (4) Ni and Cr show marked depletion with increasing differentiation in both groups, consistent with their strong partitioning into olivine and clinopyroxene (see Table 8). (5) Concentrations of moderately compatible elements such as Sc decrease with decreasing Mg value for both basalt groups, but the correlation is best for flank basalts. (6) The alkali (Na, K) and the alkaline earth metals (Rb, Sr, Ba), represented here by Sr, show a small increase with decreasing Mg value in all lavas, but substantial scatter for higher Mg values. (7) Highly incompatible elements, Zr and La, correlate poorly with Mg value in the valley lavas. This correlation is better for flank basalts, but still scattered for lavas with Mg values greater than about 0.6.

In common with other alkalic basalt suites (Irving, 1980; Frey and Prinz, 1978; Frey and others, 1978; Wass, 1980; Kay and Gast, 1973), all GVF lavas are light REE–enriched relative to heavy REE (Fig. 8). Valley lavas exhibit a limited range in rare-earth element abundances and pattern slopes on a chondrite-normalized diagram (Fig. 8C). Those lavas with the highest La (44–50 ppm) and La/Sm (6.4–7.2 ppm) have relatively low Yb (2.5–2.7 ppm) (Table 6). Conversely, samples with lower La (30–37 ppm) and La/Sm (5.0–5.6 ppm) have higher Yb (2.4–3.1 ppm). This results in a spectrum of chondrite-normalized REE

TABLE 5* (CONTINUED)

	Basanite Samples				Nepheline Hawaiite Samples		Hawaiite Samples		Mugearite Samples
	D6 V	pk-G-50 F	D12 F	pk-G-15 V	RH12 V	pk-G-66 V	pk-G-63 F	D9 F	pk-G-58 F
Oxide (wt.%)									
SiO_2	46.08	46.24	46.56	46.22	47.20	47.31	48.04	50.41	51.81
TiO_2	2.04	2.34	2.10	2.12	2.08	1.89	2.61	1.77	1.48
Al_2O_3	16.24	16.44	15.67	16.38	16.72	16.43	17.13	16.79	17.36
Fe_2O_3	4.71	0.99	4.26	3.28	4.50	n.d.	4.88	3.61	7.14
FeO	6.00	9.25	5.91	7.38	6.43	9.61	6.52	6.95	2.35
MnO	0.18	0.16	0.18	0.18	0.20	0.21	0.21	0.18	0.21
MgO	7.50	8.66	8.81	8.03	5.47	6.06	4.59	5.03	1.92
CaO	8.91	9.78	9.80	8.80	7.53	7.59	7.18	6.59	4.93
Na_2O	4.22	3.31	3.49	3.22	5.00	5.73	4.29	4.27	4.85
K_2O	1.94	1.47	1.22	2.08	2.60	2.64	2.17	2.10	3.11
P_2O_5	0.59	0.47	0.46	0.62	0.77	0.71	0.80	0.58	0.50
H_2O^+	1.04	0.64	1.81	1.54	n.d.		1.43	1.30	1.73
Total	99.45	99.75	100.27	99.85	98.53	99.25	99.85	99.58	97.39
Or	11.65	8.77	7.32	12.50	15.59	15.87	13.03	12.63	19.21
Ab	14.67	16.19	18.61	16.68	18.00	14.96	28.20	34.67	40.32
An	19.96	25.89	23.86	24.51	15.73	11.54	21.41	20.80	17.16
Ne	11.71	6.54	6.17	5.98	13.51	18.65	4.71	1.14	1.40
Di	17.52	16.48	18.49	13.21	14.54	18.54	8.31	7.46	4.55
Ol	17.20	18.98	18.56	19.74	14.85	13.74	15.37	16.68	11.16
Mt	1.68	1.65	1.59	1.69	1.71	1.58	1.79	1.67	1.48
Ilm	3.94	4.48	4.05	4.10	4.01	3.66	5.04	3.42	2.94
Ap	1.31	1.04	1.02	1.38	1.17	1.58	1.78	1.29	1.14
Total	99.64	100.01	99.69	99.79	99.66	100.11	99.63	99.75	99.36
Mg value**	0.59	0.63	0.64	0.61	0.51	0.55	0.46	0.49	0.30
DI	38.0	31.5	32.1	35.2	47.1	49.48	45.9	48.4	60.9

*Physiographic occurrence designated as V (valley) or F (flank).

**Mg value calculated as $Mg/(Mg + 0.9Fe^{tot})$.

patterns that cross near Tb. The flank lavas have lower LREE concentrations, lower LREE/HREE ratios, and HREE contents that are comparable to or higher than valley lavas. The less-evolved flank basanites and alkali olivine basalts (Fig. 8B) show greater variability in pattern slopes and REE concentrations than valley lavas for comparable states of evolution (i.e., similar Mg values, Cr and Ni concentrations). With the exception of hawaiite and mugearite samples, pk-G-63 and pk-G-58, evolved flank basalts are not exceptionally enriched in REE. Hawaiite D9 is nearly identical in REE concentrations to basanite pk-G-15, yet these lavas are markedly different in Mg values (0.49 versus 0.61) and compatible element concentrations (Table 6). Europium anomalies are negligible, even in the most evolved compositions.

The difference in relative degree of trace element enrichment between flank and valley basalts is even more pronounced for the LIL (large ion lithophile) and the HFS (high field strength) elements. Figure 9 is a plot of trace element concentrations including LIL, HFS, and rare-earth elements normalized to primordial mantle (Wood and others, 1981), in which relative degree of compatibility increases from left to right. Three representative samples are shown that encompass the range of normative Ne contents for primitive GVF basalts. Sample 50 is a flank basalt with normative Ne of 7 wt.%. Samples 55 and D7 are valley basanites that have normative nepheline contents of 12 and 16, respectively. With increasing degree of silica undersaturation, the degree of incompatible element enrichment increases for the LIL (Rb, Sr, Ba, etc.), as well as for the HFS elements Zr and Nb. Yet the elemental ratios K/Rb, Rb/Ba, Rb/Sr, and so forth, vary only slightly throughout the GVF suite (Table 7).

Zr/Nb ratios, however, are distinct for these two groups. Zr/Nb values for all GVF lavas range from 3.17 to 5.67; but within the two subgroups, flank and valley lavas, variations are more restricted. Zr/Nb in valley lavas range from 3.4 to 4.05, whereas flank lavas range from 4.5 to 5.7 (Fig. 10). The 2 σ error is within the size of the symbols and shows that this distribution is not a function of analytic precision. Other incompatible element ratios, e.g., Hf/Th, Th/Y, are also distinct (see Table 7).

In Figure 11, ratios of the incompatible elements Zr and Nb

TABLE 6. TRACE ELEMENT CONCENTRATIONS IN GVF BASALT SAMPLES

	D5	D4	pk-G-51	pk-G-22	82-1	D7	D2	QH8	pk-G-55	D6	pk-G-50	D12	pk-G-15
Trace Elements (ppm)													
Analysis													
Rb	22.7	9.3	22	48.3	43.1	51.2	45.1	32	35.4	33.9	22.8	29.3	32.3
Sr	579	650	574	795	871	732	803	741	685	761	567	553	697
Ba	367	428	453	559	585	540	573	496	470	436	360	333	393
Y	24.4	26.4	n.d.	27.4	30.2	26.8	27.2	26.5	25.7	26.3	26.1	25.5	26
Th	4.0	3.5	3.1	6.5	6.0	7.1	5.4	4.2	4.1	4.7	3.2	3.8	4.5
Hf	6.0	5.6	4.9	7.4	5.6	6.1	5.4	4.2	4.6	5.4	5.3	4.9	6.2
Zr	256	236	192	307	307	310	260	220	219	261	243	239	295
Nb	53	52	38	88	85.2	85	76.8	69	60.2	64.4	46.3	49	60
Ni	n.d.	128	148	141	70	150	127	185	200	120	158	152	129
Cr	211	203	265	240	71	373	211	274	342	167	210	286	204
V	191	160	n.d.	165	166	168	171	218	220	151	188	189	161
Sc	21.7	22.8	25.0	21.1	20.5	22.6	22.6	29.6	29.2	21.2	28.1	26.6	25.6
Rare-Earth Elements (ppm)													
La	31.4	32.8	27.7	47.7	49.1	49.7	44.1	32.3	32.0	39.5	26.6	29.7	37.5
Ce	64	66	51	90	97	91	83	66	54	83	51	53	81
Nd	34	37	28	41	53	43	44	31	36	--	28	25	35
Sm	6.59	6.38	5.92	6.97	8.11	6.90	6.93	6.43	6.27	6.67	5.75	5.97	6.89
Eu	1.99	2.10	1.66	2.25	2.46	2.18	2.29	1.63	1.59	2.08	1.47	1.65	1.90
Tb	1.02	0.76	0.90	0.97	1.12	0.93	0.80	0.98	0.85	0.99	0.93	0.84	1.21
Yb	2.2	2.6	2.7	2.6	3.0	2.7	2.5	2.8	2.4	2.4	3.0	2.9	4.3
Lu	0.35	0.38	0.42	0.48	0.45	0.45	0.45	0.46	0.41	0.44	0.34	0.37	0.51

	RH12	pk-G-66	pk-G63	D9	pk-G-58
Trace Elements (ppm)					
Analysis					
Rb	37	49	40.5	33.9	52
Sr	1040	1173	739	764	691
Ba	695	656	548	394	823
Y	n.d.	n.d.	38.3	26.9	n.d.
Th	7.3	7.3	5.0	5.0	7.8
Hf	5.4	6.0	6.1	7.3	9.5
Zr	320	286	328	325	442
Nb	95	87	66.3	60.5	85
Ni	57	96	6	77	5
Cr	83	131	10	99	5
V	115	n.d.	90	84	40
Sc	16.4	16.8	19.1	19.3	8.8
Rare-Earth Element (ppm)					
La	55.5	54.8	43.4	36.7	57.0
Ce	110	107	97	83	124
Nd	61	53	50	32	67
Sm	8.45	7.81	10.60	6.59	10.81
Eu	2.59	2.38	2.29	1.99	3.09
Tb	1.37	1.16	1.49	0.94	1.70
Yb	3.2	2.9	4.4	3.3	3.5
Lu	0.50	0.44	0.50	0.41	0.63

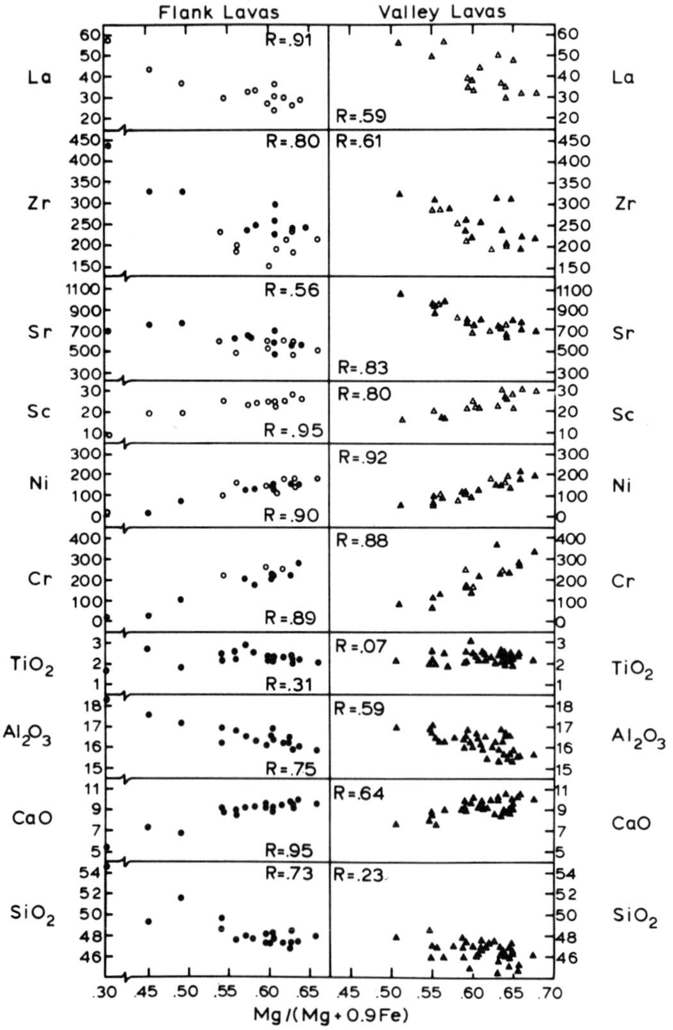

Figure 7. Variation of representative major and trace elements versus Mg value for GVF flank and valley basalts. Open symbols indicate data determined by INAA; solid symbols, data determined by XRF. Major element oxides (SiO_2, CaO, Al_2O_3, TiO_2) given in wt. %. Trace element concentrations (La, Zr, Sr, Sc, Ni, Cr) given in parts per million.

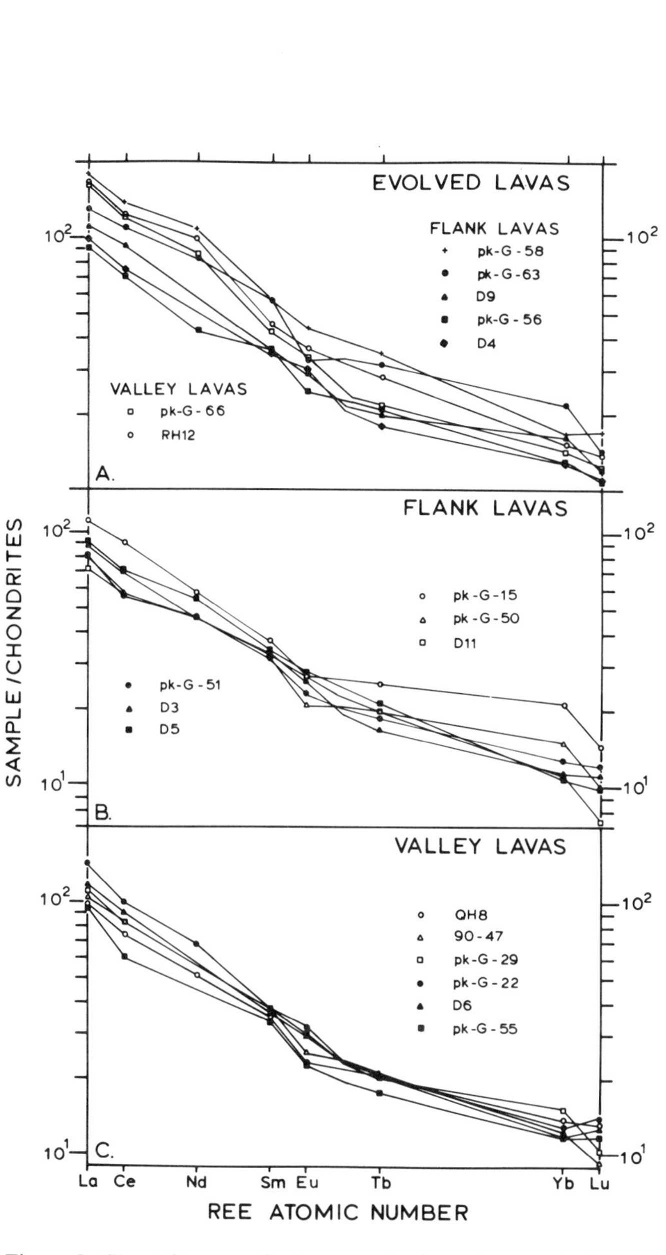

Figure 8. Chondrite-normalized rare earth element concentrations in GVF basalts versus REE atomic number. A, Lavas have Mg values <0.60; B and C, lavas have Mg values ≥0.60.

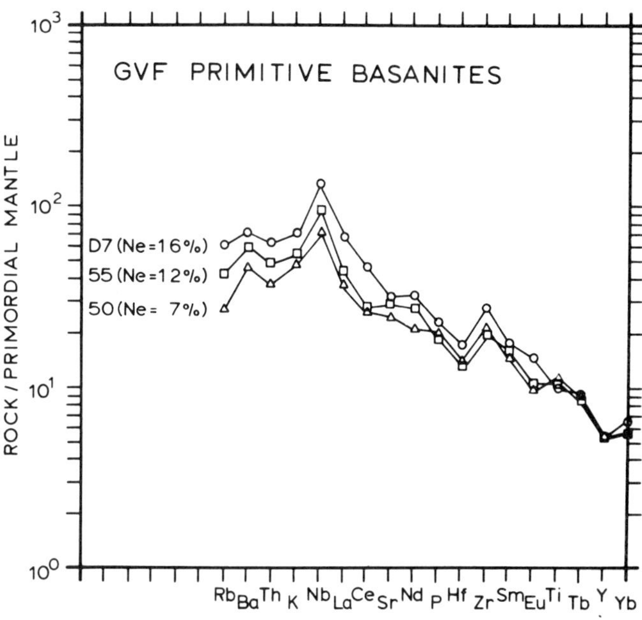

Figure 9. Incompatible trace element concentrations in GVF basalts pk-G-50, pk-G-55, and D7, normalized to primordial mantle values of Wood and others (1981). Numbers in parentheses are normative nepheline contents of each sample.

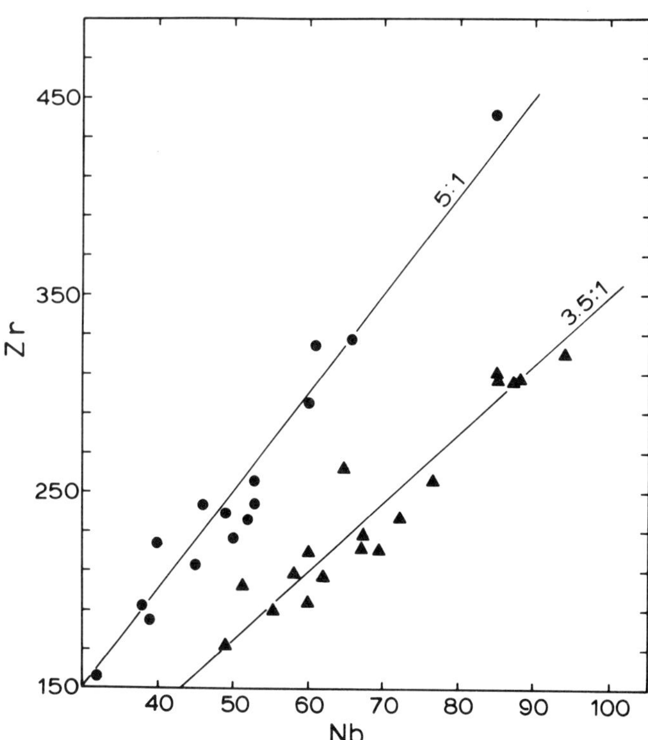

Figure 10. Zr versus Nb in GVF basalts. Symbols as in Figure 2.

TABLE 7. GVF BASALT TRACE ELEMENT RATIOS CORRECTED FOR FRACTIONAL CRYSTALLIZATION

	La/Yb*	Rb/Ba	Zr/Nb	Hf/Th	Th/Y
Flank Basalt					
Pk-G-15	8.03 (8.7)	0.082	4.90	1.4	0.16
D9	8.14 (11.2)	0.088	5.21	1.6	0.15
pk-G-50	9.01	0.065	5.25	1.7	0.12
D11	9.48 (10.8)	0.071	5.59	1.9	0.10
D4	9.57 (12.8)	0.021	4.50	1.7	0.12
D8	9.59 (11.4)	0.057	4.48	1.5	0.16
pk-G-51	10.08 (10.4)	0.049	5.01	1.6	0.12
D3	11.12 (13.1)	0.039	4.72	1.7	0.13
pk-G-56	11.91 (13.9)	0.066	4.59	1.4	0.14
pk-G-58	12.75 (17.9)	0.054	5.23	1.5	---
D5	13.31 (14.0)	0.063	4.81	1.5	0.16
Average	10.04 (11.78)	0.063	4.93	1.6	0.14
Valley Basalt					
pk-G-23B	10.80	0.059	3.23	1.3	0.15
QH8	11.39	0.064	3.17	1.0	0.16
pk-G-37	11.61	0.050	3.97	1.3	0.18
D10	11.87 (13.7)	0.063	3.28	1.4	0.13
90-47	11.91	0.073	3.29	1.1	0.18
pk-G-29	11.95 (13.9)	0.072	3.31	1.2	0.15
90-9	13.02	0.073	3.60	0.9	0.17
pk-G-55	13.46	0.074	3.51	1.1	0.16
D6	13.51 (16.3)	0.079	4.02	1.3	0.16
90-8	13.52 (15.5)	0.075	---	0.9	---
82-1	13.65 (16.3)	0.074	3.58	1.0	0.18
RH12	14.28 (16.6)	0.050	3.38	0.8	---
pk-G-66	16.46 (18.6)	0.075	3.25	0.9	0.32
pk-G-22	18.02	0.085	3.49	1.1	0.24
D2	18.06 (17.9)	0.079	3.35	1.0	0.20
D7	18.14	0.094	3.66	0.9	0.26
Average	13.86 (14.73)	0.071	3.47	1.1	0.19

*Uncorrected ratios for La/Yb given in parentheses. All other values have been corrected for fractionation based on major element mass balance models.

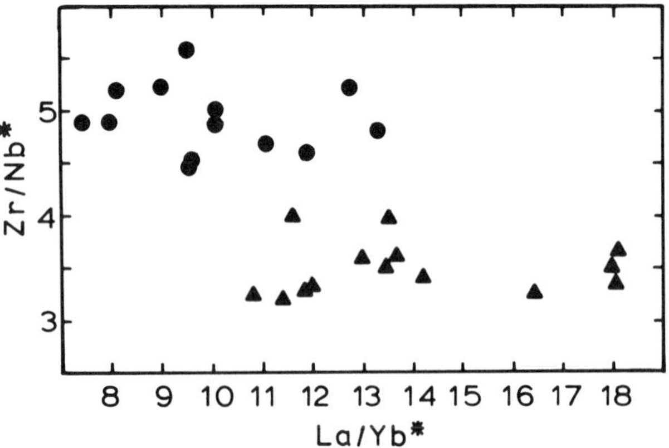

Figure 11. Zr/Nb* versus La/Yb*, corrected for effects of fractional crystallization. Circles indicate flank lavas; triangles, valley lavas.

are plotted against La/Yb for GVF basalts. The values for evolved lavas have been corrected for fractionation based on major element mass balance constraints discussed in the next section; samples so corrected are indicated with an asterisk. Flank lavas exhibit higher Zr/Nb* ratios and La/Yb* values less than or equal to the La/Yb* in the valley lavas. A similar relationship is observed between Hf/Th and Th/Y (see Table 7). Although Zr/Nb* is nearly uniform within each group, Rb/Ba* increases with increasing La/Yb* among the valley basalts ($R = 0.7$). This correlation is poorer for flank basalts, and probably reflects inadequate correction for fractionation (see Table 7).

MAJOR AND TRACE ELEMENT MODELS

Compositional variations in basaltic lavas are determined by predominantly four factors: composition and mineralogy of the mantle source region, the nature of the partial melting process, fractionational crystallization, and contamination by crustal materials. Although it has been suggested that crustal contamination played an important role in the evolution of some older flank basalts (Hoefs and others, 1984; Kempton and others, 1984b), major element mass balance calculations demonstrate that assimilation of no more than 5 to 10 percent of lower crustal granulitic material is possible, even for the most evolved basalts (Kempton, 1984). This process cannot be adequately constrained by the data presented here, but is believed to be insignificant for most GVF basalts. Within the GVF lava suite, compositional variations can be attributed to an interplay of the first three processes. In the following section, the influence of fractional crystallization on GVF basalt compositional trends is described, and probable fractionating assemblages are identified on the basis of major and trace element modeling. In the subsequent section, trace element data are used to specify some of the mineralogic and compositional characteristics of the source area for GVF lavas.

Compositional Variations Produced by Fractional Crystallization

The importance of fractional crystallization in the evolution of GVF lavas has been evaluated quantitatively through the use of major element mass balance calculations (Stormer and Nicholls, 1978). The calculations utilized the major element oxides SiO_2, Al_2O_3, TiO_2, FeO*, MgO, CaO, Na_2O, and K_2O. The validity of these models was tested by trace element calculations using the equations for Rayleigh fractional crystallization (Allegre and Minster, 1978) and the partition coefficients given in Table 8.

Potential fractionating phases include both low-pressure phenocryst compositions and megacrysts or mineral phases from Type II (Al-augite series) xenoliths believed to be the fractionation products of similar magmas at depth (Kempton, 1983; Irving, 1980; Wass, 1979a) (Table 9). All minerals selected as potential fractionating phases exhibit a range of compositions (Figs. 4 through 6), but an average composition was chosen. Based on major element mass balance and trace element calcula-

TABLE 8. PARTITION COEFFICIENTS USED IN MODEL CALCULATIONS*

Mineral Phase	Clinopyroxene		Spinel		Orthopyroxene		Amphibole		Garnet	
	Min.	Max.	Min.	Max.	Min.	Max.	Min.	Max.	Min.	Max.
Rb	0.004	0.055	0.003	0.0002	0.0006	---	0.004	0.25	0.007	0.042
Sr	0.07	0.11	0.003	0.0002	0.007	---	0.07	0.78	0.012	1.4
Ba	0.001	0.01	0.001	---	0.001	---	0.5	0.8	0.023	0.15
Zr	0.1	---	0.1	---	0.03	---	---	---	0.3	---
Nb	0.01	0.1	0.4	---	0.015	---	---	---	0.1	---
Hf	0.08	0.3	0.001	---	0.001	0.04	0.38	0.5	0.5	---
Th	0.001	0.04	0.001	---	0.001	0.13	0.038	0.07	0.001	---
Y	0.5	---	0.2	---	0.01	---	0.5	---	2.0	---
La	0.069	0.15	0.025	---	0.0005	0.006	0.1	0.3	0.001	0.02
Sm	0.3	0.45	0.052	---	0.003	0.0175	0.65	1.2	0.08	0.2
Yb	0.7	0.95	0.111	---	0.029	0.11	0.57	0.99	4.0	---
Sc	2.9	3.1	0.048	0.8	1.1	---	1.4	---	28.0	---
Cr	1.4	3.0	10.0	20.0	3.0	6.0	0.34	2.0	17.5	---
Ni	1.6	3.0	8.0	20.0	3.0	---	1.0	2.0	5.0	---

Mineral Phase	Olivine		Mica	Feldspar
	Min.	Max.		
Rb	0.0002	0.002	2.0	0.95
Sr	0.0002	0.003	1.0	2.7
Ba	0.0001	---	5.0	0.15
Zr	0.01	---	---	---
Nb	0.01	---	---	---
Hf	0.001	---	0.1	0.001
Th	0.001	---	0.04	---
Y	0.001	---	0.01	0.03
La	0.001	0.01	0.017	0.11
Sm	0.002	0.012	0.014	0.08
Yb	0.004	0.04	0.019	0.07
Sc	0.27	0.3	0.5	0.04
Cr	1.0	2.8	0.3	0.004
Ni	14.0	16.0	8.0	0.04

*Partition coefficients from:
Irving (1978, 1980)
Irving and Frey (1984)
Arth and Hanson (1975)
Frey and others (1978)
Pearce and Norry (1979)

tions, the critical phases are (in decreasing order of importance): clinopyroxene, spinel, olivine, and amphibole of moderate pressure origin; clinopyroxene of moderate to low pressure origin; and plagioclase and olivine of low pressure origin. In contrast to the results of Irving and Green (1976) for the Newer Basalts of southeast Australia, kaersutite is not the essential phase subtracted during fractionation. Amphibole may be required to produce some of the more evolved hawaiite and mugearite compositions, and may be involved in the fractionation of some more undersaturated lavas. Anorthoclase is present as megacrysts and in the groundmass of the hawaiites and mugearite lavas, but it is not a significant fractionating phase. Depletions in elements that would be generated by anorthoclase fractionation (such as Sr, Ba, Rb and Eu) are not observed (Fig. 7) in most cases. Studies by Bahat (1979) and Irving (1974, 1977) suggest that crystallization of anorthoclase is enhanced near the solidus where H_2O pressure is higher. Therefore, anorthoclase most likely crystallizes from the evolved residua of magmas trapped at depth and is accidentally incorporated by subsequent pulses of magma making their way to the surface (Bahat, 1979).

The first criterion for selecting reasonable solutions to major element mass balance calculations is that the sums of the residuals squared be less than 1.25. These calculations do not provide unique solutions, and several assemblages as well as varying proportions of phases can produce acceptably low sums of residuals squared. When necessary, three criteria were used to select a single model from among several numerically acceptable models: (1) the match between trace element concentrations calculated using the assemblage and percentage of fractionation determined by mass balance and trace element concentrations in the actual lavas; (2) the plausibility of the assemblage, that is, whether the relative abundance of fractionated phases prescribed by the model compared well with observed phenocryst or xenocryst modal abundances; and (3) when criteria 1 and 2 failed to discriminate between models, the assemblage accepted was the one with the least number of phases.

Table 10 lists the results of major element mass balance calculations for GVF basalts. These results demonstrate that low-pressure fractional crystallization alone is incapable of accounting for the compositional variations in relatively evolved lavas (Table

TABLE 9. COMPOSITIONS OF PHASES USED IN FRACTIONATION MODEL CALCULATIONS*

Analysis (wt.%)	(1)	(2)	(3)	(4)	(5)	(6)	(7)	(8)
SiO_2	48.65	50.36	41.45	37.63	38.25	0.09	40.31	50.43
TiO_2	1.54	0.76	5.17	---	---	0.54	5.51	0.10
Al_2O_3	8.69	4.56	11.68	---	---	59.09	14.50	31.23
FeO^{tot}	6.84	10.22	7.59	20.63	16.45	20.04	9.69	0.49
MgO	14.03	12.44	10.59	40.98	43.93	17.87	13.16	---
CaO	18.66	20.66	22.88	0.11	0.24	---	10.78	14.41
Na_2O	1.39	1.32	0.48	---	---	---	2.69	3.35
K_2O	---	---	---	---	---	---	1.66	0.20
Total	99.80	100.32	99.84	99.35	98.87	97.63	98.30	100.21

*Key to column heads:
1. Aluminous augite from type II xenolith.
2. Na-rich pyroxene from clinopyroxenite xenolith.
3. Titanaugite microphenocryst in basanite 90-47.
4. Olivine from type II xenolith.
5. Olivine phenocryst in basanite pk-G-50.
6. Aluminous spinel from type II xenolith.
7. Kaersutite megacryst.
8. Plagioclase phenocryst in basanite pk-G-50.

10, column 2). Although subtraction of low-pressure phenocryst phases, olivine + plagioclase + clinopyroxene, can in some cases produce low sums of residuals squared, the proportions of phases required are inconsistent with observed modes of phenocrysts and their sequence of appearance during crystallization. More than 50 percent of the subtracted assemblage is calculated to be clinopyroxene; yet clinopyroxene is the least abundant phenocryst phase in GVF lavas at low pressure. The absence of Eu anomalies in most GVF basalts indicates that plagioclase fractionation is minor as well.

Fractionation of moderate-pressure phases aluminous clinopyroxene, spinel, and olivine can account for the majority of the compositional variations produced by fractional crystallization in GVF lavas. Derivative compositions resulting from as much as 25 to 30 percent fractionation of the initial magma can be attributed to subtraction of aluminous clinopyroxene + spinel in the approximate proportion 4:1. A few moderately evolved valley basanites may also be accounted for by fractionation of kaersutite alone (Table 10, column 10). The sums of the residuals squared are slightly higher than the 1.25 cutoff adopted, but this is primarily due to the error in calculating TiO_2. Whether kaersutite or clinopyroxene + spinel (or both) constitutes the fractionating assemblage for these valley lavas is unclear because calculated trace element abundances are similar for both assemblages (Table 10, columns 13 and 14). Fractionation greater than 25 to 30 percent requires subtraction of olivine as well as clinopyroxene and spinel, and the proportions of the solid phases cpx:sp:ol shift to approximately 16:3:1. The mineral assemblages and the relative proportions of phases subtracted are generally the same for both flank and valley evolved lavas (Table 10, columns 3, 4, 9, 10), regardless of initial normative nepheline content. These modal proportions are also similar to the relative abundances of these phases in Type II xenoliths. This suggests that these model assemblages reflect cotectic proportions in this system.

Fractionation to hawaiite or mugearite compositions follows one of two general paths at GVF (see Fig. 7). Trend 1 is characterized by fractionation of kaersutite in addition to clinopyroxene, spinel, and olivine (Table 10, column 5) and results in greater enrichment in Sr, Rb, Ba, and the light REE relative to the HREE (increasing La/Yb). Trend 2 is characterized by fractionation involving plagioclase and a Na- and Fe-rich pyroxene analogous to the green megacrysts and groundmass compositions in GVF hawaiites and mugearite lavas (Table 9; Table 10, column 7). Although trend 2 can be accounted for by fractionation of the assemblage clinopyroxene + spinel + olivine in terms of major element mass balance (i.e., the calculations produce low sums of residuals), this assemblage does not reproduce the measured trace element concentrations as satisfactorily, particularly for the REE (Table 10, columns 6 and 7). Neither model reproduces the extreme depletion in compatible trace elements, Ni and Cr, observed in these lavas. Fractionation of small percentages of Al-Cr spinel, similar to the inclusions observed in some olivine phenocrysts, could produce significant depletions in these highly compatible trace elements, yet be trivial to major element mass balance calculations.

That fractionation at GVF involves moderate-pressure phases similar to those observed as megacrysts and Type II xenoliths is supported by trace element and isotopic characteristics of these inclusions. GVF lavas have $^{143}Nd/^{144}Nd$ values which range from 0.513021-0.513037 (Menzies and others, 1984). Analyses of kaersutites and Al-clinopyroxenes from GVF exhibit overlapping values: 0.513011-0.513109 (Menzies and others, 1985). Furthermore, geochemical modeling of Type II xenoliths indicates that their major and trace element compositions are consistent with fractionation from basaltic magma similar in composition to the host basanite (Irving, 1980), and that the compositional spectrum observed among xenoliths of this type is partially the result of fractionation from a spectrum of magma compositions similar to that observed at GVF (Kempton, 1984).

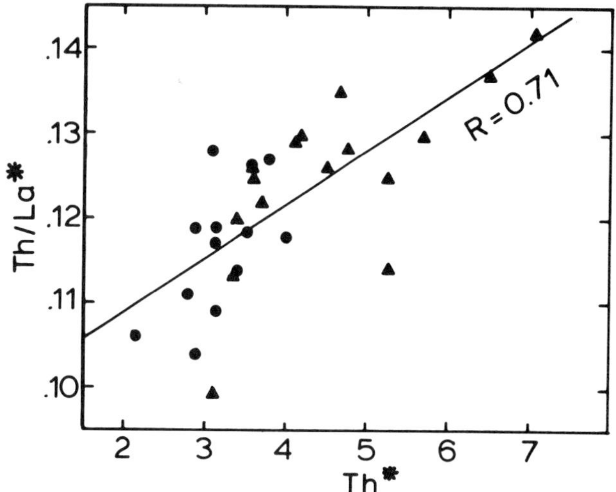

Figure 12. Th/La* versus Th*, corrected for effects of fractionation.

Compositional Variations Generated During Partial Melting

GVF basalts, particularly the younger valley basanites, exhibit considerable compositional variation that cannot be accounted for by fractionation alone (Fig. 7). Even among primitive lavas (i.e., high Mg value, Ni and Cr), major and trace elements vary markedly (normative Ne ranges from 6.6 to 17.7 wt.%; SiO_2, 44.6 to 48.0 wt.%; Al_2O_3, 15.3 to 16.6 wt.%; CaO, 8.7 to 10.7 wt.%; K_2O, 0.65 to 2.6 wt.%; La, 27 to 50 ppm; Zr 200 to 310; Rb, 23 to 51 ppm; Ba, 330 to 560 ppm; and La/Yb, 8 to 18).

Based on experimental studies of partial melting of hydrous garnet lherzolite, Green (1973) found that undersaturated compositions result from small degrees of partial melting. Subsequent experimental studies (Mysen and Kushiro, 1977; Wyllie, 1979; Olafsson, 1980; Takahashi and Kushiro, 1983) have demonstrated that several factors (pressure, temperature, vapor composition, and peridotite mineralogy or composition) may combine to influence the resultant melt composition during partial fusion. In this section we attempt to identify some of the mantle source characteristics that contribute to this diversity on the basis of trace element data.

The equation used for model calculations is that for equilibrium partial melting (Allegre and Minster, 1978):

$$C_L = C_o/[D + F(1-P)],$$

where C_L and C_o are the trace element concentrations in the liquid and the source, D is the bulk distribution coefficient with the minerals weighted by their modal proportions in the solid, P is the bulk distribution coefficient weighted according to the proportion each phase contributes to the melt, and F is the degree of partial melting. In this equation only C_L is known, but the mineralogy and composition (C_o) of the source, as well as the degree of partial melting involved can be modeled in a forward manner by assuming all of the unknown parameters and calculating a C_L that matches the observed lava composition. Unfortunately, the very low degrees of partial melting thought to be necessary for generation of alkali basalts produce a critical dependence on the accuracy of the partition coefficients used in calculations. Since most D values are not known to better than an order of magnitude (see Table 8), a very wide range of compositions can be calculated simply by varying the partition coefficients within the limits of published values. Thus, trying to reproduce GVF lava trace element abundances knowing neither the actual source concentrations nor the appropriate D values has little petrologic significance.

Modeling the inverse problem attempts to circumvent some of these assumptions, and has shown promise for some basalt suites (Allegre and Minster, 1978; Hofmann and Feigenson, 1983; Feigenson and others, 1983; Hofmann and others, 1984). But this approach assumes a source mineralogy, even if it does not assume the model proportions of phases present. It also assumes that the source areas for the individual lavas have the same mineralogy and composition, differing only in the modal proportions of phases and possibly in degree of partial melting involved. Hofmann and Feigenson (1983) have shown that if these conditions are met, ratios of highly incompatible elements (i.e., Zr/Nb, Hf/Th) and isotopic ratios must be uniform throughout the suite. In addition, the relationship between the incompatible element ratio C^H/C^i versus C^H (where C^H is a very highly incompatible element and C^i is some other generally less incompatible element) must be linear, e.g., Th/REE versus Th or Ba/Ree versus Ba. An example of such a diagram is shown for GVF data in Figure 12. The poor correlation observed between Th/La* versus Th* and the distinct elemental ratios for GVF flank and valley basalts given in Table 7 indicate that the inverse method is not applicable for GVF basalts. Instead, these results suggest that compositional variation within the GVF lava suite records a

TABLE 10. RESULTS OF FRACTIONAL CRYSTALLIZATION MODELS*
Flank Lavas

		Subtraction of Low-Pressure Phenocryst Phases	Subtraction of Moderate-Pressure Phenocryst Phases			Subtraction of Moderate and Low-Pressure Phases	
	(1) Basanite (Parent) pk-G-50	(2) Hawaiite D9	(3) Alkali Olivine Basalt D5	(4) Hawaiite D9	(5) Mugearite pk-G-58	(6) Hawaiite pk-G-63	(7)
Norm. Ne	6.5	1.1	4.0	1.1	1.4	4.7	
Mg value	0.63	0.49	0.61	0.49	0.30	0.46	
Model Results							
F		0.396	0.104	0.384	0.556	0.343	0.413
Residuals							
SiO_2		1.439	0.197	0.476	0.111	-0.037	0.278
TiO_2		-0.199	-0.125	-0.740	0.032	-0.155	-0.605
Al_2O_3		-0.634	0.147	0.446	0.204	-0.066	0.152
FeO		-0.913	0.063	-0.181	-0.782	0.322	0.388
MgO		0.362	0.112	0.377	0.366	-0.121	0.032
CaO		0.375	-0.014	0.007	-0.206	0.135	0.012
Na_2O		-0.263	-0.171	-0.224	-0.087	-0.064	-0.109
K_2O		-0.167	-0.210	-0.161	0.362	-0.015	-0.148
r^2		3.714	0.166	1.224	0.982	0.170	0.651
S.E.		0.862	0.166	0.495	0.496	0.185	0.403
D.F.		5	6	5	4	5	4
Proportions of Solids Fractionated (%)							
Clinopyroxene		52	81	80	47	81	45
Spinel			19	16	9	13	8
Olivine		22		4	3	6	17
Kaersutite					41		
Plagioclase		26					30

Trace Elements (ppm)

	Meas.	Calc.	Meas.	Calc.	Meas.	Calc.	Meas.	Calc.	Calc.	Meas.
Sc	28.1	23.9	21.7	14.8	19.3	11.9	8.8	16.0	21	19.1
Ni	158	130	133	55	77	21.5	5.3	20	13	6
Cr	210	169	211	87	99	35.5	4.7	37	58	10
Rb	23	25	23	37	34	46	45	34	33	41
Sr	584	624	590	877	773	937	691	828	616	745
Ba	356	400	367	578	394	677	823	543	594	548
La	26.6	29.2	31.4	40.4	36.7	55.0	57.0	39.3	43.5	43.45
Sm	5.75	6.2	6.6	6.8	6.6	7.7	1.08	7.9	8.9	10.60
Yb	2.9	3.1	2.2	3.3	3.3	3.3	3.2	3.6	4.2	4.4
Zr	243	267	256	374	325	499	442	354	399	328
Nb	46	51	53	72	60	99	85	68	77	66

*Notes:
F = Amount of fractionation required
r^2 = Sum of the squares of the residuals between calculated and observed values
S.E. = Standard error
D.F. = Degrees of freedom

complex interplay of variable mineralogy, composition, and degree of partial melting in the source region(s).

Although the absolute values of partition coefficients are not known well enough to quantitatively calculate absolute trace element abundances, it is not unreasonable to assume that the relative values of partition coefficients are established, and that trace element ratios can be used to define some of the mineralogic and compositional characteristics of the source area. Considering the variations of incompatible element abundances and ratios at GVF, there are several observations that must be explained. First, Zr/Nb and Hf/Th values for GVF flank and valley basalts are distinct (Fig. 10; Table 7). Because it is not possible to account for both the spectrum of concentrations and the ratios observed by partial melting alone, the mantle sources for these two basalt groups must be different. Also significant is the generally negative correlation between Zr/Nb and La/Yb for flank and valley lavas (Fig. 11). Lavas in both groups exhibit little variation in Zr/Nb for relatively large changes in La/Yb. Although fractional crystallization can increase La/Yb without changing the ratio of Zr/Nb significantly, it cannot explain the distribution of data observed. Fractionation increases the La/Yb value by only 15 percent after 25 percent crystallization of clinopyroxene and spinel, but the

TABLE 10. (CONTINUED)
Valley Lavas

Subtraction of Moderate Pressure Megacryst Phases

	(8) Basanite (Parent) D2	(9) Basanite RF26	(10) Nepheline Hawaiite RH12	(11)	(12) Basanite (Parent) pk-G-22	(13) Nepheline Hawaiite pk-G-66	(14)
Norm. Ne	15.2	13.8	13.51		17.7	18.65	
Mg value	0.61	0.58	0.51		0.65	0.55	
Model Results							
F		0.114	0.241	0.284		0.255	0.316
Residuals							
SiO_2		0.254	0.288	-0.157		0.117	-0.033
TiO_2		-0.243	-0.297	0.865		-0.339	1.000
Al_2O_3		0.236	0.380	-0.124		0.057	-0.515
FeO		0.007	0.076	0.271		0.406	0.095
MgO		0.046	-0.256	-0.242		-0.057	-0.605
CaO		0.031	0.340	-0.719		0.140	0.034
Na_2O		-0.129	-0.302	-0.030		0.206	0.324
K_2O		-0.204	-0.230	0.135		-0.154	-0.226
r^2		0.241	0.631	1.446		0.626	1.803
S.E.		0.200	0.300	0.425		0.232	0.475
D.F.		6	7	8		6	8
Proportions of Solids Fractionated (%)							
Clinopyroxene		82	85			68	
Spinel		18	15			17	
Olivine						15	
Kaersutite				100			100

Trace Elements (ppm)

	Meas. (8)	Calc. (9)	Meas. (9)	Calc. (10)	Calc. (11)	Meas. (RH12)	Meas. (12)	Calc. (13)	Calc. (14)	Meas. (pk-G-66)
Sc	22.6	19.2	---	15.1	19.8	16.4	21.1	16.7	18.1	16.8
Ni	127	103	85	60	91	57	141	57	96	96
Cr	211	168	---	91	152	83	240	150	163	131
Rb	45	51	57	58	58	30	48	62	64	49
Sr	803	897	826	1039	1090	1039	795	1019	1138	1173
Ba	573	644	---	752	675	695	559	720	678	656
La	44.1	49.2	---	55.9	55.5	55.4	47.4	59.5	62.1	54.8
Sm	6.9	7.5	---	8.5	7.8	8.4	7.0	8.3	8.0	7.8
Yb	2.5	2.6	---	2.8	2.9	3.3	2.6	2.9	3.1	2.9
Zr	260	289	248	333	349	320	307	387	434	286
Nb	77	85	71	99	106	95	88	110	129	87

*Notes:
F = Amount of fractionation required
r^2 = Sum of the squares of the residuals between calculated and observed values
S.E. = Standard error
D.F. = Degrees of freedom

range of La/Yb in valley basalts is significantly larger than this. In addition, the compositions of many of the lavas comprising this spectrum suggest minimal, if any, fractionation, i.e., high Ni, Cr, and Mg values. The third observation requiring explanation is the much greater variability in heavy REE abundances in flank lavas (Yb_{CH} = 11 to 21) relative to valley basalts (Yb_{CH} = 12 to 13), Figure 8.

Returning to equation 1, the ratio of two elements can be expressed as:

$$C_L^1/C_L^2 = C_o^1/C_o^2\,[(D^2 + F(1-P^2)/(D^1 + F(1-P^1)].$$

If the initial ratios C_o^1/C_o^2 in the source region(s) are constant, the concentration ratio in the liquid C_L^1/C_L^2 will be determined by the value of $[(D^2 + F(1\text{-}P^2)/(D^1 + F(1\text{-}P^1)]$. Figure 13 shows the variation in this parameter for La/Yb (A) versus the same parameter calculated for Zr/Nb (B) as a function of modal mineralogy and F. Partition coefficients used in these calculations are given in Table 8. Figure 13A shows the changes that result from increasing the proportion of garnet (gar) relative to clinopyroxene (cpx) for different degrees of partial melting. In all cases the percentage of change actually observed in the liquid concentration ratios depends on C_o^1/C_o^2 and the degree of melting, neither of which are known. Nonetheless, all of the B coefficients are less than 1, such that Zr/Nb decreases as B gets smaller. Similarly, the A coefficient for La/Yb is greater than 1, and La/Yb increases for those conditions that increase the A coefficient. Although the curves defining the change in A and B as a

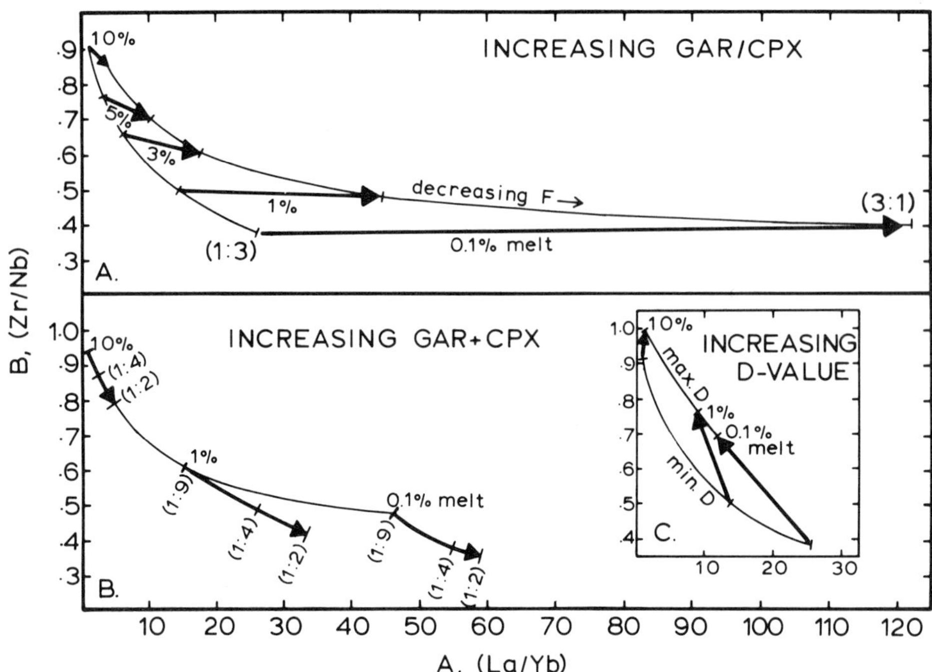

Figure 13. Model curves showing relative effects of modal variations and degree of partial mleting on La/Yb and Zr/Nb ratios of generated liquid (see text for explanation). Light curves in each figure indicate variation in F from 0.1 to 10 percent partial melting. A, Variation in A and B as function of gar/cpx ratio. Curve (1:3) = 15 percent cpx, 5 percent gar, 58 percent ol and 22 percent opx. Curve (3:1) = 5 percent cpx, 15 percent gar, 58 percent ol, and 22 percent opx. Heavy arrows indicate relative change of Zr/Nb and La/Yb, with increasing gar:cpx of 1:3 to 3:1 at same value of F. B, Variation as result of increasing gar + cpx relative to ol + opx at constant gar:cpx. Initial curve calculated for assemblage 5 percent cpx, 5 percent gar, 63 percent ol, 27 percent opx; (i.e., gar + cpx/ol + opx = 1:9); dark arrows indicate change in coefficients in A and B with increasing gar + cpx = 1:4 and 1:2 for the same value of F. C, Variation created by range of D values at same value of F.

function of F are relatively steep for higher percentages of partial melting, little variation in Zr/Nb results relative to La/Yb because the B coefficient rapidly approaches 1. For example, increasing F from 1 to 10 percent for the assemblage 15 percent cpx, 5% gar, 58% ol, and 22% opx increases Zr/Nb less than a factor of 2 while the same increase in F decreases La/Yb by a factor of 12. Increasing the gar/cpx ratio produces a much greater change in La/Yb than in Zr/Nb, particularly at low volumes of partial melting, but results in a general increase in La/Yb and decrease in Zr/Nb with increasing gar/cpx. Figure 13B shows the relative change expected for increasing the modal proportion of cpx + gar (i.e., the major phases containing incompatible trace elements) relative to ol + opx. The starting curve is for a mode of 5% cpx, 5% gar, 63% ol, and 27% opx. With increasing cpx + gar, the ratio of Zr/Nb decreases while La/Yb increases; at 1 percent partial melting, La/Yb increases by a factor of 2 while Zr/Nb decreases approximately 30%. Figure 13C emphasizes the problem created by uncertainties in D values. For the model assemblage 15% cpx, 5% gar, 58% ol, and 22% opx, the area defined by the range of D values reported in Table 8 encompasses most of the range of values created by increasing gar:cpx from 1:3 to 3:1 (compare with Fig. 13A). Nonetheless, if it is assumed that D is constant at some value in this range during partial melting, the curves will simply shift on the diagram, but relative changes produced by the remaining variables will be the same. Therefore, relative changes in each parameter may at least define relative source material differences.

Considering the GVF lavas in terms of these model curves, it appears that increasing gar/cpx or changes in F at low degrees of partial melting could explain the trends within each group. Variations between groups could result from differences in the amount of cpx + gar or from changes in F at moderate degrees of partial melting. The greater degree of silica undersasturation represented in the valley lavas relative to the older flank basalts is consistent with generation of the younger basalts at smaller degrees of partial fusion, as are major element mass balance constraints (Kempton, 1984). Although increasing the gar/cpx ratio does not alter Zr/Nb, a difference in the gar/cpx ratio for flank and valley basalts is indicated by the differences in heavy REE abundances between these two groups. The small range of chondrite-normalized Yb values in valley lavas indicates a high gar/cpx ratio, whereas the greater range in Yb values in the flank can be produced only at lower gar/cpx ratios. For example, for the assemblage gar:cpx = 3:1, Yb increases 140 percent from 10 to 1 percent partial melting, but for the assemblage gar:cpx = 1:3, Yb increases only 40 percent for the same decrease in degree of partial melting.

The presence of secondary metasomatic phases or enrichment by mantle metasomatism cannot be dismissed by this analysis. Although major element mass balance calculations appear to require the presence of mica to balance the alkali element contents, its presence would have much less of an influence on the resultant REE contents or ratios of the liquid than the cpx/gar ratio. Mantle mica is typically low in REE contents and has relatively low and uniform D values for all of the RE elements (Irving and Frey, 1984). It is unlikely that the mantle sources for GVF lavas have chondritic ratios of Zr/Nb and La/Yb; the coefficients necessary to produce GVF lavas would not be within the area defined by the model parameters in Figure 13. In particular, the B coefficient is too low (B = 0.2 to 0.4, A = 5 to 10). If the mantle sources for GVF basalts are enriched in La relative to Yb and Zr relative to Nb, the coefficients necessary to generate the lavas must approach 1, and an origin at relatively large degrees of partial melting is required. If GVF basalts are produced at smaller degrees of partial melting, the source materials must be depleted in La/Yb and Zr/Nb. Extremely small degrees of partial melting of an enriched mantle source are not permitted, according to this analysis.

Achieving reasonable results when calculating absolute concentrations of trace elements in this and previous studies (Kay and Gast, 1973; Frey and others, 1978; Fitton and Dunlop, 1985) requires exceedingly small degrees of partial melting (<1%) of a chondritic mantle. Several factors have been used to argue against such small degrees of partial melting. Mass balance constraints, for example, would indicate that GVF alkalic basalts are generated by 5 to 10 percent partial melting (Evans and Nash, 1978; Kempton, 1984). Experimental data suggest that very small degrees of partial fusion (<1–2%) would produce much more undersaturated compositions than the alkali basalts and basanites represented here (Takahashi and Kushiro, 1983; Olafsson, 1980). It has also been argued that removing such small volumes of melt from the mantle may be problematic (Arndt, 1976).

Fitton and Dunlop (1985) have recently argued that these objections need not require a metasomatic event prior to basalt genesis. Equations derived by McKenzie (1984) describing the behavior of melt migration in partially molten rock suggest that melt migrating through mantle at or near its solidus will be sufficiently rapid to restrict the amount of melt present in the mantle to less than 3 percent, even for large initial degrees of melting. Based on experimental and surface-energy considerations, Watson (1982) has suggested that melt in equilibrium with its surroundings will tend to infiltrate the dry grain boundaries of its host at a relatively rapid rate of 1 to 2 mm/day. Applying these experimental and theoretical conclusions, Fitton and Dunlop further point out that magma at depth is at least initially more likely to move toward the surface along grain boundaries than in large conduits. Elements concentrated at grain boundaries or small volumes of melt/glass remaining along grain boundaries or interstitially will be selectively incorporated into the infiltrating melt. If a magma produced by 10 percent partial melting of a depleted mantle source is allowed to percolate through 100 times the volume of mantle from which it was originally derived, it will assume the trace element characteristics of a liquid derived by significantly smaller degrees of equilibrium partial melting. Its major element composition would be buffered by equilibration with the surfaces of the silicate mantle phases so that it will still arrive at the surface as a basaltic liquid. Thus, the very small volumes of melt required by trace element models may be a reflection of the volume of mantle "processed" by the melt as it migrates through the mantle on its way to the surface.

DISCUSSION AND SYNTHESIS

Compositional variations observed within alkalic basalt suites cannot be attributed to any one igneous process. Experimental data demonstrate that, for small degrees of partial melting, melt composition varies markedly as a function of small differences in the percent of melting (temperature), pressure, and vapor composition. Heterogeneities in source mineralogy and composition can strongly influence trace element concentrations in the melt. Variability among GVF primitive lava compositions is believed to be a function of variable modal mineralogy and differing degrees of partial melting of a garnet lherzolite mantle.

Systematic variations in petrographic and compositional characteristics of GVF basalts record a temporal evolution for the volcanic field. The older flank lavas (>3.5 Ma) record generally larger degrees of partial melting at lower modal gar/cpx and lower total gar + cpx relative to ol + opx. The valley lavas (<3.5 Ma) represent, on the average, smaller degrees of partial fusion of a more fertile mantle, richer in gar/cpx and total gar + cpx. Thus, the source area for GVF basalts is not constant, but changes with time. Since the progression is one of increasing modal garnet/clinopyroxene in the source with time, this change cannot be explained as depletion of the initial source during the partial melting event. Rather, eruption of melts generated at progressively greater depths in the mantle is indicated. If the lowest degrees of partial melting calculated in trace element models reflect large enrichments during the melt migration and segregation process, the younger valley basanites probably processed a greater mantle volume, consistent with their being generated at slightly greater depths.

In both flank and valley lavas, these primary variations are overprinted by the effects of fractional crystallization, but the effects of fractionation are most pronounced in the older flank basalts. Major element mass balance calculations and analyses of included xenoliths (Kempton, 1984) indicate that fractionation of as much as 25 to 30 percent of the moderate-pressure phases, aluminous clinopyroxene + spinel (4:1), can account for the majority of the compositional variations produced by fractional crystallization in GVF lavas. Fractionation in amounts greater than 25 to 30 percent of the initial magma requires subtraction of olivine, as well in the proportions 16:3:1 of clinopyroxene, spinel, and olivine, respectively.

Substantial magma reservoirs in which fractionation is conventionally envisioned to occur, however, did not exist within the crust beneath GVF. Small magma chambers may have formed periodically throughout the evolution of the volcanic field, as suggested by the presence of megacrysts of jadeite-rich pyroxene, anorthoclase, and apatite, but the frequency and physical dimensions of these remain unknown. Based on a study of alkalic basalts from the Isle of Skye, Thompson and others (1980) have proposed a "ramifying plexus of subvolcanic conduits and fissures resembling a sponge, but without good lateral connections" (p. 247). In plumbing systems of this type, fractional crystallization takes place polybarically as liquidus phases are plated out along conduit walls by the ascending basanitic magmas (Irving, 1980).

Although it has been generally maintained that basaltic rocks with greater than 5 percent normative nepheline fractionate to more silica-undersaturated compositions (Coombs and Wilkinson, 1969; Macdonald and Katsura, 1964), the cumulative effect of fractionating the described assemblages is to decrease the degree of undersaturation in the melt. This is primarily due to crystallization of spinel. At lower pressures where plagioclase crystallizes with olivine and clinopyroxene from alkalic basalts instead of spinel, the fractionation could produce increasing silica undersaturation of the melt. However, fractionation of clinopyroxene + spinel + olivine from GVF alkali olivine basalts and basanites at moderate pressures results in decreasing undersaturation.

Derivation of some of the more evolved hawaiite and mugearite compositions at GVF requires fractionation at lower pressures of kaersutite or plagioclase + Na- and Fe-rich pyroxene, in addition to the moderate-pressure assemblages described above. Xenoliths containing these phases typically include magnetite and sometimes apatite, suggesting that these phases may have fractionated in minor amounts.

The higher apparent degree of fractional crystallization calculated for the older flank lavas is inferred to have resulted from passage to the surface through a relatively cold lithosphere during initial development of the subvolcanic conduit system. Establishment of the initial conduit system resulted in interaction between the ascending magmas and the conduit walls, at least on a scale of centimeters. Such interactions may alter the conduit walls (contact metasomatism; Menzies, 1983), and produce crosscutting intrusive relations similar to those seen in Type II/Type II composite xenoliths (Kempton and others, 1984a; Kempton, 1985). The ascending magmas probably continued to crystallize liquidus phases that attached to the conduit walls (Irving, 1980), effectively insulating subsequent magmas from the wall rock. Thus, at Geronimo Volcanic Field, penetration of the crust and upper mantle by the older flank lavas is believed to have had a "priming" effect. The younger valley lavas reached the surface having undergone less fractionation and interaction with the wallrock than did the older flank lavas.

ACKNOWLEDGMENTS

We thank Drs. C. Langmuir, W. Hart, and D. Gust for their helpful and insightful reviews. This work was largely funded by the Graduate Traineeship program of the National Aeronautics and Space Administration, a grant in support of P.D.K.'s doctoral dissertation; the manuscript was written while P.D.K. was an NRC National Research Council Fellow at NASA–Johnson Space Center. Partial funding for field work and some analytical work was provided by a Penrose grant from the Geological Society of America, by a grant to P.D.K. from the Institute for the Study of Earth and Man at Southern Methodist University, and by a grant from the National Science Foundation, EAR 8026451-01, to M.A.D. Thanks to M. L. Lake for the typing of Tables 5 and 6.

REFERENCES CITED

Allegre, C. J., and Minster, J. F., 1978, Quantitative models of trace element behavior in magmatic processes: Earth and Planetary Science Letters, v. 38, p. 1–25.

Arndt, N., 1976, The separation of magma from partially molten peridotite: Carnegie Institute of Washington Yearbook, v. 76, p. 424–428.

Arth, J. G., and Hanson, G. N., 1975, Geochemistry and origin of the early Precambrian crust of northeastern Minnesota: Geochimica et Cosmochimica Acta, v. 39, p. 325–362.

Bahat, D., 1979, Anorthoclase megacrysts; Physical conditions of formation: Mineralogical Magazine, v. 43, p. 287–291.

Basaltic Volcanism Study Project, 1981, Basaltic volcanism in the terrestrial planets: New York, Pergamon Press, 1286 p.

Batiza, R., 1978, Petrology and chemistry of Guadalupe Island; An alkalic seamount on a fossil ridge crest: Geology, v. 5, p. 760–764.

Bence, A. E., and Albee, A. L., 1968, Empirical correction factors for the electron microanalysis of silicates and oxides: Journal of Geology, v. 76, p. 382–403.

Coombs, D. S., and Wilkinson, J.F.G., 1969, Lineages and fractionation trends in undersaturated volcanic rocks from the East Otago volcanic province (New Zealand) and related rocks: Journal of Petrology, v. 10, p. 440–501.

Cooper, J. R., 1959, Reconnaissance geologic map of southeastern Cochise County, Arizona: U.S. Geological Survey Mineral Investigation Field Studies Map MF-213.

Drewes, H. D., 1981, Tectonics of southeastern Arizona: U.S. Geological Survey Professional Paper 1144, 96 pp.

Ehrenberg, S. N., 1982, Rare earth geochemistry of garnet lherzolite and megacrystalline nodules from minette of the Colorado Plateau province: Earth and Planetary Science Letters, v. 57, p. 191–210.

Evans, S. H., Jr., and Nash, W. P., 1979, Petrogenesis of xenolith-bearing basalts

APPENDIX 1. LOCATION OF SAMPLES DISCUSSED IN TEXT

Sample No.	Physiographic Distribution and Rock Type	Location	
		Position	Description
D2	Valley basanite flow	31°40'N, 109°18'W	1 mi Northwest of Krentz Ranch
D3	Flank basanite flow	31°37'N 109°19'W	Mulberry Canyon, Pedregosa Mountains
D4	Flank alkali olivine basalt	31°37'N 109°19'W	Mulberry Canyon, Pedregosa Mountains
D5	Flank alkali olivine basalt flow	31°22'N 109°05'W	Starvation Canyon, Peloncilla Mountains
D6	Valley basanite flow	31°33'N 109°06'W	Southern rim of Paramore Maar
D7	Valley basanite flow	31°29'N 109°13'W	Black Draw, 1 mi south of Mormon Tank
D8	Flank alkali olivine basalt	31°31'N 109°06'W	1 mi east of Fairchild Ranch, Peloncilla Mountains
D9	Flank hawaiite flow	31°30'N 109°06'W	Hog Canyon, 2 mi southeast of Fairchild Ranch, Peloncilla Mountains
D10	Valley basanite flow	31°37'N 109°13'W	Mangus Maar, 1 mi northeast of Hwy 80
D11	Flank alkali olivine basalt flow	31°24'N 109°24'W	Hog Canyon, Perilla Mountains
D12	Flank basanite flow	31°37'N 109°19'W	Northern rim of Mulberry Canyon, Pedregosa Mountains
pk-G-15	Flank alkali olivine basalt flow	31°29'N 109°23'W	2 mi north of bend in Hwy 80
pk-G-21	Valley cinder cone with crater (abundant mafic and ultramafic xenoliths)	31°27'N 109°17'W	Western rim of Cochise Maar
pk-G-22	Valley basanite cinder cone with pyroclasitc tuff ring (abundant mafic and ultramafic xenoliths)	31°30'N 109°17'W	2 mi east of Hwy 80
pk-G-23	Valley basanite cinder cone (abundant mafic and ultramafic xenoliths)	31°27'N 109°18'W	1 mi east of Indian Creek
pk-G-29	Valley basanite cinder cone	31°25'N 109°17'W	1 mi southeast of Cinder Hill
pk-G-37	Valley basanite cinder cone	31°29'N 109°13'W	0.5 mi southwest of Mormon Tank
pk-G-50	Flank basanite flow	31°36'N 109°21'W	East rim of Halfmoon Valley, Pedregosa Mountains
pk-G-51	Flank alkali olivine basalt flow	31°31'N 109°22'W	1 mi north of Deep Well Canyon, Pedrogosa Mountains
pk-G-56	Flank alkali olivine basalt flow	31°27'N 109°05'W	2 mi northeast of MacDonald Ranch, Peloncilla Mountains
pk-G-58	Flank mugearite flow	31°32'N 109°06'W	2 mi northeast of Fairchild Ranch, Peloncilla Mountains
pk-G-63	Flank hawaiite flow	31°33'N 109°04'W	East rim of Skeleton Canyon, Peloncilla Mountains
pk-G-66	Valley nepheline hawaiite flow	31°23'N 109°16'W	2 mi north of Geronimo Trail at BM 3828
82-1	Valley basanite flow (flow predates maar eruption)	31°33'N 109°09'W	Southwest rim of Paramore Maar
90-8	Valley basanite flow	31°28'N 109°18'W	2 mi southeast of Hwy 80
90-9	Valley basanite flow	31°26'N 109°19'W	Southwest of pk-G-23
90-47	Valley basanite flow (ultramafic xenoliths common)	31°25'N 109°19'W	Southwest of Cinder Hill
QH8	Valley basanite cinder cone with cinder quarry	31°33'N 109°16'W	4 mi west of Hwy 80
RF26	Valley basanite flow	31°40'N 109°18'W	1 mi northwest of Krentz Ranch, near locality D2
RH12	Red Hill, Valley cinder cone	31°36'N 109°13'W	0.5 mi east of Hwy 80, south of SH14
SH14	Valley basanite flow crossing Hwy 80	31°37'N 109°14'W	Southeast of Mangus Maar

from southeastern Arizona: American Mineralogist, v. 64, p. 249–268.

Feigenson, M. D., Hofmann, A. W., and Spera, F. J., 1983, Case studies on the origin of basalt; II. The transition from tholeiitic to alkalic volcanism on Kohala volcano, Hawaii: Contributions to Mineralogy and Petrology, v. 84, p. 390–405.

Fitton, J. G., and Dunlop, H. M., 1985, The Cameroon line, West Africa, and its bearing on the origin of oceanic and continental alkali basalt: Earth and Planetary Science Letters, v. 72, p. 23–38.

Frey, F. A., and Prinz, M., 1978, Ultramafic inclusions from San Carlos, Arizona; Petrologic and geochemical data bearing on their petrogenesis: Earth and Planetary Science Letters, v. 38, p. 129–176.

Frey, F. A., Green, D. H., and Roy, S. D., 1978, Integrated models of basalt petrogenesis; A study of quartz tholeiites to olivine melilitites from southeastern Australia utilizing geochemical and experimental petrologic data: Journal of Petrology, v. 19, p. 463–513.

Green, D. H., 1973, Conditions of melting of basanite magma from garnet peridotite: Earth and Planetary Science Letters, v. 17, p. 456–465.

Herd, D., and McMasters, C., 1982, Surface faulting in the Sonora, Mexico, earthquake of 1887: Geological Society of America Abstracts with Programs, v. 14, p. 172.

Hoefs, J., Kempton, P. D., Harmon, R. S., and Dungan, M. A., 1984, Oxygen isotope constraints on magmagenesis, Geronimo Volcanic Field (GVF) southeastern Arizona: Geological Society of America Abstracts with Programs, v. 16, p. 540–541.

Hofmann, A. W., and Feigenson, M. D., 1983, Case studies on the origin of basalt; I. Theory and reassessment of Grenada basalts: Contributions to Mineralogy and Petrology, v. 84, p. 382–389.

Hofmann, A. W., and White, W. M., 1982, Mantle plumes from ancient oceanic crust: Earth and Planetary Science Letters, v. 57, p. 421–436.

Hofmann, A. W., Feigenson, M. D., and Raczek, I., 1984, Case studies on the origin of basalt; III. Petrogenesis of the Mauna Ulu eruption, Kilauea, 1969–1971: Contributions to Mineralogy and Petrology, v. 88, p. 24–35.

Irving, A. J., 1974, Megacrysts from the Newer Basalts and other basaltic rocks of southeastern Australia: Geological Society of America Bulletin, v. 85, p. 1503–1514.

—— , 1977, Origin of megacryst suites in basaltic dikes near 96 Ranch, West Texas and Hoover Dam, Arizona [abs.]: EOS American Geophysical Union Transactions, v. 58, p. 526.

—— , 1978, A review of experimental studies of crystal liquid trace element partitioning: Geochimica et Cosmochimica Acta, v. 42, p. 743–770.

—— , 1980, Petrology and geochemistry of composite ultramafic xenoliths in alkali basalts and implications for magmatic processes within the mantle: American Journal of Science, v. 280A, p. 389–426.

Irving, A. J., and Frey, F. A., 1984, Trace element abundances in megacrysts and their host basalts: Contributions to Mineralogy and Petrology, v. 48, p. 1201–1222.

Irving, A. J., and Green, D. H., 1976, Geochemistry and petrogenesis of the Newer Basalts of Victoria and south Australia: Journal of the Geological Society of Australia, v. 23, p. 45–66.

Jacobs, J. W., Korotev, R. L., Blanchard, D. P., and Haskin, L. A., 1977, A well-tested procedure for instrumental neutron activation analysis of silicate rocks and minerals: Journal of Radioanalytical Chemistry, v. 40, p. 93–114.

Kay, R. W., and Gast, P. W., 1973, The rare earth content and origin of alkali-rich basalts: Journal of Geology, v. 81, p. 653–682.

Kempton, P. D., 1983, Peridotites from the Geronimo Volcanic Field; Evidence for multiple interactions between basanitic magmas and the mantle beneath southeastern Arizona: Geological Society of America Abstracts with Programs, v. 15, p. 302.

—— , 1984, I. Alkalic basalts from the Geronimo Volcanic Field; Petrologic and geochemical data bearing on their petrogenesis. II. Petrography, petrology and geochemistry of xenolithis and megacrysts from the Geronimo Volcanic Field, southeastern Arizona. III. An interpretation of contrasting nucleation and growth histories from the petrographic analysis of pillow and dike chilled margins, hole 504B, DSDP Leg 83 [Ph.D. thesis]: Dallas, Southern Methodist University, p. 275.

—— , 1985, Mineralogic and geochemical evidence for differing styles of metasomatism in spinel lherzolite xenoliths; Analogues for enriched mantle source regions?, *in* Menzies, M., and Hawkesworth, C., eds., Mantle metasomatism: London, Academic Press (in press).

Kempton, P. D., Menzies, M. A., and Dungan, M. A., 1984a, Petrography, petrology, and geochemistry of xenoliths and megacrysts from the Geronimo Volcanic Field, southeastern Arizona, *in* Kornprobst, J., ed., Vol. 2; The mantle and crust-mantle relationships: Amsterdam, Elsevier, p. 71–85.

Kempton, P. D., Blanchard, D., Dungan, M., Harmon, R., and Hoefs, J., 1984b, Petrogenesis of alkalic basalts from the Geronimo Volcanic Field; Geochemical constraints on the roles of fractional crystallization and crustal contamination, *in* Dungan, M. A., Grove, T. L., and Hildreth, W., eds., Proceedings of the ISEM Field Conference on Open Magmatic Systems: Dallas, Southern Methodist University, Institute for the Study of Earth and Man, p. 91–93.

Lynch, D. J., 1972, Reconnaissance geology of the San Bernardino volcanic field, Cochise County, Arizona [M.S. thesis]: Tucson, University of Arizona, 101 p.

—— , 1978, The San Bernardino volcanic field of southeastern Arizona: New Mexico Geological Society Guidebook, 29th Field Conference, p. 261–268.

MacDonald, G. A., and Katsura, T., 1964, Chemical composition of Hawaiian lavas: Journal of Petrology, v. 5, p. 82–133.

Maxwell, J. A., 1968, Rock and mineral analysis: New York, Interscience, p. 419–421.

McKenzie, D., 1984, The generation of compaction of partially molten rock: Journal of Petrology, v. 25, p. 713–765.

Menzies, M., 1983, Mantle ultramafic xenoliths in alkaline magmas; Evidence for mantle heterogenity modified by magmatic activity, *in* Hawkesworth, C. J., and Norry, M. J., eds., Continental basalts and mantle xenoliths: Nantwich, United Kingdom, Shiva Publishing, p. 92–110.

Menzies, M. A., and Murthy, V. R., 1980, Mantle metasomatism as a precursor to the genesis of alkline magmas-isotopic evidence: American Journal of Science, v. 280A, p. 622–638.

Menzies, M. A., Leeman, W. P., and Hawkesworth, C. J., 1984, Isotope geochemistry of Cenozoic volcanic rocks reveals mantle heterogenity below western U.S.A.: Nature, v. 303, p. 205–209.

Menzies, M., Kempton, P. D., and Dungan, M. A., 1985, Multiple enrichment events in residual MORB-like mantle below the Geronimo Volcanic Field Arizona, U.S.A., Journal of Petrology (in press).

Mysen, B. O., and Kushiro, I., 1977, Compositional variations in coexisting phases with degree of melting in peridotite in the upper mantle: American Mineralogist, v. 62, p. 843–856.

Norrish, K., and Chappell, B. W., 1967, X-ray fluorescence spectrometry, *in* Zussman, J., ed., Physical methods in determinative mineralogy: New York, Academic Press, p. 161–214.

Norrish, K., and Hutton, J. T., 1969, An accurate x-ray spectrographic method for the analysis of a wide range of geologic samples: Geochimica et Cosmochimica Acta, v. 33, p. 431–453.

Olafsson, M., 1980, Partial melting of peridotite in the presence of small amounts of volatiles with special reference to the low-velocity zone [M.S. thesis]: University Park, Pennsylvania State University, 59 p.

Pearce, J. A., and Norry, M. J., 1979, Petrogenetic implications of Ti, Zr, Y, and Nb variations in volcanic rocks: Contributions to Mineralogy and Petrology, v. 69, p. 33–47.

Ringwood, A. E., 1982, Phase transformations and differentiation in subducted lithosphere; Implications for mantle dynamics, basalt petrogenesis, and crustal evolution: Journal of Geology, v. 90, p. 611–643.

Stormer, J. C., Jr., and Nicholls, J., 1978, XLFRAC; A program for the interactive testing of magmatic differentiation models: Computers and Geosciences, v. 4, p. 143–159.

Sun, S. S., and Hanson, G. N., 1975, Evolution of the mantle; Geochemical evidence from alkali basalt: Geology, v. 3, p. 297–302.

Takahashi, E., and Kushiro, I., 1983, Melting of a dry peridotite at high pressure and basalt magma genesis: American Mineralogist, v. 68, p. 859–879.

Thompson, R. N., Gibson, I. L., Marriner, G. F., Mattey, D. P., Morrison, M. A., 1980, Trace-element evidence of multistage mantle fusion and polybaric fractional crystallization in the palaeocene lavas of Sky, N. W. Scotland: Journal of Petrology, v. 21, p. 265–293.

Wass, S. Y., 1973, The origin and petrogenetic significance of hour-glass zoning in titaniferous clinopyroxenes: Mineralogical Magazine, v. 39, p. 133–144.

—— , 1979a, Fractional crystallization of late-stage kimberlitic liquids; Evidence in xenoliths in the Kiama Area, N.S.W., Australia, *in* Boyd, F. R., and Meyer, H.O.A., eds., The mantle sample; Inclusions in kimberlites and other volcanics: Washington, D.C., American Geophysical Union, p. 366–375.

—— , 1979b, Multiple origins of clinopyroxenes in alkali basaltic rocks: Lithos, v. 12, p. 115–132.

—— , 1980, Geochemistry and origin of xenolith-bearing and related alkali basaltic rocks from the Southern Highlands, New South Wales, Australia: American Journal of Science, v. 280A, p. 639–666.

Watson, E. B., Melt infiltration and magma evolution: Geology, v. 16, p. 236–240.

Wilshire, H. G., and Trask, N. J., 1971, Structural and textural relationships of amphibole and phlogopite in peridotite inclusions, Dish Hill, California: American Mineralogist, v. 56, p. 240–255.

Wood, D. A., Tarney, J., and Weaver, B. L., 1981, Trace element variations in Atlantic Ocean basalts and Proterozoic dykes from northwest Scotland: Their bearing upon the nature and geochemical evolution of the upper mantle: Tectonophysics, v. 75, p. 91–112.

Wyllie, P. J., 1979, Magmas and volatile components. American Mineralogist, v. 64, p. 469–500.

Manuscript Accepted by the Society October 15, 1986

Printed in U.S.A.

SELECTED ABSTRACTS

The papers in this volume represent about half of the material presented at the Symposium of Alkalic Rocks and Kimberlites. In order to more fully represent the spectrum of petrology addressed by this symposium, the abstracts of all other papers presented orally or in poster session are reprinted in this appendix. They are printed in the order that they were presented at the Fayetteville meeting. These abstracts were originally published in Geological Society of America Abstracts with Programs, v. 16, no. 3, 1985.

GENERATION OF ALKALINE MAGMAS BY PRESSURE RELEASE IN WITHIN-PLATE OCEANIC AND CONTINENTAL PROVINCES

LAMEYRE, Jean, Laboratoire de Petrologie, Universite Pierre et Marie Curie, 4 place Jussieu - 75230 - Paris Cedex 05

The mantle plume hypothesis does not give a complete and coherent explanation of all the geological characters observed in within-plate provinces. The magmatic activity, mainly alkaline, occurs sporadically but may last for millions of years. It may be reactivated after several hundred millions years as in West African provinces. Moreover its location along the limits of cratons or main transcurrent faults, the shapes of provinces (georhombs, sigmoids) strongly suggest a structural lithospheric control and an activation by shear systems.

As an alternate solution to the classical question set by the lack of intersection between the geotherm and the peridotite solidus, we propose that pressure release may be triggered by fracturing of the lithosphere. The geotherm and solidus proposed by Green and Ringwood delimit a narrow interval of possible melting between lithosphere thicknesses of 90 to 125 km and offer the conditions for the generation of alkaline magmas when openings reach the deep levels of a thick lithosphere, and of tholeiitic magmas at shallow depths. The generation of kimberlite supposes thicknesses over 150 km which may be accepted in old cratons where these rocks are generally located.

When magma fills lithospheric openings, it generates pressure which inhibits further melting. Magma generation in these conditions is thus self regulated. It starts when the lithosphere is opened and stops when the magma fills the fractures. This self-sealing system may be easily related to the geological patterns of within-plate provinces. It is also consistent with the short life span of alkaline magmas indicated by the Rb/Sr isotopic studies performed on materials from Kerguelen Islands. The general change of composition in oceanic islands from tholeiitic to alkaline, and in opening continental rifts from alkaline to tholeiitic may be related to lithosphere thickening in the first case and thinning down in the second.

A COMPARATIVE STUDY OF GARNET PERIDOTITE NODULES FROM POTASSIC VOLCANIC ROCKS AND KIMBERLITE

SMITH, Douglas, Dept. Geological Sciences, University of Texas, Austin, Texas 78713

Potassic minette from The Thumb (Colorado Plateau) and kimberlite from Lesotho contain garnet peridotite xenoliths with similar bulk compositions and mineral zoning. Compositions of typical xenoliths in each suite span nearly identical ranges of MgO, CaO, Al_2O_3, and FeO*, although on the average, nodules from The Thumb are slightly richer in Fe and Cr and poorer in Na. Garnets in some fertile, porphyroclastic rocks from The Thumb and from Lesotho are zoned; one pattern is zoning from relatively Cr-rich cores to Fe and Ti-rich rims. Such zoning appears to reflect metasomatic addition of Fe, Ti, and other elements, and calculations based on diffusion rates indicate that the metasomatism occurred shortly before eruption. A reconstruction indicates increases of at least 1.2% FeO and 0.1% TiO_2 in the bulk composition of a xenolith from The Thumb. The observed zoning supports the hypothesis that the fertility of some lherzolites partly reflects metasomatism. Bulk-rock correlations between Ca and Al in both suites are typical of garnet and spinel lherzolites, however, and hence some compositional differences between fertile and sterile lherzolites likely predate metasomatism. Equilibration temperatures and pressures calculated by the preferred methods of Finnerty and Boyd (1984) for nodules from The Thumb delineate a short geotherm segment, on which most points for fertile porphyroclastic rocks plot at higher T and P than points for coarse nodules. The data are consistent with a compositional gradient with depth in the small mantle interval sampled by the minette. Though some nodules from Lesotho have much higher equilibration pressures and temperatures, they document similar mantle compositions and processes.

ARCHAEAN HARZBURGITES WITH GARNET OF DIAMOND FACIES FROM SOUTHERN AFRICAN KIMBERLITES

NIXON, P. H., Department of Earth Sciences, The University of Leeds, Leeds, U.K. LS2 9JT; BOYD, F. R., Geophysical Laboratory, Washington, D.C. 20008; and HAWKESWORTH, C., Department of Earth Sciences, The Open University, Walton Hall, Milton-Keynes, U.K.

Low-calcium garnets (LCG) similar to those occurring in diamonds are also found in rare depleted harzburgites from Liqhobong, Jagersfontein, and Premier kimberlites, southern Africa. The associated mineral suite of these xenoliths is also similar to that of the 'armoured' diamond inclusions but with evidence of some overprint, particularly further decalcification, attributed to volatile reactions. Graphite, but not diamond, was observed. The LCG in a harzburgite from Liqhobong, Lesotho has similar Nd isotopes to the Archaean-age LCG diamond inclusions (Richardson, et al., 1984) from Finsch and Kimberly. The results also demonstrate that the Archaean lithosphere extended near to the present-day boundary in Lesotho.

GLIMMERITES, MARID AND PKP XENOLITHS FROM KIMBERLEY, RSA

JONES, R. A., Earth Sciences, Leeds, LS2 9JT, UK, BOYD, F. R., and SCHULZE, D. J., Geophysical Lab., Wash. DC 20008

Abundant deformed glimmerite xenoliths from Kimberley are composed of Ti-phlogopite, Cr-diopside and ilmenite having compositions distinct from similar minerals in amphibole-bearing peridotite (PKP) and MARID xenoliths. Diopside and phlogopite separated from four glimmerites define an Rb/Sr isochron age of 89.8 ±1.3 Ma that is taken to be the age of eruption of the host kimberlites. Low initial $^{87}Sr/^{86}Sr$ (0.7035) and high estimated Rb/Sr (>10) constrain glimmerites to have formed within 4 Ma before eruption. The initial $^{87}Sr/^{86}Sr$ and $^{143}Nd/^{144}Nd$ are similar to those of group I kimberlites.

Potassic richterite from a PKP xenolith has initial $^{87}Sr/^{86}Sr$ (0.709) and ϵNd (−8) which are within the range of group II (lamprophyric) kimberlites. Minerals in a MARID xenolith have isotopic compositions intermediate between the two kimberlite groups. Together with published data, these results form arrays on isotope ratio plots which can be modelled by bulk-mixing of group I and group II fluids.

It is proposed that the glimmerites are disrupted pegmatitic segregations of group I kimberlites. MARID and PKP xenoliths may have formed in magmatic/metasomatic processes resulting from a mixing of group I magma intruding the lithosphere with fluids of group II kimberlite that originated by partial fusion of ancient, enriched lithosphere. The mixing must have occurred between limited volumes, possibly in peripheral veins.

Metasomatic xenoliths from Kimberley are interpreted to have formed as the direct result of kimberlite magmatism and may, therefore, be of little significance in discussions of global geochemical evolution.

CARBONATITES—CLUES TO MANTLE EVOLUTION

BELL, Keith and BLENKINSOP, John, Ottawa-Carleton Centre for Geoscience Studies, Dept. of Geology, Carleton University, Ottawa, Ontario, Canada K1S 5B6

A linear relationship between age and initial $^{87}Sr/^{86}Sr$ ratio of carbonatites from Ontario and Quebec (Bell et al., 1982) shows Sr derivation from a region with a constant Rb/Sr ratio between at least 2700 and 100 Ma ago. The Rb/Sr value (about 0.018) is lower than that for "bulk Earth", indicating that the Sr is from a LIL-depleted region of the mantle. The intersection of the carbonatite and "bulk Earth" lines at ca. 3000 Ma agrees well with the age of Superior Province rocks. We have proposed a model in which "bulk Earth" material differentiated about 3000 Ma ago to form an enriched fraction (the Superior crust) and a depleted remnant, the source region of carbonatite Sr. These regions were still coupled 100 Ma ago. Pb isotopic results (e.g. Tilton and Grunenfelder, 1983) are consistent with a depleted mantle source.

Initial $^{143}Nd/^{144}Nd$ ratios from a preliminary Sm/Nd study of the same samples also show a linear relationship, and confirm the existence of a depleted source. Its Sm/Nd ratio (about 0.36) is only about 10% higher than that of CHUR; a tentative value of the intersection age is about 3000 Ma. The present-day epsilon value of +7 falls below the range of normal MORB, and suggests, as do the Sr results, that the depleted source is not the MORB source. Chondrite-normalized REE patterns for the carbonatites show very strong light-REE enrichments, and imply that mantle metasomatism may have occurred just prior to intrusion of the carbonatites.

LOW-ALKALI CARBONATITES IN ALKALINE COMPLEXES: SEPARATE MANTLE SOURCES FOR CARBONATE AND ALKALIS?

TREIMAN, Allan H., Lunar and Planetary Laboratory, University of Arizona, Tuscon AZ 85721

Carbonatites are usually thought of as products of alkaline volcanism because they are associated with ijolitic or nephelinitic rocks, they contain alkali minerals, and they have caused alkaline (fenitic) alteration in wall-rocks and xenoliths. However, some carbonatites in ijolite/nephelinite complexes are not alkaline; they contain no alkali minerals except phlogopite and have not caused fenitization.

An example from the Oka carbonatite/ijolite complex, Quebec, is a calcite-monticellite-diopside±olivine carbonatite, which is present as dikes and a central plug. The carbonatite contains little K_2O (avg. 0.3%), essentially no Na_2O (avg. 0.03%) and no alkali minerals except phlogopite. The diopside averages 0.05% Na_2O, compared to 0.3% in augite from a nearby alkaline carbonatite. More important is that the monticellite-carbonatite caused no fenitization. Instead, augite and melilite in the wall-rock are replaced by monticellite! Absence of fenitization implies that the monticellite-carbonatite lacked alkali-rich fluids and that the original magma was not alkaline.

Three non-alkaline carbonatites are present in the Oka complex, and others are documented at the Magnet Cove (Arkansas) and Kerii-masi (Kenya) complexes, and possibly at Fort Portal (Uganda). Existence of non-alkaline carbonatites suggests that the carbonatite/ijolite association is related more to processes of mantle fusion or magma emplacement than to mantle or melt chemistry. Alkali enrichment and carbonate enrichment of the mantle may be distinct processes, which usually work together as precursors to normal alkaline-carbonatite/ijolite volcanism.

ALKALIC ULTRAMAFIC MAGMAS, MISSOURI BREAKS, MONTANA: THE KIMBERLITE - ALNOITE CONTINUUM

HEARN, B. Carter, 959 National Center, U.S. Geological Survey, Reston, VA 22092

The chilled magmatic components of the Missouri Breaks swarm of diatremes in north-central Montana are alnoite (AL, mellite-bearing), monticellite peridotite (MP, melilite-free), kimberlite, and rare carbonatite. Plots of major-element variation vs. MgO show that the MP field (19-31% MgO) overlaps part of the AL field (10-25% MgO) for oxides other than CaO and SiO_2. The AL and MP fields on the CaO and SiO_2 plots are close but do not overlap. The elongations of the combined AL plus MP fields (except that of P_2O_5) suggest that AL magma is derived by loss of olivine of Fo 80 to 90 from MP magma. The four Williams Ranch, Montana, kimberlite samples are within, or close to, the MP field.

Major-oxide compositions (except K_2O) of the MgO-rich range of MP are indistinguishable from the range of kimberlite compositions given by 4 cluster groups of kimberlitic rocks from Africa, averages of Lesotho and Siberian kimberlites, average basaltic and micaceous kimberlites, and averages within several individual kimberlite diatremes in South Africa. Higher K_2O content of Williams kimberlites resembles K_2O contents in phlogopite-rich kimberlites such as New Elands, South Africa. Sm contents and La/Yb of Montana MP are similar to those of Montana kimberlites and kimberlites from other localities. The compositional ranges suggest that the alkalic ultramafic magmas in Montana represent a continuum of compositions from kimberlite through MP to AL, controlled by mantle melting processes and addition or subtraction of xenocrystic and phenocrystic olivine.

ALNOITE IN THE SIERRAS SUBANDINAS, NORTHERN ARGENTINA

MEYER, Henry O. A., Department of Geosciences, Purdue University, West Lafayette, IN 47907; and VILLAR, Luisa M., Servicio Minero Nacional, Ministerio de Economia, 1060-Buenos Aires.

A new and unique intrusion of micaceous ultrabasic rock similar to alnoite is present in the Sierras Subandinas zone of the eastern Andes in northern Argentina. The single sheet-like intrusion extends for at least 10 km and was emplaced into various sedimentary rocks during the Lower Carboniferous. The major minerals in the rock consist of plates of phlogopite (TiO_2 ~ 3.5 wt.%; FeO ~ 7.0 36% mode) that enclose serpentinized oivine (Fo_{88-81} 20% mode) and minor Ti-magnetite (TiO_2 ~ 11.5 wt.%; MgO 5.0 4% mode) and apatite (1%). Other minerals are diopsidic clinopyroxene (approx. Fe_6Mg_{45}:8% mode), perovskite (10%) and calcite (5%). Accessory andradite garnet may be either Ti-rich (TiO_2 ~ 16 wt.%) or Ti poor (<0.1 wt.%). Chemical zoning is absent in the various minerals, and there is very little change in modal mineralogy throughout the rock body. The bulk chemistry is similar to that of melilite-bearing alnoite from Isle Cadieux, Quebec and to ultrabasic dikes having alnoitic affinities from Marathon, Ontario. The presence of a tinguaite (peralkaline phonolite) in the same general area of Argentina, plus the known occurrence of other alkaline rocks to the north in Bolivia may indicate the presence of an important igneous alkalic province in this region of Argentina.

ALKALINE IGNEOUS ROCKS AND CARBONATITES OF PARAGUAY

MARIANO, A. N., 48 Page Brook Rd., Carlisle, MA 01741, and DRUECKER, M. D., Geology Dept., Univ. Iowa, Iowa City, IA 52242

Regional geophysical studies of the northern area of eastern Paraguay have delineated a major NW-SE fracture system related to rift tectonics during the early Mesozoic and NE-SW linears interpreted to be regional basement structures associated with intense folding and thrusting of marginal basement sediments during the last phase of the Trans-Brasilian orogenic cycle (Cambrio-ordovician). Alkaline igneous rocks with cogenetic carbonatites have been emplaced in this structural setting.

Emphasis is placed on the prominent circular structures of Chiriguelo, Sarambi and Cerro Guazú in northeastern Paraguay. Carbonatites are known to occur in the first two while at Cerro Guazú carbonatite is inferred. In southeastern Paraguay carbonatite is also suspected at Sapucaí.

This study includes an examination and classification of fenitization, supergene REE mineralogy in Chiriguelo laterite, hydrothermal pyrochlore crystallization at Chiriguelo, and positive and negative anomaly distributions for soil and rock geochemical surveys within and around the four complexes. These data are correlated with petrology and geophysics and some comparisons are made with the carbonatites on the eastern margin of the Paraná Basin.

COMPOSITIONAL TRENDS OF SPINELS IN KIMBERLITES AND LAMPROITES

MITCHELL, Roger H., Department of Geology, Lakehead University, Thunder Bay, Ontario, Canada P7B 5E1.

Four general spinel compositional trends have so far been observed in kimberlites. 1. A macrocrystal trend consisting of TiO_2-poor magnesian aluminous chromites and aluminous magnesian chromites plotting on the base of the spinel prism. 2. Magmatic trend 1 with spinels evolving from titanian magnesian chromites (1-12% TiO_2) towards members of the magnesian ulvospinel-ulvospinel-magnetite series. The evolutionary trend is across the spinel prism from the base to the ulvospinel apex at constant mg contents. 3. Magmatic trend 2 with spinels evolving from titanium magnesian chromites to titanian chromite to ulvospinel magnetites. This trend is found in micaceous kimberlites and is along the axis of the spinel prism from the base to the magnetite or magnesioferrite apex. 4. A reaction trend resulting in the formation of pleonaste mantles upon Ti and Cr-rich spinels. Spinels in lamproites evolve from aluminous magnesian chromites to ulvospinel-magnetites. The evolutionary trend is similar to that of the kimberlite magmatic trend 2, however the spinels are relatively poor in magnesium. Lamproite spinel compositions can be used to determine the degree of evolution of any particular lamproite.

THE OLD COLD ROOT OF THE KAAPVAAL CRATON

BOYD, F. R., Geophysical Laboratory, 2801 Upton St., N.W., Washington, D.C. 20008

Occurrence of diamonds in kimberlites in southern Africa is primarily confined to the outcrop area of Archaean rocks in the Kaapvaal craton. Isotopic studies of garnet inclusions in diamonds from the Finsch and Kimberley pipes erupted within the craton have yielded Sm-Nd model ages of 3.2-3.3 b.y. (Richardson, 1984). The age of eruption of these kimberlites with their diamond xenocrysts is Late Cretaceous. Hence, these diamonds have been contained in rigid lithosphere—i.e., a craton root—for 3 b.y. under circumstances where they were protected from dispersal by asthenosphere flow. Consideration of the stability relations of diamond indicates that the root must extend to a depth of at least 150 km. Estimates of the temperatures of equilibration for twenty-nine out of thirty-one garnet-olivine pairs of inclusions in diamonds from the Finsch kimberlite have the range 900°-1150°C at an assumed pressure within the diamond stability field of 50 kb. This range corresponds to temperatures found for a continental geotherm in the range 127-175 km, calculated for a surface heat flow of 40 mW/m^2 (Pollack and Chapman, 1977). Estimates for both depths and temperatures of equilibration of two diamond-bearing garnet-lherzolite xenoliths from the Finsch kimberlite are concordant with the estimates for diamond inclusions. These data thus suggest that the Kaapvaal root has an age of over 3 billion years and that at that time it had a temperature profile comparable to that estimated for the present. This conjecture appears inconsistent with increased heat generation from radioactive sources during the Archaean, but that greater heat may possibly have been dissipated by enhanced oceanic volcanism (e.g. Burke and Kidd, 1978).

EXPERIMENTAL CONSTRAINTS ON THE GENESIS OF ALKALIC BASALT LAVAS

SACK, Richard O., Department of Geosciences, Purdue University, West Lafayette, IN 47907; CARMICHAEL, Ian S. E., Department of Geology and Geophysics, University of California, Berkeley, CA 94720

Results of 1 atmosphere (f_{O_2} = QFM) melting and crystallization experiments on alkali olivine basalt and basanite lavas define the relationships among the ratios of olivine, high-calcium pyroxene, and feldspathoid (nepheline + kalsilite + leucite) normative components, temperature, and $X_{FeO}/(X_{FeO}+X_{MgO})$ in alkali basalt liquids on the olivine + high-calcium pyroxene + plagioclase + spinel (± leucite) cotectic. Combined with a temperature-pressure calibration for these normative ratios in high pressure basaltic liquids saturated with olivine and orthopyroxene, (Stolper, 1980; Takahashii and Kushiro, 1983), these experimental constraints provide a format for determining temperatures and pressures of upper mantle source regions and recognizing the imprints of low pressure and polybaric crystallization and crystal fractionation on major-element composition trends in alkalic volcanic and hypabassal rock suites. Applications of these systematics to lavas of the ankaramite-basanite-hawaiite-mugearite-benmoreite-trachyte series from the east African rift volcano (Karisimbi (DeMulder, 1984) show that these lavas are the products of low-pressure crystallization and crystal fractionation of picritic magmas generated at temperatures and pressures of 1380±50°C and 19±3 kbars and suggest that at least some assimilation of a "granulite" component is required to interrelate the more evolved members of this series. Their application to the malignites of the northern Crazy Mountains (M. Garner, in prep.) and alkalic lavas from the oceanic islands suggests methods for determining P_{H_2O} in hypabassal alkalic suites and reveals the extensive imprint of polybaric fractionation on the latter.

THE SPEISS RIDGE—EARLY STAGES OF CONSTRUCTION OF A "HOT SPOT" VOLCANO ON THE SW INDIAN RIDGE

DICK, Henry J. B. and BRYAN, Wilfred B., Woods Hole Oceanographic Institution, Woods Hole, MA 02543

The Bouvet Hot Spot is thought to be located at Bouvet Island, an active volcano located next to the median valley of the SW Indian Ridge (Morgan 1971). Nearby, however, a major seamount (Speiss) lies astride the axis of the ridge raising the possibility that the Bouvet Hot Spot may have recently moved. The Speiss Seamount lavas are evolved ferrobasalts with isotopic and incompatible element ratios characteristic of "enriched" MORB (leRoex et al. 1982, 1983) associated with the main constructional phase of ocean island platforms above hot spots. The alkaline Bouvet lavas are more characteristic of the late stages of such volcanism, consistent with the possibility that Bouvet lies above an abandoned hotspot location. This is consistent with the absence of a median valley at the axis of the SW Indian Ridge at Speiss Seamount and the presence of a large rift valley adjacent to Bouvet Island. Median valleys are characteristic of slow spreading ridges except where they cross major hot spots.

The Speiss ferrobasalts define two distinct major element cotectic trends which correlate with subtle but consistent differences in trace element and isotope ratios. This suggests that the two basalt groups were derived by different degrees of melting of a source in which these differences have persisted for a significant period of time—for example a heterogeneous veined mantle ascending in a mantle plume.

PETROGENESIS OF BASALTS FROM THE WALVIS RIDGE, A CHAIN OF ALKALINE VOLCANOES.

DIETRICH, Volker, J., Institut für Mineralogie und Petrographie, ETH-Zentrum, CH-8092 Zürich, Switzerland, and CARMAN, Max F., Dept. of Geosciences, Univ. of Houston, Univ. Park, Houston, TX 77004

Walvis Ridge connects the islands of the Tristan da Cunha group with the Etendeka volcanic area of southwest Africa. Dredging along the Ridge and drilling across it on Legs 73 & 74 of DSDP provide samples of rocks composing the system. Cores from Hole 524A, Leg 73, can be compared with other Walvis rocks and provide a basis for deducing their origin. Hole 524A rocks include an upper suite of alkali-basaltic pillow lavas, underlain by an alkaline diabase sill dated at 65 m.y. (K-Ar). Below these is a tholeiitic diabase sill 74 m.y. old. The alkaline basalts have typical plume (P-type) chemistry and show considerable differentiation. Tholeiite of the lower sill falls within the range of N-type MORB chemically, but has slightly high alkalis, a slightly enriched, nearly flat REE pattern, and moderate evolution. Zr/Y, Zr/Nb and Y/Nb relations confirm the P- and N-designations for these rocks. Basalts from Leg 74 are *transitional* between those of Leg 73 on virtually all chemical bases. Those from dredges range from transitional to P-types and those from Tristan da Cunha are almost entirely P-type. All share a diagnostically high Ba content. Isotopic studies of Leg 74 rocks (Richardson et al., 1982) show that Nd, Sr and Pb relations cannot be explained by simple binary mixing of *typical* N-type and P-type basalts.

Considering structural, temporal, and chemical relations we believe the origin of Walvis Ridge rocks is complex, involving a deep plume source for the alkalic rocks which were extruded on top of a tholeiitic base derived from a shallow spreading ridge source. Either source may have been isotopically affected by ancient subduction prior to opening of the Atlantic Ocean. Differentiation at different levels and times can explain departures from a *simple* binary magma mixing pattern.

ISOTOPIC INTERPRETATION OF THE SUBOCEANIC MANTLE UNDER THE ONTONG JAVA PLATEAU, SOUTH WEST PACIFIC

NEAL, Clive R., Depar. of Earth Sciences, Leeds University, Leeds, U.K. and DAVIDSON, Jon P., Southern Methodist University, Dallas, TX U.S.A.

Ultrabasic mantle xenoliths from the Alnoite volcanic province in central northern Malaita, can be divided into three types, based on mineralogy, texture, and isotopic composition: 1) GARNET-SPINEL LHERZOLITES ($^{87}Sr/^{86}Sr$ = .70277 – .70417; $^{143}Nd/^{144}Nd$ = .512831 – .5132044), 2) SPINEL LHERZOLITES ($^{87}Sr/^{86}Sr$ = .70256 – .70382; $^{143}Nd/^{144}Nd$ = .513040 – .513482) and 3) SPINEL LHERZOLITES WITH SYMPLECTITE INTERGROWTHS ($^{87}Sr/^{86}Sr$ = .70271 – .70280; $^{143}Nd/^{144}Nd$ = .512847 – .512878).

It is postulated that the Alnoite volcanics have sampled an isotopically and mineralogically zoned mantle with garnet lherzolites (i.e. spinel lherzolites showing symplectite intergrowths) being derived from the deepest levels. The garnet-spinel lherzolites form a transition group, both mineralogically and isotopical isotopically, between spinel and garnet lherzolites.

Spinel lherzolites have been affected by a metasomatic fluid resulting in LREE enrichment and less radiogenic $^{87}Sr/^{86}Sr$. This correlates negatively with La/Nd ratios against Nd and also La/Nd ratios against Nd. This fluid is shown not to originate from the alnoite magma, and also thought not to be due to crustal contamination. The spinel lherzolites show time integrated LREE depletion, whereas the spinel lherzolites with symplectite intergrowths show a decoupling between trace elements and isotopic ratios (i.e. Sm/ND BULK EARTH $^{143}Nd/^{144}Nd$). The garnet-spinel-lherzolites show both time integrated LREE enrichment and depletion.

ALKALI BASALTS OF CASCADE HEAD, OREGON: SEAMOUNTS OR TRANSITION ZONE VOLCANISM?

BARNES, Melanie A. W., Department of Geosciences, Texas Tech University, Lubbock, TX 79409

Cascade Head is one of three major Late Eocene volcanic centers in the Coast Range of Oregon. 300 to 600 m of volcanic rocks are interbedded with thin-bedded, tuffaceous, brackish-water to marine siltstones of the Nestucca Fm. The Nestucca Fm. and intercalated basalts unconformably overlie the Yamhill Fm. and lie on the west limb of a broad, NE-plunging anticline. The Cascade Head volcanic sequence consists of basal submarine basaltic breccia, waterlain lapilli tuff, porphyritic olivine-pyroxene alkali basalt, aphyric submarine to subaerial alkali basalt flows that account for 75% of the volcanic pile, Hornblende dacite dikes, and uppermost basaltic sandstone.

The volcanic rocks at Cascade Head are mildly alkaline and show alkali enrichment of AFM plots. Major and trace element trends are typical of fractional crystallization of olivine, clinopyroxene, and plagioclase. REE patterns exhibit a steep negative slope ($(La/Lu)_n$ = 15-20) typical of alkali basalts.

Discrimination diagrams suggest the Cascade Head basalts were erupted in a non-oceanic, within-plate environment. Geologic and geochemical evidence suggests that the volcanic sequence originated in a transitional zone between oceanic and continental crust in a tectonic environment of extension due to rifting and is not directly related to subduction. The data are in accord with Wells' et al. (Tectonics, 1984) tectonic interpretation of a Late Eocene extensional environment due to a decrease in the rate of plate convergence.

ALKALIC VOLCANIC ROCKS FROM SW PACIFIC ISLAND ARCS: SUBDUCTION OR RIFT-RELATED?

PERFIT, Michael R., Department of Geology, University of Florida, Gainesville, FL 32611

Major element, trace element and isotopic characteristics of late Cenozoic volcanic rocks from west Melanesian island arcs define at least four different regional rock associations. The rocks range from low-K arc tholeiites to medium-K calcalkaline rocks to high-K, silica saturated and undersaturated varieties. Alkalic rocks occur on Vanuatu, the Solomon Islands and Bougainville Island and are particularly abundant in mainland Papua New Guinea and the Tabar-to-Feni Islands. High-K basalts and unique high-Na and Ti alkalic basalts have also been recovered from the New Georgia-Woodland Basin forearc region. Strongly undersaturated basanite, nephelinite, tephrite and phonolite are common only in the Tabar-to-Feni chain. Chemically, most of the alkalic rocks have arc signatures with low concentrations of TiO_2 (<1.2wt%), HFS ions and Th/U ratios and high abundances of LIL elements, Zr/Nb, Ba/La. The light-REE are moderately to highly fractionated relative to the heavy-REE, but there is no marked depletion of heavy-REE as in intraplate alkalic basalts. Sr- and Nd- isotopic values vary from .7036 to .7047 and (ϵNd) +8.7 to +2. Like many arc volcanics, these alkalic samples plot to the right of the "mantle" array" and extend the Sr-Nd trend of subalkaline volcanics in the region. Fairly systematic increases in LIL and LREE correlate with increasing $^{87}Sr/^{86}Sr$ - but each arc exhibits a moderate degree of heterogeneity. Enrichments relative to MORB-type values appear to correspond to the availability of crustal material and the tectonic history of each arc. Most of the alkalic centers are not related to present day subduction but rather are associated with tensional and strike-slip tectonics. The alkaline rocks reflect one or more previous subduction episodes that heterogeneously enriched the sub-arc mantle. Subsequently small degrees of partial melting, initiated by rifting, produced the variety of alkalic rocks erupted.

SILICA OVERSATURATED AND UNDERSATURATED MIASKITIC ALKALINE TRENDS OF KERGUELEN ARCHIPELAGO

BEAUX, J. F. and GIRET, A., Laboratoire de Petrologie T 25, E 3, Universite Pierre et Marie Curie 4, place Jussieu 75230-Paris Cedex 05

Recent petrological, geochemical and geophysical studies all point to a mantle origin for the rocks of Kerguelen Archipelago which was situated along the East Indian Paleo Ridge 42 m.y. ago. The magmatism at first tholeiitic then transitional becomes alkaline around 26 m.y. A striking feature in the last 16 m.y. is the individualization of two subprovinces characterized by alkaline rocks which are silica oversaturated to the west and undersaturated to the east of N 15° W line.

The oversaturated suite is composed of gabbros, monzosyenites quartz syenites and quartz poor granites. The undersaturated suite comprises gabbros, monzonites, nepheline-bearing and nepheline syenites. The basic rocks of both suites are very similar and the divergent evolution only appears clearly in the monzonitic types which are rare and amphibole-poor in the first suite whereas they are abundant and amphibole-rich in the second. Common features in both suites are: early emplacement of gabbros which are less abundant than the syenites; exceptionally high Fe content of the mafic minerals in the gabbros with Fe/Mg ratios frequently higher than in the corresponding minerals of the syenites; enrichment of incompatible elements is a function of the D.I. and of the activity of fluid phases with the U/Th ratio ¼ remains constant; REE proportions increase with differentiation but whereas the Eu anomaly is positive in the gabbros, it is negative in the syenites and weak or absent in the monzonites.

Models are presented to explain the divergent evolutionary trends in the two-sub-provinces.

TEMPORAL EVOLUTION OF ALKALINE MAGMATISM IN THE NORTHERN CANADIAN CORDILLERA

EICHE, G. E., FRANCIS, D. M. (Dept. of Geological Sciences, McGill University, Montreal, Canada H3A 2A7) and LUDDEN, J. N. (Dept. de Géologie, Université de Montréal, Canada H3C 3J7)

Quaternary basalts associated with a field of small eruptive centers near Alligator Lake in the southern Yukon have evolved toward increasingly silica-undersaturated compositions during their eruption history. The oldest lava flows are plagioclase-phyric, hy-normative transitional basalts with variable MgO contents (4.5-9.2 wt. %). They are overlain by relatively less differentiated, ne-normative alkaline basalts (8.5–13.7 wt. %) in which the predominant phenocryst phase is olivine. Plagioclase appears as a phenocryst only in flows with less than 10 wt. % MgO. The younger lavas are also olivine-phyric, ne-normative (10.4-14.8 wt. % MgO) alkaline basalts but are rich in LIL elements with respect to the older lavas. These LIL-rich alkaline basalts carry abundant spinel lherzolite xenoliths, megacrysts (cpx, opx, ol, and spinel), and granitoid fragments.

This temporal evolution of magmas towards more Si-undersaturated compositions is similar to that commonly documented for oceanic island volcanism. The alkaline basalts from Alligator Lake lie along a trend which has distinctly lower Si values in Al-Si space than the trend that is characteristic of the older transitional basalts. This suggests that two chemically distinct magma types may have been involved, and that younger alkaline magmas were tapped from deeper mantle sources than older transitional magmas. The LIL-rich nature of the youngest basalts, however, must reflect an additional complication such as contamination or source heterogeneity in the petrogenesis of these magmas.

THE CAMEROON LINE, WEST AFRICA, AND ITS BEARING ON THE ORIGIN OF OCEANIC AND CONTINENTAL ALKALI BASALT

FITTON, J. G., Grant Institute of Geology, University of Edinburgh, West Mains Road, Edinburgh EH9 3JW (U.K.)

The Cameroon line is a unique example of a within-plate volcanic province which straddles a continental margin. It consists of a chain of Tertiary to Recent, generally alkaline volcanoes stretching from the Atlantic island of Pagalu to the interior of the African continent. It provides an ideal area in which to compare the sub-oceanic and sub-continental mantle sources for alkali basalt.

Basaltic rocks in the oceanic and continental sectors are geochemically and isotopically indistinguishable which suggests that they have identical mantle sources. This conclusion rules out substantial lithosphere involvement in the generation of alkali basalts and therefore weakens the case for metasomatism as a necessary precursor to alkali basalt magmatism. The convecting upper mantle is a much more likely source as it will be well-stirred and unlikely to show any ocean-continent differences. The long history of Cameroon line magmatism (65 Ma) and lack of evidence for migration of volcanism with time makes a deeper mantle source unlikely.

Mid-ocean ridge basalts (MORB) also originate within the convecting upper mantle and so must share a common source with the Cameroon line alkali basalts (and, by implication, ocean island basalts and continental rift basalts). A grossly homogeneous mantle with a bulk composition depleted in large-ion lithophile elements (LILE), but containing streaks of old LILE-enriched material, provides a plausible common source. Large degree, near-surface melting of such a source would produce MORB. Smaller-degree melts produced at deeper levels would percolate upwards along grain boundaries and become enriched in LILE by leaching LILE-rich grain boundary films. The mixing of these liquids with melts from the LILE-rich streaks will produce magmas with the geochemical and isotopic features of ocean islands basalts.

IGNEOUS PETROLOGY AND STRUCTURAL GEOLOGY OF NINE POINT MESA, BREWSTER COUNTY, TEXAS

BOBECK, Patricia, Dept. of Geological Sciences, University of Texas at Austin, Austin, TX 78712

Nine Point Mesa is a Tertiary-age quartz syenite sill roughly circular in plan view, 8 km in diameter and 350 m thick, intruded into Upper Cretaceous Boquillas and Aguja Formations. Its planar upper surface is devoid of sedimentary cover. Its margin is chilled at the contact with an underlying intrusion of nepheline-normative basalt, which has a surface outcrop of 1 by 2 km and an observed thickness of 100 m. This basalt, whose preliminary K-Ar age is 47 m.y., bowed up the Boquillas as it intruded.

The northern margin of the quartz syenite still abuts against a northwest trending down-to-the-south fault that was active prior to intrusion of the syenite. Two other northwest trending faults that cut the syenite show dip-slip displacement of 30 to 70 m down to the south. The latter faults are interpreted as Basin and Range structures.

A feeder zone for the intrusion extends NE to SW across the syenite, separating the Boquillas-hosted syenite from Aguja-hosted syenite. At the feeder zone, the Pen Formation clay is absent; the Aguja rests directly on the Boquillas. This disconformity is thought to have permitted the intrusion to step up into the incompetent clays of the Aguja.

The fayalite-bearing quartz syenite is composed of 80 to 85% alkali feldspars, 5% clinopyroxene, and 5% quartz, with lesser amounts of plagioclase, olivine, apatite, zircon, magnetite, ilmenite, pyrrhotite and pyrite. No petrographic variation was detected in the sill. Its composition is similar to five other intrusions located within 55 km of Nine Point Mesa, suggesting a common magmatic source.

CHEMICAL CHARACTERIZATION OF CAMBRIAN BASALTIC LIQUIDS FROM THE SOUTHERN OKLAHOMA AULACOGEN

GILBERT, M. Charles, Dept. of Geology, Texas A&M University, College Station, TX 77843, and HUGHES, Scott S., Dept. of Chemistry and Radiation Center, Oregon State University, Corvallis, OR 97331

Within the Wichita Mountains portion of the Southern Oklahoma aulacogen three important stratigraphically distinct units of basaltic bulk chemistry crop out. INAA on representative samples from two of these three, the older Roosevelt Gabbros (RG) and the younger late diabases (LD), yields the following:

1) Incompatible elements, Ba, Rb, LREE are enriched relative to MORB, being similar to continental tholeiites.

2) Trace element patterns for Ba, REE, Sc, Ta, Th, normalized to volatile-free C1 chondrites, are fractionated and have distinctive, common shapes for all samples.

3) Patterns show positive Eu anomalies.

4) The Sandy Creek Gabbro member of the RG has the lowest abundances of REE while Mount Sheridan Gabbro member is similar to the LD.

5) Those LD occurring in a granitic/rhyolitic host show some evidence of contamination.

Conclusions are:

1) Parallelism of all patterns suggests a common source, which also implies a short time interval between them.

2) This "common source" may itself have been emplaced in the upper crust and caused partial melting leading to the granites/rhyolites.

3) The two basaltic units may have segregated from a cumulate pile during various stages of differentiation. Neither unit represents truly primary liquid.

4) Cambrian rifting here has generated basaltic liquids similar to those found in the Proterozoic Keweenawan suite rather than the alkalic types of the southern Rio Grande Rift or East Africa.

PETROLOGY OF ALKALINE ROCKS IN THE CARBONATITE COMPLEX AT IRON HILL, POWDERHORN DISTRICT, GUNNISON COUNTY, COLORADO—NEW GEOCHEMICAL AND ISOTOPIC DATA

ARMBRUSTMACHER, Theodore J. and FUTA, Kiyota, U.S. Geological Survey, MS 916, Box 25046, DFC, Denver, CO 80225

Major- and minor-element analyses and Rb/Sr and Sm/Nd isotope systematics of rocks of the carbonatite complex at Iron Hill show that pyroxenite, ijolite, uncompahgrite, and carbonatite may be comagmatic and that the rocks have been derived from alkaline magmas generated in the upper mantle. For these rocks, the initial $^{87}Sr/^{86}Sr$ ratios range from 0.70295 to 0.70334 and the initial $^{143}Nd/^{144}Nd$ ratios range from 0.511961 to 0.512133 (ϵ_{Nd} at 550 m.y. range from +0.6 to +0.4). These data suggest that the rocks were derived from a reservoir depleted in LIL elements relative to the bulk Earth. Nepheline syenite may be comagmatic with the other intrusive rocks, but isotopic and geologic evidence suggest another alternative interpretation: nepheline syenite has been influenced by fenitizing fluids passing outward from the carbonatite into the granitic country rock.

According to earlier workers, uncompahgrite originated by reaction of pyroxenite with later hydrothermal fluids. If this is true, the isotopic data require that the hydrothermal fluids were derived directly from the same alkaline magma from which the other rocks of the complex formed without the addition of significant nonmagmatic components.

Preliminary rare-earth-element (REE) analyses suggest that light REE are strongly enriched relative to heavy REE and both LREE and HREE are enriched relative to chondrites. Total REE abundances in rocks of the complex follow the sequence: magnetite-ilmenite-perovskite segregations >carbonatite >uncompahgrite >ijolite >nepheline syenite >pyroxenite.

COMBINED Rb/Sr AND Sm/Nd SYSTEMS STUDY ON THREE ALKALINE INTRUSIVE COMPLEXES IN NORTHWESTERN MONTANA

FUTA, Kiyoto, U.S. Geological Survey, Box 25046, MS 963, Denver Federal Center, Denver, CO 80225, and ARMBRUSTMACHER, Theodore J., U.S. Geological Survey, MS 916, Denver Federal Center, Denver, CO 80225

Sr and Nd isotopic ratios and trace element concentration data determined on ultramafic and syenitic rocks from three Montana complexes, Rainy Creek, Haines Point, and Skalkaho, suggest independent magma sources for the complexes although they all may be mid-Cretaceous in age. Rb/Sr data on syenites from the Haines Point complex yield an age of 107 Ma, and the Rb/Sr data on the syenites from the other two complexes lie along the same isochron line. However, no isochron age can be determined by the Sm/Nd data on the same syenites. Both the syenites and pyroxenites have enriched concentrations of LREE compared to their HREE contents. The combination of enriched LREE and high $^{143}Nd/^{144}Nd$ isotopic ratios around 0.5128 implies a recent disturbance in the Sm/Nd system. Most of the pyroxenites have lower initial $^{87}Sr/^{86}Sr$ and higher $^{143}Nd/^{144}Nd$ ratio than the syenites. The initial $^{87}Sr/^{86}Sr$ and $^{143}Nd/^{144}Nd$ ratios show a negative correlation, but the isotopic ratios do not show a correlation with their respective Sr and Nd concentrations. The initial $^{143}Nd/^{144}Nd$ ratios for most of the ultramafics have positive ϵNd values suggesting a long term residence in an environment depleted in LREE compared to HREE; whereas the syenites have ϵNd values around zero or slightly negative.

COMPLEX ZONING OF CLINOPYROXENE IN SHONKINITES AND MAFIC PHONOLITES, HIGHWOOD MTNS., MONTANA: EVIDENCE FOR PERIODIC MIXING WITH A K-RICH BASANITIC MAGMA

O'BRIEN, H. E., IRVING, A. J., AND McCALLUM, I. S., Dept. of Geological Sciences, Univ. of Washington, Seattle, WA 98195

Mafic phonolite lavas and compositionally equivalent hypabyssal shonkinites from the Highwood Mtns. contain clinopyroxene phenocrysts that show discontinuous zoning among augite (*mg* ≃76), diopside (*mg* ≃90) and low-Al augite (*mg* ≃82). Diopside occurs as cores within diopside, as distinct continuous bands ≃20-100 μm wide, and as subhedral xenocrysts within the augite-saturated shonkinitic magmas. Some diopside cores contain melt-filled cavities rimmed by augite and some diopside bands have irregular inner boundaries yet abrupt outer boundaries. These textures suggest the existence of separate augite-saturated and diopside-saturated magmas.

Olivine phenocrysts in mafic phonolite/shonkinite are $Fo_{65\text{-}76}$ whereas large olivine xenocrysts within the phonolite are ≃Fo_{87}. This latter composition is consistent with crystallization from a diopside-saturated, MgO-rich magma such as a K-rich basanite.

The existence of diopside xenocrysts compositionally identical to the diopside bands excludes an origin for the latter by an oscillatory growth process. Instead we suggest that the diopsode bands record the periodic influx of basanitic magma. The abrupt outer terminations of diopside bands may mark the point at which the densities of the basanitic and shonkinitic magmas became equal, thereby promoting rapid mixing and an abrupt change in clinopyroxene composition.

Although some low-Al augites occur as separate growth bands and rarely as phenocryst cores in shonkinites, they occur primarily as phenocrysts in augite-biotite minettes. Therefore, the minettes may represent an intermediate stage of fractional crystallization and mixing between the shonkinitic and the K-rich basanitic magmas.

Printed in U.S.A.

Index

[Italic page numbers indicate major references]

Typeset by WESType Publishing Services, Inc., Boulder, Colorado
Printed in U.S.A. by Malloy Lithographing, Inc., Ann Arbor, Michigan